Fundamentals of Machine Design
Volume II

Machine design is part of the broader discipline of Engineering Design. *Fundamentals of Machine Design* is compiled in two volumes. Vol. II is a follow-up to the first volume. Unit 1 of this volume begins by discussing fundamental concepts, types and applications of belt drives, pulleys, rope drives, chain and sprocket drive, spur gears, helical gears, bevel gears, worm gears and gear trains using simple and epicyclic gears in detail. Unit 2 discusses construction aspects, classification, material required, design procedures and selection parameters for hydrodynamic and rolling bearings. The design steps are discussed comprehensively, which helps students and teachers in practical classes. Unit 3 discusses different types and construction processes of important parts of an internal combustion engine including cylinder, piston, connecting rod, crank shaft and valve gears. The final unit 4 comprehensively discusses the design procedure, types and construction of flywheels, clutches, brakes and pressure vessels.

Pedagogical features in the book include solved examples, unsolved exercises, design problems and review questions. The text is primarily a follow-up introductory course on machine design, meant for undergraduate students of mechanical engineering. It is accompanied with teaching resources including a solutions manual for instructors.

Ajeet Singh retired as Professor and Head of the Department of Mechanical Engineering from Motilal Nehru National Institute of Technology (MNNIT), Allahabad. In addition to teaching, he worked in many administrative positions like Dean Academic, Dean Research and Consultancy etc. He has about three decades of teaching experience at undergraduate, graduate and doctoral level in India and 15 years abroad. He taught courses in machine drawing, machine design, internal combustion engines, tribology, computer aided design and engineering processes. He has been consultant to industries like BHEL, TEW etc. Besides publishing several papers in national and international journals, he has published three textbooks: *Working with AutoCAD 2000 with updates to AutoCAD 2000i* (2002), *Machine Drawing: Includes AutoCAD 2005* (2005) and *Machine Drawing; Includes AutoCAD 2010* (2012).

Fundamentals of Machine Design
Volume II

Ajeet Singh

CAMBRIDGE
UNIVERSITY PRESS

CAMBRIDGE
UNIVERSITY PRESS

University Printing House, Cambridge CB2 8BS, United Kingdom

One Liberty Plaza, 20th Floor, New York, NY 10006, USA

477 Williamstown Road, Port Melbourne, VIC 3207, Australia

314-321, 3rd Floor, Plot 3, Splendor Forum, Jasola District Centre, New Delhi - 110025, India

79 Anson Road, #06-04/06, Singapore 079906

Cambridge University Press is part of the University of Cambridge.

It furthers the University's mission by disseminating knowledge in the pursuit of education, learning and research at the highest international levels of excellence.

www.cambridge.org
Information on this title: www.cambridge.org/9781316630419

First published 2017

A catalogue record for this publication is available from the British Library

ISBN 978-1-316-63041-9 Paperback

Additional resources for this publication at www.cambridge.org/9781316630419

Dedicated to my parents, wife,

Daughters: Preety, Diljeet and Maneet

Grandchildren: Gaganjit, Karanjit, Ananya, Neha, Tanvi and Simar

Contents

Unit 1 - Drives

1. Belts and Pulleys

6. Helical Gears

9. Gear Trains and Gear Boxes

Unit 2– Bearings

10. Hydrodynamic Bearings

Unit 3 - Design of I.C. Engine Parts

12. Cylinder of an I.C. Engine

13. Pistons

17. Fly Wheels

Unit 4 – Design of Miscellaneous Parts

18. Clutches

19. Brakes

Preface

Fundamentals of machine design considers the concepts of design for each element separately such as shaft, bearing, pulley and gears. Since the number of parts is very large, the book is divided into two volumes. The language is direct and simple, so that every student can understand easily. *Volume 1* of the book describes the basic knowledge needed for designing a part. Various types of stresses that appear due to load and the analysis to calculate the size of a part, which will work satisfactorily, are given. Designs of various types of permanent and temporary joints like riveted, welded, threaded, cotter etc. and some important parts such as shafts, keys, couplings, and springs are described.

Volume 2 has four units; covering design of drives, bearings, I.C. engine parts, and miscellaneous parts such as clutches, brakes, and pressure vessels.

Unit 1 describes different types of drives. Chapter 1 gives the design of belts and pulleys of various types. Chapter 2 is on the design of ropes used for large power transmission and hoisting applications. Chapter 3 describes the design of chain drives. One of the important methods of transmitting power is using gears and is discussed in detail. Chapter 4 describes the fundamentals of the gear design. Various types of gears such as spur, helical, bevel, and worm are described separately in chapters 5 to 8. Design of gear boxes using gear trains and epicyclic gears is explained in Chapter 9.

Unit 2 gives the design of two important types of bearings used in many applications. Design of slide bearings is given in Chapter 10 and rolling element bearings in Chapter 11.

Unit 3 is on the design of I.C. engine parts. Each important part is described as a separate chapter. Chapter 12 is on the design of cylinder, Chapter 13 on piston, Chapter 14 on connecting rod, and Chapter 15 on crank shaft. These engines require valves to control flow of air and exhaust. All parts in valve gear mechanism such as cam, its follower, push rod, rocker arm, and valves are described in Chapter 16. There is large variation in torque in these engines and hence to reduce the fluctuations in speed, flywheels are required, which are described in Chapter 17.

Unit 4 describes miscellaneous parts in three chapters. Chapter 18 is on the design of various types of clutches, Chapter 19 on the design of various types of brakes, and the last Chapter 20 is on the design of pressure vessels.

Pedagogical features of the book are excellent. At the beginning of each chapter *outcomes* are given, which gives an idea as to what a student is going to learn in that chapter. Every effort has been made to explain the theory with figures. This volume contains *289 figures*. To make the book more illustrative, *122 license-free pictures* are given from the internet. Students face a lot of difficulty in solving design problems; hence a large number of *158 solved examples* are given.

At the end of each chapter a *summary* is given for quick revision of the course and formulas at the time of examination. Each chapter is followed by *theory questions*. To practice for quiz type questions, *216 multiple choice* questions have been given. To practice the design problems, *180 unsolved problems* with the answers are given. Solution to the unsolved examples shall be put in the *solution manual* on the internet in due course of time.

Competition examinations questions of past 3–4 years from Engineering services examinations and *GATE* examinations are given at the end of the chapters to help students preparing for such examinations.

After successful completion of the course, the student shall be able to understand the design process with all stresses on it and shall be able to find a size for its satisfactory working. The mastering of the course is a precondition for a successful design.

Audience: This book can be easily recommended as a *textbook* on the subject of Machine Design for undergraduate students. The book can also be used by practising engineers, students appearing for competition examinations, and for graduate admission tests.

Although every effort is made to minimize the errors, but still a human being is likely to commit mistakes. Also, there is always a possibility of improving the book. Any errors, omissions, or suggestions for the improvement of the book may please be written to the publisher or email to the email address of the author at ajeet41@yahoo.com.

Acknowledgements

First of all, I want to thank *almighty God*, who gave me patience, courage, and health even at this age to write such a voluminous text book.

The author thanks Mr. Gauravjeet Singh Reen, commissioning editor at Cambridge University Press, who had been very helpful and prompt in interaction for any of my queries or doubts.

I am thankful to the *reviewers* for giving encouraging remarks in their reviews and appreciating my effort in preparing the book. Effort has been made to incorporate all their valuable suggestions mentioned in their reviews for the book. Thanks are due to the editorial and production staff of M/S Cambridge University Press, for their cooperation and help in the publication of the book. I wish to acknowledge my gratitude to Indian Standards Institution, for the extracts of some of standards used in this book. In the last, I thank all my family members for their moral support in preparation of the book.

Many books on the subject of machine design have been consulted and the author would like to thank the publishers and authors of the books referred given in the references at the end of the book.

Belts and Pulleys

1.1 Introduction

In machines, mechanical power is transmitted from one shaft to another shaft. Various types of drive systems are available. Important ones are belt and pulleys, chains and sprockets, and gears. This chapter deals only with belt and pulleys. Pulleys are used to transmit power with the help of belts. A pulley is a circular machine element, having a hub in the center with a key way, which fits on the

shaft. Pulleys are mounted on shafts, while the belt passes over both the pulleys. The power is transmitted due to the friction between the pulley and the belt. It is suitable, if the center distance between the shafts is large. Some features towards advantages and some towards disadvantages are given below:

1.1.1 Advantages

- It is simple and inexpensive.
- The center distance between the shafts is not that critical as for gears.
- It does not require axially aligned shafts and can tolerate high misalignment.
- The distance between the shafts can be large.
- It needs no lubrication and requires minimal maintenance.
- It protects the machines from overload and jam.
- It damps and isolates noise and vibrations.
- Load fluctuations and shocks are absorbed.
- It does not increase the cost much, even if the shafts are far apart.
- Clutch action can be obtained by releasing / adjusting belt tension.
- It can operate at temperatures ranging from –35° C to 85° C.
- It can work in dusty or corrosive environment.
- Different speeds can be obtained by step or tapered pulleys.
- It offers high transmission efficiency (output / input) 90–98 per cent, usually 95 per cent.

1.1.2 Disadvantages

- Adjustment of center distance or addition of an idler pulley is needed to compensate for wear and stretch.
- Tension of the belt causes bending of the shafts and load on the bearings.
- Short life in comparison to other types of drives.
- A cover, generally of wire mesh is needed on the drive for safety purpose.

1.2 Types of Belts

Belts are of many types as shown in Figure 1.1 and pictures below. Use of a particular type of belt depends on power to be transmitted and type of service:

- Flat belt
- V belt
- Grooved belt
- Ribbed belt
- Film belt
- Circular belt

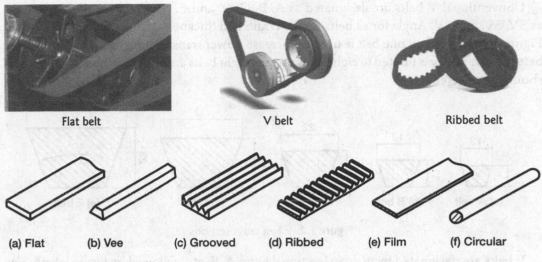

Flat belt   V belt   Ribbed belt

(a) Flat   (b) Vee   (c) Grooved   (d) Ribbed   (e) Film   (f) Circular

Figure 1.1 Types of belts

Flat and V belts are generally used and hence described below.

1.2.1 Flat belts

Plain flat belts are of rectangular cross section with no teeth or groove as shown in Figure 1.1(a). These belts are used for line shafts in factories, farming, mining applications, saw mills, flour mills, conveyors, etc. These are low cost and used on small pulleys. They need high tension resulting in high bearing loads. They are noisier than other types of belts and have low efficiency at moderate speeds. These are made of leather, fabric, rubberized fabric, nonreinforced rubber / plastic, reinforced leather, etc.

1.2.2 V belts

The V belt was developed in 1917 by John Gates. These are endless, and their cross-sectional shape is trapezoidal. They reduce the slippage and alignment problem. They provide the best combination of power transmission, speed of movement, load of the bearings, and long service life. The belt tends to wedge into the groove, as the load increases. Greater the load, more is the wedging action, thus improving the torque transmission and need lesser width and tension than flat belts. The preferred center distance is larger than the largest pulley diameter, but less than three times the sum of diameter of both the pulleys. Optimal belt speed range is 5–35 m/s.

V belts have long life (3–5 years) and offer quiet operation and low maintenance. These are most commonly used in industry and are available in standard cross-sectional sizes and lengths. They offer more speed than flat belts. The best speed for V belts is between 8 and 30 m/s. V belts are made in two sizes: conventional and narrow. Ideal speed for standard belt is 23 m/s and for narrow belts it is 50 m/s.

Conventional V belts are designated as A, B, C, D, and E. Narrow belts are designated as 3V, 5V, and 8V. Angle for all belts is 40°. Width and thickness of these belts are shown in Figure 1.2. More than one belt is used to increase power transmission capacity. Number of belts on one pulley is limited to eight. If more than eight belts are required, then larger section should be selected.

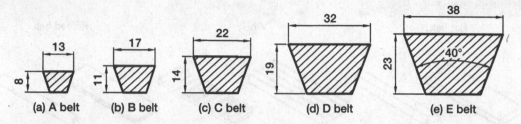

(a) A belt   (b) B belt   (c) C belt   (d) D belt   (e) E belt

Figure 1.2 V belt cross sections

V belts are designated by its cross-sectional letter A, B, etc., followed by inside length. For example, B 420 means a belt of cross section B and inside length 420. Table 1.1 shows the sizes of various belts.

Table 1.1 Size and Weight of V Belts

Belt Section	Large Width (w) (mm)	Thickness (t) (mm)	Small Width ($w2$) (mm)	Area (mm^2)	Volume per meter (V) (mm^3)	Weight per meter (N)
A	13	8	7.2	81	81,000	1.06
B	17	11	9.0	143	143,000	1.86
C	22	14	11.8	237	237,000	3.43
D	32	19	18.2	477	477,000	5.96
E	38	23	21.3	681	681,000	9.41

Small width of the of the section = Width at large size – (2 × thickness × tan 20°)

$$w2 = w - (2t \times 0.364) = w - 0.728t$$

Area of the trapezoidal cross section $A = \dfrac{\text{Large width} + \text{Small width}}{2 \times \text{Thickness}} = \dfrac{w + w2}{2t} \text{ mm}^2$

Volume of belt per meter $V = 1{,}000A$

Weight of belt per meter = Volume × density = $V \times \rho$

1.2.3 Construction of V belts

V belts are made of three layers. The outer layer is of polychloroprene as an elastic cover. The central part has load-bearing cords of polyester fabric located near the center of the section

(Figure 1.3). Since the stresses are minimum in the center, these cords have not to bear too much fatigue. In between the outer part and cords is the rubber to transmit force from cords to side walls. Thermoplastic polyurethane, and Elastomers are also used for V belts. Temperature range 0° C – 80° C.

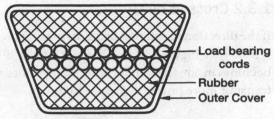

Figure 1.3 V belt construction

1.3 Types of Flat Belt Drive

The open belt drive has parallel shafts rotating in the same direction, whereas the cross-belt drive also have parallel shafts but rotate in the opposite direction. The former is far more common, and the latter is not common, because the pulleys contact both the inner and outer belt surfaces. Non-parallel shafts can be connected, if the belt's center line is aligned with the center plane of the pulley.

Three types of flat belt drives are used; open, crossed, and quarter twist depending on the direction of rotation of shafts, angular position, and distance between shafts.

1.3.1 Open belts

If the direction of rotation of the driver and driven is the same, the open belt arrangement is used as shown in Figure 1.4(a). One side of the belt is called tight side, while the other is called slack side. Tight side should be kept at the bottom as shown in Figure 1.4.

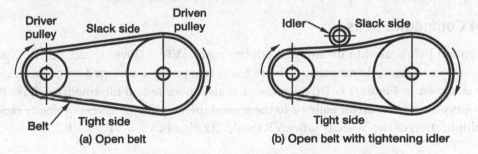

Figure 1.4 Open belt system

The center distance is an important parameter in this drive. Small center distance increases the number of turns of belt per second and hence reduces belt life. Short center distance reduces arc of contact on small pulley, which increases slip. On the other hand, large distance causes sagging of the belt. If the belt is horizontal, sagging causes swinging perpendicular to the belt. Hence, slack side is kept on the upper side so that slackness increases the arc of contact. If loose side is on the lower side, sagging reduces arc of contact and hence increases slipping.

Owing to continuous use of belt, it gets elongated. For shafts, which cannot be moved to tight the belt, an idler is used to keep the belt tight over the pulleys [Figure 1.4(b)].

1.3.2 Crossed belts

If the direction of rotation of the driven is opposite to the driver, then crossed belt arrangement is used [See Figure 1.5(a)]. The belt is twisted 180° such that outside face of belt on one pulley becomes inside face on the other pulley. Since the arc of contact is more, this arrangement can transmit more power than open belts.

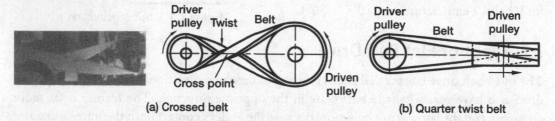

Figure 1.5 Cross and quarter twist belt arrangements

Drawback of this is that the location where two sides of the belts cross each other, rubbing between the belt faces takes place causing excessive wear and tear of the belt. To reduce this rubbing, center distance between the shafts should be about 20 times the width of the belt.

1.3.3 Quarter twist belts

If the axes of the driver and driven shafts are at 90°, quarter twist belt as shown in Figure 1.5(b) is used. Center distance between the pulleys has to be long for this arrangement. To avoid slipping of the belt off the pulley, width of pulley is kept about 1.5 times the width of the belt.

1.3.4 Compound belts

Compound belt is used to obtain high velocity ratios (VR = Driver speed / driven speed). For this type of drive, driver pulley 1 drives the driven pulley 1 mounted on an intermediate shaft as shown in Figure 1.6. Driver pulley 2 is also mounted on intermediate shaft. Belt 2 transmits power from driver pulley 2 to the second driven pulley 2. Overall velocity ratio VR is multiplication of two velocity ratios VR1 and VR2, that is, VR = VR1 × VR2.

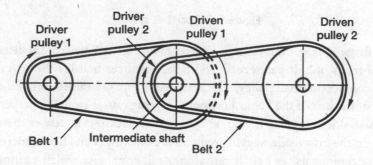

Figure 1.6 Compound belt

1.3.5 Serpentine belts

A belt can drive more than one pulley also as shown in Figure 1.7. In this figure, three pulleys are being driven from a single driver pulley. To increase the arc of contact so that the belt does not slip, idlers are used. Idler is also used to tight the belt by moving it in a direction, which tights the belt.

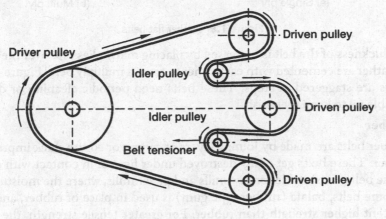

Figure 1.7 Serpentine belt

1.4 Belt Materials and Construction

The belt material is chosen depending on the use and application. Leather oak tanned belts and rubber belts are most commonly used. Plastic belts have almost twice the strength of leather belt. Fabric belts are used for temporary or short-period operations. Belt material should possess the following properties:

- High coefficient of friction for a good grip with the pulley;
- High tensile strength to bear the pull of belt and centrifugal tensions;
- High flexibility, so that it could bend easily when turning over the pulleys;
- High fatigue strength, as the outer and inner surfaces are under varying stresses; and
- High wear resistance for a long life.

Following materials are generally used for the belts:

a. **Leather**

Main advantage of leather belt is high coefficient of friction and hence high power transmitting capacity. Oak-tanned or chrome-tanned leather is used for flat belts. Leather belts of various widths are available in the form of strips up to 1.5 m length. Leather strips have two sides; one is called flash side and the other hair side. Flash side is strong, whereas the hair side is smooth and hard. Hair side is kept for contact on pulley. Tension is maximum on outside of the belt, hence flash side is kept outside. The strips are made into endless loops by making the ends tapered as shown in Figure 1.8(a). Each strip is called a ply.

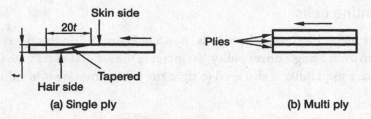

Figure 1.8 Leather flat belts

Thickness of the belt is increased by placing many plies one over the other. Strips of leather are cemented with each other to form a multiply belt [Figure 1.8(b)]. The joints are staggered in steps. These belts need periodic cleaning or dressing with suitable oil to keep them soft.

b. Rubber

Rubber belts are made by joining layers of canvas or cotton duck impregnated with rubber. These belts get easily destroyed under heat or in contact with oil or grease. These belts are suitable for saw mills and paper mills, where the moisture may exist. In some belts, balata (rubber type gum) is used in place of rubber, and it offers 25 per cent higher strength than rubber. For greater tensile strength, the rubber belts are reinforced with steel cords or nylon cords. These belts offer high load-carrying capacity and long life, and can operate at belt speed as high as 300 m/s. Their disadvantages are that they cannot be used over small pulleys and not suitable for oily environment.

c. Fabric

Fabric belts are made of canvas or woven cotton. The thickness of belt is increased either by folding or stitching each layer together. The belt thickness is built up with a number of fabric layers called plies. The plies are impregnated with filler material like linseed oil to make them water proof. These belts are cheaper and suitable for damp and warm environment. These are suitable for agricultural machines, belt conveyors, as they require very little attention. These can be easily made endless. These are manufactured by any one of the methods shown in Figure 1.9.

- **Raw edge belt** Strips are cut and placed one over the other. Edges are visible on sides and hence called raw edge belt [Figure 1.9(a)]. Edges are protected with water proof compound. Rubber is placed in between the plies. These are suitable for small pulley and high speed.

(a) Raw edge belt   (b) Folded layer belt   (c) Spiral wrapped belt

Figure 1.9 Fabric flat belts

- **Folded layer belt** It has a central ply wrapped in rectangular plies [Figure 1.9(b)]. There may not be any rubber between the plies.

- **Spiral wrapped belt** They are made by a single piece of fabric wrapped in a spiral fashion [Figure 1.9(c)]. There may not be any rubber between the plies.

d. **Plastics**

Thin plastic sheets with rubber layers are also used as flat belt material.

1.5 Properties of Belt Materials

Design strength, endurance strength, and elastic modulus are given in Table 1.2. These values are required to calculate the size of the belt. Table 1.2 also gives density of belt material, which can be used to calculate centrifugal force on the belt.

Table 1.2 Properties of Belt Materials

Material	Design Strength (MPa)	Endurance Strength (MPa)	Elastic Modulus (MPa)	Density (kg/m³)	Maximum Velocity (m/s)
Leather	2.0	6.0	30.0	1,150	50
Rubber	1.6	6.0	10.0	1,150	30
Fabric	1.5	3.0	15.0	950	25
Plastic	4.0	6.0	60.0	1,050	60

Coefficient of friction depends on pair of belt and pulley material. It is given in Table 1.3.

Table 1.3 Coefficient of Friction for Belt Materials and Different Pulleys

Belt Material	Pulley Material			
	Wood	Cast Iron / Steel		
		Dry	Wet	Greasy
Leather oak tanned	0.30	0.25	0.20	0.15
Leather chrome tanned	0.40	0.35	0.32	0.22
Rubber	0.32	0.30	0.18	–
Canvas	0.23	0.20	0.15	0.12
Woven cotton	0.25	0.22	0.15	0.12

1.6 Flat Belt Specifications

To specify a belt fully, one has to specify the following:

- Material of belt
- Number of plies for flat belt or thickness
- Maximum tensile strength

- Width of flat belt (cross-section size for V belts)
- Power rating in watts per ply per mm width
- Length at pitch diameters of the pulleys
- Timing belts in addition, require size of the teeth.

a. Material

Selection of material to be used depends on the type of service, that is, duty hours per day, smooth or shocks, type of environment; dusty, oily, moist, etc.

b. Number of ply and thickness

Number of plies is decided depending on the belt tensile strength required for a given power transmission. Standard belt thicknesses are 5, 6.5, 8, 10, and 12 mm.

c. Maximum belt stress per unit width

The belts are subjected to only tensile load. The allowable tensile load depends on the allowable stress on the belt and its cross-sectional area. It is customary to provide the belt stress value for a given belt thickness and per unit belt width.

d. Width

A designer has to select a belt thickness and then calculate the required belt width. Alternately, one can calculate the belt cross-sectional area and then adjust the belt thickness and the width from the standard widths. Values of standard belt widths are as under:

In R10 series: Width starts from 25 to 63 mm (25, 32, 40, 50, and 63).

In R20 series: Width starts from 71 to 600 mm (71, 80, 90, 100, 112, 125, 140, 160, 180, 200, 224, 250, 280, 315, 355, 400, 450, 500, 560, and 600).

The maximum belt stress also depends on the belt speed. Hence, the maximum belt stress is provided either for different belt speeds or for a specified speed.

1.7 Flat Belt Joints

Strips of the belt material are made into a loop by joining the ends of strip by different methods as described below:

Cemented Leather and fabric belts are joined with a cement by making the ends inclined. The angle is about 1 in 20 (Figure 1.10). The joint can offer strength up to 85 per cent.

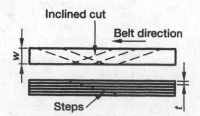

Figure 1.10 Cemented edges

Laced Ends of belt are cut squared so that they butt tightly together. Starting from each end, holes are punched. One line of holes is about 20 mm from each end as shown in Figure 1.11(a) and picture on the right. Second set of holes is about 60 mm from the end. The holes are staggered so that the belt does not become weak at the hole section. Finally, a single centered hole is punched 75 mm from the squared end for the knots. Now, similar holes are punched in the second edge as the mirror image of the first end the belt. Butted ends together will show a diamond shape of holes.

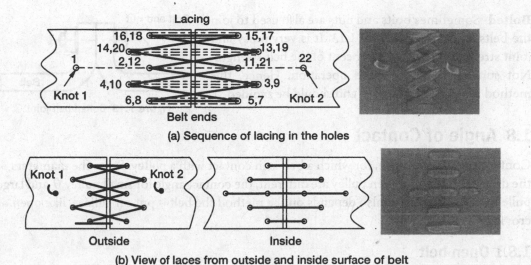

Figure 1.11 Lacing of flat belts

The far single hole is numbered 1. A knot is put at hole 1 and then lacing is done in the sequence numbered as 2, 3, 4, ... , 22 using a leather lace. Again, a knot is put at hole 22. Figure 1.11(b) shows views of lacing from outside and inside surface of the belt.

Efficiency of laced joint depends on how the joint is made. Raw hide lace gives joint efficiency of 60–70 per cent, wire laced by hand 70–80 per cent and if laced by machine, it is 75–85 per cent.

Hinge comb A metallic comb with sharp edges is pressed into the belt at the ends. The comb has a semicircular bend as shown in Figure 1.12(a). When two ends of the belts mate each other, the comb forms a circular hole, through which a pin is passed to make a hinged joint. See Figure 1.12(b).

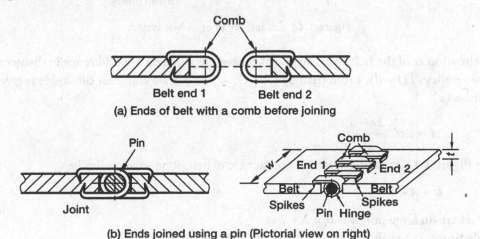

Figure 1.12 Hinged belt joint

Bolted Sometimes bolts and nuts are also used to join the belts as shown in Figure 1.13. It is very noisy and joint strength is hardly 25 per cent of the belt strength. Not suitable for high speed operation. Hence, this method is not recommended and should be avoided.

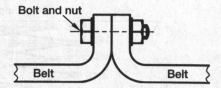

Figure 1.13 Bolted belt joint

1.8 Angle of Contact

Contact angle is the angle, for which a belt is in contact with a pulley. Since the diameters of the driver pulley and driven pulley are different, the contact angle for small pulley θ_s and large pulley θ_L are different. It also depends on the method the belt is put on pulleys, like open or crossed.

1.8.1 Open belt

A small pulley of diameter d with its center O_1 and a large pulley of diameter D with its center O_2 are shown in Figure 1.14 placed at a center distance C. Contact arc of belt, on small pulley is θ_s and on large pulley θ_L.

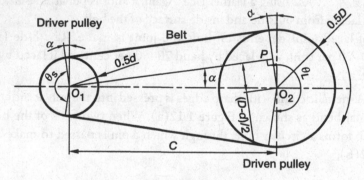

Figure 1.14 Contact arc in open belt system

Inclination α of the belt with the line of centers depends on the difference in diameters of the two pulleys $(D - d)$. From triangle $O_1 - O_2 - P$, it can be seen that this angle is given by Equation (a)

$$\alpha = \sin^{-1}\frac{D-d}{2C} \tag{a}$$

From Figure 1.14, it can be seen that contact arc of belt, θ_s on small pulley is:

$$\theta_s = \pi - 2\alpha$$

Contact arc on large pulley is $\theta_L = \pi + 2\,\alpha$.

Substituting the value of α from Equation (a) in equations above:

Angle on small pulley $\theta_s = \pi - 2\sin^{-1}\dfrac{D-d}{2C}$ \hfill (1.1)

Angle on large pulley $\theta_L = \pi + 2\sin^{-1}\dfrac{D-d}{2C}$ \hfill (1.2)

1.8.2 Cross belt

A cross belt on two pulleys is shown in Figure 1.15.

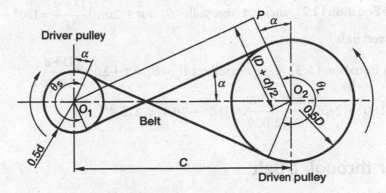

Figure 1.15 Contact arc in crossed belt system

It can be seen that angle of inclination α is now given by Equation (1.3).

$$\alpha = \sin^{-1}\frac{D+d}{2C}$$ \hfill (b)

Hence, angle of contact on both pulleys:

$$\theta_L = \theta_s = \pi + 2\alpha$$

Substituting the value of α from Equation (b):

$$\theta_L = \theta_s = \pi + 2\sin^{-1}\frac{D+d}{2C}$$ \hfill (1.3)

Example 1.1

Angle of contact of a belt over a given set of pulleys

Two pulleys of diameter 400 mm and 850 mm with center distance 2 m are driven by a belt. Calculate angle of contacts of belt for small and large pulleys for following types of drives:

a. Open belt

b. Crossed belt

Solution

Given $d = 400$ mm $D = 850$ mm $C = 2$ m

   a. Open belt

From Equation (1.1), angle on small pulley $\theta_s = \pi - 2\sin^{-1}\dfrac{D-d}{2C}$

$$\theta_s = 180° - 2\sin^{-1}\frac{850 - 400}{2 \times 2{,}000} = 180° - 13° = 167°$$

From Equation (1.2), angle on large pulley $\theta_L = \pi + 2\sin^{-1}\dfrac{D-d}{2C} = 180° + 13° = 193°$

   b. Crossed belt

From Equation (1.3), angle on pulleys $\theta_L = \theta_s = \pi + 2\sin^{-1}\dfrac{D+d}{2C}$

$$= 180° + 2\sin^{-1}\frac{850 + 400}{2 \times 2{,}000} = 180° + 36.4° = 216.4°$$

1.9 Power through a Belt

Power transmission is a function of belt tensions and belt velocity. However, increasing tension increases stress in the belt and load on bearings. The ideal belt is that, which with lowest tension does not slip at high loads.

Net force acting on the pulley $= T_1 - T_2$

Where, $T_1 =$ Belt tension in tight side

     $T_2 =$ Belt tension in slack side

The torque on pulley due to this force $T = (T_1 - T_2) \times \dfrac{d}{2}$

This force acts at pitch diameter d. Belt speed is to be taken at pitch circle diameter. Pitch line is the line between the inner and outer surfaces of the belt, which is subjected to neither tension nor compression. It is midway through the surfaces in flat belts and depends on the cross-sectional shape and size of V belts.

If N is the speed of pulley in rpm., substituting the value of T from equation above, power P in watts is given as:

$$P = \frac{2\pi N T}{60} = \frac{\pi d N (T_1 - T_2)}{60} \tag{1.4}$$

Belt speed $v = \dfrac{\pi d N}{60}$

Substituting this value in Equation (1.4): $P = (T_1 - T_2)v$ (1.5)

1.10 Belt Tensions

The belt drives primarily operate on the friction between the belt and the pulley and are responsible for transmitting power from one pulley to the other. The driving pulley gives motion to the belt and the motion of the belt is transmitted to the driven pulley. Owing to the presence of friction between the pulley and the belt surfaces, tensions on both the sides of the belt are not equal. So it

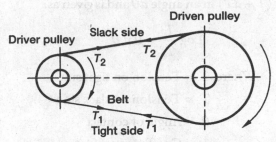

Figure 1.16 Tight and slack side of belt

is important that one has to identify the higher tension (tight) side and the lesser tension (slack) side, shown in Figure 1.16.

Equilibrium of the belt segment suggests that T_1 is higher than T_2. Here, T_1 is referred as the tight side and T_2 as the slack side. Slack side of the belt is kept in the upper side and the tight side of the belt is in the lower side as shown in Figure 1.17(a). The slack side of the belt, due to self-weight, is not in a straight line and sags and the angle of contact increases. However, the tight side does not sag to that extent. Hence, the net effect is an increase of the angle of contact or angle of wrap. It will be shown later, that due to the increase in angle of contact, the power transmission capacity of the drive increases.

On the other hand, if the slack side is on the lower side and the tight side is on the upper side as shown in Figure 1.17(b), for the same reason given above, angle of wrap decreases and the power transmission capacity decreases. Hence, in case of horizontal drive system the tight side is on the lower side and the slack side is always on the upper side. If the drive is vertical, slack side does not affect the power transmission.

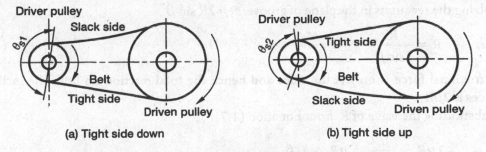

Figure 1.17 Contact arc variation with tight side down and up

1.11 Belt Tensions Ratio (Capstan Equation)

1.11.1 Flat belts

Figure 1.18 shows a belt over a small segment AB of a pulley. The equation, which relates tensions of tight and slack side, is given by Capstan Equation (1.6).

The tension in the belt increases from T to $(T + dT)$ in an angle $d\theta$ and is given as:

$$\text{Log}\, \frac{T_1}{T_2} = \mu\, d\, \theta$$

Where, T_1 = Tension on tight side

T_2 = Tension on slack side

θ = Angle of contact

μ = Coefficient of friction

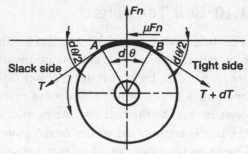

Figure 1.18 Tensions in a belt due to friction

Or, $\quad T_1 = T_2 e^{\mu\theta}$ \hfill (1.6)

1.11.2 V belts

A V belt uses a grooved pulley. Included angle of the trapezoidal section of all V belt is 40°, whereas the groove angle for the pulley varies from 32° to 38°. This reduction in angle provides more wedging action for a firm grip. Smaller the pulley, lesser is the angle. Small pulley groove angle lies between 32° and 34°, whereas for large pulley, it varies from 36° to 38°. Figure 1.19 shows an enlarged view of the groove.

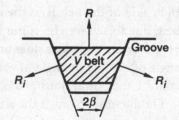

Figure 1.19 Belt reactions in a V groove

Let R = Radial reaction in the plane of groove.

R_i = Inclined reaction between the sides of groove and belt.

β = Angle of inclined face of the groove with the center line (2β is groove angle).

Resolving the reactions in the plane of groove: $R = 2R_i \sin\beta$

Or, $\quad R_i = \dfrac{R}{2\sin\beta}$ \hfill (1.7)

The frictional force is only at the sides and hence the total reaction on both the inclined surfaces is $2\mu R_i$.

Substituting the value of R_i from Equation (1.7):

$$2\mu R_i = \frac{\mu R}{\sin\beta} = \mu R \operatorname{cosec}\beta$$

Like flat belts, tension ratio for the V belt can be found as:

$$T_1 = T_2 e^{\mu\,\theta\,\operatorname{cosec}\beta}$$ \hfill (1.8)

So it is concluded that a V belt offers ($\operatorname{cosec}\beta$) times more grip then flat belt.

1.12 Initial Tension

Power transmitted by belt depends on the firmness of the belt over the pulley. To increase the grip, the belt is initially tightened either by an idler or by sliding one of the shaft to increase the center distance. Thus some tension called as initial tension T_0 exists, even when the pulleys are not rotating.

When pulleys start rotating, the driver pulley pulls the belt and the tension changes on tight side. Thus tension on tight side is $(T_1 - T_0)$.

If α is coefficient of increase in length of belt for a unit force, increase in length is given as:

$$\delta L = \alpha(T_1 - T_0)$$

Assuming the length of belt remains same, decrease in length on slack side will also be same as δL and is given by relation:

$$\delta L = \alpha(T_0 - T_2)$$

Equating both the equations:

$$\alpha(T_1 - T_0) = \alpha(T_0 - T_2)$$

or $\qquad (T_1 - T_0) = (T_0 - T_2)$

or $\qquad T_0 = \dfrac{T_1 + T_2}{2}$ $\qquad\qquad\qquad\qquad$ (1.9)

However, due to elastic properties of the belt, it has been found practically that:

$$T_0 = \left(\frac{\sqrt{T_1} + \sqrt{T_2}}{2}\right)^2 \qquad\qquad\qquad\qquad (1.10)$$

Example 1.2

Belt tensions for flat and V belts

A pulley drive transmits 5 kW at 1,440 rpm from a pulley of 300 mm diameter to another pulley of 600 mm placed at a distance of 1,700 mm. Assuming coefficient of friction 0.3 find:

a. Belt tensions, if flat belt is used.

b. Belt tensions, if V belt is used. Assume groove angle 36° and pitch circle same as for flat pulleys.

c. Initial tension required for a flat belt.

Solution

Given $\quad P = 5\,\text{kW} \qquad N1 = 1,440\,\text{rpm} \qquad d = 300\,\text{mm} \qquad D = 600\,\text{mm}$
$\qquad\qquad\quad C = 1,700\,\text{mm} \qquad \mu = 0.3 \qquad\quad 2\beta = 36°$

a. Flat belt

Angle on small pulley $\theta_s = \pi - 2\sin^{-1}\dfrac{D-d}{2C}$

Or, $\theta_s = 180° - 2\sin^{-1}\dfrac{600-300}{2 \times 1,700} = 180° - 10° = 170° = 2.965$ rad

Belt velocity $v = \dfrac{\pi d N}{60} = \dfrac{3.14 \times 0.3 \times 1,440}{60} = 22.6\,\text{m/s}$

From Equation (1.5), $P = (T_1 - T_2)\,v$ or $5,000 = (T_1 - T_2) \times 22.6$

Or, $(T_1 - T_2) = 221.2\,\text{N}$ (a)

From Equation (1.6), $T_1 = T_2 e^{\mu\theta} = T_2^{0.3 \times 2.965} = T_2 e^{0.8895} = 2.43\,T_2$
Using this value in Equation (a):

$(2.43T_2 - T_2) = 221.2\,\text{N}$ or $T_2 = 154.3\,\text{N}$

$T_1 = 2.43\,T_2 = 2.43 \times 154.3 = 375\,\text{N}$

b. V belt

Angle of contact will remain same, that is, 2.965 rad

From Equation (1.8), $T_1 = T_2 e^{\mu\theta \cosec\beta}$

$T_1 = T_2 e^{0.3 \times 2.965 \times \cosec(18°)} = T_2 e^{2.878} = 17.78T_2$

From Equation (a),

$(17.78T_2 - T_2) = 221.2\,\text{N}$ or $T_2 = 13.2\,\text{N}$

$T_1 = 17.78 \times T_2 = 17.78 \times 13.2 = 237\,\text{N}$

c. Initial tension

Using Equation (1.9), initial tension $T_0 = \dfrac{T_1 + T_2}{2} = \dfrac{375 + 154.3}{2} = 264.6\,\text{N}$

Using Equation (1.10), initial tension considering elastic properties is:

$T_0 = \left(\dfrac{\sqrt{T_1} + \sqrt{T_2}}{2}\right)^2$

$T_0 = \left(\dfrac{\sqrt{375} + \sqrt{154.3}}{2}\right)^2 = \left(\dfrac{19.36 + 12.42}{2}\right)^2 = 252\,\text{N}$

1.13 Centrifugal Tension

When pulleys start rotating, belt also rotates over them, which causes centrifugal force F_c on tight and slack sides. It may not be appreciable, if belt velocity is less than 10 m/s, but it becomes significant for higher belt velocities and cannot be neglected.

A small segment of belt AB over a pulley is shown in Figure 1.20, which extends an angle $d\theta$ at the center of pulley. Mass of this portion of belt causes centrifugal force, when the belt passes over the pulley. If r is the radius of pulley:

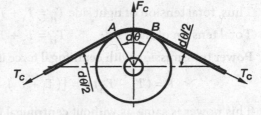

Figure 1.20 Centrifugal tensions

Length of belt AB $= r\, d\theta$

Volume of flat belt $= b\, t\, r\, d\theta$

Mass of belt $= \rho\, b\, tr\, d\theta$

Centrifugal force $= \rho\, b\, tr\, d\theta \dfrac{v^2}{r}$

Where, b = Width of belt

t = Thickness of belt

ρ = Density of belt material

For V belts, mass per unit length can be taken as m:

Mass of belt length AB $= mr\, d\theta$

Centrifugal force $\quad F_c = mr\, d\theta \dfrac{v^2}{r}$

This force T_c is balanced by tension in the belt on slack and tight side. It may be noted from Figure 1.20, that belt makes an angle $d\theta / 2$.

Horizontal component of belt tension gets cancelled, whereas vertical component provides reaction to centrifugal force. Equating these forces:

$$F_c = mr\, d\theta \frac{v^2}{r} = 2T_c \sin \frac{d\theta}{2}$$

For small angles, $\sin \dfrac{d\theta}{2} = \dfrac{d\theta}{2}$

Hence, $\quad mr\, d\theta \dfrac{v^2}{r} = 2T_c \dfrac{d\theta}{2}$

Or, $\quad T_c = mv^2$ $\qquad\qquad\qquad\qquad\qquad$ (1.11)

Where, T_c = Centrifugal force in N

m = Mass of belt in kg per meter length

v = Belt velocity in m/s.

This centrifugal tension is added to both tension of tight and slack side.

Thus, total tension of tight side $T_{t1} = T_1 + T_c$

Total tension of slack side $\quad T_{t2} = T_2 + T_c$

Power transmission with centrifugal force is:

$$P = (T_{t1} - T_{t2})\,v = [(T_1 + T_c) - (T_2 + T_c)]\,v = (T_1 - T_2)v$$

This power is same as without centrifugal force. This proves that centrifugal tension has no effect on power transmission.

Ratio of tensions given by Capstan equation (1.6) is modified with centrifugal tensions as:

$$\text{Log}\,\frac{T_{t1} - T_c}{T_{t2} - T_c} = \mu d\theta \qquad\qquad (1.12)$$

1.14 Maximum Tension

Maximum tension T_{max} that a belt can take depends on safe stress of the belt material.

If, σ = Safe stress

A = Area of belt

t = Thickness of belt

Area for a flat belt: $A = b \times t$

Area for a V belt: $\quad A = \dfrac{[w + \{w - (2t \times \tan 20°)\}]}{2} \times t$

Then, maximum tension $\qquad\qquad T_{max} = \sigma \times A$

Maximum tension with centrifugal force: $T_{max} = T_1 + T_c \qquad (1.13)$

If centrifugal force is not considered: $\quad T_{max} = T_1$

Example 1.3

Angle of wrap and belt tensions with centrifugal force in cross belt

A flat belt of width 200 mm transmits 60 kW at a belt speed of 25 m/s. Belt mass is 2 kg/m of belt length. The belt drives two pulleys in crossed arrangement of diameters 300 mm and 900 mm at a center distance of 6 m. Calculate:

a. Angle of wrap over small pulley.

b. Belt tensions, if coefficient of friction is 0.38.

Solution

Given $b = 200$ mm $P = 60$ kW $v = 25$ m/s $m = 2$ kg/m

$d = 300$ mm $D = 900$ mm $C = 6$ m $\mu = 0.38$

a. Initial tension not given, hence taken it as zero.

$$\alpha = \sin^{-1}\frac{D+d}{2C} = \sin^{-1}\frac{900 + 300}{2 \times 6,000} = 5.75° = 0.1 \text{ rad}$$

From Equation (1.3), wrap angle $\theta_L = \theta_s = \pi + 2\sin^{-1}\dfrac{D+d}{2C}$

$$= 180° + (2 \times 5.75) = 180° + 11.5° = 191.5° = 3.34 \text{ rad}.$$

b. From Equation (1.11), centrifugal force $T_c = m\,v^2 = 2 \times 25 \times 25 = 1,250$ N.

From Equation (1.12):

$$\log\frac{T_{t1} - T_c}{T_{t2} - T_c} = \mu\theta \quad \text{Or,} \quad \frac{T_{t1} - 1,250}{T_{t2} - 1,250} = e^{\mu\theta} = e^{0.38 \times 3.34} = 3.56$$

Or, $T_{t1} - 1,250 = 3.56\,T_{t2} - 4,450$ or $T_{t1} = 3.56\,T_{t2} - 3,200$ (a)

Also from Equation (1.5):

$$P = (T_{t1} - T_{t2})\,v \quad \text{Or,} \quad 60,000 = (T_{t1} - T_{t2}) \times 25 \quad \text{or} \quad (T_{t1} - T_{t2}) = 2,400 \text{ N}.$$

Substituting the values from Equation (a):

$$3.56T_{t2} - 3,200 - T_{t2} = 2,400 \quad \text{or} \quad T_{t2} = 2,187.5 \text{ N}$$

From Equation (a), $T_{t1} = (3.56 \times 2187.5) - 3,200 = 4,587.5$ N

1.15 Condition for Maximum Power

The power that can be transferred by belt depends on tensions in belt, coefficient of friction, and belt velocity. To find a condition for maximum power, use the following relations for power and tensions:

$$P = (T_1 - T_2)\,v \quad \text{and} \quad T_1 = T_2 e^{\mu\theta}$$

Substituting the value of T_2 in power equation: $\quad P = \left(T_1 - \dfrac{T_1}{e^{\mu\theta}}\right)v = Cv\,T_1$

Where, $C = \left(1 - \dfrac{1}{e^{\mu\theta}}\right)$

If T_{max} is the maximum tension, which a belt can take, $T_1 = (T_{max} - T_c)$

Hence, $\quad P = Cv(T_{max} - T_c)$

Because, $\quad T_c = mv^2$,

Hence, $\quad P = C(T_{max}v - mv^3)$ $\hfill (1.14)$

To get the condition for maximum power for a given velocity, $\dfrac{dP}{dv} = 0$

Differentiating Equation (1.14) with respect to v:

$$T_{max} - 3mv^2 = 0 \quad \text{Or,} \quad v = \sqrt{\dfrac{T_{max}}{3m}} \hfill (1.15)$$

1.16 Slip of Belt

In power transmission, it is assumed that there is a firm grip between the pulley and belt due to friction. But generally, the grip is not sufficient and there is slight slipping of the belt over the pulley. This is called slip of belt.

Let s_1 = Slip between the driver pulley and belt,

s_2 = Slip between belt and the driven pulley, and

s = Total slip between driver and driven pulley $(s_1 + s_2)$

Pitch line velocity of the driver pulley, $v = \dfrac{\pi\, dN_1}{60}$

Velocity of belt v will reduce due to slip s_1 and will be equal to v' as given below:

$$v' = \dfrac{\pi dN_1}{60} - \left[\dfrac{\pi dN_1}{60} \times \dfrac{s_1}{100}\right] \quad \text{Or,} \quad v' = \dfrac{\pi dN_1}{60}\left[1 - \dfrac{s_1}{100}\right] \hfill (1.16)$$

Velocity of belt passing over the driven pulley will be same as v' but due to slip, the driven pulley speed will reduce, that is:

$$\dfrac{\pi dN_2}{60} = v'\left[1 - \dfrac{s_2}{100}\right] \quad \text{Or,} \quad \dfrac{\pi dN_2}{60} = \dfrac{\pi dN_1}{60}\left[1 - \dfrac{s_1}{100}\right]\left[1 - \dfrac{s_2}{100}\right]$$

Or, $\quad \dfrac{N_2}{N_1} = \dfrac{d}{D}\left[1 - \dfrac{s_1}{100}\right]\left[1 - \dfrac{s_2}{100}\right]$ $\hfill (1.17)$

Since values of s_1 and s_2 are small, hence $s_1 \times s_2$ can be neglected. Therefore, velocity ratio with slip is as under:

$$\dfrac{N_2}{N_1} = \dfrac{d}{D}\left[1 - \dfrac{s_1}{100} - \dfrac{s_2}{100}\right] = \dfrac{d}{D}\left[1 - \dfrac{s_1 + s_2}{100}\right]$$

Or, $\quad \dfrac{N_2}{N_1} = \dfrac{d}{D}\left[1 - \dfrac{s}{100}\right]$ $\hfill (1.18)$

If belt is thin, its thickness t can be neglected. But, if thickness t is also considered, pitch diameters increase by belt thickness:

$$\frac{N_2}{N_1} = \frac{d+t}{D+t}\left[1 - \frac{s}{100}\right] \tag{1.19}$$

1.17 Creep of Belt

It has been described that one side of the belt has more tension called tight side and other side has less tension called slack side. Thus, when belt passes over the pulley, it has to undergo from cycles of more tension to less tension. When tension increases, the belt is stretched according to its young's modulus of elasticity E and causes some relative motion between belt and pulley. This relative motion is called creep. The net effect is that the speed of the driven pulley decreases slightly. The velocity ratio is given by the following equation:

$$\frac{N_2}{N_1} = \frac{d}{D} \times \frac{E+\sqrt{\sigma_2}}{E+\sqrt{\sigma_1}} \tag{1.20}$$

Where, N_1, N_2 = Speed of driver and driven respectively

E = Young's modulus of elasticity for the belt material

σ_1, σ_2 = Stress in tight side of the belt and slack side respectively

1.18 Length of Belt

If center distance C between the two shafts is given, pitch length of belt L_p for open and cross belts is given by the following relations:

Open belt:

Length of belt L_p = Length of belt over small pulley + length of belt over large pulley + twice the inclined length of belt between two pulleys

If θ is the angle of wrap, actual length of the belt is given by $r\,\theta$, that is,

$$L_p = \frac{D(\pi+\alpha)}{2} + \frac{d(\pi-\alpha)}{2} + 2C\cos\alpha \tag{1.21}$$

For small values of $\alpha = \sin\alpha = \dfrac{D-d}{2C}$

Hence, $\cos\alpha = 1 - 2\sin^2\left(\dfrac{\alpha}{2}\right) = 1 - \dfrac{2(D-d)^2}{4C^2}$

Substituting the value of α and $\cos \alpha$ in Equation (1.21) and rearranging the terms:

$$L_p = \frac{\pi(D+d)}{2} + \frac{(D-d)^2}{4C} + 2C\left[1 - \frac{2(D-d)^2}{4C^2}\right]$$

Or, $$L_p = 2C + \frac{\pi}{2}(D+d) + \frac{(D+d)^2}{4C} \qquad (1.22)$$

Similarly for cross belt: $L_p = 2C + \frac{\pi}{2}(D+d) + \frac{(D+d)^2}{4C} \qquad (1.23)$

If pitch length of belt L_p is known, center distance C for open belt can be calculated from Equation (1.22). Multiplying by $4C$ on both sides and rewriting Equation (1.22):

$$4CL_p = 8C^2 + 2C\pi(D+d) + (D-d)^2$$

Or, $$8C^2 - 4CL_p + 2C\pi(D+d) + (D-d)^2 = 0$$

Or, $$C^2 - 2C\left(\frac{L_p}{4} - \frac{\pi(D+d)}{8}\right) + \frac{(D-d)^2}{8} = 0$$

Let, $$A = \frac{L_p}{4} - \frac{\pi}{8}(D+d) \qquad (1.24)$$

$$B = \frac{(D-d)^2}{8} \qquad (1.25)$$

The above quadratic equation becomes:

$$C^2 - 2AC + B = 0$$

Positive value of the solution of the equation is:

$$C = A + \sqrt{A^2 - B} \qquad (1.26)$$

A V belt is specified by its section and inside length L. Its length at its neutral axis is more due to its thickness. Relation between pitch length L_p and inside length L is:

For A belt	$L_p = L + 36$ mm	For B belt	$L_p = L + 43$ mm
For C belt	$L_p = L + 56$ mm	For D belt	$L_p = L + 79$ mm
For E belt	$L_p = L + 92$ mm		

For example, a belt specified as C 1,524 has section C and inside length 1,524 mm. Its pitch length is $1,524 + 56 = 1,580$ mm. Standard length L and pitch length L_p are given in Table 1.4.

Table 1.4 Standard V Belt Inside Length L and Pitch Length L_p in mm [IS 2494–1974]

	L	610	680	711	787	830	889	914	965	991	1016	1,067	1,092
	L_p	646	696	747	828	848	925	950	1,001	1,026	1,051	1,102	1,128
A	L	1,168	1,219	1,295	1,397	1422	1473	1,524	1,600	1,626	1,651	1,727	1,778
	L_p	1,204	1,255	1,331	1,433	1,458	1,509	1,560	1,636	1,661	1,687	1,763	1,814
	L	1,905	1,981	2,032	2,057	2,159	2,286	2,438	2,667	2,845	3,048	3,251	3,658
	L_p	1,941	2,017	2,068	2,096	2,195	2,322	2,474	2,703	2,880	3,084	3,287	3,693
B	L	889	965	1,016	1,067	1,168	1,219	1,295	1,397	1,372	1,422	1,524	1,651
	L_p	9,362	1,008	1,059	1,110	1,212	1,262	1,339	1,415	1,440	1466	1,567	1,694
	L	1,727	1,778	1,905	1,981	2,057	2,159	2,286	2,464	2,540	2,667	2,845	3,048
	L_p	1,770	1,824	1,948	2,024	2,101	2,202	2,329	2,507	2,583	2,710	2,888	3,091
	L	3,251	3,658	4,013	4,115	4,394	4,572	4,953	5,334	-	-	-	-
	L_p	3,294	3,701	4,056	4,158	4,437	4,615	4,996	5,377	-	-	-	-
C	L	1,295	1,524	1,727	1,905	2,057	2,159	2,286	2,438	2,667	2,845	3,048	3,150
	L_p	1,351	1,580	1,783	1,961	2,113	2,215	2,342	24,94	2,723	2,901	3,104	3,206
	L	3,251	3,404	3,658	4,013	4,115	43,94	4,572	4,953	5,334	6,045	6,807	7,569
	L_p	3,307	3,460	3,714	4,069	4,171	4,450	4,628	5,009	5,390	6,101	6,863	7,625
	L	8,331	9,093	-	-	-	-	-	-	-	-	-	-
	L_p	8,387	9,149	-	-	-	-	-	-	-	-	-	-
D	L	3,048	3,251	3,658	4,013	4,115	4,394	4,572	4,953	5,334	6,045	6,807	7,569
	L_p	3,127	3,330	3,737	4,092	4,194	4,473	4,651	5,032	5,413	6,124	6,886	7,648
	L	8,331	9,093	9,855	10,617	12,141	13,655	15,189	16,713	-	-	-	-
	L_p	8,410	9,172	9,934	10,696	12,220	13,744	15,268	16,792	-	-	-	-
E	L	5,334	6,045	6,807	7,569	8,331	9,093	9,855	10,617	12,141	13,655	15,189	16,713
	L_p	5,426	6,137	6,899	7,661	8,423	9,168	9,947	10,709	12,233	13,757	15,281	16,805

Example 1.4

Width and length of flat belt drive for the given power

A pump is driven by an electric motor using an open type flat belt. Determine the width and length of belt for the following data:

Power transmission = 15kW

Rotational speed of the motor = 1,440 rpm

Motor pulley diameter = 250 mm

Pump pulley diameter = 500 mm

Coefficient of friction for motor pulley = 0.26

Coefficient of friction for pump pulley = 0.22

Center distance between the pulleys = 1,200 mm

Thickness of the belt = 5 mm

Density of belt material = 1,000 kg/m³

Allowable stress for the belt material = 2 MPa

Solution

Given $P = 15$ kW $N1 = 1,440$ rpm $d = 250$ mm $D = 500$ mm

$\mu1 = 0.26$ $\mu2 = 0.22$ $C = 1,200$ mm $t = 5$ mm

$\rho = 1,000$ kg/m³ $\sigma = 2$ MPa

$$\alpha = \sin^{-1}\left(\frac{D-d}{2C}\right) = \sin^{-1}\left(\frac{500-250}{2 \times 1,200}\right) = 6°$$

Arc of contact of small pulley $\theta_s = 180° - (2 \times 6) = 168° = 2.93$ rad

Arc of contact of large pulley $\theta_L = 180° + (2 \times 6) = 192° = 3.37$ rad

Velocity of belt $v = \dfrac{\pi \, dN_1}{60} = \dfrac{\pi \times 0.25 \times 1,440}{60} = 18.84$ m/s

Mass of belt per meter, $m = bt\rho = \dfrac{b}{10^3} \times \dfrac{5}{10^3} \times 1,000 = 0.005b$ kg/m

Centrifugal force $T_c = mv^2 = 0.005\, b \times 18.84^2 = 1.77\, b$ N

$\mu1 \times \theta_s = 0.26 \times 2.93 = 0.762$

$\mu2 \times \theta_L = 0.22 \times 3.37 = 0.741$

Lesser of the two values, that is, 0.741 will be chosen to be on the safe side. Using Equation (1.12),

$$\frac{T_1 - 1.77b}{T_2 - 1.77b} = e^{0.741} = 2.09 \tag{a}$$

Power is given by equation:

$$P = (T_1 - T_2) \times v$$

Hence, $\left(T_1 - T_2\right) = \dfrac{15,000}{18.84} = 796$ N $\tag{b}$

As per allowable stress $T_1 = \sigma b\, t = 2\, b \times 5 = 10b$ N $\tag{c}$

From Equations (b) and (c):

$$(10b - T_2) = 796 \quad \text{or} \quad T_2 = 10b - 796 \text{ N}$$

Substituting values of T_1 and T_2 in Equation (a):

$$\frac{10b - 1.77b}{\left(10b - 796\right) - 1.77b} = 2.09 \quad \text{Or,} \quad \frac{8.23b}{8.23b - 796} = 2.09$$

Or, $\quad b = 185$ mm

From Equation (1.22), length for open belt, $L_p = 2C + \frac{\pi}{2}(D + d) + \frac{(D-d)^2}{4C}$

Substituting the values:

$$L_p = (2 \times 1,200) + 1.57(500 + 250) + \frac{(500 - 250)^2}{4 \times 1,200}$$

Or, $\quad L_p = 2,400 + 1,178 + 13 = 3,591$ mm

Example 1.5

Length of V belt for the given center distance and size of pulleys
A pulley drive transmits powers using C belt, through two pulleys of diameters 800 mm and 300 mm, which are to be mounted on shafts approximately 1.8 m apart.
  a. Find a standard belt size for open belt system.
  b. Exact center distance for the selected belt.

Solution

Given $\quad D = 800$ mm $\quad d = 300$ mm $\quad C = 1.8$ m

Belt section $= C$ Open belt

  a. Using Equation (1.22), Pitch length $L_p = 2C + \frac{\pi}{2}(D + d) + \frac{(D-d)^2}{4C}$

Substituting the values: $L_p = (2 \times 1.8) + 1.57(0.8 + 0.3) + \frac{(0.8 - 0.3)^2}{4 \times 1.8}$

$$= 3.6 + 1.727 + 0.035 = 5.362 \text{ m} = 5,362 \text{ mm.}$$

From Table 1.4 for C belt, nearest standard length L is 5,334, for which pitch length L_p is 5,390 mm.

  b. Using Equation (1.24),

$$A = \frac{L_p}{4} - \frac{\pi}{8}(D+d) = \frac{5.390}{4} - 0.392(0.8 + 0.3) = 0.9092$$

Using Equation (1.25), $B = \dfrac{(D-d)^2}{8} = \dfrac{(0.8-0.3)^2}{8} = 0.031$

Using Equation (1.26),

Center distance $C = A + \sqrt{A^2 - B} = 0.9092 + \sqrt{0.9092^2 - 0.031} = 1{,}801$ mm

1.19 Design of Flat Belt Drive

For the design of belt drive, usually following data are given:

- Power to be transmitted (P)
- Type of service for the drive (light, medium, heavy, and working hours per day)
- Driver speed (N_1)
- Driven speed (N_2) or velocity ratio (VR $= N_1 / N_2$)

It is required to find:

- Diameter of small pulley (d) and large pulley (D)
- Width of belt (b) and its thickness (t)
- Center distance between the shafts (C)

Following design procedure can be adapted for the belt drive:

Step 1   Calculate pulley diameters

Assume a belt velocity of 18 m/s and calculate diameter of small pulley using relation:

$$d = \frac{60 \times 1{,}000\, v}{\pi N_1} \tag{1.27}$$

Alternately, diameter of small pulley can be calculated using following empirical relation:

Diameter of small pulley in millimeters, $d = (1{,}100 \text{ to } 1{,}300)\left(\dfrac{P}{N_1}\right)^{1/3}$ (1.28)

Where, P = Power in kW

N_1 = Speed of small pulley in rpm

Select preferred pulley diameter from Table 1.5.

Table 1.5 Standard Flat Pulley Diameters (mm)

20	22	25	28	32	36	40	45	50	56	63	71
80	90	100	112	125	140	160	180	200	224	250	280
315	355	400	450	500	560	630	710	800	900	1,000	1,120
1,250	1,400	1,600	1,800	2,000	2,240	2,500	2,800	3,150	3,550	4,000	5,000

Again calculate actual belt velocity v with the standard pulley diameter d : $v = \dfrac{\pi d N_1}{60}$

Calculate large pulley diameter D with slip (s) of about 2–5 per cent from Equation (1.18).

$$\frac{N_2}{N_1} = \frac{d}{D}\left[1 - \frac{S}{100}\right]$$

Step 2 Calculate design power

To consider service conditions of an application, service factor K_s is used. It is a multiplier, applied to the required power to determine the design power. Calculate design power P_d by considering a suitable service factor K_s from Table 1.6.

$$P_d = P \times K_s \tag{1.29}$$

Table 1.6 Service Factors (K_s) for Various Applications

Service Duty	K_s for Duty Hours per Day		
	<10 Hours	10 – 16 Hours	>16 Hours
Light duty	1.0 – 1.2	1.1 – 1.2	1.2 – 1.3
Medium duty	1.1 – 1.3	1.2 – 1.3	1.3 – 1.4
Heavy duty	1.2 – 1.4	1.3 – 1.5	1.4 – 1.6
Very heavy duty	1.3 – 1.5	1.4 – 1.6	1.5 – 1.8

Step 3 Calculate arc of contact

Contact arc angle θ_s for small pulley can be calculated using Equation (1.1) for open or Equation (1.3) for cross belt.

Step 4 Calculate corrected power

Power rating of the belts is given for 180° arc of contact. Actual arc of contact for smaller pulley is less than 180°. To account for this lesser angle, arc of contact factor K_a is used to modify the value. Table 1.7 gives the value of this factor for different values of arc of contact. Select value of this factor for the angle of contact calculated in step 2.

Table 1.7 Arc of Contact Factor for Contact Angles of Smaller Pulley

θ_s°	120	130	140	150	160	170	180	190	200
K_a	1.33	1.26	1.19	1.13	1.08	1.04	1.0	0.97	0.94

$$\text{Corrected power } P_c = P_d \times K_a = P \times K_s \times K_a \qquad (1.30)$$

Step 5 Calculate corrected power rating of belt

Power rating of the belts is specified per mm of belt width per ply. It depends on belt velocity. For belt speed of about 5 m/s, power rating varies between 0.012 and 0.015 kW/mm/ply.

$$\text{Calculate load rating at actual velocity} = \text{Load rating at 5 m/s} \times \left(\frac{v}{5}\right) \qquad (1.31)$$

Step 6 Calculate belt width and number of plies

$$\text{Belt width } b \times \text{number of plies } p = \frac{\text{Corrected power } P_c}{\text{Corrected power rating of belt}}$$

Use Table 1.8 to select a suitable combination of width and number of plies $(b \times p)$.

Table 1.8 Standard Width of Belts for Various Plies

No. of Plies (p)	Standard Belt Width (b) (mm)												
3	25	32	40	44	50	63	76	90	100	–	–	–	–	–
4	25	32	40	44	50	63	76	90	100	112	125	140	152	200
5	76	90	100	112	125	152	200	224	250	–	–	–	–	–
6	100	112	125	152	180	200	250	–	–	–	–	–	–	–
8	200	250	305	355	400	–	–	–	–	–	–	–	–	–

Step 7 Check pulley diameter for the number of plies

Pulley diameter should not be less than the recommended value given in Table 1.9 for the selected number of plies.

Table 1.9 Minimum Pulley Diameter (mm) for Given Speed and Number of Plies

No. of Plies	Belt Speed (m/s)				
	10	15	20	25	30
2	50	63	80	90	112
3	90	100	112	140	180
4	140	160	180	200	250
5	200	224	250	315	355
6	250	315	355	400	450
7	355	400	450	500	560
8	450	500	560	630	710
9	560	630	710	800	900
10	630	710	800	900	1,000

Example 1.6

Flat belt design for given power and center distance

15 kW motor running at 1,440 rpm drives a compressor at 600 rpm. Design an open belt for center distance of 2.5 m, service factor 1.3, and belt velocity approximately 18 m/s.

Solution

Given $P = 15$ kW $N_1 = 1,440$ rpm $N_2 = 600$ rpm $C = 2.5$ m $K_s = 1.3$

$$v = \frac{\pi d N_1}{60} \quad \text{or} \quad d = \frac{60\,v}{\pi \times N1}$$

Assuming belt velocity 18 m/s

$$d = \frac{60 \times 18,000}{\pi \times 1,440} = 239 \text{ mm}$$

Standard pulley $d = 250$ mm   (from Table 1.5)

Large pulley diameter $D = \dfrac{250 \times 1,440}{600} = 600$ mm (slip is not given, hence neglected)

Actual belt speed $v = \dfrac{\pi d N_1}{60} = \dfrac{\pi \times 0.25 \times 1,440}{60 \times 1,000} = 18.85$ m/s

Design power $P_d = P \times K_s = 15 \times 1.3 = 19.5$ kW

Arc of contact on small pulley $\theta_s = 180° - 2\sin^{-1}\left(\dfrac{D-d}{2C}\right) = 180° - (2 \times 4°) = 172°$

Interpolating arc of contact for 172° from Table 1.7: $K_a = 1.032$

Corrected power $P_c = 1.032 \times 19.5 = 20.14\,kW$

Assuming power rating varies between 0.012 kW/mm/ply.

Corrected belt rating $P_c = \dfrac{0.012 \times 18.85}{5} = 0.045$

$$b \times p = \dfrac{20.14}{0.045} = 462\,mm$$

For number of plies refer Table 1.8:

For $p = 4$, $b = 462/4 = 115.5\,mm$

For $p = 5$, $b = = 462/5 = 92.4\,mm$

So standard width 112 (which is near to 115.5) is selected.

From Equation (1.22),

belt length, $L_p = 2C + \dfrac{\pi}{2}(D + d) + \dfrac{(D - d)^2}{4C}$

Or, $\quad L_p = (2 \times 2.5) + \dfrac{\pi}{2}(600 + 250) + \dfrac{(600 - 250)^2}{4 \times 2.5} = 6{,}347\,mm$

So belt selected is 4 ply, 112 mm width and of pitch length 6,347 mm.

1.20 Center Distance

Center distance C of a drive depends on velocity ratio, that is, ratio of large and small pulley diameters; D and d given in Table 1.10.

Table 1.10 Selection of Center Distance C

D / d ratio	1	2	3	4	5	6–9
C / D ratio	1.5	1.2	1.0	0.95	0.9	0.85

Minimum center distance, $C_{min} = 0.55\,(D + d) + t$ (1.32)

Where, t = Belt thickness

Maximum center distance, $C_{max} = 2\,(D + d)$.   (1.33)

1.21 Power Rating of V Belts

The power, which a single V belt can transfer, depends on small pulley diameter d and belt velocity v. It can be calculated using the following formula:

$$P_r = v\left[\frac{a}{v^{0.09}} - \frac{b}{dK_p} - cv^2\right] \qquad (1.34)$$

Where, P_r = Power in kW

v = Belt velocity in m/s

d = Small pulley diameter in millimeters

a, b, c = Constants given in Table 1.11

K_p = Small diameter factor depending on D/d ratio is given in Table 1.12.

Table 1.11 Values of Constants for Power Rating of Belt and Bending factor

Belt Cross Section	a	b	c	Bending factor F_b
A	0.45	19.62	0.765×10^{-4}	25
B	0.79	50.8	1.32×10^{-4}	66
C	1.47	142.7	2.34×10^{-4}	185
D	3.22	506.7	4.78×10^{-4}	650
E	4.58	952	7.05×10^{-4}	1,242

Table 1.12 Values of small pulley factor K_p

D/d Ratio Range	K_p	D/d Ratio Range	K_p	D/d Ratio Range	K_p
1 – 1.019	1.00	1.11 – 1.42	1.05	1.341 – 1.429	1.10
1.02 – 1.032	1.01	1.143 – 1.178	1.06	1.430 – 1.562	1.11
1.033 – 1.055	1.02	1.179 – 1.222	1.07	1.563 – 1.814	1.12
1.056 – 1.081	1.03	1.223 – 1.274	1.08	1.815 – 2.948	1.13
1.082 – 1.109	1.04	1.275 – 1.340	1.09	2.949 and more	1.14

1.22 Life of Belts

Fatigue, more than abrasion, is the culprit for most belt problems. This wear is caused by stress from rolling around the pulleys. High belt tension, excessive slippage, adverse environmental conditions, and belt overloads caused by shock, vibration, or belt slapping all contribute to belt fatigue.

Belt life factors:

$$F_1 = T_1 + T_c + T_{b1} \qquad (1.35a)$$

$$F_2 = T_1 + T_c + T_{b2} \qquad (1.35b)$$

Where, F_1 = Total belt forces for tight side
F_2 = Total belt forces for slack side
T_{b1} = Bending force for tight side = F_b / d_1 (Value of F_b from Table 1.11)
T_{b2} = Bending force for slack side = F_b / d_2
T_1 = Belt tensions in tight side
T_2 = Belt tensions in slack side
T_c = Centrifugal force

Number of times peak forces F_1 sustained, $$M_1 = \left(\frac{Q}{F_1}\right)^x \qquad (1.36a)$$

and number of times peak forces F_2 sustained, $$M_2 = \left(\frac{Q}{F_2}\right)^x \qquad (1.36b)$$

Where, Q, x = Life factors given in Table 1.13

If N = Number of passes a belt can sustain then:

$$\frac{1}{N} = \frac{1}{M_1} + \frac{1}{M_2} \qquad (1.37)$$

Table 1.13 Values of Q and x for Different V Belt Sections

Belt Section	Q	x
A	3,060	11.089
B	5,420	10.924
C	9,260	11.173
D	19,120	11.105
E	27,550	11.10
3V	3,310	12.464
5V	7,520	12.593
8V	16,450	12.629

Belt life:
Life of belt in hours is given as:

$$L_h = \frac{NL_p}{v \times 3,600} \qquad (1.38)$$

Where, L_p = Belt length in meters
v = Belt velocity in m/s
N = Number of passes, which a belt can sustain.

Example 1.7

Life of a V belt for given application and belt length

A pulley of 0.3 m diameter transmits 6.75 kW at 1,160 rpm to another pulley of diameter 0.4 m using a V belt of cross-section A and pitch length 2.2 m. Assuming coefficient of friction 0.28 and weight of belt 1.06 N/m, calculate:

a. Center distance

b. Life of belt

V belt A-1024 has total tensions of 600 N and 100 N for tight and slack side respectively. Calculate life in number of hours, if belt velocity is 20 m/s

Solution

Given $P = 6.75 \text{ kW}$ $d = 0.3 \text{ m}$ $D = 0.4 \text{ m}$ $N_1 = 1,160 \text{ rpm}$ $C = 2.2 \text{ m}$

$\mu = 0.28$ Belt section = A $L_p = 2.2 \text{ m}$ $m = 1.06 \text{ N/m}$

a. Velocity of belt $v = \dfrac{\pi d N_1}{60} = \dfrac{\pi \times 0.3 \times 1,160}{60} = 18.21 \text{ m/s}$

From Equation (1.24) $A = \dfrac{L_p}{4} - \dfrac{\pi}{8}(D+d) = \dfrac{2.2}{4} - \dfrac{3.14}{8}(0.4 + 0.3) = 0.275 \text{ m}$

From Equation (1.25) $B = \dfrac{(D-d)^2}{8} = \dfrac{(0.4-0.3)^2}{8} = 0.00125$

From Equation (1.26): $C = A + \sqrt{A^2 - B} = 0.275 + \sqrt{(0.275)^2 - 0.00125} = 0.55 \text{ m}$

b. From Equation (1.1), angle on small pulley $\theta_s = \pi - 2\sin^{-1}\dfrac{D-d}{2C}$

Putting the values: $\theta_s = 180 - 2\sin^{-1}\dfrac{0.4-0.3}{2 \times 0.55} = 180° - 10.4° = 169.6° = 2.96 \text{ rad}$

From Equation (1.1) $T_c = mv^2 = \dfrac{1.06}{9.81} \times 18.21^2 = 35.8 \text{ N}$

From Equation (1.12) $\dfrac{T_1 - 35.8}{T_2 - 35.8} = e^{\mu \theta \csc 20°} = e^{0.28 \times 2.96 \times 2.92} = 11.28$

or $T_1 - 35.8 = 11.28 T_2 - 403.8$ or $T_1 = 11.28 T_2 - 368$ \hfill (a)

Also from Equation (1.11) $P = (T_1 - T_2) v$ Substituting the values:

$6.75 = (T_1 - T_2) \times 18.21$ or $T_1 - T_2 = 123$ \hfill (b)

From Equations (a) and (b): $T_1 = 170.8 \, N$ $T_2 = 47.8 \, N$

From Table 1.11 for A belt, $F_b = 25$

$$T_{b1} = \frac{F_b}{d} = \frac{25}{0.3} = 83.33 \, N \qquad \text{and} \qquad T_{b2} = \frac{F_b}{D} = \frac{25}{0.4} = 62.5 \, N$$

From Equation (1.35a) $F_1 = T_1 + T_c + T_{b1} = 170.8 + 35.8 + 83.33 = 289.9 \, N$

From Equation (1.35b) $F_2 = T_1 + T_c + T_{b2} = 47.8 + 35.8 + 62.5 = 269.1 \, N$

From Table 1.13 for A belt, $Q = 3,060$ and $x = 11.089$

Using Equation (1.36a), Number of times peak forces F_1 sustain, $M_1 = \left(\dfrac{Q}{F_1}\right)^x$

Putting the values, $M_1 = \left(\dfrac{3,060}{289.9}\right)^{11.089} = 10.555^{11.089} = 0.223 \times 10^{12}$

Using Equation (1.36b), Number of times peak forces F_2 sustain, $M_2 = \left(\dfrac{Q}{F_2}\right)^x$

Putting the values, $M_2 = \left(\dfrac{3,060}{269.1}\right)^{11.089} = 11.371^{11.089} = 0.5103 \times 10^{12}$

Using Equation (1.37), $\dfrac{1}{N} = \dfrac{1}{M_1} + \dfrac{1}{M_2}$ Substituting the values:

$$\frac{1}{N} = \frac{1}{0.223 \times 10^{12}} + \frac{1}{0.5103 \times 10^{12}} \quad \text{Or,} \quad N = 48.85 \times 10^{10} \text{ passes}$$

Using Equation (1.38), Life of belt in hours is given as: $L_h = \dfrac{N L_p}{v \times 3,600}$

Substituting the values: $L_h = \dfrac{48.85 \times 10^{10} \times 2.2}{18.22 \times 3600} = 16.39 \times 10^4 \text{ Hours}$

1.23 Design of V Belt Drive

1. From the design power calculated from Equation (1.29), and speed of driver, select a suitable section of V belt (A, B, C, D, or E) from Figure 1.21.

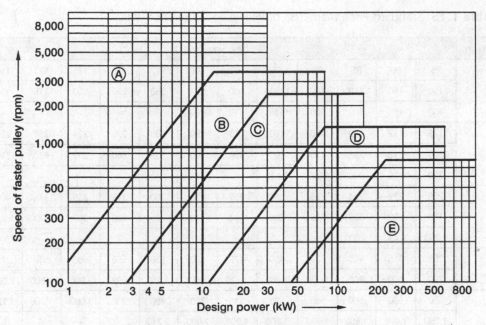

Figure 1.21 Selection of V belt cross section

2. If space is not a problem, for the selected section of the belt, pulley size in no case should be smaller than minimum pulley diameter given in Table 1.14.

Table 1.14 Power, Weight, and Pulley Dimensions

Belt Section	Power Range (kW)	Weight per Meter Length (N)	Minimum Pitch Diameter of Pulley (mm)	Recommended Pulley Size (mm)
A	0.7–3.5	1.06	75	85
B	2–15	1.89	125	185
C	7.5–75	3.43	200	325
D	20–150	5.96	355	425
E	30–350	9.41	500	700

Select a standard pulley diameter from Table 1.15.

Table 1.15 Standard Pitch Diameters of Pulleys for Different V Belts

Section	Pulley Size (mm)											
A	75	80	85	90	95	100	106	112	118	125	132	140
	150	160	170	180	190	200	224	250	280	300	315	355
	400	450	500	560	630	710	800	–	–	–	–	–
B	125	132	140	150	160	170	180	190	200	224	250	280
	300	315	355	375	400	450	500	530	560	600	630	710
	750	800	900	1,000	1,120	–	–	–	–	–	–	–
C	200	212	224	236	250	265	280	300	315	355	375	400
	450	500	530	560	600	630	710	750	800	900	1,000	1,200
	1,250	1,400	1,600	–	–	–	–	–	–	–	–	–
D	355	375	400	425	450	475	500	530	560	600	630	710
	750	800	900	1,000	1,060	1,120	1,250	1,400	1,500	1,600	1,800	2,000
E	500	530	560	600	630	710	750	800	900	1,000	1,060	1,120
	1,250	1,400	1,500	1,600	1,800	1,900	2,000	2,240	2,500	–	–	–

3. Calculate large pulley diameter D from the speed of driver pulley N_1 and speed of driven pulley N_2.

$$D = \frac{0.98 d\ N_1}{N_2} \text{ (allowing slip of 2 per cent)}$$

4. For the selected small pulley size, calculate belt speed v from the following equation:

$$v = \frac{\pi d N_1}{60}$$

Belt velocity v varies from 5 to 25 m/s, but the best velocity is 20 m/s.

If v is not near to best velocity, calculate small pulley pitch diameter assuming belt velocity of 20 m/s from the same equation, $d = \dfrac{60v}{\pi N_1}$

5. Center distance C between shafts varies from $(D + 0.5d)$ to $3(D + d)$. Small center distance reduces belt life due to more number of turns of belt in a second. Large center distance increases the cost of belt and promotes sagging of belt on loose side. Select a suitable value of center distance depending on the space available.

6. Calculate pitch length of the belt L_p from center distance and pulley diameters using Equation (1.22). If the direction of rotation of the driven shaft required is opposite to the driver shaft use Equation (1.23) for cross belt.

7. From Table 1.4, choose a standard available pitch belt length L_p and standard specified belt size L from the same table. Get belt length correction factor K_L from Figure 1.22.

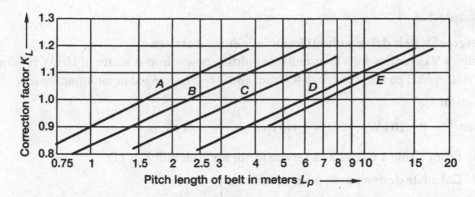

Figure 1.22 Correction factor for pitch length

8. Calculate rated power P_r using Equation (1.34).

9. Get a suitable service factor K_s from Table 1.6 for the given type of application.

10. Find $\left(\dfrac{D-d}{C}\right)$ and then value of arc contact factor K_a from Table 1.16.

Table 1.16 Arc of Contact Factor for Contact Angles of Smaller Pulley

$\dfrac{D-d}{c}$	1	0.9	0.8	0.7	0.6	0.5	0.4	0.3	0.2	0.1
θ_s°	120	127	133	139	145	151	157	163	169	174
K_a	0.82	0.85	0.87	0.89	0.91	0.93	0.96	0.97	0.98	0.99

11. Calculate number of V belts using the relation:

$$\text{Number of V belts } z = \frac{P \times K_s}{P_r \times K_L \times K_a} \tag{1.39}$$

Where, K_s = Service factor (Table 1.6)

K_L = Pitch length factor (Figure 1.22)

K_a = Contact arc factor (Table 1.16)

12. Calculate number of runs of belt per second from the belt velocity v using relation:

$$v = \frac{\pi d N_1}{60} \text{ and runs}/s = \frac{v}{L_p}$$

Runs per second should be less than 10.

Example 1.8

Design of V belt drive without knowing center distance
Design a V belt drive for a flour mill transmitting power from a motor of 10 kW rotating at 1,440 rpm. The mill rotates at 400 rpm and is to be arranged in minimum space.

Solution

Given $P = 10\,\text{kW}$ $N_1 = 1,440\,\text{rpm}$ $N_2 = 400\,\text{rpm}$

1. From Table 1.6, assume a service factor for flour mill $K_s = 1.3$.
2. Calculate design power from equation: $P_d = P \times K_s$

    Design power $P_d = 10 \times 1.3 = 13\,\text{kW}$

   From the design power 13 kW and speed of driver 1,440 rpm, from Figure 1.21 belt section B is selected.

3. Speed ratio $VR = \dfrac{N_1}{N_2} = \dfrac{1,440}{400} = 3.6$

4. From Table 1.14, minimum pulley size for B belt is 125 mm, whereas recommended is 185 mm. So initially 185 mm is selected.

5. Calculate belt speed v from the equation: $v = \dfrac{\pi\, d N_1}{60}$

$$v = \frac{\pi \times 0.185 \times 1,440}{60} = 13.9 \ \text{m/s}$$

   But the best velocity is 20 m/s. Hence, assuming $v = 20$ m/s and calculate small pulley diameter from the same equation.

$$d = \frac{60\,v}{\pi\,N_1} = \frac{60 \times 20}{\pi \times 1,440} = 0.273 \ \text{m}$$

   From Table 1.15, standard pulley diameter 280 mm is selected.

6. Assuming slip $s = 2$ per cent, calculate big pulley pitch diameter D from equation:

    $D = d(1 - s) \times VR$

       $= 0.28\,(1 - 0.020) \times 3.6 = 0.978\,\text{m say } 980\,\text{mm}$

7. Center distance C between shafts varies from $(D + 0.5d)$ to $3(D + d)$

   Since there is space limitation, hence choosing minimum center distance as:

    $C = (D + 0.5d) = [0.98 + (0.5 \times 0.28)] = 1.12\,\text{m}.$

8. Pitch length of the belt L_p from center distance and pulley diameters for the open belt is given by Equation (1.22): $L_p = 2C + \dfrac{\pi}{2}(D + d) + \dfrac{(D - d)^2}{4C}$

Substituting the values: $L_p = 2 \times 1.12 + \dfrac{\pi}{2}(0.98 + 0.28) + \dfrac{(0.98 - 0.28)^2}{4 \times 1.12} = 4.327$ m

9. From Table 1.4, choose a standard available belt pitch length near to this size, which is 4,394. Inside standard length L for this belt is 4,437 mm.

10. Calculate exact center distance C between the shafts with the new pitch length of the belt using Equations (1.24), (1.25) and (1.26).

$$A = \frac{L_p}{4} - \frac{\pi}{8}(D + d) = \frac{4.45}{4} - \frac{\pi}{8}(0.98 + 0.28) = 0.618$$

And $B = \dfrac{(D - d)^2}{8} = \dfrac{(0.98 - 0.28)^2}{8} = 0.0612$

Center distance; $C = A + \sqrt{A^2 - B} = 0.618 + \sqrt{0.618^2 - 0.0612} = 1.184$ m

11. Calculate number of runs of belt per second from the belt velocity v using relation:

$$v = \frac{\pi\, dN1}{60} = \frac{\pi \times 0.28 \times 1,440}{60} = 21.1 \text{ m/s}$$

$$\text{Runs/s} = \frac{v}{L_p} = \frac{21.1}{4.45} = 4.94. \text{ It is less than 10 / s, hence safe.}$$

12. Calculate rated power using Equation (1.34):

$$P_r = v\left[\frac{a}{v^{0.09}} - \frac{b}{d k_p} - c v^2 \right]$$

Values of a, b, and c are from Table 1.11: $a = 0.79$, $b = 50.8$, and $c = 1.32 \times 10^{-4}$

For velocity ratio 3.6, value F_b from Table 1.12 is 1.14.

Substituting the values $P_r = 21.1\left[\dfrac{0.79}{21.1^{0.09}} - \dfrac{50.8}{280 \times 1.14} - \left\{ 1.32 \times 10^{-4} \times (21.1)^2 \right\} \right]$

$P_r = 21.11(0.6 - 0.159 - 0.0587) = 8.06$ kW

13. For belt length 4,437 mm, value of K_L from Figure 1.22 is 1.15

14. $\dfrac{D - d}{c} = \dfrac{980 - 280}{1184} = 0.59$

Value of contact factor from Table 1.16 $K_a = 0.91$.

15. Calculate number of belts:

From Equation (1.39), Number of V belts $z = \dfrac{P \times K_s}{P_r \times K_L \times K_a}$

Or, $z = \dfrac{10 \times 1.3}{8.06 \times 1.15 \times 0.91} = 1.54$ say 2. So two belts B 4,437 are recommended.

Example 1.9

Design of a V belt with given center distance

Design a V belt to transmit 7.5 kW power from a pulley rotating at 1,440 rpm to a fan pulley to rotate at 480 rpm (neglect slip). The fan is to be used for 24 hours a day and space does not permit to have center distance more than 1,250 mm.

Solution

Given $P = 7.5$ kW $N_1 = 1,440$ rpm $N_2 = 480$ rpm $C = 1,250$ mm
24 hours service

From Figure 1.21, for $P = 7.5$ kW and speed 1,400 rpm selection of cross section is B.

Recommended pulley diameter for B belt is 185, selecting a standard pulley diameter from Table 1.14, of 200 mm.

Large pulley diameter $D = \dfrac{200 \times 1,440}{480} = 600$ mm

$$v = \frac{\pi d N_1}{60} = \frac{\pi \times 200 \times 1,440}{60} = 15 \text{ m/s}$$

$$L_p = 2C + \frac{\pi}{2}(D + d) + \frac{(D - d)^2}{4C}$$

Substituting the values; $L_p = (2 \times 1,250) + 1.57(600 + 200) + \dfrac{(600 - 200)^2}{4 \times 1,250} = 3,789$ mm

Choosing a standard length of L, 3,658 mm from Table 1.4, whose pitch length $L_p = 3,701$ mm exact center distance = 1205.6 mm

Velocity ratio $= \dfrac{N_1}{N_2} = \dfrac{1,440}{480} = 3$

Rated power $P_r = v \left[\dfrac{a}{v^{0.09}} - \dfrac{b}{d k_p} - c v^2 \right]$

Values of a, b, and c are from Table 1.11: $a = 0.79$, $b = 50.8$, and $c = 1.32 \times 10^{-4}$.
For velocity ratio 3, value k_p from Table 1.12 is 1.14.

Substituting the values, $P_r = 15 \times \left[\dfrac{0.79}{15^{0.09}} - \dfrac{50.8}{200 \times 1.14} - \left\{ 1.32 \times 10^{-4} \times (15)^2 \right\} \right]$

$P_r = 15 \times (0.619 - 0.223 - 0.0297) = 5.49 \text{ kW}$

For pitch belt length 3,701 mm, value of K_L from Figure 1.22 is 1.1

$$\frac{D-d}{c} = \frac{600-200}{1205.6} = 0.332$$

Value of contact arc factor from Table 1.16 for 0.332 by interpolation is, $K_a = 0.966$
From Table 1.6, for 24 hours working, service factor $K_s = 1.3$
Calculate number of belts: From Equation (1.39),

Number of V belts; $z = \dfrac{P \times K_s}{P_r \times K_L \times K_a}$

Or, $z = \dfrac{7.5 \times 1.3}{5.49 \times 1.1 \times 0.966} = 1.69$ say 2. So two V belts B 3,701 are recommended.

1.24 Types of Pulleys

Pulleys are used to transmit power with the help of belts. A pulley is a circular machine element having a hub in the center, with a key way that fits on the shaft with a key. Small pulleys can be made of forged steel, but large pulleys are made of cast iron. Aluminum pulleys are used, where light construction is required. Sometimes, wood is also used as a pulley material. Pulleys are classified in many ways as given below:

A. **According to shape:**
   - Flat (Section 1.25)
   - Stepped (Section 1.25.5)
   - Grooved pulley (Section 1.26)

B. **According to driving belt:**
   - Flat belt pulley: The driving belt is of rectangular section (Section 1.25).
   - Single V groove pulley: A single V belt having 40° angle is used (Section 1.26.1).
   - Multi V groove pulley: Many V belts are used in parallel for more power transmission (Section 1.26.2).
   - Toothed belt pulley: It has teeth on the outer periphery and is used for positive drive (Section 1.27).
   - Rope pulley: A circular rope is used in a groove of the pulley (See Chapter 2).

C. According to construction:

- Solid pulley: Pulley is one solid casting or forging (Section 1.25.1).
- Webbed pulley: A disc is provided between hub and the rim (Section 1.25.2).
- Armed pulley: It has rim for belt, hub in the center, and four or six arms joining hub and rim (Section 1.25.3).
- Built-up pulley: Large pulleys are fabricated by joining two or more pieces of rims (Section 1.25.4).

D. According to use:

- Fast pulley: The pulley is keyed to a shaft (Sections 1.26 and 1.27).
- Fast and loose pulley: Fast pulley is keyed to a shaft, whereas loose pulley has no key (Section 1.25.6).
- Idler pulley: Does not transmit any power. Generally used for providing belt tension.

1.25 Flat Belt Pulleys

Flat belt pulley has three parts. Its outer periphery is almost flat but with slight curvature and is called rim. A slight camber is given to make the outer periphery slightly convex or arched (Figure 1.23). This helps in keeping the belt positioned centrally on the pulley.

The central portion is called hub, which is mounted on the shaft using a key. Rim and hub are joined by different methods given below:

- Small-size flat pulleys are made solid and have no arm or web.
- Large-size pulleys are webbed, or have straight or curved arms.
- Very large pulleys are built up, by joining different circular segments of pulley.

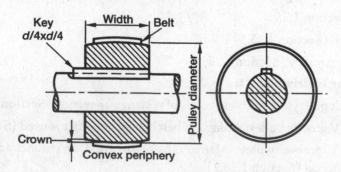

Figure 1.23 Solid pulley

Minimum pulley diameter depends on number of plies and belt speed. Higher the speed, more diameter is required. Diameter also increases, if the number of plies is increased as it can be seen in Table 1.8.

Standard diameters for V belt pulleys are given in Table 1.14. Width of pulley is kept wider than the belt width as given in Table 1.17.

Table 1.17 Width of Belt and Width of Pulley (mm) [3142 – 1993]

Width of Belt (mm)	<125	125–250	250–375	375–500
Width of Pulley Greater Than Belt Width	13	25	38	50

Crown (convex surface on periphery of rim) is provided to keep the belt centrally on the pulley. It is provided on both the pulleys. It is necessary, if pulleys are mounted on slightly non-parallel shaft or misaligned shafts. Crown is radial length in millimeters from outer edge of pulley to extreme surface in the center of pulley (Figure 1.23). Its value depends on width and diameter of pulley and is tabulated in Table 1.18 for different size of pulleys.

Table 1.18 Crown (mm) on Pulleys for Different Widths of Pulleys

Pulley Diameter (mm)	Width of Pulleys (mm)						
	<125	140 – 160	180 – 200	224 – 250	280 – 315	315 – 355	>400
40 – 112	0.3	–	–	–	–	–	–
125 – 140	0.4	–	–	–	–	–	–
160 – 180	0.5	–	–	–	–	–	–
200 – 224	0.6	0.6	–	–	–	–	–
250 – 280	0.8	0.8	–	–	–	–	–
315 – 355	1.0	1.0	1.0	1.0	1.0	–	–
400 – 450	1.0	1.2	1.2	1.2	1.2	1.2	1.2
500 – 560	1.0	1.5	1.5	1.5	1.5	1.5	1.5
630 – 710	1.0	1.5	2.0	2.0	2.0	2.0	2.0
800 – 1,000	1.0	1.5	2.0	2.5	2.5	3.0	3.0
1,120 – 1,250	1.2	1.5	2.0	2.5	3.0	3.0	3.5
1,400 – 1,600	1.5	2.0	2.5	3.0	3.5	4.0	5.0
1,800 – 2,000	2.0	2.5	3.0	3.5	4.0	5.0	6.0

1.25.1 Solid pulley

Small pulleys are made solid and its sectional view is shown in Figure 1.23. The hole is of the shaft size (d) according to the fit required and has a key way of size $d/4 \times d/4$ or a collar to fit by a screw. Values of crown are given in Table 1.18. It can be seen that belt is wider than width of the belt.

1.25.2 Webbed pulley

Medium-size pulleys up to 200 mm diameter are made webbed to reduce weight and save material (Figure 1.24). Web is a disk between the rim and hub, whose thickness is about 30 – 50 per cent of width of rim. Sometimes, the web is provided with holes to decrease the weight further and save material. Number of holes varies from 4 to 6.

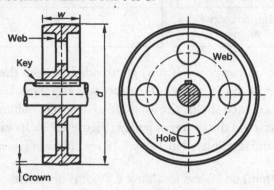

Figure 1.24 Webbed pulley

1.25.3 Armed pulley

Large-size pulleys of diameter more than 300 mm have arms joining the rim and central hub. The arms can be straight as shown in Figure 1.25(a) or curved as in Figure 1.25(b). These are made either of cast iron or of mild steel. For steel pulleys, the rim is shaped circular from a steel plate and welded or screwed with a mild steel hub using round steel bars. These pulleys are lighter than cast iron pulleys.

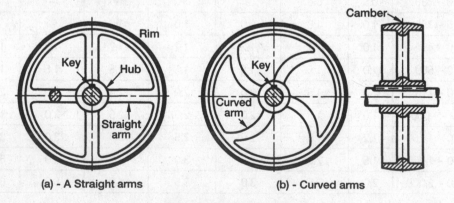

(a) - A Straight arms   (b) - Curved arms

Figure 1.25 Armed pulley

Following proportions can be initially assumed and then checked for the safe stresses.

   Hub   The hub has a hole equal to shaft diameter d and keyway of size $(0.25d \times 0.25d)$.

   Hub diameter $d_h = 1.5d + 25$ mm, but not more than $2d$

   Hub length $L_h = 0.67w$ to w

Where, w is of width of the rim.

Arm   Number of arms n = 4 for pulleys of diameter up to 400 mm, and

$$n = 6 \text{ for diameter more than } 450 \text{ mm.}$$

Cross section of the arm is generally elliptical but can be circular. Major axis a is in the plane of rotation. Width of arm near the boss can be initially assumed as:

$$a = 2.94 \sqrt[3]{\frac{wD}{4n}} \tag{1.40}$$

Where, w = Width of rim

D = Diameter of pulley

n = Number of arms

Thickness of arm $b = 0.5a$

Fillet radius of arms with boss and rim $r = 0.5b$

Arms are generally tapered with 2 to 4 mm per 100 mm of arm length.

Rim   Its radial thickness varies from $(0.0033D + 2 \text{ mm})$ to $(0.005D + 3 \text{ mm})$, where D is pulley outside diameter.

Design of arms

The section is checked for stresses, assuming it as a cantilever beam supported at hub and a concentrated load at rim.

$$\text{Length of arm} = \text{Inside radius of rim} - \text{Hub outside radius}$$

If hub diameter is not known, it can be approximated as twice the shaft diameter. It is assumed that the power is transmitted only through half the number of arms n that is $0.5n$, as the belt covers only about 50 per cent of the periphery of the rim. Thus, bending moment:

$$M = \frac{2T}{n}$$

Cross section of arm can be taken as elliptical with major axis a in the plane of rotation and minor axis b along pulley axis. Modulus of section Z for this section is:

$$Z = \frac{\pi}{32} \times b \times a^2$$

Minor axis is taken as half the major axis, that is, $b = 0.5a$. Hence,

$$Z = \frac{\pi}{64} \times a^3 \tag{1.41}$$

Then use relation $M = \sigma_b \times Z$.

Bending moment M and safe bending stress σ_b are known, calculate value of major and minor axes.

Example 1.10

Design of a flat pulley for the given power

Design a cast iron pulley mounted on a shaft of 40 mm diameter to transmit 15 kW at 480 rpm using a 200 mm wide flat belt at speed of 20 m/s. Take safe bending stress 15 MPa.

Solution

Given $d = 40$ mm $P = 15$ kW $N = 480$ rpm $v = 20$ m/s $w = 200$ mm $\sigma_b = 15$ MPa.

$$v = \frac{\pi DN}{60} \quad \text{or} \quad D = \frac{60v}{\pi N}$$

Hence, large pulley diameter, $D = \dfrac{60 \times 20}{\pi \times 480} = 0.796$ m

From Table 1.5, standard pulley diameter near to this value is 800 mm.

From Table 1.17, width of pulley = 200 + 25 = 225 mm (rim width).

Since pulley size is more than 400 mm, armed pulley is assumed.

Pulley diameter is more than 450 mm, hence $n = 6$ arms are selected.

$$T = \frac{P \times 60}{2\pi N} = \frac{15,000 \times 60}{2 \times 3.14 \times 480} = 298.6 \text{ Nm}$$

Hub diameter $d_h = 2d = 2 \times 40 = 80$ mm (d is shaft diameter)

From Equation (1.40), Width of arm near hub:

$$a = 2.94 \sqrt[3]{\frac{wD}{4n}} = 2.94 \sqrt[3]{\frac{225 \times 800}{4 \times 6}} = 58 \text{ mm}$$

Radial thickness of rim = 0.005 D + 3 mm = (0.005 × 800) + 3 = 7 mm

Length or arm = Radius of rim – Thickness of rim – Radius of hub

$$= 400 - 7 - 40 = 353 \text{ mm}$$

Taper of arm is 2 mm per 100 mm. So for 353 mm taper = 2 × 3.53 = 7 mm

Width of arm near rim $a' = a - 7$ or
$a' = 58 - 7 = 51$ mm
Minor axis of the arm $b = 0.5a = 0.5 \times 58$
$= 29$ mm
Bending moment = Torque = 298.6 Nm

$$M = 746 \times 0.4 = 298.6 \text{ Nm}$$

Checking the size for bending stress:

Out of six arms only three share moment at one time.

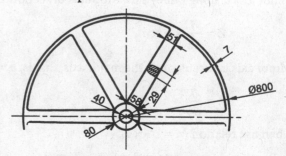

Figure 1.51 Armed pulley design

Bending moment per arm $M = \dfrac{298.6}{3} = 99.5$ Nm

For elliptical arm $Z = \dfrac{\pi}{32} \times b \times a^2 = 0.0981 \times 29 \times 58^2 = 9{,}570$ mm^3

$$\sigma = \frac{M}{Z} = \frac{99.5 \times 10^3}{9{,}570} = 10.4 \, \text{MPa (safe)}$$

It is lesser than safe bending stress 15 MPa, hence safe.
Final dimensions of the pulley are given in Figure 1.S1.
Alternate approach (calculate size from safe stress)

Modulus of section $Z = \dfrac{\pi}{32} \times b \times a^2$

Since $b = 0.5\,a$, $\qquad Z = \dfrac{\pi}{32} \times 0.5a \times a^2 = \dfrac{\pi}{64} \times a^3$

Also, $\qquad\qquad Z = \dfrac{M}{\sigma}$

Substituting the values:

$$\frac{\pi}{64} \times a^3 = \frac{99.5 \times 10^3}{15} \quad \text{or} \quad a = 51.3 \text{ mm say 52 mm}$$

$$b = 0.5; \, a = 0.5 \times 52 = 26 \text{ mm}$$

Values of a and b are lesser than calculated earlier as stress was 10.4 MPa in earlier calculations and here it is 15 MPa. Even these values will also be safe.

1.25.4 Built-up pulley

Built-up pulleys are used for very large diameters. Both hub and rim are made in two halves and then joined together with bolts and nuts (Figure 1.26). Hub and rim are provided with radial holes. Arms (spokes) are of circular cross section with outer end having a collar and step at one end and inner end is also stepped. Inner ends of the rods are shrunk fit into holes of the hub. The rim is supported by the collar at the outer end of the rod. The stepped portion of rod is riveted to the rim in a tapered hole. The rim halves are joined by a curved strap plate on the inside periphery of the rim. The strap is

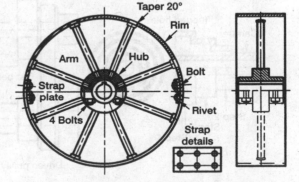

Figure 1.26 Built-up pulley in half section

riveted to one side of the rim and bolted by a counter sunk bolt at the other side. The two halves of hub are bolted on shaft with a key in between.

Spokes Following guidelines can be used for spokes:

- Pulleys having rim width less than 300 mm, provide a single row of spokes;
- Pulleys having width more than 300 mm, use two rows of spokes;
- Diameter of spoke is taken as 19 mm up to pulley of diameter 710 mm;
- Diameter of spoke is taken 22 mm for pulleys larger than 710 mm; and
- Number of spokes is decided by pulley size as given in Table 1.19.

Table 1.19 Number of Spokes for Steel Pulley Diameters (mm)

Pulley Diameter	<280	280 – 500	560 – 710	800 – 1,000	1,120	1,250	1,400	1,600	1,800	2,000
No. of Spokes	4	6	8	10	12	14	16	18	20	22

Boss Length of the boss is taken half the width of face (rim) but should not be less than 100 mm for spokes of 19 mm and 140 mm for 22 mm spokes.

Rim Most of the rims are made of 5 – 6-mm-thick plates.

1.25.5 Stepped pulley

Stepped pulley consists of many pulleys of different diameters casted or forged together as one piece (Figure 1.27). Step pulleys are used in pair and arranged in such a way that the smallest pulley drives the largest pulley and the largest drives the smallest. The size of step is calculated such that the belt length remains same, on any one set of pulleys. Speed of the driven pulley is changed by changing the position of belt from one step to another step. The steps can be three or four. More than four are not commonly used.

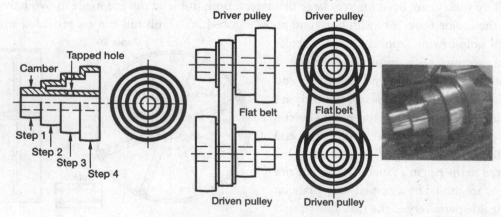

Figure 1.27 Stepped pulley

1.25.6 Fast and loose pulley

Two pulleys placed adjacent to each other are used as fast and loose pulley. One of the pulleys is keyed to the shaft called fast pulley, whereas the other is without key, called loose pulley. Loose pulley runs freely on a bush without transmitting power to shaft. Whenever the power transmission is not required, belt is made to slip from fast to loose pulley. Some machine tools use this type of power transmission to connect and disconnect power, whenever required as shown in Figure 1.28 and in picture on its right side.

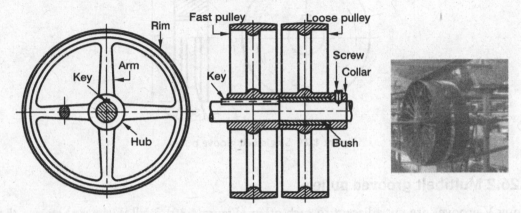

Figure 1.28 Fast and loose pulley

1.26 Grooved Pulleys

Grooved pulleys use V belts. These belts have a standard cross section A, B, C, D, and E. The section to be selected depends on the power to be transmitted. Section A is for less power and E for large power. If the power to be transmitted is more than the capacity of one belt, two or more belts can be used in parallel.

If pulley diameter is less, belt has to bend more and hence less life and chances of slipping due to less contact area. Hence minimum pulley size depends on cross section of the belt. Table 1.14 gives minimum size and preferred diameter of the pulley required for different size of belts. Standard pulley diameters are given in Table 1.15.

Although the angle of the belt is 40°, the angle of pulley groove is made lesser than 40° so that the wedging action causes more gripping force between pulley and belt. Driver pulleys (generally smaller) have included angle between 32° and 34° and large pulley between 36° and 38°.

Depth of the groove is also kept about 10 per cent more than the actual depth of belt so that when the belt gets compressed due to continuous use, it may not touch the bottom surface of the groove, else the wedging action will not take place.

1.26.1 Single belt grooved pulley

Single belt grooved pulley for V belt is shown in Figure 1.29. It may be solid or webbed or armed.

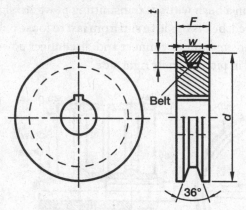

Figure 1.29 Single belt groove pulley

1.26.2 Multibelt grooved pulley

Many V grooves are cut adjacent to each other (Figure 1.30). Wall thickness between the grooves varies between 3 mm and 5 mm. Outer wall thickness is kept slightly more than inside wall thickness. The gap between grooves depends on the material also. Larger size pulleys are webbed and holes are also made to make them light.

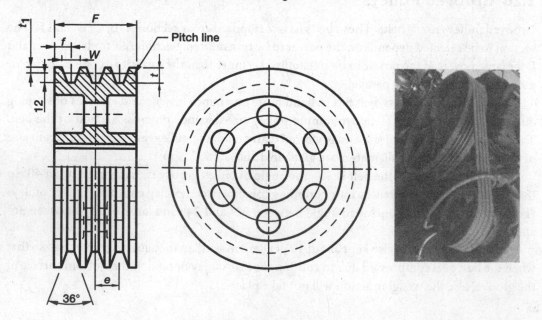

Figure 1.30 Multibelt groove pulley

Table 1.20 gives the dimensions of the pulley for various belts.

Table 1.20 Dimensions of Multi V Grooved Pulleys (mm)

Section	Pitch Width of Groove (w)	Height above Pitch Line (t1)	Height below Pitch Line (t2)	Depth of Groove (t)	Distance between Grooves (e)	Groove Center from Outer Face (f)
A	11	3.3	8.7	12	15	10
B	14	4.2	10.8	115	19	12.5
C	19	5.7	14.3	20	25.5	17
D	27	8.1	19.9	28	37	24
E	32	9.6	23.4	33	44.5	29

1.26.3 Stepped grooved pulley

Purpose of stepped grooved pulley is same as that of stepped pulley for flat belt. These pulleys are also used in pair, aligned such that belt of the smallest pulley aligns with the biggest pulley. The angle of groove is kept between 34° and 38° (Figure 1.29). The groove is kept deeper than the height of the belt. Size of groove depends on cross section of belt A, B, or C. The stepped pulley is shown in Figure 1.31 and its picture on the right side.

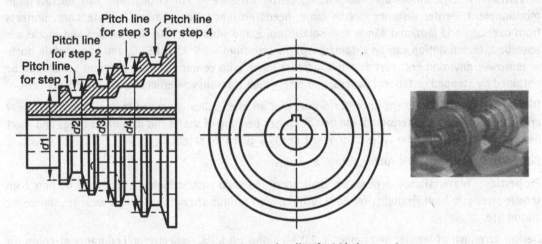

Figure 1.31 Stepped pulley for V belts

1.27 Toothed Pulley

Their use is limited to applications, where slip may affect the performance of the machine. The outer periphery has teeth that fit into the grooves of the toothed belt to check slip. Such a pulley is shown in Figure 1.32. Their use can be seen for cam shaft drive in Internal

Combustion (I.C.) engines. A collar is provided on the sides of driver to keep the belt within the width of the pulley to prevent shift of belt on sides due to slight misalignment of shafts.

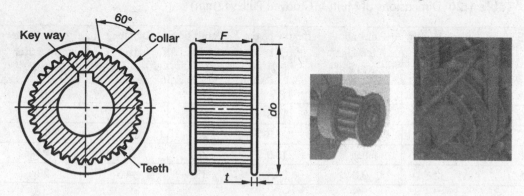

Figure 1.32 Toothed pulley and drive

Summary

Belt drive system   Mechanical power is transmitted from one shaft to another shaft using belt and pulleys, chains and sprockets, and gears. A pulley is a circular machine element, having a hub in the center with a key way, which fits on the shaft. A belt passes over pulleys to transmit the power.

Advantages   It is simple and inexpensive; center distance is not critical and can tolerate high misalignment; center distance can be large, needs no lubrication, minimal maintenance, protects from overload and jam, and damps and isolates noise and vibration; load fluctuations and shocks are absorbed; clutch action can be obtained, can operate from – 35° C to 85° C, and can work in dusty or corrosive environment; cost does not increase much with center distance; different speeds can be obtained by stepped or tapered pulleys; and offers high transmission efficiency usually 95 per cent.

Disadvantages   Adjustment of center distance or an idler pulley is needed to compensate for wear and stretch, that is, the tension of the belt. It causes bending of shafts and load on bearings and short life in comparison to other types of drives. It needs a guard for safety.

Belt materials are leather, rubber, fabric, and plastics.

Properties   Material used depends on application. It should possess high coefficient of friction, high tensile strength, high flexibility for easy turning, high fatigue strength, and high wear resistance for a long life.

Design strength of leather and rubber is 2.0 MPa and 1.6 MPa, respectively. Endurance strength for both is 6 MPa. Density of all materials is 1,150 kg/m³, except for fabric, which is 900 kg/m³. Maximum velocity for leather is 50 m/s and for rubber 30 m/s.

Coefficient of friction depends on pair of belt and pulley material and condition of surface like dry, wet, and grease. For dry surface, it varies from 0.2 to 0.35 and for wet 0.15 to 0.3. For greasy condition, it reduces to 0.12.

Types of belts are flat belt, V belt, and grooved belt.

Flat belts These are of rectangular cross section and used for line shafts in factories, farming, mining applications, saw mills, flour mills, and conveyors. These are low cost and used on small pulleys. They need high tension resulting in high-bearing loads. They are noisier than other belts and have low efficiency at moderate speeds.

V belts These are endless, and their cross-sectional shape is trapezoidal with angle 40°. Slippage is less due to wedging action and offer higher power transmission than flat belts. V belts have long life (3–5 years), and offer quiet operation and low maintenance. These are most commonly used in industry and are available in standard cross-sectional sizes and lengths.

Best speed for V belts is 8–30 m/s. V belts are made in two sizes: conventional and narrow. Ideal speed for standard belt is 23 m/s and for narrow belts 50 m/s. Conventional belts are designated as A, B, C, D, and E. Narrow belts are designated as 3V, 5V, and 8V.

Belt specifications A flat belt is specified for material, number of plies, and thickness. Standard belt thicknesses are 5, 6.5, 8, 10, and 12 mm. Maximum belt stress per unit width is used. Values of standard belt widths are: in R10 series, 25, 32, 40, 50, and 63 mm and in R20 series, width starts from 71 mm to 600 mm.

Types of flat belt drives Three types of flat belt drives are used depending on the direction of rotation of shafts, angular position, and distance between shafts.

Open belt drive is for parallel shafts rotating in the same direction. One side of the belt is called tight side, whereas the other is called slack side.

Small center distance reduces belt life and arc of contact on small pulley, which increases slip.

Large center distance causes sagging of the belt especially for horizontal belts. So slack side is kept on the upper side.

Crossed belt arrangement gives direction of rotation of driven opposite to the driver. The belt is twisted 180°, on the other pulley. Owing to more arc of contact, it can transmit more power than open belts. Its drawback is that rubbing between the belt faces causes excessive wear and tear of the belt. Hence, center distance should be about 20 times the width of the belt.

Quarter twist belt Axes of the driver and driven are at 90°. Center distance between the pulleys has to be long. To avoid slipping, width of the pulley is kept about 1.5 times the width of the belt.

Compound belt: It is used to obtain high-velocity ratios. Output of the first stage is taken as input for the second stage. Overall velocity ratio VR is multiplication of two velocity ratios VR1 and VR2, that is, VR = VR2 × VR2.

Angle of contact is the angle for which a belt is in contact with a pulley.

Open belt For small pulley, contact angle is $\theta_s = \pi - 2\sin^{-1}\dfrac{D-d}{2C}$

Contact on large pulley $\theta_L = \pi + 2\sin^{-1}\dfrac{D-d}{2C}$

Cross belt Contact angle on both pulleys: $\theta_L = \theta_s = \pi + 2\sin^{-1}\dfrac{D+d}{2C}$ Owing to the presence of friction, tensions on both the sides are not equal. Tight side tension is T_1 and T_2 as the slack side.

Power through a belt: Power P is a function of belt tensions T_1 on tight side, T_2 on slack side and belt velocity v. Power, $P = (T_1 - T_2)\, v$.

Belt tensions ratio For angle of contact θ and coefficient of friction, μ the ratio is: For flat belts: $T_1 = T_2 e^{\mu\theta}$ and for V belts $T_1 = T_2 e^{\mu\theta\, \text{cosec}\beta}$ (2β is groove angle)

Initial tension To increase grip, the belt is initially tightened to some tension, even when the pulleys are not rotating. It is given as: $T_0 = \left(\dfrac{\sqrt{T_1}+\sqrt{T_2}}{2}\right)^2$

Centrifugal tension When pulleys start rotating, centrifugal force F_c acts on tight and slack sides.

For belt velocities $v > 10$ m/s, it cannot be neglected. $F_c = mr\, d\theta \dfrac{v^2}{r}$ (m is mass per unit length). Centrifugal tension $T_c = mv^2$.

Centrifugal tension has no effect on power transmission.

Ratio of tensions with centrifugal tension by Capstan equation is $\log \dfrac{T_{t1} - T_c}{T_{t2} - T_c} = \mu\, d\theta$

Maximum tension, which a belt can take, depends on safe stress of the belt. Maximum tension with centrifugal force: $T_{max} = T_1 + T_c$

Condition for maximum power: Belt velocity $v = \sqrt{\dfrac{T_{max}}{3m}}$

Slip of belt Generally, there is slight slipping of belt over pulley. If s_1 is slip between the driver pulley and belt, and s_2 between driven pulley. Total slip $s = (s_1 + s_2)$.

With slip $\dfrac{N_2}{N_1} = \dfrac{d}{D}\left[1 - \dfrac{s_1}{100}\right]\left[1 - \dfrac{s_2}{100}\right] = \left[1 - \dfrac{s}{100}\right]$, N_1 and N_2 are driver and driven speeds respectively.

For a thick belt, thickness t is also considered and $\dfrac{N_2}{N_1} = \dfrac{d+t}{D+t}\left[1 - \dfrac{s}{100}\right]$

Creep of belt: When belt passes over the pulley, it undergoes from cycles of more tension to less tension. This causes relative motion and is called creep. The net effect is that the speed of the driven pulley decreases slightly. The velocity ratio $\dfrac{N_2}{N_1} = \dfrac{d}{D} \times \dfrac{E+\sqrt{\sigma_2}}{E+\sqrt{\sigma_1}}$. Where, E is Young's modulus of elasticity, σ_1 and σ_2 are stresses in tight and slack sides respectively.

Length of V belt If center distance C between the two shafts is given:

For open belt, length $L_p = 2C + \dfrac{\pi}{2}(D+d) + \dfrac{(D-d)^2}{4C}$

For cross belt, $L_p = 2C + \dfrac{\pi}{2}(D+d) + \dfrac{(D+d)^2}{4C}$

If pitch length of belt L_p is known, center distance C for open belt can be calculated.

If $A = \dfrac{L_p}{4} - \dfrac{\pi}{8}(D+d)$ and $B = \dfrac{(D-d)^2}{8}$ Then, $C = A + \sqrt{A^2 - B}$

A V belt is specified by its section and inside length. Its length at its neutral axis is more due to its thickness. The relation between pitch length L_p and inside length L is:

For A belt, $L_p = L + 36$ mm, B belt, $L_p = L + 43$ mm, C belt, $L_p = L + 56$ mm and so on.

Center distance of a drive depends on ratio of large and small pulley diameters D and d, respectively. For D/d ratios 1, 2, 3, 4, 5 and 6, C/D ratios are 1.5, 1.2, 1.0, 0.95, 0.9, and 0.85, respectively.

Minimum distance $C_{min} = 0.55 \ (D + d) + t$ where, $t =$ belt thickness.

Maximum center distance $C_{max} = 2 \ (D + d)$.

Rated power of V belts   Power, which a single V belt, can transfer depends on small pulley diameter d and belt velocity v.

Rated power $P_r = v \left[\dfrac{a}{v^{0.09}} - \dfrac{b}{d\,k_p} - cv^2 \right]$ Values of constants a, b, c, and F_b can be taken from Table 1.11.

Design of flat belt drive   Data given is: power P (kW), type of service, N_1 is driver speed and N_2 is driven speed in rpm or velocity ratio VR. In design find sizes of pulleys d and D, size of belt b, and center distance C between the shafts.

1. *Calculate pulley diameters*   Assume a belt velocity v of 18 m/s and calculate diameter of small pulley using $d = \dfrac{60 \times 1,000 v}{\pi N_1}$ Alternately, $d = (1,100 - 1,300) \left(\dfrac{P}{N_1} \right)^{1/3}$

    Select preferred pulley diameter (mm) 20, 22, 25, 28, 32, 36, 45, 56, and 63.

    Again calculate the actual belt velocity v with the standard diameter, $v = \dfrac{\pi d N_1}{60}$

    Calculate large pulley diameter D assuming slip s: $\dfrac{N_2}{N_1} = \dfrac{d}{D} \left[1 - \dfrac{s}{100} \right]$

2. *Calculate design power*   Assuming a service factor K_s (From Table 1.6) between 1.1 and 1.8, calculate design power $P_d = P \times K_s$

3. *Calculate arc of contact*   Contact arc angle θ_s for small pulley for open or cross belt.

4. *Calculate corrected power*   Take value of K_a from Table 1.7. Calculate corrected power $P_c = P_d \times K_a$

    Where, K_a is arc of contact factor varying from 1.33 for θ_s 120° to 0.94 for θ_s 200°.

5. *Calculate corrected power rating of belt*   For belt speed of about 5 m/s, power rating varies between 0.012 and 0.015 kW/mm/ply. Calculate:

    Load rating at actual velocity = Load rating at 5 m/s $\times \left(\dfrac{v}{5} \right)$

6. *Calculate belt width and number of plies:* Number of plies p varies from 3 to 8.

    $b \times p = \dfrac{\text{Corrected power } P_c}{\text{Corrected power rating of belt}}$ (b is width of belt)

    Select a standard belt width b (mm) 25, 32, 40, 44, 50, 63, 76, 90, 100, 112, 125, etc.

7. *Check pulley diameter for the number of plies*   It should not be less than recommended.

Life of belts   High belt tension, excessive slippage, belt overloads caused by shock, vibration, or belt slapping all contribute to belt fatigue.

Total belt force on tight side, $F_1 = T_c + T_{b1} + T_1$ and for slack side $F_2 = T_c + T_{b2} + T_1$

Where, T_c is centrifugal force and bending force is $T_{b1} = F_{b1} / d$ and $T_{b2} = F_{b2} / D$ for tight and slack side respectively

Number of times peak forces F_1 sustained $M_1 = \left(\dfrac{Q}{F_1}\right)^x$ Value of x from Table 1.13 is about 11.

Value of Q for A, B, C, D, and E belts are 3,060, 5,420, 9,260, 19,120, and 27,550, respectively.

Number of times peak forces F_2 sustained $M_2 = \left(\dfrac{Q}{F_2}\right)^x$

If N is the number of passes a belt can sustain, then $\dfrac{1}{N} = \dfrac{1}{M_1} + \dfrac{1}{M_2}$

Life of belt in hours is given as $L_h = \dfrac{NL_p}{v \times 3{,}600}$

   Where, L_p = Belt length in meters.

Design of V belt drive

1. For design power and speed of driver, select a suitable V belt from Figure 1.21.
2. Small pulley diameter should not be smaller than minimum pulley diameter given in Table 1.14. Take a standard pulley diameter from Table 1.15.
3. Calculate large pulley diameter D from speeds and N_2. $D = \dfrac{0.98d\, N_1}{N_2}$ (slip of 2 per cent)
4. For the selected small pulley size, calculate belt speed $v = \dfrac{\pi\, dN_1}{60}$

   Assume $v = 20$ m/s and calculate small pulley diameter $d = \dfrac{60v}{\pi\, N_1}$
5. Select suitable center distance C, which varies from $(D + 0.5d)$ to $3(D + d)$.
6. Calculate pitch length of the belt $L_p = 2C + \dfrac{\pi}{2}(D + d) + \dfrac{(D - d)^2}{4C}$ (for open belt)
7. Choose a standard available pitch belt length L_p and standard specified belt size L from Table 1.4. Get belt length correction factor K_L from Figure 1.22.
8. Calculate rated power $P_r = v\left[\dfrac{a}{v^{0.09}} - \dfrac{b}{d\,K_p} - cv^2\right]$
9. Get a suitable service factor K_s from Table 1.6 and contact arc factor K_a from Table 1.16.
10. Calculate number of V belts: $z = \dfrac{P \times K_s}{P_r \times K_L \times K_a}$
11. Calculate number of runs of belt per second $= \dfrac{v}{L_p}$ It should be less than 10.

Types of pulleys   Pulleys are classified in many ways:

According to shape   Flat, stepped, and grooved pulley

According to driving belt   Flat belt pulley, single or multi V groove pulley, toothed pulley.

According to construction   Solid pulley, webbed pulley, armed pulley, and built-up pulley.

According to use   Fast pulley, fast and loose pulley, and idler pulley.

Flat belt pulleys have generally three parts; outer periphery called *rim* almost circular flat. A slight *camber* is given to make slightly convex or which helps in keeping the belt centrally on the pulley. The central portion is *hub*, which is mounted on the shaft using a key. Rim and hub are joined using *arms*.

Small-size flat pulleys are made solid and has no arm.

Large-size pulleys are webbed, or have straight or curved arms.

Very large pulleys are built up using joining different circular segments of pulley.

Minimum pulley diameter depends on number of plies and belt speed.

Standard diameters for flat and V belt pulleys are given in Table 1.15.

Width of pulley is kept wider than the belt width as given in Table 1.17.

Crown (convex surface) is provided to keep the belt centrally on the pulley (Table 1.18).

Solid pulley Small pulleys are made solid. The hole is of the shaft size and has a key way of size $d/4 \times d/4$.

Webbed pulley Pulleys up to 200 mm diameter are made webbed to reduce weight and save material. Web is a disk between the rim and hub, whose thickness is about 30–50 per cent of width of rim. Sometimes, the web is provided with 4 – 6 holes to reduce weight.

Armed pulley Large-size pulleys of diameter more than 300 mm have arms (straight or curved) joining the rim and hub. These are made either of cast iron or of mild steel.

Design of pulley Assume following proportions and then checked for the safe stresses.

Hub It has hole equal to shaft diameter d and keyway of size $(0.25d \times 0.25d)$.

Hub diameter $d_h = 1.5\ d + 25$ mm but not more than $2d$.

Hub length $L_h = 0.67\ w$ to w Where, w is of width of the rim.

Arm Number of arms $n = 4$ for diameter up to 400 mm, and $n = 6$ for more than 450 mm.

Cross section of the arm is generally elliptical but can be circular. Major axis a is in the plane of rotation. Width of arm near the boss assumed as: $a = 2.94 \sqrt[3]{\dfrac{wD}{4n}}$

w is width of rim, and thickness of arm $b = 0.5\ a$

Arms are generally tapered with 2 – 4 mm per 100 mm of arm length.

Rim Its radial thickness varies from $(0.0033D + 2$ mm$)$ to $(0.005D + 3$ mm$)$.

The section is checked for stresses assuming it as a cantilever beam supported at hub and a concentrated load at rim. Thus bending moment $M = \dfrac{2T}{n}$

Modulus of section for elliptical section: $Z = \dfrac{\pi}{64} a^3$ (assuming $b = 0.5a$) $M = \sigma_b \times Z$.

Values of M and σ_b are known, calculate Z and then major axis size a.

Built-up pulleys are used for very large diameter. Both hub and rim are made in two halves and then joined together with bolts and nuts or rivets. Hub and rim are provided with radial holes in which arms of circular cross section are fixed. The two halves of hub are bolted on shaft with a key in between.

Spokes are in a single row for diameter < 300 mm and two rows for diameter > 300 mm.

Diameter of spoke is 19 mm up to pulley 710 mm and 22 mm for larger diameter.

Number of spokes is 4 for diameter < 280 mm, 6 up to 500 mm, and 8 up to 710 mm.

Boss   Length is taken half the width of face (rim) but should not be less than 100 mm for spokes of 19 mm and 140 mm for 22 mm spokes.

Rim   Most of the rims are made of 5 – 6 mm thick plates.

Stepped pulley   consists of many pulleys of different diameters casted or forged together as one piece. These are used in pair and arranged in such a way that the smallest pulley drives the largest pulley and the largest drives the smallest. The steps can be three or four.

Fast and loose pulley   Two pulleys placed adjacent to each other. One pulley is keyed to the shaft called *fast pulley*, whereas the other is without key, called *loose pulley*. Loose pulley runs freely on a bush without transmitting power to shaft. Whenever the power transmission is not required, belt is made to slip over to loose pulley.

Grooved pulleys   These pulleys use V belts of standard cross section A, B, C, D, and E. Section A is for less power and E for large power. If the power to be transmitted is more than of one belt, two or more belts are used in parallel. Use Table 1.14 for minimum size and preferred diameter. Standard pulley diameter from Table 1.15.

Angle of pulley groove is made lesser than 40° for more wedging action. Driver pulleys have angle 32°–34° and large pulley 36° – 38°. The depth of the groove is also kept about 10 per cent more than the actual depth.

Single belt grooved pulley   It may be solid or webbed or armed.

Multibelt grooved pulley   Many V grooves are cut adjacent to each other. Wall thickness between the grooves is 3 – 5 mm. Outer wall thickness is kept slightly more than inside wall thickness. The gap between grooves depends on material.

Stepped grooved pulley   Purpose is same as that of stepped pulley for flat belt. Angle of groove is kept between 34° and 38°.

Toothed pulley   Their use is limited to applications, where slip may affect performance of machine, for example, cam shaft drive in I.C. engines. The outer periphery has teeth, which fits into the grooves of the toothed belt to check slip. A collar is provided on the sides of driver to keep the belt within the width of the pulley.

Theory Questions

1. What are the advantages and disadvantages of a belt drive?
2. Why the slack side of the belt of a horizontal belt drive is preferable to place on the top side?
3. Describe the different belt arrangements over pulleys.
4. What are the various types of belts? Describe them by sketches.
5. What are the various standard cross sections of V belts? How are these designated?
6. Describe the construction of a V belt with a neat sketch.
7. What are the various belt materials? Explain their properties and use.
8. Explain the terms 'initial tension' and 'centrifugal tension'.

9. What is the effect of centrifugal tension on power transmission?

10. Discuss the various joints for joining a flat belt with neat sketches.

11. Derive the expression for calculating contact angle for small and large pulleys for open and crossed belts.

12. What is the power capacity of a belt? How is it calculated for a belt of given size?

13. Describe slip and creep in belts. .

14. Write the design procedure for a flat belt drive.

15. Derive expressions for length of an open and cross belt drive.

16. How will you calculate center distance, if lengths of belt and pulley diameters are known?

17. Write the procedure to design a V belt drive system.

18. Explain the length and arc factors for V belt design.

19. Classify pulleys. Sketch any three types.

20. Differentiate between solid and webbed pulley. What are the advantages of webbed pulley over solid pulley?

21. Describe the construction of a built-up pulley with a neat sketch.

22. What are the various types of armed pulleys? Sketch any two types.

23. What is the use of a stepped pulley? Sketch a stepped pulley and describe its construction.

24. Describe the application of a fast and loose pulley. How do they work?

25. Differentiate between a flat and grooved pulley with sketches.

26. Sketch a stepped grooved pulley and describe its use and applications.

27. Where is multi-grooved pulley used? Sketch a pulley for three V belts of B size.

28. What is the application of a toothed belt? What are its advantages over other types?

Multiple Choice Questions

1. Belt drives are suitable where:
   (a) Center distance is large
   (b) Speed ratio need not be exact
   (c) Lubrication is not possible
   (d) All given in a, b, and c

2. Which of the following is not a suitable material for belt?
   (a) Leather
   (b) Rubber
   (c) Asbestos
   (d) Plastic

3. Type of belt used for line shafts is:
   (a) Flat
   (b) V
   (c) Grooved
   (d) Ribbed

4. In an open horizontal belt system, tight side of belt should be kept at:
   (a) Top
   (b) Bottom
   (c) Can be anywhere
   (d) Not important

5. Direction of rotation of driven shaft is opposite to driver shaft in:
   (a) Open-belt system
   (b) Crossed-belt system
   (c) Quarter twist-belt system
   (d) None of above

6. Included angle for a V belt is:
   (a) 32°
   (b) 36°
   (c) 40°
   (d) 45°

7. If 2β is groove angle of a pulley, ratio of gripping capacity of V belt to flat belt is:
   (a) 0.5 cosec β
   (b) cosec β
   (c) cosec 2β
   (d) sin 2β

8. If T_1 and T_2 are tight side and slack side belt tensions, initial tension is given as:
   (a) $\left(\dfrac{\sqrt{T_1}+\sqrt{T_2}}{2}\right)^2$
   (b) $\left(\dfrac{\sqrt{T_1}-\sqrt{T_2}}{2}\right)^2$
   (c) $\dfrac{\sqrt{T_1}+\sqrt{T_2}}{2}$
   (d) $\dfrac{\sqrt{T_1}\cdot\sqrt{T_2}}{2}$

9. Centrifugal tension:
   (a) Decreases power transmission
   (b) Increases power transmission
   (c) Has no effect on power transmission
   (d) Decreases belt life

10. If m is mass per unit length of V belt and T_{max} maximum torque, then for maximum power, belt velocity should be:
    (a) $\sqrt{\dfrac{T_{max}}{4m}}$
    (b) $\sqrt{\dfrac{T_{max}}{3m}}$
    (c) $\sqrt{\dfrac{T_{max}}{2m}}$
    (d) $\dfrac{T_{max}}{4m}$

11. Creep in a belt is:
    (a) Cyclic slip on pulley surface
    (b) Permanent deformation of belt
    (c) Slip of belt due to low friction
    (d) None of above

12. Power transmission capacity of a belt depends on:
    (a) Belt tensions
    (b) Coefficient of friction
    (c) Contact angle
    (d) All given in a, b, and c

13. Life of a belt in number of hours increases if:
    (a) Speed of rotation is high
    (b) Belt length is more
    (c) Center distance is less
    (d) Its cross-sectional area is more

14. V belts of cross section A are suitable for:
    (a) Low speed low power
    (b) Medium speed low power
    (c) High speed low power
    (d) Medium speed high power

15. Large-size pulleys are made:
    (a) Using web between hub and boss
    (b) With arms of mild steel
    (c) Solid and forged
    (d) In two or three parts and then joined

16. Generally arms of a large pulley have cross section:
    - (a) Square
    - (b) Elliptical with major axis in plane of rotation
    - (c) Circular
    - (d) Elliptical with minor axis in plane of rotation

17. Fast and loose pulleys are used where:
    - (a) Speed has to vary from slow to fast
    - (b) There is no change in speed
    - (c) Power has to be stopped intermittently
    - (d) Power varies with time

18. Stepped pulleys are used where:
    - (a) Uniform speed variation is desired
    - (b) Speed varies in steps
    - (c) Power is to be transmitted vertically upwards
    - (d) Gear box cannot be used

19. Toothed pulley is used for:
    - (a) Variable speed ratio
    - (b) Escalators
    - (c) Timing gears in I.C. engine
    - (d) None of above

20. Crown on a flat pulley is used to:
    - (a) Increase belt tensions
    - (b) Easy removal of belt from the pulley
    - (c) Increase life of a belt
    - (d) Keep the belt centrally positioned on pulley

Answers to multiple choice questions

1. (d)	2. (c)	3. (a)	4. (b)	5. (b)	6. (c)	7. (b)	8. (a)	9. (c)	10. (b)
11. (a)	12. (d)	13. (b)	14. (c)	15. (d)	16. (b)	17. (c)	18. (b)	19. (c)	20. (d)

Design Problems

Flat belts

1. Design a flat belt for transmitting 20 kW at 720 rpm to a pump at 240 rpm at a belt speed of 25 m/s. Coefficient of friction = 0.4, design stress for belt is 2.2 MPa, and density of belt material 0.01 N/m².
    $$[t = 6 \text{ mm}; b = 140 \text{ mm}]$$

2. A flat belt runs over two pulleys of diameter 800 mm and 1,400 mm with center distance 2.4 m. A power of 90 kW is transmitted from 800 mm pulley. Assuming slip as 2 per cent and design stress for belt material 2.8 MPa, calculate the size of flat belt. Assume belt velocity 20 m/s, coefficient of friction 0.3. and safe stress 2.1 MPa.
    $$[t = 15 \text{ mm}; b = 250 \text{ mm}]$$

3. Select a suitable flat belt to transmit 18 kW from a motor rotating at 1,440 rpm to other shaft at 680 rpm with center distance 2.6 m. Assume belt velocity 18 m/s and coefficient of friction 0.38. Take density of belt material 0.01 N/m³, and safe stress 2.1 MPa.
    $$[t = 8 \text{ mm}; b = 100 \text{ mm}]$$

4. An electric motor of 5 kW rotating at 2,880 rpm drives a compressor at 720 rpm. Design a flat leather belt for minimum center distance.
    $$[b = 50 \text{ mm}; \text{plies} = 3; L_p = 1.992 \text{ mm}]$$

5. A power of 30 kW is transmitted from a shaft rotating at 300 rpm to another shaft rotating at 150 rpm. The driver pulley is located in the middle of two bearings 800 mm apart. The driven pulley

is over hung by 80 mm. Take coefficient of friction 0.3 and center distance between the shafts as of sum of diameter of pulleys.

(a) Select a suitable flat belt.

(b) Calculate shaft diameters, taking safe shear stress 80 MPa.

$$[(a) \ b = 250 \ mm; p = 6 \ (b) \ d1 = d2 = 50 \ mm]$$

6. Design a belt drive to transfer 110 kW through two pulleys of diameters 900 mm and 1,200 mm with center distance 3.6 m, belt velocity 20 m/s, coefficient of friction 0.28, slip of 2 per cent at each pulley, power rating of the belt 0.012 kW/mm/ply, and 20 per cent over load. $[b = 355 \ mm; p = 8]$

7. In a horizontal belt drive for a centrifugal blower, the blower is belt driven at 600 rpm by a 15 kW 1,750 rpm electric motor. The center distance is twice the diameter of the larger pulley. The density of the belt material is 1500 kg/m^3. Maximum allowable stress is 4 MPa, $\mu_1 = 0.5$ (motor pulley), $\mu_2 = 0.4$ (blower pulley), and peripheral velocity of the belt = 20 m/s. Determine the following:

(a) Pulley diameters

(b) Belt length

(c) Minimum initial tension for operation without slip

(d) Resultant force in the plane of blower, when operation with an initial tension 50 per cent greater than the minimum value.

$$[(a) \ d_1 = 218 \ mm; d_2 = 636 \ mm; (b) \ L = 3.92 \ m; A = 294 \ mm^2; (c) \ T_i = 801 \ N; (d) \ F = 1,231 \ N]$$

8. A belt 100 mm wide and 10 mm thick is transmitting power at 1,000 m/min. The net driving tension is 1.8 times the tension on the slack side. If the safe permissible stress on the belt section is 1.6 MPa. Calculate the power that can be transmitted at this speed. Assume the density of the leather as 1,000 kg/m^3. Calculate the absolute maximum power that can be transmitted by this belt and the speed at which it can be transmitted.

$$[P = 11.85 \ kW, P_{max} = 16.42 \ kW, v = 23.1 \ m/s]$$

9. A flat belt is required to transmit 30 kW from a pulley of 1.6 m effective diameter running at 400 rpm. The angle of contact is spread over one third of circumference. The coefficient of friction between the belt and pulley surface is 0.3. Taking centrifugal tensions into account, determine the width of belt required. It is given that the belt thickness is 9.8 mm. Density of material is 1,300 kg/m^3 and related permissible working stress is 2.5 MPa. $[b = 190 \ mm]$

10. Determine width of flat belt and belt length for the following particulars of a drive.

Power = 100 kW

Center distance = 6,000 mm

Pulley diameters = 1,680 mm and 420 mm

Speed of bigger pulley = 240 rpm

Leather belt thickness = 8 mm

Coefficient of friction $\mu = 0.3$

Material density $\rho = 920$ kg/m^3

Allowable stress in belt = 3.0 N/mm^2 $[b = 340 \ mm]$

11. Layout of a 6-mm-thick leather belt drive transmitting 15 kW power at 1440 rpm to other pulley at 480 rpm as shown in Figure 1.P1. The center distance between the pulleys is twice the diameter of the big pulley. The belt velocity is approximately 20 m/s and the allowable stresses for the belt is 2.25 N/mm². Taking the density of leather as 950 kg/m³ and coefficient of friction as 0.35, calculate:

    (a) Diameter of pulleys
    (b) The belt tensions
    (c) The length pitch and width of the belt

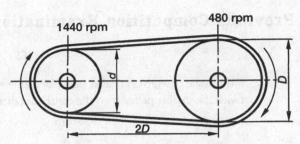

Figure I.PI A flat belt drive system

$$[(a)\ d = 270\ mm,\ D = 810\ mm;\ (b)\ T_1 = 1{,}199\ N,\ T_2 = 449\ N]$$
$$(c)\ L_p = 4{,}982\ mm,\ b = 90\ mm]$$

V Belts

12. 10 kW power is transmitted through a V belt B over a pulley of pitch circle diameter 200 mm from a shaft rotating at 2,000 rpm to another shaft at 800 rpm. Center distance is approximately 1 m and service factor may be taken as 1.2. Calculate the pitch length of standard belt and number of belts.

$$[L_p = 3{,}091\ mm,\ z = 2]$$

13. An electric motor of 25 kW rotating at 1,440 rpm drives a compressor at 900 rpm using open type V belt. Center distance is to be approximately 2.5 m. Service factor can be taken as 1.1. Design cross section and standard pitch length of a V belt, number of belts, and pulley diameters.

$$[Cross\ section\ C,\ L_p = 6{,}873\ mm,\ z = 2,\ d = 315\ mm,\ D\ 500\ mm]$$

14. Two shafts are located 1.3 m apart. Driver pulley rotating at 1000 rpm on a shaft has effective diameter of 300 mm and transfers 95 kW through a V belt of included angle 40° to driven pulley rotating at 375 rpm. Area of belt is 400 mm², and safe tensile strength 2.1 MPa. Coefficient of friction is 0.28 and density of belt material is 1,100 kg/m³. Driven pulley is overhung by 200 mm from the nearest bearing. Calculate:

    (a) Type and number of belts required.
    (b) Diameter of driven pulley shaft, if shear stress is limited to 50 MPa.
    (c) Pitch length of belt. $[(a)\ Belt\ size\ D,\ z = 10;\ (b)\ D = 70\ mm;\ (c)\ L_p = 4{,}375\ mm]$

15. Design a V belt for transmitting 7.5 kW to a blower with the following specifications:

    Speed of driver pulley = 1,440 rpm

    Speed of driven pulley = 400 rpm

    Center distance = 1,600 mm

    Service conditions = 16 hours/day   [Two V belts B-5377]

Previous Competition Examination Questions

IES

1. If the angle of wrap on smaller pulley of diameter 250 mm is 120° and diameter of larger pulley is twice the smaller pulley, then the center distance between the pulleys for an open belt drive is:

   (a) 1,000 mm   (b) 750 mm   (c) 500 mm   (d) 250 mm

   [Answer (a)] [IES 2015]

2. If the velocity ratio for an open belt drive is 8 and the speed of driving pulley is 800 rpm, then considering elastic creep of 2 per cent, the speed of driven pulley is:

   (a) 104.4 rpm   (b) 102.4 rpm   (c) 100.4 rpm   (d) 98.04 rpm

   [Answer (d)] [IES 2015]

3. In the assembly of pulley, key, and shaft:

   (a) Pulley is made the weakest   (b) Key is made the weakest

   (c) Key is made the strongest   (d) All three are designed for equal strength

   [Answer (b)] [IES 2013]

4. Considering the effect of centrifugal tension in a flat belt drive with T_1 = tight side tension and T_c = centrifugal tension, and m = mass per unit length of the belt, the velocity of the belt for maximum power transmission is given by:

   (a) $v = \sqrt{\dfrac{T_1}{3m}}$ (b) $v = \sqrt{\dfrac{T_c}{3m}}$ (c) $v = \sqrt{\dfrac{T_1 - T_c}{3m}}$ (d) $v = \sqrt{\dfrac{T_1 + T_c}{3m}}$

   [Answer (d)] [IES 2013]

GATE

5. Which of the following statement is correct regarding power transmission through V belts?

   1. V belts are used at high speed end.
   2. V belts are used at low speed end.
   3. V belts are of standard length.
   4. V angles of pulleys and belts are standardized.

   Select the correct answer using the code given below:

   (a) 1 and 3 only   (b) 2 and 4 only

   (c) 2, 3, and 4   (d) 1, 3, and 4

   [Answer (d)] [GATE 2015]

6. If with T_1 and m represnt the maximum tension and mass per unit length of a belt, then the maximum permissible speed of the belt is given by:

   (a) $v = \sqrt{\dfrac{T_1}{3m}}$ (b) $v = \sqrt{\dfrac{3T_1}{m}}$ (c) $v = \sqrt{\dfrac{2T_1}{m}}$ (d) $v = \sqrt{\dfrac{T_1}{m}}$

   [Answer (a)] [GATE 2015]

7. A flat belt has an angle of wrap of 160° on the smaller pulley. Adding an idler increases the wrap angle to 200°. The slack side tension is the same in both the cases and centrifugal force is negligible. By what percentage is the torque capacity of the belt drive increased by adding the idler? Use coefficient of friction 0.3. [Answer (35.4 per cent)] [GATE 2001]

8. The ratio of tension on the tight side to that on the slack side in a flat belt drive is:

(a) Proportional to the product of coefficient of friction and lap angle.

(b) An exponential function of product of coefficient of friction and lap angle.

(c) Proportional to lap angle.

(d) Proportional to the product of coefficient of friction.

[Answer (b)] [GATE 2000]

□□□

Rope Drives

2.1 Rope Drives

Belt drives described in Chapter 1 are useful for moderate power and center distance between the two shafts up to 8 m. Drives using a rope are used, where power is large and distance between pulleys is also more. Rope drives can be of two types:

- Fibre rope drive [Figure 2.1(a)]: For pulleys up to 50 m apart.

- Wire rope drive [Figure 2.1(b)]: For pulleys up to 150 m apart.

(a) Fiber rope   (b) Wire rope

Figure 2.1 Types of ropes

2.2 Fibre Rope Drive

2.2.1 Advantages

Fibre ropes offer the following advantages:
- High mechanical efficiency
- Smooth and quiet drive
- Little misalignment of the shafts can be easily tolerated
- Suitable for outdoor applications

2.2.2 Fibre rope materials

Ropes are available in a variety of materials. The type of material used for the rope is the main determinant of the rope's strength, abrasion resistance, ease of use, and price.

Nylon Nylon is the strongest easily available rope material. It is quite strong, but loses about 15 per cent of its strength in wet condition. So it should not be used, where the rope will be exposed to water. It is very dense and sinks in water. It is the best choice of rope for many applications like marine, general purpose, and towing.

Polypropylene It is most popular due to its price, as it is the cheapest of all synthetic fibres. It is strong for its weight, but it is not very resistant to abrasion or heat. For this reason, it is generally not a good choice for long-term applications, where the rope is exposed to sun or abrasion like in a dock line. Its characteristic, other than its excellent price, is that it floats. It is an important feature in rescue line operations.

Polyester Polyester is almost as strong as nylon, but it does not lose strength when wet. It has the highest resistance to abrasion and heat. Main difference between nylon and polyester is the elasticity. Nylon stretches much more than polyester. Polyester is also more expensive and not easy to work with. It is also more difficult to untie than a nylon rope.

Because of these factors, nylon is more appropriate for the applications, where slight stretch is desirable, such as dock lines or the core of dynamic climbing ropes. Similarly, polyester line is more appropriate for applications, where stretch is not desired, such as lifting slings and hammock guy lines.

Natural fibres   In general, natural fibres are heavier, weaker, and less resistance to all forms of abrasion than synthetic fabrics. For this reason, there are a few applications in which a natural fibre is preferred. Some specific applications, where natural fibres are preferred include those, where the roughness of the rope is beneficial (e.g. climbing rope). Natural fibre rope is also commonly used for decorative purposes, but not recommended for load bearing.

Manila   Manila fibres are rough and not much flexible. These ropes have less strength and wear internally due to sliding of fibres. To reduce wear, they are lubricated with tar or graphite. These ropes are available between 38 mm to 50 mm diameter. Their breaking strength in kilo newton can be taken as $50d^2$ (where d is rope diameter in mm).

Cotton   Cotton ropes are soft and smooth. Sometimes, these ropes are also lubricated to reduce wear. These ropes have lesser strength and are costlier than manila ropes. Maximum diameter of these ropes can be 50 mm. Breaking strength of these ropes is lesser than manila ropes and it can be taken in kilo newton as $35d^2$ (where d is rope diameter in mm).

2.3 Sheave for Fiber Ropes

Pulley is used for flat and V belts, and the pulley used for ropes is called a sheave. It is the rotating part of the pulley system. Number of grooves can be one or more than one as shown in Figure 2.2. Groove angle is made 45°. The bottom of the circular rope does not touch the bottom of the groove and is squeezed between the inclined sides of the groove. Sheave diameter is kept between 35 to 40 times the diameter of rope, to reduce bending stresses.

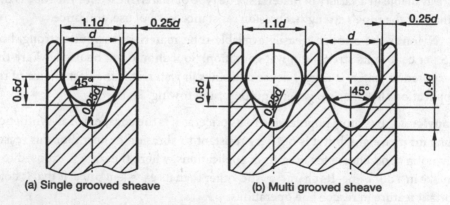

(a) Single grooved sheave   (b) Multi grooved sheave

Figure 2.2 Sheave for fibre ropes

2.4 Design of Rope Drive

Fibre rope drive is also designed as for V belts, except that the angle is 45° instead of 40° for V belts. Ratio of tight and slack side tensions T_1 and T_2, respectively, is given as:

$$\frac{T_1}{T_2} = e^{\mu\theta\cosec 22.5°} = e^{2.63\,\mu\theta} \tag{2.1}$$

Where, θ = Angle of wrap over the sheave

μ = Coefficient of friction

If centrifugal force T_c is to be considered: $T_c = \dfrac{wv^2}{g}$

And tension ratio is: $\dfrac{T_1 - T_c}{T_2 - T_c} = e^{2.63\,\mu\theta}$ (2.2)

Where, T_c = Centrifugal force (N)

w = Weight of rope per meter length (N/m)

v = Rope velocity (m/s)

For maximum power, speed of belt / rope is taken as: $v = \sqrt{\dfrac{T}{3m}}$ (2.3)

Where, T = Maximum tension in rope (N)

m = Mass of rope (kg/m)

Weight of the rope is given in Tex, which depends on its density. Tex is the weight of rope in kilo newton for 1 km length of rope. Weight and breaking load for some of the cotton ropes are given in Table 2.1.

Table 2.1 Weight and Breaking Load for Cotton Ropes (IS 2452–1985))

Diameter of Rope d, (mm)	Weight per meter, w (N/m))	Breaking Load, W (kN)
19	2.90	10.34
22	3.70	15.21
25	4.60	19.64
28	6.05	24.63
32	7.30	32.17
38	10.80	45.17
44	14.20	60. 83
50	17.35	78.55

If Table 2.1 is not available, for cotton rope of diameter d in mm, following empirical relations can be used to estimate approximate values as under:

Weight of rope (N/m), $w = 0.0072d^2$ (2.4)

Breaking load (kN), $W = 0.0315d^2$ (2.5)

- Factor of safety is taken quite high for rope drives. It varies from 35 to 40.
- Rope velocity varies from 15 m/s to 35 m/s, but most economical is 20 m/s to 25 m/s.

Breaking load is taken as the maximum tension, that is, tension on tight side T_1.

Example 2.1

Power capacity with fiber rope of given size

Calculate maximum power, which a 25-mm fibre rope can transmit at a rope speed of 20 m/s. Breaking load can be taken 19 kN, factor of safety 40, angle of wrap over smaller sheave 160°, and coefficient of friction 0.15. Assume weight of rope $0.007d^2$ N/m.

Solution

Given $d = 25$ mm $v = 20$ m/s $W = 19$ kN $FOS = 40$ $\theta = 160°$

$\mu = 0.15$ $w = 0.007d^2$

Weight of rope, $w = 0.007d^2 = 0.007 \times 25^2 = 4.375$ N/m

Centrifugal tension $T_c = \dfrac{wv^2}{g} = \dfrac{4.375 \times 20^2}{9.81} = 178.6$ N

Contact angle $\theta = \dfrac{160 \times 3.14}{180} = 2.79$ rad

From Equation (2.2): $\dfrac{T_1 - T_c}{T_2 - T_c} = e^{2.63\mu\theta}$

Substituting the values: $\dfrac{T_1 - 178.6}{T_2 - 178.6} = e^{2.63 \times 0.15 \times 2.79} = e^{1.1} = 3$

Or, $T_1 = 3\,T_2 - 357.2$

Safe load $= \dfrac{W}{FOS} = \dfrac{\text{Breaking load}}{40} = \dfrac{19,000}{40} = 475$ N

Maximum tension in belt should not be more than this load. That is, $T_1 = 475$

Hence, $3T_2 - 357.2 = 475$ N   or $T_2 = 277.4$ N

Power (kW) $= \dfrac{(T_1 - T_2)v}{1,000} = \dfrac{(475 - 277.4) \times 20}{1,000} = 3.95$ kW

Example 2.2

Diameter of a rope for given power

A power of 60 kW is to be transmitted using fibre ropes through a set of pulleys with speed ratio 4:1. Rope speed is limited to 18 m/s and number of ropes is limited to 6. Calculate the size of rope with factor of safety 35, coefficient of friction 0.18, weight of load per $0.007\,d^2\,\text{N/m}$, and breaking load in kilo newton $35d^2$ (d is rope diameter in mm). Minimum sheave diameter can be assumed $40\,d$.

Solution

Given $P = 60\,\text{kW}$ $v = 18\,\text{m/s}$ $FOS = 35$ $N_1 / N_2 = 4$
$\qquad\qquad W = 35d^2$ $\mu = 0.18$ $z = 6$ $w = 0.007\,d^2$
$\qquad\qquad D_1 = 40\,d$

Minimum sheave diameter $D_1 = 40\,d$

Bigger sheave diameter $D_2 = 4 \times 40\,d = 160\,d$

Minimum center distance between shafts:

$$C = (D_2 - D_1) + 1.5\,D_2 = (160\,d - 40\,d) + (1.5 \times 160\,d) = 360\,d$$

$$\theta = \pi - 2\sin^{-1}\left(\frac{D_2 - D_1}{2C}\right) = 180 - 2\sin^{-1}\left(\frac{160\,d - 40\,d}{2 \times 360\,d}\right) = 161°$$

Contact angle in radians $\theta = \dfrac{161 \times 3.14}{180} = 2.81\,\text{rad}$

Centrifugal tension $T_c = \dfrac{wv^2}{g} = \dfrac{0.007d^2 \times 18^2}{9.81} = 0.231d^2\,\text{N}$

With centrifugal tension $\dfrac{T_1 - T_c}{T_2 - T_c} = e^{2.63\mu\theta} = e^{2.63 \times 0.18 \times 2.81} = e^{1.32} = 3.74$

Substituting the values:

$$\frac{T_1 - 0.231d^2}{T_2 - 0.231d^2} = 3.74$$

Safe load $= \dfrac{\text{Breaking load}}{FOS} = \dfrac{35d^2}{35} = d^2\,\text{N}$

Maximum tension in belt should not be more than this load. That is, $T_1 = d^2$

Hence,

$$\frac{d^2 - 0.231d^2}{T_2 - 0.231d^2} = 3.74$$

Or, $0.769d^2 = 3.74\,T_2 - 0.864d^2$ Or, $T_2 = 0.437d^2$

Power for z number of ropes (kW) $P = \dfrac{(T_1 - T_2)zv}{1,000} = \dfrac{(d^2 - 0.437d^2) \times 6 \times 18}{1,000}$

Substituting the values: $60 = \dfrac{0.563d^2 \times 6 \times 18}{1,000}$ Or, $d = 31.4\,\text{mm}$

Selecting a standard diameter, $d = 32$ mm.

2.5 Wire Ropes

Wire rope is a flexible cable, which consists of several strands of metal wires. Earlier the wires were made of wrought iron, but now steel wires are used, which are twisted to form a helix. Wire ropes of diameter less than 9.5 mm are called cables. Use of wire ropes is quite old since 1830s, when these were used for mining hoist applications. Typical use of wire ropes can be seen in many applications like:

- Lifting cranes, lifts, and hoisting applications;
- In mechanisms, like clutch and gear wires, connections from airplane cockpit using levers and pedals to different parts;
- In pre-stressed concrete bridges, use of static steel wire ropes can be seen; and
- Suspension bridges use steel wire rope between two towers at extreme ends.

2.5.1 Advantages of wire ropes

Wire ropes offer the following advantages over fiber ropes:

- Lighter than fiber ropes for the same strength
- Can withstands shocks
- More reliable and economical
- Higher efficiency
- Longer life than fiber ropes

2.5.2 Wire rope materials

A wire rope can be made of fabric or of metallic wires. Materials for each are given below: Stranded wire rope and cable are made from various grades of stainless steel or carbon steel.

Stainless steel (S/S) Type 302/304 S/S is the most common non-magnetic stainless steel (18 parts chromium and 8 parts nickel). Type 305 S/S cable is commonly available in 1.5 mm and 0.75 mm diameters. This type of steel is generally used for sensitive instrumentation or other systems, which can be affected by magnetism. Type 316 S/S is popular in many outdoor saltish air environments. It has excellent corrosion protection and is the preferred choice of the marine industry.

Monel It is an uncommon grade of stainless steel used in applications, where resistance to corrosive substances and liquids is required. The main drawback is its less breaking strength, which is about 30 per cent less than that of Type 302/304 S/S. It is not readily available.

Carbon steel wire It is available in different grades, sometimes referred as plow steel, improved plow steel, or even extra improved plow steel. Wire rope manufactured from uncoated wire is commonly referred as Bright wire rope and is available in sizes 6 mm and above.

Galvanized carbon steel wire It is frequently used to manufacture wire rope, as these provide good corrosion protection. In sizes less than 6 mm, it is used when stainless steel is not specified.

2.6 Construction of Wire Ropes

A wire rope consists of a core in the center, around which a number of multi-wired strands are laid helically. The most popular arrangement is six strands around the core. This combination gives the best balance. The number of wires per strand may vary from 3 to 91. Most of the wire ropes use 7-wires, 19-wires, or 37-wires strands.

2.6.1 Core

Purpose of the core is to provide support and maintain the position of the outer strands during operation. Any number of multi-wired strands may be laid around the core. There are two general types of cores for wire rope.

a. **Fibre core** It is made from natural or synthetic fibres. Fibre cores are the most flexible and elastic, but have the disadvantage of getting crushed easily.

b. **Wire strand core** The wire strand core can be of any one type given below:
   - Independent wire rope core (IWRC) It is the most durable in all types of environments.
   - Strand core (SC) It is made up of one additional strand of wire and is typically used for suspension.

2.6.2 Wires

Steel wires for wire ropes are normally made of carbon steel with 0.4 – 0.95 per cent carbon. High strength of the wires enables wire ropes to bear large tensile loads and runs over sheaves with relatively small diameters without exceeding the safe bending stress. The most common steel wire grades are:
   - Improved plow steel (IPS)
   - Extra improved plow steel(EIP)
   - Stainless steels and other special grades are provided for special applications.

Most wire ropes are made with round wires. Triangular and other shaped wires are also used for special constructions. Higher the strength of the wire, lower is its ductility.

2.6.3 Strands

In a strand, wires of the different layers cross each other in different ways; parallel lay (Figure 2.3) and crossed lay, which are seldom used.

Parallel lay strands   Lay length of all the wire layers is equal and the wires of any two superimposed layers are parallel, resulting in linear contact. It is used most commonly. The wire of the outer layer is supported by two wires of the inner layer. These wires are neighbors along the whole length of the strand.

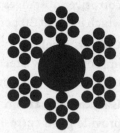

Figure 2.3 Strands in a wire rope

Parallel lay strands are made in one operation. Parallel lay strands with two wire layers are called as Filler, Seale, or Warrington (See Section 2.8). Endurance strength of wire ropes with parallel lay is much higher than cross lay strands.

2.7 Lay of Wire Ropes

Wire rope is identified not only by its component parts but also by its construction, that is, the way the wires have been laid to form strands, and by the way the strands have been laid around the core. The lay of a wire rope is a description of the arrangement; wires and strands that are placed during construction. A lay is specified by its direction of strands.

- **Right lay**   Strands pass from left to right across the rope. The lay direction of the strands in the rope can be right (symbol Z) or left (symbol S); and
- **Left lay**   Strands pass from right to left. Lay direction of the wires can be right (symbol z) or left (symbol s). Lowercase characters are used.

The way wires are placed within each strand is given in the next section.

2.7.1 Regular lay

Wires in the strands are laid opposite in direction to the lay of the strands. It results in individual wires running parallel to the longitudinal axis of the rope [Figure 2.4(a) and (b)].

2.7.2 Lang lay

Wires are laid in the same direction as the lay of the strands [Figure 2.4(c)]. Lang lay rope is one, in which the direction of the wires of the individual strand is the same as that of the strands in the rope resulting in the individual wires running diagonally across the longitudinal axis of the rope. Lang lay ropes are more flexible and greater wearing surface per wire than regular lay ropes. It is recommended for excavating, construction, mining applications, and hoist lines. These are more prone to the abuses of bending over small diameter sheaves, pinching in undersize sheave grooves, and crushing, while winding on drums. It has the greatest usage in oil fields on rod and tubing lines and blast hole rigs.

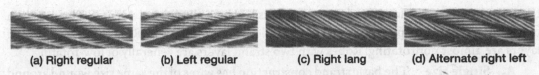

| (a) Right regular | (b) Left regular | (c) Right lang | (d) Alternate right left |

Figure 2.4 Lay of wire ropes

Most of the wire ropes are right and regular lay. This combination of lay is used by most of the applications and meets the requirements. Other lay specifications are considered exceptions and have to be specially ordered.

Lang lay ropes are to be recommended by a qualified wire rope engineer. Their use is limited for some applications only. These are advantageous over regular lay ropes in the following properties:

- Greater wearing surface, result in less wear on sheave;
- Greater bending fatigue resistance; and
- Not as stable as regular lay ropes from crushing resistance point of view.

2.7.3 Ordinary lay

The lay direction of the wires in the outer strands is in the opposite direction to the lay of the outer strands.

2.7.4 Alternate right and left lay

The wire rope strands are alternately regular lay and lang lay [Figure 2.4(d)]. This construction is seldom used, its principal application is for conveyors.

2.8 Types of Wire Ropes

2.8.1 Spiral ropes

Spiral ropes are round strands with assembly of layers of wires laid helically over a center, with at least one layer of wires in the opposite direction to that of the outer layer. Spiral ropes can be dimensioned in such a way that they are non-rotating, which means that under tension the rope torque is nearly zero. The open spiral rope consists of only round wires. The half-locked coil rope and the full-locked coil rope always have a center made of round wires. The locked coil ropes have one or more outer layers of profile wires. Advantages of spiral rope are:

- Its construction prevents the penetration of dirt and water to a greater extent;
- It also protects from loss of lubricant; and
- Ends of a broken outer wire cannot leave the rope.

2.8.2 Stranded ropes

Most types of stranded ropes have one strand layer over the core. Stranded ropes are of several strands laid helically in one or more layers around a core. Multi-strand ropes are less resistant to rotation and have at least two layers of strands laid helically around a center. The direction of the outer strands is opposite to that of the underlying strand layers. Ropes with three strand layers are almost nonrotating. Ropes with two strand layers are generally of low-rotating.

A. Single size The basic strand construction has wires of the same size wound around a center as shown in Figure 2.5(a).

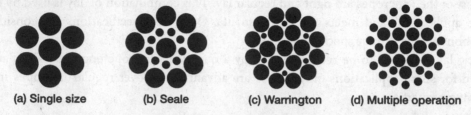

(a) Single size   (b) Seale   (c) Warrington   (d) Multiple operation

Figure 2.5 Types of ropes

B. Seale Large outer wires with the same number of smaller inner wires around a core wire [Figure 2.5(b)] provide an excellent abrasion resistance but less fatigue resistance. When used with an IWRC, it offers excellent crush resistance over drums.

C. Warrington Outer layer of alternately large and small wires provides good flexibility and strength, but low abrasion and crush resistance [Figure 2.5(c)].

D. Multiple operation One of the above strand designs may be covered with one or more layers of uniform-sized wires [Figure 2.5(d)].

Finish

- Bright finish is suitable for most applications.
- Galvanized finish is available for corrosive environments.
- Plastic jacketing is also available on some constructions.

2.9 Designation of Ropes

Wire ropes are designated as: $S \times W$

Where, S = Number of strands

W = Number of wires in each strand

For example, the most commonly used wire rope 6×7 means, that it has six strands and seven wires in each strand.

2.10 Classification of Wire Ropes

A. 6×7 Classification These ropes have six strands and number of wires may vary from 3 to 14. Rope with seven wires is most common. Outside wires are not more than nine. These are available in diameter from 6 mm to 38 mm. Lay may be left or right, and regular or lang.

Characteristics:

- Give long service, where ropes are dragged along the ground or over rollers;
- Excellent abrasion resistance;
- Less bending fatigue resistance; and
- Larger sheaves and drums are required to avoid breakage from fatigue.

Applications

Dragging and haulage in mines and in tramways.

B. 6 × 19 Classification   These ropes have six strands and number of wires may vary from 15 to 26. Outside wires are not more than 12. Rope with 19 wires is most common.

Characteristics

- Provide a balance between fatigue and wear resistance;and
- Give excellent service with sheaves and drums of moderate size.

C. 6 × 37 Classification   These ropes contain six strands and wire numbers vary from 27 to 49 wires. Outside wires are not more than 18. Each strand contains many wires of small diameter. Flexibility increases with the number of wires. These are available in diameter from 6 mm to 125 mm. These are available in right or left lay and regular or lang lay. Characteristics of these wire ropes is that it is more flexible but less abrasion resistant than the 6 × 19 classification.

Typical applications

Overhead cranes and mobile crane hoist ropes.

2.10.1 Classification according to usage

Wire ropes can be classified according to their usage:

Running ropes   The rope turns over a pulley called sheave to lift a load. The stresses are tensile and bending due to turn over sheave.

Stationary ropes   These ropes carry tensile stresses due to static or varying loads.

Track ropes   Used for aerial ropeways. Rollers of the cabin move over the rope. Radius of the roller does not change the shape of rope to that of roller as in running ropes but causes a free bending radius. This radius decreases as the load increases and vice versa.

Wire rope slings   Used to pack various types of goods. The rope undergoes tensile stress and also bending stress at the corner of the goods.

2.11 Wire Rope Terminations

End of a wire rope tends to fray (wires hang and become loose), hence it has to be prevented from fraying. The most common method is to make a loop by turning it back and putting a wire clip. Different methods of terminating the end are; using a flemish eye or swaging (Figure 2.6).

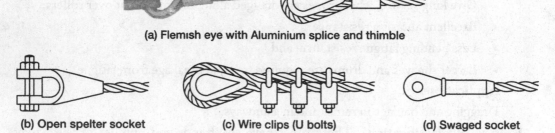

(a) Flemish eye with Aluminium splice and thimble

(b) Open spelter socket   (c) Wire clips (U bolts)   (d) Swaged socket

Figure 2.6 Terminations of wire rope

The most common and useful type of end fitting for a wire rope is to turn the end back to form a loop as shown in Figure 2.6(a). Efficiency (strength of the terminated end to the strength of wire rope) of flemish eye and other types is 90 per cent.

a. **Eye splice** An eye splice may be used to terminate the loose end of a wire rope, when forming a loop. The strands of the end of a wire rope are unwound a certain distance, and plaited back into the wire rope, forming the loop, or an eye, called an eye splice.

b. **Thimbles** When the wire rope is terminated with a loop, there is a risk that it will bend too tightly. A thimble can be installed inside the loop to preserve the natural shape of the loop and protect the cable from pinching and abrading on the inside of the loop as shown in Figure 2.6(a). The use of thimbles in loops is the best practice. The thimble prevents the load from coming into direct contact with the wires.

c. **Spelter socket** Efficiency as high as 100 per cent can be obtained with spelter socket [Figure 2.6(b)].

d. **Wire rope clamps / clips** A wire rope clamp, also called a clip, is used to fix the loose end of the loop back to the wire rope. It usually consists of a U shaped bolt, a forged saddle, and two nuts. This is shown in Figure 2.6(c).

e. **Swaged terminations** Purpose of swaging wire rope fittings is to connect two wire rope ends together, or to otherwise terminate one end of wire rope to something else. A mechanical or hydraulic swager is used to compress and deform the fitting, creating a permanent connection as in Figure 2.6(d). There are many types of swaged fittings: threaded studs, ferrules, sockets, and sleeves. Swaging with fibre core ropes is not recommended.

f. **Wedge sockets** A wedge socket termination is useful when the fitting needs to be replaced frequently. For example, if the end of a wire rope is in a high-wear region, the rope may be periodically trimmed, requiring the termination hardware to be removed and reapplied.

The end loop of the wire rope enters a tapered opening in the socket, wrapped around a separate component called the wedge. The arrangement is knocked in place, and load gradually eased onto the rope. As the load increases on the wire rope, the wedge becomes more secure, gripping the rope tighter.

2.12 Selection of Wire Ropes

Several factors are to be considered while selecting the most suitable type of wire rope for an application. Following requirements will help select a rope, for optimum performance and long life of the wire rope.

- Breaking strength
- Bending fatigue strength
- Crushing strength
- Strength for abrasion
- Fatigue strength for vibrations

It is impossible for a single wire rope to have qualities that will maximize performance for each selection criterion. A high rating in one area lowers rating in another. Requirements for an application must be carefully analyzed so that a rope satisfies the most important criteria and sacrifices the least important. Generally, a small number of large wires are more abrasion resistant and less fatigue resistant than a large number of small wires.

2.12.1 Selecting a type of wire rope center

There are three general types of wire rope centers.

Strand center   It is usually used in stationary ropes, such as guys, suspension bridge cables, and in ropes of small diameter, such as aircraft cable. It is not suitable, where severe crushing may exist.

Fibre center   This center is made of pre-lubricated plastic fibres, usually polypropylene. These fibres are extremely hard-laid, which can withstand high pressures of rope service. These are used only when normal operating loads do not break the fibres. The polypropylene center is generally recommended for acidic environment.

Independent wire rope center

This rope center increases the strength by 7 per cent and weight by 10 per cent, but decreases flexibility slightly. It greatly increases crushing resistance. It is recommended, where severe loads are placed on ropes on sheaves or drums. Unless required for one or more of the following three properties, the use of IWRC should be avoided.

- Increased strength
- Greater resistance to crushing
- Resistance to excessive heat

2.13 Stresses in Wire Rope

When a wire rope passes over a pulley or wound on a drum, many types of stresses are developed as given below:

a. **Tensile stress** This stress occurs due to axial load coming on the rope and also due to the own weight of the rope.

Let, W = Load to be lifted

w = Weight of the rope

A = Cross-sectional area of rope

The tensile stress: $\sigma_t = \dfrac{W + w}{A}$
$\hspace{4cm}$ (2.6)

b. **Bending stress** When a rope is wound over a drum or it passes over the sheave, it gets bend, which causes bending stress. Smaller the diameter of drum / sheave, higher is the bending stress. This stress depends on:

- Rope construction
- Diameter of wire (d_w)
- Diameter of sheave (D)

Table below gives the recommendations for values of wire and sheave diameters.

Table 2.2 Ratio of sheave/wire diameters for different rope constructions

Rope construction	Minimum (D / d_w) ratio
6 × 26 Seale	30
6 × 31 Seale	26
6 × 36 Seale	23
6 × 50 Seale	22

Reuleaux suggested a relation for the approximate bending stress as under:

Bending stress $\sigma_b = \dfrac{E_r d_w}{D}$
$\hspace{4cm}$ (2.7)

Where, E_r = Modulus of elasticity of the rope

$\hspace{1.5cm} E_r$ = 83 GPa for steel wires

$\hspace{1.5cm} E_r$ = 77 GPa for Wrought Iron (W.I.) wires

$\hspace{0.3cm} d_w$ = Diameter of wire (mm)

$\hspace{0.3cm} D$ = Sheave diameter (mm)

Load on the whole rope due to bending

$W_b = \sigma_b \times n \times \dfrac{\pi}{4}(d_w)^2$
$\hspace{4cm}$ (2.8)

Where, n = Number of wires

σ_b = Bending stress

Alternately, it can also be calculated using the relation:

$$W_b = \frac{E_r A d_w}{D} \qquad (2.9)$$

Where, A = Area of wire (mm^2)

D = Sheave / drum diameter (mm)

c. **Stresses due to change in speed**   When the load is lifted from rest, it gets accelerated causing additional stress, which depends on the load and value of acceleration. It can be calculated from the time, it takes to attain a velocity.

   Acceleration, $a = \dfrac{v_2 - v_1}{t}$

Where, v_1 = Initial velocity in m/s. If starting from rest it is taken as zero

v_2 = Final velocity in m/s

t = Time in seconds

Additional load due to acceleration is:

$$W_a = \frac{(W + w) \times a}{g} \qquad (2.10)$$

Where, a = Acceleration (m/s^2)

g = Acceleration due to gravity (m/s^2)

d. **Stress due to sudden stopping**   While lowering, if a hoisting drum is suddenly stopped, the kinetic energy of the moving masses becomes zero all of a sudden. This energy has to be absorbed by the rope, which is many more times the static tensile stress due to load. The rope gets stretched during this operation and impact stresses are to be calculated.

e. **Impact stress due to slack in rope**   If the rope has a slack of h, it has to be overcome first before the rope becomes tight and starts lifting the load. This causes impact stress in the rope. Impact load at the time of starting W_s is given by the relation:

$$W_s = (W + w)\left[1 + \sqrt{1 + \frac{2ahE_r}{\sigma_t L g}}\right] \qquad (2.11)$$

Where, h = Slack of rope (mm)

L = Length of rope (mm)

a = Acceleration of load and rope (m/s^2)

E_r = Modulus of elasticity of the rope (MPa)

Acceleration a is calculated from the velocity of rope V_r when the rope becomes tight after moving a height h using relation $V_r = \sqrt{2ah}$.

If there is no slack, that is, $h = 0$, Equation (2.11) becomes: $W_s = 2(W+w)$

Hence, the starting stress $\sigma_s = \dfrac{2(W+w)}{A}$ \hfill (2.12)

f. **Effective stress** Sum of direct tensile and bending stress is called effective stress, which acts on rope during normal working, that is, $\sigma_t + \sigma_b$. Effective total stress σ_{total} during acceleration becomes:

$$\sigma_{total} = \sigma_t + \sigma_b + \sigma_s \hfill (2.13)$$

Effective loads under different conditions are:

During normal working $= W + w + W_b$

During starting $= W_s + W_b = 2(W+w) + W_{sb}$

During acceleration $= W + w + W_b + W_a$

2.14 Drum and Sheave Arrangement

Drum is a cylindrical part, on which the rope is wound or unwound (Figure 2.7). Small-and medium-size drums are made plain with no grooves, whereas large-size drums are provided with helical grooves of cross section less than semicircle on the periphery as shown in Figure 2.8. Grooves are of radius 0.53 d of depth 0.25 d. Pitch between the grooves is kept 1.15 d.

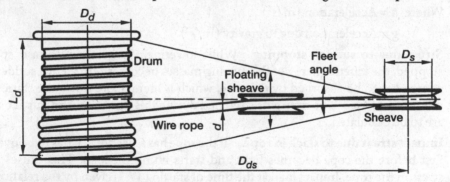

Figure 2.7 Drum and sheave arrangement

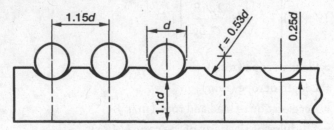

Figure 2.8 Large-size drum with grooved surface

Sheave is the rotating part of the pulley system; it is a grooved wheel, into which a rope fits. The rope from the drum passes over a sheave as shown in Figure 2.7. Sometimes a floating shave is also provided in between the drum and sheave. This can slide axially to adjust according to the angle of rope. Maximum angle of rope from the sheave to drum is called fleet angle. The distance between drum and sheave has to be such that the fleet angle does not exceed 2°.

Wire ropes operating over sheaves and drums are subjected to bending stresses, which cause fatigue in wire rope. The wires on the top side of rope elongate and wires on underside get compressed. Severity of these stresses is related to the rope load and the ratio of sheave diameter to rope diameter (D_s / d). This continual shifting causes the inner wires to move against one another, when the rope is under load. So select a sheave diameter from Table 2.3.

Table 2.3 Sheave Diameter D_s for Wire Ropes in Terms of Rope Diameter d

Type of Rope	Sheave Diameter D_s		Application
	Minimum	**Recommended**	
6 × 7	42d	72d	Mines
6 × 19	20d	30d	Elevators, well drilling
	30d	45d	Hoisting rope
	60d	100d	Cargo cranes, mines
6 × 37	18d	27d	High-speed elevators
8 × 19	21d	31d	Flexible hoisting rope

If the sheave or drum is too small, it causes severe bend in the rope, hence wire movement is adversely affected, because the wires cannot easily move to compensate for the bend. Larger the sheave or drum diameter, better is the rope's fatigue performance. To maximize rope life, the equipment should have properly designed and maintained sheaves.

2.14.1 Construction of sheave

Outer periphery of the sheave has a groove of a semicircular cross section as shown in Figure 2.9. Proportions of the groove in terms of rope diameter are indicated in this figure. If d is the rope diameter, groove radius is 0.53d. The flared groove is called throat and it makes an angle of 45°. Small sheaves can be made solid but large sheaves are provided with a web of thickness 0.75d.

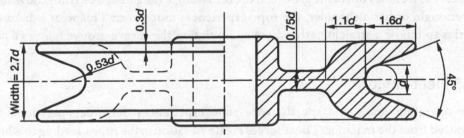

Figure 2.9 Proportions of groove sheave

An effective sheave must have:

- High D_s / d ratio, that is, sheave diameter should be much bigger than rope diameter;
- Minimum surface area possible for the rope or cable; and
- Must be resistant to abrasions or warping as this causes the rope to fray.

2.14.2 Groove size

Sheave and drum grooves are under constant pressure during normal working. A tight sheave causes high groove pressures at the point of contact. A rope ultimately may become of same size as the groove, but it decreases rope life.

A too large groove also has a negative effect on performance. An oversized groove does not support the rope and allows it to flatten, thus unbalancing the rope. An unbalanced rope develops internal stresses that exceed design factors and lead to early fatigue of wires, wire breakage, and rope crushing. A right sheave can be determined using a sheave gauge. It is a set of thin steel blades with semicircular ends of different radii.

The best groove is which gives maximum support to the rope. Greater the groove support, lower is the sheave bearing pressures and the longer rope life.

When regrooving a sheave or drum, the recommended contact to rope contour is 150°. It should lie between 146° and 157°. The groove contours must be checked when ropes are changed.

2.14.3 Groove hardness

The hardness of sheaves and drums is often neglected. Wire rope is very hard having Brinell of 430–500. Owing to this hardness of rope, the pressures exerted from the rope on the groove cause groove corrugations, if it has not been flame hardened. Mines typically alternate rope lays to combat corrugations. Once this condition occurs, the groove needs to be reconditioned and surface hardened. Groove hardness should be in the range of 475–540 Brinell to prevent the premature wear of the grooves.

2.14.4 Throat angle

On leaving the sheave, the rope forms an angle with the sheave equal to the fleet angle (See Figure 2.7) as it comes out of the groove. If the throat angle (angle between the groove flanges) is not enough, or has a shoulder, the rope experiences more wear. This wear is because the rope tries to follow a straight path, where as a point on the sheave groove follows a circular path.

2.14.5 Fleet angle

Under operating loads and speeds, wire rope experiences dynamic operational forces transmitted from the machine. These forces create vibration in the ropes, leading to whipping. To minimize the resultant damage, the fleet angle should be minimized. The angle should not exceed 1.5° for a plain drum and 2° for a grooved drum. Excessive fleet angle results in abrasion between the rope and sheave groove flange.

2.14.6 Sheave alignment

If the sheave is not properly aligned, the wire rope comes in constant contact with the sheave flange. This creates an additional stress area on the rope, which leads to abrasive damage and wire fatigue breaks at the point of contact.

2.15 Design Procedure for Wire Rope Drive

1. Select type of rope for the given application from Table 2.4 given below:

Table 2.4 Selection of Wire Rope for a Given Application

Application	Type of Rope	Range of Wire Rope Diameter (mm)
Excavators and cranes	6 × 19	8–40
	6 × 37	8–56
Haulage in mines	6 × 7	8–35
	6 × 19	13–38
Lifts, elevators, hoists	6 × 19	6–25
	8 × 19	8–25
Oil well drilling	6 × 7	10–25
	6 × 19	13–38
	6 × 37	13–38
	8 × 19	13–29

2. For the application, select a factor of safety from Table 2.5.

Table 2.5 Factor of Safety for a Given Application

Application	Factor of Safety	Application	Factor of Safety
Guys	3.5	Mine hoists: depth > 900 m	5
Track cables	4	600 m–900 m	6
General hoists	5	300 m–600 m	7
Haulage ropes	6	<300 m	8
Small air hoists	7	Slings and cranes	8

3. Calculate the design load by multiplying the given load by factor of safety = $W \times FOS$.

4. Find tensile strength S in terms of d^2 for the selected tensile strength of wire from Table 2.6.

Table 2.6 Average Weight and Tensile Strength of Wire and Wire Rope of Diameter (d)

Type of Rope	Diameter d (mm)	Average Weight w (N/m)	Tensile Strength S_t of Wire (MPa)	Tensile Strength S of Rope (N)
6 × 7	10 – 25	0.037 d^2	1,600 – 1,800	550 d^2
			1,800 – 2,000	610 d^2
6 × 7	8 – 35	0.0347 d^2	1,600	530 d^2
			1,800	600 d^2
6 × 19	6 – 25	0.0383 d^2	1,100 – 1,250	385 d^2
			1,250 – 1,400	445 d^2
6 × 19	13 – 38	0.0363 d^2	1,600	530 d^2
			1,800	595 d^2
6 × 19	13 – 38	0.037 d^2	1,600 – 1,800	510 d^2
			1,800 – 2,000	570 d^2
			2,000 – 2,250	630 d^2
6 × 19	8 – 40	0.0375 d^2	1,600 – 1,750	540 d^2
			1,750 – 1,900	590 d^2
6 × 37	13 – 38	0.037 d^2	1,600 – 1,800	490 d^2
			1,800 – 2,000	540 d^2
			2,000 – 2,250	600 d^2
6 × 37	8 – 56	0.038 d^2	1,600 – 1,750	510 d^2
			1,750 – 1,900	550 d^2
8 × 19	8 – 25	0.034 d^2	1,100 – 1,250	355 d^2
			1,250 – 1,400	445 d^2
8 × 19	13 – 29	0.0338 d^2	1,800 – 2,000	530 d^2

5. Find wire rope diameter d from the design load from the relation: $S = W \times FOS$
   Where, S = Strength given in the last column of Table 2.6 in terms of d^2.

6. Select a standard wire diameter from Table 2.7.

Table 2.7 Standard Wire Diameter (mm) of Different Wire Ropes

Wire Rope	Standard Wire Diameters (mm)
6 × 7	8, 9, 10, 11, 12, 13, 14, 16, 18, 19, 20, 21, 22, 24, 25, 26, 27, 28, 29, 31, 35
6 × 19	8, 9, 10, 11, 12, 13, 14, 16, 18, 19, 20, 21, 22, 24, 25, 26, 27, 28, 29, 31, 32, 36, 38, 40

6 × 37	8, 9, 10, 11, 12, 13, 14, 16, 18, 19, 20, 21, 22, 24, 25, 26, 27, 28, 29, 31, 32, 36, 38, 40, 44, 48, 52, 56
8 × 19	8, 9, 10, 12, 13, 14, 16, 18, 20, 22, 25, 29

7. Find wire diameter d_w and rope area A from Table 2.8.

Table 2.8 Wire Diameter (mm) and Area (mm²) of Different Wire Ropes

Type of Rope	6 × 7	6 × 19	6 × 37	8 × 19
Wire Diameter (d_w)	0.106d	0.063d	0.045d	0.05d
Area of Rope (A)	0.38d²	0.38d²	0.38d²	0.35d²

8. Knowing rope diameter d, calculate weight of rope w per meter from Table 2.6.
9. Select a sheave diameter from Table 2.2. It can be taken between minimum and recommended value.
10. Calculate bending stress from Equation (2.7): $\sigma_b = \dfrac{E_r d_w}{A}$

And then bending load from equation: $W_b = \sigma_b \times A$
11. Calculate additional load due to acceleration a from given data and Equation (2.8).

$$W_a = \frac{(W+w)\times a}{g}$$

12. Calculate stress at the time of starting from Equation (2.12): $W_s = 2(W+w)$
13. Calculate effective load at the time of starting: $W_{total} = W_b + W_s$
14. Check for effective load during acceleration: $W_{total} = W + w + W_b + W_a$

Example 2.3

Design of wire rope of a lift for given load and height
Select a suitable wire for a lift to carry a load of 10 persons with an average weight of 70 kg. Weight of the lift can be assumed 300 kg. The lift has to move 20 floors each of 3 m height with maximum acceleration of 0.1 g. Assume Young's modulus of elasticity for rope as 85×10^3 MPa. For the selected rope, calculate factor of safety during acceleration and starting.

Solution

Given $W = (70 \times 10) + 300 = 1,000 \text{ kg} = 10 \text{ kN}$ $h = 20 \times 3 = 60 \text{ m}$
$a = 0.1 g$ $E_r = 85 \times 10^3 \text{ MPa}$

Initially as the wire details are not known, neglect weight of wire.

Step 1 Select a wire rope for the application from Table 2.4. For lifts type of wire rope: 6 × 19 with range of rope diameters 6 to 25 mm.

Step 2 Select factor of safety for the application from Table 2.5: Selected 5.

Step 3 Calculate the design load = $W \times FOS = 10,000 \times 5 = 50,000$ N.

Step 4 From Table 2.6 select wire for type of rope: Selecting a wire of strength S_t 1,100 to 1,250 MPa

Step 5 Find tensile strength of rope S for the selected wire in terms of d^2 from Table 2.6 : $S = 385d^2$

Equating it to the load on wire: $50,000 = 385d^2$ or $d = 11.4$ mm

Step 6 Select standard wire rope from Table 2.7: Standard wire rope is of 12 mm.

Step 7 Find wire diameter: From Table 2.8, $d_w = 0.063\ d = 0.063 \times 12 = 0.76$ mm.

Step 8 Find weight of wire: From Table 2.6, $w = 0.0383\ d^2 = 0.0383 \times (12)^2 = 5.51$ N/m.

Height is 60 m, hence weight of the wire rope $w = 60 \times 5.51 = 331$ N.

Step 9 Select a sheave diameter (Table 2.3) recommended diameter $45\ d = 45 \times 12 = 540$ mm.

Step 10 Calculate bending stress from Equation (2.8):

$$\sigma_b = \frac{E_r d_w}{D} = \frac{85 \times 10^3 \times 0.76}{540} = 119 \text{ MPa}$$

From Table 2.6, area of wire $A = 0.38\ d^2 = 0.38 \times (12)^2 = 49.4$ mm^2

Bending load, $W_b = \sigma_b \times A = 119 \times 49.4 = 5,879$ N

Step 11 Additional load due to acceleration:

From Equation (2.10): $W_a = \dfrac{(W+w) \times a}{g} = \dfrac{(10,000+331) \times 0.1g}{g} = 1,033$ N

Step 12 Additional load at time of starting from Equation (2.11):

$W_s = 2(W + w) = 2 \times (10,000 + 331) = 20,662$ N

Effective load during starting = $2(W+w) + W_b = 20,662 + 5,879 = 26,541$.

Step 13 Effective load during acceleration

$W_{total} = W + w + W_b + W_a$

$W_{total} = 10,000 + 331 + 5,879 + 1,033 = 17,243$ N

Step 14 Factor of safety during acceleration: $FOS = \dfrac{50,000}{17,243} = 2.9$

Factor of safety during starting: $FOS = \dfrac{50,000}{26,541} = 1.88$

Example 2.4

Size of wire rope for a crane with given load and height
Calculate wire rope size for a crane with the following data:

Wire rope type 6 ×19

Maximum load = 30 kN

Maximum height of crane = 20 m

Drum diameter = 45d (where d is rope diameter)

Maximum acceleration = 1.5 m/s²

Weight of hook = 10 kN

Ultimate strength of wire = 1,600 MPa

Young's modulus of elasticity = 82GPa

Diameter of wire = 0.063d

Weight of rope per meter length = $0.0375d^2$

Area of wire rope = $0.38d^2$

FOS = 5

Solution

Given $W = 30$ kN $h = 20$ m $a = 1.5$ m/s² $w' = 10$ kN

$S_{ut} = 1,600$ MPa $E_r = 82$GPa $d_w = 0.063d$ $w = 0.0375d^2$ N/m

$A = 0.38d^2$ FOS = 5

Bending stress, $\sigma_b = \dfrac{E_r d_w}{D} = \dfrac{82 \times 10^3 \times 0.063\,d}{45\,d} = 114.8$ MPa

Equivalent bending load, $W_b = \sigma_b \times A = 114.8 \times 0.38d^2 = 43.6d^2$ N

Weight of wire, $w = 20 \times 0.0375d^2 = 0.75d^2$ N

Load due to acceleration, $W_a = \dfrac{(W + w' + w) \times a}{g}$

$= \dfrac{(30,000 + 10,000 + 0.75d^2) \times 1.5}{9.81} = 6116 + 0.115d^2$

$W_{total} = W + w' + w + W_b + W_a = 30,000 + 10,000 + 0.75d^2 + 6,116 + 0.115d^2$

$= 46,116 + 0.865\,d^2$

Allowable stress, $\sigma_t = \dfrac{S_{ut}}{FOS} = \dfrac{1,600}{5} = 320$ MPa

Allowable load, $W_{total} = \sigma_t \times A = 320 \times 0.38d^2 = 121.6d^2$ N

Equating this with total load:

$$46,116 + 0.865d^2 = 121.6d^2 \quad \text{Or,} \quad d^2 = 463.6 \text{ mm}^2 \quad \text{Or,} \quad d = 21.5 \text{ mm}$$

Select a standard wire rope diameter = 22 mm.

Example 2.5

Size of wire rope for a trolley for given inclined distance and load

A trolley carrying coal with total weight of 50 kN is pulled up 400 m on an inclined rail having a slope of 25° with horizontal as shown in Figure 2.S1. Friction of trolley on rail can be taken 30 N/kN of trolley weight. Maximum acceleration is 1.5 m/s². Calculate diameter (d) of 6 × 19 wire rope with factor of safety 6. Drum diameter is 60 times the rope diameter. Properties of 6 × 19 wire rope are as under:

Weight of the rope per meter length = $0.038d^2$ N/m
Tensile strength of rope = $385d^2$ N
Wire diameter = $0.063d$ mm
Equivalent modulus of elasticity of rope = 82 GPA

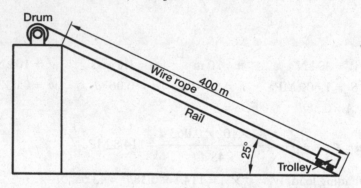

Figure 2.S1 A trolley on an inclined rail

Solution

Given $W_T = 50$ kN $\quad L = 400$ m $\quad \alpha = 25°$ $\quad FOS = 6$
$\quad\quad\quad F_f = 30$ N/kN $\quad a = 1.5$ m/s² $\quad d_w = 0.063d$ $\quad D = 60\ d$
$\quad\quad\quad w = 0.038d^2$ N/m $\quad S = 385d^2$ $\quad E_r = 82 \times 10^3$ MPa

Load due to friction = $30 \times 50 = 1,500$ N

Load on rope $W = (50,000 + 1,500) \sin \alpha = 51,500 \times \sin 25° = 21,765$ N

Component of weight of wire along slide $w = (400 \times 0.0375d^2) \times \sin 25° = 6.4\ d^2$

$$W + w = 21,765 + 6.4d^2$$

Bending stress $\sigma_b = \dfrac{E_r d_w}{D} = \dfrac{82 \times 10^3 \times 0.063\ d}{60\ d} = 86.1$ MPa

Area of a wire $\dfrac{\pi}{4}d_w^2 = 0.785 \times (0.063d)^2 = 0.00311d^2 \ \text{mm}^2$

Number of wires in 6×19 rope $= 6 \times 19 = 114$

Area of all wires $A = 0.00311d^2 \times 114 = 0.355\ d^2$

Equivalent bending load $W_b = \sigma_b \times A = 86.1 \times 0.355d^2 = 30.5d^2\ \text{N}$

Load due to acceleration $W_a = \dfrac{(W+w) \times a}{g}$

$$= \dfrac{(21{,}765 + 6.4d^2) \times 1.5}{9.81} = 3{,}328 + 0.98d^2$$

$$W_{total} = W + w + W_b + W_a = 21{,}765 + 6.4d^2 + 30.5d^2 + 3328 + 0.98d^2$$

$$= 25{,}093 + 37.9d^2$$

Allowable load $\dfrac{S}{FOS} = \dfrac{385d^2}{6} = 64.16d^2$

Equating this with total load:

$$64.16d^2 = 25{,}093 + 37.9d^2 \quad \text{Or,} \quad d^2 = 955\ \text{mm}^2 \quad \text{Or,} \quad d = 31\ \text{mm}$$

Select a standard diameter from Table 2.7, $d = 31$ mm

Summary

Rope drives are used, where power is large and distance between pulleys is also more. Rope drives can be of two types:

- Fibre rope drive: For pulleys up to 50 m apart.
- Wire rope drive: For pulleys up to 150 m apart.

Fibre rope drive

Advantages: High mechanical efficiency, smooth and quiet drive, little misalignment of the shafts can be easily tolerated, and suitable for outdoor applications.

Fibre rope materials

The type of material used for the rope is the main determinant of strength of the rope, abrasion resistance, ease of use, and price.

Nylon is the strongest easily available rope material. It is quite strong but loses about 15 per cent of its strength in wet condition.

Polypropylene is the cheapest of all synthetic fibres. It is strong for its weight, but it is not resistant to heat and abrasion.

Polyester is almost as strong as nylon, but it does not loose strength when wet. It has highest resistance to abrasion and heat.

Natural fibres are heavier, weaker, and less resistance to all forms of abrasion than synthetic fabrics.

Manila fibres are rough and not much flexible. These ropes have less strength and wear internally due to sliding of fibres. To reduce wear, they are lubricated with tar or graphite.

Cotton ropes are soft and smooth. These ropes are also lubricated to reduce wear.

Sheave Pulley used for ropes is called a sheave. Number of grooves can be one or more than one. Groove angle is made 45°. Sheave diameter is kept between 35 and 40 times the rope diameter to reduce bending stresses.

Design of rope drive

Ratio of tight and slack side tensions T_1 and T_2 is: $\dfrac{T_1}{T_2} = e^{\mu\,\theta\cosec 22.5°} = e^{2.63\mu\,\theta}$

Centrifugal force; $T_c = \dfrac{wv^2}{g}$

Tension ratio is; $\dfrac{T_1 - T_c}{T_2 - T_c} = e^{2.63\,\mu\,\theta}$

For maximum power, $v = \sqrt{\dfrac{T}{3m}}$

Where, T = Maximum tension

m = Mass of rope (kg/m)

Breaking load for some of the cotton ropes are given in Table 2.1. If no table then, weight of rope of diameter d in N/m can be calculated using $w = 0.0072d^2$.

Breaking load $W = 0.0315d^2$ in kN and d in mm for cotton ropes.

- Factor of safety is taken quite high for rope drives. It varies from 35 to 40.
- Rope velocity varies from 15 m/s to 35 m/s, but most economical is 20 m/s to 25 m/s.

Braking load is taken as the maximum tension, that is, T_1.

Wire rope is a flexible cable, which consists of several strands of metal wires.

Wire ropes of diameter less than 9.5 mm are called cables. These are used for:

- Lifting cranes, lifts, and hoisting applications.
- In mechanisms, like clutch and gear wires, connections from airplane cockpit.
- In pre-stressed concrete used in bridges, use of static steel wire ropes can be seen.
- Suspension bridges use steel wire rope between two towers at extreme ends.

Advantages of wire ropes: Lighter than fibre ropes, can withstands shocks, more reliable, more economical, higher efficiency, and longer life.

Wire rope materials

Stainless steel (18 parts chromium and 8 parts nickel). Type 305 S / S cable is commonly available in 1.5 mm and 0.75 mm diameters. This type of steel is generally used for sensitive instrumentation or other systems, which can be affected by magnetism.

Monel is an uncommon grade of stainless, used in applications where added resistance to corrosive substances and liquids is required.

Carbon steel wire is available in different grades. Wire rope manufactured from uncoated wire is commonly referred as Bright.

Galvanized carbon steel wire is used to manufacture wire rope as they provide good corrosion protection.

Construction of wire ropes A wire rope consists of a core in the center, around which a number of multi wired strands are laid helically. The most popular arrangement is six strands around the core. Most of wire ropes use 7-wires, 19-wires, or 37-wires strands.

Core is to provide support and maintain the position of the outer strands during operation. There are two general types of cores: fibre cores and wire strand cores.

Wires Steel wires for wire ropes are normally made of carbon steel with a carbon 0.4–0.95 per cent. Most wire ropes are made with round wires.

Strands In a strand, wires of the different layers cross each other in different ways; parallel lay and crossed lay that is seldom used.

Parallel lay strands Lay length of all the wire layers is equal and the wires of any two superimposed layers are parallel, resulting in linear contact. It is used most commonly. Endurance of wire ropes with parallel lay is much higher than cross lay strands.

Lay of wire ropes It is a description of the arrangement; wires and strands are placed during construction. A lay can be specified by its direction of strands.

Right lay Strands pass from left to right across the rope.

Left lay Strands pass from right to left.

The way wires are placed within each strand is given below:

Regular lay Wires in the strands are laid opposite in direction to the lay of the strands.

Lang lay Wires are laid in the same direction as the lay of the strands. These ropes are more flexible and greater wearing surface per wire than regular lay ropes.

Most of the wire ropes are right and regular lay.

Ordinary lay The lay direction of wires in the outer strands is in opposite direction to the lay of the outer strands.

Alternate right and left lay The wire rope strands are alternately regular lay and lang lay.

Types of wire ropes

Spiral ropes are round strands with assembly of layers of wires laid helically over a center with at least one layer of wires in the opposite direction to that of the outer layer.

The open spiral rope consists only of round wires.

Advantages of spiral rope are that it prevents the penetration of dirt and water, protects from loss of lubricant, and ends of a broken outer wire cannot leave the rope.

Stranded ropes are of several strands laid helically in one or more layers around a core. Multistrand ropes are less resistant to rotation and have at least two layers of strands laid helically around a center.

  a. *Single size* has wires of the same size wound around a center.

  b. *Seale* Large outer wires with the same number of smaller inner wires around a core wire. Provides excellent abrasion resistance but less fatigue resistance.

  c. *Warrington* Outer layer of alternately large and small wires provides good flexibility and strength, but low abrasion and crush resistance.

  d. *Multiple operation* One of the above strand designs may be covered with one or more layers of uniform-sized wires.

Finish

  – Bright finish is suitable for most applications.

  – Galvanized finish is available for corrosive environments.

  – Plastic jacketing is also available on some constructions.

Designation of ropes Wire ropes are designated as: $S \times W$

Where, S = Number of strands and W = Number of wires in each strand.

Classification of wire ropes

6×7 *Classification* These ropes have six strands and number of wires may vary from 3 to 14. Seven wires are most common. Used in dragging and haulage in mines, and in tramways.

6×19 *Classification* These ropes have six strands and number of wires may vary from 15 to 26. Outside wires are not more than 12. Rope with 19 wires is most common.

6×37 *Classification* These ropes contain six strands and number of wires may vary from 27 to 49 wires. Outside wires are not more than 18. These wire ropes are more flexible but less abrasion resistant than the 6×19 classification.

Classification according to usage

Running ropes The rope turns over a sheave to lift a load.

Stationary ropes These ropes carry tensile stresses due to static or varying loads.

Track ropes Used for aerial ropeways. Rollers of the cabin move over the rope.

Wire rope slings Used to pack various types of goods.

Wire rope terminations End of a wire rope tends to fray, hence it has to be prevented from fraying. The most common method is to make a loop by turning it back. Different methods of terminating the end are: using a flemish eye or swaging.

  a. *Eye splice* An eye splice may be used to terminate the loose end of a wire rope, when forming a loop. The strands of the end of a wire rope are unwound to a certain distance, and plaited back into the wire rope, forming the loop, or an eye, called an eye splice.

  b. *Thimbles* A thimble is installed inside the loop to preserve the natural shape of the loop and protect the cable from pinching and abrading on the inside of the loop.

  c. *Spelter socket* Efficiency as high as 100 per cent can be obtained with spelter socket.

  d. *Wire rope clamps / clips* It usually a U shaped bolt, a forged saddle, and two nuts.

e. *Swaged terminations* Purpose of swaging wire rope fittings is to connect two wire rope ends together, or to otherwise terminate one end of wire rope to something else. Swaging with fibre core ropes is not recommended.

f. *Wedge sockets* It is useful, when the fitting needs to be replaced frequently. The end loop of the wire rope enters a tapered opening in the socket, wrapped around a separate component called the wedge.

Selection of wire ropes Several factors must be considered to select a suitable wire rope for a job. Following requirements will help selection:

- Breaking strength, bending fatigue strength,crushing strength, strength for abrasion, and fatigue strength for vibrations.

Selecting a wire rope center: There are three general types of wire rope centers.

- *Strand center* It is usually used in stationary ropes, such as guys, suspension bridge cables, and in ropes of small diameter, such as aircraft cable. It is not suitable, where severe crushing may exist.

- *Fibre center* This center is made of either pre-lubricated Java sisal fibers or plastic fibres, usually polypropylene. These fibres are extremely hard-laid, which can withstand high pressures of rope service. These are used only when normal operating loads do not break the fibres.

- *Independent wire rope center* This rope center increases the strength by 7 per cent and weight by 10 per cent, but decreases flexibility slightly. It greatly increases crushing resistance. It is recommended, where severe loads are placed on ropes on sheaves or drums.

Stresses in wire rope

a. *Tensile stress* Owing to axial load *W* coming on the rope and also due to the own weight *w* of the rope also.

The tensile stress $\sigma_t = \dfrac{W + w}{A}$

b. *Bending stress* Owing to bending of wires of diameter d_w rope over the drum of diameter *D* or sheave.

Bending stress $\sigma_b = \dfrac{E_r d_w}{D}$

E_r = 83 GPa for steel wires

If there are *n* wires, load on the whole rope, due to bending is: $W_b = \sigma_b \times n \times \dfrac{\pi}{4}(d_w)^2$

c. *Stresses due to acceleration* It is calculated from $a = \dfrac{v_2 - v_1}{t}$

Additional load due to acceleration is: $W_a = \dfrac{(W + w) \times a}{g}$

d. *Stress due to sudden stopping* This energy has to be absorbed by rope, which is many more times the static tensile stress due to load.

e. *Impact stress due to slack in rope* If the rope has a slack of *h*, it has to be overcome first before the rope becomes tight. Impact load at the time of starting is

$$W_s = (W + w)\left[1 + \sqrt{1 + \frac{2ahE_r}{\sigma_t Lg}}\right]$$

If there is no slack, that is, $h = 0$, Then $W_s = 2(W + w)$

Hence the starting stress $\sigma_s = \dfrac{2(W + w)}{A}$

f. *Effective stress* During normal working, that is, $\sigma_t + \sigma_b$.

Effective loads under different conditions are

During normal working = $W + w + W_b$

During starting = $W_s + W_b = 2(W + w) + W_b$,

During acceleration = $W + w + W_b + W_a$

Drum and sheave arrangement Drum is a cylindrical part, on which the rope is wound or unwound. Large-size drum are provided with helical grooves of cross section less than semi-circle on the periphery. Grooves are of radius $0.53d$ and depth $0.25d$. Pitch between the grooves is kept $1.15d$.

Sheave is the rotating part of the pulley system as a grooved wheel, into which a rope fits. A floating sheave can slide axially to adjust according to angle of rope. Maximum angle of rope from the sheave to drum is called fleet angle. It should be $< 2°$. Severity of these stresses is related to the rope load and the ratio of sheave diameter to rope diameter (D_s / d). So select this ratio from Table 2.2.

Construction of sheave

Groove size Outer periphery of the sheave has a groove of a semicircular cross section. If d is the rope diameter, groove radius is $0.53d$. Small sheaves can be made solid, but large sheaves are provided with a web of thickness $0.75d$.

Groove hardness Wire rope has Brinell hardness of 430–500. This causes groove corrugations, if it is not been flame hardened. Groove hardness should be in the range of 475–540 Brinell to prevent premature wear of the grooves.

Throat angle On leaving the sheave, rope forms an angle with the sheave, which is equal to fleet angle. If the throat angle is not enough, or has a shoulder, the rope experiences more wear.

Fleet angle Wire rope experiences dynamic operational forces transmitted from the machine. To minimize the resultant damage, angle should be $< 1.5°$ for a plain drum and $2°$ for a grooved drum.

Sheave alignment If the sheave is not properly aligned, it creates an additional stress area on the rope, which leads to abrasive damage and wire fatigue.

Design procedure

1. Select type of rope for the given application from Table 2.4.
2. For the application, select a factor of safety from Table 2.5.
3. Calculate the design load by multiplying the given load by factor of safety = $W \times FOS$.
4. Find tensile strength S in terms of d^2 for the selected wire from Table 2.6.
5. Find wire rope diameter d from the design load from the relation: $S = W \times FOS$.

Where, S = Strength given in the last column of Table 2.5 in terms of d^2.

6. Select a standard wire diameter from Table 2.7.

7. Find wire diameter d_w and rope area A from Table 2.8.

8. Knowing rope diameter d, calculate weight of rope w per meter from Table 2.5.

9. Select a sheave diameter from Table 2.3.

10. Calculate bending stress from $\sigma_b = \dfrac{E_r d_w}{D}$

    Bending load $W_b = \sigma_b \times A$.

11. Calculate load due to acceleration a $W_a = \dfrac{(W+w) \times a}{g}$.

12. Calculate stress at the time of starting from $W_s = 2(W + w)$.

13. Calculate effective load at the time of starting $W_{total} = W_b + W_s$.

14. Check for effective load during acceleration $W_{total} = W + w + W_b + W_a$.

Theory Questions

1. What are the applications of wire ropes and the advantages over other drives?

2. Describe the various materials used for fiber ropes.

3. What are the different materials used for wire ropes?

4. Describe construction of a wire rope with a sketch.

5. What do you mean by strands and lays in a wire rope? What are the standard lays? Show them with a freehand sketch.

6. Explain with sketches different types of wire ropes.

7. How a wire rope is designated? Give examples.

8. Write typical applications for (6 × 7), (6 × 19), and (6 × 37) wire ropes.

9. Discuss the various stresses developed in a wire rope.

10. How will you calculate the size of a wire for an application?

11. Write the design procedure to get the size of a rope for a given application.

12. Differentiate between drum, pulley, and sheave. Show their use with a sketch.

13. Sketch a sheave with suitable proportions in terms of rope diameter.

14. Why a minimum size of sheave is important? Give reasons.

15. Write a short note on grooves on a drum and sheave.

16. Describe the terms: throat, throat angle, fleet angle, and pitch circle diameter.

17. What is meant by terminators? What are the different types of terminators? Show them by sketches?

Multiple Choice Questions

1. Which one is not a suitable material for a fabric rope?
   (a) Nylon   (b) Manila
   (c) Polyester   (d) Asbestos

2. Groove angle of a rope sheave is
   (a) 25°   (b) 40°   (c) 45°   (d) 50°

3. Lay in a wire rope is the
   (a) Way the wires have been laid to form strands
   (b) Gap between two adjacent wires
   (c) Way in which the core is placed
   (d) Orientation of the wires

4. Which of the following is not a standard wire rope classification?
   (a) 6 × 7   (b) 6 × 15
   (c) 6 × 19   (d) 6 × 37

5. Surface of a large drum is
   (a) Flat   (b) Semicircular grooves
   (c) Grooves of arc less than a circle   (d) Any shape

6. Diameter of a sheave for 6 × 7 wire rope of diameter d varies
   (a) $5d - 10d$ (b) $15d - 20d$
   (c) $25d - 30d$ (d) $40d - 60d$

7. Fleet angle is the angle, which a rope makes
   (a) Between pulley and extreme ends of the drum
   (b) Between center of the drum and extreme of the drum
   (c) Between two pulleys
   (d) None of above

8. Average weight of the wire ropes of diameter d varies from
   (a) $0.005d^2 - 0.008d^2$ (b) $0.016d^2 - 0.025d^2$
   (c) $0.034d^2 - 0.038$ (d) $0.364d^2 - 0.41d^2$

9. Strength of the wire ropes of diameter d varies from
   (a) $2,100d - 3,200d$ (b) $125d^2 - 280d^2$
   (c) $350d^2 - 600d^2$ (d) $0.07d^3 - 0.09d^3$

Answers to multiple choice question

1. (d)   2. (c)   3. (a)   4. (b)   5. (c)   6. (d)   7. (a)   8. (c)   9. (c)

Design Problems

1. Power of 50 kW is to be transmitted from a pulley of 800 mm diameter with groove angle 45° rotating at 500 rpm to another shaft at a distance of 3 m. Ratio of driver to driven pulley is three. Breaking load in kilo newton can be taken $W = 0.0315d^2$. Coefficient of friction can be assumed 0.28. Neglect centrifugal forces. Assuming a factor of safety 5, calculate

   (a) Rope diameter if only one rope is to be used.

   (b) Number of ropes, if diameter is not to exceed 19 mm. [(a) $d = 22$ mm; (b) $n = 2$]

2. A wire rope pulley system has the following data

   Power being transferred = 85 kW

   Driver pulley diameter = 500 mm

   Driven pulley diameter = 1,200 mm

   Center distance = Twice the larger pulley diameter

   Driver speed = 200 rpm

   Number of ropes = 6

   Coefficient of friction = 0.3

   Groove angle = 45°

   Mass of rope = $0.0072d^2$ N/m

   Allowable load(N) = $3d^2$

   Calculate diameter of the rope and tension in tight side. [$d = 32$ mm; $T_1 = 3,072$ N]

3. A construction crane is lifting building materials having maximum weight 6 kN for a height of 30 m. Sheave diameter can be taken as 40 times the rope diameter. Wire diameter 0.063 times the diameter of rope in millimeters. If acceleration is not to exceed 1.3 m/s². Calculate diameter of 6 × 19 wire rope with a factor of safety as 8. [$d = 31$ mm]

4. A 6 × 7 wire rope of diameter $d = 15$ mm is hoisting a lift weighing 20 kN for a height of 10 m with maximum acceleration 1 m/s². Calculate number of ropes required for a factor of safety 8. Take the weight of rope $0.37d^2$ and breaking load = $550d^2$. Area of wires in rope = $0.38d^2$.

   [$Z = 1.4$ say 2]

5. A lift carries a load of 12 kN to a height of 50 m. Maximum acceleration of the lift is limited to 2 m/s². Take tensile strength of rope in N as $355d^2$ for diameter d in millimeters for 8 × 19 wire rope. Diameter of wire is $0.05d$ and weight of rope in N/m is $0.034d^2$. Area of rope in millimeters² = $0.35d^2$. Young's modulus of elasticity for the rope 84 GPa. Take sheave diameter 40 times the rope diameter. Calculate

   (a) Rope diameter with a factor of safety 5.5

   (b) Bending stress

   (c) Effective load during acceleration [(a) $d = 14$ mm; (b) 105 MPa; (c) 19,787 N]

6. Recommend a suitable diameter of 8 × 9 wire rope for a trolley carrying a total weight of 80 kN pulled up 200 m on an inclined rail having a slope of 30° with horizontal. Drum diameter is 50 times the rope diameter. Following are the data for the trolley

Maximum acceleration = 1.6 m/s²

Friction of trolley on rail can be taken 0.02 of trolley weight

Friction on the guide sheave 0.04 of trolley weight

Factor of safety = 5   [d = 35 mm]

Properties of 8 × 19 wire rope for d in mm are:

    Weight of the rope per meter length = $0.037d^2$ N

    Tensile strength of rope = $450d^2$ N.

    Area of wires in rope = $0.38d^2$ mm².

    Wire diameter = $0.063d$ mm

    Young's modulus of elasticity = 82 GPa   [d = 36 mm]

Previous Competition Examination Questions

IES

1. Given W = weight of the load handled, w_r = weight of the rope, and f = acceleration, the additional load in ropes of a hoist during starting is given by

    (a) $F_a = \left(\dfrac{w - w_r}{g}\right)f$ (b) $F_a = \left(\dfrac{w + w_r}{g}\right)f$

    (c) $F_a = \left(\dfrac{w}{g}\right)f$ (d) $F_a = \left(\dfrac{w_r}{g}\right)f$

[Answer (b)] [IES 1997]

2. Effective stress in wire ropes during normal working is equal to the stress due to

    (a) Axial load plus stress due to bending

    (b) Acceleration / retardation of masses plus stress due to bending

    (c) Axial load plus stress due to acceleration / retardation

    (d) Bending plus stress due to acceleration / retardation

[Answer (a)] [IES 1996]

3. When compared to a rod of same diameter and material, a wire rope

    (a) Is less flexible

    (b) Has a much smaller load carrying capacity

    (c) Does not provide much warning before failure

    (d) Provides much greater time for remedial action before failure

[Answer (d)] [IES 1994]

GATE

4. A wire rope is designatedas 6 × 19 standard hosting. The number 6 × 19 represent

   (a) Diameter in millimeter × length in meter

   (b) Diameter in centimeter × length in meter

   (c) Number of strands × number of wires in each strand

   (d) Number of wires in each strand × number of strands

   [Answer (c)] [GATE 2003]

   ❑❑❑

Chain Drives

3.1 Introduction

Chains and gears offer a non-slip drive. Belt drives are associated with slip (except ribbed belt) and hence cannot be used where a positive drive is required. A chain is a reliable machine component, which transmits power by means of tensile forces in its links and is used primarily for power transmission and conveyor systems.

A chain consists of many metallic links hinged using a pin joint at ends to provide flexibility, so that it could wrap over a wheel called sprocket (Figure 3.1). These wheels have teeth, which fit in the recess of the chain links at a pitch p to offer non-slip drive. The center distance between the two sprockets can be as large as 8 m and power capacity up to 100 kW.

The velocity ratio (VR) of the driver and driven shaft depends on number of teeth on the sprockets and can go up to 10.

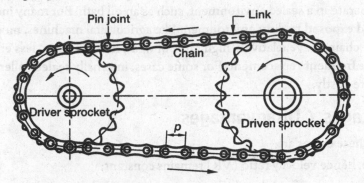

Figure 3.1 Chain drive

Chain drive did not become a prominent technology until the late nineteenth century. Roller chain using sprockets are found extensively in many industries. Some of typical applications are given below (See pictures also):

- Bicycles / motor cycles use roller chains to connect their road wheels to pedals or gear box;
- Sprockets are used in film projectors, but use perforated film stock instead of roller chains;
- Fork-lifts provide vertical lift, with sprockets and roller chains;
- Chains are used as a conveyor to move articles using a hook or otherwise;
- Wind mills and water mills use chain drives to connect machinery to the power source;
- Largest version of a roller chain transmission can be seen as main drive of an army tank; and earth digging machines;
- Chains are used for bucket elevators in conveying system.

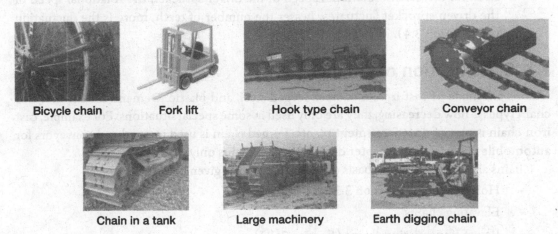

Bicycle chain   Fork lift   Hook type chain   Conveyor chain

Chain in a tank   Large machinery   Earth digging chain

Many driving chains operate in clean environments, for example, a camshaft inside an internal combustion engine, and thus the wearing surfaces like pins and bushings are safe from grit. Some chains operate in a sealed environment, such as an oil bath. But many industrial chains run unprotected exposed to dirty environment like agricultural machines, motor cycles, and bicycles. These chains have relatively high rates of wear, more friction, less efficiency, more noise, and more frequent replacement. For some cases, internally sealed roller chains can be used but they are costly.

3.2 Advantages / Disadvantages

Advantages of these drives are:

- No slip, hence velocity ratio (VR) remains constant;
- Chain are made of metals, which are stronger than belt materials, hence lesser width is required for a given power;
- Chains can work in a smaller space (small center distance), whereas belts require large center distance;
- Transmission efficiency up to 98 per cent can be achieved with chains;
- Since there are no initial tension as in belts, load on shafts / bearings is less;
- One single chain can drive many sprockets with same efficiency;
- High velocity ratios even up to 10 can be used in a single stage;
- Environmental conditions like oily or moisture do not affect the drive; and
- These are capable of transmitting more power than belts.

Disadvantages of this drive are:

- Costlier than belt drives;
- Requires correct mounting;
- Needs continuous maintenance and lubrication;
- Even for a constant rotational speed of the driver sprocket, the rotational speed of the driven sprocket fluctuates. Lesser the number of teeth, more is the fluctuation (See Section 3.4).

3.3 Classification of Chains

Chains are made of cast iron, cast steel, forged steel, and plastic. Demand for the first two chain types is now decreasing; they are only used in some special situations. For example, cast iron chain is part of water-treatment plants. Forged chain is used in overhead conveyors for automobile factories. This chapter deals with steel chains only.

Chains are classified on the basis of the application as given below:

- Hoisting chains (Section 3.3.1)
- Elevating chains
- Power transmission chains (Section 3.3.3)

On the basis of construction, these can be classified as under:

- Link chain: Oval or rectangular
- Hook type chain
- Roller chain (See Section 3.3.4)
- Silent chain (See Section 3.17)

3.3.1 Hoisting chains

Hoisting chains are used to lift loads, such as in chain pulley blocks (Figure 3.2). The chain is received by a special type of pulley in which the link fits into the recess, which does not allow to slip the chain. These chains can be used up to a speed of 0.25 m/s.

Figure 3.2 Hoisting chain

3.3.2 Conveyor chains

Conveyor chains are shown in Figure 3.3. These can be used up to 2 m/s.

Figure 3.3 Conveyor chains

3.3.3 Power transmission chains

Roller chain is generally used to transmit power from one shaft to other. These chains can operate at speed up to 15 m/s. It consists of many parts as shown in Figure 3.4 which are described in section 3.3.4.

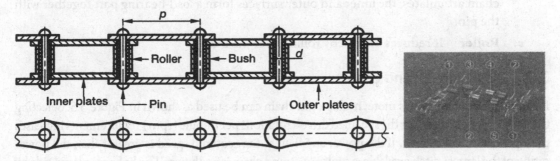

Figure 3.4 Components of a roller chain

3.3.4 Roller chains

Parts of roller chain and their functions

Earlier chains used inner and outer plates joined by pins, which were directly contacting the sprocket. These types of chains were wearing very rapidly, hence roller chains were developed, in which the pin supports a roller with a bush inside. The use of roller in bushed chains increased not only its life (as the sliding friction was changed to rolling friction) but also the transmission efficiency.

Parts of chain should have not only great static tensile strength but also good fatigue strength to bear dynamic forces of load and shock. Further, they should meet the environmental resistance for corrosion, abrasion, etc.

a. **Link**   This component bears the tension coming in the chain. In a roller chain, there are alternating links of two types. The first set is of inner links, having two inner plates held together by two bushes upon which two rollers rotate. Two outer plates are held together by pins passing through the bushings of the inner links.

   The bushless roller chain is similar in operation of chain with bushed roller chain, but not in construction. Instead of separate bushes holding the inner plates together, the plate has a tube stamped into it, protruding from the hole. This has the advantage of removing one link in assembly of the chain. There are two plates in a link; one on each side of the roller (Figure 3.4). This pair of plates is called inner plates. The other pair of plates put outside the inner plates is called outer plates. For such chains, the number of links has to be even.

b. **Offset link**   It is used when an odd number of chain links is required. It is 35 per cent lower in fatigue strength than the chain itself. The pin and two plates are slip fit. Two-pitch offset link is also available.

c. **Pin** It is subjected to shearing and bending forces transmitted by the plate. It is a load-bearing part, with the bushing, when the chain flexes during sprocket engagement. Therefore, the pin needs high tensile, shear, fatigue strength and resistance to bending, and wear.

d. **Bush**   Bush is put between the roller and pin to act as bearing. In addition, when the chain articulates, the inner and outer surfaces form a load-bearing part together with the pin.

e. **Roller**   It reduces friction by rolling over the bush.

3.3.5 Multiple-strand chains

If the power transmitted is more, multistrand chain can be used as shown in Figure 3.5. Selecting the smallest pitch chain for the specified drive's rotations per minute (rpm) requirements increases the options for the chains. If the power supplied at the required speed is greater than the power rating of the largest pitch available, a multiple-strand chain is a solution. The links are placed side by side with a long pin passing through all the bushes. Number of strands can go as high as 10.

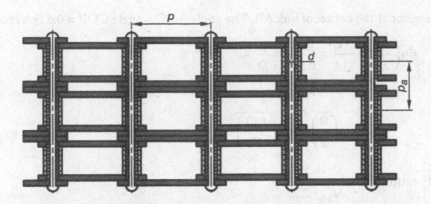

Figure 3.5 Multistrand roller chain (Three strands)

3.4 Pitch and Pitch Circle Diameter

Pitch of a chain (p) is the distance between hinge center of one link to the center of adjacent link (Figure 3.6). Pitch circle diameter (D) is the diameter of a circle, on which the center of chain links lie, when it wraps over the sprocket. Since the link being rigid cannot take arc shape, pitch length is a chord joining two centers and not the arctual length between the two centers.

In Figure 3.6, A and B are the two centers of a link, which subtend an angle θ at the center O. If a sprocket has Z number of teeth, angle between two adjacent teeth:

$$\theta = \frac{360}{Z} \qquad (3.1)$$

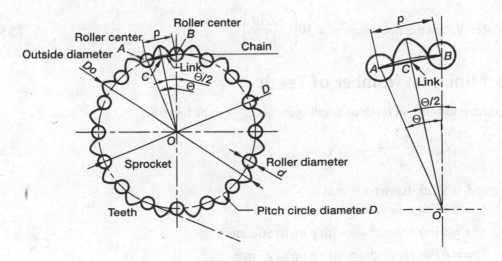

Figure 3.6 Pitch of a chain

Let C be a point at the center of link AB. The angle AOC = angle COB = 0.5 θ, hence:

$$\sin\left(\frac{\theta}{2}\right) = \frac{AC}{OA} = \frac{0.5p}{0.5D} = \frac{p}{D}$$

Or,
$$D = \frac{p}{\sin\left(\dfrac{\theta}{2}\right)} = \frac{p}{\sin\left(\dfrac{180}{Z}\right)} \tag{3.2}$$

Velocity ratio:
$$VR = \frac{N_1}{N_2} = \frac{Z_2}{Z_1} \tag{3.3}$$

Where, N_1, N_2 = Speed of driver and driven shafts respectively

$\quad\quad$ Z_1, Z_2 = Number of teeth of driver and driven sprocket respectively

Average chain velocity v is given by the relation:

$$v = \frac{\pi D_1 N_1}{60} = \frac{\pi D_2 N_2}{60} \tag{3.4}$$

Where, D_1, D_2 = Pitch circle diameter of driver and driven sprocket respectively

The pitch of a chain depends on the power to be transmitted, speed of the chain, and rotational speed of the driver. Power capacity of different roller chains is given in Table 3.6 In section 3.11. If center distance C is known, pitch p can be approximated using the relation:

$$p = \frac{C}{30 \text{ to } 60} \tag{3.5a}$$

If center distance is not given:
$$p \leq 10\left(\frac{3{,}640}{N_1}\right)^{2/3} \tag{3.5b}$$

3.5 Minimum Number of Teeth

Minimum number of teeth on small sprocket Z_{min} can be taken as:

$$Z_{min} = \frac{4d_1}{p} + x \tag{3.6a}$$

Where, d_1 = Shaft diameter in mm

$\quad\quad$ $x = 4$ For roller chain with $p \geq 25$ mm

$\quad\quad$ $x = 5$ For roller chain with p up to 50 mm

$\quad\quad$ $x = 6$ For silent chain with p up to 25 mm

If pitch is not known, use the following relation:

$$Z_{min} = 31 - \left(2 \times VR\right) \tag{3.6b}$$

Minimum number of teeth depends on velocity ratio. Table 3.1 gives number for different velocity ratios with different types of chains. Velocity ratio more than seven is not recommended for chains.

Table 3.1 Minimum Number of Teeth for Different Velocity Ratios

Velocity Ratio		1	2	3	4	5	6
Type of Chain	Bushed Roller	31	27	25	23	21	17
	Silent	40	35	31	27	23	19

In general, minimum number of teeth is 13–15, if chain velocity is less than 2 m/s. Although 17 teeth are the recommended minimum for a drive sprocket, but as little as seven teeth are also available. When the maximum bore of a 17 tooth sprocket cannot accommodate a required shaft, a sprocket with more teeth has to be used. Hardened teeth are recommended for sprockets with 25 teeth or less. For velocity more than 2 m/s, this number is increased to 19.

3.6 Chordal Action

In spite of constant speed of driver sprocket, chain velocity does not remain constant. Variation in its speed depends on the number of teeth of the sprocket. See Figure 3.7(a) in which end 1 of link (1–2) is shown at its maximum distance from the center. At this position, chain velocity is maximum and is given as:

$$v_{max} = \frac{0.5\pi DN}{60} \tag{a}$$

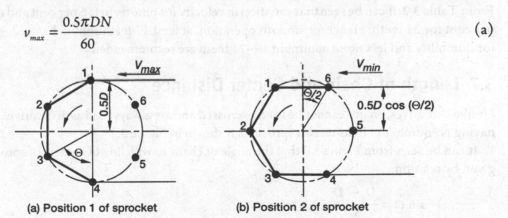

(a) Position 1 of sprocket   (b) Position 2 of sprocket

Figure 3.7 Change in chain velocity due to chordal action

Let θ be the angle between adjacent teeth at the center. When the sprocket rotates by an angle 0.5θ, highest position is now by the link (1–6) as shown in Figure 3.7(b). This position will be at the lowest distance from the center and hence gives minimum velocity v_{min}, which is given as:

$$v_{min} = \frac{0.5\pi D\cos\left(\theta/2\right)N}{60} \qquad \text{(b)}$$

This means chain velocity varies from v_{max} to v_{min} in each revolution as number of times as the number of teeth. Variation chain velocity is given by:

$$\Delta V = v_{max} - v_{min}$$

Substituting the values from Equations (a) and (b):

Variation in chain velocity, $\quad \Delta V = \dfrac{0.5\pi DN}{60} - \dfrac{0.5\pi D\cos(\theta/2)N}{60}$

Or, $$\Delta V = \frac{\pi DN}{120}\left[1 - \cos\left(\theta/2\right)\right] \qquad (3.7)$$

If number of teeth is increased, value of θ decreases and lesser is the variation of ΔV by $\left[1 - \cos\left(\theta/2\right)\right]$ times as tabulated below in Table 3.2:

Table 3.2   Speed Variation with Number of Teeth

No. of Teeth	7	9	11	13	15	17	19	21	23	25
Speed Variation (%)	9.9	6.0	4.0	2.9	2.2	1.7	1.4	1.1	1.0	0.8

From Table 3.2, it can be seen that variation in velocity for nine teeth is 6 per cent and only 1.1 per cent for 21 teeth. Hence for smooth operation, at least 17 teeth should be used, whereas for durability and less noise minimum 19–21 teeth are recommended.

3.7 Length of Chain and Center Distance

Unlike belt drives, chains cannot be used as crossed and are always used as open drive. A chain having N_L number of links on two sprockets is shown in Figure 3.8.

It can be seen from Figure 3.8 that the angle of chain α with line of centers of sprockets is given by relation:

$$\sin \alpha = \frac{D_2 - D_1}{2C}$$

Where, D_1, D_2 = Pitch circle diameters of small and large sprocket respectively.

$\qquad C$ = Center distance between the two shafts.

Length of inclined portion of chain on both sides is:

$$= 2 \times \sqrt{C^2 + \left(\frac{D_2 - D_1}{2}\right)^2}$$

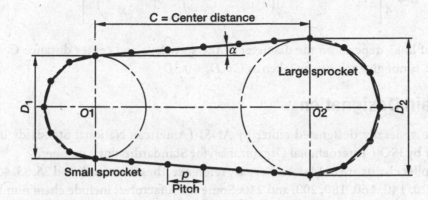

Figure 3.8 Length of chain

Length of chain warpped over small and large sprockets is $\frac{\pi}{2}(D_1 + D_2)$.

Hence, length of chain,

$$L = \frac{\pi}{2}(D_1 + D_2) + \left(2 \times \sqrt{C^2 + \left(\frac{D_2 - D_1}{2}\right)^2}\right) \tag{3.8}$$

Length of chain has to be multiple of number of links N_L. If p is distance between centers of the link, length of chain L is given as:

$$L = N_L \times p \qquad \text{Or,} \qquad N_L = \frac{L}{p}$$

Because $D_1 = \frac{pZ_1}{\pi}$ and $D_2 = \frac{pZ_2}{\pi}$

Substituting these values in equation above and value of L from Equation (3.8):

$$\text{Number of links: } N_L = \frac{Z_2 + Z_1}{2} + \frac{2C}{p} + \frac{p}{C}\left(\frac{Z_2 - Z_1}{2\pi}\right)^2 \tag{3.9}$$

Where, Z_1, Z_2 = Number of teeth of small and big sprocket respectively

C = Center distance (mm)

p = Pitch (mm)

Center distance from number above equation is:

$$C = \frac{p}{4}\left[N_L - \frac{Z_1 + Z_2}{2} + \sqrt{\left(N_L - \frac{Z_2 + Z_1}{2} \right)^2 - 8\left(\frac{Z_2 - Z_1}{2\pi} \right)^2} \right] \qquad (3.10)$$

Number of links depend on the diameter of the sprockets and center distance C. If center distance C, is not given, it can be taken as $C = D_2 + 0.5D_1$.

3.8 Chain Designation

Chains are generally designated either by ANSI (American National Standards Institute) number or by ISO (International Organization for Standardization) number.

ANSI roller chains are in 14 sizes. For easy reference, these are numbered 25, 35, 40, 50, 60, 80, 100, 120, 140, 160, 180, 200, and 240. Some manufacturers include chain numbers 320 and 400 also to the list of standard chains.

The last numeral '0' stands for standard chains, '1' for narrow chains and '5' for bushing chains, which do not have rollers.

Next number from the right indicates the pitch in 1 / 8th of an inch. To calculate the pitch, multiply this number by 3.175 mm (25.4 / 8). For example, for chain 100, last 0 on right side is for standard chains, next number 10 is for pitch, which is calculated as:

$$\text{Pitch}, p = \left(10 \times \frac{1}{8} \right) = 1.25" \quad \text{or in mm it is } 10 \times 3.175 = 31.75 \text{ mm}$$

ISO number represents pitch in 1 / 16th of an inch followed by a letter 'B', which indicates adherence to British standards. For example, 16B means a pitch $= \left(16 \times \frac{1}{16} \right) = 1"$, that is, 25.4 mm. B stands for British standard.

The number in both the standards may have a hyphen followed by a number of chain strands. For example, 8B-2 means, sprocket will have a pitch of half inch and it is a double-strand chain.

There are many other standards like:

- JIS: Japanese standards are limited to Japan
- KS: Korean standard
- DIN (Deutsche Institute fuer Normung): Codes are for German industries. They produce chain in dimensions very similar to both British and American standards.

Table 3.3 gives the dimensions and breaking load for different chains.

Table 3.3 Dimensions (mm) and Breaking loads of Chains [IS 2403–1991]

ISO No.	ANSI No.	Pitch p	Roller Diameter d	Strand Spacing p_a	Breaking Load W (kN)		
					No. of Strands		
					1	2	3
05B	–	8.00	5.00	5.64	4.6	8.0	11.4
06B	–	9.52	6.35	10.24	8.9	16.9	24.9
08A	40	12.7	7.95	14.38	13.8	27.6	41.4
08B	–	12.7	8.51	13.92	17.8	31.1	44.5
10A	50	15.88	10.16	18.11	21.8	43.6	65.4
10B	–	15.88	10.16	16.59	22.2	44.5	66.7
12A	60	19.05	11.91	22.78	31.1	62.3	93.4
12B	–	19.05	12.07	19.46	28.9	57.8	86.7
16A	80	25.40	15.88	29.29	55.6	111.2	166.8
16B	–	25.40	15.88	31.88	42.3	84.5	126.8
20A	100	31.75	19.05	35.76	86.7	173.5	260.2
20B	–	31.75	19.05	36.45	64.5	129.0	193.5
24A	120	38.10	22.23	45.44	124.6	249.1	373.7
24B	–	38.10	25.40	48.36	97.9	195.7	293.6
28A	140	44.45	25.40	48.87	169.0	338.1	507.1
28B	–	44.45	27.94	59.56	129.0	258.0	387.0
32A	160	50.80	28.58	58.55	222.4	444.8	667.2
32B	–	50.80	29.21	58.55	169.0	338.1	507.1
40A	200	63.50	39.68	71.55	347.0	693.9	1,040.9
40B	–	63.50	39.37	72.29	262.4	524.9	787.3
48A	240	76.20	47.63	87.83	500.4	1,000.8	1,501.3
48B	–	76.20	48.26	91.21	400.3	1,423.4	1,201.0

ISO number of a chain is an approximate number of roller diameter of the chain in mm. Table 3.4 gives thickness of plates, pin size, and weight of roller chains.

Table 3.4 Size (mm) and Weight of Chains (IS-2403–1991)

b_1 = width between inner plates

b_2 = width between outer plates, d_p = Pin diameter and t = Thickness of plates

ISO No.	ANSI No.	Roller Diameter	d_p	b_1	b_2	t	Weight (w) (N/m) No. of Strands		
							1	2	3
05B	–	5.00	2.31	3.0	4.9	0.95	1.8	3.3	5.2
06B	–	6.35	3.27	5.72	8.66	1.3	4.1	7.7	10.9
08A	40	7.95	3.98	7.92	10.92	1.5	5.91	11.76	18.68
08B	–	8.51	4.45	7.75	11.43	1.6	6.75	13.21	21.88
10A	50	10.16	5.09	9.53	13.59	2.03	9.97	19.58	29.19
10B	–	10.16	5.08	9.65	13.41	1.7	8.32	16.26	25.58
12A	60	11.91	5.96	12.7	17.12	2.42	13.58	26.65	39.95
12B	–	12.07	5.72	11.63	15.75	1.85	11.47	22.73	63.55
16A	80	15.88	7.94	15.88	22.38	3.25	25.79	49.95	75.22
16B	–	15.88	8.28	17.02	25.88	4.1	26.09	51.72	77.10
20A	100	19.05	9.53	19.05	27.05	4.00	38.80	75.74	112.86
20B	–	19.05	10.19	19.56	29.14	4.5	36.45	72.1	107.89
24A	120	22.23	10.10	25.4	35.00	4.80	55.96	107.36	161.07
24B	–	25.40	16.63	25.4	37.40	6.0	69.08	135.68	199.19
28A	140	25.40	12.70	29.4	40.60	5.60	74.39	144.62	216.33
28B	–	27.94	15.90	30.99	46.71	6.0	87.21	184.19	274.60
32A	160	28.58	14.27	31.75	44.55	6.40	100.37	194.95	291.55
32B	–	29.21	17.81	34.99	45.70	7.5	97.97	193.99	290.30
40A	200	39.68	19.65	38.1	54.10	8.00	166.95	327.55	486.9
40B	–	39.37	22.89	38.10	55.88	8.5	158.27	314.30	470.33
48A	240	47.63	26.74	44.68	63.68	9.50	236.82	462.87	690.77
48B	–	48.26	29.24	45.72	70.69	12.0	244.23	484.96	786.53

3.9 Forces on Chain

There are three forces acting on chain: Force due to pull, centrifugal force, and force due to sagging. Each can be calculated as given below:

a. Force due to pull on tight side $F_p = \dfrac{\text{Power in kilowatts} \times 1,000}{v} = \dfrac{1,000\,P}{v}$ (3.11)

On slack side, it is zero.

b. Centrifugal force $F_c = \dfrac{w\,v^2}{g}$ (3.12)

Where, w = Weight of chain (N/m; given in Table 3.4)
      v = Chain velocity (m/s)

c. Force due to sagging $F_s = C\,w\,K_a$ (3.13)

Where, C = Center distance (m)
      K_a = Coefficient for angle of chain drive (line joining the centers).
        Vertical orientation, $K_a = 1$
        Inclined 45°, $K_a = 2$
        Horizontal orientation, $K_a = 4$

Total force on the chain $F = F_p + F_c + F_s$ (3.14)

3.10 Breaking Load and Factor of Safety

Breaking load W for a chain is given in Table 3.3. This has to be more than the total forces F on the chain. Factor of safety (FOS) is given as:

$$FOS = \frac{\text{Breaking load}}{\text{Force on chain}} = \frac{W}{F}$$

FOS for the chains is given in Table 3.5 for different small sprocket speeds.

Table 3.5   Factor of Safety for Bush Roller Chains

Pitch of Chain (p) (mm)	Speed of Small Sprocket (rpm)							
	≤ 50	200	400	600	800	1,000	1,200	1,600
12.7–15.875	7	7.8	8.55	9.35	10.2	11.0	11.7	13.2
19.5–25.4	7	8.2	9.35	10.3	11.7	12.9	14.0	16.3
31.75–38.1	7	8.55	10.2	13.2	14.8	16.3	19.5	–

3.11 Power Capacity of Chains

Power rating P of single strand roller chain in kilowatts for 17 teeth driving sprocket for a life of 15,000 hours is given in Table 3.6. Power capacity increases with speed, but after some speed, it starts decreasing. Maximum power is shown dark shaded.

Table 3.6 Power Capacity in Kilowatts of Single Strand Roller Chains with 17 Teeth

Speed (rpm)	Chain No.										
	06B	08A	08B	10A	10B	12A	12B	16A	16B	20A	20B
50	0.14	0.28	0.34	0.53	0.64	0.94	1.07	2.06	2.59	4.29	5.36
100	0.25	0.53	0.64	0.98	1.18	1.74	2.01	4.03	4.83	7.96	9.95
200	0.47	0.98	1.18	1.83	2..19	3.40	3.75	7.34	8.94	14.31	17.53
300	0.61	1.34	1.70	2.68	3.15	4.56	5.43	11.63	13.06	21.47	23.99
500	1.09	2.24	2.72	4.34	5.01	7.69	8.53	16.99	20.57	34.00	41.14
700	1.48	2.95	3.66	5.91	6.71	10.73	11.63	23.26	27.73	44.73	53.32
1,000	2.03	3.94	5.09	8.05	8.97	14.32	15.65	28.63	34.89	39.37	47.98
1,400	2.73	5.28	6.81	11.18	11.67	14.32	18.15	18.49	38.47	34.00	70.76
1,800	3.44	6.98	8.10	8.05	13.03	10.44	19.85	–	–	–	–
2,000	3.80	6.26	8.67	7.16	13.49	8.55	20.57	–	–	–	–

Power given in Table 3.6 is for single strand. To increase power, number of strands is increased to 2 or 3 or 4 maximum. Then, the rated power becomes $P_r \times K_s$. Value of strand factor K_s is given in Table 3.7.

Table 3.7 Strand Factor K_s

No. of Strands	1	2	3	4
Strand Factor (K_s)	1.0	1.7	2.5	3.3

Example 3.1

Power transmitted by a specified chain

A 17 teeth sprocket transmits power at 1,500 rpm using 08A chain. Find:

   a. Chain velocity

   b. What maximum power it can transmit, with single and dual strands?

   c. What maximum load it can take, with double strand?

Solution

Given $Z_1 = 17$ $N_1 = 1,500$ rpm   Chain = 08 A

a. From Table 3.3, pitch of 08A chain is 12.7 mm.

Circumference at pitch circle diameter = $p \times Z_1 = 12.7 \times 17 = 215.9$ mm

$$\text{Chain velocity } v = \frac{215.9 \times 1,500}{60 \times 1,000} = 5.4 \,\text{m/s}$$

b. From Table 3.6, power rating for single strand after interpolation for 1,500 rpm is 5.63 kW. From Table 3.7 for double strand, strand factor $K_s = 1.7$.

Hence power for dual strand is:

$P = 5.63 \times 1.7 = 9.57$ kW

c. From Table 3.3, breaking load for chain 08A for two strands is $W = 27.6$ kN.

3.12 Design Power and Corrected Power

Rated power P_r given in Table 3.6 is for 17 teeth, with single strand.

Corrected power of the chain is $P_c = P_r \times K_s$ (K_s is strand factor from Table 3.7)

Design power P_d for an application depends on power P to be transmitted, which needs correction due to following correction factors.

3.12.1 Tooth correction factor

Power rating of chains in Table 3.6 is given for 17 teeth of sprocket. If number of teeth of driver sprocket Z_1 is other than 17, tooth correction factor K_t as given in Table 3.8 is to be applied.

Table 3.8   Tooth Correction Factor K_t on Number of Teeth of Driver Sprocket Z_1

Z_1	11	12	13	14	15	16	17	18	19	20	21	22
K_t	0.53	0.62	0.70	0.78	0.85	0.95	1.0	1.05	1.11	1.18	1.26	1.29
Z_1	23	24	25	27	29	31	35	40	45	50	55	60
K_t	1.35	1.41	1.46	1.57	1.68	1.77	1.95	2.15	2.37	2.51	2.66	2.80

3.12.2 Load factor

If the power is transmitted to applications with shocks, load correction factor as given in Table 3.9 is to be applied.

Table 3.9   Load Factor K for Power Transmitting Roller Chains

Type of Output Load	Electric Motor Drive	I. C. Engine Drive
Smooth	1.0	1.2
Moderate shock	1.2–1.3	1.4–1.5
Heavy shock	1.4	1.7

3.12.3 Service factor

Power rating of chains in Table 3.6 is for 8 hours of service. If its service is required for more than 8 hours, then service factor as given below is to be applied:

8 hours per day $K_h = 1$

16 hours per day $K_h = 1.25$

24 hours per day $K_h = 1.5$

3.12.4 Lubrication factor

Method of lubrication also affects power transmission capacity. Following values of lubrication factor K_l can be used:

Continuous lubrication $K_l = 0.8$

Drop lubrication $K_l = 1.0$

Periodic lubrication $K_l = 1.5$

From all the correction factors selected from Tables 3.8 and 3.9, calculate design power.

$$\text{Design power } P_d = \frac{P \times K \times K_l \times K_h}{K_t} \tag{3.15}$$

Example 3.2

Design power and *FOS*

A two strand roller chain 12B-2 is used to transmit power for 16 hours a day with heavy shocks between 13 teeth driving sprocket rotating at 300 rpm to 52 teeth driven sprocket.

  a. How much power this drive can transmit with continuous lubrication?

  b. How much is factor of safety? Neglect centrifugal and sagging forces.

Solution

Given   Chain: 12B-2 $Z = 13$ $N_1 = 300$ rpm $Z_1 = 13$ $Z_2 = 52$

          16 hours service K = Heavy shocks K_l = Continuous lubrication

a. **Design power** From Table 3.8, tooth correction factor for 13 teeth $K_t = 0.7$

For 16 hours per day $\qquad K_h = 1.25 \qquad$ (Section 3.12.3)

For continuous lubrication $\quad K_l = 0.8 \qquad$ (Section 3.12.4)

Load factor for heavy shocks $K = 1.5 \qquad$ (Table 3.9)

Design power: $P_d = \dfrac{P \times K \times K_l \times K_h}{K_t} = \dfrac{P \times 1.5 \times 0.8 \times 1.25}{0.7} = 2.14\,P$ kW

Corrected power From Table 3.3, pitch of 12-B chain is 19.05 mm.

Rated power from Table 3.6 for single strand for 17 teeth is $P_r = 5.43$ kW

For two strands, strand factor from Table 3.7 is $K_s = 1.7$

Corrected power with two strands: $P_c = P_r \times K_s = 5.43 \times 1.7 = 9.23$ kW

Equating corrected power capacity of chain with the design power:

$\qquad 2.14P = 9.23 \qquad$ Or, Power capacity, $\quad P = 4.3$ kW

b. **FOS** Circumference at pitch circle diameter $= p \times Z_1 = 19.05 \times 13 = 247.65$ mm

Chain velocity: $\quad v = \dfrac{247.65 \times 300}{60 \times 1,000} = 1.24$ m/s

Force due to pull $F_p = \dfrac{\text{Power (kW)} \times 1,000}{v} = \dfrac{9.23 \times 1,000}{1.24} = 7{,}443$ N $= 7.443$ kN

From Table 3.3, breaking load for chain 12B with two strands $W = 57.8$ kN

$\qquad FOS = \dfrac{\text{Breaking load}}{\text{Force on chain}} = \dfrac{W}{F_p} = \dfrac{57.8}{7.443} = 7.76$

Example 3.3

Chain size and length for a given application

An I.C. engine transmits 7.5 kW at 800 rpm through a shaft driven by belt with velocity ratio 4 and transmission efficiency 95 per cent. This shaft has a sprocket having 17 teeth and drives a power press at 100 rpm giving heavy shocks. Assume continuous lubrication and 8 hours duty.

a. Select a suitable chain

b. Calculate length of chain $\qquad$ c. Number of links, and center distance

Solution

Given $\quad Z_1 = 17 \qquad P = 7.5$ kW $\qquad N = 800$ rpm $\qquad N_2 = 100$ rpm $\qquad$ VR1 = 4

$\eta = 95$ per cent $\quad Z_1 = 17 \quad$ Heavy shocks $\quad$ 8 hours a day $\quad$ Continuous lubrication

$$VR2 = \frac{800}{4 \times 100} = 2$$

a. **Chain selection** $VR1 = 4 = \dfrac{N}{N_1} = \dfrac{800}{N_1}$ or Speed of small sprocket, $N_1 = 200\,\text{rpm}$

Power to the sprocket shaft $= P \times \eta = 7.5 \times 0.95 = 7.125\,\text{kW}$

For heavy shocks, service factor from Table 3.9: $K = 1.5$

Tooth correction factor for 17 teeth from Table 3.8: $K_t = 1.0$

For 8 hours per day $K_h = 1.0$ (Section 3.12.3)

For continuous lubrication $K_l = 0.8$ (Section 3.12.4)

Design power $P_d = \dfrac{P \times K \times K_l \times K_h}{K_t} = \dfrac{7.125 \times 1.5 \times 0.8 \times 1.0}{1} = 8.55\,\text{kW}$

From Table 3.6, for single strand chain 16B power rating is 8.94.

It is more than design power, that is, 8.55 kW, hence chain 16B is safe.

b. **Chain length** Pitch for chain 16B is $p = 25.4$

Pitch circle diameter of small sprocket $D_1 = \dfrac{Z_1 \times p}{\pi} = \dfrac{17 \times 25.4}{3.14} = 137.5\,\text{mm}$

Pitch circle diameter of big sprocket $D_2 = 2 \times D_1 = 2 \times 137.5 = 275\,\text{mm}$

Minimum center distance $C_{min} = D_2 + 0.5\,D_1 = 275 + (0.5 \times 137.5) = 344\,\text{mm}$

From Equation (3.8) length of chain $L = \dfrac{\pi}{2}\left(D_1 + D_2\right) + 2\left(\sqrt{C^2 + \left(\dfrac{D_2 - D_1}{2}\right)^2}\right)$

$$L = 1.57 \times \left(137.5 + 275\right) + 2 \times \left(\sqrt{\left(344\right)^2 + \left(\dfrac{275 - 137.5}{2}\right)^2}\right)$$

Or,

$$L = 647.625 + \left(2 \times \sqrt{1,23,060}\right) = 1,349\,\text{mm}$$

c. **Number of links** $N_L = \dfrac{L}{p} = \dfrac{1,349}{25.4} = 53.1$ say 54 links (select even number)

Exact length of chain $= N_L \times p = 54 \times 25.4 = 1,371.6\,\text{mm}$

Exact center distance from Equation (3.10):

$$C = \frac{p}{4}\left[N_L - \frac{Z_1 + Z_2}{2} + \sqrt{\left(N_L - \frac{Z_2 + Z_1}{2} \right)^2 - 8\left(\frac{Z_2 - Z_1}{2\pi} \right)^2} \right]$$

$$\text{Or, } C = \frac{25.4}{4}\left[54 - \frac{17+34}{2} + \sqrt{\left(54 - \frac{34+17}{2} \right)^2 - 8\left(\frac{34-17}{2\times3.14} \right)^2} \right]$$

$$\text{Or, } C = 6.35 \times \left[54 - 25.5 + \sqrt{1{,}190.25 - 58.62} \right] = 394.6 \text{ mm}$$

The chain gets extended due to pull, hence exact center distance $C = 394$ mm

3.13 Maximum Number of Teeth

Maximum number of teeth depends on VR, $Z_2 = Z_1 \times VR$. If this number is large, chain gets stretched and pitch gets increased by Δp. If this is more, chain may come out of the sprocket or can even break. Ratio $\Delta p/p$ is limited to 1.25 per cent only. Hence with this condition, following is recommended:

For roller chains, $Z_{max} = 120$ and for silent chains $Z_{max} = 140$

Preferred numbers are 38, 57, 76, 95, and 114.

3.14 Maximum Chain Speed

Maximum speed, which a chain can be allowed, depends on number of teeth on small sprocket. It is given in Table 3.10.

Table 3.10 Maximum Small Sprocket Speed (Thousands of RPM) for Roller Chains

No. of Teeth	25	35	40	50	60	80	100	120	160
	04C	06C	8A	10A	12A	16A	20A	24A	32A
11–13	10	10	8.0	5.8	4.0	2.8	1.8	1.5	1.2
14–16	10	10	7.5	5.4	4.0	2.6	1.8	1.4	1.1
17–19	10	10	7.5	5.0	4.0	2.6	1.8	1.4	1.1
20–22	10	10	7.5	5.0	3.8	2.4	1.6	1.6	1.0
23–25	10	10	7.0	4.5	3.5	2.4	1.6	1.2	0.9
26–32	10	10	6.0	4.0	3.0	2.4	1.4	1.1	0.8
35–45	10	7.5	4.5	3.5	2.6	1.8	1.2	1.0	0.7

3.15 Bearing Pressures

Force F on the chain causes bearing pressure, which is given as force per unit projected area of the pin.

Bearing pressure, $p_b = \dfrac{F \times K_h \times K_l}{d_p b_2}$ (3.16)

Where, K_h = Service hours: 8 hours = 1.0, 16 hours = 1.25, and 24 hours = 1.5

d_p = Diameter of roller pin from Table 3.4.

b_2 = Length of roller pin from Table 3.4.

K_l = Lubrication factor (See Section 3.12.4)

The allowable bearing pressures are given in Table 3.11 for different roller chains.

Table 3.11 Allowable Bearing Pressures (N/mm²)

Chain Size	Smaller Sprocket Speed (RPM)				
	50	200	400	800	1,200
08B–10B	34.3	30.9	28.1	23.7	20.6
12B–16B	34.3	29.4	25.7	20.6	17.2
20B–24B	34.3	28.1	23.7	18.1	14.7
28B–32B	34.3	25.7	20.6	14.7	–

3.16 Design of Chain Drive

For the design problems, the data, given are, power to be transmitted, speed of driver and of driven or velocity ratio, service conditions, any constraint on space available, etc. Follow the steps given below:

Step 1 Select number of teeth on sprockets

If velocity ratio is not given, calculate from the speeds of driver N_1 and driven N_2:

$$VR = \frac{N_1}{N_2}$$

Maximum speed ratio for roller chain is 7 and if it is more than 8, a compound drive, that is, chain drive in two stage is recommended. For silent chains, maximum speed ratio is 12 for one stage.

From Table 3.1, for the given velocity ratio, find minimum number of teeth Z_{min}. If this table is not available, it can be calculated using the formula:

$$Z_{min} = 31 - (2 \times \text{Velocity ratio})$$

Number of sprocket teeth is selected odd number.

For values of velocity ratio more than 7, multiple strands should be used or the drive can be in two stages, that is, with one intermediate shaft. If the chain wrap angle is less than 120° around a sprocket, an idler sprocket is recommended to increase the grip between the sprocket and chain.

Step 2 Select various correction factors

a. If number of teeth of driver sprocket Z_1 is other than 17, tooth correction factor K_t as given in Table 3.8.

b. Find lubrication factor K_l as given in Section 3.12.4.

c. Find load factor K for power transmitting chains as given in Table 3.9.

d. Find service factor K_h as given in Section 3.12.3.

Step 3 Calculate design power

From all the correction factors calculated in step 2, calculate design power. Initially, start with single strand.

$$\text{Design power} = \frac{P \times K \times K_l \times K_h}{K_t}$$

Step 4 Select chain

From Table 3.6, select a suitable chain, which should have rated power P_r more than design power. If it is less, increase the number of strands to 2 or 3 or 4 as maximum. Then, the rated power becomes $P_r \times K_s$. Value of strand factor K_s is given in Table 3.7.

Step 5 Check pitch

For the selected chain, see the pitch p from Table 3.3.

$$\text{Pitch} \leq \frac{2,330}{N^{2/3}}$$

Where, N = Speed of small sprocket (rpm)

Step 6 Calculate pitch line velocity

$$\text{Velocity}, \ v = \frac{p N_1 Z_1}{60,000}$$

For normal roller chain properly covered and lubricated chains, velocity should be up to 8 m/s. For exposed roller chains, 15–20 m/s; and for silent chains, 20–40 m/s.

Step 7 Calculate center distance

Center distance between the sprockets should be $> D_2 + 0.5 D_1$. If VR is greater than 3, the center distance must be $> D_2 + D_1$.

Longer center distances give better chain wrap around a sprocket. For common applications, a center distance of 30–50 chain pitches is recommended for best results. For transmissions with variable loads, 20–30 pitches may be suitable. For transmissions longer than 80 chain pitches in length, idler sprockets should be used to support chain.

Step 8 Determine load on chain (See Section 3.9)

Using Equation (3.11), force due to pull on tight side $F_p = \dfrac{1{,}000P}{v}$

Determine weight of chain per unit length w from Table 3.4.

Using Equation (3.12), calculate centrifugal force $F_c = \dfrac{wv^2}{g}$

Using Equation (3.13), calculate force due to sagging $F_s = CwK_a$

Using Equation (3.14), calculate total force $F = F_p + F_C + F_s$

Step 9 Calculate *FOS*

From Table 3.3, find the breaking load W_B for the selected chain.

Calculate *FOS* using following equation:

$$FOS = \frac{\text{Breaking load of chain}}{\text{Total force on chain}} = \frac{W_B}{F}$$

Step 10 Find number of links

Calculate N_L using Equation (3.9).

Step 11 Calculate exact center distance

Use Equation (3.10) to calculate exact center distance.

Step 12 Check for wear

Using Equation (3.16) calculate bearing pressure, $p_b = \dfrac{F \times K_h \times K_l}{d_p b_2}$

It should be lesser than allowable bearing pressure given in Table 3.11.

Example 3.4

Design of a chain drive

An internal combustion engine developing 7.5 kW running at 1,000 rpm drives a compressor at 350 rpm for 16 hours per day. The chain drive is horizontal and center distance cannot be more than 500 mm due to space limitations. Design a roller chain.

Solution

Given $P = 7.5$ kW $N_1 = 1{,}000$ rpm $N_2 = 350$ rpm $C < 500$ mm

         Horizontal drive,   Duty = 16 h / day

Step 1 Velocity ratio, $VR = \dfrac{N_1}{N_2} = \dfrac{1{,}000}{350} = 2.857$

$Z_{min} = 31 - (2 \times VR) = 31 - (2 \times 2.857) = 25.29$

Selecting an odd number $Z_1 = Z_{min} = 25$

Number of teeth of driven sprocket $= Z_1 \times VR = 25 \times 2.857 = 71.4$.

Selecting an odd number $Z_2 = 71$

Step 2 From Table 3.8 for 25 teeth, tooth correction factor $K_t = 1.46$

Assuming continuous lubrication. From section 3.12.3, $K_l = 0.8$

From Table 3.9, Load factor $K = 1.5$

Step 3 Design power $= \dfrac{P \times K \times K_l \times K_h}{K_t} = \dfrac{7.5 \times 1.5 \times 0.8 \times 1.25}{1.46} = 7.7 \text{ kW}$

Step 4 From Table 3.6, for 1,000 rpm suitable chain is 10B (single strand) with capacity 8.97 kW.

From Table 3.3, its pitch is 15.88 mm.

Step 5 Check pitch:

$$\text{Pitch} \leq \frac{2,330}{N^{2/3}} \leq \frac{2,330}{(1,000)^{2/3}} = 23.3 \text{ mm}$$

It should be less than 23.3 mm, hence 15.88 mm is safe.

Step 6 Pitch line velocity $v = \dfrac{p N_1 Z_1}{60,000} = \dfrac{15.88 \times 1,000 \times 25}{60,000} = 6.6 \text{ m/s}$

Less than 8 m/s hence safe.

Step 7 Minimum center distance can be taken as:

$$C = D_2 + 0.5 D_1 = \frac{pZ_2}{\pi} + 0.5 \frac{pZ_1}{\pi}$$

Or, $C = \dfrac{15.88 \times 71}{\pi} + 0.5 \times \dfrac{15.88 \times 25}{\pi} = 359 + 63 = 422 \text{ mm}$

It should vary between 30p and 50p, that is, 476 mm to 794 mm

To account for addendum for outside diameters, $C = 480$ mm (0.48 m) is taken.

It is less than maximum given center distance 500 mm, hence accepted.

Step 8 a. Force due to pull on tight side $F_p = \dfrac{P \times 1,000}{v} = \dfrac{7.7 \times 1,000}{6.6} = 1,167 \text{ N}$

Weight of chain per unit length w from Table 3.4 for single strand is 8.32 N/m

b. Calculate centrifugal force $F_c = \dfrac{w v^2}{g} = \dfrac{8.32 \, (6.6)^2}{9.81} = 37 \text{N}$

c. Calculate force due to sagging $F_s = C w K_a$ (for horizontal drive $K_a = 4$)

$F_s = 0.4 \times 8.32 \times 4 = 13.3 \, \text{N}$

Total force $F = F_p + F_c + F_s = 1{,}167 + 37 + 13.3 = 1{,}217.3 \, \text{N}$

Step 9 From Table 3.3, for chain 10B breaking load for single strand is 22.2 kN

Find factor of safety: $FOS = \dfrac{W_B}{F} = \dfrac{22.2 \times 10^3}{1{,}217.3} = 18.23$

From Table 3.5, FOS should be 11. It is more than that, hence safe.

Step 10 Number of links from Equation (3.9): $N_L = \dfrac{Z_2 + Z_1}{2} + \dfrac{2C}{p} + \dfrac{p}{C}\left(\dfrac{Z_2 - Z_1}{2\pi}\right)^2$

Putting the values: $N_L = \dfrac{71 + 25}{2} + \dfrac{2 \times 480}{15.88} + \dfrac{15.88}{480}\left(\dfrac{71 - 25}{2 \times 3.14}\right)^2$

Or, $N_L = 48 + 60.5 + 1.8 = 110.3$

Number of links are kept even, hence 110 is selected.

Length of chain $L = 15.88 \times 110 = 1{,}746.25 \, \text{mm}$

Step 11 Exact center distance: From Equation (3.10):

$$C = \frac{p}{4}\left[N_L - \frac{Z_1 + Z_2}{2} + \sqrt{\left(N_L - \frac{Z_2 + Z_1}{2}\right)^2 - 8\left(\frac{Z_2 - Z_1}{2\pi}\right)^2}\right]$$

Or, $C = \dfrac{15.88}{4}\left[110 - \dfrac{25 + 71}{2} + \sqrt{\left(110 - \dfrac{71 + 25}{2}\right)^2 - 8\left(\dfrac{71 - 25}{2 \times 3.14}\right)^2}\right]$

Or, $C = 3.968\left[110 - 48 + \sqrt{3{,}844 - 429}\right] = 477.9 \, \text{mm}$

Decrease center distance by 1–2 mm to allow extension of chain so $C = 477 \, \text{mm}$

Step 12 Bearing pressure $p_b = \dfrac{F \times K_h \times K_l}{d_p b_2} = \dfrac{1{,}217.3 \times 1.25 \times 1.0}{5.08 \times 13.41} = 22.3 \, \text{N/mm}^2$

(Take values of pin diameter and length from Table 3.4)

From Table 3.11, allowable bearing pressure for 1,000 rpm after interpolation is 22.2. It is almost same, hence OK.

Example 3.5

Chain for a motor cycle

An internal combustion engine of 7.5 kW at 3,000 rpm drives a motor cycle with speed ratio in top gear 1:1 with the following data:

Maximum speed of motor cycle = 100 km/h

Tyre outside diameter 600 mm.

Transmission efficiency = 75 per cent

Sprocket size at wheel D_2 = 186 mm

· Shock factor = 1.4

$FOS = 4$

  a. Calculate speed reduction by chain and sprockets.

  b. Recommend a suitable chain.

  c. Diameter and number of teeth of small sprocket.

Solution

Given $P = 7.5$ kW $N_1 = 3,000$ rpm $S = 100$ km/h

          $D_w = 600$ mm $\eta = 75$ per cent $D_2 = 186$ mm

            $K = 1.4$ $FOS = 4$

a. Speed of motor cycle, $S = \dfrac{100 \times 1,000}{3,600} = 27.78$ m/s

Also linear speed of wheel, $S = \dfrac{\pi \times D_w \times N_2}{60}$

Or, $N_2 = \dfrac{60S}{\pi \times D_w} = \dfrac{60 \times 27.78}{3.14 \times 0.6} = 885$ rpm

Speed reduction by chain and sprockets $= \dfrac{N_1}{N_2} = \dfrac{3,000}{885} = 3.39$

b. Torque $T = \dfrac{60,000\,P}{2\,\pi N} = \dfrac{60,000 \times 7.5}{6.28 \times 3,000} = 23.9$ Nm

Torque available at driver shaft $T = 23.9 \times 0.75 = 17.9$ Nm

Diameter of driving sprocket $D_1 = 186 / 3.39 = 54.86$ mm $= 0.05486$ m

Force on chain $F = \dfrac{T}{0.5 D_1} = \dfrac{17.9}{0.5 \times 0.05486} = 652.6$ N

Breaking load capacity of the chain required:

$$F \times K \times FOS = 652.6 \times 1.4 \times 4 = 3,654 \text{ N}$$

From Table 3.3, the single strand chain, which can bear this load is 05B and whose breaking load is 4,400 N will be safe.

c. From Table 3.3, pitch for 05B chain is 8 mm.

Diameter of the large sprocket $186 = \dfrac{8 \times Z_2}{\pi}$ or $Z_2 = 73$

Number of teeth driver sprocket $Z_1 = \dfrac{73}{3.39} = 21.5$ say 21

Exact diameter of small sprocket, $D_1 = \dfrac{p}{\sin\left(\dfrac{180}{z_1}\right)} = \dfrac{8}{\sin\left(\dfrac{180}{21}\right)} = 53.67$ mm

---

Example 3.6

Chain driving three sprockets

A small sprocket A shown in Figure 3.S1(a) having 15 teeth, rotates at 500 rpm and delivers 2.5 kW. It drives two other sprockets B and C at 250 rpm and 150 rpm and transfers 1 kW and 1.5 kW power, respectively. Distance between centers of B and C is 600 mm, whereas A is 300 mm above D, which is the center of BC. Calculate the tensions in the chains in different segments and reactions at the shafts at A, B, and C.

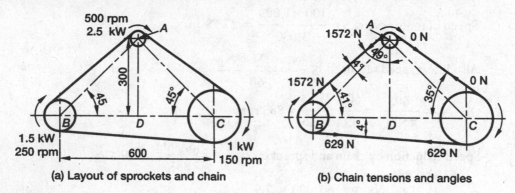

(a) Layout of sprockets and chain   (b) Chain tensions and angles

Figure 3.S1 A chain between three sprockets

Solution

Given $Z_A = 15$ $P_A = 2.5$ kW $P_B = 1.5$ kW $P_C = 1$ kW
               AD = 300 mm $N_A = 500$ rpm $N_B = 250$ rpm $N_C = 150$ rpm
               BC = 600 mm

From Table 3.6 for 2.5 kW at 500 rpm, suitable chain is 08B whose pitch from Table 3.3 is 12.7 mm

Chain velocity $v = \dfrac{p Z_A N_A}{60,000} = \dfrac{12.7 \times 15 \times 500}{60,000} = 1.59$ m/s

Calculate chain tensions

Chain tension of sprocket A on tight side, $F_A = \dfrac{1,000P}{v} = \dfrac{1,000 \times 2.5}{1.59} = 1,572$ N

Chain tension on slack side of sprocket A is 0 N as this is the driver sprocket.

Difference in tight and slack side tensions of B sprocket, $B = \dfrac{1,000 \times 1.5}{1.59} = 943$ N

Hence, tension on slack side of sprocket B = 1,572 − 943 = 629

Chain tension of sprocket C on tight side $F_C = \dfrac{1,000 \times 1}{1.59} = 629$ N and on slack side 0 N.

Calculate angles of chains with X axis

Number of teeth of B sprocket $Z_B = Z_A \left(\dfrac{N_A}{N_B} \right) = \dfrac{15 \times 500}{250} = 30$

Number of teeth of C sprocket $Z_C = Z_A \left(\dfrac{N_A}{N_B} \right) = \dfrac{15 \times 500}{150} = 50$

From Equation (3.2) diameter of A sprocket $D_A = \dfrac{p}{\sin\left(\dfrac{180}{Z_A} \right)} = \dfrac{12.7}{\sin\left(\dfrac{180}{15} \right)} = 61.1$ mm

Diameter of B sprocket $D_B = \dfrac{12.7}{\sin\left(\dfrac{180}{30} \right)} = 121.4$ mm

Diameter of C sprocket $D_C = \dfrac{12.7}{\sin\left(\dfrac{180}{50} \right)} = 202.2$ mm

Distance $AB = AC = 300\sqrt{2} = 424$ mm and $BC = 600$ mm $\left[\text{See Figure 3.S1(b)} \right]$

Angle of chain between sprockets A and B with line $AB = \dfrac{D_B - D_A}{2AB} = \dfrac{121.4 - 61.1}{2 \times 424} = 4°$

(For small angles, sin of angle = Angle in radians)

Angle of chain between sprockets B and C with line $BC = \dfrac{D_C - D_B}{2BC} = \dfrac{202.2 - 121.4}{2 \times 600} = 4°$

Calculate horizontal, vertical, and resultant reactions

Since the tension on one side of the sprockets A and C is zero, the tension of the tight side is the reaction at the sprockets A and C. That is

Reaction at A, $R_A = 1,572$ N at an angle of 41° with horizontal and

Reaction at C, $R_C = 629$ N at an angle of − 4° with horizontal

Reaction at B in X direction, $R_{BX} = 1{,}572 \cos 41° + 629 \cos 4° = 1{,}814$ N

Reaction at B in Y direction, $R_{BY} = 1{,}572 \sin 41° - 629 \sin 4° = 986$ N

Resultant reaction at B, $R_B = \sqrt{1{,}814^2 + 986^2} = 2{,}065$ N

3.17 Silent Chains

The first commercial application of silent chain occurred in 1,843, with the launch of the SS Great Britain in a propeller driven, iron steamship. In the twentieth century, improvements in material quality and chain design increased the load and speed capacity of silent chain. Its use in a gear box is shown in the picture on right side.

Silent chains are made up of stacked rows of flat and tooth-shaped driving links that mesh with sprockets having compatible tooth spaces similar to a rack and pinion mesh. Washers or spacers are used in some chain constructions. All these components are held together by riveted pins located in each chain joint.

Chains are always used on compatible sprockets. When considering different silent chain designs, it is essential that sprocket compatibility is also to be considered.

3.17.1 Comparison with roller chains

- Higher speed and power capacity than roller chains
- Longer sprocket life
- Greater efficiency (99 per cent)
- Less velocity variation
- Reduced noise and vibration
- More uniform wear characteristics
- Less affected by chordal action

3.17.2 Use of silent chains

Silent chains are used for high speed applications. These chains operate more smooth and noiseless than roller chains. However, they are heavier and expensive and hence use is limited.

Silent power transmission chains are capable of transmitting loads and speeds are more than the capacity of all other chains and belts. This is because of increased load and speed capability of specialized chain and sprocket designs. These can be used to 2,000 kW, with speeds 35 m/s, and mechanical efficiency of 99 per cent.

Silent conveying chains are available in a variety of standard widths and offer a smooth, durable, flat conveying surface, and less velocity variation than other conveying chains. Chains are manufactured from hardened steel, which last long, and can work at elevated temperatures.

3.17.3 Construction of silent chains

It consists of a series of toothed link-plates assembled on pin connectors, as shown in Figure 3.9, permitting smooth joint articulation. Many links are joined in parallel with a common pin to increase its width. Pitch is the distance between the teeth of driving links.

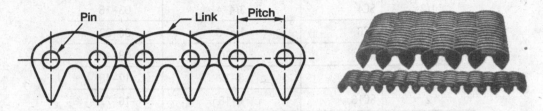

Figure 3.9 Silent chain

Driving Links   Silent chains have tooth-shaped driving links that engage sprockets with similar shaped tooth spaces, similar to a rack and pinion. The shape and size of the tooth on each link can vary, depending on the type and manufacturer as shown in Figure 3.10(a)–(c).

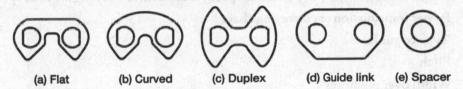

(a) Flat   (b) Curved   (c) Duplex   (d) Guide link   (e) Spacer

Figure 3.10 Silent chain

Guide links   Some chains also contain guide links that maintain proper tracking of the chain on sprockets. Guide links are in the middle for center guide chains and on the sides for double-sided guide chain [Figure 3.10(d)].

Pins and joints   Pin in each joint holds the chain together and allows it to swivel. There may be a single pin in each joint or two pins, depending on the type of chain.

Spacers or bushings   Chains for conveying are provided with spacers [Figure 3.10(e)] between link plates to reduce weight, increase airflow, and allow releasing of debris.

Silent chains pass over the face of a sprocket that looks like a spur gear. The sprocket teeth do not protrude through the chain as in roller chains. The chain meshes with the sprocket by means of teeth extending across the width of the chains underside. The links have no sliding action on teeth, thus providing a quiet and smooth rolling operation.

3.17.4 Standard widths

Chains of different pitches are available in standard sizes as given in Table 3.12.

Table 3.12 Standard Widths of Silent Chains

Pitch (Inches)	Chain No.	Chain Width (Inches)	Description
3/16	SC03	5/32–31/32*	05–31
3/8	SC3	1/2–3	02–12
1/2	SC4	3/4–4	03–16
3/4	SC6	1½–5	06–32
1	SC8	1–10	08–40
1½	SC12	3–12	12–48
2	SC16	4–18	16–72

* All chains are with increment of 1 / 4" except SC03 whose increment is 1 / 32.

* See section 3.17.7 for description

3.17.5 Specifying a chain

For replacing a silent chain, it is important to specify it correctly. Following are to be specified:

- Power transmission or conveying chain
- Type of guides
- Pitch
- Width over pin heads
- Width between guides
- Number of pins: one pin or two pins
- Simple or duplex chain

3.17.6 Selection and design tips

Follow the tips given below while selecting / designing a silent chain:

- Minimize cost by selecting a standard chain rather than special size.
- Use smaller pitch chains and sprockets of large numbers of teeth to reduce noise and vibration. Sprockets should have a minimum of 21 teeth.
- Velocity ratios even of 12:1 are possible, but more than 8:1 should be avoided by using two-stage drive.
- Provide 1 per cent of center distance for adjusting shaft centers to allow for chains.
- Use a chain wider than calculated size to reduced chain stresses and longer life.
- Center distances should not exceed 60 pitches.

- Center distance between the shafts should be sufficient such that the wrap on small sprocket is at least 120°.
- Chain length should be an even number of pitches. An odd number of pitches requiring the use of an offset section, weakens the chain.
- An adjustable idler sprocket should be used to maintain chain tension on drives, if the shafts are fixed.
- For maximum life and safety, enclose the drive with proper lubrication.
- Preferred drive orientation between shaft centers is horizontal or inclined, but the angle should be less than 45°.
- If one shaft is located over the other shaft vertically, chains are to be re-tensioned as it wears. Positioning off the vertical is better.
- Slack side should be on the bottom, if center distance is large.

3.17.7 Chain designation

Silent chains are designated with two letters SC followed by one or two digits giving the chain pitch in increments of $1/8$" and then followed by two or three digits that give the chain width in increments of $1/4$". For example, chain number SC420 is a chain with pitch $4/8" = \frac{1}{2}"$ (12.7 mm) and width $20/4" = 5"$ (127 mm). For complete specifications, the following sequence is used:

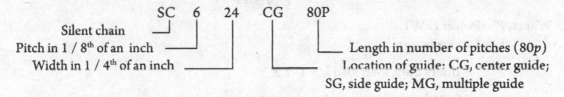

$$\text{SC} \quad 6 \quad 24 \quad \text{CG} \quad 80\text{P}$$

Silent chain ⌐
Pitch in $1/8^{th}$ of an inch ——
Width in $1/4^{th}$ of an inch ———— Length in number of pitches (80p)
Location of guide: CG, center guide; SG, side guide; MG, multiple guide

The above given specification is for a silent chain with pitch $p = \frac{3}{4}"$, width $w = 6"$, center guide and length $L = 80$ pitches, that is, 60".

3.17.8 Factor of safety

FOS for silent chains depends on pitch and speed of small sprocket. It is given in Table 3.13.

Table 3.13 Factor of Safety for Silent Chains

Pitch of Chain (p)(mm)	Speed of Small Sprocket (RPM)							
	≤50	200	400	600	800	1,000	1,200	1,600
12.7–15.875	20	22	24.4	28.7	29.0	31.0	33.0	37.8
19.5–25.4	20	23.4	26.7	30.0	33.4	36.8	40.0	46.5
31.75	20	25	32	36.5	41	46	51	–

3.17.9 Power capacity

Power capacity $P1$ of a silent chain for 1 mm of chain width can be found by an empirical relation given below:

$$\text{Power } P1 = \frac{pv}{231}\left[1 - \frac{v}{2.16(Z_1 - 8)}\right] \tag{3.17}$$

Where, $P1$ = Power of chain width (kW/mm)

$\quad v$ = Chain velocity (m/s)

$\quad Z_1$ = Number of teeth of small sprocket

$\quad p$ = Pitch (mm)

Width of chain can be from p to $10\,p$, normally $2p$ to $6p$, but should not be more than 0.75 m.

3.17.10 Maximum speed of silent chain

For silent chains, maximum speed of small sprocket N_1 (rpm) is:

$$\text{Maximum permissible speed } N_1 = 9{,}550\sqrt[3]{\frac{P \times k_1 k_2}{b\,p\,z_1 p_b}} \tag{3.18}$$

Where, P = Power (kW)

$\quad k_1$ = Service factor

$\qquad$ 8 hours duty, even load k_1 = 1–1.2

$\qquad$ 16–24 hours duty, even load k_1 = 1.2–1.4

$\qquad$ 16–24 hours duty, heavy load k_1 = 1.5–1.7

$\quad k_2$ = Lubrication method:

$\qquad$ For continuous lubrication k_2 = 1.0

$\qquad$ Periodic lubrication k_2 = 1.3

$\quad b$ = Width of silent chain. It can be taken 3–5 times the pitch

$\quad Z_1$ = Number of teeth of driver sprocket

$\quad p_b$ = Allowable bearing pressure: for silent chains (See Table 3.14)

$\qquad P_b$ = 20 N/mm², if speed is less than 50 rpm

$\qquad P_b$ = 8 N/mm², if speed is more than 50 rpm

Table 3.14 Allowable Bearing Pressures for Silent Chain (N/mm²)

Pitch(mm)	Smaller Sprocket Speed (RPM)				
	50	200	400	800	1,200
12.7–15.8	19.6	17.6	16.1	13.7	11.8
19.05–25.4	19.6	16.7	14.7	11.87	9.81
30–38.1	19.6	16.1	13.7	10.3	8.43

Table 3.15 Maximum Speed (RPM)

Pitch	Inches	3/8	1/2	5/8	3/4	1	1½
	mm	10	12.7	16	19	25.4	38.1
Speed from		2,100	1,600	1,250	1,100	900	650
Speed up to		5,200	3,200	2,700	2,100	1,600	1,100

Example 3.7

Roller and silent chain for a centrifugal pump

A chain drive is to be used to connect a 960 rpm motor to a centrifugal pump rated at 4,000 lpm against a head of 9 m at 480 rpm of pump. Efficiency of pump is 70 per cent and of chain drive 97 per cent. Duty hours are 10 hours per day. Select:

a. A roller chain

b. A silent chain

Solution

Given $P = 7.5$ kW $N_1 = 960$ rpm $h = 9$ m $N_2 = 480$ rpm
$Q = 4,000$ lpm $\eta_{pump} = 70$ per cent $\eta_{chain} = 97$ per cent

Duty hours = 10 hours / day

Water power $= \dfrac{mgh}{60,000} = \dfrac{4,000 \times 9.81 \times 9}{60,000} = 5.88$ kW

Power to be transmitted, $P = \dfrac{5.88}{0.7 \times 0.97} = 8.66$ kW

Velocity ratio $VR = \dfrac{N_1}{N_2} = \dfrac{960}{480} = 2$

a. **Roller chain**

Minimum number of teeth $Z_1 = Z_{min} = 31 - (2 \times VR) = 31 - 4 = 27$

For 27 teeth from Table 3.8, $K_t = 1.57$

Design power $P_d = \dfrac{P \times K \times K_l \times K_h}{K_t} = \dfrac{8.66 \times 1.0 \times 0.8 \times 1.25}{1.57} = 5.5 \text{ kW}$

Number of teeth of driven sprocket $Z_2 = 27 \times 2 = 54$

From Equation (3.5b):

$$p \leq 10 \left(\frac{3{,}640}{N_1}\right)^{2/3} \quad \text{or} \quad \leq 10 \left(\frac{3{,}640}{960}\right)^{2/3} \quad \text{or} \quad \leq 24.3 \text{ mm}$$

So selecting a standard pitch of 15.8 mm. Chain with this pitch is 10B.

From Table 3.6, power capacity of single strand chain 10B is 8.9 kW, hence safe.

Diameter of small sprocket $D_1 = \dfrac{p Z_1}{\pi} = \dfrac{15.88 \times 27}{3.14} = 136.5 \text{ mm}$

Diameter of driven sprocket $D_2 = D_1 \times VR - 136.5 \times 2 = 273 \text{ mm}$

Center distance, $C = D_2 + 0.5 \times D_1 = 273 + (0.5 \times 136.5) = 341.2 \text{ mm}$

Using Equation (3.8), length of chain, $L = \dfrac{\pi}{2}(D_1 + D_2) + 2\left(\sqrt{C^2 + \left(\dfrac{D_2 - D_1}{2C}\right)^2}\right)$

$$L = 1.57 \times (136.5 + 273) + 2 \times \left(\sqrt{(341.2)^2 + \left(\frac{273 - 136.5}{2 \times 341.2}\right)^2}\right)$$

Or, $L = 642.9 + \left(2 \times \sqrt{118{,}164}\right) = 1{,}327 \text{ mm}$

Number of links, $N_L = \dfrac{L}{p} = \dfrac{1{,}327}{15.88} = 83.56 \text{ say } 84 \text{ links} \left(\text{select even number}\right)$

Exact length of chain, $L = N_L \times p = 84 \times 15.88 = 1{,}334 \text{ mm}$

b. **Silent chain**

Velocity of the chain $v = \dfrac{p N_1 Z_1}{60{,}000} = \dfrac{15.8 \times 960 \times 27}{60{,}000} = 6.826 \text{ m/s}$

From Equation (3.17),

$$P1 = \frac{pv}{231}\left[1 - \frac{v}{2.16(Z-8)}\right]$$

$$P1 = \frac{15.8 \times 6.826}{231}\left[1 - \frac{6.826}{2.16(27-8)}\right] = 0.4669 \times 0.8337 = 0.389 \text{ kW/mm}$$

Power is 8.66 kW, hence width of chain required $= \dfrac{8.66}{0.389} = 22.3$ mm say 25 mm

Maximum allowable speed from Table 3.15 for this pitch is 1,250 rpm to 2,700 rpm. Here, it is only 960 rpm, hence safe.

3.18 Lubrication of Chains

It is important for long life of chain that it is lubricated properly. A sheet metal cover should be provided to protect from dust for a longer life. There are four types of lubrication as given below. Anyone can be chosen depending on chain velocity as given in Table 3.16.

Type I : Periodic oil supply

Type II : Drip lubrication

Type III : Oil with maintained oil level

Type IV : Oil supply by pump

Table 3.16   Type of Lubrication for Maximum Chain Speed (m/s)

Lubrication Type	Chain Pitch (mm)											
	6.25	9.5	12.7	15.9	25.4	31.8	38.1	44.5	50.8	62.5	76.2	101.6
I	2.5	1.85	1.5	1.25	1.1	0.85	0.75	0.65	0.575	0.5	0.425	0.375
II	12.5	8.5	6.5	5	4.25	3.25	2.6	2.15	1.85	1.65	1.3	1.1
III	17.5	14	11.5	10	9	7.5	6.5	6	5.5	5	4.5	4
IV	Up to maximum permissible speed											

A chain is lubricated efficiently by oil, but for very slow chain speeds, grease lubrication can also be done.

3.19 Sprockets

Sprockets usually are forged and then milled or, but may be flame cut. Large-size sprockets may be produced in halves or segments to aid in transportation. Metal is most common sprocket materials, but nonmetallic plastic and hybrid versions, like metal wheels with plastic teeth, are also used. Sprocket wheels are made of carbon steels,which are heat treated to increase hardness up to 180 BHN. Surface hardness up to 300 BHN is also used for high-speed applications.

Sprockets are specified by following information:

- Pitch
- Number of strands
- Number of teeth
- Bore diameter
- Style of body (Section 3.19.1)

3.19.1 Body styles

Sprocket body can be any of the the following types:

Plate body A plated body provides stronger structure and is used for small-size sprockets. See picture on right side.

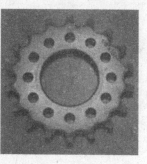

Arms Where large-size sprockets are needed, armed sprocket is used. Use of arms lowers weight, and therefore inertia, and cost.

Split body By segmenting the sprocket body in two halves, it is easier to assemble and disassemble the sprocket from a shaft.

Flanged With a short extension on the sprocket rim, the chain's plates rest on the sprocket flange. Some load from the chain's pins is transferred to flange. These are also called 'chain-saver sprockets' because they increase life of chains.

3.19.2 Sprocket mounting

Sprockets of any design are mounted in many ways as given below:

- **Set screws** Screws are tightened through the sprocket into the shaft. It is used for very small torques.
- **Simple bore** It is used in conjunction with press-fit or adhesive mounting as shown in Figure 3.11(a). It is used for small sprockets up to 50 mm.
- **Keyway** Sprockets bigger than 100 mm are made with a boss, which can be on one side or both sides. One or more keyways align with the shaft, transferring torque through keys in the hub as shown in Figure 3.11(b).
- **Shear pin** A shear pin is put between the shaft and the sprocket Figure 3.11(c). If the chain drive is overloaded or jammed, the shear pin snaps, preventing damage to the machinery. It just acts as a fuse in electric connections.
- **Bolted** Sprocket is drilled axial holes on a pitch circle and bolted to the flange mounted on the shaft. This type of mounting is shown in Figure 3.11(d).

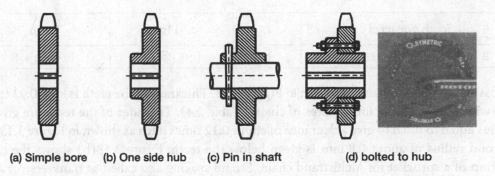

| (a) Simple bore | (b) One side hub | (c) Pin in shaft | (d) bolted to hub |

Figure 3.11 Methods of sprocket mounting

3.19.3 Sprocket proportions

Enlarged profile of tooth of a sprocket is shown in Figure 3.12. Arc 3–4 forms the seat of roller of diameter d. It is semicircle of radius r_s, which is slightly bigger than roller radius. Radius of flanks arcs 1-2-3 or 4-5-6 is r_f. Difference between outside diameter D_o and pitch circle diameter D is addendum a, which is kept $0.5p - r$. Root diameter D_i is pitch circle diameter minus roller diameter, that is, $D_i = D - d$.

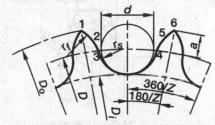

Figure 3.12 Tooth profile

Table 3.17 gives the minimum and maximum proportions of the sprocket wheel for single and multistrand chains in terms of roller diameter d, pitch circle diameter D, and number of teeth Z. For a multistrand chain, sprockets are put side by side at a distance called axial pitch p_a.

Table 3.17 Maximum and Minimum Proportions of Sprocket [IS 2403–1991]

S. No.	Dimension	Notation	Minimum Value	Maximum Value
1	Roller seating radius	r_s	$r = 0.5d$	$0.5d + 0.07\sqrt[3]{d}$
2	Tooth flank radius	r_f	$0.24\, r\, (Z + 2)$	$0.016\, r\, (Z_2 + 180)$
3	Height of tooth above Pitch Circle Diameter (PCD)	A	$0.5\, p - r$	$0.625\, p - r + (0.8\, p\,/\, Z)$
4	Outside diameter	D_o	$D + p\left(1 - \dfrac{1.6}{Z}\right) - 2r$	$D + 1.25P - 2r$
5	Root diameter	D_i	$D - r_s$	$D - r_s$
6	Tooth width	t	$0.93b$	$0.95b$
7	Tooth side radius	r_t	p	p

| 8 | Tooth side relief | t_1 | $0.1p$ | $0.15p$ |
| 9 | Width over n teeth | t_n | $(n-1)p_a + t$ | $(n-1)p_a + t$ |

Cross section of a sprocket is shown in Figure 3.13. Thickness of the teeth is kept 0.93 times the width between the inner plates of chain (Table 3.4). The sides of the teeth are given a radius equal to pitch to give a clearance of about 0.12 times pitch as shown in Figure 3.13(a). Shroud radius of about 0.8 mm is given below the teeth. Figure 3.13(b) shows the cross section of a sprocket for multistrand chain. Strand spacing also called as transverse or axial pitch p_a depends on chain size and is given in Table 3.3.

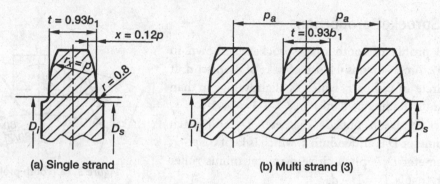

(a) Single strand   (b) Multi strand (3)

Figure 3.13 Rim profile

Summary

Chains and gears offer a nonslip drive. A **chain** consists of many metallic links hinged together using a pin joint at ends to provide flexibility so that it could wrap over a wheel called **sprocket**. It is used primarily for power transmission and conveyor systems.

The center distance between the two sprockets can be as large as 8 m and power capacity up to 100 kW. The velocity ratio (VR) can go up to 10.

Applications of chains are in bicycles / motorcycles, I.C. engines, fork-lifts, wind mills, water mills, agricultural machines, main drive of army tank, and bucket elevators.

Advantages No slip, lesser width is required for a given power, can work in a smaller space, high transmission efficiency up to 98 per cent, load on shafts/bearings is less as no initial tension is needed as in belts, one single chain can drive many sprockets, and high velocity ratios. Environmental conditions do not affect the drive and more power transmission capacity than belts.

Disadvantages Costlier than belt drives, requires correct mounting, needs continuous maintenance and lubrication, and rotational speed of the driven sprocket fluctuates.

Material for chains is cast iron, cast steel, forged, steel, and plastic.

Classification on **application basis** Hoisting chains, elevating chains, and power transmission chain.

On the basis of **construction** Link chain, hook type chain, roller chain, and silent chain.

Parts of roller chain

Link This component bears the tension coming in the chain. In a roller chain, there are alternating links of two types. The first set is of inner links, having two inner plates held together by two bushes upon which two rollers rotate.

Offset link is used when an odd number of chain links is required.

Pin is subjected to shearing and bending forces transmitted by the plate. It is a load-bearing part, with the bushing, when the chain flexes during sprocket engagement.

Bush is put between the roller and pin to act as bearing.

Roller reduces friction by rolling over the bush.

Multiple-strand chains used for more power. The links are placed side by side with a long pin passing through all the bushes. Number of strands can go as high as 10.

Pitch of a chain is the distance between hinge centers of one link to the center of adjacent link. **Pitch circle diameter** is the diameter of a circle, on which the center of chain links lie, when it wraps over the sprocket. **Pitch length** is a chord joining two centers and not the arctual length between two centers.

Angle subtended by link at the center $\theta = \dfrac{360}{Z}$

Where, Z is number of teeth on sprocket.

Pitch circle diameter $D = \dfrac{p}{\sin\left(\dfrac{180}{Z}\right)}$

Where, p is pitch of the chain.

Velocity ratio $VR = \dfrac{N_1}{N_2} = \dfrac{Z_2}{Z_1}$

Average chain velocity $v = \dfrac{\pi D_1 N_1}{60} = \dfrac{\pi D_2 N_2}{60}$

If center distance C is known, pitch p can be approximated as $p = \dfrac{C}{30-60}$

If center distance is not given: $p \leq 10\left(\dfrac{3{,}640}{N_1}\right)^{2/3}$

Minimum number of teeth on small sprocket $Z_{min} = \dfrac{4d_1}{p} + x$ (x varies from 4 to 6)

If pitch is not known use $Z_{min} = 31 - (2 \times$ Velocity ratio$)$

In general minimum number of teeth is 13–15, if chain velocity is less than 2 m/s. Although 17 teeth are the recommended minimum for a drive sprocket, but as little as seven teeth are also available.

Chordal action causes speed variation of the driven shaft in spite of constant speed of driver sprocket. Variation in speed depends on the number of teeth of the sprocket. Chain velocity does not remain constant.

Variation in velocity $\Delta V = \dfrac{0.5\pi DN}{60}\left[1 - \cos(\theta/2)\right]$

Variation in velocity for nine teeth is 6 per cent and only 1.1 per cent for 21 teeth.

Length of chain is: $L = \dfrac{\pi}{2}(D_1 + D_2) + 2\left(\sqrt{C^2 + \left(\dfrac{D_2 - D_1}{2C} \right)^2} \right)$

Number of links: $N_L = \dfrac{Z_2 + Z_1}{2} + \dfrac{2C}{p} + \dfrac{p}{C}\left(\dfrac{Z_2 - Z_1}{2\pi} \right)^2$

Center distance $C = \dfrac{p}{4}\left[N_L - \dfrac{Z_1 + Z_2}{2} + \sqrt{\left(N_L - \dfrac{Z_2 + Z_1}{2} \right)^2 - 8\left(\dfrac{Z_2 - Z_1}{2\pi} \right)^2} \right]$

Center distance, if not given can be taken as $C = D_2 + 0.5D_1$.

Chain designation is done either by ANSI number or by ISO number.

ANSI roller chains are in 14 sizes. For easy reference, these are numbered 25, 35, 40, 50, 60, 80, 100, 120, 140, 160, 180, 200, and 240. The last numeral '0' stands for standard chains, '1' for narrow chains, and '5' for bushing chains, which do not have rollers. Next number from the right indicates the pitch in 1/8th of an inch. Letter A follows at the end of chain number.

ISO number represents pitch in 1/16th of an inch followed by a letter 'B', which indicates to British standards. For example, 16B means a pitch $= \left(16 \times \dfrac{1}{16} \right) = 1''$, that is, 25.4 mm. A hyphen followed by a number gives number of strands. 8B-2 means double-strand chain.

Other standards: JIS stands for Japan. Korean standard **KS**, and Germany's **DIN**.

Forces on chain

a. Force due to pull on tight side $F_p = \dfrac{1{,}000P}{v}$ On slack side, it is zero.

b. Centrifugal force $F_C = \dfrac{wv^2}{g}$

Where, w = Weight of chain (N/m)

v = Chain velocity.

c. Force due to sagging $F_s = C\,w\,K_a$

Where, C = Center distance (m)

For vertical orientation $K_a = 1$;   Inclined at angle $45° K_a = 2$;   Horizontal orientation $K_a = 4$.

Total force on the chain $F = F_p + F_c + F_s$

Breaking load W for a chain has to be more than the total forces F on the chain.

Factor of safety $FOS = W / F$

Power rating P of single strand roller chain in kilowatts for 17 teeth driving sprocket for a life of 15,000 hours (ISO Type A) is to be found (Table 3.6). To increase power, increase the number of strands to 2 or 3 or 4 as maximum. Then, the rated power becomes $P_r \times K_s$. Value of strand factor K_s is given in Table 3.7.

Rated power P_r given in Table 3.6 is for 17 teeth, with single strand.

Corrected power of the chain is $P_c = P_r \times K_s$ (K_s from Table 3.7)

Design power P_d for an application depends on power P to be transmitted, which needs correction due to following correction factors.

  a. *Tooth correction factor* K_t For sprocket Z_1 other than 17, use tooth correction factor as given in Table 3.8.

  b. *Load factor K* For power with shocks, load correction factor is applied (Table 3.9).

  c. *Service factor* K_h Power rating for 8 h is in Table 3.6. For more than 8 hours, use service factor as: for 16 hours a day, $K_h = 1.25$ and for 24 hours a day $K_h = 1.5$.

  d. *Lubrication factor* K_l Continuous $K_l = 0.8$, drop lubrication $K_l = 1.0$, and periodic lubrication $K_l = 1.5$.

From all the correction factors, calculate design power $P_d = \dfrac{P \times K \times K_l \times K_h}{K_t}$

Maximum number of teeth For roller chains, $Z_{max} = 120$ and for silent chains $Z_{max} = 140$. Preferred numbers are 38, 57, 76, 95, and 114 (Table 3.6).

Maximum chain speed It is given in Table 3.10.

Bearing pressure $P_b = \dfrac{F \times K_h \times K_l}{d_p b_2}$; $K_h =$ Service factor

$d_p =$ Diameter of roller pin, $b_2 =$ Length of roller pin, and $K_l =$ Lubrication factor. The allowable bearing pressures are given in Table 3.11.

Design procedure Follow the steps given to design a chain drive:

  1. Calculate velocity ratio, $VR = N_1 / N_2$. For VR, < 7 select roller chain and <12 silent chains. For >12 use two stage.

      Minimum number of teeth $Z_{min} = 31 - (2 \times VR)$ or use Table 3.1. Select odd number.

  2. Select various correction factors like *tooth correction factor* K_t as given in Table 3.8. *Lubrication factor* K_l, *load factor* from Table 3.9, and *service factor* K_h.

  3. Calculate design power: $P_D = \dfrac{P \times K \times K_L \times K_h}{K_t}$. Initially, start with single strand.

  4. Select chain from Table 3.6, with rated power P_r more than corrected power. If it is less, increase the number of strands to 2 or 3 or 4 as maximum. Then, the rated power becomes $P_r \times K_s$. Value of strand factor K_s is given in Table 3.7.

  5. Check pitch p from Table 3.3 for the selected chain. It should be $\leq \dfrac{2,330}{N^{2/3}}$ where, $N =$ Speed of small sprocket (rpm)

  6. Calculate pitch line velocity $v = \dfrac{pN_1 Z_1}{60,000}$. Velocity should be up to 8 m/s. For exposed roller chains, 15–20 m/s and silent chains, 20–40 m/s.

  7. Calculate center distance $C = D_2 + 0.5D_1$. For $VR > 3$, C should be $> D_2 + D_1$.

      For common applications, C is 30–50 chain pitches. For transmissions with variable loads, $C = 20p$ to $30p$. For transmissions longer than 80 p, use idler sprocket.

8. Determine force due to pull on tight side $F_p = 1,000 \, P / v$,

   Centrifugal force $F_c = wv^2 / g$

   Where, w is weight of chain per unit length (Table 3.4).

   Force due to sagging $F_s = C \, w K_a$

9. From Table 3.3, find breaking load for the selected chain. $FOS = W_B / F$

10. Find number of links N_L using Equation (3.9).

11. Using Equation (3.10) calculate exact center distance.

12. Check bearing pressure $p_b = \dfrac{F \times K_h \times K_l}{d_p b_2}$. It should be lesser than the bearing pressure given in Table 3.11

Silent chains are used for high-speed application as they operate more smooth and noiseless than roller chains. These are heavier and expensive and hence use is limited.

$$\text{Power in kW/mm of chain width } P1 = \frac{pv}{231}\left[1 - \frac{v}{2.16(Z - 8)}\right]$$

Where, v = chain velocity (m/s), Z = Number of teeth of small sprocket, and p = Pitch (mm).

Width of chain can be p to $10p$, normally $2p$ to $6p$ but should not be more than 0.75 m. Maximum permissible speed for chains in thousands of rpm (See Table 3.14).

$$\text{For silent chains: } p = 60\sqrt[3]{\frac{P \times k_1 k_2}{b\,\omega_1\, z_1\, p_b}}$$

Where, P = Power (kW) k_1 = Service factor

   k_2 = Lubrication factor b = Width of silent chain.

It can be taken $3p$ to $5p$, ω_1 = Angular velocity of driver, Z_1 = Number of teeth of driver procket, p_b = Allowable bearing pressure: For silent chains: P_b = 20 N/mm², if N < 50 rpm or 8 N/mm², if N > 50 rpm

Chain lubrication For very slow chain speeds use grease. There are four types of lubrication. Select as per chain velocity given in Table 3.15.

Type I   : Periodic oil supply   Type II : Drip lubrication

Type III : Oil with maintained oil level   Type IV : Oil supply by pump.

Sprockets These are usually forged and then milled or may be flame cut. Large-size sprockets may be produced in halves or segments to aid in transportation. Metal like carbon steel is most common, which is heat treated to increase hardness up to 180 BHN. For high-speed applications hardness up to 300 BHN is also used. Non-metallic plastic are also used.

Specification of sprockets is by pitch, number of strands, number of teeth, bore diameter, and style of body.

Sprocket body features

Plate body   Used for small-sized sprockets.

Arms   For large size sprocket. Arms lower weight, and therefore inertia, and cost.

Split body   sprocket body in two halves, it is easier to assemble and disassemble.

Flanged   With a short extension on sprocket rim, chain's plates rest on the sprocket flange.

Sprocket mounting is in many ways as given below:

Set screws   Screws is used for very small torques. Tightened through the sprocket.

Simple bore   It is used in conjunction with press-fit or adhesive mounting. It is used for small sprockets up to 50 mm.

Keyway   Sprockets bigger than 100 mm are made with a boss, which can be on one side or both sides. One or more keyways align with the shaft, through keys in the hub.

Shear pin   A shear pin is put between the shaft and the sprocket. If the chain drive is overloaded or jammed, the shear pin snaps preventing damage to the machinery.

Bolted   Sprocket is drilled axial holes on a pitch circle and bolted to the flange on the shaft.

Sprocket proportions

Radius of flanks r_f varies from 0.24 $r(Z+2)$ to 0.016 $r(Z_2+180)$

Addendum $a=$ outside diameter; D_o, Pitch circle diameter $D=0.5\ p-r$.

Root diameter $D_i=$ Pitch circle diameter, D–Roller diameter d.

Thickness of the teeth $t=0.93\times$ width between the inner plates of chain.

Sides of the teeth are given a radius equal to pitch to give a clearance of about 0.12p.

Shroud radius of about 0.8 mm is given below the teeth.

Strand spacing or axial pitch p_a depends on chain size and is given in Table 3.3.

Theory Questions

1. Describe the construction of a roller chain with the help of a sketch.
2. What are the different types of chains? Give a sketch of each type and its application.
3. What are the advantages and disadvantages of chain drive over other types of drives?
4. Differentiate between a simple roller chain and bushed roller chain.
5. Explain the speed variation in chain drive due to chordal action.
6. Derive an expression for the length of chain in a chain drive.
7. Describe ISO and ANSI designations of a chain.
8. What are the forces coming on a chain? Write an expression for each.
9. Describe breaking load and *FOS* for a chain drive.
10. How a power capacity of a chain is defined? What are the standard parameters for this capacity? How it is modified for a multistrand chain?
11. What is meant by design power and how it is calculated? Describe the various factors affecting the power.
12. Discuss tooth correction factor.
13. Explain lubrication factor and bearing pressures in a chain.

14. Describe the construction of a silent chain with the help of a neat sketch.

15. Discuss the construction and specifications for a sprocket.

16. Sketch a tooth profile of a sprocket and give the various proportions.

Multiple Choice Questions

1. A chain drive is preferred where:
   (a) Positive drive is desired
   (b) Cost has to be less
   (c) High power is to be transmitted
   (d) Constant speed of driven is desired

2. If driver sprocket rotates at constant speed, driven speed is:
   (a) 98 per cent of driver speed
   (b) Same as driver speed
   (c) More than driver speed
   (d) Variation depends on number of teeth.

3. Pitch circle diameter is related with pitch as:
   (a) $\dfrac{p}{\sin\left(\dfrac{180}{Z}\right)}$
   (b) $\dfrac{p}{\sin\left(\dfrac{360}{Z}\right)}$
   (c) $\dfrac{\sin\left(\dfrac{360}{Z}\right)}{p}$
   (d) $\sin\left(\dfrac{180}{Z}\right)$

4. Minimum number of teeth is important as it affects:
   (a) Velocity ratio
   (b) Life of chain
   (c) Speed variation
   (d) Cost

5. Chain with oval links is used for:
   (a) Cranes
   (b) Conveyor applications
   (c) Slow speed hoists
   (d) Where noise level is important

6. For a roller chain of diameter d, outside sprocket diameter is related to pitch circle diameter D as:
   (a) $D + 0.5\,d$
   (b) $D + 0.8\,d$
   (c) $D + d$
   (d) $D + 1.2\,d$

7. Root of the sprocket tooth is:
   (a) Elliptical
   (b) Semicircle of roller radius
   (c) Arc of any radius
   (d) Cycloid

8. A silent chain has
   (a) Rollers of high precision
   (b) Rollers with rubber bushing.
   (c) A cover with a sound proof casing
   (d) Inverted teeth

9. Pitch of a roller chain is proportional to the power P as:
   (a) $P^{1.5}$
   (b) P^2
   (c) $P^{2/3}$
   (d) $\sqrt[3]{P}$

Answers to multiple choice questions

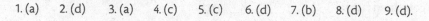

1. (a)   2. (d)   3. (a)   4. (c)   5. (c)   6. (d)   7. (b)   8. (d)   9. (d).

Design Problems

1. A single strand 10A chain is used to drive a 21 teeth sprocket at 1,200 rpm. Calculate the power, which it can transfer with 8 hours duty per day and good lubrication. [P = 115 kW]

2. A jaw crusher with heavy shocks needs 20 kW power at 150 rpm. The driver sprocket has 17 teeth and rotates at 500 rpm. Assuming 16 hours per day working and periodic lubrication, suggest a suitable chain for the drive. [Chain 16A-4 or Chain 20A-2]

3. An electric motor running for 8 hours per day transmits 3 kW at 1,000 rpm to an application giving moderate shocks. The driver sprocket has 19 teeth and driven sprocket 38 teeth. Approximate center distance between the shafts can be taken twice the driven sprocket. Select a suitable chain, number of links, and exact center distance with drop lubrication. [08A, 78 links, C = 311.9 mm]

4. A roller chain is to transmit 20 kW from a 17 teeth sprocket at a speed of 300 rpm to 34 teeth sprocket. Load is with moderate shock ($K = 1.3$) and adverse service conditions like poor lubrication, dirty surrounding, cold temperature, etc. Equipment is supposed to operate for 16 hours a day. Specify the size and length of chain for center distance of 25 pitches.

[20B-2, Two strands, Pitch = 31.75 mm, L = 2.413 m, C = 796.8 mm]

5. An electric motor of 10 kW running at 1,440 rpm drives a compressor at 400 rpm with horizontal roller chain for 8 hours per day with periodic lubrication. Use minimum center distance due to space limitations.

   (a) Recommend a suitable chain for the drive.

   (b) Pitch circle diameters of sprockets and center distance.

   (c) Total force coming on chain for design power.

   (d) Bearing pressure

   [(a) 10A-2 strands; (b) D_1 = 116.3 mm, D_2 = 418.7 mm, C = 480 mm;
   (c) 1,521 N; (d) p_b = 16.5 MPa]

6. Find the kilowatts rating of 08B four strand roller chain with 15 teeth on the driver sprocket at 1,400 rpm. Assume good lubrication and shock factor 1.2. [P = 22 kW]

7. A power of 50 kW at 500 rpm is transferred using a 19 teeth sprocket to drive a shaft at about 200 rpm with moderate shocks and 16 h per day with poor lubrication conditions. Center distance between the shafts can be taken 30 times the pitch of chain. Assume lubrication factor 1.2 and service factor 1.2. Suggest a suitable four stranded roller chain. Also find the number of links and length of chain.

[Chain 16B, N_L = 95, L = 2,413 mm]

8. A 12B chain transmits power from 1,000 rpm motor to a shaft at 360 rpm. Driving sprocket has 21 teeth. Assuming mild shocks and 8 h of service with good lubrication conditions, find:

   (a) Rated power required for the chain

   (b) Factor of safety in braking load [(a) 15.17 kW; (b) FOS = 12.7]

9. An I.C. engine gives power output of 25 kW at 700 rpm and delivers to a belt conveyor approximately at 200 rpm. Recommend a suitable two strand roller chain with good lubrication if driver sprocket has 17 teeth. Assume center distance 40p and shock factor 1.3. Calculate the diameter of the sprockets, number of links, and chain length. [Chain 16A, N_L = 118, L = 2,997 mm]

10. An electric motor of 10 kW running at 1,440 rpm drives a compressor at 400 rpm for 8 hours per day using a horizontal roller chain. Assume minimum center distance and periodic lubrication:

(a) Recommend a suitable chain with two strands.

(b) Number of teeth and diameters of the sprockets.

(c) Total force in chain considering its weight and centrifugal force also.

(d) Bearing pressure at the pin.

$$[(a) \text{ Chain 10A-2; (b) } Z_1 = 23, D_1 = 116 \text{ mm}, Z_2 = 83, D_2 = 418.6 \text{ mm};$$
$$(c) \text{ 1,331 N; (d) 14.43 MPa]}$$

Previous Competition Examination Questions

IES

1. Sources of power loss in a chain drive are given below:

1. Friction between chain and sprocket teeth.

2. Overcoming the chain stiffness.

3. Overcoming the friction in shaft bearing.

4. Frictional resistance to the motion of chain in air or lubricant.

The correct sequence of descending order of power loss due to these sources is:

(a) 1 2 3 4

(b) 1 2 4 3

(c) 2 1 3 4

(d) 2 1 4 3

[Answer (a)] [IES 1995]

◻◻◻

Gear Fundamentals

Outcomes

➤ Learn terms used for different types of gears. Their comparison with other types of drives

➤ Advantages and disadvantages of gear drives

➤ Different pitches used for gears and tooth proportions

➤ Cycloid and involute tooth profiles. Law of gearing

➤ Involute tooth system, base circle, contact ratio, arc and path of contact

➤ Interference and parameters affecting it

➤ Minimum number of teeth to avoid interference

➤ Maximum addendum radius

➤ Largest gear with specified pinion

➤ Slide velocity between teeth of mating gears

4.1 Introduction

Belts and pulleys transmit power with some slippage. If this slippage cannot be tolerated, gears are used. See Figure 4.1(a) in which two wheels, A and B are mounted on parallel shafts and are made to touch. Now, if A is rotated in anticlockwise direction, wheel B starts rotating in clockwise direction due to friction at their periphery. If slight load is put on the driven shaft, the wheel starts slipping. To avoid this slipping, teeth are cut on their periphery and such a wheel is called gear as shown

in Figure 4.1(b). Speed can be increased or decreased by selecting the diameters in the proportion of speed reduction. Gear ratios (driver to driven speed ratio) up to 5 are common. A gear is never used alone, but always used in pair; one is called driver gear and the other driven gear. The size depends on the power to be transmitted and center distance between the shafts.

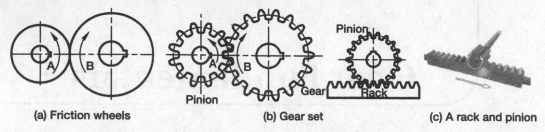

(a) Friction wheels   (b) Gear set   (c) A rack and pinion

Figure 4.1 Gear concepts and drives

A gear of small diameter is called pinion, whereas a gear of infinite diameter is called a rack, that is, it is straight as shown in Figure 4.1(c). For speed ratio more than 5, gear trains (gear sets in series described in Chapter 9) or worm gears (Chapter 8) are used.

A gear is a circular wheel having teeth on its periphery. The teeth may be external as shown in Figure 4.2(a) or internal, inside a ring shown in Figure 4.2(b). The gear with internal teeth is called a ring gear or an annular gear. Direction of rotation of both the gears is same.

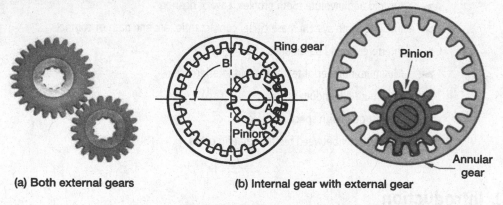

(a) Both external gears   (b) Internal gear with external gear

Figure 4.2 An external and internal gear

4.2 Gear Drives versus Other Drives

Drives are used to transmit power from one shaft (generally of a prime mover) to the other shaft of a machine. The prime mover may be an engine or an electric motor, etc., which may run at speed different than the speed required for the machine. Hence, a drive is used to transmit power and at the same time reduce or increase the speed as required by the machine. Many types of drives are used such as belt and pulleys, chain and sprockets, and a gear set. Gear drive is compared with other drives as under:

- Belt drive slips, whereas gear drive does not slip and offers a positive drive.

- Chain drive does not slip, but there is slight speed variation in the driven shaft speed depending on the number of teeth of the driver sprocket (Chapter 3), whereas in gear drive, there is no speed variation. Speed is constant for any number of teeth.

- Belt and chain drives are preferred, if center distance between the shafts is more,whereas gears are not suitable for large center distances.

- In belt drive, if there is any locking on the driven shaft, it starts slipping and acts as a fuse, whereas in gear drive the teeth may break.

- Belts do not transfer the shocks of driver shaft, whereas in gear drives shocks are also transferred.

4.3 Advantages and Disadvantages of Gear Drives

Advantages

- Gears can be used for small to large powers.
- The drive is positive and gives exact velocity ratio.
- Transmission efficiency is high.
- Offers reliable service.
- Maintenance is inexpensive. If properly lubricated, these drives offer the longest service life in comparison to other drives.
- Occupies less space, hence compact design.

Disadvantages

- Suitable for small center distances only.
- Costly due to special tools required for manufacturing.
- Can cause vibrations due to error in manufacturing.
- Requires proper lubrication for satisfactory working.

4.4 Types of Gear Drives

Gears are of many types. These are classified according to the inclination of their teeth as follows.

4.4.1 Spur gears

Spur gear drive is the simplest type of drive and is most commonly used. They have straight teeth, whose axis is parallel to the axis of the gear [See Figure 4.3(a)].

4.4.2 Helical gears

Helical gears have their teeth inclined at helix angle a as shown in Figure 4.3 (b and c). Owing to this inclination, these gears give rise to axial thrust. This thrust can be cancelled using double helical gears called Herringbone gears as shown in Figure 4.3 (d and e).

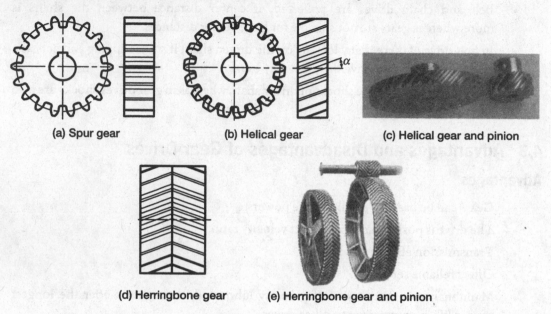

(a) Spur gear   (b) Helical gear   (c) Helical gear and pinion

(d) Herringbone gear   (e) Herringbone gear and pinion

Figure 4.3 Types of gears

4.4.3 Bevel gears

Bevel gears are used, where power is to be transmitted at an angle. The gear blank takes the form of a frustum of a cone instead of a cylinder (Figure 4.4). Thickness of teeth and height also decrease towards center. All teeth point towards the center.

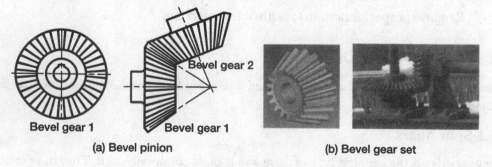

Bevel gear 1   Bevel gear 2
                       Bevel gear 1

(a) Bevel pinion   (b) Bevel gear set

Figure 4.4 Bevel gears

4.4.4 Worm and worm wheel

Worm and worm gears are used, when large speed reduction is required. Ratio up to 20 or even more can be got in one stage only. Worm is in the form of a threaded screw of trapezoidal section, with large pitch, that matches with the worm wheel having mating grooves of the same profile (Figure 4.5).

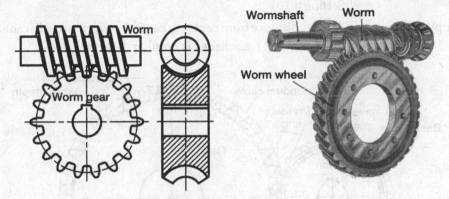

Figure 4.5 Worm and worm gear

4.5 Terminology

Following terms used for a gear are shown in Figure 4.6:

Pitch circle diameter: An imaginary circle, on which rolling takes place without slip. This diameter is called pitch circle diameter (D) in short PCD also.

(D_p) PCD of pinion.

(D_g) PCD of gear.

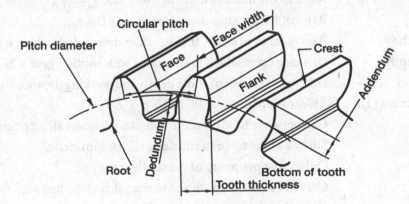

Figure 4.6(a) Gear terminology

Addendum (a) — Radial distance from tip of tooth to PCD.

Dedendum (d) — Radial distance from PCD to root of tooth.

Addendum circle (D_a) — Outermost circle diameter $D_a = D + 2a$.

Dedendum circle (D_d) — Circle passing through the roots of teeth $D = D - 2d$.

Base circle (D_b) — Imaginary circle on which involute tooth profile is generated (See Section 4.10)

Circular pitch (p_c) — Arctual distance from center to center between two adjacent teeth at PCD. $p_c = \pi D / z$, where z is number of teeth.

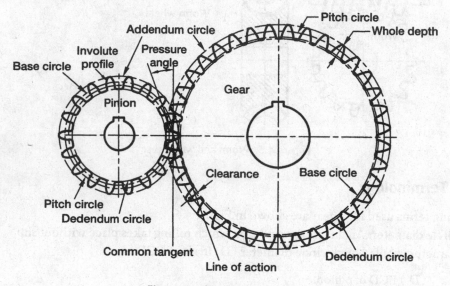

Figure 4.6(b) Pinion and gear

Diametrical pitch (p_d) — Ratio of the number of teeth z to PCD, $p_d = z / D$.

Module (m) — Ratio of PCD to number of teeth, $m = D / z$.

Whole depth (h) — Radial depth of tooth, that is, addendum + dedendum, $h = a + d$.

Working depth — Distance up to which tooth mates with another gear $= 2a$.

Clearance (c) — Difference between whole depth and working depth, $c = d - a$.

Tooth thickness (t) — Thickness of tooth at PCD, $t = p_c / 2$.

Backlash — Gap between tooth space and tooth thickness along pitch circle.

Top land — Surface at the tip of tooth along addendum circle.

Crest — Outermost periphery of the tooth at the tip.

Face — Curved surface of tooth above the pitch circle up to its tip.

Flank — Curved surface of tooth below the pitch circle up to root.

Face width (b) — Width of the gear along the axis of shaft.

Root radius (r)	Radius at the bottom between tooth flank and root circle.
Pitch point	Common point between two pitch circle diameters of gear and pinion.
Pressure angle (Φ)	Angle between a line tangent to the base circles and common normal to the two gears at the contact point. Standard pressure angles are 14.5°, 20° and 25°
Path of contact	Path traced by the contact point, from beginning to end of contacting of teeth.
Arc of contact	Path traced by a point on pitch circle, from beginning to end of contacting of teeth.
Arc of approach	Portion of path of contact from beginning to the pitch point.
Arc of recess	Portion of path of contact from pitch point to the end of engagement.
Contact ratio	It is the ratio of arc of contact to circular pitch. It indicates number of teeth in contact.

4.6 Types of Pitches

For gears, pitch is given by any one of the three different methods given below:

- Circular pitch (p_c),
- Diametral pitch (p_d), and
- Module (m).

If D is PCD and Z is number of teeth, these are related to each other as under:

$$p_c = \frac{\pi \times D}{Z} \tag{4.1a}$$

$$p_d = \frac{Z}{D} \tag{4.1b}$$

$$m = \frac{D}{Z} \tag{4.1c}$$

These pitches are related to each other. If one is known, the other can be calculated from the relations given below:

$$p_c = \pi m = \frac{\pi}{p_d} \tag{4.2a}$$

$$p_d = \frac{1}{m} = \frac{\pi}{p_c} \tag{4.2b}$$

$$p_c \times p_d = \pi \tag{4.2c}$$

4.7 Gear Tooth Proportions and Standard Modules

The relations in terms of module and PCD for pinion or gear (D) are given as under:

Circular pitch $p_c = \pi m$

Outside gear diameter $D_o = D + (2 \times \text{Addendum}) = D + 2a$

Root diameter $D_r = D - (2 \times \text{Dedendum}) = D - 2d$

Fillet radius $r = 0.4\, m$

Tooth thickness $t = 0.5 \times p_c = 0.5\, \pi m = 1.57\, m$

Base circle diameter $D_b = D \times \cos \Phi$

Where, Φ is pressure angle

Angle between adjacent teeth, $\theta = \dfrac{360}{Z}$

The standard module of gear can be selected from Table 4.1 given below:

Table 4.1 Standard Modules [IS 2535:1978 Revised 2001]

Increment	Standard Modules										
0.125	1	1.125	1.25	1.375	1.5	–	–	–	–	–	–
0.25	1.5	1.75	2	2.25	2.5	2.75	3	3.25	3.5	3.75	4
0.5	4	4.5	5	5.5	6	6.5	7	–	–	–	–
1	7	8	9	10	11	12	–	–	–	–	–
2	12	14	16	18	20	–	–	–	–	–	–

Proportions of gear tooth for addendum a, dedendum d, whole depth, and clearance for different tooth profiles are given in Table 4.3 in Section 4.9.

Example 4.1

Gear tooth proportions for a given module

A gear has 20 teeth of module 5 and pressure angle 20°. Calculate:

  a. Addendum and dedendum

  b. Pitch circle, outside, base circle, and root diameters

  c. Tooth thickness and fillet radius

  d. Circular and diametrical pitch

Solution

Given $Z = 20$ $m = 5$ $\Phi = 20°$

a. Addendum $\qquad a = 1\,m = 5$ mm

   Dedendum $\qquad d = 1.25\,m = 1.25 \times 5 = 6.25$ mm

b. Pitch circle diameter $D = m\,Z = 5 \times 20 = 100$ mm

   Outside diameter $\quad D_o = D + 2m = 100 + (2 \times 5) = 110$ mm

   Root diameter $\qquad D_r = D - (2 \times 1.25\,m) = 100 - (2 \times 1.25 \times 5) = 87.5$ mm

   Base circle diameter $D_b = D \times \cos \Phi = 100 \times \cos 20° = 93.9$ mm

c. Tooth thickness $\qquad t = 1.57m = 1.57 \times 5 = 7.855$ mm

   Fillet radius $\qquad r = 0.4\,m = 0.4 \times 5 = 2$ mm

d. Circular pitch $\qquad p_c = \dfrac{\pi D}{Z} = \dfrac{\pi \times 100}{20} = 15.7$ mm

   Diametrical pitch $\quad p_d = \dfrac{Z}{D} = \dfrac{20}{100} = 0.2$

4.8 Tooth Profiles

There are two standard tooth profiles: cycloid and involute. Each type is described below:

4.8.1 Cycloid profile

When a circular wheel rolls on a flat surface, the locus generated by a point on the circumference of the wheel is called a cycloid. The locus shown in Figure 4.7 is for a point '0' on the wheel. Line 0-0 is the circumference of the wheel. Points 1, 2, 3, etc., are eight segments of circumference. Points 1', 2', 3', etc., are the centers of the wheel when it rolls 1 / 8th of the revolution. Points 1", 2", etc., are the points, where the arc of radius of wheel R cuts the vertical heights projected from each position. The curve joining the points 1", 2", 3", etc., is a cycloid.

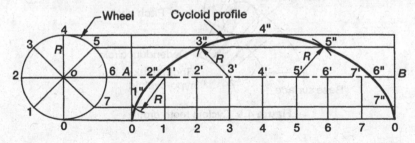

Figure 4.7 Cycloid profile

If the bottom surface is convex instead of a straight surface, the path generated is epicycloid. It is drawn in a similar way as the cycloid, except that the horizontal lines are replaced by arcs parallel to the bottom surface and vertical lines are replaced by radial lines as shown in Figure 4.8(a).

If the bottom surface is concave and circular, the curve generated is hypocycloid. Rest of the procedure is same as that for epicycloid. It is shown in Figure 4.8(b).

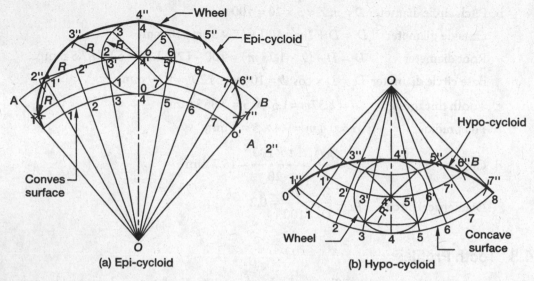

Figure 4.8 Epicycloid and hypocycloid profiles

A cycloid tooth is epicycloid above pitch circle and its flank is hypocycloid below pitch circle (Figure 4.9). This profile requires exact center distance between shafts, which is difficult to attain; hence cycloid profile is generally not used and majority of the gears have involute profile described in the next section 4.8.2.

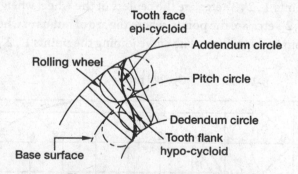

Figure 4.9 Cycloid tooth profile

4.8.2   Involute profile

If a string is wound around a cylinder and then unwound keeping the end stretched, the locus of free end of string is involute, which forms flank and face of the gear tooth. Other side of the tooth is generated by changing the direction of wound of string from clockwise to anticlockwise.

The profile shown in Figure 4.10 is only for left side of the tooth. Points 1, 2, 3, etc., are on the base circle described in the next section 4.10. Tangents are drawn on these points and intercepts equal to the actual length from point P are cut on these lines to give points 1', 2', 3', etc. These tangential lines are called generating lines. The curve passing through these points is the left face of the tooth. Right face is the mirror image of this profile about a radial line passing through the center of tooth thickness.

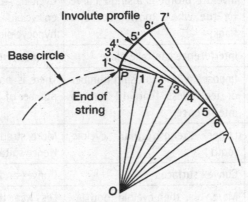

Figure 4.10 Involute profile

4.8.3 Properties of involute teeth

a. Shape of involute depends on base circle diameter.
b. Angular motion remains uniform even, if center distance is changed slightly.
c. Contact point during engagement of mating teeth moves along a straight line, which is called line of action.
d. The point, where line of action intersects line of centers is the pitch point. The pitch circle radii are measured from this point to the respective center.
e. Ratio of pitch circles and base circles for the driven and driver gear remains the same.
f. Angle between the line of action and a common tangent at the pitch point is called pressure angle. This angle varies, if the center distance is changed.
g. Profile of a rack meshing with involute pinion is a straight line.

4.8.4 Involute versus cycloid profile

Advantages and disadvantages of both the tooth profiles are compared in Table 4.2.

Table 4.2 Involute versus Cycloid Profile

S.No.	Parameter	Involute Profile	Cycloid Profile
1.	Pressure angle	Remains constant during contact period	Maximum at the beginning of contact, reduces to zero at pitch circle, and again increases to maximum at the end of contact

2	Center distance	Can be slightly varied without affecting velocity ratio	Needs exact center distance for satisfactory working
3	Manufacturing	Simple to manufacture the tooth shape of an involute rack is straight line	Difficult to manufacture. Teeth are cut with special machine tools
4	Profiles	Involute profile is a single curve for the whole tooth (face and flank)	Cycloid tooth has two curves: epicycloid for the face and hypocycloid for the flank
5	Interference*	Interference can take place	No interference
6	Minimum number of teeth	Imposes a limit in using number of teeth less than 12 due to interference	There is no such limit on minimum number of teeth
7	Strength per pitch	Lesser strength than cycloid tooth	More strength than involute tooth. Hence suitable for cast teeth
8	Contact surfaces	Convex surfaces	Convex and concave surfaces
9	Wear	More wear than cycloid tooth	Less wear than involute tooth

* Interference is the under cutting at the root of teeth of one gear, due to tip of the mating gear, when the number of teeth are less (See Section 4.16).

4.9 Involute Gear Tooth Systems

Following systems of involute gear tooth proportions are used:

a. **14.5° composite** Pressure angle is 14.5°. The tooth has cycloid profile at top and bottom and involute at the middle of the tooth. The teeth are cut by formed milling cutters. This system is used for general purpose gears and does not offer interchangeability.

b. **14.5° full depth** Pressure angle is 14.5°. The complete gear tooth has involute profile. These teeth can be cut by hob cutters.

c. **20° full depth** Pressure angle is 20°, which makes the teeth stronger than the teeth of 14.5° pressure angle.

d. **20° stub** Pressure angle is 20° but due to smaller height of cantilever tooth, it becomes strong and suitable for heavy loads.

Gear tooth proportions for various systems of teeth are given in Table 4.3.

Table 4.3 Gear Tooth Proportions for Various Systems of Teeth

S.No.	Parameter	14.5° Composite / Full Depth	20° Full Depth	20° Stub
1	Addendum	$1m$	$1m$	$0.8m$
2	Dedendum	$1.157m$	$1.25m$	$1m$

3	Working depth	2m	2m	1.6m
4	Total depth	2.157m	2.25m	1.8m
5	Clearance	0.157m	0.25m	0.2m
6	Minimum teeth*	23	17	14
7	Fillet radius	0.4m	0.4m	0.4m

* To avoid interference (See Section 4.16.1).

4.10 Base Circle

Involute profile of the gear tooth is generated on the base circle. Its center is same as of pitch circle. Figure 4.11 shows two mating gears with their pitch circles touching each other at common point P called as pitch point. Line AB is tangent to both the base circles at this point P at line of centers O1-O2 at pressure angle Φ.

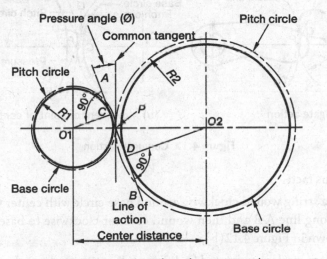

Figure 4.11 Base circle and pressure angle

This line is called line of action. From the centers O1 and O2, radial lines O1C and O2D are drawn, which are perpendicular to this line of action. Circles drawn tangent to the line of action are called base circles. These circles pass through points C and D, at pinion and gear respectively. Value of the radius of base circle (O1C) can be calculated by formulas:

$$O1C = R_1 \cos \Phi \quad \text{and} \quad O2D = R_2 \cos \Phi$$

Where, R_1 and R_2 are half of PCD for the respective gears.
Pressure angle is chosen according to the load coming on gear tooth. For light loads, pressure angle is 14.5°, for medium loads 20°, and for heavy loads 25°.

4.11 Law of Gearing

Tooth profile of the gears should be such that, when driver rotates at constant angular speed, the driven gear should also rotate at constant angular velocity. This is called law of gearing. Such an action is called conjugate action.

Refer Figure 4.12(a), where two curved surfaces rotate about their centers O_1 and O_2. One surface pushes the other surface at point C. Line AB is the pressure line, which intersects the line $O_1 - O_2$ (line joining the centers of the rotating bodies) at P. Velocity ratio of the driver and driven gear is inversely proportional to the ratio of distances of point P from the centers. To have constant angular velocity ratio, the common normal at point C of contact between a pair of surfaces (gear teeth) must pass through point P lying on PCD. For involute profile, the point of contact moves along the pressure line AB, which passes through point P, hence constant angular velocity.

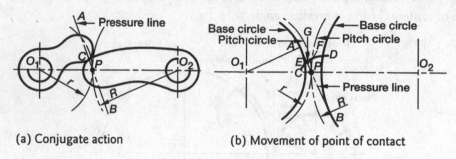

(a) Conjugate action (b) Movement of point of contact

Figure 4.12 Conjugate action

To understand this fact:

- Imagine a string wound clockwise around base circle with center O_1 and is stretched tightly along line A-B and then wound counter-clockwise to base circle with center O_2 as shown in Figure 4.12(b).

- Now, if base circles are rotated in different directions to keep the string tight a point G on the string will trace out involute C-D on gear 2 and E-F on gear 1.

- Tracing point G therefore represents point of contact, whereas potion of string A-B is generating line. Point of contact moves along the generating line.

- Since generating line is always perpendicular to involutes at point of contact, the condition for the uniform angular velocity is satisfied.

4.12 Velocity Ratio

For a gear set, velocity ratio or gear ratio G is defined as the ratio of the rotational speed of the driver gear on input shaft to the speed of driven gear on output shaft. A value more than one indicates speed reduction and lesser than one increase in speed. Mathematically, it is given by the following relation:

$$G = \frac{N_p}{N_g} = \frac{Z_g}{Z_p} = \frac{D_g}{D_p} \qquad (4.3)$$

Where, N_g, N_p = Rotational speed of gear and pinion respectively

Z_g, Z_p = Number of teeth on gear and pinion respectively

D_g, D_p = PCD of gear and pinion respectively

A single gear set can be used satisfactorily for a velocity ratio of 5. If this ratio is more than 5, then a gear train is used for which output shaft of the first gear set becomes the input shaft for the second set and so on. Thus, a gear train is gear sets joined in series. If suffixes 1, 2, etc., represent the number of gear sets, then the velocity ratio G becomes:

$$G = \frac{N_{p1}}{N_{g1}} \times \frac{N_{p2}}{N_{g2}} = \frac{Z_{g1}}{Z_{p1}} \times \frac{Z_{g2}}{Z_{p2}} = G_1 \times G_2$$

For gear sets more than two, velocity ratio for a gear train can be calculated as:

$$G = G_1 \times G_2 \times G_3 \qquad (4.4)$$

Since speed of the pinion for the second set N_{p2} is same as the speed of gear for the first set N_{g1} (being on the same shaft), the above equation for a gear train of two gear sets reduces to:

$$G = \frac{N_{p1}}{N_{g2}} \qquad (4.5)$$

4.13 Path of Contact

A pinion with center O_p meshes with a gear with center O_g as shown in Figure 4.13. Their pitch circles touch at point P. When tooth tip of the driven gear touches the line of action, contact of the tooth starts. After that there exists combination of sliding and rolling motion and the point of contact moves along pressure line A-B. The pressure line A-B is at pressure angle Φ with the normal to the line of centers at point P.

When the gears rotate, the contact starts at point C, when the addendum circle of gear touches the line AB and remains up to a point D. Point C is on addendum circle of gear and point D on addendum circle of pinion. Thus, distance O_pD is radius of addendum circle (r_a) of pinion and O_gC is radius of addendum circle (R_a) of gear.

Upper case symbols are used for gear and lower case for pinion for the derivation.

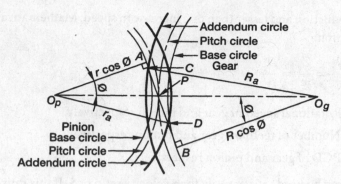

Figure 4.13 Path of contact

The total length of path of contact CD (L) can be separated as CP called as path of approach L_1 and PD as path of recess L_2.

That is, CD = CP + PD   or $L = L_1 + L_2$ $\qquad\qquad$ (a)

From Figure 4.14, $L_1 = CP = (CB - PB) = \sqrt{(O_g C)^2 - (O_g B)^2} - PB$

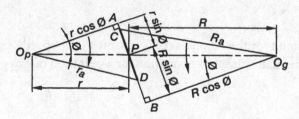

Figure 4.14 Path of approach and recess

Or, $\qquad L_1 = \sqrt{R_a^2 - (R\cos\Phi)^2} - R\sin\Phi$ $\qquad\qquad$ (b)

Similarly, $L_2 = PD = (DA - AP) = \sqrt{(O_p D)^2 - (O_p A)^2} - AP$

Or, $\qquad L_2 = \sqrt{r_a^2 - (r\cos\Phi)^2} - r\sin\Phi$ $\qquad\qquad$ (c)

Substituting the values from Equations (b) and (c) in Equation (a)
Length of path of contact is given as: $L = L_1 + L_2$

Length of contact, $L = \sqrt{R_a^2 - (R\cos\Phi)^2} - R\sin\Phi + \sqrt{r_a^2 - (r\cos\Phi)^2} - r\sin\Phi$ $\qquad$ (4.6)

Maximum length of contact $L_{max} = R\sin\Phi + r\sin\Phi = (R + r)\sin\Phi$ $\qquad\qquad$ (4.7)

4.14 Arc of Contact

It is the path traced by a point on the PCD of any gear from beginning to end of engagement. In Figure 4.15, engagement of the tooth starts at point C. At this position, point P1 is on the pitch circle of gear tooth face. Point 1 is the intersection of tooth flank and base circle. When the gears rotate, the contact point moves to point P2 on pitch circle at the line of centers. Point 1 reaches at point 2. The arc between the points P1-P2 is called arc of approach.

When the gears rotate further, the contact remains till point D. P3 is a point on pitch circle at this moment at the end of contact and point 3 on base circle. The arc P2-P3 is called arc of recess.

Arc of contact is sum of these two arcs, that is:

Arc of contact = Arc of approach + Arc of recess = (P1 – P2) + (P2 – P3) = (P1 – P3)

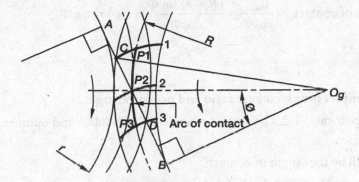

Figure 4.15 Arc of contact

The time taken t for the contact point to move from C to P2 at angular velocity of ω is same as the time taken t by the tooth face from P1 to P2.

$$\text{Hence, Arc P1} - \text{P2} = \omega R t$$

Multiplying and dividing by cos Φ in numerator and denominator:

$$\text{Arc } P1\text{-}P2 = \frac{\omega (R \cos \Phi) t}{\cos \Phi} = \frac{\omega (\text{Distance } O_g B) t}{\cos \Phi}$$

$$= \frac{\text{Tangential velocity at point B} \times t}{\cos \Phi}$$

Time taken t to move from C to P2 is same as from point 1 to 2.

$$\text{Hence, Arc } P1\text{-}P2 = \frac{\text{Arc} 1\text{-}2}{\cos \Phi} = \frac{(\text{Arc B-1}) - (\text{Arc B-2})}{\cos \Phi}$$

According to the definition of the involute: Arc B1 = Line BC and Arc B2 = Line BP2

Arc of approach: Arc $P1-P2 = \dfrac{BC - BP2}{\cos\Phi} = \dfrac{CP2}{\cos\Phi} = \dfrac{\text{Path of approach}}{\cos\Phi}$

Similarly, arc of recess Arc $P2-P3 = \dfrac{P2D}{\cos\Phi} = \dfrac{\text{Path of recess}}{\cos\Phi}$

Hence, Arc of contact $= \dfrac{CP2 + P2D}{\cos\Phi} = \dfrac{CD}{\cos\Phi}$

Arc of contact $= \dfrac{\text{Path of contact}}{\cos\Phi}$ 　　　　　　　　　　(4.8)

Maximum arc of contact $= \dfrac{L_{max}}{\cos\Phi} = \dfrac{(R + r)\sin\Phi}{\cos\Phi} = (R + r)\tan\Phi$ 　　(4.9)

Example 4.2

Length of contact for given gear ratio and pressure angle

A gear set of gear ratio 1:2.5 has pressure angle 20°, module 5 and number of teeth on pinion 24.

    a. What will be the length of contact?

    b. Angle turned by the pinion, while in contact with the gear.

Solution

Given　　$G = 2.5$　　　$\Phi = 20°$　　　$m = 5$　　　$Z_p = 24$

    a.　$G = \dfrac{Z_g}{Z_p}$　　or　　$2.5 = \dfrac{Z_g}{24}$　　or　　$Z_g = 60$

PCD of pinion, d = Number of teeth × Module = 24 × 5 = 120 mm

Pitch circle radius of pinion, $r = 0.5\,d = 60$ mm

PCD of gear, D = Number of teeth × Module = 60 × 5 = 300 mm

Pitch circle radius of gear, $R = 0.5D = 150$ mm

Addendum, $a = 1\,m = 5$ mm

Addendum radius of pinion, $r_a = r + a = 60 + 5 = 65$ mm

Addendum radius of gear,　$R_a = R + a = 150 + 5 = 155$ mm

From Equation (4.6) Length of path of contact is:

$$L = \sqrt{R_a^2 - (R\cos\Phi)^2} - R\sin\Phi + \sqrt{r_a^2 - (r\cos\Phi)^2} - r\sin\Phi$$

$$= \sqrt{155^2 - (150\cos 20°)^2} - 150\sin 20° + \sqrt{65^2 - (60\cos 20°)^2} - 60\sin 20°$$

$$= \sqrt{24{,}025 - 19{,}868} - 51.3 + \sqrt{4{,}225 - 3{,}179} - 20.5$$

$$= 64.5 - 51.3 + 32.3 - 20.5 = 25 \text{ mm}$$

b. From Equation (4.7):

$$\text{Length of arc of contact} = \frac{L}{\cos\Phi} = \frac{25}{\cos 20°} = 26.6 \text{ mm}$$

$$\text{Angle turned by pinion} = \frac{\text{Arc of contact} \times 360}{\text{Circumference}} = \frac{26.6 \times 360}{2\pi \times 60} = 25.4°$$

4.15 Contact Ratio

Contact Ratio (CR) is defined as the ratio of arc of contact and circular pitch, that is,

$$CR = \frac{\text{Arc of contact}}{\text{Circular pitch}} = \frac{\text{Arc of contact}}{\pi m}$$

Value of CR equal to one means that when one tooth is ending contact, the next tooth is just at the instant of making contact. This results only one tooth in contact during contact period. Following may be kept in mind about CR:

- In no case, value of CR should be less than one, otherwise the contact will be lost.
- Its value exactly one is dangerous, as due to some manufacturing errors, the contact may not exist for some time. Hence, it should not be less than 1.2.
- It is kept always more than one, to ensure that the next tooth surely makes a contact before the one ends contact.
- Its value 1.6 means that full contact is for one pair of teeth and with two teeth for 60% of arc of contact.
- Its value 2 means that the contact is for two teeth all the time.
- A value of CR between 1.5 and 1.7 is satisfactory.

Example 4.3

Line of contact arc and contact ratio for a given gear set

A gear set of pinion with 22 teeth meshes with a gear of 42 teeth with module 5 and pressure angle 20°. Find:

   a. Length of approach, length of recess, length of contact.

   b. Length of arc of contact.

   c. Contact ratio.

   d. Angle turned by pinion during contact period.

Solution

Given $z = 22$ $Z = 42$ $m = 5$ $\Phi = 20°$

a. PCD of pinion, $d = m\,z = 5 \times 22 = 110$ mm

   Pitch circle radius of pinion, $r = 0.5d = 0.5 \times 110 = 55$ mm

   Addendum radius of pinion, $r_a = r + m = 55 + 5 = 60$ mm

   PCD of gear, $D = m\,Z = 5 \times 42 = 210$ mm

   Pitch circle radius of gear, $R = 0.5D = 0.5 \times 210 = 105$ mm

   Addendum radius of gear, $R_a = R + m = 105 + 5 = 110$ mm

   Approach length, $L_1 = \sqrt{R_a^2 - (R\cos\Phi)^2} - R\sin\Phi$

   Or, $L_1 = \sqrt{110^2 - (105 \times \cos 20°)^2} - 105 \times \sin 20° = 12.66$ mm

   Recess length, $L_2 = \sqrt{r_a^2 - (r\cos\Phi)^2} - r\sin\Phi$

   Or, $L_2 = \sqrt{60^2 - (55 \times \cos 20°)^2} - 55 \times \sin 20° = 11.65$ mm

   Length of contact $= L_1 + L_2 = 12.66 + 11.65 = 24.3$ mm

b. Arc of contact $= \dfrac{\text{Length of contact}}{\cos\Phi} = \dfrac{24.3}{\cos 20°} = 25.86$ mm

c. Contact ratio $CR = \dfrac{\text{Arc of contact}}{\pi m} = \dfrac{25.86}{3.14 \times 5} = 1.64$

d. Angle turned by pinion during contact period

$$CR = \frac{CR \times 360}{2\pi r} = \frac{1.64 \times 360}{2 \times 3.14 \times 55} = 26.9°$$

4.16 Interference

An involute tooth has involute profile above the base circle. For small number of teeth, dedendum is so large that it lies below the base circle, which is not an involute. In this condition, addendum of involute tooth of one gear interferes with the noninvolute profile of the other gear below the base circle. The contact area of nonconjugate teeth is called interference. During interference, tip of the tooth face tries to dig into the flank of the mating gear. This occurs when the number of teeth is less than minimum recommended number of teeth, which depends on pressure angle. Figure 4.16 shows two mating gears.

Points A and B are on the pressure line, where it touches the base circle. Points C and D are the point of engagement and disengagement of teeth, respectively. It can be seen that the contact starts at point C before the point A and ends at D after point B.

It should be noted that contact at point C starts, where addendum circle of the driven gear cuts the pressure line. Contact ends at point D, where addendum circle of the driver gear cuts the pressure line.

If points of contact are outside the points A and B, interference occurs. Hence, for no interference, contact points should be within the points of tangency on base circle and not outside.

It can also be noted that to have no interference, maximum length of approach L_{1max} is given as: $L_{1max} = r \sin \Phi$.

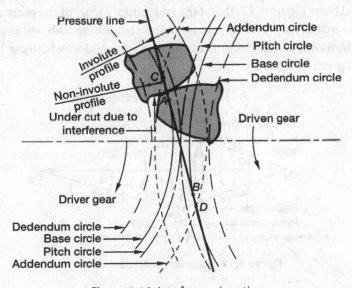

Figure 4.16 Interference in action

When the teeth are cut by generation process, interference is automatically eliminated by cutting some extra portion of the tooth flank. This is called under cut and it weakens the tooth. Hence, number of teeth should not be less than a specified number, which depends on pressure angle. This minimum number of teeth can be calculated by the analysis given in Section 4.18.

4.16.1 Parameters affecting interference

- **Number of teeth** is most important parameter affecting interference. But increasing number of teeth, increases size of the gear and hence the drive is not compact. The weight, size, and cost also increases.

- **Pressure angle** also affects interference. Increase in pressure angle increases radial load and reduces tangential load, which results in less power transmission.

- **Stubbed tooth** has 0.8 times addendum than the standard value. So when addendum is reduced, it does not cross the root circle, but it reduces the contact ratio. This may lead to shock load and noise.

- **Center distance** is increased slightly so that tooth tip does not dig the mating space.

Each method of reducing interference has some disadvantage. The option of using stubbed teeth is generally used.

4.17 Maximum Addendum Radius

In Section 4.14, Figure 4.15 shows a pinion meshing with a gear. The pressure line A-B touches the base circles. Points C and D are the points of engagement and disengagement of teeth, respectively. The points are within the points A and B and hence interference will not take place.

It can be noted from Figure 4.17, that if the addendum radius of the pinion r_a is increased, point D shifts towards B. Any further increase in this radius beyond B will cause interference. Hence, O_p B is the maximum addendum radius r_{amax} for the pinion, which can be used without interference. It is given by Equation (4.10).

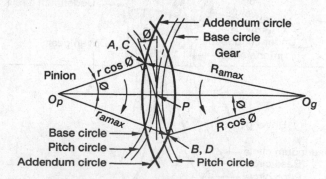

Figure 4.17 Maximum addendum radius

From triangle $O_p AB$, $(O_p B)^2 = (O_p A)^2 + (AB)^2 = (O_p A)^2 + (AP + PB)^2$

Or, $\quad r_{amax}^2 = \left(r \cos \Phi \right)^2 + \left(r \sin \Phi + R \sin \Phi \right)^2$

Or, $\quad r_{amax} = \sqrt{\left(r \cos \Phi \right)^2 + \left(r \sin \Phi + R \sin \Phi \right)^2}$

Because $R = G \times r$

$$r_{amax} = \sqrt{r^2 + \sin^2 \Phi (R^2 + 2Rr)} = \sqrt{r^2 + r^2 \sin^2 \Phi (G^2 + 2G)}$$

$$r_{amax} = r \sqrt{1 + G(G + 2)\sin^2 \Phi} \qquad (4.10)$$

Maximum addendum for the pinion $a_{pmax} = r_{amax} - r$

Similarly, if the addendum radius of the gear R_a is increased, point C shifts towards point A. Any further increase in this radius will cause interference. Hence, the maximum addendum radius of the gear is $O_g A = R_{amax}$ is given by Equation (4.11).

From triangle $O_g AB$, $(O_g A)^2 = (O_g B)^2 + (AB)^2 = (O_g B)^2 + (AP + PB)^2$

$$R_{amax}^2 = (R\cos\Phi)^2 + (r \sin \Phi + R \sin \Phi)^2$$

Or, $R_{amax} = \sqrt{(R\cos\Phi)^2 + (r \sin \Phi + R \sin \Phi)^2}$

Or, $R_{amax} = R \sqrt{1 + \frac{1}{G}\left(\frac{1}{G} + 2\right)\sin^2 \Phi} \qquad \left[G = \frac{R}{r}\right] \qquad (4.11)$

Maximum addendum for the gear $a_{gmax} = R_{amax} - R$

For a gear set having equal number of teeth, that is, equal radii $R = r$, Equations (4.10) and (4.11) will be as under:

$$r_{amax} = r\sqrt{1 + 3\sin^2 \Phi}$$

Since $r = R$, hence $r_{amax} = R_{amax} = r\sqrt{1 + 3\sin^2 \Phi} = R\sqrt{1 + 3\sin^2 \Phi} \qquad (4.12)$

4.18 Minimum Number of Teeth to Avoid Interference

Minimum number of teeth to avoid inference for a pinion and gear depend on pressure angle and the gear ratio. These are calculated as described below.

4.18.1 Minimum teeth for a pinion

Maximum addendum radius for pinion r_{amax} is given by Equation (4.10).

$$r_{amax} = \frac{mZ_p}{2}\sqrt{\left[1 + G(G + 2)\sin^2 \Phi\right]} \left(\text{Because, } r = \frac{mZ_p}{2}\right)$$

If k is defined as addendum depth factor, by which standard module of 1 m is to be multiplied to addendum of pinion to avoid interference is given as:

$$k = \frac{\text{Ratio of actual addendum}(a)}{\text{Standard addendum}(1m)} = \frac{a}{m}$$

Maximum addendum radius r_{amax} = Pitch circle radius of pinion (r) + Addendum = $r + m\,k$

Using Equation(4.10); $(r_{amax} - r) \leq \dfrac{mZ_p}{2}\sqrt{\left[1 + G\left(G + 2\right)\sin^2\Phi\right]} - \dfrac{mZ_p}{2}$

To avoid interference, addendum $(r_{amax} - r)$ should be less than $(m\,k)$.

Or, Addendum, $mk \leq \dfrac{Z_p}{2}\left[\sqrt{\left[1 + G(G+2)\sin^2\Phi\right]} - 1\right]$ 　　　　　(4.13)

From Equation (4.13), minimum number of teeth on pinion Z_{pmin} is given as:

$$Z_{pmin} = \frac{2k}{\sqrt{1 + G\left(G+2\right)\sin^2\Phi} - 1}$$ 　　　　　(4.14)

For a gear set having equal number of teeth on pinion and gear, that is, $G = 1$, minimum number of teeth to avoid interference is given as:

$$Z_{pmin} = \frac{2\,k}{\sqrt{\left(1 + 3\sin^2\Phi\right)} - 1}$$ 　　　　　(4.15)

For a full depth pinion, that is, $k = 1$ and pressure angle 14.5°, minimum number of teeth from Equation (4.15) can be calculated as 23. For pressure angle 20°, it is 12.3 say 13.

4.18.2 Minimum teeth for a gear wheel

Maximum addendum radius for gear R_{amax} is given by Equation (4.11):

Substituting $R = \dfrac{mZ_g}{2}$,

Maximum addendum, $R_{amax} = \dfrac{mZ_g}{2}\sqrt{1 + \dfrac{1}{G}\left(\dfrac{1}{G} + 2\right)\sin^2\Phi} = R + \left(k \times m\right)$

For no interference, maximum addendum radius of the gear R_{amax} is given as:

Maximum addendum, $\left(k \times m\right) = \dfrac{mZ_g}{2}\sqrt{1 + \dfrac{1}{G}\left(\dfrac{1}{G} + 2\right)\sin^2\Phi} - \dfrac{mZ_g}{2}$

Or,　　$k = \dfrac{Z_g}{2}\left[\sqrt{\left[1 + \dfrac{1}{G}\left(\dfrac{1}{G} + 2\right)\sin^2\Phi\right]} - 1\right]$

Or, Minimum number of teeth on gear, $Z_{gmin} = \dfrac{2k}{\sqrt{1 + \dfrac{1}{G}\left(\dfrac{1}{G} + 2\right)\sin^2\Phi} - 1}$

$$\text{Or,} \quad Z_{gmin} = \frac{2Gk}{\sqrt{G^2 + (1+2G)\sin^2 \Phi} - G} \tag{4.16a}$$

This equation is used in the form given by Equation (4.16b) also.

Multiplying Equation (4.16a) by $\left[\sqrt{G^2 + (1+2G)\sin^2 \Phi} + G\right]$ in numerator and denominator:

$$Z_{gmin} = \frac{2Gk}{\sqrt{G^2 + (1+2G)\sin^2 \Phi} - G} \times \frac{\sqrt{G^2 + (1+2G)\sin^2 \Phi} + G}{\sqrt{G^2 + (1+2G)\sin^2 \Phi} + G}$$

$$\text{Or,} \quad Z_{gmin} = \frac{2Gk\left[\sqrt{G^2 + (1+2G)\sin^2 \Phi} + G\right]}{(1+2G)\sin^2 \Phi} \tag{4.16b}$$

Note In a gear set addendum of both the gears is same. Maximum addendum for no interference is more for the gear than that of the pinion. Hence the tooth of the gear enters the non-involute profile of the pinion, whereas tooth of pinion will not interfere with that of the gear. Therefore minimum number of teeth for pinion is calculated first for the gear and then is divided by the gear ratio G (Equation 4.16c) (See Example 4.4).

$$Z_{pmin} = \frac{2k}{G\left[\sqrt{1 + \frac{1}{G}\left(\frac{1}{G} + 2\right)\sin^2 \Phi} - 1\right]} \tag{4.16c}$$

4.18.3 Minimum teeth for a pinion with a rack

Refer Figure 4.18 showing a rack and a pinion mating each other at pitch line.

If k is defined as addendum depth factor, by which standard module of $1\ m$ is to be multiplied to the addendum of a rack to avoid interference is given as:

Addendum of rack $a = k \times m = CD = PC \sin \Phi = (O_pP \sin \Phi) \sin \Phi$

$$\text{Or,} \quad k \times m = O_pP\sin^2 \Phi = \frac{mZ_p}{2}\sin^2 \Phi$$

$$\text{Or,} \quad k = \frac{Z_p}{2}\sin^2 \Phi$$

With a rack the smallest pinion without interference will have minimum number of teeth:

$$Z_{pmin} = \frac{2k}{\sin^2 \Phi} \tag{4.17}$$

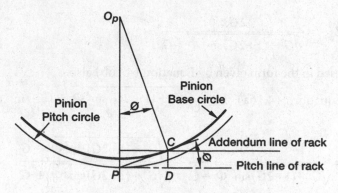

Figure 4.18 Minimum teeth for a pinion with a rack

Example 4.4

Minimum number of teeth for a given addendum size

A pair of gear set with gear ratio 4 has pressure angle of 20°. Calculate minimum number of teeth on pinion and gear with:

   a. Addendum = 0.8 m

   b. Addendum = 1 m   c. Gear ratio = 1 and addendum = 1 m

Solution

Given $G = 4$ $\Phi = 20°$ (a) $k = 0.8$ (b) $k = 1$ (c) $G = 1$ and $a = 1\,m$

   a. Using Equation (4.16a) minimum number of teeth on gear:

$$Z_{gmin} = \frac{2Gk}{\sqrt{G^2 + (1+2G)\sin^2 \Phi} - G} = \frac{2 \times 4 \times 0.8}{\sqrt{4^2 + \left[1 + (2 \times 4)\right]\sin^2 20°} - 4} = 49.4 \text{ say } 50$$

Number of teeth on pinion $Z_p = \dfrac{50}{4} = 12.5 \text{ say } 13$

Hence number of teeth on gear $Z_g = 13 \times 4 = 52$

   b. Using Equation (4.16a) with $k = 1: Z_{gmin} = \dfrac{49.4}{0.8} = 61.75 \text{ say } 62$

Number of teeth on pinion $Z_p = \dfrac{62}{4} = 15.5 \text{ say } 16$

Hence number of teeth on gear $Z_g = 6 \times 4 = 64$

   c. Using Equation (4.16a): $Z_{gmin} = \dfrac{2 \times 1 \times 1}{\sqrt{1^2 + \left[1 + (2 \times 1)\right]\sin^2 20°} - 1} = 12.3 \text{ say } 13$

Number of teeth on pinion and gear $Z_p = Z_g = 13$

Note   Equation (4.15) can also be used directly for $G = 1$.

Example 4.5

Interference with a given size of gears

Check, if a set of gears of module 6 having full length addendum with pressure angle 20° will have interference or not? Number of teeth on pinion = 14, and number of teeth on gear = 63.

Solution

Given $m = 6$ $k = 1$ $\Phi = 20°$ $Z_p = 14$ $Z_g = 63$

$$G = \frac{Z_g}{Z_p} = \frac{63}{14} = 4.5$$

Minimum number of teeth on pinion for no interference is by Equation (4.14):

$$Z_{pmin} = \frac{2k}{G\left[\sqrt{1+\frac{1}{G}\left(\frac{1}{G}+2\right)\sin^2\Phi}-1\right]}$$

$$Z_{pmin} = \frac{2 \times 1 \times 6}{4.5\left[\sqrt{1+\frac{1}{4.5}\left(\frac{1}{4.5}+2\right)\sin^2 20°}-1\right]} = 15.6 \text{ say } 16$$

Since there are 14 teeth on pinion, while minimum teeth required are 16, hence there will be interference.

Example 4.6

Length of approach and interference for given gears

A pair of gears has full depth with 14 and 22 teeth of module 12 and pressure angle 20°. Find:

   a. Length of approach.

   b. Will interference be there or not?

Solution

Given $a = 1m$ $Z_p = 14$ $Z_g = 22$ $m = 12$ $\Phi = 20°$

   a. Gear ratio $G = \dfrac{Z_g}{Z_p} = \dfrac{22}{14} = 1.57$

      Pitch circle radius of pinion $r = \dfrac{mZ_p}{2} = \dfrac{12 \times 14}{2} = 84$ mm

      Pitch circle radius of gear $\quad R = \dfrac{mZ_g}{2} = \dfrac{12 \times 22}{2} = 132$ mm

For full depth teeth addendum is equal to $(1m)$, that is, $a = 12$ mm

Addendum radius of gear $R_a = R + a = 132 + 12 = 144$ mm

Approach length $L_1 = \sqrt{R_a^2 - (R\cos\Phi)^2} - R\sin\Phi$

$$= \sqrt{144^2 - (132\cos 20°)^2} - 132\sin 20° = 34.5 \text{ mm}$$

b. Maximum allowable length of approach $= r \sin\Phi = 84 \times \sin 20° = 28.7$ mm

Since the approach length is more than the allowable length, hence there will be interference.

Example 4.7

Minimum number of teeth and approach length

A gear of module 8 and pressure angle 20° and 40 teeth meshes with a pinion of 20 teeth. Calculate:

a. Length of approach.

b. Check, if there is interference with this pinion.

c. Minimum number of teeth so that there is no interference.

Solution

Given $\quad m = 8 \qquad \Phi = 20° \qquad Z_g = 40 \qquad Z_p = 20$

PCD of pinion $d = m\,z = 8 \times 20 = 160$ mm

Pitch circle radius of pinion $r = 0.5 \times d = 0.5 \times 160 = 80$ mm

PCD of gear $D = m\,Z_g = 8 \times 40 = 320$ mm

Pitch circle radius of gear $R = 0.5 \times D = 0.5 \times 320 = 160$ mm

Addendum diameter of gear $D_a = D + 2m = 320 + (2 \times 8) = 336$ mm

Addendum radius of gear $R_a = 0.5 \times D_a = 0.5 \times 336 = 168$ mm

a. Approach length $L_1 = \sqrt{R_a^2 - (R\cos\Phi)^2} - R\sin\Phi$

$$\text{Or,} \quad L_1 = \sqrt{168^2 - (160 \times \cos 20°)^2} - 160 \times \sin 20° = 20.13 \text{ mm}$$

b. Maximum length of approach of pinion with 20 teeth is:

$$L_{1max} = r \times \sin\Phi = 80 \times \sin 20° = 27.36 \text{ mm}$$

Since L_{1max} is more than approach length L_1, hence, no interference.

c. For minimum number of teeth, approach length $= r \times \sin\Phi$

Approach length L_1, $20.13 = r \times \sin 20°$ or $r = 58.5$ mm

Minimum diameter of pinion $= 2r = 2 \times 58.5 = 117.7$ mm

Minimum number of teeth $Z_{min} = \dfrac{d}{m} = \dfrac{117.7}{8} = 14.7$ say 15 teeth.

4.18.4 Largest gear with a specified pinion

The largest gear, which can be used with a specified pinion without interference, is:

$$Z_{gmax} = \frac{Z_p^2 \sin^2 \Phi - 4k^2}{4k - 2Z_p \sin^2 \Phi} \tag{4.18}$$

For example, for a pinion of full depth (i.e., $k = 1$) 16 teeth with pressure angle 20°, the largest gear, which will mesh with this gear without interference, is:

$$Z_{gmax} = \frac{(16^2 \times \sin^2 20°) - 4}{4 - (2 \times 16 \times \sin^2 20°)} = 99 \text{ teeth}$$

Similarly, for 14 and 13 teeth on pinion, maximum number of teeth on gear is 24 and 16, respectively.

4.19 Slide Velocity

Slide velocity v_s is defined as the difference between the rolling velocities of the surfaces of gear and pinion teeth. Figure 4.19 shows meshing teeth with K as the point of contact on the line of action. Line KQ is the line of action, which is tangent to the base circles.

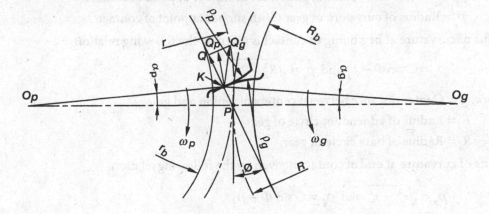

Figure 4.19 Slide velocity

KQ_p and KQ_g represent the tangential velocities of pinion and gear about their centers O_p and O_g, respectively, at this point. QQ_p and QQ_g represent tangential velocities of the surfaces of pinion and gear, respectively:

Tangential velocity of pinion $v_{pT} = \omega_p \times r_K = \dfrac{2\pi r_K N_p}{60}$

Tangential velocity of gear $\quad v_{gT} = \omega_g \times R_K = \dfrac{2\pi R_K N_g}{60}$

Where, r_K = Radial distance of point K from center of pinion O_p

$\quad\quad R_K$ = Radial distance of point K from center of gear O_g

When two involute teeth mesh at a point, there is pure rolling of tooth surfaces at pitch circle, but some sliding of the surfaces also takes place, as the point of contact moves along line of action and makes an angle α. Rolling is a product of rotational speed ω and radius of curvature ρ. Because $\rho = r \sin \alpha$, these velocities are as under:

Rolling velocity of pinion, $\quad v_{pR} = \dfrac{2\pi \, \rho_p N_p}{60} = v_{pT} \times \sin\alpha_p$

Rolling velocity of gear, $\quad v_{gR} = \dfrac{2\pi \, \rho_g N_g}{60} = v_{gT} \times \sin\alpha_g$

Slide velocity, $\quad\quad\quad\quad v_s = v_{pR} - v_{gR}$ $\hfill$ (4.19a)

Where, α_p = Angle between line of centers and radial line from point K to the center O_p.

$\quad\quad \alpha_g$ = Angle between line of centers and radial line from point K to the center O_g.

$\quad\quad \rho_p$ = Radius of curvature of pinion tooth surface at point of contact

$\quad\quad \rho_g$ = Radius of curvature of gear tooth surface at point of contact

Radius of curvature at beginning of contact is given by the following relation:

$$\rho_p = C \sin\Phi - \rho_g \text{ and } \rho_g = \sqrt{R_a^2 - R_b^2}$$

Where, C = Center distance between centers of pinion and gear

$\quad\quad R_a$ = Radius of addendum circle of gear

$\quad\quad R_b$ = Radius of base circle of gear

Radius of curvature at end of contact is given by the following relation:

$$\rho_p = \sqrt{r_a^2 - r_b^2} \text{ and } \rho_g = C \sin \Phi - \rho_p$$

Where, r_a = Radius of addendum circle of pinion

$\quad\quad r_b$ = Radius of base circle of pinion

Slide velocity in terms of radius of curvature, $v_s = \dfrac{2\pi\, \rho_p N_p}{60} - \dfrac{2\pi\, \rho_g N_g}{60}$ (4.19b)

Slide velocity v_s is the difference between the rolling velocities of the gear and pinion. Since pinion and gear rotate in opposite directions hence:

Slide velocity in terms of rotational speeds, $v_s = s\,(\omega_p + \omega_g)$ (4.19c)

Where, s = Distance of any point P from the PCD along line of action.

Slide velocity will be maximum v_{smax}, when the point K is at the tip of the tooth, that is, s_{max} = Bigger of approach length L_1 or recess length L_2.

$$v_{smax} = s_{max}\,(\omega_p + \omega_g)$$ (4.20)

Following points should be noted about the slide velocity:

- When point P is on PCD ($s = 0$), pure rolling takes place and there is no slide velocity, that is, $v_s = 0$. At any other location rolling velocities are different and therefore sliding takes place.
- Slide velocity increases as the point of contact P shifts towards addendum or dedendum.
- Slide velocity is maximum when the point P is on the tip of tooth.
- Direction of slide velocity is such that it tries to compress the tooth at pitch point and tries to elongate the tooth at dedendum circle.

Example 4.8

Addendum and slide velocity for a given gear set

A spur gear set has involute profile, pressure angle 20° and module 8. Pinion has 20 teeth and gear 45 teeth. Addendum on each wheel is such that path of approach and recess are equal and half of the maximum possible length for contact without interference. Find:

a. Addendum for pinion and gear.

b. Length of arc of contact.

c. Maximum sliding velocity, if pinion rotates at 300 rpm.

Solution

Given $\Phi = 20°$ $m = 8$ $Z_p = 20$ $Z_g = 45$ $L_1 = L_2 = 0.5 L_{max}$

$N_p = 300$ rpm

a. Pitch circle radius of pinion $r = \dfrac{m Z_p}{2} = \dfrac{8 \times 20}{2} = 80$ mm

Pitch circle radius of gear $R = \dfrac{m Z_g}{2} = \dfrac{8 \times 45}{2} = 225$ mm

$L_1 = L_2 = 0.5 \times L_{max} = 0.5(R + r)\sin\Phi = 0.5 \times (225 + 80)\sin 20° = 52.15$ mm

Approach length $L_1 = \sqrt{R_a^2 - (R\cos\Phi)^2} - R\sin\Phi$

Or, $52.15 = \sqrt{R_a^2 - (225 \times \cos 20°)^2} - 225 \times \sin 20°$

Or, $(52.15 + 76.95)^2 = R_a^2 - 44,703$ or $R_a = 247.73$ mm

Addendum $a = R_a - R = 247.73 - 225 = 22.73$ mm

Recess length $L_2 = \sqrt{r_a^2 - (r\cos\Phi)^2} - r\sin\Phi$

Or, $52.15 = \sqrt{r_a^2 - (80 \times \cos 20°)^2} - 80 \times \sin 20°$

Or, $(52.15 + 13.68)^2 = r_a^2 - 5651.3$ or $r_a = 99.92$ mm

Addendum $= r_a - r = 99.92 - 80 = 19.92$ mm

b. Path of contact $L = L_1 + L_2 = 52.15 + 52.15 = 104.3$ mm

Arc of contact $= \dfrac{\text{Path of contact}}{\cos\Phi} = \dfrac{104.3}{\cos 20°} = 111$ mm

c. Rotational speed of pinion $\omega_p = \dfrac{2\pi N_p}{60} = \dfrac{2 \times 3.14 \times 300}{60} = 31.4$ rad/s

Rotational speed of gear $\omega_g = \dfrac{31.4}{G} = \dfrac{31.4 \times 20}{45} = 13.95$ rad/s

$s_{max} = $ Bigger of L_1 or L_2 (here both are equal)

Maximum slide velocity $v_{smax} = s_{max}(\omega_p + \omega_g)$

$$= 52.15 \times 10^{-3} \times (31.4 + 13.95) = 2.36 \text{ m/s}$$

Summary

Belt and pulleys transmit power with some slippage. Gears are used, where slipping cannot be tolerated. Gear ratios (driver to driven speed ratio) up to 5 are common. Gear of small diameter is called pinion, whereas a gear of infinite diameter is called rack. For speed ratio more than 5, worm gears (Chapter 8) or gear sets in series (Chapter 9) are used. The teeth on a gear can be external or internal inside a ring. The gear with internal teeth also is called a ring gear or an annular gear and direction of rotation of both the gears is same.

Gear drive is a positive drive without slip. Unlike chain drive, speed is constant for any number of teeth. Gears are not suitable for large center distances. In case of overloading, the teeth of gear may break, whereas a belt starts slipping and acts as a fuse. These drives transfer shock also.

Advantages Used for small to large powers, gives exact velocity ratio, transmission efficiency is high, offers reliable service, maintenance is inexpensive, and occupy less space. If properly lubricated, offer longest service life in comparison to other drives.

Disadvantages Suitable for small center distances, costly manufacturing, cause vibrations due to error in manufacturing, and requires proper lubrication.

Classification according to the inclination of their teeth:

Spur gear most commonly used. Have straight teeth, axis is parallel to the axis of the gear.

Helical gears have inclined teeth at helix angle α. There exists axial thrust also.

Herringbone gears called double helical gears. Axial thrust gets cancelled by double helical.

Bevel gears transmits power at an angle. The gear blank is of the form of a frustum of a cone. Teeth thickness and height decrease towards center.

Worm and worm gears are used for large speed reduction. Ratio up to 20 or even more can be got in one stage only. Worm is in the form of a threaded screw of trapezoidal section, with large pitch that matches with the worm wheel.

Important terms

PCD (D) is an imaginary circle on which non-slip rolling occurs.

Addendum (a) Is radial distance from tip of tooth to PCD.

Dedendum (d) is radial distance from PCD to root of tooth.

Addendum circle (D_a) is outermost circle diameter, $D_a = D + 2a$.

Dedendum circle (D_d) passes through the roots of teeth, $D_d = D - 2d$.

Base circle (D_b) is an imaginary circle on which the tooth profile involute is generated.

Circular pitch (p_c) is arctual distance between centers of two adjacent teeth at PCD.

Diametrical pitch (p_d) is the ratio of number of teeth z to PCD, $p_d = z / d$.

Module (m) is the ratio of PCD to number of teeth, $m = d / z$

Whole depth (h) is the radial depth of tooth, that is, addendum + dedendum, $h = a + d$.

Clearance (c) is the difference between whole depth and working depth, $c = d - a$.

Tooth thickness (t) is thickness of tooth at PCD $t = p_c / 2$.

Pressure angle (Φ) is angle between a line tangent to base circles and common normal.

Various pitches: Circular pitch $p_c = \dfrac{\pi \times D}{Z}$, diametral pitch $p_d = \dfrac{Z}{D}$, and module $m = \dfrac{D}{Z}$

Tooth proportions Circular pitch $p_c = \pi m$, outside gear diameter $D_o = D + 2a$

Root diameter $D_r = D - 2d$, fillet radius $r = 0.4 \, m$, tooth thickness $t = 1.57 \, m$

Base circle diamete $D_b = D \times \cos \Phi$, Angle between adjacent teeth $\theta = \dfrac{360}{Z}$

Tooth profiles: Two standard profiles are; cycloid and involute.

When a circular wheel rolls on a flat surface, the locus generated by a point on the circumference of the wheel is called a cycloid. If the bottom surface is convex instead of a straight surface, the path generated is epicycloid. If the bottom surface is concave and circular, the curve generated is hypocycloid.

Cycloid tooth is epi-cycloid above pitch circle and its flank is hypo-cycloid below pitch circle. This profile requires exact center distance hence is generally not used.

Involute profile: If a string is wound around a cylinder and then unwound keeping the end stretched, the locus of free end of string is involute, which forms flank and face of the gear.

Properties of involute teeth Angular motion remains uniform, even if center distance is changed slightly, and contact point during engagement of mating teeth moves along a straight line, called line of action. The point, where line of action intersects line of centers is the pitch point. Ratio of pitch circles and base circles for the driven and driver gear remains the same. Angle between the line of action and a common tangent at the pitch point is called pressure angle. This angle varies, if the center distance is changed. Profile of a rack meshing with involute pinion is a straight line.

Involute gear tooth systems

14.5° composite Pressure angle is 14.5°. The tooth has cycloid profile at top and bottom and involute in the middle of the tooth.

14.5° full depth Pressure angle is 14.5°. The complete gear tooth has involute profile. These teeth can be cut by hob cutters.

20° full depth Pressure angle is 20°, which makes the teeth stronger than teeth with 14.5° pressure angle.

20° stub Pressure angle is 20° but due to smaller height of cantilever tooth, it becomes strong and suitable for heavy loads.

Fillet radius for all systems is 0.4 m.

Involute profile of the gear tooth is generated on the base circle. Base circle for pinion, BC1 = $R_1 \cos \Phi$ and Base circle for gear, BC2 = $R_2 \cos \Phi$

Where, R_1 and R_2 are radii of pinion and gear, respectively.

For light loads, pressure angle is 14.5°, for medium loads 20° , and for heavy loads 25°.

Law of gearing Tooth profile of gears should be such that when one gear rotates at constant angular speed, the driven gear should also rotate at constant angular velocity.

Velocity ratio or gear ratio G is defined as the ratio of the rotational speed of the driver gear on input shaft to the speed of driven gear on output shaft.

$$G = \frac{N_p}{N_g} = \frac{Z_g}{Z_p} = \frac{D_g}{D_p}$$

A gear train is gear sets joined in series. $G = \dfrac{N_{p1}}{N_{g1}} \times \dfrac{N_{p2}}{N_{g2}} = \dfrac{Z_{g1}}{Z_{p1}} \times \dfrac{Z_{g2}}{Z_{p2}} = G_1 \times G_2$

For gear sets more than two, velocity ratio is: $G = G_1 \times G_2 \times G_3$

Path of contact is divided as path of approach L_1 and path of recess L_2.

$$L_1 = \sqrt{R_a^2 - (R\cos\Phi)^2} - R\sin\Phi \text{ and } L_2 = \sqrt{r_a^2 - (r\cos\Phi)^2} - r\sin\Phi$$

Maximum length of contact $L_{max} = (R + r)\sin\Phi$

Arc of contact: It is the path traced by a point on the PCD of any gear from beginning to end of engagement. Arc of contact = Arc of approach + Arc of recess

$$\text{Arc of contact} = \frac{\text{Path of contact}}{\cos\Phi}$$

$$\text{Maximum arc of contact} = \frac{L_{max}}{\cos\Phi} = \frac{(R + r)\sin\Phi}{\cos\Phi} = (R+r)\tan\Phi$$

$$\text{Contact ratio CR} = \frac{\text{Arc of contact}}{\pi m}$$

Value of CR equal to one means that when one tooth is ending contact, the next tooth is just at the instant of making contact. In no case, value of CR should be less than one. A value of CR between 1.5 and 1.7 is satisfactory.

The contact area of non-conjugate teeth is called interference. During interference, tip of the tooth face tries to dig into the flank of the mating gear. This occurs when the number of teeth is less than minimum recommended number of teeth, which depends on pressure angle. For no interference, maximum length of approach $L_{1max} = r\sin\Phi$.

Parameters affecting interference

Number of teeth Increasing number of teeth decreases interference but increases the size of the gear. The weight, size, and cost increases.

Pressure angle Increase in pressure angle increases radial load and reduces tangential load, which results in less power transmission.

Stubbed tooth has 0.8 times standard addendum. So when addendum is reduced, it reduces the contact ratio leading to shock load and noise.

The last option of using stubbed teeth is generally used.

Maximum addendum radius r_{amax} for the pinion, without interference is:

$$r_{amax} = r\sqrt{1 + G(G + 2)\sin^2\Phi}$$

Maximum addendum for the pinion $a_{pmax} = r_{amax} - r$

$$\text{Maximum addendum radius of gear } R_{amax} = R\sqrt{1 + \frac{1}{G}\left(\frac{1}{G} + 2\right)\sin^2\Phi}$$

Maximum addendum for the gear $a_{gmax} = R_{amax} - R$

For a gear set having equal number of teeth $r = R$,

$$r_{amax} = R_{amax} = r\sqrt{1 + 3\sin^2\Phi} = R\sqrt{1 + 3\sin^2\Phi}$$

Minimum number of teeth for a pinion to avoid inference depends on pressure angle and the gear ratio. Gear ratio $G = R / r$. If k is defined as addendum depth factor, by which standard module of $1m$ is multiplied to addendum (a) of pinion to avoid interference that is $k = a / m$:

To avoid interference, addendum ($r_{amax} - r$) should be less than ($m\,k$).

$$Z_{pmin} = \frac{2k}{G\left[\sqrt{1 + \frac{1}{G}\left(\frac{1}{G} + 2\right)\sin^2\Phi} - 1\right]}$$

For $G = 1$, $Z_{pmin} = \dfrac{2k}{\sqrt{(1 + 3\sin^2\Phi)} - 1}$

Minimum number of teeth for a gear:

$$Z_{gmin} = \frac{2Gk\left[\sqrt{G^2 + (1 + 2G)\sin^2\Phi} + G\right]}{(1 + 2G)\sin^2\Phi} = \frac{2Gk}{\sqrt{G^2 + (1 + 2G)\sin^2\Phi} - G}$$

Minimum number of teeth for a pinion with a rack $Z_{pmin} = \dfrac{2k}{\sin^2\Phi}$

Slide velocity v_s is defined as the difference between the rolling velocities at the surfaces of teeth of the gear and pinion. $v_s = s\,(\omega_p - \omega_g)$ where $s =$ distance contact point P from the PCD. It is maximum at the commencement of engagement or leaving whichever is greater, that is, length of approach. Maximum slide velocity, $v_{smax} = (L_1 \text{ or } L_2) \times (\omega_p + \omega_g)$. (plus sign as gear rotates in opposite direction).

Theory Questions

1. Describe the advantages and disadvantages of the gear drive over other types of drives.

2. What are the various types of gears? Explain by giving neat sketches.

3. Define addendum, dedendum, pitch radius, base circle, and pressure angle.

4. Differentiate between circular pitch, diametrical pitch, and module.

5. What are the standard gear tooth proportions?

6. Differentiate between involute and cycloid profile using sketches.

7. Compare the advantages and disadvantages of involute and cycloid profiles.

8. What do you mean by conjugate action? Describe the law of gearing.

9. What is velocity ratio? How is it calculated?

10. Using a neat sketch, define base circle, pressure angle, and line of action.

11. Define path of contact. Derive an equation to calculate its value.

12. Define arc of approach, arc of recess, and arc of contact. How does it relate with path of contact.

13. Define contact ratio. Explain what happens if this ratio is less than one, equal to one, and more than one.

14. Define the phenomenon of interference and the parameters affecting it.

15. How do you calculate minimum number of teeth?

16. How do you calculate maximum number of teeth for a given size of pinion and gear ratio?

17. What is meant by slide velocity in spur gears? Derive an expression for its calculation.

Multiple Choice Questions

1. Pitch of a gear is defined by:
   (a) Circular pitch
   (b) Diametrical pitch
   (c) Module
   (d) Any one of the above

2. If D is PCD and Z number of teeth, module is defined as:
   (a) $\dfrac{D}{Z}$
   (b) $\dfrac{Z}{D}$
   (c) $\dfrac{\pi D}{Z}$
   (d) $\dfrac{Z}{\pi D}$

3. Tooth profile of a gear which can accommodate slight variation in center distance is:
   (a) Cycloid
   (b) Epi-cycloid
   (c) Involute
   (d) Circular

4. Base circle of a gear is:
   (a) Same as pitch circle
   (b) Circle on which involute is generated
   (c) Above pitch circle
   (d) Generally between addendum circle and pitch circle

5. Satisfactory contact ratio of a gear drive varies from:
   (a) 0.8 to 1.0
   (b) 1.0 to 1.2
   (c) 1.5 to 1.7
   (d) 1.9 to 2.5

6. For no interference, contact points should be:
   (a) Within the points of tangency on base circles and not outside
   (b) Outside the tangency line
   (c) Any where on the tangency line
   (d) None of the above

7. For no interference, minimum number of teeth on a pinion depends on:
   (a) Gear ratio
   (b) Pressure angle
   (c) Number of teeth on driver
   (d) All given above

8. Slide velocity is defined as difference between the:
   (a) Rotational speeds of gear and pinion
   (b) Difference in rolling velocities of the gear and pinion teeth surfaces
   (c) Backlash between teeth
   (d) Not applicable for spur gears

9. Shape of teeth of a rack is:

(a) Cycloid

(b) Involute

(c) Arc

(d) Straight line

Answers to multiple choice questions

1. (d)   2. (a)   3. (c)   4. (b)   5. (c)   6. (a)   7. (d)   8. (b)   9. (d).

Design Problems

1. A pinion having 30 teeth of involute profile with pressure angle 20° and module 3 meshes with a gear of 90 teeth. Calculate:

(a) Addendum and dedendum.

(b) Pitch circle, outside, and root diameters of pinion.

(c) Pitch circle and outside diameters of gear.

(d) Center distance.

(e) Tooth thickness.

[(a) 3 mm, 3.75 mm; (b) 90 mm, 96 mm, 82.5 mm;

(c) 270 mm; 276 mm (d) 180 mm; (e) 4.71 mm]

2. A pinion and gear are meshing to have center distance 240 mm and give a gear ratio of 3. Pressure angle is 20° and module 4. Calculate:

(a) Number of teeth for pinion and gear.

(b) Pitch circle diameters.

(c) Outside diameters.

[(a) 30, 90; (b) 120 mm, 360 mm; (c) 128 mm, 368 mm]

3. A pinion with 28 teeth drives a gear of 70 teeth cut with module 6 cutter of pressure angle 20°. Find:

(a) Length of contact, length of arc of contact.

(b) Contact ratio.

(c) Angle turned by pinion during contact period.

[(a) 30.5 mm, 32.8 mm; (b) 1.74; (c)22.4°]

4. A gear set of standard full depth involute teeth has pressure angle 20° and module 6. The pinion has 20 teeth and gear 50 teeth. Find:

(a) The number of teeth in contact at a time.

(b) Thickness and height of tooth and radial clearance while meshing.

[(a) 1.65; (b) $t = 9.42$ mm, $h = 13.5$ mm, $c = 1.5$ mm]

5. A gear set has a gear ratio of 5:1 with full depth involute profile of pressure angle 20°. Find:
   (a) Minimum number of teeth, if no interference is desired.
   (b) Module, if approximate center distance is 300 mm. [(a) 16; (b) 6]

6. Two gears transmitting power through involute teeth of pressure angle 20° of module 4 have center distance of 320 mm with gear ratio of 4. Driver speed is 600 rpm. Calculate:
   (a) Number of teeth of pinion and gear,
   (b) Base circle diameters, and
   (c) Outside diameters of gears.

   [(a) 32, 128; (b) 120.3, 481.1; (c) 136 mm; 520 mm]

7. A gear of module 6 and 60 teeth is driven by a pinion of 20 teeth with pressure angle 20°. Find:
   (a) Length of contact,   (b) Arc of contact,
   (c) Contact ratio,   (d) Angle turned by gear during contact.

   [(a) 29.67 mm; (b) 31.6; (c) 1.68; (d)10°]

8. A pinion drives a gear with gear ratio of 4 and pressure angle 14.5°. Find:
   (a) Number of pinion teeth without interference,
   (b) Length of contact in terms of circular pitch

   [(a) 15; (b) $L = 18.6\, p_c$];

9. A gear set of module 8 has number of teeth 14 and 35 for pinion and gear, respectively, and pressure angle 20°. Find, if this gear set will have interference or not? What is the ratio of contact length and module?.

   [Yes interference ; 4. 64]

10. A gear set of module 8, pressure angle 20°, and addendum $1m$ has a gear ratio of 2.8. Find number of teeth on pinion and gear to avoid interference. [15, 42]

11. A rack drives a pinion of 20 teeth, pressure angle 20°, and module 6. Calculate addendum to avoid interference and contact ratio. [$a = 7$ mm, CR = 2.04]

Previous Competition Examination Questions

IES

1. In an interference fit between a shaft and a hub, the state of stress or the shaft due to interference fit is:
   (a) Only compressive radial stress
   (b) A tensile radial stress and compressive tangential stress
   (c) A tensile tangential stress and a compressive radial stress
   (d) A compressive tangential stress and a compressive radial stress

   [Answer (a)] [IES 2015]

2. (i) State the law of gearing to maintain the condition for constant velocity ratio between a pair of toothed wheels. Name two types of gear tooth profiles to satisfy these.

(ii) A pinion having 36 teeth drives a gear having 96 teeth. The profiles of the gears are involute with 20° pressure angle, 10 mm module, and 10 m addendum. Find the length of path of contact and contact ratio   [Answer L = 52.76 mm, CR = 1.79] [IES 2013]

3. The arc of contact and the path of contact are respectively 27 mm and 25.4 mm, in a pair of involute spur gears. The pressure angle would be:

   (a) 14.5°   (b) 17.2°

   (c) 19.8°   (d) 20.5°

[Answer (c)] [IES 2013]

GATE

4. It is desired to avoid interference in a pair of spur gears having a 20° pressure angle. With increase of pinion to gear speed ratio, the minimum number of teeth on the pinion:

   (a) Increases   (b) Decreases

   (c) First increases and then decreases   (d) Remains unchanged

[Answer (b)] [GATE 2014]

5. Tooth interference in an external involute spur gear pair can be reduced by:

   (a) Decreasing center distance between gear pair   (b) Decreasing module

   (c) Decreasing pressure angle   (d) Increasing number of gear teeth

[Answer (a)] [GATE 2010]

6. Backlash in spur gear is:

   (a) Desirable in gear drive.

   (b) Gap between tooth space and tooth gear at pitch circle.

   (c) For silent operation.

   (d) To make manufacturing easy.

[Answer (b)] [GATE 2010]

7. Form factor of a spur gear tooth depends on:

   (a) Number of teeth and addendum

   (b) Number of teeth and pressure angle

   (c) Pressure angle and circular pitch

   (d) Number of teeth, pressure angle, and addendum modification coefficient

[Answer (d)] [GATE 2010]

8. The minimum number of teeth on the pinion to operate without interference in standard full height involute teeth gear mechanism with 20° pressure angle is:

   (a) 14   (b) 12

   (c) 18   (d) 32

[Answer (c)] [GATE 2010]

❏❏❏

Spur Gears

5.1 Introduction

Spur gears shown in picture on right side are used to transmit power between two parallel shafts. In Chapter 4, geometrical properties of gears, like pitch / module, tooth profiles, contact ratio, interference, path of approach, recess, etc., have been described. This chapter describes selection of materials and the design method for spur gears for strength, for static and dynamic loads, and wear. In spur gears, tooth contact is over one straight line over the tooth face at one instant and then it changes from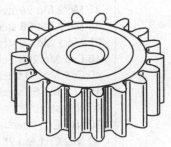
one tooth to next tooth abruptly. Therefore, gear loads are applied suddenly, developing high impact stresses and noise. In comparison to other types of gears, these are easy to manufacture due to straight teeth.

5.2 Gear Materials

Gears are made of metallic materials such as cast iron, steel as well as nonmetallic like nylon to reduce noise. Cast iron is commonly used material for big gears, due to its resistance to wear. Casting can easily produce complicated shapes, but needs machining for smooth operation. Plain carbon steels and alloy steels offer high strength. These gears are heat treated to impart hardness and toughness. Table 5.1 shows some materials for gears with their tensile strength and Brinell hardness.

Table 5.1 Properties of Gear Materials

S. No.	Material	Condition	Minimum Tensile Strength (MPa)	Brinell Hardness No.
1.	Malleable cast iron White Black	– –	280 320	217 (Maximum) 149 (Maximum)
2.	Cast iron Grade 20 Grade 25 Grade 35	As cast As cast As cast	200 250 350	179 (Minimum) 197 (Minimum) 207 (Minimum)
3.	Cast steel grade 1	–	550	145 (Minimum)
4.	Forged carbon steel 0.3% carbon 0.35% carbon 0.4% carbon 0.55% carbon	H.T.* Normalized H.T. H.T.	600 720 600 700	152 201 179 223
5.	Alloy steel C 55 Mn 75 steel 55 Cr 70 steel 27 Mn 2 steel 37 Mn 2 steel 35 Mn 2 Mo 28 steel 45 Mn 2 Mo 28 steel 35 Mn 2 Mo 45 steel 40 Cr 1 Mo 28 steel 40 Cr 1 Mo 60 steel 15 Cr 1 Mo 55 steel 40 Ni 3 steel 30 Ni 4 Cr.1 steel 40 Ni 2 Cr 1 Mo 28 steel 31 Ni 3 Cr 56 Mo 55 steel	H.T. H.T. H.T. H.T. H.T. H.T. H.T. H.T. H.T. H.T. H.T. H.T. H.T. H.T.	700 900 600 700 800 700 800 700 900 900 800 1,540 1,550 1,000–1,550	223 225 (Minimum) 170 (Minimum) 201 (Minimum) 229 (Minimum) 201 (Minimum) 229 (Minimum) 201 (Minimum) 248 (Minimum) 255 (Minimum) 229 (Minimum) 444 (Minimum) 444 (Minimum) 311–444

6.	*Surface hardened steel*			**Core**	**Case**
	C 40 steel	–	551	145	460
	C 55 steel		709	200	520
	C 55 Cr steel	–	866	250	500
	1% Cr steel	–	709	250	500
	3 % Ni steel		709	200	500
	1.5% Ni 1% Cr steel	–	866	250	500
7.	*Case hardened steel*				
	0.15% Carbon steel	–	504	140	600
	0.2% Carbon steel	–	504	140	640
	3% Ni steel	–	709	200	620
	5% Ni steel	–	866	250	600

* H.T. stands for hardened and tempered.

5.3 Gear Design Considerations

Gear design must meet the following requirements:

- Sufficient strength for high starting torque.
- Strength to withstand static and dynamic loading under normal running conditions.
- If shafts are long, misalignment of the gears due to deflected shaft should be considered.
- Minimum noise. Spur gears make more noise than helical gears (Chapter 6).
- No overheating of teeth under normal running conditions.
- Minimum wear qualities for a long life.
- Should occupy minimum space.
- Satisfactory lubrication system.
- Minimum cost.

5.4 Gear Tooth Strength

Calculation for the maximum stress in a loaded tooth is complicated due to variation in magnitude and direction of the load, during contact period of the teeth. Shape of the tooth also varies near the root due to fillet. Owing to change in cross section of the tooth, stress concentration is also to be considered.

5.4.1 Loads on gear tooth

The load F_n acts normal to the tooth profile at point of contact. The normal force changes because:

- The point of contact goes on changing as the gear rotates, therefore the arm of bending changes.

- Number of teeth in contact vary depending on contact ratio, hence load per tooth changes.
- Manufacturing inaccuracies in gear tooth spacing and tooth profile may cause dynamic forces. These are cyclic and can be even more than the static forces.

The normal force F_n can be resolved in two components; (a) tangential and (b) radial as shown in Figure 5.1.

Tangential force F_t is calculated from the given torque T and gear diameter d using the following relation:

$$T = 0.5d \times F_t \tag{5.1}$$

Tangential load F_t causes bending moment on the tooth at the point of contact. It is related with normal force F_n and pressure angle Φ as under:

$$F_t = F_n \cos \Phi \tag{5.2}$$

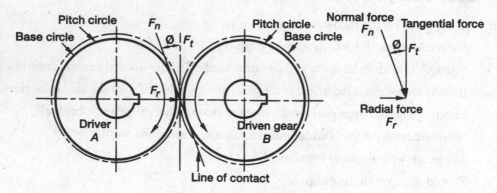

Figure 5.1 Components of the forces

Radial load F_r causes compressive stress. This being eccentric from the center line of the tooth is nonuniform over the surface. This tries to separate the gears and hence also called as separating force F_s. This can be calculated from the following relations:

$$F_r = F_n \sin \Phi \tag{5.3a}$$

Or, $$F_r = F_t \tan \Phi \tag{5.3b}$$

Example 5.1

Gear tooth forces for given power and module
A 20 teeth pinion transmits 5 kW at 1,440 rpm to a spur gear having module 5 mm.
Calculate:
  a. Driving force
  b. Separating force

c. Force on mounting shaft

d. Pitch line velocity

Solution

Given $Z_p = 20$ $P = 5\ kW$ $N_p = 1{,}440\ rpm$ $m = 5$

a. $p = \dfrac{2\pi NT}{60}$ or $5 \times 10^3 = \dfrac{2\pi \times 1{,}440 \times T}{60}$ or $T = 33.17\ Nm$

$d_p = m \times Z_p = 5 \times 20 = 100\ mm$

Driving force, $\qquad F_t = \dfrac{T}{0.5 d_p} = \dfrac{33.17 \times 10^3}{0.5 \times 100} = 663\ N$

b. Radial separating force, $\quad F_r = F_t \tan \Phi = 663 \times \tan 20° = 241\ N$

c. Force on mounting shaft, $F_n = \dfrac{F_t}{\cos \Phi} = \dfrac{663}{\cos 20°} = 705\ N$

d. Pitch line velocity, $\qquad v = \dfrac{\pi d_p N_p}{60} = \dfrac{\pi \times 100 \times 1{,}440}{60} = 7.54\ m/s.$

5.4.2 Lewis equation

Owing to varying forces and complexity of the geometry of the tooth discussed above, Wilfred Lewis gave an equation in 1,892 to simplify the problem. He made some assumptions given below:

- Neglect the radial load.

- Consider only the tangential load causing bending.

- Tooth profile is assumed parabolic as shown in Figure 5.2(a). Width of parabola is taken as thickness t of the tooth at the root. This offers uniform strength at all sections.

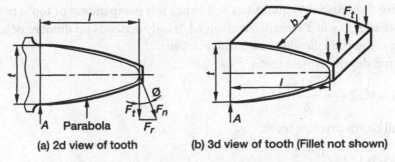

(a) 2d view of tooth (b) 3d view of tooth (Fillet not shown)

Figure 5.2 Parabolic approximation of tooth profile

- Normal load is taken at the center of the tooth.
- Load acts at the tip of parabola Figure 5.2(b).

Since the actual width of the tooth is more than parabola except at point A, this is the critical section. Therefore, stress is calculated at this section and is taken as the basis of design.

Modulus of section for the tooth $Z = \dfrac{bt^2}{6}$

Where, b = Face width of the gear tooth

t = Thickness of the tooth

Bending stress $\sigma = \dfrac{M}{Z} = \dfrac{6F_t l}{bt^2}$ (because $M = F_t l$)

Where, l = length of gear tooth as a cantilever beam

Multiplying and dividing the above equation by circular p_c:

$$\sigma = \frac{6F_t l p_c}{bt^2 p_c} = \frac{F_t}{b p_c} \times \frac{6l p_c}{t^2}$$

Using a Lewis form factor $y = \dfrac{t^2}{6 l p_c}$ $\qquad$ (5.4)

The above equation becomes: $\sigma = \dfrac{F_t}{b y p_c}$ or $F_t = \sigma b y p_c$

Thus the tooth strength is given by Lewis equation,

$$F_t = \sigma b y p_c \qquad\qquad (5.5)$$

If module m is to be is used instead of circular pitch p_c, Equation (5.5) becomes:

$$F_t = \sigma b Y m \qquad\qquad (5.6)$$

Where, $Y = \pi y = \dfrac{\pi t^2}{6 l p_c}$ and $m = \dfrac{P_c}{\pi}$

Both y or Y are dimensionless quantities and hence it is independent of tooth proportions. If tooth size is changed, y or Y remains unchanged. It only depends on number of teeth and can be calculated using the following empirical relations:

For 14.5° full depth involute teeth

$$y = 0.124 - \frac{0.684}{Z} \qquad\qquad (5.7a)$$

For 20° full depth involute teeth

$$y = 0.154 - \frac{0.912}{Z} \qquad\qquad (5.7b)$$

For 20° stub involute teeth

$$y = 0.175 - \frac{0.841}{Z} \qquad (5.7c)$$

Values of form factor Y (πy) based on these relations are tabulated in Table 5.2.
Face width b is generally taken as under:

$b = 2$ to 5 times the circular pitch P_c, but preferable value is $3p_c$ to $4p_c$.

If module is considered, face width b is taken as $6m$ to $15m$.

Table 5.2 Values of Lewis Form Factor (Y)

No. of Teeth	External Gears			Internal Gears	
	14.5° Full Depth	20° Full Depth	20° Stub	Pinion	Gear 20° Full Depth
12	0.210	0.245	0.311	0.327	–
13	0.223	0.264	0.324	0.329	–
14	0.236	0.276	0.339	0.330	–
15	0.245	0.289	0.349	0.331	–
16	0.255	0.295	0.360	0.333	–
17	0.264	0.302	0.368	0.342	–
18	0.270	0.308	0.377	0.348	–
20	0.283	0.320	0.393	0.364	–
22	0.292	0.330	0.400	0.373	–
24	0.302	0.337	0.411	0.382	–
26	0.308	0.344	0.421	0.391	–
28	0.314	0.352	0.430	0.399	0.691
30	0.318	0.358	0.437	0.404	0.678
32	0.322	0.364	0.443	0.407	0.672
35	0.327	0.373	0.442	0.412	0.661
37	0.330	0.389	0.454	0.415	0.655
40	0.336	0.389	0.459	0.419	0.645
45	0.340	0.399	0.468	0.427	0.628
50	0.346	0.408	0.474	0.437	0.612
55	0.352	0.415	0.480	0.441	0.604
60	0.355	0.421	0.484	0.445	0.597
65	0.358	0.425	0.488	0.447	0.594
70	0.360	0.429	0.496	0.449	0.587
75	0.361	0.433	0.496	0.452	0.581

80	0.363	0.436	0.499	0.454	0.575
90	0.366	0.442	0.503	0.457	0.570
100	0.368	0.446	0.506	0.460	0.565
150	0.375	0.458	0.518	0.468	0.550
200	0.378	0.463	0.524	0.472	0.547
300	0.382	0.471	0.534	0.477	0.534
Rack	0.390	0.484	0.540	-	-

Example 5.2

Module and face width using Lewis equation for given power

A pinion of width 12 m with 24 full depth teeth of 20° pressure angle transmits 30 kW at 1,200 rpm. The form factor for 20° full depth involute teeth is given as: $y = 0.154 - \dfrac{0.912}{Z}$.

Using Lewis equation, find the module and face width of the pinion. Assume safe bending stress 200 MPa.

Solution

Given $b = 12\,m$ $Z_p = 24$ $\Phi = 20°$ $P = 30\,kW$ $N_p = 1,200\,rpm$ $\sigma = 200\,MPa$

$$P = \frac{2\pi N T}{60}$$

Substituting values: $30 \times 10^3 = \dfrac{2\pi \times 1,200 \times T}{60}$ or $T = 239\,Nm$

$$d_p = m \times Z_p = m \times 24 = 24m \text{ mm}$$

Tangential force, $F_t = \dfrac{T}{0.5 d_p} = \dfrac{239 \times 10^3}{0.5 \times 24m} = \dfrac{19,910}{m}$

Form factor, $y = 0.154 - \dfrac{0.912}{Z} = 0.154 - \dfrac{0.912}{24} = 0.116$

Lewis equation is: $F_t = \sigma b y p_c$

Substituting the values: $\dfrac{19,910}{m} = 200 \times (12m) \times 0.116 \times (\pi \times m)$ [Because $P_c = \pi \times m$]

Or, $m^3 = \dfrac{19,910}{200 \times 12 \times 0.116 \times \pi} = 22.7$ Hence module, $m = 2.83$

Using standard module, $m = 3$ mm

Face width, $b = 12\,m = 12 \times 3 = 36$ mm.

5.5 Dynamic Loads

In a gear drive, dynamic loads occur due to the following reasons:

- Inaccuracy of tooth profiles
- Errors in tooth spacing
- Deflection of teeth when loaded
- Misaligned bearings
- Elasticity of parts
- Inertia of rotating parts on the shaft

To account for the dynamic loading, a factor called velocity factor K_v is used (Section 5.5.1). It does not give accurate value of dynamic loads, hence this is used for a preliminary design and then final design is done using Buckingham's equation (Section 5.6).

5.5.1 Velocity factor

Strength of gear depends on the pitch line velocity. This is considered using a velocity factor K_v developed by Barth, which depends on, as to how the teeth are cut. Values of velocity factor are given in Table 5.3 for different manufacturing methods.

Table 5.3 Relation for Velocity Factor (K_v) for Different Pitch Line Velocities

Type of Teeth	Casted Teeth	Milled Teeth	Hobbed / Shaped	Ground
Velocity (m/s)	<10	10–15	15–20	>20
Velocity factor, K_v	$\dfrac{3}{3+v}$	$\dfrac{6}{6+v}$	$\dfrac{50}{50+\sqrt{200v}}$	$\sqrt{\dfrac{78}{78+\sqrt{200v}}}$

Dynamic load is then calculated as: $F_d = \dfrac{F_t}{K_v}$

Hence considering dynamic loads, Lewis equation becomes:

$$F_t = \sigma \, b \, Y K_v \, m \tag{5.8}$$

Although velocity factor is easy to use and has been used for many years with satisfactory results, but it does not give good results if:

- Pitch line velocity is outside a limited range.
- Gear materials have too different modulus of elasticity.

Example 5.3

Approximate dynamic load using velocity factor

An electric motor running at 960 rpm drives a compressor at 200 rpm using a spur gear set with milled teeth. The center distance between the two shafts is 450 mm. A torque of 4,000 Nm is transmitted and the starting torque is 30 per cent higher than the average torque. Calculate the dynamic load using a suitable value of the velocity factor.

Solution

Given $N_p = 960$ rpm $\quad N_g = 200$ rpm $\quad C = 450$ mm $\quad T_{av} = 4{,}000$ Nm $\quad T_{max} = 1.3\,T_{av}$

$$\frac{d_g}{d_p} = \frac{N_p}{N_g} = \frac{960}{200} = 4.8 \quad \text{or} \quad d_g = 4.8\,d_p \qquad\qquad\text{(a)}$$

Also $\quad d_p + d_g = 2C = 2 \times 450 = 900$ mm $\qquad\qquad\qquad\qquad\qquad$ (b)

Substituting the value of d_g from equation (a) in equation (b)

$$d_p + 4.8\,d_p = 900 \quad\quad \text{or} \quad\quad d_p = 155 \text{ mm}$$

Hence, $\qquad\qquad d_g = 4.8 \times 155 = 745$ mm

Pitch line velocity, $v = \dfrac{\pi d N_p}{60} = \dfrac{\pi \times 0.155 \times 960}{60} = 7.8$ m/s

From Table 5.3, for milled teeth velocity factor is given as:

$$K_v = \frac{6}{6+v} = \frac{6}{6+7.8} = 0.435$$

$$T_{max} = 1.3\,T_{av} = 1.3 \times 4{,}000 = 5{,}200 \text{ Nm}$$

$$F_t = \frac{T}{0.5d_p} = \frac{5{,}200 \times 10^3}{0.5 \times 155} = 67{,}010 \text{ N}$$

Dynamic load, $F_d = \dfrac{F_t}{K_v} = \dfrac{67{,}010}{0.435} = 154{,}123$ N.

5.6 Buckingham's Equation for Dynamic Load

5.6.1 Dynamic load

Size of the tooth is fixed using Lewis equation. Then Buckingham's equation is used for determining dynamic loads and strength in wear. Once the gear dimensions are known using velocity factor, final design can be done with a better value of dynamic load using Buckingham's equation. According to this equation incremental dynamic load F_i is given as:

$$F_i = \frac{21v(Ceb + F_t)}{21v + \sqrt{(Ceb + F_t)}}$$ (5.9a)

Where, v = Pitch line velocity (m/s)

C = Deformation factor (N/mm²) (See Section 5.6.3)

e = Sum of errors of two meshing gears (mm) (See Section 5.6.4)

b = Face width (mm)

F_t = Tangential force as per rated torque (N)

If $A = (Ceb + F_t)$. Equation (5.9a) can be written as:

$$F_i = \frac{21vA}{21v + \sqrt{A}}$$ (5.9b)

Incremental load and dynamic load are shown in Figure 5.3.

The dynamic load F_d can be calculated as:

$$F_d = F_t + F_i$$ (5.10)

For a design to be safe, dynamic load should be lesser than beam strength (Section 5.6.2).

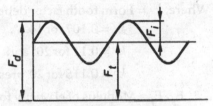

Figure 5.3 Incremental load and dynamic load

5.6.2 Beam strength

A gear tooth experiences varying load as described above, hence the endurance strength S_e is to be considered instead of bending strength in the Lewis equation. Thus the beam strength for σ_e as safe endurance stress becomes:

$$S_b = \sigma_e b y p_c = \sigma_e b y \pi m$$ (5.11)

See Table 5.4 for the values of endurance strength for different materials.

Table 5.4 Endurance Strength S_e in (MPa) for Different Hardnesses

Material	Brinell Hardness No. (BHN)										
	100	150	160	200	240	280	300	320	350	360	> 400
Steel	–	252	–	350	420	490	525	560	595	630	700
Grey cast iron	–	–	84	–	–	–	–	–	–	–	–
Phosphor Bronze	168	–	–	–	–	–	–	–	–	–	–

Note   For steels, endurance strength can be calculated using the relation $S_e = 1.75$ BHN.

To ensure that the gear tooth does not break, the beam strength has to be more than the dynamic load. Buckingham recommends the following:

For steady loads $\quad S_b \geq 1.25 \, F_d$

For pulsating loads $S_b \geq 1.35 \, F_d$

For shock loads $\quad S_b \geq 1.5 \, F_d$

5.6.3 Deformation factor

The deformation factor C used in Buckingham's equation is calculated from the values of modulus of elasticity of the pinion and gear materials using the equation given below:

$$C = \frac{k}{\left[\dfrac{1}{E_p} + \dfrac{1}{E_g}\right]} \qquad (5.12)$$

Where, k = Form tooth factor depending on pressure angle as:

$\quad k = 0.107$ for 14.5°,

$\quad k = 0.11$ for 20° and

$\quad k = 0.115$ for 20° pressure angle stub teeth.

E_p, E_g = Modulus of elasticity for pinion and gear materials respectively

Values of C are given in Table 5.5 for different combinations of pinion and gear, and Young's modulus of elasticity in Table 5.6.

Table 5.5 Values of Deformation Factor C (N / mm²)

Material Combination		Type of Teeth and Pressure Angle		
Pinion	Gear	Full Depth 14.5°	Full Depth 20°	Stub 20°
Grey cast iron	Grey cast iron	5,500	5,700	5,900
Steel	Grey cast iron	7,600	7,900	8,100
Steel	Steel	11,000	11,400	11,900

Table 5.6 Values of Modulus of Elasticity (GPa)

Material	Malleable iron	Nodular iron	Cast iron	Steel	Aluminum bronze
Modulus of Elasticity (E)	170	170	150	200	120

5.6.4 Errors in gears

The error e in Equation (5.9a) depends on the method, by which the gears are manufactured. The quality of gear manufacture is divided in 12 grades of precision. Grade 1 is of highest quality and needs grinding and grade 12 is most rough. Following grades can be expected from different manufacturing methods:

Surface grinding	Grades 1–2
Gear shaving	Grades 3–4
Hobbing and rough grinding	Grades 5–7
Fine hobbing	Grades 8–10
Casting	Grades 11–12

The tolerance or the maximum error e for adjacent pitch is given in Table 5.7, which depends on the module m and a tolerance factor f. The factor f is given as:

$$f = m + \sqrt{d} \qquad (5.13)$$

Where, m = Module

d = Pitch circle diameter of pinion or gear (mm)

Table 5.7 Tolerance e (μm) on Adjacent Pitch for Different Grades

Grade	Tolerance	Grade	Tolerance	Grade	Tolerance	Grade	Tolerance
1	$0.80 + 0.06f$	4	$3.2 + 0.25f$	7	$11.0 + 0.90f$	10	$32.0 + 2.50f$
2	$1.25 + 0.1f$	5	$5.0 + 0.40f$	8	$16.0 + 1.25f$	11	$45.0 + 3.55f$
3	$2.0 + 0.16f$	6	$8.0 + 0.63f$	9	$22.0 + 1.80f$	12	$63.0 + 5.00f$

The total error in Equation (5.9a) is the algebraic sum of errors in microns of pinion e_p and of gear e_g and is given as:

$$e = e_p + e_g \qquad (5.14)$$

5.6.5 Maximum allowable error

The maximum error e_{max}, which can be tolerated, depends on pitch line velocity v and is tabulated in Table 5.8 for well-cut commercial gears.

Table 5.8 Maximum Allowable Error e_{max} (μm) for Different Velocities (m/s)

Velocity	Max. Error	Velocity	Max. Error	Velocity	Max. Error	Velocity	Max. Error
1.25	92.5	7.5	47.5	15	22.5	22.5	15.0
2.5	80.0	10	37.5	17.5	17.5	25	12.5
5	60.0	12.5	30.0	20	15.0	> 25	12.0

Example 5.4

Dynamic load for a gear set using Buckingham's equation

A steel pinion of width 80 mm, module 8 and 20 teeth meshes with a cast iron gear of 50 teeth. Values of young's modulus of elasticity of pinion and gear are 200 and 150 GPa, respectively. A tangential force of 65 kN is transmitted through pinion rotating at 600 rpm. Teeth are cut stub with pressure angle 20°. Take tooth form factor $k = 0.115$. Calculate:

a. Deformation factor.

b. Maximum error, if gears are made by hobbing and then rough grinding.

c. Dynamic load capacity using Buckingham's equation.

Solution

Given $b = 80$ mm $m = 8$ $Z_p = 20$ $Z_g = 50$ $E_p = 200$ GPa

$E_g = 150$ GPa $F_t = 65$ kN $N_p = 600$ rpm $\Phi = 20°$ $k = 0.115$

a. From Equation (5.12) deformation factor,

$$C = \frac{1}{\left[\dfrac{1}{E_p} + \dfrac{1}{E_g}\right]} = \frac{0.115}{\left[\dfrac{1}{200 \times 10^3} + \dfrac{1}{150 \times 10^3}\right]} = 9{,}860 \text{ N/mm}^2$$

b. For hobbed and rough ground gears, accuracy grade varies from 5 to 7. Taking it as 6. From Table 5.7 for grade 6 tolerance is: $e = 8 + 0.63f$

The tolerance factor f is given by Equation (5.13) is given as: $f = m + \sqrt{d}$

Pinion diameter, $d_p = m \times Z_p = 8 \times 20 = 160$ mm

Gear diameter, $d_g = m \times Z_g \times = 8\,0 \times 50 = 400$ mm

For pinion error, $e_p = 8 + 0.63\left(8 + \sqrt{160}\right) = 21$ μm

For gear error, $e_g = 8 + 0.63(8 + \sqrt{400}) = 25.6$ μm

Total error, $e = e_p + e_g = 21 + 25.6 = 46.6$ μm

Pitch line velocity, $v = \dfrac{\pi d N}{60} = \dfrac{\pi \times 0.160 \times 600}{60} = 5$ m/s

From Table 5.8, maximum allowable error for this velocity is 60 μm, hence safe.

$$A = (C\,e\,b + F_t) = [9{,}860 \times (46.6 \times 10^{-3}) \times 80] + 65{,}000 = 101{,}758 \text{ N}$$

Incremental load as per Buckingham's Equation (5.9b) is given as:

$$F_i = \frac{21vA}{21v + \sqrt{A}} = \frac{21 \times 5 \times 101,758}{(21 \times 5) + \sqrt{101,758}} = 25,200 \text{ N}$$

Dynamic load according to Equation (5.10):

$$F_d = F_t + F_d = 65,000 + 25,200 = 90,200 \text{ N}$$

5.7 Gear Design for Wear Strength

Gear tooth may fail due to beam strength lesser than the allowable strength and many other reasons also as given below:

- Surface failure due to wear.
- Scoring due to lack of lubrication.
- Abrasion due to presence of foreign particles in oil or environment.
- Pitting failure due to high contact stresses. When the surface stress exceeds the endurance limit of the material, pitting takes place. It causes small pits on the surface, which leads to tooth failure due to wear. Hence the hardness of the surface has to be sufficient so that this failure does not occur.

The maximum load that a gear tooth can carry without premature wear depends on:

- Radius of curvature of tooth profiles,
- Elasticity, and
- Surface fatigue limit of material.

5.7.1 Hertz stresses on tooth surface

When two cylinders of diameters $2r_1$ and $2r_2$ are pressed against each other, contact stresses called hertz stresses are developed at the contact line. Owing to elastic behavior of the surfaces, the contact area becomes an ellipse instead of a line. The major axis of this stressed area is equal to the length of the cylinder and the minor axis is of width $2x$. The maximum hertz stress is given by Equation (a)

$$\sigma_H = p_{max} = \frac{2F_n}{\pi b x} \tag{a}$$

Where, x = Semi-minor axis given by Equation (b).

Half width of minor axis $x = \sqrt{\dfrac{2F_n\left(\dfrac{1 - \mu_p^2}{E_p} + \dfrac{1 - \mu_g^2}{E_g}\right)}{\pi b\left(\dfrac{1}{r_p} + \dfrac{1}{r_g}\right)}}.$ \hfill (b)

Where, σ_H = Maximum hertz stress

F_n = Normal force to the surface

E_p, E_g = Modulus of elasticity for pinion and gear materials respectively

μ_p, μ_g = Poison's ratio of pinion and gear materials respectively

In case of gears, the tangential force F_t is related with normal force F_n as under:

$$F_n = \frac{F_t}{\cos \Phi} \qquad (c)$$

Substituting the value of x from Equation (b) and F_n from Equation (c) in Equation (a):

$$\sigma_H = \sqrt{\frac{2F_t}{\pi b \cos \Phi} \left[\frac{\dfrac{1}{r_p} + \dfrac{1}{r_g}}{\dfrac{1-\mu_p^2}{E_p} + \dfrac{1-\mu_g^2}{E_g}} \right]} \qquad (d)$$

Wear of gear tooth is generally maximum at the pitch line, when only one pair of teeth is taking the entire load. Hence radii of curvature r_1 and r_2 of surfaces of tooth profiles for pinion and gear, respectively, are taken at this point. These are related with pressure angle Φ as under:

$$r_p = \frac{d_p}{2} \sin \Phi \quad \text{and} \quad r_g = \frac{d_g}{2} \sin \Phi$$

Hence, $\dfrac{1}{r_p} + \dfrac{1}{r_g} = \dfrac{2}{\sin \Phi}\left(\dfrac{1}{d_p} + \dfrac{1}{d_g} \right) = \dfrac{2}{\sin \Phi}\left(\dfrac{d_p + d_g}{d_p \times d_g} \right)$

Because, $d_g = G \times d_p$

Hence, $\dfrac{1}{r_p} + \dfrac{1}{r_g} = \dfrac{2}{\sin \Phi}\left(\dfrac{d_p + Gd_p}{d_p \times Gd_p} \right) = \dfrac{2}{\sin \Phi}\left(\dfrac{1+G}{Gd_p} \right)$

Let, $I = \dfrac{\sin \Phi \cos \Phi}{2}\left(\dfrac{Gd_p}{G+1} \right)$ \qquad (5.15a)

For internal gears, $I = \dfrac{\sin \Phi \cos \Phi}{2}\left(\dfrac{Gd_p}{G-1} \right)$ \qquad (5.15b) and elastic coefficient be $C_p = \sqrt{\dfrac{1}{\pi \left(\dfrac{1 - \mu_p^2}{E_p} + \dfrac{1 - \mu_g^2}{E_g} \right)}}$ (5.16)

Substituting the values from Equations (5.15a) and (5.16) in Equation (d):

So maximum hertz stress, considering velocity factor K_v for inaccuracies in manufacturing

$$\sigma_H = C_p \sqrt{\dfrac{F_t}{K_v \, b d_p I}}$$ (5.17)

Units of C_p are $\sqrt{\text{MPa}}$ and its values are given in Table 5.9.

Table 5.9 Values of C_p in $\sqrt{\text{MPa}}$ for Different Pinion and Gear Materials

Pinion Material	E (GPa)	C_p for Gear Material					
		Steel	M.I	N.I	C.I.	Al. Br.	Tin Br.
Steel	200	191	181	179	174	162	158
Malleable iron (M.I)	170	181	174	172	168	158	154
Nodular iron (N.I)	170	179	172	170	166	156	152
Cast Iron (C.I.)	150	174	168	166	163	154	149
Aluminum Bronze	120	162	158	156	154	145	141
Tin Bronze	110	158	154	152	149	141	137

Example 5.5

Hertz stresses on tooth surface for given power through a gear set

A steel pinion of pitch circle diameter 75 mm and pressure angle 20° meshes with a cast iron spur gear having gear ratio 3 and transmits 2 kW power at 360 rpm. Gear width is 12 times the module. These have the following material properties:

Modulus of elasticity for pinion material = 200 GPa
Modulus of elasticity gear material = 174 MPa
Poison's ratio of pinion material = 0.29
Poison's ratio of gear material = 0.21

Calculate the Hertz stresses at the surface of the tooth.

Solution

Given $d_p = 75$ mm $G = 3$ $\Phi = 20°$ $P = 2$ kW $N_p = 360$ rpm
             $b = 12m$ $E_p = 200$ GPa $E_g = 174$ MPa $\mu_p = 0.29$ $\mu_g = 0.21$

Module $m = \dfrac{d_p}{Z_p} = \dfrac{75}{15} = 3$

$$P = \dfrac{2\pi NT}{60}$$

Substituting the values: $2 \times 10^3 = \dfrac{2\pi \times 360 \times T}{60}$ or $T = 53.1\,\text{Nm}$

Tangential force, $F_t = \dfrac{T}{0.5 d_p} = \dfrac{53.1 \times 10^3}{0.5 \times 75} = 1{,}415\,\text{N}$

Pitch line velocity, $v = \dfrac{\pi d_p N}{60} = \dfrac{\pi \times 75 \times 10^{-3} \times 360}{60} = 1.4\,\text{m/s}$

For velocity less than 5 m/s $K_v = \dfrac{3}{3+v} = \dfrac{3}{3+1.4} = 0.682$

$$I = \dfrac{\sin \Phi \cos \Phi}{2}\left(\dfrac{G d_p}{G+1}\right) = \dfrac{\sin 20^\circ \; \cos 20^\circ}{2}\left(\dfrac{3 \times 75}{3+1}\right) = 9.04$$

Elastic coefficient $C_p = \sqrt{\dfrac{1}{\pi\left(\dfrac{1-\mu_p^2}{E_p} + \dfrac{1-\mu_g^2}{E_g}\right)}} = \sqrt{\dfrac{1}{\pi\left(\dfrac{1-0.29^2}{200 \times 10^3} + \dfrac{1-0.21^2}{174 \times 10^3}\right)}} = 178.5$

Maximum hertz stress $\sigma_H = C_p \sqrt{\dfrac{F_t}{K_v b d_p I}}$

Substituting the values:

$$\sigma_H = 178.5 \sqrt{\dfrac{1{,}415}{0.682 \times (12 \times 3) \times 75 \times 9.04}} = 52\,\text{MPa}$$

5.7.2 Buckingham equation for wear

Buckingham gave an equation for estimating maximum wear load F_w for satisfactory performance of the gear. For the design to be safe, wear load F_w should be less than the dynamic load F_d.

$$F_w = d_p\, b\, Q\, K \tag{5.18}$$

Where, d_p = Pitch circle diameter of pinion (mm)
$\quad\; b$ = Face width (mm)

$$Q = \text{Ratio factor} = \frac{2Z_g}{Z_g + Z_p} = \frac{2G}{G + 1}$$

here, G = Gear ratio $= \dfrac{Z_g}{Zp}$

K = Load stress factor, which is given by Equation (5.19):

$$K = \frac{(S_{es})^2 \sin\Phi \cos\Phi}{1.4}\left(\frac{1}{E_p} + \frac{1}{E_g}\right) \tag{5.19}$$

Where, S_{es} = Surface endurance strength (MPa)

Φ = Pressure angle

E_p = Modulus of elasticity for pinion material

E_g = Modulus of elasticity gear material

Surface endurance strength S_{es} for steels in MPa is given by the relation given in Equation (5.20):

$$S_{es} = 2.76\,\text{BHN} - 70 \tag{5.20}$$

Where, BHN = Brinell Hardness Number of the softer material

Values of surface endurance strength for different materials are given in Table 5.10.

Table 5.10 Endurance / Wear Strength (MPa) for Different Gear Materials

Material	Core Hardness (BHN)	Endurance Strength	Wear Strength
Grey cast iron	160	84	630
Semi steel	200	125	630
Soft cast steel	110	178	–
Medium cast steel	120	206	–
Steel	150	260	350
Steel	200	345	490
Steel	280	480	721
Steel	300	515	770
Steel	400	686	1,050

A life of 10^8 stress cycles has been assumed. American Gear Manufacturing Association (AGMA) recommends that this surface fatigue strength should be modified as corrected surface fatigue strength as under:

$$S_H = \frac{C_L C_H}{C_T C_R} S_{es} \tag{5.21a}$$

Where, S_H = Corrected surface fatigue strength

C_L = Life factor (From Table 5.11)

C_H = Hardness factor = 1 for spur gears

C_T = Temperature factor, C_T = 1, if temperature is less than 120° C

C_R = Reliability factor (From Table 5.11)

Thus for spur gears and temperature lesser than 120° C, the above equation reduces to:

$$S_H = \frac{C_L}{C_R} S_{es}$$ (5.21b)

Values of life and hardness factors can be taken from Table 5.11 given below:

Table 5.11 Values of Life and Reliability Factors (C_L and C_R)

Life in Cycles	C_L	Reliability	C_R
10^4	1.5	Up to 0.99	0.8
10^5	1.3	Up to 0.999	1.0
10^6	1.1	0.9999 and more	1.25
10^8	1.0	–	–

Example 5.6

Strength in wear for a gear set for given power

A spur gear set has pressure angle 20° full depth involute 14 teeth transmits 10 kW power through a pinion rotating at 450 rpm to a gear with gear ratio 3.5. Width can be taken as 13 times the module. Safe allowable stress for the pinion and gears can be taken as 100 MPa and 85 MPa, respectively. Young's modulus of elasticity for pinion and gear are 200 GPa and 170 MPa, respectively. Overload factor can be taken 1.25. Velocity factor for velocity up to 10 m/s can be taken as $K_v = \dfrac{3}{3+v}$ Calculate:

   a. Module.

   b. Factor of safety in wear, if endurance strength is 550 MPa.

Solution

Given $\Phi = 20°$ $P = 10\,\text{kW}$ $Z_p = 14$ $N_p = 450\,\text{rpm}$
            $G = 3.5$ $b = 13\,m$ $K_o = 1.25$ $\sigma_p = 100\,\text{MPa}$
            $\sigma_g = 85\,\text{MPa}$ $\sigma_e = 550\,\text{MPa}$ $E_p = 200\,\text{GPa}$ $E_g = 170\,\text{GPa}$

   a. Pitch line velocity, $v = \dfrac{\pi\, d_p\, N_P}{60{,}000} = \dfrac{\pi m Z_p N_P}{60{,}000} = \dfrac{3.14 m \times 14 \times 450}{60{,}000} = 0.33 m\ \text{m/s}$

Tangential force $F_t = \dfrac{P \times K_o}{v} = \dfrac{10{,}000 \times 1.25}{0.33m} = \dfrac{37{,}879}{m}$

For 20° full depth involute teeth from Equation (5.7b): $y = 0.154 - \dfrac{0.912}{Z}$

Form factor for pinion $y_p = 0.154 - \dfrac{0.912}{Z_p} = 0.514 - \dfrac{0.912}{14} = 0.088$

Form factor for gear $y_g = 0.154 - \dfrac{0.912}{Z_g} = 0.154 - \dfrac{0.912}{3.5 \times 14} = 0.135$

For pinion $\sigma_p \times y_p = 100 \times 0.088 = 8.8$

For gear $\sigma_g \times y_g = 85 \times 0.135 = 11.475$

Pinion is weaker than gear, hence design will be based on pinion's strength.

According to Lewis equations: $F_t = \sigma_p\, k_v\, b\, \pi\, m\, y_p = (\sigma_p \times y_p) \times k_v \times (13m) \times \pi\, m$

Substituting the values:

$$\frac{37{,}879}{m} = 8.8 \times \left(\frac{3}{3 + 0.33m}\right) \times 13m \times 3.14 \times m$$

Or, $\qquad \dfrac{37{,}879}{m} = \dfrac{1{,}007.6\, m^2}{3 + 0.33\, m} \qquad$ or $\qquad 3 + 0.33m = 0.0285\, m^3$

By trial error module is between 5 and 6. Selecting a higher standard value $m = 6$.

Pitch line velocity, $v = 0.33\, m = 0.33 \times 6 = 1.98$ m/s

It is lesser than 10 m/s, hence the value of velocity factor assumed is satisfactory.

Diameter of pinion, $d_p = m \times Z_p = 6 \times 14 = 84$ mm

Diameter of gear, $d_g = m \times Z_g = 6 \times 3.5 \times 14 = 294$ mm

Ratio factor, $Q = \dfrac{2G}{G+1} = \dfrac{2 \times 3.5}{3.5 + 1} = 1.555$

Load stress factor is given by Equation (5.19): $K = \dfrac{(\sigma_{es})^2 \sin \Phi \cos \Phi}{1.4}\left(\dfrac{1}{E_p} + \dfrac{1}{E_g}\right)$

$$K = \frac{(550)^2 \sin 20° \times \cos 20°}{1.4}\left(\frac{1}{200 \times 10^3} + \frac{1}{170 \times 10^3}\right) = 0.755 \text{ MPa}$$

b. Using Equation (5.18) for safe wear load:

$$F_w = d_p\, b\, Q\, K$$

Substituting the values $F_w = 84 \times (13 \times 6) \times 1.555 \times 0.755 = 7,692$ N

Tangential force $F_t = \dfrac{37,879}{6} = 6,313$ N

FOS in wear strength $\dfrac{F_w}{F_t} = \dfrac{7,692}{6,313} = 1.22$

5.8 Factors Affecting Gear Design

5.8.1 Overload factor (K_o)

Shocks coming on the driver and driven gears are considered using an overload factor K_o. Its values are given in Table 5.12 for different level of shocks on driver and driven gear.

Table 5.12 Values of Overload Factor (K_o)

Driver Gear	Driven Gear		
	Uniform	Moderate Shock	Heavy Shock
Uniform	1.0	1.25	1.75
Light shock	1.25	1.5	2.0
Medium shock	1.5	1.75	2.25

5.8.2 Load distribution factor (K_l)

The shaft on which a gear is mounted gets deflected depending on its stiffness and type of mounting in bearings. Hence effect of load distribution on gear tooth depends on, how accurate is mounting of the shaft. Its effect depends on face width. More the face width, higher is the effect of mounting. Values of load distribution factor K_l are given in Table 5.13 for different face width and type of mounting.

Table 5.13 Values of Load Distribution Factor (K_l)

Mounting Accuracy	Face Width (mm) (b)*			
	0–50	51–150	151–225	226–400
Accurate minimum deflection	1.3	1.4	1.5	1.8
Less rigid, less accurate	1.6	1.7	1.8	2.2
Poor mounting	> 2.2	> 2.2	> 2.2	> 2.2

*Assume face width b as under for the given center distance C.

For ordinary gears $b = 0.3 \, C$

For improved gears $b = 0.4 \, C$

For hardened gears $b = 0.25 \, C$

5.8.3 Mounting factor (K_m)

The location of gear mounting over the shaft affects the effect of shaft deflection causing misalignment in meshing the gears. This factor is considered by considering a mounting factor K_m given in Table 5.14.

Table 5.14 Values of Mounting Factor (K_m)

Type of Mounting	Face Contact	
	Partial	Full
Symmetrical mounting between bearings	1.0	1.1
Near one end of bearing with rigid shaft	1.1	1.3
Near one end of bearing with flexible shaft	1.2	1.5
Cantilever	1.3	2.0

5.8.4 Surface finish factor (K_f)

Wear life of a gear depends on surface finish of the gear teeth. Surface finish factor K_f depends on ultimate strength of the gear material. Its values are given in Table 5.15.

Table 5.15 Values of Surface Finish Factor (K_f)

Ultimate Strength (MPa)	500	600	700	800	900	1,000	1,000	1,200	1,400
Surface Finish Factor	0.8	0.74	0.72	0.7	0.68	0.67	0.65	0.63	0.62

5.8.5 Rotation factor (K_n)

In most of the applications, gear rotates in one direction, but in some applications, its direction of rotation may change. If it reverses, a rotation factor K_n is to be considered as given below:

For one side rotation $K_n = 1.4$

For reversed rotation $K_n = 1.0$

5.8.6 Reliability factor (K_r)

Depending on type of application, a reliability factor K_r is to be considered as given in Table 5.16.

Table 5.16 Values of Reliability Factor (K_r)

Reliability	0.5	0.9	0.95	0.99	0.999	0.9999
Reliability Factor	1.08	0.897	0.868	0.814	0.753	0.702

5.8.7 Size factor (K_s)

Size of gear tooth depends on module m. Bigger the tooth, that is, higher the value of module, the tooth has tendency to bend. It is considered using a size factor K_s given in Table 5.17.

Table 5.17 Values of Size Factor (K_s)

Module	1	2	2.5	3	4	5	6	8
Size Factor	1	0.99	0.974	0.956	0.930	0.910	0.894	0.870
Module	10	12	16	20	25	32	40	50
Size Factor	0.850	0.836	0.813	0.796	0.779	0.76	0.744	0.728

If $m < 1$, then $K_s = 1$
If $m > 1$, then use Table 5.17 or calculate using the following relation:

$$K_s = 1.189 \times (\pi m)^{-0.097} \qquad (5.22)$$

5.8.8 Temperature factor (K_t)

If the operating temperature is more than 70°C, then only the temperature factor is to be considered as given below:

If temperature $t \leq 70°C$, then $K_t = 1$

If temperature $t > 70°C$, then $K_t = \dfrac{344}{273 + t}$ $\qquad (5.23)$

5.9 Design Procedure

For spur gear design, following input parameters are usually given:

- Power P (kW)
- Speeds of driver N_1 and driven gear N_2 or speed of any one gear and gear ratio G
- Service conditions
- Center distance C between the two gears may be given or not given. The design procedure changes, if the center distance is not given. Both the procedures are given in subsequent Sections 5.9.1 and 5.9.2, respectively.

5.9.1 Design procedure with given center distance

A. Calculate stress for static load

1. Calculate torque T from the relation:

$$T = \frac{60,000\,P}{2\pi N_p} \tag{5.24}$$

Where, T = Torque (Nm)

P = Power (kW)

N_p = Speed of the driver gear (rpm)

2. Calculate gear ratio $G = \dfrac{N_p}{N_g}$ (It is usually more than 1. N_g is speed of driven gear).

3. Calculate pinion and gear diameters d_p and d_g, respectively, from the center distance C using the relations. $d_p = \dfrac{2C}{1+G}$ and $d_g = G\,d_p$.

4. Calculate tangential force F_t on the tooth of pinion, from the torque T in Nm and diameter d_p in mm from the following equation:

$$F_t = \frac{2T \times 10^3}{d_p}\,N \tag{5.25}$$

5. Calculate pitch line velocity v (m/s) from the pinion diameter d_p (mm) from relation:

$$v = \frac{\pi\,d_p\,N_p}{60,000} \tag{5.26}$$

6. Select a suitable manufacturing method according to the application. Calculate velocity factor K_v for the pitch line velocity v from Table 5.3.

7. From the given service conditions for shock, find overload factor K_o from Table 5.12.

8. Assuming a suitable mounting method depending on application, find load distribution factor K_l from Table 5.13.

9. Find mounting factor K_m from Table 5.14 depending on type of mounting.

10. Select pressure depending on type of load; 14.5° for light loads, 20° for moderate loads, and 25° for heavy loads.

11. Find minimum number of teeth for the gear / pinion Z_{min} depending on gear ratio G and pressure angle Φ to avoid interference from the equation (4.16c) in Chapter 4.

12. Calculate module from equation

$$m = \frac{d_p}{Z_{min}} \tag{5.27}$$

Choose a standard value of module from Table 5.18 given below.

Table 5.18 Preferred Modules (*m*)

First Choice	Next Choice	First Choice	Next Choice	First Choice	Next Choice
1	1.125	4	4.5	16	18
1.25	1.375	5	5.5	20	22
1.5	1.75	6	7	25	28
2	2.25	8	9	32	36
2.5	2.75	10	11	40	45
3	3.5	12	14	50	55

13. Calculate the following coefficients, whichever is applicable:

Surface finish factor K_f from Table 5.15

Reliability factor K_r from Table 5.16

Rotation factor K_n (Section 5.8.5)

Size factor K_s from Table 5.17

Temperature factor K_t from Equation (5.23)

14. From all the coefficients calculated above, calculate factor K as given below:

$$K = K_f \times K_r \times K_n \times K_s \times K_t \tag{5.28}$$

15. Take values of K_v, K_o, K_l and K_m from Tables 5.3, 512, 5.13 and 5.14 respectively. Calculate allowable static stress σ_o from the equation given below:

$$\sigma_o = \frac{F_t \left(K_o K_l K_m \right)}{mYbKK_v} \tag{5.29}$$

This stress should be lesser than allowable stress.

B. Check for dynamic load

16. Select a suitable gear manufacturing process or the grade. Calculate tolerance factor for pinion and gear using Equation (5.13): $f = m + \sqrt{d}$

17. From the tolerance formula in Table 5.7 for the selected grade, calculate error for pinion e_p and for gear e_p. Then calculate total error e from Equation (5.14): $e = e_p + e_g$.

18. Calculate deformation factor C from Equation (5.12):

$$C = \frac{k}{\left[\dfrac{1}{E_P} + \dfrac{1}{E_g} \right]}$$

19. Calculate A from the relation $A = Ceb + F_t$

20. Calculate incremental load F_i from Equation (5.9): $F_i = \dfrac{21vA}{21v + \sqrt{A}}$

21. Calculate dynamic load F_d using Equation (5.10): $F_d = F_t + F_i$

22. Check that this value of error e is lesser than the maximum allowable error e_{max} given in Table 5.8. If not, increase accuracy level and recalculate.

    Beam strength S_b should be more than dynamic load F_d.

C. Check for hertz stress

23. Calculate I from Equation (5.15a):

$$I = \frac{\sin \Phi \cos \Phi}{2} \left(\frac{Gd_p}{G+1} \right)$$

24. Calculate elastic coefficient C_p from Equation (5.16):

$$C_p = \sqrt{\frac{1}{\pi \left(\dfrac{1 - \mu_p^2}{E_p} + \dfrac{1 - \mu_g^2}{E_g} \right)}}$$

25. Calculate hertz stress σ_H from Equation (5.17): $\sigma_H = C_p \sqrt{\dfrac{F_t}{K_v b d_p l}}$

    Value of strength S_H should be more than hertz stress σ_H to be safe.

D. Check for wear

26. Calculate Ratio factor Q from the relation:

$$Q = \frac{2G}{G+1}$$

27. Calculate load stress factor K from the equation:

$$K = \frac{(S_{es})^2 \sin \Phi \cos \Phi}{1.4} \left(\frac{1}{E_p} + \frac{1}{E_g} \right)$$

28. Calculate surface fatigue strength S_{es} from Equation (5.20): $S_{es} = \left(2.76 \, BHN - 70 \right)$

29. Calculate corrected surface fatigue strength from Equation (5.21):

$$S_H = \frac{C_L C_H}{C_T C_R} S_{es}$$

    Take values of C_L and C_H from Table 5.11.

30. Calculate load capacity in wear F_w from Equation (5.18): $F_w = d_p b Q K$

    Wear load capacity F_w should be more than dynamic load F_d. If not, then change either the material or hardness and recalculate the value till it is safe.

Example 5.7

Calculate module for a drive with known center distance

A bronze pinion rotating at 360 rpm transmits 12.5 kW power to a cast iron gear through 20° involute gear, which rotates at 240 rpm. Center distance between shafts may be taken as 225 mm and face width as 12 times the module. If safe stress for pinion is 80 MPa and for gear 90 MPa, calculate:

a. Number of teeth for each gear

b. Module and face width

Solution

Given
$$N_p = 360 \text{ rpm} \quad P = 12.5 \text{ kW} \quad \Phi = 20° \quad N_g = 240 \text{ rpm}$$
$$C = 225 \text{ mm} \quad b = 12\,m \quad \sigma_p = 80 \text{ MPa} \quad \sigma_g = 90 \text{ MPa}$$

a. Gear ratio $G = \dfrac{N_p}{N_g} = \dfrac{360}{240} = 1.5$

Therefore $d_g = 1.5\,d_p$

Center distance $C = \dfrac{d_g + d_p}{2}$

Substituting values: $225 = \dfrac{1.5d_p + d_p}{2}$ or $d_p = 180 \text{ mm}$

Hence gear diameter $d_g = 1.5 \times d_p = 1.5 \times 180 = 270 \text{ mm}$

Pitch line velocity $v = \dfrac{\pi d N}{60} = \dfrac{\pi \times 0.18 \times 360}{60} = 3.39 \text{ m/s}$

From Table 5.3, $K_v = \dfrac{3}{3+v} = \dfrac{3}{3+3.39} = 0.47$

From Equation (4.16a) of Chapter 4: $Z_{g\,min} = \dfrac{2\,G\,k}{\sqrt{G^2 + \left(1 + 2G\right)\sin^2 \Phi} - G}$

Assuming full depth tooth $(k = 1)$

$$Z_{gmin} = \dfrac{2 \times 1.5 \times 1}{\sqrt{1.5^2 + \left[1 + \left(2 \times 1.5\right) \times 0.117\right]} - 1.5} = 21 \text{ teeth}$$

Number of teeth on pinion $Z_p = \dfrac{21}{1.5} = 14$

b. From Equation (5.7b), Lewis factor $y = 0.154 - \dfrac{0.912}{Z} = 0.154 - \dfrac{0.912}{14} = 0.0888$

$$Y = \pi y = 3.14 \times 0.0888 = 0.278$$

From Table 5.2, for 14 teeth and $\Phi = 20°$ value of Y can also be taken as 0.276

From Table 5.2 for gear having 21 teeth $Y = 0.3225$

For pinion: $\sigma_p \times Y = 80 \times 0.278 = 22.08$

For gear: $\sigma_g \times Y = 90 \times 0.325 = 29.25$

Stress for pinion is lesser than gear, hence design is based on pinion strength

Tangential force $F_t = \dfrac{P}{v} = \dfrac{12,500}{3.39} = 3,687$ N

Substituting the values in Lewis Equation (5.8) $F_t = \sigma K_v b\, Y m$

$$3,687 = 80 \times 0.47 \times (12m) \times 0.278 \times m \quad \text{or} \quad m = 5.4$$

Select a standard module from Table 5.18 $m = 6$

Hence face width $b = 12\, m = 12 \times 6 = 72$ mm

5.9.2 Design procedure when center distance is not given

1. Calculate torque T from Equation (5.24): $T = \dfrac{60,000 P}{2\pi N_1}$

2. Assume pressure angle and then calculate minimum number of teeth Z_{min} from Equation (4.14) given in Chapter 4. Number of teeth on pinion $Z_p = Z_{min}$

3. Calculate teeth on gear from the speeds $Z_g = Z_p \dfrac{N_p}{N_g}$

4. Assume face width $b = 10\, m$.

5. Calculate form factor Y from Equation (5.7 a, b, c) as applicable.

6. Select velocity factor K_v depending on gear manufacturing from Table 5.3.

7. Calculate module from the equation: $\sigma_o = \dfrac{2T}{(10m)(mZ_p)mY K_v}$

   Find the nearest standard module from Table 5.18.

8. Calculate pinion and gear diameters from equations: $d_p = m Z_p$ and $d_g = m Z_g$

9. Calculate pitch line velocity from equation $v = \dfrac{\pi d_p N_p}{60,000}$

10. Calculate exact velocity factor K_v.

11. Calculate coefficients K_o, K_p and K_m.

12. Calculate coefficients K_f, K_r, K_n, K_s, K_t. From all these coefficients, calculate factor

$$K = K_f \times K_r \times K_n \times K_s \times K_t$$

13. Calculate exact face width b from the equation:

$$b = \frac{F_t \left(K_o K_l K_m \right)}{\sigma_o m Y K K_v} \tag{5.30}$$

Check, if this face width b lies between $9.5\,m$ to $12.5\,m$ but should not be $< 7m$ or $>15m$.

14. Calculate surface fatigue strength S_{es} from Equation (5.20): $S_{es} = 2.76\,\text{BHN} - 70$

15. Calculate corrected hertz strength S_H from Equation (5.21) $S_H = \dfrac{C_L C_H}{C_T C_R} S_{es}$

16. Calculate elastic coefficient C_p from values of E and μ.

17. Calculate value of I from the value of pressure angle Φ and gear ratio G.

18. Calculate maximum hertz stress from Equation (5.17): $\sigma_H = C_p \sqrt{\dfrac{F_t}{K_v b d_p I}}$

19. Check that $\sigma_H > S_H$. If not, re-iterate.

Example 5.8

Forces on tooth and radial force on bearings

A spur gear set having gear ratio 5 and pressure angle 25° with approximate center distance 360 mm transmits 40 kW power at 1,500 rpm through pinion. Find:

 a. Module without any interference.

 b. Pitch circle diameters of pinion and gear.

 c. If gear width is $7\,m$, gear tooth pressure/mm width of gear.

 d. Total load on bearings.

Solution

Given $G = 5$ $\Phi = 25°$ $C \cong 360\,\text{mm}$ $P = 40\,\text{kW}$ $N_p = 1,500\,\text{rpm}$ $b = 7m$

 a. Minimum number of teeth for no interference is given by the following equation (4.16c) (Chapter 4). For $k = 1$

$$Z_{pmin} = \frac{2k}{G\sqrt{1 + \dfrac{1}{G}\left(\dfrac{1}{G} + 2\right) \sin^2 \Phi} - G} = \frac{2}{5\sqrt{1 + \dfrac{1}{5}\left(\dfrac{1}{5} + 2\right) \sin^2 25} - 5} = 10.3 \text{ say } Z_p = 11$$

Hence number of teeth on gear $Z_g = G \times Z_p = 5 \times 11 = 55$

Center distance, $C = \dfrac{d_g + d_p}{2}$

Substituting the values: $225 = \dfrac{5d_p + d_p}{2}$ Or, $d_p = \dfrac{450}{6} = 120\,mm$

Module, $m = \dfrac{d_p}{Z_p} = \dfrac{120}{11} = 10.9$ selecting a standard module $m = 11$

b. Pinion diameter based on 11 module, $d_p = 11 \times 11 = 121\,mm$

Hence gear diameter, $d_g = 5 \times d_p = 5 \times 121 = 605\,mm$

c. Torque $T = \dfrac{P \times 60{,}000}{2\pi N} = \dfrac{40 \times 60{,}000}{2 \times 3.14 \times 1{,}500} = 254.77\,Nm = 254{,}770\,Nmm$

Tangential force $F_t = \dfrac{T}{0.5 d_p} = \dfrac{254{,}770}{0.5 \times 121} = 4{,}211\,N$

Normal force $F_n = \dfrac{F_t}{\cos\Phi} = \dfrac{4{,}211}{\cos 25°} = 4{,}646\,N$

Gear width $b = 7\,m = 7 \times 11 = 77\,mm$

Pressure/mm width, $p = \dfrac{4{,}646}{77} = 60\,N/mm$

d. Radial force $F_r = F_n \sin\Phi = 4{,}646 \times \sin 25° = 1{,}964\,N$

Example 5.9

Power transmission capacity of a given gear set

A spur gear pinion having 16 full depth involute teeth of module 6 mm and pressure angle 20° rotates at 1,000 rpm transmitting power to a mating gear having 64 teeth and width 10 m. If safe bending strength for pinion and gear are 100 MPa and 80 MPa, respectively, find the maximum power, which this set can transmit on the basis of strength. Take form factor $y = 0.154 - \dfrac{0.912}{Z}$.

Solution

Given $Z_p = 16$ $m = 6\,mm$ $\Phi = 20°$ $N_p = 1{,}000\,rpm$
             $Z_g = 64$ $b = 10\,m$ $\sigma_p = 100\,MPa$ $\sigma_g = 80\,MPa$

Pitch circle diameter of pinion, $d_p = m \times Z_p = 6 \times 16 = 96\,mm$

Pitch line velocity $v = \dfrac{\pi d_p N_p}{60} = \dfrac{\pi \times 96 \times 10^{-3} \times 1{,}000}{60} = 5.02\,m/s$

Velocity factor $k_v = \dfrac{3}{3+v} = \dfrac{3}{3+5.02} = 0.374$

Form factor for pinion $y_p = 0.154 - \dfrac{0.912}{Z_p} = 0.154 - \dfrac{0.012}{16} = 0.097$

Form factor for gear $y_g = 0.154 - \dfrac{0.912}{Z_g} = 0.154 - \dfrac{0.912}{64} = 0.14$

For pinion $\sigma_p \times y_p = 100 \times 0.097 = 9.7$

For gear $\sigma_g \times y_g = 80 \times 0.14 = 11.2$

Pinion is weaker than gear, hence design will be based on pinion's strength.

$$Y = \pi y = 3.14 \times 0.097 = 0.305$$

Lewis equation is: $F_t = \sigma k_v b \, Y m$ Substituting the values:

$$F_t = 100 \times 0.374 \times (10 \times 6) \times 0.305 \times 6 = 4{,}106 \text{ N}$$

Power, $P = F_t \times v = 4{,}106 \times 5.02 = 20{,}612 \text{ W} = 20.6 \text{ kW}$

Example 5.10

Module for a gear drive with given approximate center distance

A spur gear set transmits 20 kW power through a pinion rotating at 300 rpm to a gear with gear ratio 3. Stub teeth with pressure angle 20° are used and approximate center distance is 500 mm. Width can be taken as 9 times the module. Safe stress for both the gears can be taken as 80 MPa. Overload factor can be taken 1.2. Calculate:
   a. Module
   b. Exact center distance

Solution

Given $P = 20 \text{ kW}$ $N_p = 300 \text{ rpm}$ $G = 3$ $\Phi = 20°$

$C = 500 \text{ mm}$ $b = 9\,m$ $\sigma_p = 80 \text{ MPa}$ $\sigma_g = 80 \text{ MPa}$

$K_o = 1.2$

a. Approximate center distance, $C = \dfrac{d_g + dp}{2} = \dfrac{(3 \times d_p) + d_p}{2} = 2d_p$

Hence, approximate pinion diameter $d_p = 0.5 \times C = 0.5 \times 500 = 250 \text{ mm}$

Approximate gear pitch circle diameter, $d_g \cong G \times d_p = 3 \times 250 \cong 750 \text{ mm}$

Pitch line velocity $v = \dfrac{\pi d_p N_p}{60} = \dfrac{\pi \times 250 \times 10^{-3} \times 300}{60} = 3.925 \text{ m/s}$

Velocity factor from Table 5.3: $k_v = \dfrac{3}{3+v} = \dfrac{3}{3+3.925} = 0.433$

Number of teeth on pinion $Z_p = \dfrac{250}{m}$

For stub teeth from Equation (5.7c), form factor is:

$y_p = 0.175 - \dfrac{0.814}{Z_p} = 0.175 - \dfrac{0.841 \times m}{250} = 0.175 - 0.0036\, m$

Tangential force, $F_t = \dfrac{P \times K_o}{v} = \dfrac{20,000 \times 1.2}{3.925} = 6,115\,\text{N}$

Since material is same for pinion and gear, $(\sigma_p \times y_p)$ will be smaller for pinion than gear. Hence, the design is done on the basis of pinion strength.

Using Lewis equations: $F_t = \sigma_p\, b\, \pi\, m\, y_p$ Substituting the values:

$6,115 = 80 \times 0.433 \times 9\, m \times 3.14\, m \times (0.175 - 0.0036\, m)$

Or, $6,115 = 171.3 m^2\, 3.29 m^3$ or $m^3 - 52 m^2 + 1,848 = 0$

Solving above equation by trial and error:

For $m = 5$, value of $m^3 - 52 m^2 + 1,848 = 673$

For $m = 6$, value of $m^3 - 52 m^2 + 1,848 = 192$

For $m = 6.5$ value of $m^3 - 52 m^2 + 1,848 = -74.4$

It means its value is in between 6 and 6.5. Selecting next higher standard value of module as $m = 7$.

Number of teeth on pinion, $Z_p = \dfrac{250}{7} = 35.7$ say 36

Hence, exact diameter of pinion: $d_p = m \times Z_p = 7 \times 36 = 252\,\text{mm}$

Exact diameter of gear, $d_g = G \times d_p = 3 \times 252 = 756\,\text{mm}$

b. Exact center distance $c = \dfrac{d_g + d_p}{2} = \dfrac{756 + 252}{2} = 504\,\text{mm}$

5.10 Internal Gears

Internal gears are also called as annular gear or ring gear as their shape is like a ring [Figure 5.4(a)]. The teeth are cut on the internal surface of the annulus. External gear is generally a driver and annulus gear is driven. The tooth and tooth space of these gears are proportioned like addendum and dedendum of standard gears in reverse direction. The size is kept slightly bigger than the standard calculations. Tooth space of internal gear is concave and almost similar to the tooth of external gear and tooth of internal gear is almost similar to tooth space of external gear mating with it. Tooth surface of internal gear is concave, whereas for external gear it is convex as shown in Figure 5.4(b).

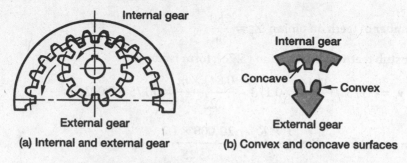

Figure 5.4 Contact of internal and external gears

5.10.1 Advantages and disadvantages

Advantages

- Teeth of internal gear wraps over the teeth of external gear, which reduces noise. It also decreases the impact effect on the tooth.
- Contact ratio is high as more than one tooth is engaged, hence higher load capacity.
- Surface stress per tooth also decreases due to many teeth in contact.
- Owing to lesser stress intensity, life of these gears is more.
- The drive is compact as the center distance is small in comparison to external gears; hence, used in planetary gear drives described in Chapter 9.
- Direction of rotation on both the gears is same, hence sliding velocity is less. This gives less wear and higher transmission efficiency.
- Outer surface of the gear is cylindrical, which acts as a guard over the meshing teeth and thus offers more safety.

Disadvantages

- Manufacturing of internal gears is difficult than external gears, hence costlier.
- Assembly of this gear set is difficult.

5.10.2 Interference in internal gears

In these gears, interference is of two types. The first one is similar to external gears in which, involute profile of one gear meshes with non-involute profile of the mating gear. Tip of the tooth of the internal gear extends into the base circle of the driver gear, where there is no involute profile. This is eliminated by modifying the shape of the tooth.

Second type of interference called "fouling" also occurs, if difference between internal gear diameter and external gear is less. It is there, because the tooth surface of the internal gear is concave and external gear convex. This can be avoided by using the generating method of manufacturing of these gears. Alternately to avoid interference, difference of number of teeth between internal and external gear is kept as under:

For 14.5° pressure angle with full depth $Z_g - Z_p = 12$

For 20° pressure angle with full depth $Z_g - Z_p = 10$

For 20° pressure angle with stub tooth $Z_g - Z_p = 8$

A thumb rule can be taken that, pitch circle diameter of external gear should not be more than 65 per cent of the internal gear pitch circle diameter.

Or, $Z_g - 1.2 Z_p = 10$.

To avoid interferences, two methods are employed. The first one is to decrease the addendum circle of internal gear and increase of external driving pinion. Addendum is changed from 1 m to 0.6m for internal gear and 1.25 m for external gear. Thus:

Addendum circle diameter (tip diameter) of internal gear is taken: $d_{ag} = m Z_g - 1.2 m$.

Addendum circle diameter of the pinion $d_{ap} = m Z_p + 2.5 m$.

The second method is to reduce the addendum of internal gear, if the difference in number of teeth is more than 10. The addendum is reduced to 0.6 m to 0.95 m, depending on number of teeth of driver pinion.

Minimum number of teeth for the pinion can be taken as given below:

Slow speed light load $Z_p = 10$

Medium speed medium load $Z_p = 12$

High speed heavy load $Z_p = 16$

5.10.3 Design of internal gears

Since the tooth of internal gear is stronger than external gear, hence the module calculated for the driver pinion is safe for the internal gear also. In design, the difference is in the ratio factor Q of Equation (5.18) for external gears is $F_w = d_p b Q K$

For internal gears, Ratio factor $\quad Q = \dfrac{2G}{G-1}$ (5.31)

Example 5.11

Design of internal gear for given power

5 kW power is to be transmitted from an external pinion rotating at 960 rpm to rotate an annular gear at 480 rpm using 20° involute teeth. Face width can be taken 10 times the module. Elastic coefficient for the gear materials is 190. If safe stress for both pinion and is 90 MPa, calculate:

a. Number of teeth for each gear

b. Module and face width

Solution

Given
$$P = 5 \text{ kW} \qquad N_p = 960 \text{ rpm} \qquad \Phi = 20° \qquad b = 10m$$
$$C_p = 190 \qquad \sigma_p = 90 \text{ MPa} \qquad \sigma_g = 90 \text{ MPa} \qquad N_g = 480 \text{ rpm}$$

Torque $T = \dfrac{60,000\,P}{2\pi N_1} = \dfrac{60,000 \times 5}{2 \times 3.14 \times 960} = 49.76 \text{ Nm}$

Assuming number of teeth for drive pinion 12 as the power is less and speed is high.

Teeth on gear from the speeds $Z_g = Z_p = \dfrac{N_p}{N_g} = 12 \times \dfrac{960}{480} = 24$

$$Z_g - Z_p = 24 - 12 = 12$$

It is more than 10, hence OK.

For 20° full depth involute teeth $y = 0.154 - \dfrac{0.912}{Z} = 0.154 - \dfrac{0.912}{12} = 0.078$

Pitch circle diameter of pinion $d_p = m\,Z_p = 12\,m$

Tangential force $F_t = \dfrac{T}{0.5 d_p} = \dfrac{49.76 \times 10^3}{0.5 \times 12m} = \dfrac{8,293}{m} \text{N}$

Velocity $v = \dfrac{\pi d_p N_p}{60,000} = \dfrac{\pi \times 12m \times 960}{60,000} = 0.6m \text{ m/s}$

Assuming milled teeth, velocity factor $K_v = \dfrac{6}{6+v} = \dfrac{6}{6+0.6m}$

Dynamic load, $F_d = \dfrac{F_t}{K_v} = \dfrac{8,293(6+0.6m)}{m \times 6} = \dfrac{8,923 + 829.3m}{m} \text{N}$

Beam strength $S_b = \sigma b\,y\,\pi m = 90 \times 10\,m \times 0.078 \times 3.14 \times m = 220.43\,m^2 \text{ N}$

Equating dynamic load with beam strength: $\dfrac{8,923 + 829.3m}{m} = 220.43m^2$

Or, $40.48 + 4.048\,m = m^3$

Solving by trial and error: $m = 3.8$

Selecting a standard module $m = 4$

Gear diameters from equations: $d_p = 4 \times 12 = 48 \text{ mm}$ and $d_g = 4 \times 24 = 96 \text{ mm}$

Check for wear:

$$I = \frac{\sin\Phi\cos\Phi}{2}\left(\frac{Gd_p}{G-1}\right) = \frac{\sin 20°\cos 20°}{2}\left(\frac{2\times 48}{2-1}\right) = 96 \text{ mm}$$

Tangential force $F_t = \dfrac{T}{0.5d_p} = \dfrac{49.76\times 10^3}{0.5\times 48} = 2{,}073.3 \text{ N}$

$$K_v = \frac{6}{6+v} = \frac{6}{6+0.6m} = \frac{6}{6+(0.6\times 4)} = 0.714$$

Maximum hertz stress $\sigma_H = C_p\sqrt{\dfrac{F_t}{K_v bd_p I}} = 190\sqrt{\dfrac{2{,}073.3}{0.714\times 40\times 48\times 96}} = 23.85 \text{ MPa}$

It is quite less, hence safe in wear.

To avoid interference, addendum is modified 1.1 m for internal gear and 1.25 m for external gear. Hence sizes are as under:

Width of both the gears $b = 10\ m = 40$ mm

External gear:

Outside diameter $= d_p + 2.5\ m = 48 + 10 = 58$ mm

Root diameter $= d_{pr} - 2.4\ m = 48 - 9.4 = 38.6$ mm

Internal gear:

Tip diameter $= d_g - 1.2\ m = 96 - 4.8 = 91.2$ mm

Root diameter $= d_{gr} - 2.4\ m = 96 - 9.4 = 86.6$ mm

Center distance $C = \dfrac{d_g - d_p}{2} = \dfrac{96 - 48}{2} = 24$ mm

5.11 Non-circular Gears

Circular gears are mainly used for transmitting power with a fixed gear ratio, whereas non-circular gears (NCGs) are used. The gear ratio varies as the driver rotates. Two simple examples are elliptical and square gears shown in Figure 5.5. In elliptical gears shown in Figure 5.5(a), speed ratio increases and then decreases twice in one revolution. Speed ratio changes as the radius of driver and driven varies with angular rotation. Their use can be seen for giving feed to the grinder of internal grinding machines. In square gears [Figure 5.5(b)], the speed variation occurs four times in one revolution. The pitch circle profile is so adjusted that the sum of radial distances of the driver and driven gear remains same equal to center distance in both the cases. In case of spiral gears [Figure 5.5(c)] for constant speed of driver, the speed of driven constantly increases in one revolution.

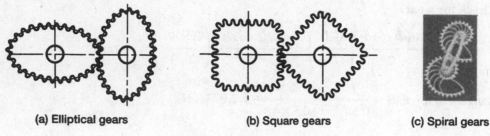

(a) Elliptical gears   (b) Square gears   (c) Spiral gears

Figure 5.5 Noncircular gears

Summary

Spur gears have straight teeth and are easy to manufacture but noisy.

Gear materials are cast iron, steel as well as non-metallic like nylon to reduce noise. Cast iron is commonly used material for big gears, due to its resistance to wear. Plain carbon steels and alloy steels offer high strength. These gears are heat treated to impart hardness and toughness.

Tensile strength of different gear materials is given in Table 5.1.

Gear design requirements Sufficient strength at static loading, strength to withstand dynamic loading under normal running conditions, minimum wear qualities for a long life, should occupy minimum space, minimum cost, minimum noise, satisfactory lubrication system, and no overheating of teeth under normal running conditions. If shafts are long, misalignment of the gears due to deflected shaft should be also considered.

Loads on gear tooth The load F_n acts normal to the tooth profile at point of contact. It can be resolved in two components; tangential and radial.

Tangential force F_t is calculated from torque $T = 0.5\ d \times F_t$.

Tangential load causes bending moment on the tooth. It is related as $F_t = F_n \cos \Phi$

Radial load F_r causes compressive stress. This being eccentric from the center line of the tooth is non-uniform over the surface. $F_r = F_n \sin \Phi$. Also $F_r = F_t \tan \Phi$

Owing to change in cross section of the tooth, stress concentration is also to be considered.

Wilfred Lewis gave an equation to simplify the varying forces and geometry of the tooth. He made some assumptions and gave the formula for tooth strength: $S_b = \sigma\, b\, y\, \pi\, m$ Where, b = Face width of tooth ($6m$ to $15\ m$), y = Lewis factor, ($Y = p\, y$ and $p_c = pm$)

Both y and ($Y = \pi y$) are dimensionless quantities and hence if size is changed, y or Y remain unchanged. It only depends on number of teeth and is calculated by the relations given below:

For 14.5° full depth involute teeth $\quad y = 0.124 - \dfrac{0.684}{Z}$

For 20° full depth involute teeth $\quad y = 0.154 - \dfrac{0.912}{Z}$

For 20° stub involute teeth $\qquad y = 0.175 - \dfrac{0.841}{Z}$

Values of form factor Y based on these relations are tabulated in Table 5.2.

Dynamic loads occur due to inaccuracy of tooth profiles, errors in tooth spacing, deflection of teeth when loaded, misaligned bearings, elasticity of parts. To account for all this, velocity factor K_v is used. It does not give accurate value of dynamic loads, hence used for a preliminary design and then final design is done using Buckingham's equation.

Velocity factor   Strength of gear depends on pitch line velocity. This is considered using a velocity factor K_v. Values of velocity factor for different pitch line velocity are:

$$v < 10 \text{ m/s and } \quad K_v = \frac{3}{3+v}, \qquad\qquad v = 10 - 15 \text{ m/s and } K_v = \frac{6}{6+v},$$

$$v = 10 - 15 \text{ m/s and } K_v = \frac{50}{50+\sqrt{200v}}; \text{ and} \qquad v > 20 \text{ m/s and } K_v = \sqrt{\frac{78}{78+\sqrt{200v}}}$$

Approximate dynamic load: $F_d = \dfrac{F_t}{K_v}$

Considering dynamic loads Lewis equation: $F_t = \sigma\, b\, Y K_v\, m$

Buckingham's equation is used for determining dynamic loalds and the hardness. An incremental

dynamic load $\;F_i = \dfrac{21v\,(C\,e\,b + F_t)}{21v + \sqrt{(C\,e\,b + F_t)}}$

Where, v = Pitch line velocity; C = Deformation factor; e = Sum of errors of two meshing gears; b = Face width in mm; and F_t = Tangential force.

If $A = (C\,e\,b + F_t)$, $\;F_i = \dfrac{21v\,A}{21v + \sqrt{A}}$

Then dynamic load $F_d = F_t + F_i$

For a design to be safe, dynamic load should be lesser than beam strength.

Beam strength (S_b)   A gear tooth experiences varying load, hence the endurance strength S_e is to be considered instead of bending strength in the Lewis equation. $S_b = \sigma_e\, b\, y\, \pi\, m$

Values of endurance strength S_e are given in Table 5.4.

For steels, endurance strength can be calculated using the relation $S_e = 1.75$ BHN

To ensure that the gear tooth does not break Buckingham recommends the following:

For steady loads $S_b \geq 1.25\, F_d$. For pulsating loads $S_b \geq 1.35\, F_d$. For shock loads $S_b \geq 1.5\, F_d$

Deformation factor C used in Buckingham's equation is calculated from the values of modulus of elasticity of the pinion and gear materials using the equation given below:

$$C = \frac{k}{\left[\dfrac{1}{E_p} + \dfrac{1}{E_g}\right]}$$

Where, k = Form tooth factor; k = 0.11 for 20°; and k = 0.115 for 20° stub teeth.

E_p, E_g = Modulus of elasticity for pinion and gear material, respectively.

Values of C are given in Table 5.5 for different material combinations, E_p, E_g in Table 5.6.

Error e in gears depends on the method, by which these are manufactured. The quality of gear manufacture is divided in 12 grades of precision. Grade 1 is of highest quality and needs grinding and grade 12 is most rough. The tolerance or the maximum error e for adjacent pitch is given in Table 5.7, which depends on the module m and a tolerance factor f. The factor f is given as: $f = m + \sqrt{d}$

The total error e of pinion e_p and of gear e_g and is $e = e_p + e_g$

Maximum allowable error e_{max}, which can be tolerated depends on pitch line velocity v and is tabulated in Table 5.8 for well-cut commercial gears.

Gear design for wear strength   Gear tooth may fail due to beam strength lesser than the allowable strength and many other reasons such as surface failure of due to wear, scoring due to lack of lubrication, abrasion due to presence of foreign particles in oil or environment, and pitting failure due to high contact stresses. Hence the hardness of the surface has to be sufficient so that this failure does not occur.

Hertz stresses on tooth surface

When two cylinders of diameters $2r_1$ and $2r_2$ are pressed against each other, contact stresses called hertz stresses are developed at the contact line.

Maximum hertz stress $\sigma_H = \sqrt{\dfrac{2F_t}{\pi b \cos\Phi}\left[\dfrac{\dfrac{1}{r_p}+\dfrac{1}{r_g}}{\dfrac{1-\mu_p^2}{E_p}+\dfrac{1-\mu_g^2}{E_g}}\right]} = C_p\sqrt{\dfrac{F_t}{K_v\, b d_p\, I}}$

Where, for external gears $I = \dfrac{\sin\Phi\,\cos\Phi}{2}\left(\dfrac{G d_p}{G+1}\right)$

For internal gears $I = \dfrac{\sin\Phi\,\cos\Phi}{2}\left(\dfrac{G d_p}{G-1}\right)$

and elastic coefficient be $C_p = \sqrt{\dfrac{1}{\pi\left(\dfrac{1-\mu_p^2}{E_p}+\dfrac{1-\mu_g^2}{E_g}\right)}}$

Units of C_p are $\sqrt{MPa}$ and its values are given in Table 5.9.

E_p and E_g = Modulus of elasticity for pinion and gear material, respectively.

Wear load   Buckingham gave an equation for maximum wear load $F_w = d_p\, b\, Q\, K$.

Where, d_p = Pitch circle diameter of pinion; b = Face width; Q = Ratio factor $= \dfrac{2G}{G+1}$

Load stress factor $K = \dfrac{(S_{es})^2 \sin\Phi\cos\Phi}{1.4}\left(\dfrac{1}{E_p}+\dfrac{1}{E_g}\right)$

Where, S_{es} = Surface endurance strength

For the design to be safe, wear load F_w should be less than the dynamic load F_d

Surface fatigue strength S_{es} for steels in megapascals is $S_{es} = 2.76 \, BHN - 70$

Where, BHN = Brinell Hardness Number of the softer material.

Values of surface endurance strength for different materials are given in Table 5.10.

A life of 10^8 stress cycles has been assumed. AGMA recommends that this surface fatigue strength should be modified as corrected surface fatigue strength as $S_H = \dfrac{C_L C_H}{C_T C_R} S_{es}$

C_L = Life factor; C_H = Hardness factor; C_T = Temperature factor; and C_R = Reliability factor.

Values of life and hardness factors can be taken from Table 5.11.

Factors affecting gear design

Overload factor (K_o)   Considers shocks coming on the driver and driven gears (Table 5.12).

Load distribution factor (K_l)   Considers change in load distribution due to misalignment (Table 5.13).

Mounting factor (K_m)   It considers the effect of misalignment in meshing the gears (Table 5.14).

Surface finish factor (K_f)   Wear life of a gear depends on surface finish also (Table 5.15).

Rotation factor (K_n)   For one side rotation $K_n = 1.4$. For reversed rotation $K_n = 1.0$

Reliability factor (K_r)   Reliability factor K_r is to be considered (Table 5.16).

Size factor (K_s)   It is considered using a size factor K_s given in Table 5.17.

If $m < 1$ then $K_s = 1$. If $m > 1$ then $K_s = 1.189 \times (\pi m)^{-0.097}$

Temperature factor (K_t)   For temperature $\leq 70°$ C, $K_t = 1$. For $> 70°$ C, $K_t = \dfrac{344}{273 + t}$

Design procedure with given center distance

A.   Calculate stress for static load

1.   Calculate torque $T = \dfrac{60,000 P}{2 \pi N_p}$; N_p = Speed of the driver gear (rpm).

2.   Calculate gear ratio $G = N_p / N_g$

3.   Calculate pinion and gear diameters d_p and d_g. $d_p = \dfrac{2C}{1 + G}$ and $d_g = G \, d_p$

4.   Calculate tangential force $F_t = \dfrac{2T \times 10^3}{d_p}$

5.   Calculate pitch line velocity $v = \dfrac{\pi d_p N_p}{60,000}$

6.   Calculate velocity factor K_v for the pitch line velocity v using suitable formula from Table 5.3.

7.   From the given service conditions for shock, find overload factor K_o from Table 5.12.

8.   Find load distribution factor K_l from Table 5.13, mounting factor K_m from Table 5.14.

9. Select Φ; 14.5° for light loads, 20° for moderate loads, and 25° for heavy loads.

10. Find minimum number of teeth on pinion $Z_{pmin} = \dfrac{2k/G}{\sqrt{G^2 + (1 + 2G)\sin^2 \Phi} - 1}$

11. Calculate module $m = \dfrac{d_p}{Z_{min}}$

12. Choose a standard value of module from Table 5.18 given below.

13. Calculate Surface finish factor K_f from Table 5.15, reliability factor K_r from Table 5.16, rotation factor K_n, size factor K_s from Table 5.17, and temperature factor K_t.

14. Calculate factor $K = K_f \times K_r \times K_n \times K_s \times K_t$

15. Calculate stress from relation: $\sigma_o = \dfrac{F_t \left(K_o\,K_i K_m\right)}{m\,Y\,b\,K\,K_v}$.

B. Check for dynamic load

16. For $K = 0.107$ to 0.115 depending upon pressure angle, calculate deformation factor $C = \dfrac{k}{\left[\dfrac{1}{E_p} + \dfrac{1}{E_g}\right]}$

17. Calculate $A = Ceb + F_t$

18. Calculate incremental load $F_i = \dfrac{21v\,A}{21v + \sqrt{A}}$

19. Calculate dynamic load $F_d = F_t + F_i$

20. Calculate tolerance factor for pinion and gear using $f = m + \sqrt{d}$

21. Calculate error for pinion e_p and gear e_g. Then total error: $e = e_p + e_g$

22. This value of error e should be lesser than the maximum allowable error e_{max} given in Table 5.8

C. Check for hertz stresses

23. Calculate $I = \dfrac{\sin\Phi\,\cos\Phi}{2}\left(\dfrac{G\,d_p}{G + 1}\right)$

24. Calculate elastic coefficient $C_p = \sqrt{\dfrac{1}{\pi\left(\dfrac{1 - \mu_p^2}{E_p} + \dfrac{1 - \mu_g^2}{E_g}\right)}}$

25. Calculate hertz stress $\sigma_H = C_p\sqrt{\dfrac{F_t}{K_v\,b\,d_p\,I}}$. Strength S_H should be more than σ_H.

D. Check for wear

26. Calculate ratio factor Q from the equation: $Q = \dfrac{2G}{G + 1}$

27. Calculate load stress factor $K = \dfrac{(S_{es})^2\,\sin\Phi\cos\Phi}{1.4}\left(\dfrac{1}{E_p} + \dfrac{1}{E_g}\right)$

28. Calculate surface fatigue strength $S_{es} = (2.76\ \text{BHN} - 70)$

29. Calculate corrected surface fatigue strength: $S_H = \dfrac{C_L C_H}{C_T C_R} S_{es}$, C_L and C_H – (Table 5.11).

30. Calculate load capacity in wear $F_w = d_p\, b\, Q\, K$. F_w should be more than F_d.

Design procedure when center distance is not given

1. Calculate torque $T = \dfrac{60{,}000\,P}{2\pi N_1}$

2. Assume pressure angle and then calculate minimum number of teeth $Z_{min} = Z_p$

3. Calculate teeth on gear from the speeds $Z_g = Z_p \dfrac{N_p}{N_g}$

4. Assume face width $b = 10\ m$.

5. Calculate form factor Y from equations as applicable.

6. Select velocity factor K_v depending on gear manufacturing from Table 5.3.

7. Calculate module from: $\sigma_o = \dfrac{2T}{(10m)(mZ_p)mY K_v}$. Take standard module.

8. Calculate pinion and gear diameters $d_p = m\,Z_p$ and $d_g = m\,Z_g$

9. Calculate pitch line velocity from equation $v = \dfrac{\pi d_p N_p}{60{,}000}$

10. Calculate exact velocity factor K_v

11. Calculate coefficients K_o, K_i and K_m.

12. Calculate coefficients K_f, K_r, K_n, K_s, K_t and then $K = K_f \times K_r \times K_n \times K_s \times K_t$

13. Calculate exact face width $b = \dfrac{F_t (K_o K_i K_m)}{\sigma_o m Y K K_v}$

    Check if this face width b lies between $9.5\ m$ and $12.5\ m$ but should not be $< 7m$ or $> 15m$.

14. Calculate surface fatigue strength $S_{es} = 2.76\ \text{BHN} - 70$

15. Calculate corrected hertz strength $S_H = \dfrac{C_L C_H}{C_T C_R} S_{es}$

16. Calculate elastic coefficient C_p from values of E and μ.

17. Calculate value of I from the value of pressure angle Φ and gear ratio G.

18. Calculate maximum hertz stress $\sigma_H = C_p \sqrt{\dfrac{F_t}{K_v\, b d_p\, I}}$

19. Check that $\sigma_H > S_H$. If not, reiterate.

Internal gears are also called as annular gear or ring gear. The teeth are cut on the internal surface of the annulus. The size is kept slightly different than the standard calculations. **Tooth space** of internal gear is almost similar to the tooth of external gear. Tooth space of internal gear is concave, whereas for external gear is convex.

Advantages of internal gears are greater contact ratio, which results in reduced noise, compact drive, less stress on teeth, longer life, and smooth working.

Disadvantages are that cost of manufacturing is more than external gears.

Interference is avoided by modifying the tooth addendum and keeping the difference between gear and pinion number of teeth more than 10–16 depending on pressure angle.

Design procedure for the internal gears is same as for external gears as teeth of internal gears for the same module are stronger than external gear. The difference is only for ratio factor, which is $Q = \dfrac{2G}{G-1}$ for these gears.

Noncircular gears are used where the velocity ratio has to vary with rotation of the driver gear. The pitch profile can take the shape of an ellipse or square with rounded corners. The profile is so adjusted that the sum of radial distance of the driver and driven remains equal to the center distance. Spiral gears are used for continuous change in speed.

Theory Questions

1. Describe the various gear materials and their approximate hardness and strength.
2. What parameters are considered, while designing a gear?
3. Describe the various loads coming on a gear tooth. How are these related with the tangential force coming on tooth?
4. Describe the assumptions made in deriving the Lewis equation? Discuss all the terms used in this equation?
5. Explain Lewis form factor. What are the parameters, which affect its value?
6. What are the parameters causing dynamic loads on gear tooth? How are these considered in gear design?
7. Discuss 'velocity factor' and the various parameters affecting its value.
8. Discuss Buckingham's equation for dynamic loads.
9. Describe 'deformation factor' and the parameters affecting this factor.
10. Differentiate between the gear design on strength basis and design for wear.
11. How the hertz stresses are considered in gear design for wear?
12. What are the various correction factors considered in gear design for strength?
13. Describe stepwise the method used for gear design when center distance is given.
14. If the center distance is not given, how will you design a spur gear?
15. Write construction of an internal gear. What are the advantages and disadvantages of these gears over external gears?
16. Describe interference in an internal gear drive. What measures are taken to avoid interference?
17. Describe construction of noncircular gears with sketches.

Multiple Choice Questions

1. Normal force on a gear tooth is normal to the:
   (a) Line of gear centers
   (b) Surface of pinion tooth only
   (c) Surface of gear tooth only
   (d) Surface of both the teeth

2. Lewis equation for gears is applicable for strength of:
   (a) Pinion only
   (b) Gear only
   (c) Weaker of the gear set
   (d) Stronger of the gear set

3. For a given safe stress, strength of a gear tooth depends on:
   (a) Width, velocity, module, and number of teeth of pinion
   (b) Rotational speed, circular pitch, and number of teeth of gear
   (c) Pressure angle, width, velocity, and module
   (d) Diametrical pitch, width, velocity, and addendum

4. Form factor of a gear tooth depends on:
   (a) Only pressure angle
   (b) Only module
   (c) Pressure angle and number of teeth
   (d) None of above

5. Deformation factor is a function of:
   (a) Modulus of elasticity for pinion
   (b) Modulus of elasticity of both the gears and form tooth factor
   (c) Form tooth factor and allowable safe tensile stress
   (d) Safe stress and pressure angle

6. Pitting on gear tooth is due to:
   (a) Abrasion over tooth
   (b) Lack of lubrication
   (c) High contact stresses
   (d) Wear of surface

7. Buckingham's equation is used to find:
   (a) Wear load
   (b) Dynamic load
   (c) Strength of tooth
   (d) Life of gear

8. For a gear ratio G and center distance C, diameter of pinion is:
   (a) $\dfrac{2C}{1+G}$
   (b) $\dfrac{2C}{G}$
   (c) $\dfrac{G}{C+1}$
   (d) $2C + G$

9. To avoid interference for internal gear drive, addendum in terms of module is kept:
   (a) $0.9\,m$
   (b) $0.8\,m$
   (c) $0.7\,m$
   (d) $0.6\,m$

Answers to multiple choice questions

1. (d) 2. (c) 3. (a) 4. (c) 5. (b) 6. (c) 7. (b) 8. (a) 9. (d)

Design Problems

1. A pinion of pressure angle is 20° with 18 milled involute teeth transmits 15 kW at 1,440 rpm. The driven gear has 45 teeth with thickness 12 m. Ultimate strength of pinion and gear are 410 MPa and 200 MPa respectively. Design the gears assuming factor of safety as 1.5 and starting torque of the driver as 1.5 times the normal running torque.　　$[m = 5, d_p = 90\,mm, d_g = 225\,mm, b = 60\,mm]$

2. A power of 10 kW is being transmitted through a full depth 15 involute teeth rotating at 600 rpm driving a gear with steady load, and gear ratio 3. Face width can be taken 12 times the module and pressure angle 20°. Allowable strengths for pinion and gear are 100 MPa and 80 MPa respectively. Endurance strength is 600 MPa. Yong's modulus of elasticity for pinion and gear are 220 GPa and 100 GPa respectively. Velocity factor can be taken as: $\dfrac{4.5}{4.5 + v}$. Design the gears and factor of safety

   in wear.　　$[\,m = 5, d_p = 90\,mm, d_g = 225\,mm, b = 60\,mm, FOS = 1.28]$

3. A gear set with gear ratio 2.5 transmits 30 kW through a pinion having casted 14 teeth with pressure angle 20° at 900 rpm with a load factor 1.2. Allowable strengths for pinion and gear are 125 MPa and 95 MPa respectively. Assuming face width as 12 m, find the module, effective load and the center distance.　　$[\,m = 7, F_e = 20.47\,kN, C = 171.5]$

4. Design a pair of gear set of straight teeth with pressure angle 20° has to transmit 12 kW at 1,400 rpm with a speed reduction of 3. Starting torque of the driver can be assumed 140 per cent of normal torque. Use a FOS 3 for the Lewis equation. Both pinion and gear are made of carbon steel having ultimate strength 570 N/mm². Suggest a suitable hardness number.
   $$[m = 4, Z_p = 15, Z_g = 45, d_p = 60\ mm, d_g = 180\ mm, b = 40\ mm, BHN = 270]$$

5. A pinion of pressure angle is 20° with 18 milled involute teeth transmits 15 kW at 1,440 rpm. The driven gear has 45 teeth. Ultimate strength of pinion and gear are 410 MPa and 200 MPa, respectively. Design the gears assuming FOS as 1.2 and starting torque of the driver as 1.3 times the normal running torque.　　$[\,m = 5, d_p = 90\ mm, d_g = 225\ mm, b = 50\ mm]$

6. A gear set of 20 and 45 teeth, having full depth with pressure angle 20° is used to transmit 10 kW power at 1,000 rpm to drive a compressor. Service factor can be taken as 1.5. Module is 5 mm and width is 8 times the module. If ultimate strength of pinion / gear material is 580 MPa, calculate:

   (a) FOS on the basis of strength.　　　　(b) Suitable hardness.

   $[(a)\ 1.62\ (b)\ BHN = 298]$

7. A gear set of steel has allowable strength 200 MPa and BHN 410 MPa with the following data:

Service factor	1.6
Number of teeth on pinion	22
Number of teeth on gear	48
Module	4
Pressure angle	20°
Speed of pinion	1,440 rpm
Face width	36 mm

   Calculate the power, which can be transmitted safely with FOS for dynamic loads 1.5.

   $[P = 7.9\ kW]$

8. Design gears of a gear box driven by an electric motor at 1,440 rpm to drive a compressor at 360 rpm. The compressor requires 4.75 kW of power. Starting torque is 1.5 times the normal torque. The center distance can be taken approximately 200 mm. Ultimate strength of material is 690 MPa and take a FOS 3. Deformation factor $C = 11,400$ and width 12 times the module. Form factor initially to calculate module can be assumed 0.35. Tolerance grade can be taken 5. What is the dynamic load? Design the gear set with factor of safety in strength as 1.8.

$$[m = 3, b = 36 \text{ mm}, Z_p = 27, Z_g = 108, d_p = 81 \text{ mm}, d_g = 324 \text{ mm}, F_d = 4,266 \text{ N}]$$

Previous Competition Examination Questions

IES

1. Two spur gears of 20° full depth involute system are transmitting motion with a gear ratio of 2. The ratio of the base circle radii of the gears would be:

   (a) 0.5

   (b) 0.68

   (c) 0.72

   (d) 1.0

   [Answer (a)] [IES 2013]

GATE

2. Gear 2 rotates at 1,200 rpm in counter clockwise direction and engages with gear 3. Gear 3 and 4 are mounted on the same shaft. Gear 5 engages with gear 4. The number of teeth on gears 2, 3, 4, and 5 are 20, 40, 15, and 30, respectively. The angular speed of gear 5 is.

   (a) 300 rpm counter-clockwise

   (b) 300 rpm clockwise

   (c) 4,800 rpm clockwise

   (d) 4,800 rpm counter-clockwise

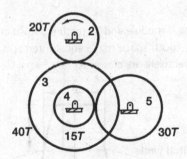

[Answer: (a)] [GATE 2014]

3. A spur pinion of pitch diameter 50 mm rotates at 200 rad/s and transmits 3 kW power. The pressure angle of the tooth of the pinion is 20°. Assuming that only one teeth is in contact, the total force (in newton) exerted by the tooth of pinion on the tooth of gear is _____

   [$F = 638.5$ N] [GATE 2014]

4. A pair of spur gear with module 5 mm and a center distance of 450 mm is used for a speed reduction of 5:1. The number of teeth on pinion is _____

   [$Z_p = 30$] [GATE 2014]

5. Which one of the following is used to convert a rotational motion into translation motion?

    (a) Bevel gears   (b) Double helical gears

    (c) Worm gears   (d) Rack and pinion

    [Answer (d)] [GATE 20014]

6. For the given statements:

    I.   Mating spur gear teeth is an example of higher pair.

    II. A revolute joint is an example of lower pair.

    Indicate the correct answer:

    (a) Both I and II are false   (b) I is true and II is false

    (c) II is true and I is false   (d) Both I and II are true

    [Answer (d)] [GATE 20014]

Linked Answer Questions 7 and 8

A 20° full depth involute spur pinion of 4 mm module and 21 teeth is to transmit 15 kW at 960 rpm. Its face with is 25 mm.

7. The tangential force transmitted (in newton) is:

    (a) 3,552   (b) 2,611

    (c) 1,776   (d) 1,305

    [Answer (a)] [GATE 2009]

8. Given that the geometry factor is 0.32 and the combined effect of the dynamic load and ailed factors intensifying the stress is 1.5, the minimum allowable stress in megapascals for the gear is:

    (a) 242.0   (b) 166.5

    (c) 121.0   (d) 74.0

    [Answer (b)] [GATE 2009]

9. One tooth of a gear having 4 module and 32 teeth is shown in Figure 5.E1. Assume that the gear tooth and corresponding tooth space make equal intercepts on the pitch circumference. The dimensions 'a' and 'b', respectively, are closest to:

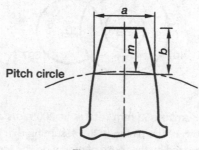

Figure 5.E1

    (a) 6.08 mm, 4 mm   (b) 6.48 mm, 4.2 mm

    (c) 6.28 mm, 4.3 mm   (d) 6.28 mm, 4.1 mm

    [Answer (d)] [GATE 2008]

10. A spur gear has a module of 3 mm, number of teeth 16, a face width of 36 mm, and a pressure angle of 20°. It is transmitting a power of 3 kW at 20 rev/s. Taking a velocity factor of 1.5 and a form factor of 0.3, the stress in the gear tooth is about:

   (a) 32 MPa

   (b) 46 MPa

   (c) 58 MPa

   (d) 70 MPa

   [Answer (b)] [GATE 2008]

Linked Answer Questions 11–13

A gear set has a pinion with 20 teeth and a gear with 40 teeth. The pinion runs at 30 rev/s and transmits a power of 20 kW. The teeth are on the 20° full depth system and have a module of 5 mm. The length of line of action is 19 mm.

11. The center distance for the above gear set in millimeters is:

   (a) 140

   (b) 150

   (c) 160

   (d) 170

   [Answer (b)] [GATE 2007]

12. The contact ratio of the contacting tooth is:

   (a) 1.21

   (b) 1.25

   (c) 1.29

   (d) 1.33

   [Answer (c)] [GATE 2007]

13. The resultant force on the contacting gear tooth in newton is:

   (a) 77.23

   (b) 212.0

   (c) 2,258.1

   (d) 289.43

   [Answer (c)] [GATE 2007]

14. 20° full depth involute profiled 19 teeth pinion and 37 teeth gear are in mesh. If the module is 5, the center distance between the gears pair will be:

   (a) 140 mm

   (b) 150 mm

   (c) 280 mm

   (d) 300 mm

   [Answer (a)] [GATE 2006]

15. Two mating spur gears have 40 and 120 teeth, respectively. The pinion rotates at 1,200 rpm and transmits a torque of 20 Nm. The torque transmitted by gear is:

   (a) 6.6 Nm

   (b) 20 Nm

   (c) 40 Nm

   (d) 60 Nm

   [Answer (d)] [GATE 2004]

Common Data Questions 16 and 17

The overall gear ratio in a two stage reduction gear box (with all spur gears) is 12. The input and output shafts of the gear box are co-linear. The counter shaft which is parallel to the input shaft and output shaft has a gear (Z_2 teeth) and pinion ($Z_3 = 15$ teeth) to mesh with pinion ($Z_1 = 16$ teeth) on the input shaft and gear (Z_4 teeth) on the output shaft, respectively. It was decided to use a gear ratio 4 with 3 module in the first stage and 4 module in the second stage.

16. Z_2 and Z_4 are:
    (a) 64 and 45
    (b) 45 and 64
    (c) 48 and 60
    (d) 60 and 48

[Answer (a)] [GATE 2003]

17. The center distance in the second stage is:
    (a) 90 mm
    (b) 120 mm
    (c) 160 mm
    (d) 240 mm

[Answer (b)] [GATE 2003]

❏❏❏

Helical Gears

Outcomes

➤ Learn about load capacity, noise, and hand of helical gears

➤ Terms used for helical gears and types of helical gears

➤ Face width, overlap, normal module,and gear tooth proportions

➤ Equivalent number of teeth

➤ Forces on gear tooth and design of helical gears

➤ Effective and dynamic load

➤ Wear strength of helical gears

6.1 Introduction to Helical Gears

Like spur gears, helical gears shown in picture on right side are also used for transmitting power between two parallel shafts. In helical gears, teeth are cut in the form of a helix on their pitch cylinder at an angle called helix angle. The helix can be right hand or left hand. Helix is called right hand, if its tooth advances in clockwise direction. In left hand helix, its tooth advances in anticlockwise direction. These gears can be used for high speeds even up to 50 m/s. These gears require  accurate machining and better lubrication for the progressive contact of the mating surfaces and increased sliding action. Salient features of helical gears over spur gears are as under:

- **Load capacity**   In spur gears, entire tooth face comes into contact with the mating gear, which causes sudden load, that is, impact loading. In helical gears, contact with the mating gear starts at the leading edge and then extends gradually along the tooth. This results in lesser stresses in helical gears than spur gears and hence they can take greater loads.

- **Noise** Owing to gradual engagement, these gears offer silent operation, whereas spur gears are noisy due to sudden contact of the teeth.
- **Axial loads** Owing to inclination of teeth, helical gear causes axial loads and hence needs thrust bearings to take these loads. Spur gears do not cause any axial load.
- **Hand of helix** Spur gears mesh, if they have the same module, whereas for helical gears to mesh, they should have the same module but hand of helix of driver should be opposite to the driven, that is, if driver has right hand helix, driven gear should have left hand helix.

6.2 Terminology for Helical Gears

Pitch circle diameter (PCD) of a helical gear depends on helix angle also. Addendum, dedendum, and other proportions remain same as that for spur gears.

Most of the terms used for spur gears are also used for helical gears, but some terms given below are related to helical gears:

Helix angle (α)
: Teeth are in the form of a helix on their pitch cylinder at an angle called helix angle. It is the angle made by helix with the axis of rotation.

Hand of helix
: These gears have a hand of helix; right hand and left hand.

Module (m)
: For helical gears, it is called transverse module and is the ratio of number of teeth to PCD. That is $m = Z / d$.

Normal module (m_n)
: Owing to helix angle α, normal module is defined as:

$$m_n = m \cos \alpha \tag{6.1}$$

Normal pressure angle (Φ_n)
: It is the angle of load, normal to the tooth surface. It is related with pressure angle Φ by the following relation:

$$\tan \Phi_n = \tan \Phi \cos \alpha \tag{6.2}$$

Face width (b)
: It is the width of gear along the axis and not along the tooth. Its value varies from $12\,m$ to $20\,m$.

Tooth length
: Unlike spur gears, it is not equal to face width. It is the length of a tooth along the tooth and it is equal to $b \sec \alpha$.

Circular pitch (p_c)
: It is the distance between two adjacent faces of teeth at PCD measured at any one face of the gear. It is same as transverse pitch. $P_c = \pi m$

Normal pitch (p_n)
: It is the distance between two adjacent faces of teeth at PCD measured normal to teeth at any one face of the gear (Figure 6.1). Normal pitch and circular pitch are related as under:

$$P_n = p_c \cos \alpha = \pi m \cos \alpha \tag{6.3}$$

Axial pitch (p_a)   It is the distance along the axis of the gear from its face, to the point where the tooth axis cuts an axis parallel to axis of gear (See Figure 6.1).

$$p_a = \frac{\pi m}{\tan \alpha} \qquad (6.4)$$

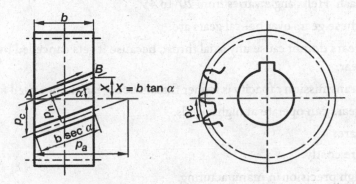

Figure 6.1 Various pitches in a helical gear

6.3 Types of Helical Gears

a. **Parallel helical gears**   These gears are used for shafts, whose axes are parallel. Helix of the driver is opposite to the driven [Figure 6.2(a)]. Helix angle α varies from 15° to 30°. Pictorial view of these gears is also shown in Figure 6.2(b).

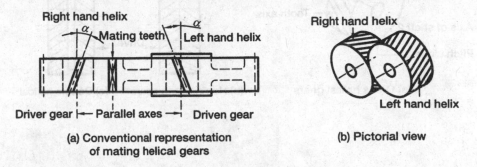

(a) Conventional representation
of mating helical gears

(b) Pictorial view

Figure 6.2 Parallel axis helical gears

b. **Crossed helical gears**   These gears are used for shafts that are crossed, that is, not parallel. Their teeth may have same or different hand of helix depending on the angle between the axes of the shafts [See Figure 6.3 (a) and picture also for crossed helical gears]. For 90° angle between shafts, helix angle of both the gears is same as 45° and use the same hand. Salient features for these gears are:

Contact area   Owing to curvature of the contacting surfaces, tooth does not have line contact, but it is a point contact.

Load capacity   Owing to small contact area of these gears, load capacity is low.

Wear rate   Owing to small contact area and high loads, wear is also high.

c. **Herringbone gears**   Two helical gears of opposite and equal value of hand of helix and module are joined face to face as shown in Figure 6.3(b) and in its picture also. Both the halves are integral and cut by two cutters reciprocating 180° out of phase to avoid clash. Helix angle varies from 20° to 45°.

Advantages of these gears over helical gears are:

- These gears do not cause any axial thrust, because it gets cancelled by the other half of the gear.
- Power transmission capacity is higher than helical gears, hence used for ship drives.
- These gears can operate at high speeds.

Disadvantages are:

- These are costly.
- Need high precision in manufacturing.

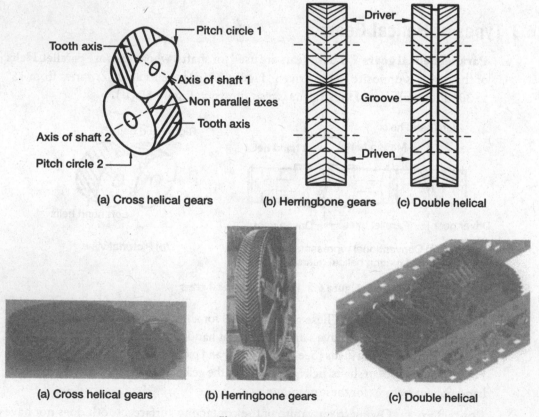

(a) Cross helical gears   (b) Herringbone gears   (c) Double helical

(a) Cross helical gears   (b) Herringbone gears   (c) Double helical

Figure 6.3 Types of helical gears

d. **Double helical gears** These are similar to herringbone gears, but there is a groove between the two halves as shown in Figure 6.3(c). Helix angle varies from 23° to 30°.

6.4 Face Width and Overlap

An inclined tooth of a helical gear is shown in Figure 6.1 in section 6.2. Arrow shows the direction of contact movement. Point A on the tooth comes into contact with the mating gear first and is called the leading edge. As the pinion rotates, the point of contact advances and it just leaves the contact at point B at the other face of the gear called the trailing edge. The distance from point A to point B in the direction of rotation is called overlap X. It is the distance along the circumference from one end of the tooth to the other end. This should be more than the circular pitch, to ensure that the other tooth comes in contact before one leaves. It indicates how many teeth will be in contact.

$$\text{Overlap}, X = b \tan \alpha \tag{6.5}$$

Where, b = Face width

α = Helix angle

From Equation (6.5) overlap increases with increase in helix angle, but increase in helix angle causes more axial thrust. Hence a compromise is to be made between the two values.

Therefore it should be about 15 per cent of circular pitch, that is, overlap = 1.15 p_c.

Thus face width,

$$b = \frac{1.15 p_c}{\tan \alpha} = \frac{1.15 \pi m}{\tan \alpha} \tag{6.6}$$

Generally, face width varies from 12 m to 20 m.

6.5 Gear and Tooth Proportions

$$\text{Pitch circle diameter}, d = \frac{Z \times p_n}{\pi \times \cos \alpha} = \frac{Z m_n}{\cos \alpha} = Zm \tag{6.7}$$

Where, m_n = Normal module

Z = Number of teeth

Other dimensions are:

Addendum $a = m_n$

Dedendum $d = 1.25\, m_n$

Outside diameter = PCD + 2 m_n

Root diameter = PCD − 2.5 m_n

Face width = 12 m to 20 m

Fillet radius = 0.35 m_n

Example 6.1

Tooth proportions of a helical gear for given module

A helical pinion having 18 teeth drives a pairing helical gear having 45 teeth. Normal module is 4, normal pressure angle is 20°, and helix angle 30°. Calculate:

a. Axial, normal and transverse pitch

b. Pitch circle diameters

c. Center distance

d. Outside diameters

e. Root diameters

Solution

Given $Z_p = 18$ $Z_g = 45$ $m_n = 4$ $\Phi_n = 20°$ $\alpha = 30°$

a. Transverse module, $m = m_n \sec \alpha = 4 \times \sec 30° = 4 \times 1.154 = 4.62$ mm

Transverse pressure angle $\tan \Phi = \tan \Phi_n \times \cos \alpha = \tan 20° \times \cos 30°$

Or, $\tan \Phi = 0.364 \times 0.866 = 0.315$ Or, $\Phi = 17.5°$

Axial pitch $p_a = \dfrac{\pi m}{\tan \alpha} = \dfrac{3.14 \times 4}{\tan 30°} = \dfrac{3.14 \times 4}{0.577} = 21.77$ mm

Transverse pitch $p_t = p_c = \pi m = 3.14 \times 4.62 = 14.5$ mm

Normal pitch $p_n = p_t \cos \alpha = 14.5 \times \cos 30° = 12.56$ mm

b. PCD of pinion $d_p = \dfrac{Z_p m_n}{\cos \alpha} = \dfrac{18 \times 4}{\cos 30°} = 83.1$ mm

PCD of gear $d_g = \dfrac{Z_g m_n}{\cos \alpha} = \dfrac{45 \times 4}{\cos 30°} = 207.8$ mm

c. Center distance $C = \dfrac{d_p + d_g}{2} = \dfrac{83.1 + 207.8}{2} = 145.5$ mm

d. Addendum $a = m_n = 4$ mm

Outside diameter of pinion $d_{op} = d_p + 2a = 83.1 + 8 = 91.1$ mm

Outside diameter of gear $d_{og} = d_g + 2a = 207.8 + 8 = 215.8$ mm

e. Dedendum $d = 1.25 \times m_n = 1.25 \times 4 = 5$ mm

Root diameter of pinion $d_{rp} = d_p - 2d = 83.1 - 10 = 73.1$ mm

Root diameter of gear $d_{rg} = d_g - 2d = 207.8 - 10 = 187.8$ mm

6.6 Equivalent Number of Teeth of Helical Gears

When a helical gear is cut, the gear blank is turned by helix angle on the machine. Thus the profile of the involute cutter is cut at a plane A–B normal to the helix angle. Intersection of pitch cylinder and plane A–B produces an ellipse, whose major axis is shown by dashed line in Figure 6.4.

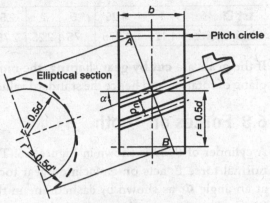

Figure 6.4 Formulative gear

Radius r' is given as:

$$r' = \frac{d}{2\cos^2 \alpha} \qquad (6.8)$$

Where, d = PCD

r' = Pitch circle radius of imaginary circle

α = Helix angle

The number of teeth on this imaginary cylinder is called equivalent or virtual formulative teeth denoted by Z'. These are related as:

$$Z' = \frac{2\pi r'}{p_n} = \left(\frac{2\pi}{p_n}\right) \times \left(\frac{d}{2\cos^2 \alpha}\right) = \frac{d}{m_n \cos^2 \alpha} \qquad (6.9)$$

Substituting the value of $d = m\,Z$

$$Z' = \frac{mZ}{m_n \cos^2 \alpha}$$

Substituting the value of m_n from Equation (6.1):

$$Z' = \frac{mZ}{(m\cos\alpha)\cos^2 \alpha} = \frac{Z}{\cos^3 \alpha} \qquad (6.10)$$

6.7 Normal Modules

The module of the helical gears is generally considered as the normal module m_n. If the gears are **cut by standard hobs**, normal pressure angle Φ_n and normal module m_n will apply and hence these standard values should be used in design. The value has to be selected from the first choice of modules preferably, otherwise take second choice given in Table 6.1.

Value of module m in the plane of rotation will have fractions, and hence PCD will also be in fractions. Selection of hob cutter has to be on the basis of equivalent number of teeth given by Z'.

Table 6.1 Normal Module for Helical Gears

1st Choice	1	1.25	1.5	2	2.5	3	4	5	6	8	10	12	16	20
2nd Choice	1.125	1.375	1.75	2.25	2.75	3.5	4.5	5.5	7	9	11	14	18	22

If the gears are **cut by gear shaping**, the module m and pressure angle Φ will apply to the plane of rotation and hence the standard values of m should be used.

6.8 Forces on Tooth

A cylinder of PCD is shown in Figure 6.5. The normal force F_n acts on an inclined gear tooth at an angle Φ_n as shown by dashed line in this figure. This force can be resolved to give three main components; that is, (1) tangential along the direction of rotation at contact point, (2) radial towards center of the gear, and (3) axial along axis of the gear. These are related to each other as given in Equations (6.11), (6.12), and (6.13).

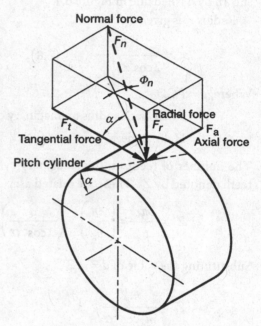

Figure 6.5 Components of forces on helical gear tooth

$$\text{Tangential force } F_t = \frac{2T}{d_p} \qquad (6.11)$$

$$\text{Radial force } F_r = F_t \left(\frac{\tan \Phi_n}{\cos \alpha} \right) \qquad (6.12)$$

$$\text{Axial force } F_a = F_t \tan \alpha \qquad (6.13)$$

Where, T = Torque (Nm)

d_p = PCD of pinion (m)

α = Helix angle

Φ = Pressure angle

The load calculated from the given torque is tangential. The normal force can be calculated using the relation:

$$F_t = F_n \cos \Phi_n \cos \alpha$$

$$\text{Or,} \quad F_n = \frac{F_t}{\cos \Phi_n \cos \alpha} \qquad (6.14)$$

Example 6.2

Tooth forces on helical gear for given power
A helical gear set has 22 teeth pinion, which transmits 7.5 kW at 600 rpm to its mating gear. Normal module is 6 mm and normal pressure angle 20°, helix angle 35°. Calculate tooth forces in tangential, axial, and radial directions. What is the normal force on the tooth?

Solution

Given $Z_p = 22$ $P = 7.5$ kW $\Phi_n = 20°$ $\alpha = 35°$ $N_p = 600$ rpm $m_n = 6$ mm

Torque, $T = \dfrac{60p}{2\pi N} = \dfrac{60 \times 7.5 \times 10^3}{2 \times 3.14 \times 600} = 119.4$ Nm

From Equation (6.7), Pitch circle diameter of pinion, $d_p = \dfrac{Z_p m_n}{\cos\alpha} = \dfrac{22 \times 6}{\cos 35°} = 161.1$ mm

From Equation (6.11), Tangential force, $F_t = \dfrac{2T}{d_p} = \dfrac{2 \times 119.4 \times 10^3}{161.1} = 1,482$ N

From Equation (6.13), Axial force, $\quad F_a = F_t \tan\alpha = 1,482 \times \tan 35° = 1,037.6$ N

From Equation (6.12), Radial force, $\quad F_r = F_t\left(\dfrac{\tan\Phi_n}{\cos\alpha}\right) = 1,482\left(\dfrac{\tan 20°}{\cos 35°}\right) = 658.4$ N

From Equation (6.14), Normal force, $\quad F_n = \dfrac{F_t}{\cos\Phi_n \cos\alpha} = \dfrac{1,482}{\cos 20° \cos 35°} = 1,860$ N

6.9 Design of Helical Gears

The design procedure for helical gears is similar to that for spur gears, but following points are to be kept in mind:

- Minimum number of teeth is to be replaced by equivalent number of teeth.
- Module is replaced by normal module.
- Helix angle should be between 8° and 20° but not more than 30° to reduce side thrust.

The gear design is done on the basis of the following three considerations:

a. On the basis of **strength** of tooth using Lewis equation (Section 6.10).
b. On the basis of **dynamic** loads (Sections 6.11 and 6.12).
c. Design for **wear** considerations (Section 6.13).

6.10 Lewis Equation for Helical Gears

Tooth of a helical gear is considered as a cantilever beam of length equal to height of the tooth (addendum + dedendum). Beam strength of a tooth is the maximum force, which it can take without causing failure in bending. The strength is calculated for the formulative imaginary gear of diameter and number of teeth.

Following assumptions are made in the Lewis equation:

- Tangential component is uniformly distributed over the entire face of the tooth.
- Effect of radial component causing compressive stresses is neglected.
- Point of contact moves, which causes change in resultant force. This change is neglected.
- Although number of teeth that are in contact is more than one, but it is assumed that only one pair of teeth takes the entire load.
- Analysis is valid, when gears are stationary or rotating at very slow speed.
- Effect of the dynamic forces is neglected in strength calculations.
- Effect of stress concentration is neglected.

Beam strength for helical gears S_b is given by Lewis equation as under:

$$S_b = \sigma_b \pi m_n b y' \tag{6.15}$$

Where, σ_b = Bending strength (N/mm²)

If not given, it can be taken as one-third of ultimate tensile strength.

m_n = Normal module (mm)

b = Face width (mm)

y' = Lewis factor based on equivalent number of teeth

Equation (6.15) is known as Lewis equation for helical gears. It gives the beam strength in the plane of rotation. It gives the maximum tangential force, which a gear tooth can take without failure in bending.

Beam strength normal to the tooth is given as:

$$S_{bn} = \sigma_b m_n \sec \alpha\, b\, Y' \qquad (Y' = \pi y') \tag{6.16}$$

Where, S_{bn} = Normal beam strength

α = Helix angle

Thus, $S_{bn} = S_b \sec \alpha$ \hfill (6.17)

For a design to be safe in strength, its value should be more than tangential force, that is, $S_b > F_t$. Torque T is calculated using the relation:

$$P = \frac{2\pi N T}{60,000} \quad \text{or} \quad T = \frac{60,000 P}{2\pi N} \tag{6.18}$$

Where, P = Power (kW)

N = Speed of input shaft (rpm)

T = Torque (Nm)

Tangential force F_t is calculated using the relation:

$$F_t = \frac{T}{0.5d} \qquad (6.19)$$

Where, d = Pitch circle diameter of driver pinion / gear

6.11 Effective Load

Tangential force on a gear tooth is calculated from the given power and speed, assuming that the load is constant. In case of prime movers like Internal Combustion engines, there may be mild to heavy shocks. The maximum torque may be more than the average torque and is called as dynamic load. This effect is considered by considering two factors:

- Service factor
- Velocity factor

6.11.1 Service factor

Value of service factor depends on the type of driver machine and type of driven machine. The torque varies from maximum to average and is given by relation:

$$K_s = \frac{\text{Maximum torque}}{\text{Average torque}} = \frac{T_{max}}{T_{av}} \qquad (6.20)$$

Table 6.2 can be used to select a suitable value of service factor.

6.11.2 Velocity factor

Manufacturing and assembly inaccuracies, circumferential velocity, and elastic bending of tooth also increase the actual load on gear tooth. The manufacturing process decides the value of velocity factor as given below:

Commercial cut gears using form cutters with low helix angle and velocity $v < 5$ m/s:

$$k_v = \frac{4.6}{4.6 + v} \qquad (6.21a)$$

Table 6.2 Values of Service Factor K_s for Different Types of Loads

| Type of Driver | | Type of Driven Machine | | |
| | | Uniform | Medium Shock | Heavy Shock |
Type of Load	Type of Machine	Machine Tools, Generators, Conveyors, Turbo Blowers, and Hoists	Cranes, Feed Pump, and Heavy Elevators	Press, Rolling Mills, Well Drilling Machines, and Power Shovel
Uniform load	Electric motors Turbines	1.0	1.25	1.75
Light shock	Multi cylinder I.C. engine	1.25	1.5	2.0
Medium shock	Single cylinder I.C. engine	1.5	1.75	2.25

Carefully cut gears using form cutters with velocity between 5 m/s and 10 m/s:

$$k_v = \frac{6}{6+v} \tag{6.21b}$$

Accurately cut gears using hob cutters with velocity between 10 m/s and 20 m/s:

$$k_v = \frac{15}{15+v} \tag{6.21c}$$

Precisely cut with shaving and ground / lapped gears and velocity > 20 m/s:

$$k_v = \frac{5.6}{5.6+\sqrt{v}} \tag{6.21d}$$

The effective load F_e is calculated from the selected value of k from Table 6.2 and velocity factor k_v using a suitable equation (Equation 6.21). The relation is given in Equation 6.22:

$$F_e = F_t\left(\frac{K_s}{K_v}\right) \tag{6.22}$$

The velocity factor is used only to calculate approximate dynamic load in preliminary calculations. Exact calculation is done using Buckingham's equation described in Section 6.12.

In the preliminary design of gears, module, number of teeth, and face width are not known, so module is calculated equating effective load using Equation (6.22) with the beam strength calculated from Lewis equation (6.15).

$$F_e = F_t\left(\frac{K_s}{K_v}\right) = \sigma_b \pi m_n b y' \tag{6.23}$$

6.12 Dynamic Load

In final design, when gear size is approximately known, the maximum dynamic load can be calculated more accurately using equation given below:

$$F_d = (k_s F_t) + F_i \qquad (6.24)$$

Where, F_d = Maximum design load

F_i = Incremental additional load due to the dynamic conditions

F_t = Tangential load

k_s = Service factor

Note Velocity factor is not used in Equation (6.24), as it will be considered more precisely in incremental load described below.

Earle Buckingham gave the following relation to calculate this additional load:

$$F_i = \frac{21 v B \cos \alpha}{21v + \sqrt{B}} \qquad (6.25)$$

Where, $B = C e b \cos^2 \alpha + F_t (\text{N})$

α = Helix angle (°)

C = Deformation factor, which depends on Young's modulus of elasticity

e = Total error in manufacturing = Error of pinion e_p + Error of gear e_g (mm)

b = Face width of gear along the axis of rotation (mm)

v = Pitch line velocity (m/s)

Note Value of error for pinion or gear depends on accuracy grade given in Table 5.7 of Chapter 5. It is dependent on number of teeth, hence errors for pinion and gear are to be calculated separately and then summed up to get total error:

$$e = e_p + e_g \qquad (6.26)$$

Deformation factor C can be calculated from the following relation:

$$C = \frac{K_t}{\left(\dfrac{1}{E_p} + \dfrac{1}{E_g} \right)} \qquad (6.27)$$

Where, K_t = A constant depending on type of tooth for deformation factor (Table 6.3)

K_t = 0.107 for 14.5° full depth tooth

K_t = 0.11 for 20° full depth tooth

K_t = 0.115 for 20° stub tooth

E_p = Young's modulus of elasticity for pinion material (N/mm²)

E_g = Young's modulus of elasticity for gear material (N/mm²)

Table 6.3 Values of C

Material		Type of Teeth		
Pinion	Gear	14.5°	20 ° Full Teeth	20°Stub Teeth
Grey C.I.	Grey C.I.	5,500	5,700	5,900
Steel	Grey C.I.	7,600	7,900	8,100
Steel	Steel	11,000	11,400	11,900

Calculate dynamic load F_d using Equation (6.24). Its value should be lesser than the beam strength S_b. The factor of safety (FOS) for dynamic load is given as:

$$FOS = \frac{S_b}{F_d}$$ (6.28)

6.13 Wear Strength of Helical Gears

Wear strength of a gear is the maximum tangential force, which a tooth can take without pitting. For helical gears, teeth are longer than face width, hence wear strength is calculated for helical gears using the following Buckingham's wear strength equation:

$$S_w = \frac{d_p b Q K_w}{\cos^2 \alpha}$$ (6.29)

Where, S_w = Wear strength (N/mm²)

α = Helix angle (°)

d_p = PCD of pinion (mm)

b = Face width along the axis of gear (mm)

Q = Ratio factor given by relation: $Q = \frac{2Z_g'}{Z_g' + Z_p'}$

Z_p', Z_g' =Number of teeth on virtual pinion and on virtual gear, respectively.

Dividing by $\cos^3 \alpha$ in numerator and denominator: $Q = \frac{2Z_g}{Z_g + Z_p}$

Z_p, Z_g = Number of teeth on pinion and on gear, respectively.

Dividing by Z_p in numerator and denominator and using: $G = \frac{Z_g}{Z_p}$

$$Q = \frac{2G}{G+1}$$ (6.30)

K_w = A constant given by relation:

$$K_w = \frac{S_{es}^2 \sin\Phi_n \cos\Phi_n \left(\dfrac{1}{E_p} + \dfrac{1}{E_g}\right)}{1.4} \tag{6.31}$$

Φ_n = Normal pressure angle (°)

S_{es} = Surface endurance strength. If BHN is given, it can be calculated from:

$$S_{es} = 2.76\,BHN - 70 \tag{6.32}$$

E_p, E_g = Young's modulus of elasticity for pinion and gear material, respectively.

If BHN is given, value of K_w can be approximated, using the relation:

$$K_w = 0.156\left(\frac{BHN}{100}\right)^2 \tag{6.33}$$

For a design to be safe, wear strength S_w should be more than the effective load F_e or dynamic load F_d. The FOS for wear is given as:

$$FOS = \frac{S_w}{F_d} \tag{6.34}$$

Example 6.3

Module and beam strength of a helical gear for given power

A helical gear with full depth 20° pressure angle involute teeth transmits 10 kW with helix angle 40°. The pinion of 76 mm diameter rotating at 3,000 rpm drives the gear with gear ratio 4. The material both for pinion and gear has allowable bending strength of 95 MPa and endurance strength 580 MPa. Take gear width 15 times the module.

a. Find the module from strength point of view and design the gear, if service factor is 1.25.

b. Beam strength.

c. FOS in wear, if Young's modulus of elasticity for both gears is 190 GPa.

Solution

Given $P = 10\,kW$ $\Phi = 20°$ $\alpha = 40°$ $d_p = 76\,mm$ $N_p = 3,000\,rpm$

$G = 4$ $\sigma_b = 95\,MPa$ $S_{es} = 580\,MPa$ $b = 15\,m$ $K_s = 1.25$

$E_p = E_g = 190\,GPa$

a. Module

$$T = \frac{60,000\,P}{2\pi N} = \frac{60,000}{2 \times 3.14 \times 3,000} = 31.78\,Nm$$

$$F_t = \frac{T}{0.5d_p} = \frac{31.78 \times 10^3}{0.5 \times 76} = 836\,N$$

$$Z_p = \frac{d_p}{m} = \frac{76}{m}$$

Equivalent number of teeth $Z'_p = \dfrac{Z_p}{\cos^3 \alpha} = \dfrac{76/m}{\cos^3 40°} = \dfrac{169}{m}$

Since material for gear and pinion is same, pinion is always weaker than gear. So calculate form factor y'_p for the pinion.

Form factor for 20° is: $y'_p = 0.154 - \dfrac{0.912}{Z'_p}$

Or, $y'_p = 0.154 - \dfrac{0.912m}{169} = 0.154 - 0.0054m$

Pitch line velocity $v = \dfrac{\pi d N}{60} = \dfrac{3.14 \times 76 \times 10^{-3} \times 3,000}{60} = 11.86$ m/s

For velocity more than 10 m/s, $k_v = \dfrac{15}{15+v} = \dfrac{15}{15+11.86} = 0.558$

From Equation (6.23), effective load $F_e = F_t \left(\dfrac{K_s}{K_v} \right) = 836 \left(\dfrac{1.25}{0.558} \right) = 1,873$

From Lewis equation: $F_e = \sigma_b b \pi m_n y'$

Substituting the values: $1,873 = 95 \times (15m) \times 3.14 \times (m \cos 40°) \times (0.154 - 0.0054m)$

Or, $1,873 = 3427 \, m^2 (0.154 - 0.0054 \, m) = 528 \, m^2 - 18.5 \, m^3$

Or, $m^3 - 28.5 \, m^2 + 101.2 = 0$

Solving above equation by trial and error: $m \cong 2$ say 2

Number of teeth of pinion, $Z_p = \dfrac{d_p}{m} = \dfrac{76}{2} = 38$ mm

Number of teeth of gear, $Z_g = Z_p \times m = 38 \times 2 = 152$ mm

Width $b = 15$ m $= 15 \times 2 = 30$ mm

$y'_p = 0.154 - 0.0054 \, m = 0.143$

b. Beam strength

$S_b = \sigma_b \pi m_n b y' = 95 \times 3.14 \times (2 \cos 40°) \times 30 \times 0.143 = 1,960.2$ N

Beam strength is more than the effective load 1,873 N, hence the design is safe.

c. *FOS* in wear

Ratio factor $Q = \dfrac{2G}{G+1} = \dfrac{2 \times 4}{4+1} = 1.6$

$\tan \Phi_n = \tan \Phi \times \cos \alpha = \tan 20° \times \cos 40° = 0.278$. Hence $\Phi_n = 15.6°$

$$K_w = \frac{S_{es}^2 \sin\phi_n \cos\phi_n \left(\dfrac{1}{E_p} + \dfrac{1}{E_g}\right)}{1.4}$$

Substituting the values: $K_w = \left(\dfrac{(580)^2 \sin 15.6° \times \cos 15.6°}{1.4}\right) \left(\dfrac{1}{190} + \dfrac{1}{190}\right) = 0.655$

Wear strength, $S_w = \dfrac{d_p b Q K_w}{\cos^2 \alpha} = \dfrac{76 \times 30 \times 1.6 \times 0.655}{\cos^2 40°} = 4,018 \text{ N}$

FOS in wear, $FOS = \dfrac{S_w}{F_e} = \dfrac{4,018}{1,873} = 2.15$

Example 6.4

Wear strength and power capacity of a given gear set

A parallel axis helical gear set has a 20 teeth pinion and 80 teeth gear of width 50 mm. Pinion rotates at 900 rpm. Gear material has ultimate strength 570 MPa with BHN 280. Normal module is 5 mm and normal pressure angle 20°, helix angle 30°. Service factor can be taken as 1.5 and FOS 2.5. Calculate:

 a. Beam strength.

 b. Check, if this gear set is safe in wear strength.

 c. Safe power capacity.

Solution

Given $Z_p = 20$ $Z_g = 80$ $b = 50$ mm $N_p = 900$ rpm $m_n = 5$ mm $\Phi = 20°$ $\alpha = 30°$

$S_{ut} = 570$ MPa   BHN = 280   FOS = 2.5 $K_s = 1.5$

 a. Formulative number of teeth, $Z_p' = \dfrac{Z_p}{\cos^3 \alpha} = \dfrac{20}{\cos^3 30°} = 30.8$ say 31

Form factor for 20° is: $y_p' = 0.154 - \dfrac{0.912}{Z_p'} = 0.154 - \dfrac{0.912}{31} = 0.124$

Allowable bending stress: $\sigma_b = \dfrac{S_{ut}}{2.5} = \dfrac{570}{2.5} = 228$ MPa

From Lewis equation: $S_b = \sigma_b\, b\, \pi m_n y_p'$. This should be at least equal to tangential force.

Substituting the values:

Tangential force, $F_t = 228 \times 50 \times 3.14 \times 5 \times 0.124 = 22,193$ N

b. Gear ratio $G = \dfrac{Z_g}{Z_p} = \dfrac{80}{20} = 4$

Ratio factor $Q = \dfrac{2G}{G+1} = \dfrac{2 \times 4}{4+1} = 1.6$

Pinion PCD: $d_p = Z_p \times m = \dfrac{Z_p m_n}{\cos \alpha} = \dfrac{20 \times 5}{\cos 30°} = 115\text{ mm}$

Approximate value of K_w is: $K_w = 0.156 \left(\dfrac{BHN}{100} \right)^2 = 0.156 \left(\dfrac{280}{100} \right)^2 = 1.22$

Wear strength, $S_w = \dfrac{d_p b Q K_w}{\cos^2 \alpha} = \dfrac{115 \times 50 \times 1.6 \times 1.22}{\cos^2 30} = 14{,}965\text{ N}$

Wear strength is lesser than tangential force, hence the gear will fail in pitting.

c. Pitch line velocity, $v = \dfrac{\pi d N}{60} = \dfrac{3.14 \times 115 \times 10^{-3} \times 900}{60} = 5.4\text{ m/s}$

For velocity 5–10 m/s: $k_v = \dfrac{6}{6+v} = \dfrac{6}{6+5.4} = 0.526$

Wear strength required, $S_w = \dfrac{(F_t \times FOS) \times K_S}{k_v} = \dfrac{(F_t \times 2.5) \times 1.5}{0.526} = 7.13 F_t$

Equating wear strength: $14{,}965 = 7.13\, F_t$ or $F_t = 2{,}099\text{ N}$

Torque $T = F_t \times (0.5\, d_p) = 2{,}099 \times 0.5 \times 115 = 120{,}692\text{ Nm}$

Power $P = \dfrac{2\pi N T}{60{,}000}$

Substituting the values: $P = \dfrac{2 \times 3.14 \times 900 \times 120{,}692}{60{,}000} = 11{,}369\text{ watts say } 11.4\text{ kW}$

Example 6.5

Power capacity of a gear set with given error

Two helical gears having 25 and 35 teeth, pressure angle 20°, normal module 5, and with width 50 mm are mounted on parallel axis shafts having center distance is 190 mm. Pinion rotates at 750 rpm. BHN is 300, maximum error is 0.1 mm. Safe bending strength is 150 MPa. Young's modulus of elasticity for both the gears is 200 GPa. Calculate:

a. Helix angle.

b. Wear strength.

c. Dynamic load assuming C = 11,400 MPa and service factor = 1.2.

d. Safe power capacity.

e. Total error, if accuracy is changed to grade 8.

Solution

Given $Z_p = 25$ $Z_g = 35$ $\Phi = 20°$ $m_n = 5$ mm $b = 50$ mm

$c = 190$ mm $N_p = 750$ rpm BHN = 300 $e = 0.1$ mm $\sigma_b = 150$ MPa

$E_p = E_g = 200$ GPa $C = 11,400$ $K_s = 1.2$

a. Center distance $190 = \dfrac{d_p + d_g}{2} = \dfrac{dp\left(1 + \dfrac{35}{25}\right)}{2} = 1.2 d_p$. Hence, $d_p = 158.3$ mm

$$d_g = \frac{35}{25} d_p = \frac{35}{25} \times 158.3 = 190 \text{ mm}$$

$$m = \frac{d_p}{Z_p} = \frac{158.3}{25} = 6.33$$

$m_n = m \cos \alpha$ or $5 = 6.33 \cos \alpha$. Hence $\alpha = 37.8°$

b. $\tan \Phi_n = \tan \Phi \times \cos \alpha = \tan 20° \times \cos 37.8°$. Hence $\Phi_n = 16°$

Formulative number of teeth, $Z'_p = \dfrac{Z_p}{\cos^3 \alpha} = \dfrac{25}{\cos^3 37.8} = 51$

Form factor for 20° is: $y'_p = 0.154 - \dfrac{0.912}{Z'_p} = 0.154 - \dfrac{0.912}{51} = 0.136$

From Lewis equation: $S_b = \sigma_b b \pi m_n y'_p$

Substituting the values: $S_b = 150 \times 50 \times 3.14 \times 5 \times 0.136 = 16,014$ N

Endurance strength Equation (6.32) $S_{es} = 2.76 \text{BHN} - 70 = (2.76 \times 300) - 70 = 758$ MPa

From Equation (6.31): $K_w = \dfrac{S_{es}^2 \sin \Phi_n \cos \Phi_n \left(\dfrac{1}{E_p} + \dfrac{1}{E_g}\right)}{1.4}$

Substituting the values: $K_w = \dfrac{(758)^2 \sin 16° \times \cos 16°}{1.4}\left(\dfrac{1}{200} + \dfrac{1}{200}\right) = 1.08$

Ratio factor: $Q = \dfrac{2Z_g}{Z_g + Z_p} = \dfrac{2 \times 35}{35 + 25} = 1.167$

Wear strength, $S_w = \dfrac{d_p b Q K_w}{\cos^2 \alpha} = \dfrac{158.3 \times 50 \times 1.167 \times 1.08}{\cos^2 37.8^\circ} = 15{,}986 \text{ N}$

c. $B = C\,e\,b\cos^2\alpha + F_t$

Substituting the values: $B = 11{,}400 \times 0.1 \times 50 \times \cos^2 37.8^\circ + 16{,}014 = 51{,}602 \text{ N}$

Pitch line velocity, $v = \dfrac{\pi D N}{60{,}000} = \dfrac{3.14 \times 158.3 \times 750}{60{,}000} = 6.21 \text{ m/s}$

From Equation (6.25) $F_i = \dfrac{21v B \cos \alpha}{21v + \sqrt{B}}$

Substituting the values:

$$F_i = \dfrac{21 \times 6.21 \times 51{,}602 \times \cos 37.8^\circ}{(21 \times 6.21) + \sqrt{51{,}602}} = 14{,}877 \text{ N}$$

From Equation (6.24) $F_d = (k_s F_t) + F_i = (1.2 \times 16{,}014) + 14{,}877 = 34{,}094 \text{ N}$

d. Since wear strength is lower than dynamic load, hence safe power is calculated on the basis of this strength.

Power capacity (kW), $P = \dfrac{S_w \times v}{1{,}000} = \dfrac{15{,}986 \times 6.21}{1{,}000} = 99.3 \text{ kW}$

e. From Table 5.7 of Chapter 5, for accuracy grade 8, $e = 16 + 1.25 f$

For pinion: $\quad f_p = m_n + 1.25\sqrt{d_p} = 5 + 1.25\sqrt{158.3} = 20.73$

For gear: $\quad f_g = m_n + 1.25\sqrt{d_g} = 5 + 1.25\sqrt{190} = 22.23$

Error for pinion, $e_p = 16 + 1.25 f_p = 16 + (1.25 \times 20.73) = 41.9\,\mu$

Error for gear, $\quad e_g = 16 + 1.25 f_g = 16 + (1.25 \times 22.23) = 43.8\,\mu$

Total error: $\quad e = e_p + e_g = 41.9 + 43.8 = 85.7\,\mu = 0.0857 \text{ mm}$

Example 6.6

Module and BHN of a gear set for given power

Two parallel shafts mounted with precisely cut helical gears transmit 10 kW at 3,000 rpm through 20 teeth pinion of width 14 times the module to a gear rotating at 1,000 rpm. Stub teeth are cut with 20° pressure angle in diametrical plane and helix angle 30° with accuracy grade 6. Allowable beam stress is 100 MPa. Young's modulus of elasticity for both the gears can be taken 200 GPa. Calculate :

(a) Module (b) Dynamic load

(c) Hardness for safe working in wear.

Solution

Given

$P = 10\,kW$	$N_p = 3,000\,rpm$	$Z_p = 20$	$\Phi = 20°$	$b = 14\,m$
$N_g = 1,000\,rpm$	$\alpha = 30°$	$S_{ut} = 570\,MPa$		Grade 6
$\sigma_b = 100\,MPa$	$E_p = E_g = 200\,GPa$			

a. Pitch line velocity, $v = \dfrac{\pi d N}{60,000} = \dfrac{\pi \times (20m) \times 3,000}{60,000} = 3.14m \text{ m/s}$

Tangential force, $F_t = \dfrac{P}{v} = \dfrac{10,000}{3.14m} = \dfrac{3,185}{m}$

Formulative number of teeth, $Z'_p = \dfrac{Z_p}{\cos^3 \alpha} = \dfrac{20}{\cos^3 30°} = 30.8$ say 31

Form factor for 20° stub teeth, $y'_p = 0.175 - \dfrac{0.814}{Z'_p} = 0.175 - \dfrac{0.841}{31} = 0.147$

Velocity factor for precise teeth: $k_v = \dfrac{5.6}{5.6+v} = \dfrac{5.6}{5.6+\sqrt{3.14m}}$

Effective load, $F_e = F_t \left(\dfrac{K_s}{K_v}\right)$ or $F_t = F_e \times K_v$ (assuming $k_s = 1$)

From Lewis equation: $S_b = \sigma_b b \pi m_n y'_p$. This should be at least equal to F_e.

Hence, $F_t = \sigma_b b \pi m_n y'_p k_v$

Substituting the values:

$\dfrac{3,185}{m} = 100 \times 14m \times \pi \times (m\cos 30°) \times 0.147 \times \dfrac{5.6}{5.6+\sqrt{3.14m}}$

Simplifying, $3,185 = \dfrac{3,134 m^3}{5.6 + \sqrt{3.14m}}$

By trial and error $m = 2$

Width of gear, $b = 14\,m = 28\,mm$

b. Pitch line velocity, $v = 3.14\,m = 6.28\,m/s$

Tangential force, $F_t = \dfrac{3,185}{m} = \dfrac{3,185}{2} = 1,592.5\,N$

From Table 5.7 of Chapter 5, for accuracy grade 6, $e = 8 + 0.63f$

For pinion: $f_p = m_n + 0.63\sqrt{m \times Z_p} = 2\cos 30° + 0.63\sqrt{2 \times 20} = 5.72$

For gear: $f_g = m_n + 0.63\sqrt{m \times Z_g} = 2\cos 30° + 0.63\sqrt{2 \times 30} = 6.61$

Error for pinion, $e_p = 8 + 0.63\,f_p = 8 + (0.63 \times 3.58) = 11.6\,\mu$

Error for gear, $e_g = 8 + 0.63\,f_g = 8 + (0.63 \times 4.73) = 12.16\,\mu$

Total error: $e = e_p + e_g = 11.6 + 12.16 = 23.76\,\mu$

From Equation (6.27) $C = \dfrac{K_t}{\left(\dfrac{1}{E_p} + \dfrac{1}{E_g}\right)}$ $\quad (K_t = 0.115$ for stub teeth)

Substituting the values: $C = \dfrac{0.115 \times 10^3}{\left(\dfrac{1}{200} + \dfrac{1}{200}\right)} = 11{,}500$

$B = C\,e\,b\cos^2\alpha + F_t = [11{,}500 \times (23.76 \times 10^{-6}) \times 28 \times \cos^2 30°] + 1{,}592.5 = 1{,}598\,\text{N}$

From Equation (6.25) Incremental load $F_i = \dfrac{21vB\cos\alpha}{21v + \sqrt{B}}$

Substituting the values:

$F_i = \dfrac{21 \times 6.28 \times 1{,}598 \times \cos 30°}{(21 \times 6.28) + \sqrt{1{,}598}} = \dfrac{182{,}464}{172} = 1{,}061\,\text{N}$ $\quad (v = 3.14m = 6.28)$

From Equation (6.24): $F_d = (k_s F_t) + F_i = (1 \times 1{,}592.5) + 1{,}061 = 2{,}653.5\,\text{N}$

Beam strength, $S_b = \sigma_b b\,\pi\,m_n y'_p$

Substituting the values: $S_b = 100 \times 28 \times 3.14 \times (2\cos 30°) \times 0.147 = 2{,}238\,\text{N}$

Since beam strength is lesser than the dynamic load, increase the strength either by increasing module or the easiest way by increasing the width. Hence width is increased from 28 mm to 35 mm. This will give beam strength as 2,798 N. Then the design will be safe.

c. Gear ratio $G = \dfrac{Z_g}{Z_p} = \dfrac{N_p}{N_g} = \dfrac{3{,}000}{1{,}000} = 3$

Ratio factor $\quad Q = \dfrac{2G}{G+1} = \dfrac{2 \times 3}{3+1} = 1.5$

Pinion PCD: $d_p = Z_p \times m = 20 \times 2 = 40\,\text{mm}$

$\tan \Phi_n = \tan \Phi \times \cos \alpha = \tan 20° \times \cos 30° = 0.315$. Hence $\Phi_n = 17.5°$

For design to be safe, wear strength S_w should be at least equal to dynamic load 2653.5 N

Wear strength, $S_w = \dfrac{d_p b Q K_w}{\cos^2 \alpha}$

Substituting the values: $2{,}653.5 = \dfrac{40 \times 35 \times 1.5 \times K_w}{\cos^2 30}$ or $K_w = 0.95$

From Equation (6.31) $K_w = \dfrac{S_{es}^2 \sin \Phi_n \cos \Phi_n \left(\dfrac{1}{E_p} + \dfrac{1}{E_g} \right)}{1.4}$

Substituting the values: $0.95 = \dfrac{S_{es}^2 \sin 17.5° \cos 17.5° \left(\dfrac{1}{200} + \dfrac{1}{200} \right)}{1.4 \times 10^3} = 2.04 \times 10^{-6} S_{es}^2$

Or, $S_{es} = 682$ MPa

From Equation (6.32): $S_{es} = 2.76$ BHN $- 70$

Substituting the values: $682 = 2.76$BHN -70. Hence BHN $= 272$

Summary

Helical gears have teeth in the form of a helix on their pitch cylinder at an angle called helix angle. Helix is called right hand, if its tooth advances in clockwise direction. These gears can be used for high speeds up to 50 m/s. These gears require accurate machining and better lubrication because of increased sliding action.

Salient features of helical gears are:

Addendum, dedendum, and other proportions remain same as that for spur gears.

PCD depends on helix angle also.

Load capacity   Contact with the mating gear starts at the leading edge and then extends gradually along the tooth. This results less stresses in helical gears than spur gears.

Noise   Owing to gradual engagement, these gears offer silent operation

Axial loads   Owing to inclination of teeth, helical gear causes axial loads.

Hand of helix   Hand of helix of driver pinion is opposite to the driven gear.

Helix angle (α)   It is the angle made by helix with the axis of rotation.

Module (m)   For helical gears, it is called transverse module and is $m = Z/d$.

Normal module (m_n)   Owing to helix angle α, normal module is defined as $m_n = m \cos \alpha$.

Normal pressure angle (Φ_n)   Angle of load, normal to tooth surface. $\tan \Phi_n = \tan \Phi \cos \alpha$.

Face width (b) Width of gear along the axis. It varies from 12 m to 20 m.

Tooth length It is the length of a tooth along the tooth and it is equal to $b \sec \alpha$.

Circular pitch (P_c) It is same as transverse pitch. $P_c = \pi m$.

Normal pitch (P_n) Distance between two adjacent faces of teeth at PCD measured normal to teeth. Normal pitch and circular pitch are related as $P_n = P_c \cos \alpha = \pi m \cos \alpha$.

Axial pitch (P_a) Distance along the axis of the gear from its face, to the point, where the tooth axis cuts an axis parallel to axis of $P_a = \pi m / \tan a$

Types of helical gears

Parallel helical gears Used for shafts whose axes are parallel. Helix angle α is from 15° to 30°.

Crossed helical gears These gears are used for shafts, which are not parallel. Their teeth may have the same or different hand of helix depending on the angle between the axes of the shafts. For 90° angle between shafts, helix angle of both the gears is same as 45° and use the same hand. Salient features for these gears are:

- Owing to curvature of tooth surfaces, it does not have line contact, but a point contact.
- Owing to small contact area of these gears, load capacity is low.
- Owing to small contact area and high loads, wear is also high.

Herringbone gears Two helical gears of opposite and equal value of hand of helix and module are joined face to face. Helix angle varies from 20° to 45°.

Advantages of these gears over helical gears are:

- These do not cause any axial thrust, because it gets cancelled by other half of the gear.
- Power transmission capacity is higher than helical gears, hence used for ship drives.
- These gears can operate at high speeds.

Disadvantages: These are costly and need high precision in manufacturing.

Double helical gears Similar to herringbone gears, but there is a groove between two halves. Helix angle varies from 23° to 30°.

Overlap Distance along the circumference from one end of the tooth to the other end. This should be more than the circular pitch, to ensure that the other tooth comes in contact before one leaves. It indicates how many teeth will be in contact.

Overlap, $X = b \tan \alpha$

where, b = Face width and α = Helix angle

Overlap increases with increase in helix angle, but increase in helix angle causes more axial thrust. Hence a compromise is to be made between the two values.

Therefore it should be about 15 per cent of circular pitch, that is, overlap = 1.15 P_c.

Face width, $b = 1.15 \, P_c / \tan \alpha = 1.15 \, \pi m / \tan \alpha$. Generally, it varies from 12 m to 20 m.

Gear and tooth proportions

PCD $d = \dfrac{Z \times p_n}{\pi \times \cos\alpha} = \dfrac{Zm_n}{\cos\alpha} = Zm$

Addendum $a = m_n$

Outside diameter = PCD + 2 m_n

Root diameter = PCD – 2.5 m_n

Dedendum $d = 1.25\ m_n$

Face width = 12 m to 20 m

Fillet radius = 0.35 m_n

Equivalent number of teeth   When a helical gear is cut, the gear blank is turned by helix angle on the machine. Thus the profile of the involute cutter is cut at a plane A–B normal to the helix angle. Intersection of pitch cylinder and plane A–B produces an ellipse, whose radius r' is given as

$$r' = \dfrac{d}{2\cos^2\alpha}$$

where, d = PCD, r' = Pitch circle radius of imaginary circle, and α = Helix angle.

The number of teeth on this imaginary cylinder Z' are called equivalent or virtual formulative teeth.

These are related as $Z' = \dfrac{Z}{\cos^3\alpha}$

Selection of *hob cutter* has to be on the basis of equivalent number of teeth given by Z'.

Normal module   The module of the helical gears is generally considered as the normal module m_n. If the gears are cut by *standard hobs*, normal pressure angle Φ_n and normal module m_n will apply and hence these standard values should be used in design.

Table 6.1 gives first choice of standard modules preferably, otherwise take second choice.

If the gears are cut by *gear shaping*, the module m and pressure angle Φ will apply to the plane of rotation and hence the standard values should be used.

Forces on tooth   The load calculated from the given torque is tangential $F_t = \dfrac{2T}{d_p}$

Radial force $F_r = F_t \left(\dfrac{\tan\Phi_n}{\cos\alpha} \right)$ Axial force $F_a = F_t \tan\alpha$

The normal force F_n acts on an inclined gear tooth at an angle Φ_n. $F_n = \dfrac{F_t}{\cos\Phi_n \cos\alpha}$

Design of helical gears   Gear design is done on the basis of following considerations:

   a.   On the basis of *strength* of tooth using Lewis equation.

   b.   On the basis of *dynamic loads*.

   c.   For *wear* considerations.

Lewis equation for helical gears

The strength is calculated for the formulative imaginary gear of diameter and number of teeth. Beam strength for helical gears S_b is given by Lewis equation is $S_b = \sigma_b \pi m_n b\, y'$.

Where, σ_b = Bending strength (MPa). If not given, take $\sigma_b = S_{ut}/3$

      m_n = Normal module (mm)

b = Face width (mm)

y' = Lewis factor based on equivalent number of teeth

This gives the beam strength in the plane of rotation. Beam strength normal to the tooth is: $S_{bn} = \sigma_b m_n \sec \alpha \, b \, y' = S_b \sec \alpha$. For a design to be safe in strength, $S_b > F_t$.

Effective load   The maximum torque may be more than the average torque and is called as dynamic load. This is considered by (a) service factor and (b) velocity factor.

Service factor   Value of service factor depends on the type of driver machine and type of driven machine.

The torque varies from maximum to average and is given by relation: $K_s = \dfrac{\text{Maximum torque}}{\text{Average torque}} = \dfrac{T_{max}}{T_{av}}$. Table 6.2 gives values of service factor.

Velocity factor (K_v) The manufacturing process decides the value of velocity factor as:

Commercial cut gears using form cutters velocity < 5 m/s: $K_v = \dfrac{4.6}{4.6 + v}$

Carefully cut gears using form cutters with velocity between 5 m/s and 10 m/s: $K_v = \dfrac{6}{6 + v}$

Accurately cut gears using hob cutters with velocity between 10 m/s and 20 m/s: $K_v = \dfrac{15}{15 + v}$

Precisely cut with ground / lapped gears and velocity > 20 m/s: $K_v = \dfrac{5.6}{5.6 + \sqrt{v}}$

Effective load (F_e) It is calculated from values of K_s and K_v with relation $F_e = F_t \left(\dfrac{K_s}{K_v} \right)$

The velocity factor is used only to calculate approximate dynamic load in preliminary calculations. Exact calculation is done using Buckingham's equation.

In the preliminary design module is calculated equating effective load using Equation $F_e = F_t \left(\dfrac{K_s}{K_v} \right)$ with the beam strength calculated from Lewis $F_e = F_t \left(\dfrac{K_s}{K_v} \right) = \sigma_b \pi m_n b y'$

Dynamic load   In final design, when gear size is approximately known, the maximum *dynamic load* can be calculated more accurately using equation $F_d = (K_s F_t) + F_i$

F_d = Maximum design load, K_s = Service factor, F_i = Incremental additional load

Buckingham gave the relation to calculate this additional load: $F_i = \dfrac{21 v \, B \cos \alpha}{21 v + \sqrt{B}}$

$B = [C \, e \, b \cos^2 \alpha + F_t]$ Where, C = Deformation factor, which depends on E.

Total error e = Error of pinion e_p + Error of gear e_g (mm) = $e_p + e_g$

Deformation factor C can be calculated from the relation: $C = \dfrac{K_t}{\left(\dfrac{1}{E_p} + \dfrac{1}{E_g} \right)}$

K_t = A constant depending on type of tooth for deformation factor (Table 6.3).

For 14.5° full depth K_t = 0.107; for 20° full depth K_t = 0.11; for 20° stub teeth K_t = 0.115.

E_p, E_g = Young's modulus of elasticity for pinion and gear material, respectively.

Calculate dynamic load F_d. Its value should be lesser than the beam strength S_b.

The FOS for dynamic load is given as: $FOS = S_b / F_d$

Wear strength of a gear is the maximum tangential force, which a tooth can take without pitting.

It is calculated using this Buckingham's wear strength: $S_w = \dfrac{d_p\, b\, Q\, K_w}{\cos^2 \alpha}$

b = Face width along the axis of gear and d_p = Pitch circle diameter of pinion

Ratio factor: $Q = \dfrac{2Z_g'}{Z_g' + Z_p'} = \dfrac{2G}{G+1}$

K_w = A constant given by relation: $K_w = \dfrac{S_{es}^2 \sin\Phi_n \cos\Phi_n \left(\dfrac{1}{E_p} + \dfrac{1}{E_g}\right)}{1.4}$

Surface endurance strength S_{es} = 2.76 BHN – 70

E_p, E_g = Young's modulus of elasticity for pinion and gear material, respectively.

Value of K_w can be approximated, if BHN is given using: $K_w = 0.156\left(\dfrac{BHN}{100}\right)^2$

For a design to be safe, wear strength $S_w > F_e$ or F_d FOS for wear is given as: S_w / F_d

Theory Questions

1. Compare the working of a spur and helical gear.
2. Explain the terms, helix angle, hand of helix, normal module, and overlap.
3. What are the various types of helical gears? Explain by sketches.
4. Write a short note on face width and overlap.
5. How the outside diameter and root diameter of a helical gear are related to PCD?
6. Explain the term equivalent number of teeth. How is it related with the actual number of teeth?
7. Describe the various components of forces coming on a tooth of a helical gear with the help of a neat sketch.
8. Describe the beam strength of a helical gear tooth for dynamic loads.
9. Explain service factor and velocity factor. How are these considered in calculating beam strength?
10. What is the effective load and how can it be calculated?
11. How is the wear strength of a helical gear calculated? Write the governing equations.

Multiple Choice Questions

1. Helical gears are superior to spur gears because of:
   - (a) Low cost
   - (b) Easily manufactured
   - (c) Silent operation and higher strength
   - (d) Easy meshing

2. Helix angle is the angle which a tooth makes with:
   - (a) The axis of rotation
   - (b) Normal to the axis of rotation
   - (c) Line of tangents to base circles
   - (d) Pressure angle line

3. Axes of a helical gear set:
   - (a) Have to be necessarily parallel
   - (b) Necessarily at right angles to each other
   - (c) Can be at any angle
   - (d) None of the above

4. If b is width of gear and α helix angle, then the overlap is given by the relation:
   - (a) $b \tan \alpha$
   - (b) $b \cos \alpha$
   - (c) $b \sin \alpha$
   - (d) $b \cot \alpha$

5. Equivalent number of teeth for a helical gear with helix angle α having Z number of teeth is given as:
   - (a) $Z \cos^2 \alpha$
   - (b) $Z \cos^3 \alpha$
   - (c) $Z \operatorname{cosec}^2 \alpha$
   - (d) $Z \sec^3 \alpha$

6. Helix angle varies from:
   - (a) $5°-10°$
   - (b) $15°-20°$
   - (c) $20°-35°$
   - (d) $40°-50°$

7. Distance between two adjacent faces at PCD of a helical gear normal to teeth is called:
   - (a) Axial pitch
   - (b) Normal pitch
   - (c) Diametrical pitch
   - (d) Module

8. For a helical gear having helix angle α wear strength is proportional to:
   - (a) $\sec^2 \alpha$
   - (b) $\operatorname{cosec}^2 \alpha$
   - (c) $\cos^2 \alpha$
   - (d) $\cot^2 \alpha$

Answers to multiple choice questions

1. (c)   2. (a)   3. (c)   4. (a)   5. (d)   6. (c)   7. (b)   8. (a)

Design Problems

1. A helical pinion with 30 teeth meshes with a gear of 60 teeth. Normal module is 5. If center distance is 250 mm, calculate helix angle. $[\alpha = 25.8°]$

2. A helical pinion having 25 full depth involute teeth drives a parallel axis helical gear with 75 teeth. Normal module is 4 and helix angle is 25°. Calculate PCDs of pinion and gear.

   $[d_p = 110.3 \text{ mm}, d_g = 331 \text{ mm}]$

3. A 20 teeth helical pinion meshes with a gear of 70 teeth. Normal module is 4, normal pressure angle is 20°, and helix angle is 20°. Calculate transverse module, transverse pressure angle, and axial pitch. $[m = 4.26 \text{ mm}, \Phi = 21.1°, p_a = 36.75 \text{ mm}]$

4. A pair of cast steel helical gears transmits 30 kW at 1,440 rpm. The drive gear has 20 teeth and full depth involute teeth with pressure angle 20°. Face width can be assumed three times the normal pitch. Helix angle is 30°. Allowable stress can be taken 60 MPa. Calculate module and axial thrust.

$$[m = 6 \text{ mm}, F_a = 1,915 \text{ N}]$$

5. A helical 20° normal pressure angle full depth teeth pinion driven by an I.C. engine rotating at 1,000 rpm drives a parallel axis helical gear with speed reduction 2.5 transmits 50 kW. Assume 24 hours operation with medium shocks and width as 12 times the module. Find:

   (a) Module, if allowable stress is 175 MPa and helix angle 30°.

   (b) Face width and number of teeth in each gear assuming pitch line velocity 12 m/s.

$$[m_n = 4, m = 4.62, Z_p = 50, Z_g = 125, b = 48 \text{ mm}]$$

6. A helical pinion with 24 teeth drives a helical gear to give a speed reduction of 5 with 20° full depth involute teeth with 6 mm module. Pinion rotates at 2,800 rpm, face width is 72 mm, helix angle 20°, service factor 1.25, and allowable stress is 200 MPa. Calculate the power capacity of this gear set.

$$[P = 288 \text{ kW}]$$

7. Two parallel shafts 0.5 m apart are mounted with helical gears of gear ratio 2. Teeth are involute 20° stub, power transmitted is 40 kW at 1,000 rpm. Allowable stress is 150 MPa. Overload factor is 1.75 and service factor 1.25. Face width can be taken as 12 times the module and helix angle 30°. Find the module and face width.

$$[m = 6, b = 60 \text{ mm}]$$

8. Calculate safe power, which can be transmitted through two parallel axis helical gears having 20° full depth teeth with helix angle 25° and normal module 6, face width 80 mm. Pinion speed is 1,500 rpm and has 24 teeth with gear ratio 4. Gear material has BHN = 200 and allowable stress is 185 MPa. Fatigue factor can be taken as 1.5 and overload factor 1.2.

$$[79.2 \text{ kW}]$$

9. A pair of 20° helical gears mounted on parallel shafts at 300 mm apart transmits 20 kW at 1,000 rpm. Normal module is 6 mm, and number of teeth on pinion 28 and on gear 56. Calculate helix angle and tooth forces at the center of the tooth.

$$[a = 32.8°, F_t = 1,852 \text{ N}, F_r = 674 \text{ N}, F_a = 1,193 \text{ N}]$$

10. A helical pinion has PCD 145 mm and face width 250 mm with helix angle 25° and pressure angle in the diametrical plane 20°. Gear ratio is 5 and BHN is 200. Calculate wear load capacity. [39 kN]

11. A pair of parallel axis helical gears with full depth involute teeth transmits 16 kW at 900 rpm. Number of teeth on pinion are 40 and on gear 80. Center distance is 325mm, normal module is 5, face width is 50 mm, and normal pressure angle is 20°. Ultimate strength of the gear material is 570 MPa and surface hardness is 280 BHN. Gears are cut with grade 6 accuracy. Service factor is 1.2. Calculate:

   (a) Helix angle,
   (b) Beam strength in bending,
   (c) Wear strength,
   (d) Dynamic load,
   (e) FOS in bending, and
   (f) FOS for pitting.

$$[\alpha = 22.6°, S_b = 20,284 \text{ N}, S_w = 21,260 \text{ N}, F_d = 10,518 \text{ N}, FOS_{bending} = 1.93, FOS_{wear} = 2.02]$$

12. A pair of helical gears is to transmit 15 kW. The teeth are 20° stub in diameter plane and have a helix angle 45°. The pinion runs at 1,000 rpm and has 80 mm pitch diameter. The gear has 320 mm pitch diameter. If the gears are made of cast steel having allowable strength of 100 MPa, determine suitable module and face width from static strength considerations and check the gears for wear. Given:

σ_{es} = 618 MPa, Face width b = 12.5 m, E_p = E_g = 200 kN/m^2

$$C_v = \frac{0.75}{0.75 + \sqrt{v'}}, \text{ Lewis form factor } y = 0.175 - \frac{0.841}{T_e}$$

[m = 2.5 mm, b = 45 mm, S_w = 5,554 N]

13. A helical gear transmits 20 kW with pinion of 20 teeth and module 5 running at 1,200 rpm. Pressure angle is 20° and helix angle 25°. Determine:

    (a) Tangential force

    (b) Radial force

    (c) Axial thrust

    Assume suitable positions of gear and direction of rotation accordingly draw a force component diagram. [F_t = 3,185 N, F_r = 1,107 N, F_t = 1,485 N]

14. A pair of helical gears consists of a 20 teeth pinion meshing with a 100 teeth gear. The pinion rotates at 720 rpm. The normal pressure angle is 20°. The face width is 40 mm and the normal module is 4 mm. The pinion as well as gear are made of steel having ultimate tensile strength of 600 MPa and heat treated to a surface hardness of 300 BHN. The service factor and FOS are 1.5 and 2, respectively. Assume that the velocity factor accounts for the dynamic load, calculate the power transmitting capacity of the gear. [P = 59 kW]

Previous Competition Examination Questions

IES

1. Consider the following statements:

   In case of helical gears teeth are cut at an angle to the axis of rotation of the gear.

   1. Helix angle introduces another ratio called axial contact ratio.

   2. Transverse contact ratio is equal to axial contact ratio in helical gears.

   3. Large transverse contact ratio does not allow multiple teeth to share the load.

   4. Large axial contact ratio will cause larger axial force component.

   Which of the above statements are correct?

   (a) 1 and 2   (b) 2 and 3   (c) 1 and 4   (d) 3 and 4

   [Answer (c)] [IES 2015]

GATE

2. The following data are the data for two crossed helical gears used for speed reduction.

   Gear I: PCD in plane of rotation 80 mm and helix angle 30°.

   Gear II: Pitch circle diameter in plane of rotation 120 mm and helix angle 22.5°.

   If the input speed is 1,440 rpm, the output speed in rpm is:

   (a) 1,200   (b) 900   (c) 875   (d) 720

   [Answer (b)] [GATE 2012]

□□□

Bevel Gears

Outcomes

➤ Learn about the shape of bevel gears and their terminology.

➤ Types of bevel gears

➤ Pitch angle and gear ratio for these gears

➤ Properties of bevel gears

➤ Formulative number of teeth

➤ Forces on gear tooth

➤ Strength of a bevel gear tooth

➤ Dynamic and wear strength of these gears

➤ Spiral bevel gears

7.1 Introduction

Bevel gears are used to transmit power from generally intersecting shafts, but axes are inclined to each other. Generally, the angle between the shafts is 90° but they can transmit power at any angle. The teeth are cut on a conical surface. The apex of the cone lies on the shaft axis and is known as vertex. The cross section of the tooth decreases from outer radius towards apex side. Use of these gears in a differential of an automobile is shown in the picture on the right side.

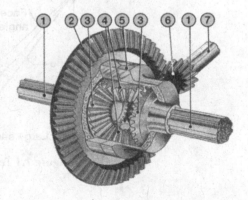

Spur gears of same module can be used with any other gear, but bevel gear cannot be used with any other bevel gear and hence these are always designed in pairs. The sum of pitch cone angles has to be equal to angle between the axes of the shafts.

7.2 Terminology

Figure 7.1 illustrates various terms used for bevel gears.

Large end	A bevel gear is in the form of a frustum of a cone. The end containing the largest radius of the cone is called large end.
Cone center	Apex of the cone is called cone center.
Pitch cone	Like other gears, pitch cone is an imaginary cone, whose surface contains the pitch line of the teeth.
Cone distance (L_o)	It is the distance from the apex to pitch line at large end. It is also called pitch cone radius.
Pitch line	A straight line joining the cone center and a point on large end of pitch circle diameter (PCD).
Pitch diameter (d)	It is the diameter of the pitch cone at the large end.
Pitch angle (γ)	Angle between pitch line and axis of cone.

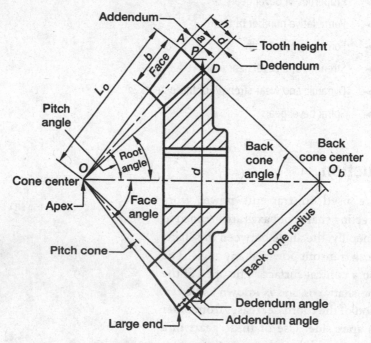

Figure 7.1 Terms used with bevel gears

Addendum (a)	Height of the tooth at the large end from the pitch line.
Addendum angle	It is the angle between outside face of tooth and pitch angle.
Dedendum (d)	Depth of the tooth at the large end, below the pitch line.
Dedendum angle	It is the angle between root of the tooth and pitch angle.
Tooth depth (h)	Sum of addendum and dedendum at the large end.
Face width (b)	Inclined length of the tooth.
Face angle	It is the angle between face of tooth with the axis of the cone.
Root angle	It is the angle between root of tooth with the axis of the cone.
Back cone	It is also a frustum of a cone on the backside of the cone containing teeth. Its inclined face is perpendicular to the pitch line.
Back cone angle	It is the angle between face of the back with the axis of the cone.
Back cone radius	It is the inclined distance from the apex of the back cone to pitch line at large end.

7.3 Types of Bevel Gears

Bevel gears are classified on the basis of the shape of their teeth as under:

a. **Straight bevel gears**   Straight teeth converge towards the apex of the cone [Figure 7.2(a)]. These gears are noisy like spur gears. These are easy to manufacture than other types of bevel gears.

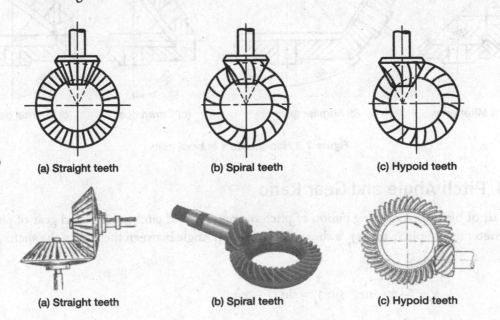

| (a) Straight teeth | (b) Spiral teeth | (c) Hypoid teeth |

| (a) Straight teeth | (b) Spiral teeth | (c) Hypoid teeth |

Figure 7.2 Types of bevel gears

b. **Spiral bevel gears** Teeth are curved, but axes of the shafts still intersect [Figure 7.2(b)]. These gears operate at lesser noise than straight teeth gears, because contact of the teeth is gradual. These are difficult to manufacture. Their use can be seen in the differential of heavy vehicles.

c. **Hypoid bevel gears** Teeth are curved, but the axes of the shafts do not intersect [Figure 7.2(c)]. There is some offset between the axes. These gears also operate silently and difficult to manufacture. Their use can be seen in the differential of cars.

Bevel gears are named on the angle between the intersecting shafts as described below:

a. **Miter gears** When two bevel gears of equal size mesh with shafts at right angle to each other as shown in Figure 7.3(a), they are called miter gears. The pitch angle γ of each gear is 45°. There is no change in speed of shafts as both have equal number of teeth.

b. **Angular gears** The angle between the shafts is of any other value except 90° [See Figure 7.3(b)]. Pitch angle γ of one of the gears is less than 45°.

c. **Crown gears** The angle between the axes of the shafts is more than 90° as shown in Figure 7.3(c). Pitch angle γ of one of the gears is 90°.

d. **Internal gears** The angle between the axes of the shafts is more than 90° [Figure 7.3(d)]. Pitch angle γ of one of the gears is greater than 90°.

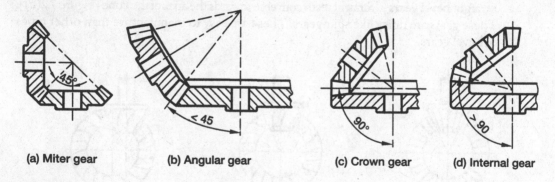

(a) Miter gear   (b) Angular gear   (c) Crown gear   (d) Internal gear

Figure 7.3 Nomenclature of bevel gears

7.4 Pitch Angle and Gear Ratio

A pair of bevel gears having pinion of pitch diameter d_p and pitch angle γ_p and gear of pitch diameter d_g and pitch angle γ_g is shown in Figure 7.4. Angle between the axes of the shafts Γ:

$$\Gamma = \gamma_p + \gamma_g$$

Or, $\qquad \gamma_g = \Gamma - \gamma_p \quad$ or $\quad \sin \gamma_g = \sin (\Gamma - \gamma_p)$

Or, $\qquad \sin \gamma_g = \sin \Gamma \cos \gamma_p - \cos \Gamma \sin \gamma_p$ $\hfill$ (7.1)

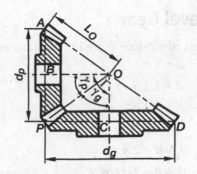

Figure 7.4 Pitch angle of bevel gears

Gear ratio $G = \dfrac{d_g}{d_p} = \dfrac{2 \times \text{OP} \sin \gamma_g}{2 \times \text{OP} \sin \gamma_p} = \dfrac{\sin \gamma_g}{\sin \gamma_p}$

Or, $\quad \sin \gamma_g = G \times \sin \gamma_p$ $\qquad\qquad$ (7.2)

Equating Equations (7.1) and (7.2):

$$G \times \sin \gamma_p = \sin \Gamma \cos \gamma_p - \cos \Gamma \sin \gamma_p$$

Dividing both sides by $\cos \gamma_p$:

$$G \times \tan \gamma_p = \sin \Gamma - \cos \Gamma \tan \gamma_p$$

Or, $\quad \tan \gamma_p = \dfrac{\sin \Gamma}{G + \cos \Gamma}$ $\qquad\qquad$ (7.3)

Similarly, $\quad \tan \gamma_g = \dfrac{\sin \Gamma}{\left(\dfrac{1}{G}\right) + \cos \Gamma}$ $\qquad\qquad$ (7.4)

7.5 Cone Distance

Referring to Figure 7.4, cone distance (L_o) is the distance from apex to pitch circle at large end, that is, OA.

$$\tan \gamma_p = \frac{\text{PB}}{\text{BO}} = \frac{\text{PB}}{\text{PC}} = \frac{0.5 \times d_p}{0.5 \times d_g} = \frac{d_p}{d_g} = \frac{m \times Z_p}{m \times Z_g} = \frac{Z_p}{Z_g}$$

$$\tan \gamma_g = \frac{\text{PC}}{\text{CO}} = \frac{\text{PC}}{\text{PB}} = \frac{0.5 \times d_g}{0.5 \times d_p} = \frac{d_g}{d_p} = \frac{m \times Z_g}{m \times Z_p} = \frac{Z_g}{Z_p}$$

$$L_0 = \sqrt{\text{AB}^2 + \text{BO}^2} = \sqrt{\left(0.5 d_p\right)^2 + \left(0.5 d_g\right)^2} \qquad\qquad (7.5)$$

7.6 Proportions of Bevel Gear

The values given below in terms of module m are to be measured at large end.

Addendum	$a = 1\,m$
Dedendum	$d = 1.2\,m$
Tooth depth	$h = 2.2\,m$
Tooth thickness	$t = 1.571\,m$
Outside diameter of cone	$d_o = d + 2\,a\cos\gamma$
Root diameter of cone	$d_o = d - 2\,d\cos\gamma$

7.7 Formulative Number of Teeth

The teeth of bevel gear converge towards apex and hence their depth and width goes on decreasing towards apex of the cone. Thus, the involute profile does not lie on a straight plane but lies on a surface of a sphere. The beam strength and wear of gear tooth are based on tooth profile. So the profile is approximated accurately using Tredgold's approximation. According to this approximation, a tangent to the sphere at the pitch point can be considered as the surface for a small distance (Figure 7.5). Hence the back cone can be used as a plane for generating involute profile at large end. Radius of the back cone r_b is taken an imaginary pitch circle radius for generating involute profile. Let Z' be the number of teeth on this pitch circle. The gear with Z' number of teeth and PCD of $2r_b$ is called formative gear. The number Z' is called virtual or formative or equivalent number of teeth. Thus:

$$Z' = \frac{2r_b}{m} \tag{a}$$

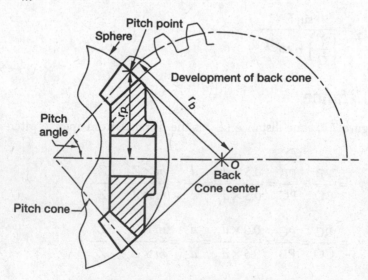

Figure 7.5 Formulative teeth for bevel gears

If m is the module and d is PCD, then the number of teeth Z is given as:

$$Z = \frac{d}{m} \tag{b}$$

From Equations (a) and (b):

$$\frac{Z'}{Z} = \frac{2r_b}{d} \tag{c}$$

Since surface of back cone is perpendicular to the large end, d is related as:

$$d = 2r_b \cos \gamma \tag{7.6}$$

Substituting this value of d in Equation (c):

$$Z' = \frac{Z}{\cos \gamma} \tag{7.7}$$

The action of bevel gears is same as that of equivalent teeth on a helical gear. Because Z' is always bigger than Z, hence larger contact ratios for bevel gears make them run more smooth than with spur gears of same number of actual teeth.

7.8 Forces on Gear Tooth

7.8.1 Forces on tooth of pinion

While considering forces on gear tooth, it is assumed that the resultant force is a concentrated load at the midpoint P of the face on pitch cone (Figure 7.6). The resultant force F comprises of three forces:

F_t = Tangential force transmitting torque, shown by arrow AP

F_r = Radial force towards center, shown by arrow BP

F_a = Axial force causing axial thrust in the shaft / bearings, shown by arrow CP

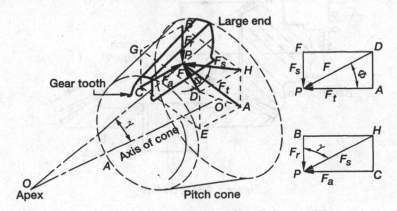

Figure 7.6 Forces on a gear tooth

$$\tan \Phi = \frac{AD}{AP} = \frac{F_s}{F_t}$$

Where, F_s = Separating force which tries to separate the gears

$$F_s = F_t \tan \Phi \tag{7.8a}$$

$$F_r = F_s \cos \gamma \tag{7.8b}$$

$$F_a = F_s \sin \gamma \tag{7.8c}$$

In the design data, only tangential force F_t is known, which is calculated from the power and speed of rotation:

$$P = \frac{2 \pi NT}{60,000} = \frac{2 \pi N F_t r_m}{60,000}$$

Or,
$$F_t = \frac{60,000 P}{2 \pi N r_m} \tag{7.9}$$

Where, γ_m = Mean radius of bevel gear at pitch circle in the middle of its inclined face (mm)

P = Design power (kW)

N = Speed of driver shaft (rpm)

Mean radius γ_m can be calculated from the simple geometry of the gear as (See in Figure 7.7):

$$r_m = \left[\frac{d}{2} - \frac{b \sin \gamma}{2} \right] \tag{7.10}$$

Where, b = Length of face

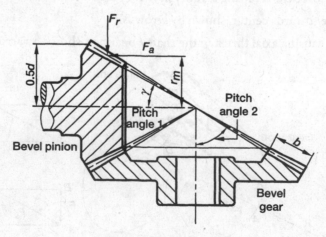

Figure 7.7 Mean radius

Converting Equations (7.8b) and (7.8c) in terms of tangential force, using Equation (7.8a). From above equations:

$$F_r = F_t \tan \Phi \cos \gamma \qquad\qquad (7.11)$$

$$F_a = F_t \tan \Phi \sin \gamma \qquad\qquad (7.12)$$

7.8.2 Forces on tooth of bevel gear

Direction of rotation of driven bevel gear is opposite to the direction of driver gear. Hence, direction of tangential force in a gear is same as its direction of rotation.

- **Radial** component acts towards the center of the gear.
- **Axial force** on the gear is equal and opposite to the radial force on the pinion, if the angle between the axes of the shafts is 90°. This force tends to separate the gears.

Example 7.1

Forces on tooth of a bevel gear for given power

A bevel gear set transmits 5 kW power at 400 rpm through 20° pressure angle full cut teeth. PCDs of pinion and gear are 200 mm and 300 mm, respectively. Face length of the teeth is 25 per cent of the cone distance. Assuming angle between the shafts as 90°, calculate the tangential, radial, and axial forces on the pinion and gear.

Solution

Given $P = 5\ kW$ $N1 = 400\ rpm$ $d_p = 200\ mm$ $d_g = 300\ mm$

$\qquad\qquad \Phi = 20°$ $b = 0.25\ Lo$ $\Gamma = 90°$

Forces on pinion:

Torque $T = \dfrac{60,000 P}{2 \pi N} = \dfrac{60,000 \times 5}{2 \times 3.14 \times 400} = 119.4\ Nm$

$\qquad \tan \gamma_p = \dfrac{d_p}{d_g} = \dfrac{200}{300} = 0.667$ or $\gamma_p = 33.7°$

Cone distance $L_0 = \sqrt{\left(0.5 d_p\right)^2 + \left(0.5 d_g\right)^2} = \sqrt{100^2 + 150^2} = 180\ mm$

Face length $b = 0.25 \times 180 = 45\ mm$

$$r_m = \left[\dfrac{d}{2} - \dfrac{b \sin \gamma_p}{2}\right] = \left[\dfrac{200}{2} - \dfrac{45 \sin 33.7°}{2}\right] = 100 - 12.5 = 87.5\ mm$$

$$F_t = \dfrac{T}{r_m} = \dfrac{119.4 \times 1,000}{87.5} = 1,365\ N$$

$F_r = F_t \tan \Phi \cos \gamma_p = 1,365 \times \tan 20° \times \cos 33.7° = 1,365 \times 0.364 \times 0.83 = 413.4\ N$

$F_a = F_t \tan \Phi \sin \gamma_p = 1,365 \times \tan 20° \times \sin 33.7° = 1,365 \times 0.364 \times 0.555 = 275.7\ N$

Forces on gear:

Tangential component of the force will remain the same as on pinion = 1,365 N

Radial component will be equal to axial component of the pinion = 275.7 N

Axial component will be equal to radial component of the pinion = 413.4 N

7.9 Strength of Bevel Gear Tooth

The tooth thickness of a bevel gear varies along its length, hence its strength is calculated at the large end considering and it is equivalent to formulative spur gear in a plane perpendicular to the tooth axis at pitch circle. Figure 7.8 shows a gear tooth with a point P at which a load equal to beam strength of the tooth S_b acts tangentially.

Lewis equation for the tapered tooth of bevel gear is modified as:

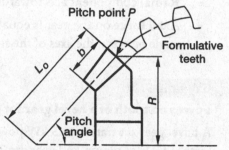

Figure 7.8 Beam strength of a bevel gear tooth

$$S_b = m\, b\, \sigma_b\, y\left[1 - \frac{b}{L_0} + \frac{b^2}{3L_0^2}\right] \qquad (7.13a)$$

Where, S_b = Safe beam strength (N)

m = Module at large end (mm)

b = Face width (mm)

σ_b = Permissible bending stress (MPa)

L_o = Cone distance (mm)

Permissible stress σ_b is usually taken as one-third of ultimate bending strength ($S_{ut}/3$). Face width b is not kept more than ($L_o/3$). Hence the terms with higher order of b are much smaller than L_o and can be neglected. The terms in the parenthesis in Equation (7.13) are:

$$\left[1 - \frac{b}{L_0} + \frac{b^2}{3L_0^2}\right] \cong \left[1 - \frac{b}{L_0}\right]$$

The factor $\left[1 - \dfrac{b}{L_0}\right]$ is called as bevel factor. Hence the module can be calculated from Equation (7.13b).

$$S_b = k_v\, m\, b\, \sigma_b\, y\left[1 - \frac{b}{L_0}\right] \qquad (7.13b)$$

Example 7.2

Module and size of gears for given power

A compressor is driven at 600 rpm using 15 kW motor at 1,000 rpm using a 90° bevel gears with pressure angle 20°. Pinion has 24 form cut teeth with face width 25 per cent of cone length. Bending strength of the material for both the gears is 120 MPa. Assume service factor 1.25. Calculate the following:

  a. Cone angles

  b. Module

  c. Size of pinion and gear

Solution

Given $P = 15\,\text{kW}$ $N1 = 1,000\,\text{rpm}$ $N2 = 600\,\text{rpm}$ $\Gamma = 90°$ $\Phi = 20°$

$Z_p = 24$ $b = 0.25\,L_o$ $\sigma_b = 120\,\text{MPa}$ $K_s = 1.25$

a. Velocity ratio $G = \dfrac{1,000}{600} = 1.667$

  Number of teeth on gear: $Z_g = 24 \times 1.667 = 40$

  Pitch angle of pinion $\gamma_p = \tan^{-1}\left(\dfrac{1}{G}\right) = \tan^{-1}\left(\dfrac{1}{1.667}\right) = 31°$

  Pitch angle of gear $\gamma_g = \Gamma - \gamma_p = 90° - 31° = 59°$

b. Formulative number of teeth on pinion $= Z_p \sec \gamma_p = 24 \sec 31° = 28$

  Formulative number of teeth on gear $= Z_g \sec \gamma_g = 40 \sec 59° = 77.7$

$$y_p = 0.154 - \frac{0.912}{Z'_p} = 0.154 - \frac{0.912}{28} = 0.121$$

$$y_g = 0.154 - \frac{0.912}{Z'_g} = 0.154 - \frac{0.912}{77.7} = 0.142$$

Since material for pinion and gear has same strength and y_p for pinion is lesser so the design will be based on strength of pinion.

Torque on pinion $T = \dfrac{60,000P}{2\pi N} = \dfrac{60,000 \times 15}{2\pi \times 1,000} = 143.3\,\text{Nm} = 143,300\,\text{Nmm}$

PCD of pinion $d_p = m \times Z_p$

$$F_t = \frac{2T}{d_p} = \frac{2 \times 143,300}{m \times Z_p} = \frac{2 \times 143,300}{m \times 24} = \frac{5,970}{m}$$

$$v = \frac{\pi d_p N_p}{60,000} = \frac{\pi m \times Z_p \times N_p}{60,000} = \frac{3.14m \times 24 \times 1,000}{60,000} = 1.256m \text{ m/s}$$

$$L_o = \frac{d_p}{2 \sin \gamma_p} = \frac{24m}{2 \sin 31°} = 23.3m \text{ mm}$$

$$b = 0.25 \quad L_o = 0.25 \times 23.3m = 5.825m \text{ mm}$$

$$k_v = \frac{3}{3+v} = \frac{3}{3+1.256m}$$

Dynamic beam strength $S_b = \sigma_b \dfrac{k_v}{K_s} b \pi m y_p \left[1 - \dfrac{b}{L_o}\right]$

Substituting the values:

$$\frac{5,970}{m} = 120 \times \frac{3}{3+1.256m} \times \frac{1}{1.25} \times 5.825m \times 3.14 \times m \times 0.121 \times \left[1 - 0.25\right]$$

From the above equation: $m^3 - 13.2\, m - 31.6 = 0$

By trial and error, m is between 4 and 5. Selecting standard value of module = 5

c. PCD of pinion $d_p = 24\, m = 24 \times 5 = 120$ mm

Outside diameter $d_{op} = 24\, m + 2\, m \cos \gamma_p = 120 + 10 \cos 31° = 128.6$ mm

Face width $b = 5.825\, m = 5.825 \times 5 = 29$ mm

PCD of gear $d_g = 40\, m = 40 \times 5 = 200$ mm

Outside diameter $d_{og} = 40\, m + 2\, m \cos \gamma_g = 200 + 10 \cos 59° = 205.2$ mm

---

Example 7.3

Power capacity of a bevel gear drive set

A bevel pinion with 25 full depth teeth set of module 5 and pressure angle 20° rotating at 2,000 rpm transmits power to a bevel gear at 90° with speed reduction of 8. Face width is 50 mm and allowable bending strength 200 MPa. Calculate the power, which this gear set can transmit?

Solution

Given $N_p = 2,000$ rpm $G = 8$ $\Gamma = 90°$ $\Phi = 20°$ $Z_p = 25$

     $b = 50$ mm $\sigma_b = 200$ MPa $m = 5$

Diameter of pinion $d_p = m \times Z_p = 5 \times 25 = 125$ mm

$$\tan \gamma_p = \frac{d_p}{d_g} = \frac{125}{8 \times 125} = 0.125 \quad \text{or} \quad \gamma_p = 7.1°$$

Formulative number of teeth on pinion $Z'_p = Z_p \sec \gamma_p = 25 \times \sec 7.1° = 25.2$

$$y_p = 0.154 - \frac{0.912}{Z'_p} = 0.154 - \frac{0.912}{25.2} = 0.118$$

$$L_o = 0.5\sqrt{d_p^2 + d_g^2} = 0.5\sqrt{(125)^2 + (1,000)^2} = 504 \text{ mm}$$

$$S_b = m\,b\,\sigma_b\,y_p\left[1 - \frac{b}{L_o}\right] = 5 \times 50 \times 200 \times 0.118 \times \left[1 - \frac{50}{504}\right] = 5,315 \text{ N}$$

Tangential force F_t can be equal to beam strength of tooth, that is, $F_t = 5,315$ N

Torque = Force × radius = $5,315 \times (0.5 \times 0.125) = 332$ Nm

$$\text{Power } P = \frac{2\,\pi\,N\,T}{60,000} = \frac{2 \times 3.14 \times 2,000 \times 332}{60,000} = 69.5 \text{ kW}$$

7.10 Dynamic Load

Dynamic load on a bevel gear occurs due to inaccuracy in the profile of gear tooth while manufacturing. The load capacity is calculated in similar way as for spur gears except that velocity has to be calculated at the pitch radius at large end. The tangential force should also be based on this velocity. The dynamic load capacity is given by Buckingham equation (7.14).

$$F_d = F_t + \frac{20.7v\left(Ceb + F_t\right)}{20.7v + \sqrt{Ceb + F_t}} \tag{7.14}$$

Where, F_d = Dynamic load

C = Dynamic load factor $= \dfrac{K}{\left[\dfrac{1}{E_p} + \dfrac{1}{E_g}\right]}$

K = A constant depending upon pressure angle;

$\quad K = 0.107$ for 14.5°

$\quad K = 0.111$ for 20° full depth

$\quad K = 0.115$ for 20° stub

E_p, E_g = Young's modulus of elasticity for pinion and gear materials, respectively

$\quad v$ = Velocity at pitch circle at the large end (m/s)

$\quad b$ = Face width (mm)

$\quad e$ = Error as given in Table 7.1:

For a safe design, the endurance beam strength should be more than dynamic loads.

Table 7.1   Maximum Error between Meshing Gears

Module	Commercial Gears Class I	Gears Cut with Care Class II	Ground and Lapped Gear Class III
1–4	0.05	0.025	0.0125
5	0.056	0.025	0.0125
6	0.064	0.03	0.015
7	0.072	0.035	0.017
8	0.08	0.038	0.019
9	0.085	0.041	0.0205
10	0.09	0.044	0.022

7.11 Wear Strength

Wear strength is defined as the maximum tangential force at the large end of a bevel gear, which can be transmitted without any pitting. This limiting load is calculated using the following equation:

$$S_w = d' \, b \, Q \, K_w \tag{7.15a}$$

Where, S_w = Wear strength (MPa)

d' = PCD of formulative pinion (mm) $= \dfrac{d}{\cos \gamma}$

b = Face width (mm)

Since one bevel gear is overhung it deflects. Therefore, assuming only 75 per cent of its width is contributing in sharing load, Equation (7.15) becomes:

$$S_w = \frac{0.75 \, b \, d \, Q \, K_w}{\cos \gamma} \tag{7.15b}$$

$$Q = \text{Tooth ratio factor} \ = \frac{2Z'_g}{Z'_g + Z'_p} = \frac{2Z_g}{Z_g + Z_p \tan \gamma_p} \tag{7.16}$$

$$K_w = \text{Material constant} = \frac{\sigma_c^2 \sin \Phi \cos \Phi \left[\dfrac{1}{E_p} + \dfrac{1}{E_g} \right]}{1.4} \tag{7.17}$$

For pressure angle $\Phi = 20°$ and same materials for pinion and gear as steel, K_w is simplified as:

$$K_w = 0.16 \left(\frac{\text{BHN}}{100} \right)^2 \tag{7.18}$$

For a safe design, wear strength S_w should be more than the tangential load at large end.

Example 7.4

Safe dynamic and wear load for given power

An electric motor rotating at 1,440 rpm transmits 7.5 kW of power with speed reduction 2.5 times through 90° class II bevel gears set. Pinion has 20 teeth generated at 14.5° pressure angle. Allowable stress for pinion made of steel is 80 MPa and for cast iron gear 50 MPa. Face width is 28 per cent of cone length. Assume service factor 1.35. Calculate the following:

  a. Outside and pitch circle diameters of pinion and gear

  b. Safe wear load, if Young's modulus of elasticity for pinion and gear are 210 GPa and 105 GPa, respectively. Endurance strength 500 MPa

  c. Safe dynamic load.

  d. Factor of safety against bending and wear.

Solution

Given $P = 7.5$ kW $N_p = 1,440$ rpm $\Gamma = 90°$ Class II cut $G = 2.5$ $Z_p = 20$

         $\Phi = 14.5°$ $\sigma_{bp} = 80$ MPa $\sigma_{bg} = 50$ MPa $b = 0.28\,L_o$ $K_s = 1.35$

         $E_p = 210$ GPa $E_g = 110$ GPa $S_e = 500$ MPa

  a. Size of gears: Number of teeth on gear $Z_g = G \times Z_p = 2.5 \times 20 = 50$

Pitch angle of pinion $\gamma_p = \tan^{-1}\left(\dfrac{1}{G}\right) = \tan^{-1}\left(\dfrac{1}{2.5}\right) = 21.8°$

Pitch angle of gear $\gamma_g = \Gamma - \gamma_p = 90 - 21.8 = 68.2°$

Formulative number of teeth on pinion $Z'_p = Z_p \sec \gamma_p = 20 \sec 21.8° = 21.5$

Formulative number of teeth on gear $Z'_g = Z_g \sec \gamma_g = 50 \sec 68.2° = 134.6$

For $\Phi = 14.5°$ pinion, $y'_p = 0.124 - \dfrac{0.684}{Z'_p} = 0.154 - \dfrac{0.684}{21.5} = 0.0922$

For gear $y'_g = 0.124 - \dfrac{0.684}{Z'_g} = 0.154 - \dfrac{0.684}{134.6} = 0.1189$

$\sigma_{bp} \times y'_p = 80 \times 0.0922 = 7.37$ MPa

$\sigma_{bg} \times y'_g = 50 \times 0.1189 = 5.946$ MPa

From these two values, strength of gear is less, hence the design will be based on strength of gear.

Speed of gear $N_g = \dfrac{N_p}{G} = \dfrac{1,440}{2.5} = 456$ rpm

Torque on gear $T = \dfrac{60,000P}{2\pi N} = \dfrac{60,000 \times 7.5}{2\pi \times 456} = 157\ \text{Nm} = 157,000\ \text{Nmm}$

PCD of gear $d_g = m \times Z_g = 50\ m$

$F_t = \dfrac{2T}{d_p} = \dfrac{2 \times 157,000}{50m} = \dfrac{6,280}{m}\ \text{N}$

$v = \dfrac{\pi d_g N_g}{60 \times 1,000} = \dfrac{\pi \times 50m \times 456}{60 \times 1,000} = 1.1932m\ \text{m/s}$

$k_v = \dfrac{6}{6+v} = \dfrac{6}{6+1.1932m}$

$L_o = \dfrac{mZ_g}{2\sin\gamma_g} = \dfrac{50m}{2\sin 68.2°} = 26.92m\ \text{mm}$

$b = 0.28\ L_o = 0.28 \times 26.92\ m = 7.54\ m\ \text{mm}$

Bevel factor $\left[1 - \dfrac{b}{L_o}\right] = (1 - 0.28) = 0.72$

Design force $F_t = \sigma_{bg}\dfrac{k_v}{K_s}\ b\,\pi\,m\,y_g\left[1 - \dfrac{b}{L_o}\right]$

Substituting the values:

$\dfrac{6,280}{m} = 50 \times \dfrac{6}{6+1.1932m} \times \dfrac{1}{1.35} \times 7.54m \times 3.14 \times m \times 0.1189 \times 0.72$

Or, $\dfrac{6,280}{m} = \dfrac{450.4m^2}{6 + 1.1932m}$

Simplifying the equation: $\qquad m^3 - 16.63m - 83.66 = 0$

Solve by trial and error method:

$\qquad m = 5 \quad \text{LHS} = 125 - 83.15 - 83.66 = -41.3$

$\qquad m = 6 \quad \text{LHS} = 216 - 99.78 - 83.66 = 32.56$

So approximate value of $m = 5.5$

Selecting a standard module $m = 6$ is selected.

Pinion PCD $\quad\quad\quad\quad d_p = Z_p \times m = 20 \times 6 = 20 \times 6 = 120$ mm

Outside diameter of pinion $d_{op} = 120 + 2m \cos \gamma_p = 120 + 12 \cos 21.8° = 131.14$ mm

PCD of gear $\quad\quad\quad\quad d_g = Z_g \times m = 50 \times 6 = 300$ mm

Outside diameter $\quad\quad d_{og} = 300 + 2m \cos \gamma_g = 300 + 12 \cos 68.2° = 304.45$ mm

Face width $\quad\quad\quad\quad b = 7.54 \, m = 7.54 \times 6 = 45.2$ mm

b. Wear strength: $K_w = \dfrac{S_e^2 \sin \Phi \cos \Phi}{1.4}\left[\dfrac{1}{E_p} + \dfrac{1}{E_g}\right]$

Substituting the values: $K_w = \dfrac{500^2 \sin 14.5° \cos 14.5°}{1.4}\left[\dfrac{1}{210,000} + \dfrac{1}{105,000}\right]$

Or, $\quad K_w = \dfrac{250,000 \times 0.25 \times 0.97 \times 3}{1.4 \times 210,000} = 0.62$

$$Q = \frac{2Z_g'}{Z_g' + Z_p'} = \frac{2 \times 134.6}{134.6 + 21.8} = 1.72$$

Wear strength: $S_w = \dfrac{0.75bdQ K_w}{\cos \gamma_p} = \dfrac{0.75 \times 45.2 \times 120 \times 1.72 \times 0.62}{\cos 21.8°} = 4{,}673$ N

Tangential force $F_t = \dfrac{2T}{d_g} = \dfrac{2 \times 157,000}{300} = 1{,}046.7$ N (Tangential force on gear

and pinion will remain same)

c. Safe dynamic load:

For pressure angle 14.5° value of $K = 0.107$

$$C = \frac{K}{\left[\dfrac{1}{E_p} + \dfrac{1}{E_g}\right]} = \frac{0.107}{\left[\dfrac{1}{210,000} + \dfrac{1}{105,000}\right]} = 7{,}490 \text{ Nmm}$$

From Table 7.1 for $m = 6$ for class II cut gears maximum error is $e = 0.03$ mm

$C e b + F_t = (7,490 \times 0.03 \times 45.2) + 1,046.7 = 11,203$

From Equation (7.14), $\quad F_d = F_t + \dfrac{20.7v\left(Ceb + F_t\right)}{20.7v + \sqrt{Ceb + F_t}}$

Or, $F_d = 1,046.7 + \dfrac{20.7 \times (1.19 \times 6) \times 11,203}{(20.7 \times 1.19 \times 6) + \sqrt{11,203}} = 1,046.7 + 6,542 = 7,588.7$ N

d. Factor of Safety

Effective load $F_e = F_d + (K_s \times F_t) = 7,588.7 + (1.35 \times 1,046.7) = 9,002$ N

Beam strength $S_b = \sigma_{bg} \, b \, \pi m y_g \left[1 - \dfrac{b}{L_o} \right]$

Substituting the values $S_b = 50 \times 45.2 \times 3.14 \times 6 \times 0.1189 \times (1 - 0.28) = 3,645$ N

FOS in bending $= \dfrac{\text{Beam strength}}{\text{Effective load}} = \dfrac{3,645}{9,002} = 0.41$ (It is less than 1 hence fail in strength)

FOS in wear $= \dfrac{\text{Wear strength}}{\text{Effective load}} = \dfrac{4,673}{9,002} = 0.52$ (It is less than 1 hence fail in wear)

Example 7.5

Factor of Safety in strength and wear for given power
A mill is driven by a pair of bevel gears having shafts at 90°. Pinion is driven at 1,440 rpm using 2 kW motor. Involute teeth of module 4 are carefully generated with pressure angle 20°. Pinion has 18 and gear 27 teeth. Face width is 30 per cent of cone length. Bending strength of the material for both for pinion and gears is 50 MPa and endurance strength 500 MPa. Value of E for both materials is 82 GPa. Assume service factor 1.2. Calculate FOS in strength and wear.

Solution

Given $P = 2\,kW$ $N1 = 1,440\,rpm$ $\Gamma = 90°$ $m = 4$ $\Phi = 14.5°$ $Z_p = 18$ $Z_g = 27$

$b = 0.3\,L_o$ $\sigma_b = 50\,MPa$ $S_e = 500\,MPa$ $K_s = 1.2$ $E_p = E_g = 82\,GPa$

Gear ratio, $G = \dfrac{27}{18} = 1.5$

Pitch angle of pinion $\gamma_p = \tan^{-1}\left(\dfrac{1}{G}\right) = \tan^{-1}\left(\dfrac{1}{1.5}\right) = 33.7°$

Pitch angle of gear $\gamma_g = \Gamma - \gamma_p = 90 - 33.7 = 56.3°$

Formulative number of teeth on pinion $Z'_p = Z_p \sec \gamma_p = 18 \sec 33.7° = 21.6$

Formulative number of teeth on gear $Z'_g = Z_g \sec \gamma_g = 27 \sec 56.3° = 48.7$

Since material for pinion and gear has same strength and y_p for pinion is lesser, the design will be based on strength of pinion.

$$y'_p = 0.154 - \frac{0.912}{Z'_p} = 0.154 - \frac{0.912}{21.6} = 0.112$$

$$d_p = m \times Z_p = 4 \times 18 = 72 \text{ mm}$$

$$d_g = m \times Z_g = 4 \times 27 = 108 \text{ mm}$$

$$v = \frac{\pi d_p N_p}{60 \times 1,000} = \frac{\pi \times 72 \times 1,440}{60 \times 1,000} = 5.246 \text{ m/s}$$

$$K_v = \frac{6}{6+v} = \frac{6}{6+5.246} = \frac{6}{11.246} = 0.525$$

$$L_o = \frac{d_p}{2 \sin \gamma_p} = \frac{72}{2 \sin 33.7^\circ} = 65 \text{ mm}$$

$$b = 0.3\, L_o = 19.5 \text{ mm}$$

Torque on pinion $T = \dfrac{60P}{2\pi N} = \dfrac{60 \times 2,000}{2\pi \times 1,440} = 13.27 \text{ Nm} = 13,270 \text{ Nmm}$

Tangential force $F_t = \dfrac{2T}{d_p} = \dfrac{2 \times 13,270}{72} = 368.6 \text{ N}$

Effective Load on tooth $F_e = \dfrac{F_t \times K_s}{K_v} = \dfrac{368.6 \times 1.2}{0.525} = 842.5 \text{ N}$

Beam strength $S_b = \sigma_b\, b\, \pi\, m\, y_p \left[1 - \dfrac{b}{L_o}\right]$

Substituting the values: $S_b = 50 \times 19.5 \times 3.14 \times 4 \times 0.112 \times (1 - 0.3) = 960 \text{ N}$

FOS based on strength $= \dfrac{960}{842.5} = 1.14$

$$K_w = \frac{S_e^2 \sin \Phi \cos \Phi}{1.4}\left[\frac{1}{E_p} + \frac{1}{E_g}\right] = \frac{500^2 \sin 14.5^\circ \cos 14.5^\circ}{1.4}\left[\frac{1}{82,000} + \frac{1}{82,000}\right] = 1.06$$

$$Q = \frac{2Z_g'}{Z_g' + Z_p'} = \frac{2 \times 48.7}{48.7 + 21.6} = 1.385$$

Strength in wear, $S_w = \dfrac{d_p\, b\, Q\, K_w}{\cos \gamma_p} = \dfrac{72 \times 19.5 \times 1.385 \times 1.06}{\cos 33.7^\circ} = 2,477 \text{ N}$

FOS based on wear $= \dfrac{2,477}{960} = 2.58$

Example 7.6

Bearing reactions in a bevel gear drive for given power

A bevel gear set with pressure angle 20° transmits 3 kW at 600 rpm. Bevel pinion diameter at large end is 150 mm and gear 250 mm with face width 30 mm. Calculate the reactions for bearing A and B, which are 260 mm apart as shown in Figure 7.S1.

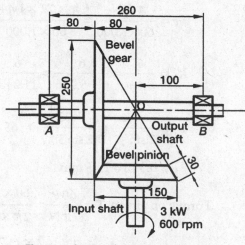

Figure 7.S1 Bevel gear set supported on shafts

Solution

Given $P = 3\,\text{kW}$ $N_p = 600\,\text{rpm}$

$\Gamma = 90°$ $\Phi = 20°$

$d_p = 150\,\text{mm}$ $d_g = 250\,\text{mm}$

$b = 30\,\text{mm}$ $L = 260\,\text{mm}$

Torque, $T = \dfrac{60P}{2\pi N} = \dfrac{60 \times 3,000}{2\pi \times 600} = 47.77\,\text{Nm}$

For bevel pinion: $\tan \gamma_p = \dfrac{d_p}{d_g} = \dfrac{150}{250} = 0.6$ or $\gamma_p = 31° \left(\text{Figure 7.S1A}\right)$

$$r_{mp} = \left[\dfrac{d_p}{2} - \dfrac{b \sin \gamma_p}{2}\right] = \left[\dfrac{150}{2} - \dfrac{30 \times \sin 31°}{2}\right] = 75 - \left(15 \times 0.488\right) = 67.3\,\text{mm}$$

$$F_t = \dfrac{T}{r_m} = \dfrac{47.77 \times 1,000}{67.3} = 710\,\text{N}$$

$F_r = F_t \tan \Phi \cos \gamma_p = 710 \times \tan 20° \times \cos 31° = 710 \times 0.364 \times 0.857 = 221.4\,\text{N}$

$F_a = F_t \tan \Phi \sin \gamma_p = 710 \times \tan 20° \times \sin 31° = 710 \times 0.364 \times 0.515 = 133.1\,\text{N}$

For bevel gear: $\gamma_g = 90 - 31 = 59°$

Refer Figure 7.S1B showing the forces at PCD and reactions at bearings.

$$r_{mg} = \left[\dfrac{d_g}{2} - \dfrac{b \sin \gamma_g}{2}\right] = \left[\dfrac{250}{2} - \dfrac{30 \times \sin 59°}{2}\right] = 125 - \left(15 \times 0.488\right) = 112.1\,\text{mm}$$

Tangential force F_t remains the same but direction becomes opposite.

F_r for gear $= F_a$ of pinion $= 133.1\,\text{N}$ and F_a for gear $= F_r$ of pinion $= 221.4\,\text{N}$

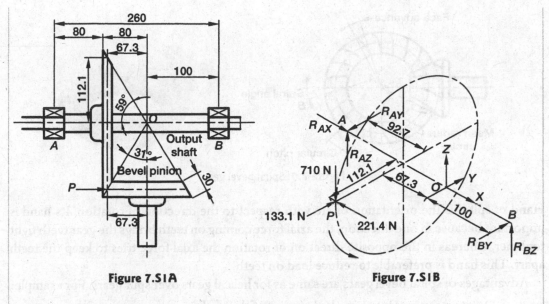

Figure 7.S1A

Figure 7.S1B

Forces in *XY* plane: Take moments about bearing A:

$$(R_{BY} \times 260) + (133.1 \times 92.7) = 221.4 \times 112.1 \qquad \text{or} \qquad R_{BY} = 48 \text{ N}$$

From equilibrium of forces: $R_{AY} = 133.1 \text{ N} + 48 \text{ N} = 181.1 \text{ N}$

Forces in *XZ* plane: Take moments about bearing A:

$$R_{BZ} \times 260 = 710 \times 92.7 \qquad \text{or} \qquad R_{BZ} = 253.1 \text{ N}$$

From equilibrium of forces: $R_{AZ} + R_{BZ} = 710 \text{ N} \qquad \text{or} \qquad R_{AZ} = 710 - 253.1 = 456.9 \text{ N}$

7.12 Spiral Bevel Gears

Bevel gears have straight teeth, whereas spiral bevel gears have teeth in the form of a spiral on pitch cone of both the gears. Minimum number of teeth is usually 12 but should not be less than six teeth. The teeth are generally ground for these gears. Following terms are used for these gears. Spiral angle ψ is the angle of teeth at the mean pitch radius and not at the large end of the gear (Figure 7.9). P is a point on the tooth on circle of mean radius. AB is a line tangent to the tooth at point P. The angle between line AB and OX is called spiral angle. It is generally kept 35°, whereas the pressure angle is same for other gears, that is, 14.5° or 20°.

Face advance is the angle formed at the apex from the start of arc at small end to the end of the arc at large end.

Face contact ratio is the face advance to the circular pitch at large end (Figure 7.9). This ratio is kept more than 1.3.

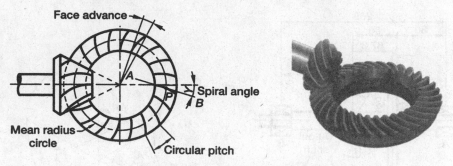

Figure 7.9 Spiral bevel gear

Hand of spiral is the orientation of arc with respect to the direction of rotation. Its hand is important because in one direction, the axial force coming on teeth brings the gear teeth tight together, whereas in the opposite direction of rotation the axial force tries to keep the teeth apart. This hand is preferable to reduce load on teeth.

Advantages of spiral bevel gears are same as for helical gears over spur gears. For example:

- Quiet operation due to gradual contact of the teeth.
- Offer more load capacity as more than one tooth is in contact.
- These can be used at higher speeds than straight bevel gears.

$$\text{Dynamic load } F_d = F_t + \frac{20.7v\left(C\,e\,b\cos^2\psi + F_t\right)\cos\psi}{20.7v + \sqrt{C\,e\,b\cos^2\psi + F_t}} \tag{7.19}$$

Where, ψ = Spiral angle. All other variables are defined in Section 7.10.

Summary

Bevel gears are used to transmit power from generally intersecting shafts, but axes are inclined to each other. Generally, the angle between the shafts is 90° but they can transmit power at any angle. The teeth are cut on a conical surface. Apex of the cone lies on the shaft axis and is known as *Vertex*. The cross section of the tooth decreases from outer radius towards apex side. Sum of the pitch cone angles of both the gears is equal to the angle between the axes of the shafts.

Terminology

Large end   A bevel gear is in the form of a frustum of a cone. The end containing the largest radius of the cone is called large end.

Cone center   Apex of the cone is called cone center.

Pitch cone   It is an imaginary cone, whose surface contains the pitch line of the teeth.

Cone distance (L_o)   It is the distance from the apex to pitch line at large end. It is also called *pitch cone radius*.

Pitch line   A straight line joining the vertex and a point on large end of PCD

Pitch diameter (d)   It is the diameter of the pitch cone at the large end.

Pitch angle (γ)   Angle between pitch line and axis of cone.

Addendum (a)   Height of the tooth at the large end from the pitch line.

Addendum angle   It is the angle between outside face of tooth and pitch angle.

Dedendum (d)   Depth of the tooth at the large end below the pitch line.

Dedendum angle   It is the angle between root of the tooth and pitch angle.

Tooth depth (h)   Sum of addendum and dedendum at the large end.

Face width (b)   Inclined length of the tooth.

Face angle   It is the angle between face of tooth with the axis of the cone.

Root angle   It is the angle between root of tooth with the axis of the cone.

Back cone   It is also a frustum of a cone on the backside of the cone containing teeth.

Back cone angle   It is the angle between face of the back with the axis of the cone.

Back cone radius   It is the inclined distance from the apex of back cone to pitch line at large end.

Classification of bevel gears   On the basis of the *shape of their teeth*

Straight bevel gears   Straight teeth converge towards the apex of the cone. These gears are noisy like spur gears. These are easy to manufacture than other types of bevel gears.

Spiral bevel gears   Teeth are curved, but axes of the shafts still intersect. These gears operate at lesser noise than straight teeth gears, because contact of the teeth is gradual.

Hypoid bevel gears   Teeth are curved but the axes of the shafts do not intersect. There is some offset between the axes.

Name of bevel gears is given on the angle between the intersecting shafts:

Miter gears   When two bevel gears of equal size mesh with shafts at right angle to each other, they are called miter gears. The pitch angle γ of each gear is 45°.

Angular gears   The angle between the shafts is of any other value except 90°. Pitch angle γ of one of the gears is less than 45°.

Crown gears   The angle between the axes of the shafts is more than 90°. Pitch angle γ of one of the gears is 90°.

Internal gears   The angle between the axes of the shafts is less than 90°. Pitch angle γ of one of the gears is greater than 90°.

Pitch angle and gear ratio   The angle between the axes of the shafts $\Gamma = \gamma_p + \gamma_g$

$$\text{Gear ratio } G = \frac{d_g}{d_p} = \frac{\sin \gamma_g}{\sin \gamma_p}$$

$$\tan \gamma_g = \frac{\sin \Gamma}{\left(\dfrac{1}{G}\right) + \cos \Gamma}$$

Cone distance   Distance from apex to pitch circle at large end

$$L_o = \sqrt{(0.5d_p)^2 + (0.5d_g)^2}$$

Proportions of bevel gear measured at large end.

Tooth thickness = 1.571 m Addendum $a = 1\ m$

Dedendum $d = 1.2\ m$ Tooth depth $h = 2.2\ m$

Outside diameter of cone $d_o = d + 2a \cos \gamma$

Root diameter of cone $d_o = d - 2d \cos \gamma$

Formulative number of teeth   Involute profile does not lie on a straight plane but lies on a surface of a sphere. So the profile is approximated accurately using Tredgold's approximation. According to this, back cone can be used as a plane for generating involute profile at large end. Radius of back cone r_b is taken an imaginary pitch circle radius for generating involute profile. Let Z' be the number of teeth on this pitch circle. The gear with Z' number of teeth and PCD of $2r_b$ is called *formative gear*.

Thus: $Z' = \dfrac{2r_b}{m}$ Also $Z = \dfrac{d}{m}$ Hence $\dfrac{Z'}{Z} = \dfrac{2r_b}{d}$

Since surface of back cone is perpendicular to the large end, $d = 2\ r_b \cos \gamma$

$$Z' = \frac{Z}{\cos \gamma}$$

Because Z' is always bigger than Z, larger contact ratios for bevel gears make them to run more smooth than with spur gears of same number of actual teeth.

Forces on tooth of pinion: Tangential force $F_t = \dfrac{60P}{2\pi N r_m}$

r_m = Mean radius of bevel gear at pitch circle in the middle of its inclined face.

Mean radius $r_m = \left[\dfrac{d}{2} - \dfrac{b \sin \gamma}{2}\right]$

Where, b = Length of face

The resultant force comprises three forces: F_t = Tangential force transmitting torque

Radial force towards center $F_r = F_t \tan \Phi \cos \gamma$

Axial force causing axial thrust in the shaft / bearings $F_a = F_t \tan \Phi \sin \gamma$

Separating force that tries to separate the gears $F_s = F_t \tan \Phi$

Forces on tooth of bevel gear

Direction of rotation of driven bevel gear is opposite to the direction of driver gear. Hence, direction of *tangential force* in a gear is the same as its direction of rotation.

Radial component acts towards the center of the gear.

Axial force on the gear is equal and opposite to the radial force on the pinion, if the angle between the axes of the shafts is 90°. This force tends to separate the gears.

Dynamic load occurs due to inaccuracy in the profile of gear tooth while manufacturing.

Dynamic load capacity $F_d = F_t + \dfrac{20.7v(Ceb + F_t)}{20.7v + \sqrt{Ceb + F_t}}$

v = Velocity at pitch circle at the large end

b = Face width

e = Error given in Table 7.1.

K = A constant depending upon pressure angle as given below:

$K = 0.107$ for $14.5°$, $K = 0.111$ for $20°$ full depth   and $K = 0.115$ for $20°$ stub

C = Dynamic load factor = $\dfrac{K}{\left[\dfrac{1}{E_p} + \dfrac{1}{E_g}\right]}$

E_p, E_g = Young's modulus of elasticity for pinion and gear materials, respectively.

For a safe design, the endurance beam strength should be more than dynamic load.

Wear strength is the maximum tangential force at the large end of a bevel gear, which can be transmitted without any pitting. Wear strength $S_w = d'bQK_w$

d' = PCD of for mulative pinion = $\dfrac{d}{\cos\gamma}$ and b = Face width

Since one bevel gear is overhung it deflects. Therefore, assuming only 75 per cent of its width is contributing in sharing load,

$$S_w = \dfrac{0.75\,b\,d\,Q\,K_w}{\cos\gamma}$$

Tooth factor $Q = \dfrac{2Z_g'}{Z_g' + Z_p'} = \dfrac{2Z_g}{Z_g + Z_p\tan\gamma_p}$

Material constant $K_w = \dfrac{\sigma_c^2 \sin\varPhi \cos\varPhi \left[\dfrac{1}{E_p} + \dfrac{1}{E_g}\right]}{1.4}$

For pressure angle $\varPhi = 20°$ and same materials for both gears as steel, $K_w = 0.16\left(\dfrac{BHN}{100}\right)^2$

For a safe design, wear strength $S_w >$ Tangential force at large end.

Spiral bevel gears have teeth in the form of a spiral on pitch cone of both the gears. Minimum number of teeth is 12 but should not be less than six teeth. For these gears:

Spiral angle ψ is the angle of teeth at the mean pitch radius and not at the large end of the gear. It is generally kept $35°$, whereas the pressure angle is $14.5°$ or $20°$.

Face advance is the angle formed at the apex from the start of arc at small end to the end of the arc at large end.

Face contact ratio is the face advance to the circular pitch at large end. This ratio is kept more than 1.3.

Hand of spiral is the orientation of arc with respect to the direction of rotation. Its hand is important because in one direction, the axial force coming on teeth brings the gear teeth tight together, whereas in opposite direction of rotation the axial force tries to keep the teeth apart. This hand is preferable to reduce load on teeth.

Advantages: Quiet operation due to gradual contact of the teeth offer more load capacity and can be used at higher speeds than straight bevel gears.

Dynamic load $\quad F_d = F_t + \dfrac{20.7v\left(Cebcos^2\,\psi + F_t\right)cos\,\psi}{20.7v + \sqrt{Cebcos^2\,\psi + F_t}}$

where, ψ = Spiral angle.

Theory Questions

1. Differentiate between spur gear, helical gear, and bevel gear.
2. What are the various types of bevel gears? Describe them with a neat sketch.
3. Define the following terms:

   a   Pitch angle   b.   Cone angle

   c.   Back cone angle   d.   Face angle

   e.   Pitch angle   f.   Pitch diameter

4. Write a short note on proportions of bevel gear.
5. What do you mean by formulative number of teeth? Derive an expression to calculate it.
6. Discuss the various forces coming on a bevel gear tooth with the help of a sketch.
7. What is dynamic load? How is it calculated?
8. Discuss the wear strength of a bevel gear. What are the different parameters affecting it?
9. What is a spiral bevel gear? How does it differ from a normal bevel gear?
10. Define back cone and cone distance of bevel gears.

Multiple Choice Questions

1. Pitch angle is the angle between axis of:

   (a) Cone and pitch line   (b) Pinion with axis of tooth

   (c) Bevel pinion and bevel gear   (d) Two adjacent teeth

2. Two bevel gears with intersecting axes and curved teeth are called:

   (a) Miter gear   (b) Crown gear

   (c) Spiral gear   (d) Hypoid gear

3. A miter gear is a gear set with:

   (a) Two unequal bevel gears with axes at 45°   (b) Two equal bevel gears with axes at 45°

   (c) Two equal bevel gears with axes at 90°   (d   Two unequal bevel gears with axes at 135°

4. For a bevel gear having Z teeth and pitch angle γ, formulative number of teeth is:

   (a) $Z\cos^2\gamma$ (b) $Z\cos ec^2\gamma$

   (c) $Z\cos\gamma$ (d) $Z\sec\gamma$

5. For a bevel gear of width b and cone distance L_o, bevel factor for calculating strength is:

   (a) $1 - (b / L_o)$ (b) b / L_o (c) $1 + (b / L_o)$ (d) $b + L_o$

6. For a miter bevel gear set radial force on a pinion is equal to:

   (a) Tangential force on gear   (b) Axial force on gear

   (c) Radial force on gear   (d) None of the above

7. If Φ is pressure angle, γ pitch angle, m module and a addendum, outside diameter of a bevel gear cone with PCD d is:

   (a) $d + a$ (b) $d + m$

   (c) $d + 2m \cos \gamma$ (d) $d + a \sin \Phi$

8. PCD for a bevel gear is measured at:

   (a) Large end   (b) Small end

   (c) Middle of large and small end   (d) 75 per cent of face width from small end

9. Dynamic load for a bevel gear of a given face width depends upon:

   (a) Young's modulus of elasticity   (b) Error in manufacturing

   (c) Pressure angle   (d) All given above

10. Spiral angle ψ is the angle of teeth with the radial line at the:

    (a) Large end   (b) Mean pitch radius

    (c) Small end.   (d) None of the above

Answers to multiple choice questions

1. (a)   2. (c)   3. (c)   4. (d)   5. (a)   6. (b)   7. (c)   8. (a)   9. (d)   10. (b)

Design Problems

1. A bevel gear has 12 teeth on pinion and 28 teeth on gear, which are at right angle to each other. The pinion is driven by an electric motor of 1 kW at 3,000 rpm. Find the output torque and speed of bevel gear. $[T = 46.67$ Nm $, N_g = 1{,}286$ rpm$]$

2. A 20° pressure angle bevel pinion transmits 5 kW at 500 rpm to a bevel gear at 90°. PCD of pinion is 200 mm and of gear 300 mm. Face width is 30 mm. Calculate the forces in tangential, radial, and axial direction. $[F_t = 1{,}041$ N$, F_r = 315.6$ N$, F_a = 208$ N$]$

3. A 30 kW motor running at 1,440 rpm drives a 25 teeth bevel pinion. The bevel gear meshing with the pinion is at right angle runs at 900 rpm. Pressure angle is 14.5°, face with is one-fourth of slant height. Calculate module, face width, slant height, addendum, dedendum and outside diameter of pinion. $[m = 7$ mm$, b = 42$ mm$, L_o = 165.2$ mm$, a = 7$ mm$, d = 8.4$ mm$, d_{op} = 165.2$ mm$]$

4. Two kilowatts power is transmitted at 1,000 rpm through a bevel gear set at right angle. PCD of cast iron bevel pinion and bevel gear are 60 mm and 80 mm, respectively. Teeth are cut with pressure angle 14.5° in a face width, which is one-third of the slant height. Allowable stress is 65 MPa and surface endurance strength 600 MPa. Young's modulus of elasticity for both pinion and gear can be taken 85 GPa. Calculate module from strength point of view and wear strength. $[m = 4, S_w = 3{,}204$ N$]$

5. An electric motor of 3 kW drives a 20° pressure angle bevel pinion at 300 rpm. Starting torque of the motor is 30 per cent more than the average torque. Pinion with 18 teeth of module 5 and face width 25 mm meshes with a 28 teeth bevel gear at right angle to the axis of the pinion. Ultimate strength can be taken 700 N/mm² and BHN = 380. Maximum inaccuracy in gear cutting can be assumed as 0.015 mm. Calculate:

   (a) Beam strength.   (b) Wear strength.
   (c) Effective dynamic load.   (d) FOS in bending and pitting (wear).

   $$[S_b = 7,115 \text{ N}, \ S_w = 6,531 \text{ N}, \ F_e = 4,479 \text{ N}, \ FOS_b = 1.59, \ FOS_w = 1.46]$$

6. An electric motor of 10 kW transmits power to a bevel pinion rotating at 1,440 rpm. Starting torque of the motor is 140 per cent of average torque. The bevel pinion has 20 teeth of pressure angle 20° and bevel gear 30 teeth. Face width is 30 per cent of the slant height. The material used for gears has allowable strength 250 MPa. The gears are cut with accuracy up to 0.012 mm, FOS = 2 for strength, and wear and C = 11,400 N/mm²

   (a) Design the gears for strength.   (b) BHN for safe in pitting.

   $[$(a) $m = 5, b = 27$ mm, $d_p = 100$ mm, $d_{op} = 108.3$ mm, $d_g = 150$ mm, $d_{op} = 155.5$ mm (b) BHN = 372$]$

Previous Competition Examination Questions

IES

1. Consider the following statements:

   The axes of the spiral bevel gear are nonparallel and intersecting.

   1. The most common pressure angle for the spiral bevel gear is 20°.
   2. The most common spiral angle for the spiral bevel gear is 35°.
   3. Spiral bevel gears are generally interchangeable.
   4. Spirals are noisy and recommended for low speeds of 10 m/s.

   Which of the above statements are correct?

   (a) 1 and 4   (b) 1 and 2   (c) 2 and 3   (d) 3 and 4

   [Answer (b)] [IES 2015]

2. In case the number of teeth on two bevel gears in mesh is 30 and 60, respectively, then the pitch cone angle of the gear will be:

   (a) $\tan^{-1} 2$ (b) $\dfrac{\pi}{2} + \tan^{-1} 2$ (c) $\dfrac{\pi}{2} - \tan^{-1} 0.5$ (d) $\tan^{-1} 0.5$

   [Answer (d)] [IES 2015]

3. In skew bevel gears, the axes are:
   (a) Nonparallel and nonintersecting teeth are curved.
   (b) Nonparallel and nonintersecting teeth are straight.
   (c) Intersecting and teeth are curved and oblique.
   (d) Intersecting and teeth are curved and can be ground.   [Answer (d)] [IES 2015]

4. When two shafts are neither parallel nor intersecting, power can be transmitted using:
   (a) A pair of spur gears.   (b) · A pair of helical gears.
   (c) An Oldham's coupling.   (d) A pair of spiral gears.

   [Answer (c)] [IES 1998]

GATE

5. A bevel gear can transmit power for an intersecting shaft at:

    (a) 90° only   (b) 90° –135°   (c) 90° –180°   (d) Any angle

    [Answer (d)] [GATE 2013]

6. Match the type of gears with their most appropriate description:

List I	List II
Type of Gear	**Description**
P. Helical	1. Axes nonparallel and nonintersecting.
Q. Spiral bevel	2. Axes parallel and teeth are inclined to the axis.
R. Hypoid	3. Axes parallel and teeth are parallel to the axis.
S. Rack and pinion	4. Axes are perpendicular and intersecting and teeth are inclined to the axis.
	5. Axes are perpendicular and used for large speed reduction.
	6. Axes are parallel and one of the gears has infinite radius.

Codes:

	P	Q	R	S
(a)	2	4	1	6
(b)	1	4	5	6
(c)	2	6	4	2
(d)	6	3	1	5

[Answer (a)] [GATE 2008]

□□□

Worm Gears

8.1 Worm and Worm Wheel

Worm and worm wheels shown in picture on right side are generally used for reducing speed with high speed reduction in a compact space. Gear ratio can be as high as 100:1 in a single stage. Input and output shafts are generally nonintersecting at right angles, but can be designed for any angle. Power generally flows from the worm to worm gear. If the gear ratio is high, it is an irreversible type of transmission, that is, power flows from worm to worm gear and cannot be from worm wheel to worm. It can be made reversible using small gear ratios.

The driver shaft has a worm, which is in the form of a cylindrical screw with trapezoidal threads with large pitch, that matches with the worm wheel having mating grooves of the same profile as shown in Figure 8.1 and picture on the right side.

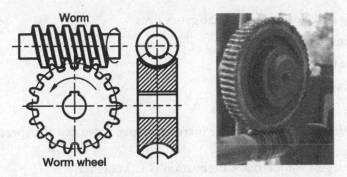

Figure 8.1 A worm and worm gear

The threads can be left hand or right hand. Single start threads are most common for high gear ratios but multi-start threads can also be used for lower gear ratios.

Driven shaft has a wheel called worm wheel, which has inclined teeth at the same angle as the helix angle of worm. Its outer periphery is generally made of concave shape according to the radius of the worm (right side of Figure 8.1). The top threaded portion of the worm gear envelopes the teeth on the worm. Diameter of the worm is based on strength calculations to transmit the required torque.

8.2 Advantages / Disadvantages of the Drive

Advantages

- Occupies lesser space in comparison to geared reduction boxes.
- High speed reduction in single stage.
- Self-locking for higher speed reductions, that is, power can flow only from worm to worm wheel and not from worm wheel to worm. It is a desirable quality especially for applications such as cranes and lifting devices.
- Silent operation.

Disadvantages

- It can be used for low and medium power transmission (up to 100 kW)
- Low efficiency in comparison to other drives.
- A lot of heat is generated due to rubbing surfaces. It needs fins at the outside of casing to dissipate heat.
- Costly, as the worm wheel is made of phosphorous bronze or bronze.

8.3 Applications

Worm and worm wheel drives are generally used as speed reducing devices. Some applications are given below:

- Opening and closing of large size gate valves
- Small hoists for lifting loads
- Machine tools
- Elevators
- Steering system of automobiles

In some applications, it can be used for increasing speed also, but the speed ratio has to be low.

- Superchargers for Internal Combustion (I.C.) engines
- Hand blowers

8.4 Terminology

Terms used with worm and worm gear are explained below and shown in Figure 8.2.

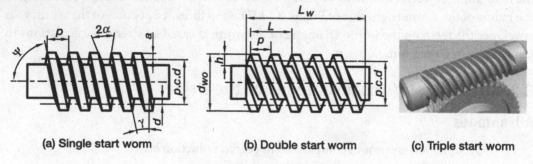

| (a) Single start worm | (b) Double start worm | (c) Triple start worm |

Figure 8.2 Worm nomenclature

Pitch circle diameter (PCD)	It is an imaginary circle, about which peripheral velocities are calculated. d_w is for worm and d_g for worm gear
Module (m)	It is the ratio of PCD of worm wheel d_g to the number of teeth Z_g. That is $m = d_g / Z_g$.
Addendum (a)	Radial distance between PCD and outside diameter of worm.
Dedendum (d)	Radial distance between PCD and root diameter.
Outside diameter (d_{wo})	Outside diameter of worm measured over the threads. $$d_{wo} = d_w + 2a$$
Axial pitch (p_a)	It is the distance measured from a point on one thread at the PCD of worm to the corresponding point on the adjacent thread in the axial direction.

Circular pitch (p_c)	It is the actual distance at PCD from a face of thread to adjacent corresponding thread.

$$p_c = \pi \frac{d_g}{Z_g} = \pi m$$

Throat diameter (d_{gt})	It is the outside diameter at the center line of the worm wheel.
Diameter quotient (q)	It is the ratio of PCD of the worm to module.

$$q = \frac{d_w}{m} \tag{8.1}$$

Number of start (n)	It is a number, indicating as to how many threads start from a plane at right angle to the axis. Figure 8.2(b) is a double start worm. Its selection depends on velocity ratio as given below:

Table 8.1 Number of Start and Gear Ratio

No. of Start (n)	1	2	3	4	6
Gear Ratio (G)	> 20	12–35	8–12	6–10	4–8

Thread angle (α)	It is half of the included angle between the two faces of the thread.
Lead (L)	It is the product of axial pitch and number of start, that is, $L = n \times p_a$.
Lead angle (γ)	It is the angle, which a tangent on thread of worm at PCD makes with the plane normal to the axis of the worm (Figure 8.3).
Helix angle (γ)	For a single start, thread helix angle is same as the lead angle.

$$\tan \gamma = \frac{L}{\pi d_w} = \frac{n \times p}{\pi d_w} = \frac{n \times \pi m}{\pi d_w} = \frac{n \times m}{d_w}$$

$$\text{or} \quad \tan \gamma = \frac{n}{q} \tag{8.2}$$

For a worm gear, lead = Circular pitch p_c of the gear

$$p_c = \pi m = \frac{\pi d_g}{Z_g} \tag{8.3}$$

$$n = \text{Number of start}$$

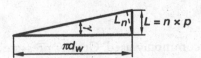

Figure 8.3 Lead angle of a worm

Length of worm (L_w)	Threaded length of worm along its axis.
Pressure angle (Φ)	It is the angle measured in the plane containing axis of the worm and is equal to half of included angle of the thread.
Face width (b)	It is the width of worm wheel at PCD of the worm.
Gear ratio (G)	Number of teeth on worm wheel / Number of start of worm

$$G = \frac{Z_g}{n} \tag{8.4}$$

8.5 Diameter Quotient

Diameter quotient q is defined as the ratio of PCD of worm d_w to module m. Worm diameter d_w is arbitrary, with the chosen number of start n and the known module m, unless some value of lead angle is taken. In the effort towards the less number of screw cutters for gears, the manufacturers recommend choosing diameter quotient q depending on the size of standardized module m. Table 8.2 gives values of m and recommended diameter quotients. Considering worm rigidity in bending, greater values of q are taken for lower values of m.

Table 8.2 Diameter Quotient for a Given Module

Module (m)	2	3	4–6	8	10–12	16–20	25
Diameter Quotient (q)	16	14–12	14–9	12–8	10–8	8	6

If the center distance is known, Table 8.3 can be used to select a suitable value of diameter quotient for a known number of teeth on gear. In the use of this table, it is assumed that the speed of the gear is more than 300 rpm. If the speed is less than 300 rpm, increase the value to 1.5 times.

Table 8.3 Preferred Values of $(n / Z_g / q / m)$ for Different Gear Ratios and Center Distances

Center Distance (mm)	Gear Ratios				
	20	25	30	40	50
100	2 / 40 / 10 / 4	–	1 / 30 / 10 / 5	1 / 40 / 10 / 4	–
125	2 / 40 / 10 / 5	2 / 50 / 10 / 4	1 / 30 / 10 / 6	1 / 40 / 10 / 5	1 / 50 / 10 / 4
160	–	2 / 50 / 10 / 5	1 / 30 / 10 / 8	–	1 / 50 / 10 / 5
200	2 / 40 / 10 / 8	–	1 / 30 / 10 / 10	1 / 40 / 10 / 8	–
250	2 / 40 / 10 / 10	2 / 50 / 10 / 8	–	1 / 40 / 10 / 10	1 / 50 / 10 / 8

8.6 Pressure Angle

Pressure angle of 20° is most commonly used. Greater pressure angle results in higher safety against fatigue break and lower danger of gear tooth undercut. On the other hand, greater pressure angle reduces the number of engaged teeth, and increases bearing load and the worm deflection load. Broadly it is taken as:

$\Phi = 14.5°$ for single and double start worms and,

$\Phi = 20°$ for triple and quadruple start worms.

Minimum number of gear teeth Z_g or lead angle γ dependence on pressure angle can be seen from Table 8.4.

Table 8.4 Pressure Angle Dependence on Minimum Number of Teeth on Worm Gear and Lead Angle

Pressure Angle ($\Phi°$)	14.5°	17.5°	20°	22.5°	25°	27.5°	30°
Minimum (Z_g)	40	27	21	17	14	12	10
Lead Angle (γ)	0–10	11–15	16–20	21–25	26–30	31–35	36–40

8.7 Types of Worms and Worm Wheels

8.7.1 Types of worms

Worms have different shapes at the cylindrical surface as described below:

a. **Cylindrical worm**   Worm is of cylindrical shape [Figure 8.4(a)]. It wraps partially over the worm wheel resulting in a line contact.

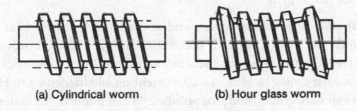

(a) Cylindrical worm   (b) Hour glass worm

Figure 8.4 Types of worms

b. **Hour glass worm**   It is so called because its shape is like an hour glass used for measuring time. The shape of the worm is modified as in Figure 8.4(b) so that it wraps more number of teeth of the worm wheel. It is also called double envelope worm.

8.7.2 Types of worm gears

Worm gear teeth are also cut in different ways as described below:

a. **Straight face**   It looks more like a helical gear [Figure 8.5(a)]. The contact area with the worm is very less, hence used for light service only.

b. **Straight hobbed face**   Hob cutter is used to make gears. It is of the shape of a worm, having cutting teeth at its cylindrical surface. When it is brought near the face of the worm wheel and pushed radially inwards, it cuts teeth of the worm wheel. The outer face is an arc and the root is also arc [Figure 8.5(b)].

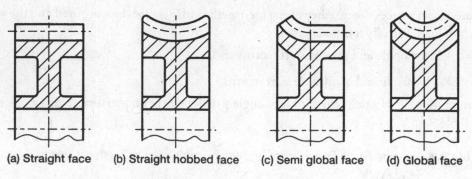

(a) Straight face   (b) Straight hobbed face   (c) Semi global face   (d) Global face

Figure 8.5 Types of worm wheel drives

c. **Semiglobal face**   It is a combination of hobbed and straight face [Figure 8.5(c)]. Advantage of the straight face is that the worm can be positioned just by sliding from one side. Owing to curved surface enveloping the worm, contact area is more, hence used for medium duty applications.

d. **Global face**   Its outer face is concave and envelops the worm from both side to give maximum contact area for more power transmission [Figure 8.5(d)]. It is used for heavy duty applications.

8.8 Material Selection

8.8.1 Materials for worm

Selection of materials is based on rigidity and stress level. It should bear the fluctuating stresses and should be tough to absorb maximum energy. Therefore worms are made of case-hardened steel. Depth of case hardness can vary from 1 mm to 4 mm. Carbon or alloy steels allow surface hardening to HRC 45-50, cementing and hardening to HRC 56-62, and nitriding. Tooth sides are ground and / or polished. When nitrided, the material need not be ground and only polishing is enough. Heat-treated or normalized steel worms are used only for lower outputs and lower peripheral speeds. Table 8.5 lists some materials used for worm with their typical use and properties.

Table 8.5 Worm Materials with Their Applications and Properties

Material of Worm	Applications	Properties
Acetal / Nylon	Toys, domestic appliances, instruments	Low cost, low duty
Cast iron	Not used frequently in modern machinery	Excellent machinability, medium friction
Carbon steel	Power gears with medium rating	Low cost, reasonable strength
Hardened steel	Power gears with high rating for extended life	High strength, good durability

Following is the chemical composition of the materials used for worms:

Carbon steels   40C8, 55C8
Case-hardened steels   10C4, 14C6
Alloy steels   16Ni80Cr60, 20Ni2Mo25,
Nickel chromium steels   13Ni3Cr80, 15Ni4Cr1
For high speeds   35SiMn, 40Cr, 40CrNi,
For heavy loads   20Cr, 15Cr, 20CrMnTi
For low speed and power   C40 hardened and tempered BHN = 220–300

8.8.2 Materials for worm gear

The desirable properties of material for worm gear should have less friction coefficient, gluing resistant to pair up worm and good running in. The basic material for worm gear is bronze, less frequently used is cast iron or brass. Plastic gears are used for lower powers to absorb shocks and provide lower noise.

For worm and worm gear, dissimilar materials are used as the coefficient of friction is less. Although stresses are same on worm gear as on worm, but number of cycles is reduced by gear ratio. Tooth profile of the gear in manufacturing is not accurate; hence, it is generally finished in the initial run out period by the hardened worm. Hence worm gear material selected is which should be soft and conformable. Bronze gears are manufactured as composite for economical reasons, that is, a bronze rim is put on a steel or cast-iron wheel.

Phosphorus bronze is most commonly used. Surface hardness varies from 90 BHN to 120 BHN.

Tin bronze with high tin content of 10–12 per cent has excellent friction properties, high resistance to seizure and good running-in, but is too expensive. Their application is justified in loaded transmissions with sliding speed more than 10 m/s.

Sn–Ni bronze, with lower tin content (5–6 per cent) or **antimony–bronze** can be used for sliding speed 4 m/s–10 m/s.

Aluminum or lead bronze, or brass is less expensive; bronze without tin is suitable for sliding speeds less than 4 m/s. These are relatively hard and strong, but are less resistant to seizure and are not so good for running-in. The mating worm, therefore, must have high surface hardness (HRC > 45).

Grey cast iron mated with a steel worm can be used for low outputs, quiet load, and low peripheral speed up to 2 m/s. Other materials are:

    High-speed transmission: ZCuSn10Pl, ZCuSnPb5Zn5
    Low-speed but heavy loads: ZCuAl10Fe3

Table 8.6 lists some materials used for worm gears with their typical use and properties.

Table 8.6 Worm Gear Materials with Their Applications and Properties

Material of Worm Gear	Applications	Properties
Acetal / Nylon	Toys, domestic appliances, instruments	Low cost, low duty
Phosphorus bronze	Normal material for worm gears with reasonable efficiency	Reasonable strength, low friction, and good compatibility with steel
Cast iron	Used frequently in modern machinery	Excellent machinability, medium friction

Table 8.7 Permissible Bending Fatigue Stress (MPa)

Material of Gear	Centrifugal cast Cu–Sn bronze	Al–Si alloy	Cast iron	Zn alloy
Fatigue Strength in Bending	23.5	11.3	11.8	7.5

8.9 Drive Proportions

8.9.1 Worm proportions

Proportions of the worm in terms of axial pitch p (it is equal to circular pitch p_c of the worm gear) or module m are given in Table 8.8 (Refer Figure 8.6):

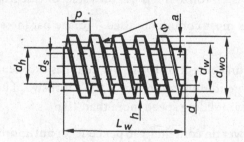

Figure 8.6 Proportions of worm

Table 8.8 Proportions of Worm [IS 3734-1966]

Dimension	Symbol	In Terms of Axial Pitch p		In Terms of Module m	
		1 or 2 Start	3 or 4 Start	1 or 2 Start	3 or 4 Start
PCD-Integral with shaft	d_w	$2.35\,p + 10^*$	$2.35\,p + 10$	$7.4\,m + 10$	$7.4\,m + 10$
PCD-Bored to fit on shaft	d_w	$2.40\,p + 28$	$2.40\,p + 28$	$7.6\,m + 28$	$7.6\,m + 28$
Outside diameter	d_{wo}	$d_w + 0.64\,p$	$d_w + 0.57\,p$	$d_w + 2\,m$	$d_w + 1.8\,m$
Hole size for shaft	d_s	$p + 13.5$	$p + 13.5$	$3.14\,m + 13.5$	$3.14\,m + 13.5$
Hub diameter	d_h	$1.7\,p + 25$	$1.7\,p + 25$	$5.2\,m + 25$	$5.2\,m + 25$

Length of worm	L_w	$4.5p + 0.02np$	$4.5p + 0.02np$	$14m + 0.06nm$	$14m + 0.06nm$
Addendum	a	$0.318\,p$	$0.286\,p$	$1.0\,m$	$0.9\,m$
Dedendum	d	$0.368\,p$	$0.337p$	$1.17\,m$	$1.06\,m$
Tooth depth	h	$0.686\,p$	$0.623\,p$	$2.17\,m$	$1.96\,m$
Pressure angle	Φ	$14.5°$	$20°$	$14.5°$	$20°$

* All dimensions and numbers are in mm.

* n = Number of start

If **center distance** c (mm) is given:

$$d_w = \frac{c^{0.875}}{1.467} \tag{8.5}$$

If circular pitch (p_c) or axial pitch (p) is known:

$$d_w = 3 \times p_c \quad \text{or} \quad d_w = 3p \tag{8.6}$$

Range of PCD of worm can be taken from the given center distance in mm from the following relations:

$$d_w = \frac{c^{0.875}}{1.467} \quad \text{to} \quad d_w = \frac{c^{0.875}}{2.5} \tag{8.7}$$

8.9.2 Worm gear proportions

The main dimensions of a worm gear in terms of circular pitch p_c and module m are given in Table 8.9.

Table 8.9 Proportions of Worm Gear (Refer Figure 8.7)

Dimension	Symbol	In Terms of Circular Pitch p_c		In Terms of Module m	
		1 or 2 Start	3 or 4 Start	1 or 2 Start	3 or 4 Start
Pressure angle	Φ	$14.5°$	$20°$	$14.5°$	$20°$
PCD	d_g	$d_w \times Z_g$	$d_w \times Z_g$	$d_w \times Z_g$	$d_w \times Z_g$
Outside diameter	d_{go}	$d_g + 1.0135\,p_c$	$d_g + 0.8903\,p_c$	$d_g + 3.2\,m$	$d_g + 2.8\,m$
Throat diameter	d_{gt}	$d_g + 0.636\,p_c$	$d_g + 0.572\,p_c$	$d_g + 2\,m$	$d_g + 1.8\,m$
Face width	b	$2.4\,p_c + 7^*$	$2.15\,p_c + 5$	$7.5\,m + 7$	$6.8\,m + 5$
Root diameter	d_r	$d_g - 0.74\,p_c$	$d_g - 0.67\,p_c$	$d_g - 2.3\,m$	$d_g - 2.1\,m$
Radius of gear face	R_f	$0.88\,p_c + 14$	$0.91\,p_c + 14$	$2.8\,m + 14$	$2.9\,m + 14$
Radius of gear rim	R_r	$2.2\,p_c + 14$	$2.1\,p_c + 14$	$6.9\,m + 14$	$6.6\,m + 14$
Radius of face edge	R_c	$0.25\,p_c$	$0.25\,p_c$	$0.8\,m$	$0.8\,m$

* All dimensions and numbers are in millimeters.

Face width b with respect to worm diameter d_w should be about 0.73 times. That is,

$$b = 0.73 d_w \qquad (8.8)$$

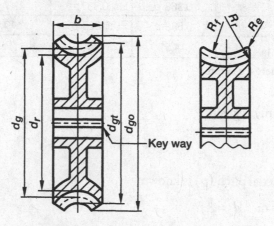

Figure 8.7 Proportions of worm gear

8.10 Drive Designation

A worm gear drive is designated by the following four terms in the sequence:

- Number of start of worm, n
- Number of teeth on worm gear, Z_g
- Diametral quotient, q
- Module, m

Drive designation is in the following sequence: $n\,/\,Z_g\,/\,q\,/\,m$

8.11 Center Distance

Center distance C between the shafts of worm and worm gear is equal to the sum of pitch circle radii of both, that is,

$$C = \frac{d_w + d_g}{2} \qquad (8.9)$$

PCD of worm d_w and lead L are related to lead angle γ as:

$$\tan\gamma = \frac{L}{\pi d_w} \quad \text{or} \quad d_w = \frac{L}{\pi\tan\gamma} = \frac{L\cot\gamma}{\pi} \qquad (a)$$

Circumference of worm gear at pitch circle = πd_g which is also equal to $G \times L$
Where, G is the gear ratio.

Hence, $\quad d_g = \dfrac{G \times L}{\pi} \qquad (b)$

Substituting values of (a) and (b) in Equation (8.9):

$$C = \frac{L}{2\pi}(\cot\gamma + GL) \tag{c}$$

Lead and normal lead are related as:

$$L = \frac{L_N}{\cos\gamma} \tag{d}$$

Substituting the value of L from Equation (d) in Equation (c):

$$C = \frac{L_N}{2\pi}\left(\cosec\gamma + G\sec\gamma\right) \quad \text{or} \quad \frac{C}{L_N} = \frac{1}{2\pi}\left(\cosec\gamma + G\sec\gamma\right) \tag{8.10}$$

Left side of Equation (8.9) is a dimensionless center distance, whereas the right side is a function of lead angle γ and gear ratio G. Values of right side for various values of G are given in Table 8.10. The minimum value of the variables is shaded. It can be seen that for higher values of gear ratio, lead angle reduces.

Table 8.10 Values of (C / L_N) for Different Lead Angles γ and Gear Ratios G

γ	Gear Ratio G														
	2	4	6	8	10	12	14	16	18	20	25	30	40	50	60
10	1.24	1.56	1.89	2.21	2.53	2.86	3.18	3.50	3.83	4.15	4.96	5.77	7.39	9.00	10.62
15	0.95	1.27	1.60	1.93	2.26	2.59	2.92	3.25	3.58	3.91	4.74	5.56	7.21	8.86	10.51
20	0.80	1.14	1.48	1.82	2.16	2.50	2.84	3.18	3.52	3.85	4.70	5.55	7.24	8.94	10.63
25	0.73	1.08	1.43	1.78	2.13	2.49	2.84	3.19	3.54	3.89	4.77	5.65	7.40	9.16	10.92
30	0.69	1.05	1.42	1.79	2.16	2.52	2.89	3.26	3.63	4.00	4.91	5.83	7.67	9.51	11.35
35	0.67	1.06	1.44	1.83	2.22	2.61	3.00	3.39	3.78	4.16	5.14	6.11	8.05	10.00	11.94
40	0.66	1.08	1.49	1.91	2.33	2.74	3.16	3.57	3.99	4.40	5.44	6.48	8.56	10.64	12.72
45	0.68	1.13	1.58	2.03	2.48	2.93	3.38	3.83	4.28	4.73	5.85	6.98	9.23	11.48	13.73
50	0.70	1.20	1.69	2.19	2.68	3.18	3.67	4.17	4.66	5.16	6.40	7.64	10.11	12.59	15.06

For a compact design, value of C has to be minimum. To get this minimum value, differentiate the right side of the Equation (8.10) with respect to γ and equate it to zero.

$$-\frac{\sin\gamma}{\cos^2\gamma} + \frac{G\cos\gamma}{\sin^2\gamma} = 0 \quad \text{or} \quad \cot^3\gamma = G \tag{8.11}$$

Lead angle based on the relation given by Equation (8.11) is tabulated. The value of this lead angle will give minimum center distance for a given gear ratio. Table 8.11 gives the values of optimum lead angle for various gear ratios.

Table 8.11 Optimum Lead Angle for Various Gear Ratios

Lead Angle	10	11	12	13	14	15	16	17	18	19	20	25	30
Gear Ratio	182.8	136.4	104.4	81.4	64.6	52.1	42.5	35.1	29.2	24.5	20.8	9.9	5.2

Example 8.1

Drive proportions and designation for given module

A worm drive has 40 teeth on worm gear, which is driven by a single start worm, integral with shaft having module 6 mm. Calculate the following:

- a. Speed reduction
- b. Size of worm
- c. Size of worm gear
- d. Center distance
- e. Specify the drive

Solution

Given $Z_g = 40$ $\quad n = 1$ $\quad m = 6$

a. Speed reduction $G = \dfrac{Z_g}{n} = \dfrac{40}{1} = 40$

b. From Table 8.8, PCD of worm with integral shaft:

$$d_w = 7.4\,m + 10 = (7.4 \times 6) + 10 = 54.4 \text{ say } 60 \text{ mm}$$

c. PCD of worm gear:

$$d_g = G \times m = 40 \times 6 = 240 \text{ mm}$$

d. Center distance between the shafts

$$C = \frac{d_w + d_g}{2} = \frac{60 + 240}{2} = 150 \text{ mm}$$

e. Diametral quotient

$$q = \frac{d_w}{m} = \frac{60}{6} = 10$$

Drive is specified as: $n / Z_g / q / m$. Hence the drive is: 1 / 40 / 10 / 6

Example 8.2

Drive proportions with given center distance

A worm gear reducing box is to be designed with approximate center distance 200 mm and helix angle 20° with single start worm. Find worm diameter, normal module, worm gear diameter, and width.

Solution

Given $C = 200$ $\quad n = 1$ $\quad \gamma = 20°$

a. From Equation (8.5), worm diameter,

$$d_w = \frac{c^{0.875}}{1.467} = \frac{(200)^{0.875}}{1.467} = 69.8 \text{ say } 70 \text{ mm}$$

From Equation (8.6), $d_w = 3 \times p_c$ or $p_c = \frac{d_w}{3} = \frac{70}{3} = 23.33 \text{ mm}$

b. Axial module, $m_a = \frac{p_c}{\pi} = \frac{23.33}{3.14} = 7.43 \text{ mm}$

Normal module, $m_n = m_a \cos \gamma = 7.43 \times \cos 20° = 6.98 \text{ mm}$

c. From Equation (8.9) $C = \frac{d_w + d_g}{2}$

Substituting the values: $200 = \frac{70 + d_g}{2}$ or $d_g = 330 \text{ mm}$

d. From Equation (8.8), width $b = 0.73 \, d_w = 0.73 \times 70 = 51.1$ say 51 mm

Example 8.3

Axial module and lead angle with given center distance

A hardened steel worm and gear set has approximate center distance 500 mm and gear ratio 25. Find axial module and lead angle.

Solution

Given $C \approx 500$ $G = 25$

PCD of worm, $d_w = \frac{c^{0.875}}{1.467} = \frac{(500)^{0.875}}{1.467} = 156.7 \text{ mm}$

From Equation (8.9): $C = \frac{d_w + d_g}{2}$ or $500 = \frac{156.7 + d_g}{2}$

Or, $d_g = 1,000 - 156.7 = 843.3 \text{ mm}$

From Equation (8.6), $p_c = \frac{d_w}{3} = \frac{156.7}{3} = 52.33 \text{ mm}$

Axial module, $m_a = \frac{p_c}{\pi} = \frac{52.33}{3.14} = \mathbf{16.63 \, mm}$

Gear ratio $G = \frac{Z_g}{n}$ or $n = \frac{Z_g}{G} = \frac{d_g}{G m_a}$ (a)

Lead angle, $\tan \gamma = \frac{p}{\pi d_w} = \frac{\pi m_a n}{\pi d_w} = \frac{m_a n}{d_w}$

Substituting the value of n from equation (a):

$$\tan \gamma = \frac{m_a}{d_w} \times \frac{d_g}{G \times m_a} = \frac{d_g}{G \times d_w} = \frac{843.3}{25 \times 156.7} = 0.1387 \quad \text{Or,} \quad \gamma = 7.9°$$

8.12 Force Analysis

In the force analysis, it is assumed that the worm is driver and worm gear is driven member. Forces acting on the worm are shown in Figure 8.8.

- Axial force along the axis of worm (F_{wa}). When the worm rotates, it rotates the worm gear due to sliding of tooth. The torque transmitted depends on this force. Equal and opposite axial force is exerted on the worm gear also.

- Tangential force in the circumferential direction at pitch radius (F_{wt}). This component of the force is due to the friction between the rubbing surfaces of worm and mating worm gear.

- Radial force from outer diameter to the center (F_{wr}). Owing to inclination of thread, the force acts at the pressure angle, which imparts radial force towards the center of the worm.

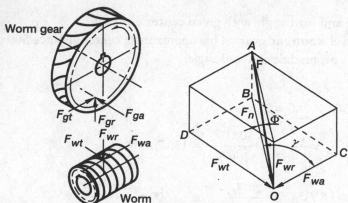

Figure 8.8 Forces on the teeth of worm and worm gear

Forces on the worm tooth are shown on right side of Figure 8.8 at an enlarged scale for more clarity. The resultant force F shown by thick line AO acts at right angle to the flank of the tooth. This force is doubly inclined as given below:

It is inclined at the angle of tooth Φ ($\angle AOB$). Its component in the plane OCBD is F_n shown by the line BO and is called normal force.

Normal force, $\qquad F_n = F \cos \Phi$ $\hfill (8.12)$

Radial force on worm, $F_{wr} = F \sin \Phi$ $\hfill (8.13)$

The force F is also inclined at lead angle γ ($\angle$BOC). Components of normal force are F_{wa} and F_{wt} shown by the lines CO and DO, respectively. These are given by equations:

Axial force on worm, $\qquad F_{wa} = F_n \cos \gamma$ $\hfill$ (8.14)

Tangential force on worm $\;F_{wt} = F_n \sin \gamma \;$ or $\; F_n = \dfrac{F_{wt}}{\sin \gamma}$ $\hfill$ (8.15)

Substituting the value of F_n from Equation (8.15) in Equation (8.14):

Axial force on worm, $\;F_{wa} = \dfrac{F_{wt}}{\sin \gamma} \cos \gamma = \dfrac{F_{wt}}{\tan \gamma}$ $\hfill$ (8.16)

Substituting value of F from Equation (8.12), that is, $\;F = \dfrac{F_n}{\cos \Phi}\;$ in Equation (8.13):

Radial force on worm, $\;F_{wr} = F \sin \Phi = \dfrac{F_n}{\cos \Phi} \sin \Phi$

Substituting value of F_n from Equation (8.15) in the above equation:

Radial force on worm, $\;F_{wr} = \dfrac{F_{wt}}{\sin \gamma \cos \Phi} \sin \Phi = \dfrac{F_{wt} \tan \Phi}{\sin \gamma}$ $\hfill$ (8.17)

Referring Figure 8.8 forces acting on the worm gear are in opposite direction to the worm, hence:

Tangential force on worm gear, $F_{gt} =$ Axial force on worm, F_{wa}

Radial force on worm gear, $\qquad F_{gr} =$ Radial force on worm, F_{wr}

Axial force on worm gear, $\qquad F_{ga} =$ Tangential force on worm, F_{wt}

Because of sliding friction between teeth of worm and gear, the frictional force is considerable, which is given as μF, where μ is coefficient of friction. Its components:

Friction in axial direction $= -\mu F \sin \gamma$ ($-$ due to opposite direction of F)

Friction in tangential direction $= \mu F \cos \gamma$

There is no friction in radial direction $= 0$

From Equations (8.12) and (8.15):

$$F = \dfrac{F_n}{\cos \Phi} = \dfrac{F_{wt}}{\sin \gamma \cos \Phi} \hfill \text{(a)}$$

Substitute the value of F from Equation (a) in equation (8.16), and get the total forces with frictional forces as under:

Total force in axial direction:

$$F_{wat} = \dfrac{F_{wt}}{\tan \gamma} - \mu F \sin \gamma = \dfrac{F_{wt}}{\tan \gamma} - \mu \dfrac{F_{wt}}{\sin \gamma \cos \Phi} \sin \gamma$$

Or,
$$F_{wat} = F_{wt}\left(\frac{1}{\tan\gamma} - \frac{\mu}{\cos\Phi}\right) \tag{8.18}$$

Total force considering friction in tangential direction:

$$F_{wtt} = F_{wt} + \mu F \cos\gamma = F_{wt} + \mu\frac{F_{wt}}{\sin\gamma\cos\Phi}\cos\gamma$$

Or,
$$F_{wtt} = F_{wt}\left(1 + \frac{\mu}{\tan\gamma\cos\Phi}\right) \tag{8.19}$$

There is no friction in radial direction, hence, total force in radial direction:

$$F_{wrt} = F_{wr} + 0 = \frac{F_{wt}\tan\Phi}{\sin\gamma} \tag{8.20}$$

Similarly forces acting on the worm gear are in opposite direction to the worm.

- Tangential force in the circumferential direction at pitch radius of worm gear is equal to axial force of worm, that is,

  Total tangential force on worm gear: $F_{gtt} = F_{wat} = F_{wt}\left(\frac{1}{\tan\gamma} - \frac{\mu}{\cos\Phi}\right)$ $\tag{8.21}$

- Radial force for worm gear is equal and opposite to radial force of worm, that is,

  Total radial force on worm gear: $F_{grt} = F_{wrt} = \frac{F_{wt}\tan\Phi}{\sin\gamma}$ $\tag{8.22}$

- Axial force along the axis of worm gear is equal to tangential force of the worm.

  Total axial force on worm gear: $F_{gat} = F_{wtt} = F_{wt}\left(1 + \frac{\mu}{\tan\gamma\cos\Phi}\right)$ $\tag{8.23}$

Example 8.4

Forces on worm and worm gear for given power

A worm drive transmits 2 kW of power at 1,440 rpm and reduces to 60 rpm. The worm is bored and has triple start teeth of module 5 mm. Lead and pressure angle both are 20°. If coefficient of friction is 0.1, calculate the forces on the worm and worm gear.

Solution

Given $P = 2$ kW $N_w = 1,440$ rpm $N_g = 60$ rpm $n = 3$
$m = 5$ mm $\gamma = 20°$ $\Phi = 20°$ $\mu = 0.1$

Gear ratio $G = \dfrac{N_w}{N_g} = \dfrac{1,440}{60} = 24$

Lead angle, $\tan\gamma = \dfrac{n}{q}$ or $\tan 20° = \dfrac{3}{q}$ or $q = \dfrac{3}{0.364}$

Hence, diameter quotient $q = 8.24$

PCD of worm, $d_w = q\,m = 8.24 \times 5 = 41.2$ mm

Outside diameter of worm, $d_{wo} = d_w + 2\,m = 41.2 + (2 \times 5) = 51.2$ mm

$$P = \frac{2\pi NT}{60} \quad \text{or} \quad 2{,}000 = \frac{2\pi \times 1{,}440 \times T}{60}$$

Hence, torque on worm $T_w = \dfrac{2{,}000 \times 60}{2\pi \times 1{,}440} = 13.263$ Nm

Tangential force on worm $F_{wt} = \dfrac{2T_w}{d_w} = \dfrac{2 \times 13.263}{41.2} = 644$ N

From Equation (8.20), Total radial force on worm $F_{wrt} = \dfrac{F_{wt}\tan\Phi}{\sin\gamma}$

Substituting the values: $F_{wrt} = \dfrac{644 \times \tan 20°}{\sin 20°} = \dfrac{234.4}{0.342} = 685$ N

From Equation (8.18), total axial force on worm

$$F_{wat} = F_{wt}\left(\frac{1}{\tan\gamma} - \frac{\mu}{\cos\Phi}\right) = 644\left(\frac{1}{\tan 20°} - \frac{0.1}{\cos 20°}\right)$$

Substituting the values: $\qquad F_{wat} = 644 \times (2.75 - 0.106) = 1{,}703$ N

Forces of worm gear from Equations (8.21, 8.22, and 8.23) are:

Tangential force on worm gear $\qquad F_{gtt} = F_{wat} = 1{,}703$ N

Radial force on worm gear $\qquad F_{grt} = F_{wrt} = 685$ N

Axial force on worm gear $\qquad F_{gat} = F_{wtt} = 644$ N

8.13 Strength of Worm Gear Tooth

Like spur and helical gears, worm gears are also checked for the following strengths:

 a. Strength in bending and
 b. Strength in wear to avoid pitting

8.13.1 Strength in bending

Since the material of the worm gear is generally weaker than the worm, to check the strength of worm gear is safe. Although the number of teeth of gear meshing with the worm is more than one, but to be on the safe side, it is assumed that only one tooth takes the load. According to Lewis equation:

$$S_b = \sigma_b k_v b \pi m y \tag{8.24}$$

Where, S_b = Strength of tooth in bending

σ_b = Bending stress (S_{yt} / FOS) (Table 8.12)

k_v = Velocity factor

b = Face width of worm gear

m = module and

y = Form factor or Lewis factor

Table 8.12 Safe Bending Stress σ_b (MPa)

Worm Gear Material	Phosphorous bronze—CC*	Phosphorous bronze—CH	Phosphorous bronze—SC	Grey cast iron	0.4 per cent carbon steel	0.5 per cent carbon steel	Case Hd. carbon steel
Safe stress	69	63	49	40	138	173	276

* CC - Centrifugal Cast; CH - Chilled cast; SC - Sand Cast; Hd - Hardened.

Velocity factor k_v depends on as to how the teeth for worm gear are cut.

For teeth cut by form cutter: $k_v = \dfrac{3}{3+v}$ \hfill (8.25a)

For generated teeth $\quad\quad k_v = \dfrac{6}{6+v}$ \hfill (8.25b)

Lewis form factor y depends on pressure angle. It is calculated in the same way as for spur gears using the following relations:

For 14.5° involute teeth, $y = 0.124 - \dfrac{0.684}{Z_g}$ \hfill (8.26a)

For 20° involute teeth, $\quad y = 0.154 - \dfrac{0.912}{Z_g}$ \hfill (8.26b)

8.13.2 Endurance strength

It is the same as for spur gears, maximum tangential force F_{ts}, which a tooth can bear:

$$F_{ts} = S_e b \pi m y \tag{8.27}$$

Where, S_e = Endurance strength, which depends on material as under:

For phosphorous bronze $\quad S_e = 160$ MPa

For cast iron $\quad\quad\quad\quad\quad S_e = 80$ MPa

Dynamic strength is considered for spur and helical gears, whereas it need not be considered for worm gears as the load is transferred from worm to worm wheel by sliding.

Example 8.5

Power capacity of a drive from strength considerations

A worm gear set with generated teeth has gear ratio 12. The worm speed is 3,000 rpm. Assume 14.5° pressure angle and center distance less than 200 mm. Calculate:

a. Power based on strength, if bending stress is not to exceed 60 MPa.

b. Check, if the design is safe in endurance strength. Assume endurance stress 160 MPa.

Solution

Given $G = 12$ $N_w = 3,000\,\text{rpm}$ $\Phi = 14.5°$ $C < 200\,\text{mm}$ $\sigma_b = 60\,\text{MPa}$ $\sigma_e = 160\,\text{MPa}$

a. From Equation (8.5)

$$d_w = \frac{c^{0.875}}{1.467} = \frac{(200)^{0.875}}{1.467} = 70.3\,\text{mm say } 70\,\text{mm}$$

From Equation (8.6), circular pitch, $p_c = \dfrac{d_w}{3} = \dfrac{70}{3} = 23.33\,\text{mm}$

Axial module $m_a = \dfrac{p_c}{\pi} = \dfrac{23.33}{3.14} = 7.43\,\text{mm say } 8\,\text{mm}$

From Table 8.1, for gear ratio 12, number of start for worm is taken as $n = 3$

Number of teeth on gear, $Z_g = n \times G = 3 \times 12 = 36$

PCD of gear, $d_g = m \times Z_g = 8 \times 36 = 288\,\text{mm}$

From Equation (8.9)

Center distance $C = \dfrac{d_w + d_g}{2} = \dfrac{70 + 288}{2} = 179\,\text{mm}$ (less than 200)

From Equation (8.11)

$$\cot^3 \gamma = G \quad \text{or} \quad \tan \gamma = \sqrt[3]{\frac{1}{G}} = \sqrt[3]{\frac{1}{12}} = 0.436 \quad \text{or} \quad \gamma = 23.6°$$

From Equation (8.10)

$$\frac{C}{L_N} = \frac{1}{2\pi}(\operatorname{cosec}\gamma + G\sec\gamma) \qquad \text{Substituting the values:}$$

$$\frac{179}{L_N} = \frac{1}{2 \times 3.14}[\operatorname{cosec} 23.6° + (12 \times \sec 23.6°)] = 2.149$$

Hence normal lead $L_N = \dfrac{179}{2.149} = 83.3$ mm

Lead, $L = \dfrac{L_N}{\cos\gamma} = \dfrac{83.3}{\cos 23.6°} = 90.9$ mm

Hence module $m = \dfrac{L}{3\pi} = \dfrac{90.9}{3\pi} = 9.6$ mm say 10 mm

Gear diameter, $d_g = m \times Z_g = 10 \times 36 = 360$ mm

With this diameter, center distance $C = \dfrac{d_w + d_g}{2} = \dfrac{70 + 360}{2} = 250$ mm

It is > 200 mm; hence module is taken as 9 mm

Gear diameter, $d_g = m \times Z_g = 9 \times 36 = 324$ mm

Center distance $C = \dfrac{d_w + d_g}{2} = \dfrac{70 + 324}{2} = 197$ mm It is < 200 mm, hence accepted.

Width of gear $b = 0.73 \, d_w = 0.73 \times 70 = 49$ mm say 50 mm

Pitch line velocity $v = \dfrac{\pi d_g N_g}{60,000} = \dfrac{3.14 \times 324 \times (3,000/12)}{60,000} = 4.24$ m/s

Velocity factor from Equation (8.25b) for generated teeth,

$$k_v = \dfrac{6}{6+v} = \dfrac{6}{6+4.24} = 0.586$$

For pressure angle 14.5° from Equation (8.26a), form factor $y = 0.124 - \dfrac{0.684}{Z_g}$

Substituting the values: $y = 0.124 - \dfrac{0.684}{36} = 0.105$

From Equation (8.24)

beam strength $S_b = \sigma_b k_v b \, \pi m y$ ⠀⠀⠀Substituting the values:

$S_b = 60 \times 0.586 \times 50 \times 3.14 \times 9 \times 0.105 = 5,216$ N

Beam strength has to be equal to tangential force, that is, $F_t = 5,216$ N

⠀⠀⠀Power, $P = F_t \times v = 5,216 \times 4.24 = 22,118$ W $= 22.12$ kW

b. Endurance strength, $S_e = \sigma_e b \, \pi m y = 160 \times 50 \times 3.14 \times 9 \times 0.105 = 23,738$ N

It is much more than the tangential force, that is, 5,216 N; hence the design is safe.

8.13.3 Strength in wear

Maximum wear load F_{tw} is given by the relation:

$$F_{tw} = d_g b K \qquad (8.28)$$

Where, d_g = PCD of the worm gear

b = Width of gear

K = Load stress factor, which depends on the materials used for worm and worm gear and lead angle (See Table 8.13 for the value of K).

Table 8.13 Values of Load Stress Factor K (MPa)

Material Combination		Lead Angle		
Worm	Worm Gear	<10°	10°–25°	>25°
Hardened steel	Cast iron	0.345	0.43	0.52
Steel BHN 250	Phosphorous bronze	0.415	0.52	0.62
Hardened steel	Phosphorous bronze	0.550	0.69	0.82
Hardened steel	Antimony bronze	0.830	1.04	1.25
Hardened steel	Chilled phosphorous bronze	0.830	1.04	1.25

The torque T on the gear for the given power P is given by:

$$\text{Torque on gear } T_g = \frac{P \times 60}{2\pi N}$$

$$\text{Tangential force } F_{gt} = \frac{T_g}{0.5 d_g}$$

The tangential force is modified using three factors; velocity factor, form factor, and load factor as:

$$F_{gtmax} = \frac{T_g K_L}{0.5 d_g k_v y} \qquad (8.29)$$

Where, F_{gtmax} = Maximum tangential force

k_v = Velocity factor (Equation 8.25)

y = Form factor (Equation 8.26)

K_L = Load factor, for light shocks K_L = 1.2 and for heavy shocks, K_L = 2

Example 8.6

Power capacity from wear considerations

For the data of the gear set given in Example 8.5, what is the power, which can be transmitted from wear point of view? Assume worm of steel BHN 250 and gear of phosphorous bronze.

Solution

From Table 8.13, for materials given in example, for lead angle 23.6°, value of $K = 0.52$

From Equation (8.28), safe wear load, $F_{tw} = d_g b K = 324 \times 50 \times 0.52 = 8,424$ N

Power from wear point of view, $P = F_{tw} \times v = 8,424 \times 4.24 = 35,718$ W $= 35.72$ kW

8.14 Friction in Worm Drives

Friction coefficient in worm drives depends on rubbing / sliding velocity V_s between worm and worm gear teeth. The slide velocity can be calculated using the following equation:

$$V_s = \frac{\pi d_w N_w}{60 \cos\gamma} \tag{8.30}$$

The variation of friction coefficient μ is nonlinear; however, an approximate curve fitting can be done for the two ranges as given below:

Rubbing velocity V_s (m/s) between 0.2 and 2.8

$$\mu = \frac{0.0422}{(V_s)^{0.28}} \tag{8.31a}$$

Rubbing velocity V_s more than 2.8 m/s

$$\mu = 0.025 + 0.0033 \, V_s \tag{8.31b}$$

8.15 Efficiency of Worm Drive

Efficiency of a worm gear is the ratio of output torque with friction f to the output torque with zero losses, that is; $f = 0$. It can also be defined as the ratio of output power to input power, that is,

$$\eta = \frac{\text{Output power}}{\text{Input power}} = \frac{P_o}{P_i}$$

Output power $P_o = F_{gtt} \, (0.5 d_g) \, n_g$ (n_g is speed in rps for worm gear)

Input power $\quad P_i = F_{wtt} \, (0.5 d_w) n_w$ (n_w is speed in rps for worm)

Hence, $\eta = \dfrac{F_{gtt} \, (0.5 d_g) n_g}{F_{wtt} \, (0.5 d_w) n_w}$ \hfill (a)

$$\frac{d_g}{d_w} = \frac{m Z_g}{mq} = \frac{Z_g}{q} = \frac{Z_g/n}{q/n} = G \tan\gamma$$

$$\frac{n_g}{n_w} = \frac{N_g}{N_w} = \frac{1}{G} \quad (N_g \text{ and } N_w \text{ are speeds of worm gear and worm in rpm respectively}) \quad \text{(b)}$$

Substituting the values of Equation (b) in Equation (a):

$$\eta = \frac{F_{gtt} \times G \times \tan\gamma}{F_{wtt} \times G} = \frac{F_{gtt}}{F_{wtt}} \tan\gamma$$

Substituting the values of F_{gtt} from Equation (8.21) and F_{wtt} from Equation (8.19):

Efficiency, $\eta = \dfrac{\cos\Phi - \mu \tan\gamma}{\cos\Phi + \mu \cot\gamma}$ (8.32)

The major losses are due to friction between worm and worm gear; hence the efficiency can be approximated from the following simple relation:

Efficiency, $\eta = \dfrac{0.95 \tan\gamma}{\tan(\gamma + \Phi_1)}$ (8.33)

Where, Φ_1 = Friction angle given as $\tan^{-1}\mu$

Plot of efficiency with a lead angle of the worm is shown in Figure 8.9.

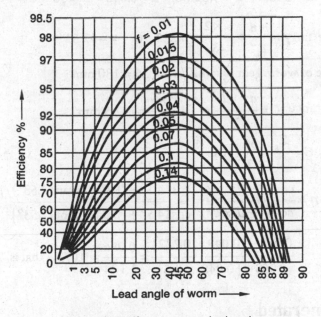

Figure 8.9 Efficiency versus lead angle

From this figure, it can be seen that efficiency depends on value of coefficient friction. More the friction f, lesser is the efficiency. Also efficiency first increases with lead angle, reaches to a maximum value, and then starts decreasing.

For maximum efficiency, relation between lead angle and coefficient of friction is given as:

$$\tan\gamma = \sqrt{1+\mu^2} - \mu$$ (8.34)

Self-locking drive

Normally a worm gear drive is irreversible, that is, a worm can drive the gear but the gear cannot drive the worm. If the lead angle is high, the drive may become reversible. A worm drive is called self-locking, when the worm wheel cannot drive the worm. This condition occurs when the lead angle of the worm is between 2° and 8°.

Example 8.7

Efficiency of worm gear drive

A worm drive has double start worm of module 5 mm. Worm has PCD 45 mm. If worm gear has 36 teeth with pressure angle 14.5°, coefficient of friction is 0.048, find the following:

- a. Lead angle of worm
- b. Center distance
- c. Gear ratio
- d. Efficiency of the drive

Solution

Given $n = 2$ $m = 5$ mm $d_w = 45$ mm $\mu = 0.048$ $Z_g = 36$ $\Phi = 14.5°$

a. Lead angle $\tan\gamma = \dfrac{mn}{d_w} = \dfrac{5 \times 2}{45} = 0.222$ or $\gamma = 12.53°$

b. Pitch circle of worm gear $d_g = mZ_g = 5 \times 36 = 180$ mm

   Center distance, $C = \dfrac{d_w + d_g}{2} = \dfrac{45 + 180}{2} = 112.5$ mm

c. Gear ratio, $G = \dfrac{Z_g}{n} = \dfrac{36}{2} = 18$

d. Efficiency $\eta = \dfrac{\cos\Phi - \mu\tan\gamma}{\cos\Phi + \mu\cot\gamma} = \dfrac{\cos 14.5° - (0.048 \times \tan 12.53°)}{\cos 14.5° + (0.048 \times \cot 12.53°)}$

$= \dfrac{0.968 - (0.048 \times 0.222)}{0.968 + (0.048 \times 4.5)} = \dfrac{0.086}{1.184} = 0.073$, that is, 73 per cent

---

8.16 Heat Generated

Efficiency of the worm gear is much less than spur or bevel gear drives. The loss of efficiency is due to sliding friction between worm and worm gear. The power lost in friction transforms, into heat generation. Thus, power lost in friction P_f is given as:

$$P_f = P(1 - \eta)$$

Where, P = Power transmitted by the drive, and

η = Efficiency of the drive.

Thus heat generated,

$$H_g = P_f = P(1 - \eta) \tag{8.35}$$

The heat generated is dissipated through oil to the casing of the drive by conduction, convection, and radiation, and then to atmosphere. Radiation can be neglected as the temperature is not very high. The heat dissipated H_d depends on:

- Overall heat transfer coefficient $K(W/m^2/{}^\circ C)$
- Outside surface area A of the gear box
- Temperature difference Δt between oil temperature t_2 and atmosphere temperature t_a, that is $(t_2 - t_a)$.

Heat dissipated,

$$H_d = KA(t_2 - t_a) = KA\Delta t \tag{8.36}$$

Overall heat transfer coefficient K is described in Section 8.16.1.

To estimate the surface area A, PCD sizes are increased to account for the addendum and clearance between gear and casing. Following proportions can be assumed in terms of PCDs:

Length L can be taken as 1.25 times the PCD of the worm gear, that is, $L = 1.25d_g$

Width W as 1.5 times the PCD of the worm, that is, $W = 1.5\,d_w$

Height H as 1.25 times the sum of PCD of worm and gear, that is, $H = 1.25\,(d_w + d_g)$

Thus the surface area of the gear casing is:

$$A = 2\left[(L \times W) + (W \times H) + (L \times H)\right]$$

Substituting the values as described above, area can be approximated as:

$$A = 4d_w^2 + 3d_g^2 + 10d_w d_g \tag{8.37}$$

For thermal equilibrium, heat generated is equal to heat dissipated, that is, $H_g = H_d$. Substituting the values from Equations (8.34) and (8.35):

$$P(1 - \eta) = KA(t_2 - t_a) \tag{8.38}$$

Knowing the power input and transmission efficiency and other parameters, temperature difference $(t_2 - t_a)$ can be calculated. The temperature difference varies between 27 °C and 38 °C. It should not be more than 38 °C. Outlet oil temperature should not exceed 95 °C.

While designing a worm drive, heat generated is to be limited, so that oil temperature does not increase more than 60 °C. American Gear Manufacturing Association (AGMA) recommends that maximum power, which can be transmitted for worm gears up to 2,000 rpm, can be checked from the following relation:

$$P = \frac{0.029\,C^{1.7}}{G+5} \tag{8.39}$$

Where, P = Input power (kW)

C = Center distance (mm) (If C is in meters, then replace 0.029 by 3,650)

G = Gear ratio

8.16.1 Heat transfer coefficient

Heat transfer coefficient depends on the gear box operation environment (ventilation, room size), gear box size, slide velocity, worm speed, temperature, etc. The use of a fan can increase the value of the coefficient. Any precise calculation of the coefficient is difficult; therefore, a value between 5 W/m²/°C and 50 W/m²/°C is selected. Recommended values for box without fan are:

Small unventilated rooms	10 W/m²/°C–20 W/m²/°C
Well ventilated rooms	12 W/m²/°C–22 W/m²/°C
With fan over the fins on casing	15 W/m²/°C–30 W/m²/°C

The values of heat transfer are modified depending on the following effects:

- **Size effect**   Small gear boxes have the coefficient up to 50 per cent higher than big ones.
- **Temperature effect**   The coefficient may increase up to 15 per cent with the increase of difference of ambient temperature and oil temperature.
- **Fan effect**   Use of a fan may increase the heat transfer coefficient up to 50 per cent.
- **Speed effect**   The coefficient increases with increase in worm speed as shown in Figure 8.10.

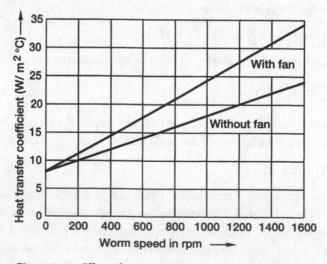

Figure 8.10 Effect of worm speed on heat transfer coefficient

8.16.2 Use of oil cooler

In gear boxes, where the heat generated is high due to higher power or lower efficiency, natural cooling is often insufficient and additional oil cooling is to be used in the form of an external oil cooler. Hot oil is taken out from the gear box using a pump and cooled in a cooler and sent back to the gear box.

Example 8.8

Power capacity with thermal considerations

A worm drive transmits 4 kW of power at 1,500 rpm. The worm is single start of module 8 mm, pressure angle 14.5°, and lead angle 10°. If surface area of the gear box is 1.2 m² and heat transfer coefficient is 20 W//°C. Calculate:

   a. Oil temperature, if atmospheric temperature is 30 °C.

   b. Power transmission capacity based on allowable temperature rise of oil as 60 °C.

Solution

Given $P = 4$ kW $N_g = 1,500$ rpm $n = 1$ $m = 8$ mm $\Phi = 14.5°$

           $\gamma = 10°$ $A = 1.2$ m² $K = 20$ W/m²/°C $t_a = 30$ °C

a. $\tan\gamma = \dfrac{mn}{d_w}$ or $\tan 10° = \dfrac{8 \times 1}{d_w}$ or, $d_w = 45.4$ mm $= 0.0454$ m

   Sliding velocity, $v_s = \dfrac{\pi d_w N_w}{\cos\gamma} = \dfrac{\pi \times 0.0454 \times 1,500}{60 \times 0.985} = 3.62$ m/s

   Velocity is more than 2.8 m/s, hence from Equation (8.31b):

     $\mu = 0.025 + 0.0033 v_s = 0.025 + 0.0033 \times 3.62 = 0.037$

   Efficiency,

$$\eta = \frac{\cos\Phi - \mu\tan\gamma}{\cos\Phi + \mu\cot\gamma} = \frac{\cos 14.5° - \left(0.037 \times \tan 10°\right)}{\cos 14.5° + \left(0.037 \times \cot 10°\right)}$$

$$= \frac{0.968 - 0.00652}{0.968 + 0.21} = \frac{0.9616}{1.7778} = 0.816 \quad \text{or} \quad 81.6\text{per cent}$$

   Heat generated, $H_g = P_f = P\,(1 - \eta) = 4,000\,(1 - 0.816) = 736$ W

   Heat dissipated, $H_d = KA\,(t_2 - t_a) = 20 \times 1.2\Delta t = 24\,\Delta t$

   Equating heat dissipated with heat generated, $24\,\Delta t = 736$ or $\Delta t = 36.8$ °C

   That is, $t_2 - t_a = 36.8$ or $t_2 = 36.8 + 30 = 66.8$ °C

b. Allowable temperature rise $\Delta t = 60 - 30 = 30$ °C

   Using Equation (8.38): $P\,(1 - \eta) = KA\,\Delta t$

   Substituting the values: $P\,(1 - 0.816) = 20 \times 1.2 \times 30$

   Or, $P = \dfrac{720}{0.184} = 3,913$ W $= 3.913$ kW

8.17 Design of Worm and Worm Wheel Drive

Initial information generally required: power or torque, input speed, output speed or gear ratio, type of service. Center distance may be given some times, due to space constraints.

8.17.1 Approximate center distance given

1. Select materials for worm and worm gear from Tables 8.5 and 8.6, get the safe allowable bending stress σ_b from Table 8.7.
2. Calculate gear ratio G from input and output speeds. If G is given, calculate output speed.
3. From the given center distance and gear ratio G, choose n / Z_g / q / m from Table 8.3.
4. Choose pressure angle 14.5° for $n = 1$ or 2 and 20° for $n = 3$ or 4 or use Table 8.4.
5. Using Equation (8.10), $\dfrac{C}{L_N} = \dfrac{1}{2\pi}\operatorname{cosec}\gamma + G\sec\gamma$ calculate normal lead angle γ. You

   can use Table 8.10 also for selecting optimum lead angle. It can also be calculated using Equation (8.11) $\cot^3\gamma = G$.
6. Calculate normal lead L_N from step 5 and lead L from the relation $L = \dfrac{L_N}{\cos\gamma}$
7. Calculate axial pitch p_a by dividing lead L by number of start, that is, L / n.
8. Calculate module from the relation $m = p_a$ / π. Round of the number to a standard module whole number. Re-calculate axial pitch with this revised module.
9. Calculate pitch circle worm diameter from the relation $d_w = \dfrac{m\,n}{\tan\gamma}$. It should be

   between $d_w = \dfrac{C^{0.875}}{1.467}$ to $d_w = \dfrac{C^{0.875}}{2.5}$.
10. Calculate number of teeth of gear with the relation $Z_g = G \times n$ and then gear diameter, $d_g = Z_g \times m$.
11. Calculate exact center distance with relation $C = \dfrac{d_w + d_g}{2}$. It should be near to the

    approximate given center distance. If more than given distance, reduce module and if less, increase module and recalculate gear diameter.
12. Calculate proportions of worm and worm gear from the relations given in Table 8.8 and 8.9, respectively. Face width of gear should not be more than $0.73d_w$.

Checking the design for beam strength

13. Calculate torque on worm gear $= \dfrac{60,000\ P}{2\pi \times N_g}$

    Where, P = Power in kilowatts, N_g = Speed of worm gear in rpm
14. Calculate tangential force on worm gear $F_t = \dfrac{2T}{d_g}$

15. Calculate pitch line velocity of gear $v = \dfrac{\pi N_g d_g}{60}$

16. Calculate velocity factor $k_v = \dfrac{6}{6+v}$ (Equation 8.25a or Equation 8.25b).

17. Calculate tooth form factor $y = 0.154 - \dfrac{0.912}{Z_g}$ (Equation 8.26a or Equation 8.26b).

18. Calculate beam strength using Lewis Equation (8.24) $S_b = \sigma_b k_v b\, \pi m\, y$. It should be more than tangential force for the design to be safe in strength.

19. Calculate endurance strength using Equation (8.27) $F_{t_s} = S_e b\, \pi m\, y$

Checking the design for wear strength

20. Calculate maximum wear load F_{tw} using Equation (8.28) $F_{tw} = d_g b\, K$.
    It should be more than tangential force for the design to be safe in wear.

Checking the design from thermal considerations

21. Calculate slide velocity from Equation (8.30) $V_s = \dfrac{\pi N_w d_w}{60 \cos\gamma}$

22. Calculate friction coefficient μ using Equation (8.31a) or Equation (8.31b).

23. Calculate efficiency using Equation (8.32) $\eta = \dfrac{\cos\Phi - \mu \tan\gamma}{\cos\Phi + \mu \cot\gamma}$

24. Calculate heat generated, Equation (8.35) $H_g = P(1 - \eta)$

25. Calculate heat dissipated, Equation (8.36) $H_d = K A\, \Delta t$

26. Equate heat generated and dissipated and calculate temperature rise Δt. For the design to be safe temperature rise should not be more than 38 °C.

Example 8.9

Design of worm drive with given center distance

Design a worm gear drive to transmit 5 kW power with worm rotating at 1,440 rpm with a reduction of speed by 20 times. Teeth may be assumed involute with pressure angle 20° and approximate center distance 200 mm.

Solution

Given $P = 5$ kW $\quad N_w = 1,440$ rpm $\quad G = 20 \quad \Phi = 20° \quad C = 200$ mm

For maximum efficiency $\cot^3 \gamma = G$

Hence $\gamma = 20.2°$ say 20°.

Using Table 8.10, center distance for gear ratio 20 and pressure angle 20° is $3.85 L_N$.

Alternately, using Equation (8.10):

$$\frac{C}{L_N} = \frac{1}{2\pi}(\operatorname{cosec}\gamma + G\sec\gamma)$$

$$C = \frac{1}{2\pi}\left(\operatorname{cosec}20° + 20\sec 20°\right)L_N = 3.85 L_N$$

Or, $200 = 3.85\ L_N$ Or, $L_N = 51.9$ mm

Lead, $L = \dfrac{L_N}{\cos\gamma} = 55.23$ mm

Hence, axial pitch $p_a = \dfrac{\text{Lead}}{\text{No. of start}} = \dfrac{55.23}{2} = 27.615$

Module, $m = \dfrac{p_a}{\pi} = 8.79$ (selecting a standard module 9).

From Table 8.1 for speed reduction 20 double start thread is selected, that is, $n = 2$

Hence number of teeth on gear $Z_g = G \times n = 20 \times 2 = 40$

$$\tan\gamma = \frac{mn}{d_w} \text{ or } \tan 20° = \frac{mn}{d_w} \quad \text{or} \quad 0.364 = \frac{m \times 2}{d_w} \text{ or } d_w = 49.45 \text{ mm}$$

It should be between $d_w = \dfrac{c^{0.875}}{1.467}$ and $\dfrac{c^{0.875}}{2.5}$, that is, 75 mm to 44 mm

Hence 49.45 mm is accepted.

Actual axial pitch, $p_a = \pi m = 28.26$ mm

Actual lead, $L = p_a n = 28.26 = 56.52$ mm

Gear diameter, $d_g = \dfrac{L \times G}{\pi} = \dfrac{56.52 \times 20}{\pi} = 360$ mm

Center distance, $C = \dfrac{d_w + d_g}{2} = \dfrac{49.45 + 360}{2} = 204.7$ mm   (Almost near to 200)

Proportions of worm (for $m = 9$)

Pitch circle diameter of worm, $d_w = 49.45$ mm

Outside diameter $d_{w0} = d_w + 2m = 49.45 + (2 \times 9) = 67.45$ mm

Length of worm $Lw = 1.25(14m + 0.06nm) = 1.25[(14 \times 9) + (0.06 \times 2 \times 9)] = 159$ mm (more by 25 per cent)

Addendum $a = 1\,m = 9$ mm

Dedendum $d = 1.17\,m = 10.53$ mm

Tooth depth $h = 2.17\,m = 19.53$ mm

Proportions of worm gear

Outside diameter $d_{go} = d_g + 3.2m = 360 + (3.2 \times 9) = 388.8$ mm

Throat diameter $d_{gt} = d_g + 2m = 360 + (2 \times 9) = 378$ mm

Root diameter $d_{gt} = d_g - 2.3m = 360 - (2.3 \times 9) = 339.3$ mm

Radius of gear face $R_f = 2.8m + 14 = (2.8 \times 9) + 14 = 39.2$ mm

Radius of gear rim $R_r = 6.9m + 14 = (6.9 \times 9) + 14 = 76.1$ mm

Radius of face edge $R_e = 0.8m = 0.8 \times 9 = 7.2$ mm

Face width $b = 7.5m + 7 = (7.5 \times 9) + 7 = 74.5$ mm

Face width should not be more than $0.73\,d_w = 0.73 \times 49.45 = 36.1$ mm; hence $b = 36$ mm is taken.

Checking the design

Speed or worm gear $N_g = \dfrac{1{,}440}{20} = 72$ rpm

Hence, torque on worm gear $T = \dfrac{60{,}000P}{2\pi \times N_g} = \dfrac{60{,}000 \times 5}{2\pi \times 72} = 663.5$ Nm

Tangential force on worm gear $F_t = \dfrac{2T}{d_g} = \dfrac{2 \times 663.5}{0.36} = 3{,}686$ N

Pitch circle velocity of gear $v = \dfrac{\pi N_g d_g}{60} = \dfrac{\pi \times 72 \times 0.360}{60} = 1.35$ m/s

Velocity factor $k_v = \dfrac{6}{6+v} = \dfrac{6}{6+1.35} = 0.816$

Tooth form factor $y = 0.154 - \dfrac{0.912}{Z_g} = 0.154 - \dfrac{0.912}{40} = 0.131$

Assuming material for worm gear as phosphorus bronze, from Table 8.6. From Table 8.12 allowable bending strength for this material $\sigma = 69$ MPa. Beam strength as per Lewis equation:

$$S_b = \sigma\, k_v\, b\, \pi\, m\, y$$

Or, $S_b = 69 \times 0.816 \times 36 \times 3.14 \times 9 \times 0.131 = 7{,}504$ N

Actual load on tooth as calculated above is 3,686 N. It is less than beam strength 7,504 N, hence safe.

Checking the design for wear strength

From Equation (8.28) maximum wear load, $F_{tw} = d_g bK$

From Table 8.13 for the materials selected $K = 0.52$

Hence $F_{tw} = 360 \times 36 \times 0.52 = 6,739$ N

It is more than tangential force 3,686 N; hence the design is safe in wear.

Checking the design from thermal considerations

From Equation (8.30), slide velocity $V_s = \dfrac{\pi N_w d_w}{60,000 \cos \gamma}$

Substituting the values $\quad V_s = \dfrac{3.14 \times 1440 \times 49.5}{60,000 \times 0.94} = 3.97$ m/s

From Equation (8.31b), friction coefficient $\mu = 0.025 + 0.0033 \times V_s = 0.038$

From Equation (8.32): Efficiency $\eta = \dfrac{\cos \Phi - \mu \tan \gamma}{\cos \Phi + \mu \cot \gamma}$

Substituting the values: $\eta = \dfrac{0.94 - (0.038 \times 0.36)}{0.94 + (0.038 \times 2.75)} = \dfrac{0.926}{1.0445} = 0.885 \quad$ or $\quad 88.5$ per cent

From Equation (8.34), heat generated, $H_g = P(1-\eta) = 5,000(1-0.885) = 572$ W

From Equation (8.37): $\quad A = 4d_w^2 + 3d_g^2 + 10\, d_w\, d_g$

$A = (4 \times 49.5^2) + (3 \times 360^2) + (10 \times 49.5 \times 360) = 576,800$ mm^2

From Figure 8.10, form worm speed 1,440 rpm, value of K can be taken as 23 W/m^2/°C.

Heat dissipated $H_d = KA\Delta t$ (Equation 8.35)

Equating it to heat generated

$\quad 572 = 23 \times 0.5768 \times \Delta t \quad$ or $\quad \Delta t = 43.1$ °C

It should not be more than 38 °C; hence the box needs a fan, for which value of K from Figure 8.10 can be taken as 32 W/m^2/°C. Using this value of K,

$\quad 572 = 32 \times 0.5768 \times \Delta t \quad$ or $\quad \Delta t = 28.3$ °C

This temperature is less than 38 °C; hence the gear box with a fan is selected for the design to be safe.

8.17.2 Center distance not given

1. Estimate center distance from the given power and gear ratio using Equation (8.39):

$$P = \frac{0.029C^{1.7}}{G+5} \quad (C \text{ in mm and } P \text{ in kW})$$

2. Once the center distance is estimated, rest of the procedure is same as described in Section 8.17.1.

Example 8.10

Design of worm drive without given center distance

Design a reduction worm gear to convert 2 kW power from 1,400 rpm to 70 rpm. Assume 20° involute teeth, safe strength in bending 90 MPa, and endurance strength 170 MPa. Load stress factor $K = 0.8 \text{ N/mm}^2$.

Solution

Given $P = 2 \text{ kW}$ $\quad N_w = 1,400 \text{ rpm}$ $\quad N_g = 70 \text{ rpm}$ $\quad \Phi = 20°$

$\sigma_b = 90 \text{ MPa}$ $\quad \sigma_e = 170 \text{ MPa}$ $\quad K = 0.8 \text{ N/mm}^2$

$$G = \frac{N_w}{N_g} = \frac{1,400}{70} = 20$$

From Table 8.1 start of worm $n = 2$

Using Equation (8.39), $P = \dfrac{0.029C^{1.7}}{G+5}$

Substituting the values: $2 = \dfrac{0.029C^{1.7}}{20+5}$ $\quad$ or $\quad C^{1.7} = \dfrac{2 \times 25}{0.029}$ $\quad$ or $\quad C = 80 \text{ mm}$

Worm diameter $d_w = \dfrac{C^{0.875}}{1.467} = 32.7 \text{ say } 35 \text{ mm}$

Worm gear diameter $d_g = 2C - d_w = 160 - 35 = 125 \text{ mm}$

Number of teeth on gear $Z_g = n \times G = 2 \times 20 = 40$

Module $m = \dfrac{d_g}{Z_g} = \dfrac{125}{40} = 3.1 \text{ say } 4 \text{ mm}$

Hence, actual gear diameter $d_g = m \times Z_g = 4 \times 40 = 160 \text{ mm}$

Modified center distance $C = \dfrac{d_w + d_g}{2} = \dfrac{35 + 160}{2} = 97.5 \text{ mm}$

Width of worm gear $b = 0.73d_w = 0.73 \times 35 = 26$ mm

Pitch circle velocity of gear $v = \dfrac{\pi d_g N_g}{60,000} = \dfrac{\pi \times 160 \times 70}{60,000} = 0.596$ m/s

Velocity factor $k_v = \dfrac{6}{6+v} = \dfrac{6}{6+0.59} = 0.91$

Tooth form factor $y = 0.154 - \dfrac{0.912}{Z_g} = 0.154 - \dfrac{0.912}{40} = 0.13$

Beam strength, $S_b = \sigma_b k_v b \pi m y$

or, $S_b = 90 \times 0.91 \times 26 \times 3.14 \times 4 \times 0.13 = 3,477$ N

Tangential force on gear, $F_t = \dfrac{P}{v} = \dfrac{2,000}{0.59} = 3,390$ N

Beam strength is more than tangential force; hence design is safe in strength.

Endurance strength, $S_e = \sigma_e b \pi m y = 170 \times 26 \times 3.14 \times 4 \times 0.13 = 7,217$ N

This strength is more than the tangential force; hence design is safe in strength.

From Equation (8.28), maximum wear load, $F_{tw} = d_g b K$

Substituting the values: $F_{tw} = 160 \times 26 \times 0.8 = 3,328$ N

It is less than tangential force 3,413 N; hence the design is unsafe in wear. So increase width from 26 to 27 mm.

New wear strength is $F_{tw} = 160 \times 27 \times 0.8 = 3,456$ N (more than F_t hence safe).

Summary

Applications Worm and worm wheels are used generally for reducing speed with high speed reduction in a compact space. Gear ratio can be as high as 100:1 in a single stage. If the gear ratio is high, it is an irreversible type of transmission. It can be made reversible using small gear ratios. Used for speed decreasing devices like opening and closing of large-size gate valves, small hoists for lifting loads, machine tools, elevators, and steering of automobiles. It can be used for increasing speed also like superchargers for I.C. engines and hand blowers.

Construction The driver shaft has a worm with trapezoidal threads on cylindrical surface with large pitch that matches with the worm wheel having mating grooves of the same profile. The threads can be left hand or right hand. Single start threads are most common for high gear ratios but multi-start threads can also be used for lower gear ratios.

Advantages Occupies lesser space, high speed reduction in single stage, and silent operation. It is irreversible, which is a desirable quality especially for applications such as cranes and lifting devices.

Disadvantages   It can be used only for low and medium power transmission, low efficiency, a lot of heat is generated due to rubbing surfaces, needs fins to dissipate heat, and costly as the worm wheel is made of phosphorous bronze or bronze.

Terminology   Terms used with worm and worm gear are:

Pitch circle diameter (PCD)   It is an imaginary circle about which peripheral velocities are calculated. d_w is for worm and d_g for worm gear.

Module   Ratio of PCD of worm wheel d_g to the number of teeth Z_g. That is, $m = d_g / Z_g$.

Outside diameter   Outside diameter of worm $d_{wo} = d_w + 2a$

Axial pitch (p_a) The distance measured from a point on one thread at PCD of worm to the corresponding point on adjacent thread in the axial direction.

Circular pitch (p_c) Actual distance at PCD from a face of thread to adjacent corresponding thread. $p_c = \pi d_g / Z_g = \pi m$

Throat diameter (d_{gt}) Outside diameter at the center line of the worm wheel.

Number of start (n) Number indicating how many threads start from a plane at right angle to the axis. $n = 1$ for gear ratio, $G > 20$, $n = 2$ for G 12–35, $n = 3$ for G 8–12, and $n = 4$ for G 6–10.

Thread angle (α) It is half of the included angle between the two faces of the thread.

Lead (L) It is product of axial pitch and number of start, that is, $L = n \times p_a$.

Helix angle (γ) Angle made by a tangent on thread at PCD with axis of the worm.

Lead angle (γ) Angle made by a tangent on thread of worm at PCD with the normal to the axis of the worm. $\tan \gamma = \dfrac{n}{q}$

Length of worm (L_w) Threaded length of worm along its axis L_w.

Gear ratio (G): Number of teeth on worm wheel / Number of start of worm, $G = Z_g / n$

Diameter quotient (q) is defined as the ratio of PCD of worm to module m, that is, d_w / m. Table 8.2 gives q if the center distance is known. Table 8.3 can be used to select q for a known number of teeth on gear. If the speed is less than 300 rpm, increase the value to 1.5 times.

Pressure angle (Φ) of 14.5° is used for single and double start worms and $\Phi = 20°$ for triple and quadruple start worms. Minimum number of gear teeth Z_g or lead angle γ dependence on pressure angle can be taken from Table 8.4.

Types of worms:

    a.   *Cylindrical worm*   It is of cylindrical shape and wraps partially over the worm.

    b.   *Hour glass worm*   Its shape is like an hour glass. It wraps more number of teeth of the worm wheel. It is also called double envelope worm.

Types of worm wheel teeth:

    a.   *Straight face*   It looks more like a helical gear. The contact area is very less, hence used for light service only.

    b.   *Straight hobbed face*   Outer face and the root are arcs, hence more contact area.

c. *Semi-global face*   It is a combination of hobbed and straight face. It can be positioned just by sliding from one side. Contact area is more, hence used for medium duty.

d. *Global face*   Its outer face is concave and envelops the worm from both side to give maximum contact area for more power transmission. It is used for heavy duty.

Materials for worm   Worms are made of case-hardened steel. Depth of case hardness can vary from 1 to 4 mm. Carbon or alloyed steels allows surface hardening to HRC 45-50, cementing and hardening to HRC 56-62, and nitriding. Table 8.5 lists some materials used for worm with their typical use and properties. Chemical composition of some materials for worms:

Carbon steels	40C8, 55C8; Case-hardened steels 10C4, 14C6
Alloy steels	16Ni80Cr 60, 20Ni 2 Mo25; Ni–Cr. steels 13Ni 3 Cr 80, 15Ni 4 Cr1
For heavy loads	20Cr, 15Cr, 20Cr Mn Ti; For high speeds 35Si Mn, 40Cr, 40Cr Ni
For low speed and power	C40 hardened and tempered BHN = 220–300

Materials for worm gear: They should have less friction coefficient, gluing resistant to pair up worm and good running in. The basic material for worm gear is bronze; less frequently used are cast iron or brass. Plastic gears are used for lower powers to absorb shocks and provide lower noise.

Phosporous bronze is most commonly used. Surface hardness varies from 90 BHN to 120 BHN.

Tin bronze with high tin content of 10 per cent–12 per cent has excellent friction properties, high resistance to seizure and good running-in, but are too expensive.

Sn–Ni bronze with lower tin content (5 per cent–6 per cent) or *Antimony–bronze* can be used for sliding speed 4 m/s–10 m/s.

Aluminum or lead bronze, or brass is less expensive; bronze without tin is suitable for sliding speeds less than 4 m/s.

Grey cast iron mated with a steel worm can be used for low outputs, quiet load, and low peripheral speed up to 2 m/s.

Table 8.6 lists some materials used for worm gears with their typical use and properties.

Table 8.7 gives permissible bending fatigue stress in megapascals.

Worm proportions for 1 and 2 start in terms module m are (Table 8.8):

PCD of integral worm $d_w = 7.4\,m + 10$	Bored to fit $d_w = 7.4\,m + 28$
Outside diameter $d_{wo} = d_w + 2\,m$	Hole size for shaft $d_s = 3.14\,m + 13.5$
Hub diameter $d_h = 5.2\,m + 25$	Length of worm $L_w = 14\,m + 0.06\,nm$
Addendum $a = 1.0\,m$	Dedendum $d = 1.17\,m$
Tooth depth $h = 2.17\,m$	Pressure angle $\Phi = 14.5°$

If circular pitch (p_c) or axial pitch (p) is known: $d_w = 3 \times p_c$ or $3\,p$

Range of PCD of worm can be taken from $d_w = \dfrac{c^{0.875}}{1.467}$ to $d_w = \dfrac{c^{0.875}}{2.5}$

Worm gear proportions for 1 or 2 start in terms of module m are (Table 8.9):

PCD of integral worm $d_g = Z_g \times d_w$ Pressure angle $\Phi = 14.5°$

Outside diameter $d_{go} = d_g + 3.2\ m$

Throat diameter $d_{gt} = d_g + 2\ m$

Face width $b = 7.5\ m + 7$

Root diameter $d_r = d_g - 2.3\ m$

Radius of gear face $R_f = 2.8\ m + 14$

Radius of gear rim $R_r = 6.9\ m + 14$

Radius of face edge $R_e = 0.8m$

Drive designation: $n\ /\ Z_g\ /\ q\ /\ m$

Center distance $C = 0.5(d_w + d_g) = \dfrac{L_N}{2\pi}(\operatorname{cosec}\gamma + G\sec\gamma)$, Where, γ = Lead angle and L_N = Normal lead

For C to be minimum, $\cot^3\gamma = G$

Force analysis   Forces acting on the worm are:

Axial force along the axis of worm $F_{wa} = F_n\cos\gamma$

Tangential force is in the circumferential direction at pitch radius $F_{wrt} = F_n\sin\gamma$

Radial force from outer diameter to the center $F_{wr} = F\sin\Phi$ and $F_n = F\cos\Phi$

The force F is also inclined at lead angle γ. In terms of tangential force, these components are:

$$F_{wa} = \frac{F_{wt}}{\tan\gamma} \text{ and } F_{wr} = \frac{F_{wt}\tan\Phi}{\sin\gamma}$$

Forces acting on the **worm gear** are in **opposite direction** to the worm, hence: $F_{gt} = F_{wa}$

$F_{gr} = F_{wr}$ and $F_{ga} = F_{wt}$

Because of sliding friction between teeth of worm and gear, the frictional force is μF, where μ is coefficient of friction. Its components:

Friction in axial direction $= -m\,F\sin\gamma(\ -$ due to opposite direction of $F)$

Friction in tangential direction $= \mu\,F\cos\gamma$

There is no friction in radial direction $= 0$

$$F = \frac{F_n}{\cos\Phi} = \frac{F_{wt}}{\sin\gamma\cos\Phi}$$

Substituting the value of F when the **normal** and **frictional** forces are added together, total forces are given as:

Total force in axial direction: $F_{wat} = F_{wt}\left(\dfrac{1}{\tan\gamma} - \dfrac{\mu}{\cos\Phi}\right)$

Total force in tangential direction: $F_{wtt} = F_{wt}\left(1 + \dfrac{\mu}{\tan\gamma\cos\Phi}\right)$

Total force in radial direction: $F_{wrt} = F_{wr} + 0 = \dfrac{F_{wt}\tan\Phi}{\sin\gamma}$

Similarly, forces acting on the worm gear are in opposite direction to the worm.

$$F_{gtt} = F_{wat} = F_{wt} \left(\frac{1}{\tan \gamma} - \frac{\mu}{\cos \Phi} \right) \text{ and } F_{grt} = F_{wrt} = \frac{F_{wt} \tan \Phi}{\sin \gamma}$$

$$F_{gat} = F_{wtt} = F_{wt} \left(1 + \frac{\mu}{\tan \gamma \cos \Phi} \right)$$

Strength in bending   According to Lewis equation with velocity factor: $S_b = \sigma_b k_v b \pi m y$

S_b = Strength of tooth in bending

k_v = Velocity factor

b = Face width of worm gear

Safe bending stress σ_b for phosphorous bronze varies from 49 MPa to 69 MPa, C.I. 40 MPa, 0.4 C steel 138 MPa, 0.5 C steel 17 MPa and case-hardened steel 276 MPa.

For teeth cut by form cutter $k_v = \dfrac{3}{3+v}$; for generated teeth $k_v = \dfrac{6}{6+v}$;

Lewis form factor y depends on pressure angle. It is calculated in the same way as for spur gears using the following relations:

For 14.5° involute teeth, $y = 0.124 - \dfrac{0.684}{Z_g}$

For 20° involute teeth, $y = 0.154 - \dfrac{0.912}{Z_g}$

Endurance strength: Maximum tangential force, which tooth can take: $F_{ts} = S_e b \pi m y$

S_e = Endurance strength; for phosphorous bronze = 160 MPa; for C.I. = 80 MPa

Dynamic strength: Not be considered for worm gears (load is transferred by sliding).

Strength in wear: Maximum wear load $F_{tw} = d_g b K$,

K = Load stress factor (Table 8.13)

Torque on gear $T_g = \dfrac{P \times 60}{2 \pi N_g}$

Tangential force $F_{gt} = \dfrac{T_g}{0.5 d_g}$

The tangential force is modified using three factors; velocity factor, form factor, and load factor to give

maximum tangential force $F_{gtmax} = \dfrac{T_g K_L}{0.5 d_g k_v y}$; K_L = Load factor.

For light shocks, K_L = 1.2 and foe heavy shocks, K_L = 2

Friction in worm drives: Depends on rubbing / sliding velocity $v_s = \dfrac{\pi N_w d_w}{60 \cos \gamma}$

Friction coefficient for V_s 0.2 to 2.8 m/s, $\mu = \dfrac{0.0422}{(v_s)^{0.28}}$

and For $V_s > 2.8$ m/s, $\mu = 0.025 + 0.0033\, v_s$

Efficiency of the worm: Ratio of output power P_0 to input power P_i.

Output power $P_0 = F_{gtt}\,(0.5 d_g)n_g$ and Input power $P_i = F_{wtt}\,(0.5 d_w)n_w$

Because, $\dfrac{d_g}{d_w} = G\tan\gamma$ and $\dfrac{n_g}{n_w} = \dfrac{1}{G}$

Efficiency $\eta = \dfrac{F_{gtt}}{F_{wtt}}\tan\gamma = \dfrac{\cos\varPhi - \mu\tan\gamma}{\cos\varPhi + \mu\cot\gamma}$

Efficiency can be approximated as, $\eta = \dfrac{0.95\tan\gamma}{\tan(\gamma + \varPhi_1)}$, where $\varPhi_1 = \tan^{-1}\mu$

Efficiency first increases with lead angle, reaches to a maximum value, and then starts decreasing. For maximum efficiency, $\tan\gamma = \sqrt{1 + \mu^2} - \mu$

Self-locking means irreversible. If lead angle is high, the drive may become reversible.

Self-locking When the worm wheel cannot drive the worm. It occurs when the lead angle of the worm is between 2° and 8°.

Heat generated Power lost in friction, transforms into heat. Thus $H_g = P_f = P\,(1 - \eta)$

Heat dissipated is through oil to the casing by conduction, convection, and radiation, and then to atmosphere. Radiation can be neglected.

Heat dissipated $H_d = K A\,(t_2 - t_a) = K A \Delta t$, where K = Heat transfer coefficient, A = surface area, Δt = Temperature difference between oil temperature t_2 and atmosphere temperature t_a

For thermal equilibrium, $H_g = H_d$ or $P\,(1 - \eta) = K A\,(t_2 - t_a)$

The temperature difference varies between 27 °C and 38 °C. It should not be more than 38 °C. Outlet oil temperature should not exceed 95 °C.

AGMA recommends for worm gears up to 2,000 rpm, maximum power $p = \dfrac{0.029 c^{1.7}}{G + 5}$, where P = Input power (kW), C = Center distance (mm), and G = Gear ratio

Heat transfer coefficient depends on gear box ventilation, room size, gear box size, slide velocity, worm speed, temperature, etc. Use of a fan can increase the value of the coefficient. Its value in small unventilated room 10 W/m²/°C–20 W/m²/°C, well ventilated room 12 W/m²/°C–22W/m²/°C, with fan over the fins 15 W/m²/°C–30W/m²/°C.

Use of oil cooler Where heat generated is high, additional oil cooling is used with an external oil cooler. Hot oil is taken out from the gear box using a pump and cooled in a cooler and sent back to the gear box.

Design procedure when approximate center distance is given:

1. Select materials for worm and worm gear to get the safe bending stress σ_b.

2. Calculate gear ratio G from input and output speeds.

3. For the given center distance and gear ratio G, choose n / Z_g / q / m from Table 8.3.

4. Choose pressure angle 14.5° for $n = 1$ or 2 and 20° for $n = 3$ or 4 or use Table 8.4.

5. Using $\dfrac{C}{L_N} = \dfrac{1}{2\pi}(\operatorname{cosec} \gamma + G \sec \gamma)$ calculate normal lead angle γ.

6. Calculate normal lead L_N from step 5 and lead L from the relation $L = \dfrac{L_N}{\cos \gamma}$

7. Calculate axial pitch p_a by dividing lead L by number of start, that is, L / n.

8. Calculate module from the relation $m = p_a / \pi$. Round of the number to a standard module whole number. Recalculate axial pitch with this m.

9. Calculate pitch circle worm diameter between $d_w = \dfrac{C^{0.875}}{1.467}$ and $d_w = \dfrac{C^{0.875}}{2.5}$.

10. Calculate number of teeth of gear $Z_g = G \times n$ and then gear diameter, $d_g = m Z_g$.

11. Calculate exact center distance with relation $C = (d_w + d_g) / 2$.

12. Calculate proportions of worm and worm gear from the relations given in Table 8.8 and 8.9, respectively. Face width of gear should not be more than 0.73 d_w.

Checking the design for beam strength

13. Calculate torque from P in kilowatts and N_g in revolutions per minute of worm gear
$T = 60{,}000\, P / 2\, \pi\, N_g$

14. Calculate tangential force on worm gear $F_t = 2T / d_g$

15. Calculate pitch line velocity of gear $v = \pi N_g\, d_g / 60$

16. Calculate velocity factor $k_v = 6 / (6 + v)$

17. Calculate tooth form factor $y = 0.154 - (0.912 / Z_g)$

18. Calculate beam strength using Lewis $S_b = \sigma_b\, k_v\, b \pi m\, y$. Check if it is $> F_t$

19. Calculate endurance strength $F_{ts} = S_e\, b \pi m y$

Checking the design for wear strength

20. Calculate maximum wear load $F_{tw} = d_g\, b\, K$. It should be $>$ tangential force.

Checking the design from thermal considerations

21. Calculate slide velocity $V_s = \dfrac{\pi N_w\, d_w}{60 \cos \gamma}$

22. Calculate friction coefficient $\mu = \dfrac{0.0422}{(V_s)^{0.28}}$; For $V_s > 2.8$ m/s, $\mu = 0.025 + 0.0033\, V_s$

23. Calculate efficiency $\eta = \dfrac{\cos \Phi - \mu \tan \gamma}{\cos \Phi + \mu \cot \gamma}$

24. Calculate heat generated $H_g = P(1 - \eta)$

25. Calculate heat dissipated $H_d = KA\Delta t$

26. Equate $H_g = H_d$ and calculate temperature rise Δt. It should be < 38 °C.

Design procedure if center distance is not given

Estimate center distance (mm) from the given power (kW) and gear ratio: $P = \dfrac{0.029\, C^{1.7}}{G+5}$

Rest of the procedure is same as given above.

Theory Questions

1. Sketch a worm and worm gear set. Describe its working and describe the conditions in which this type of drive is suitable.

2. Write advantages and disadvantages of a worm gear reduction drive in comparison to other methods. Name some typical applications where this is used.

3. Describe what is meant by diameter quotient, lead angle, normal lead, and axial pitch.

4. Describe different types of worm gear drives with the help of suitable sketches.

5. Describe the materials used for worm and worm gear and their typical applications.

6. Write the important dimensions of worm and worm gear in terms of PCD and module.

7. How is a worm gear drive designated? Explain by examples.

8. Write a short note on center distance. Derive an expression for center distance, normal lead and lead angle. What is the condition for minimum center distance?

9. Discuss the different components of forces on a worm and worm gear with the help of suitable sketches. How these forces are affected by friction.

10. Describe the Lewis equation for the beam strength of a worm gear tooth. Discuss the parameters included in this equation.

11. Explain the wear strength of a worm gear.

12. Describe friction coefficient and how its value varies with slide velocity.

13. Define efficiency of a worm drive. Derive an expression for the efficiency. How does it vary with lead angle? Show it with a sketch. What is meant by self-locking drive?

14. Describe the thermal considerations for the design of worm gear set.

15. Write a note on heat transfer coefficient. What are the parameters affecting its value?

16. Describe the procedure to design a worm gear drive, if center distance is given.

17. How do you design a worm gear reduction set, if center distance is not given?

Multiple Choice Questions

1. If PCD of worm is d_w and module is m. Diameter quotient is:

    (a) d_w / m (b) m / d_w (c) $\pi m / d_w$ (d) $m / \pi d_w$

2. For a worm gear drive with n number of start, module m, and PCD, lead angle is:

    (a) $\dfrac{m}{n \times d_w}$ (b) $\dfrac{n \times m}{d_w}$ (c) $\dfrac{n \times m}{\pi d_w}$ (d) $\dfrac{n + m}{d_w}$

3. For gear ratio more than 20, number of start used is:

    (a) One   (b) Two   (c) Three   (d) Four

4. If m is module, n number of start, Z number of teeth of gear, q diameter quotient, a worm drive is designated as:

    (a) $m/n/Z/q$ (b) $q/m/n/Z$ (c) $Z/q/m/n$ (d) $n/Z/q/m$

5. Ratio of center distance to normal lead is:

    (a) Constant for a gear ratio   (b) Depends on only lead angle

    (c) Depends on lead angle and gear ratio   (d) None of the above

6. Angle between tangent to the thread of a multi-start worm at pitch cylinder, and normal to the axis of worm is called:

    (a) Helix angle   (b) Lead angle   (c) Pressure angle   (d) Friction angle

7. If F_{wt} is tangential force on worm, Φ pressure angle, and γ lead angle, then axial force on a worm is:

    (a) $F_{wt} \cot \gamma$ (b) $F_{wt} \tan \gamma$ (c) $F_{wt} \tan \Phi$ (d) $F_{wt} \tan \Phi \tan \gamma$

8. Efficiency of a worm drive:

    (a) Increases with lead angle

    (b) Decreases with lead angle

    (c) First increases and then decreases with lead angle

    (d) Lead angle has no effect

9. If μ is friction coefficient and γ lead angle, maximum efficiency is when:

    (a) $\tan\gamma = \sqrt{1+\mu^2}$ (b) $\tan\gamma = \sqrt{1+\mu^2} - \mu$

    (c) $\tan\gamma = \sqrt{1+\mu^2} + \mu$ (d) $\tan\gamma = 1+\mu^2$

10. If L_N is normal lead, then lead L is given as:

    (a) $L_N \cot \gamma$ (b) $L_N \tan \gamma$ (c) $L_N \cos \gamma$ (d) $L_N \sec \gamma$

11. If G is gear ratio, lead angle γ then for maximum efficiency:

    (a) $\cot^3 \gamma = G$ (b) $\cos^3 \gamma = G$ (c) $\tan^3 \gamma = G$ (d) $\sin \gamma = G$

Answers to multiple choice questions

1. (a)   2. (b)   3. (a)   4. (d)   5. (c)   6. (b)   7. (a)   8. (c)   9. (b)   10. (d)   11. (a)

Design Problems

1. A single start worm gear set has 35 teeth of module 8 and diametrical quotient 12. Calculate:

    (a) Speed reduction

    (b) Center distance

    (c) Pitch circle outside and throat diameters of worm and worm gear

    (d) Face width

$$[G = 35, c = 188 \text{ mm}, d_w = 96 \text{ mm}, d_{wo} = 112 \text{ mm}, d_{wt} = 76.92 \text{ mm}, d_g = 280 \text{ mm},$$
$$d_{go} = 295.9 \text{ mm}, d_{gt} = 260.82 \text{ mm}, b = 19.6 \text{ mm}]$$

2. A worm gear set transmits 5 kW at 1,440 rpm to reduce the speed 10 times. Design the gear set assuming involute teeth of 20° pressure angle. Take allowable safe bending stress 85 MPa and K for wear 0.5 N/mm². $[d_w = 36 \text{ mm}, d_g = 180 \text{ mm}, b = 45 \text{ mm}, m = 6]$

3. A 10 kW power is to be transmitted with a speed reduction of 8 from 1,160 rpm through 20° involute teeth. Design the gear and worm, if approximate center distance is 200 mm. Assume 25 per cent overload. $[d_w = 80 \text{ mm}, d_g = 400 \text{ mm}, b = 67 \text{ mm}, m = 10]$

4. A gear set is to be designed to transmit 1 kW from a worm rotating at 1,500 rpm to a gear having speed reduction 25 with 20° involute teeth. Gear is of cast iron and worm of hardened steel.

$$[d_w = 30 \text{ mm}, d_g = 125 \text{ mm}, b = 21 \text{ mm}, m = 2.5]$$

5. A worm gear drive transmits 5 kW at 1,000 rpm with gear ratio 15. The worm has triple start threads and PCD 65 mm with module 5, pressure angle 20, and coefficient of friction 0.1. Calculate:

    (a) Tangential, axial, and radial forces on worm

    (b) Tangential, axial, and radial forces on worm gear

    (c) Efficiency of the drive

    [Worm forces: $F_{wtt} = 2,034 \text{ N}$, $F_{wat} = 250.8 \text{ N}$, $F_{wrt} = 2,034 \text{ N}$,
    Gear forces: $F_{gtt} = 250.8 \text{ N}$, $F_{gat} = 2,034 \text{ N}$, $F_{grt} = 2034 \text{ N}$, $\eta = 70.1$ per cent]

6. A hardened steel worm and worm gear set transmits power with gear ratio 16 and approximate center distance 300 mm. Find module and lead angle. $[m = 12 \text{ mm}, \gamma = 12.9°]$

7. A hardened steel worm transmits power to a phosphorous bronze gear at 1,440 rpm with gear ratio 20 and approximate center distance 1,440 rpm, pressure angle 14.5°. Find power capacity in all considerations. Assume $K = 0.6$ and safe bending strength 60 MPa.

    [Power based on strength = 11.3 kW, Power based on wear = 8.76 kW,
    Power based on heat generation = 13.4 kW, Minimum is safe, that is, 8.76 kW]

8. A 30 kW power is to be transmitted through a worm gear set, which reduces the speed from 750 rpm to 30 rpm using 20° involute teeth. Assuming safe bending strength 65 MPa and $K = 0.7$, design the gear. $[c = 438 \text{ mm}, n = 2, d_w = 145 \text{ mm}, d_g = 800 \text{ mm}, Z_g = 50, b = 102 \text{ mm}, m = 16]$

Previous Competition Examination Questions

IES

1. A worm gear set is designed to have pressure angle 30°, which is equal to the helix angle. The efficiency of the worm gear set at an interface friction of 0.5 is:

    (a) 87.9 per cent   (b) 77.9 per cent   (c) 67.9 per cent   (d) 57.9 per cent

    [Answer (a)] [IES 2015]

2. The lead angle of a worm is 22.5°; its helix angle will be:

    (a) 22.5°   (b) 45°   (c) 67.5°   (d) 90°

    [Answer (c)] [IES 1994]

GATE

3. Which one of the following is used to convert into a translational motion?

   (a) Bevel gear   (b) Double helical gear

   (c) Worm gear   (d) Rack and pinion gears

                                                                  [Answer (d)] [GATE 2014]

4. Match the following:

   Type of gears:

   P—Bevel gears   Q—Worm gears   R—Herringbone gears   S—Hypoid gears

   Arrangement of shafts:

   1.   Nonparallel offset shafts

   2.   Nonparallel intersecting shafts

   3.   Nonparallel nonintersecting shafts

   4.   Parallel shafts

       (a)   P—4, Q—2, R—1, S—3

       (b)   P—2, Q—3, R—4, S—1

       (c)   P—3, Q—2, R—1, S—4

       (d)   P—1, Q—3, R—4, S—2

                                                                    [Answer (b)] [GATE 2004]

5. Large speed reductions (greater than 20) in one stage of a gear train are possible through:

   (a) Spur gear   (b) Worm gear   (c) Bevel gear   (d) Helical gear

                                                                   [Answer (b)] [GATE 2002]

6. A 1.5 kW motor is running at 140 rev/min. It is to be connected to a stirrer running at 36 rev/min. The gearing arrangement for this application is:

   (a) Differential gear (b) Helical gear   (c) Spur gear   (d) Worm gear

                                                                   [Answer (d)] [GATE 2000]

❑❑❑

Gear Trains and Gear Boxes

9.1 Function of a Gear Box

A gear box is a box containing gears, which is used to change the speed (generally decrease), and hence increase torque. If required, change the direction of rotation between input and output shafts. The speed is decreased using a set of gears; hence the torque automatically increases in the same ratio in which the speed decreases. A gear box, in which the speed is reduced, is called a reducer. A pair of gear changes the direction of rotation, that is, if input shaft rotates clockwise (CW), output shaft will rotate counter clock wise (CCW) direction. To keep the direction same, an intermediate shaft is also used, which is called a lay shaft. When the power is transferred from the lay shaft to output shaft, the direction of rotation again becomes the same as that of the input shaft (See picture on right side). The prime mover is attached to one end of the input shaft of the gear box. The point of contact of the gear teeth is well lubricated with gear oil. The gear oil must be very clean and free of abrasive materials to avoid wearing of the gears.

A transfer case also uses gear trains, but for one input shaft, there are two output shafts. It is used for four wheel drives in automobiles. One output shaft drives rear wheels and the other the front wheels.

9.2 Applications

A gear box is used in many applications. Some of them are listed below:

- Automobile gear boxes provide different speeds (generally four or five) and also change direction of rotation for reverse. In differential to change the direction by 90° and also to reduce the speed further.

- Machine tools gear boxes provide different speeds for the spindle and also for the automatic feeds. The number of speeds may vary between 6 and 20.

- In elevators, escalators, and rolling mills, gear boxes are used as a speed reducer only at one speed.

- In many applications, where the speed is reduced such as turbines and earth moving machines.

- In multiblade windmills, the rotor rotates at slow speed; a gear box is used to increase the speed.

9.3 Construction

Inner construction of a gear box is very simple. In a single speed gear box, there are only two gears coupled with one another. In case of speed reduction, the diameter of the driver pinion and shaft is smaller than the gear of output shaft. A gear box may have one or more gear pairs. The gear pairs may be on parallel or nonparallel axes and on intersecting or non intersecting shafts. If it has more than two pairs, the setup is called a gear train (Section 9.4). Generally, they permit higher speed ratios in smaller packages, which is not feasible with a single pair of gears.

9.4 Gear Trains

A gear train comprises of two or more pairs of gears. These are of different types:

9.4.1 Simple gear train

In a gear set, if the driver rotates in CW direction, the driven gear rotates in CCW direction. If same direction of rotation is required, then an idler can be used. Idler gear does not affect the gear ratio. Such a gear train shown in Figure 9.1. is called simple gear train. Helical gears are used, if the speed is high.

9.4.2 Compound gear train

When one stage of gear pair cannot give the desired speed ratio, a compound gear train is used. Thus a compound gear train is of two or more pairs of gear sets. Driven gear of the first set is mounted on the same shaft as the pinion

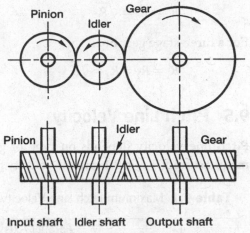

Figure 9.1 Simple gear train

of the second set as shown in Figure 9.2. The overall speed ratio is multiplication of the gear ratios of each pair. If R_1 is the gear ratio of first pair and R_2 of the second pair, then the overall gear ratio $R = R_1 \times R_2$. For a train of three pairs $R = R_1 \times R_2 \times R_3$. Picture on the right side shows a three-stage gear box. The first stage is using bevel gears, while the second and the third using helical gears.

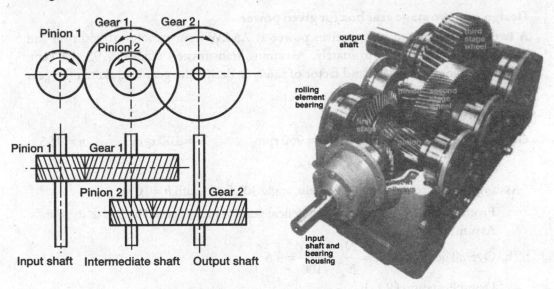

Figure 9.2 Compound gear train

In multistage gear box, speed ratio at each stage should not exceed six. Intermediate speeds are calculated on the basis of geometrical progression (GP). If R is the total reduction and R_1 and R_2 are reductions for stage 1 and stage 2, respectively then:

For a two-stage gear box:

$$R_1 = R_2 = \sqrt{R} \qquad\qquad (9.1a)$$

For a three-stage gear box:

$$R_1 = R_2 = R_3 = \sqrt[3]{R} \qquad\qquad (9.1b)$$

9.5 Pitch Line Velocity

Pitch line velocity depends on type of gear and the accuracy by which they have been manufactured. Table 9.1 gives its velocity for different types of gears.

Table 9.1 Maximum Pitch Line Velocity (m/s) for Commercial Grade Gearing

	Parallel Axis				Axes at an Angle			
	Ground		Lapped		Ground		Lapped	
	Spur	Helical	Spur	Helical	Bevel	Spiral	Bevel	Spiral
Maximum pitch line velocity	7	10	15	25	5	10	10	25

Example 9.1

Design of a two stage gear box for given power

A two-stage reduction gear receives power at 7.5 kW from a motor at 960 rpm and delivers at 100 rpm approximately. Assuming transmission efficiency 95 per cent, bending strength 200 MPa and factor of safety 2, design the gear box, sketch the gear train with gear dimensions.

Solution:

Given $P = 7.5$ kW $N_i = 960$ rpm $N_o = 100$ rpm $\eta = 0.95$

   $S_b = 200$ $FOS = 2$

Assumptions: Pressure angle 20°, helix angle 30°, face width $b = 10\,m$

a. From Table 9.1, for commercial helical gears velocity should be less than 10 m/s. Assuming 5 m/s.

b. Overall gear ratio: $R = \dfrac{N_i}{N_o} = \dfrac{960}{100} = 9.6$

   Using Equation (9.1a):

   $$R_1 = R_2 = \sqrt{R} = \sqrt{9.6} = 3.098$$

For pressure angle 20°, minimum teeth for the pinion to avoid interference is 18.

Number of teeth of the driven gear of first stage $= 18 \times 3.098 = 55.77$ say 56 teeth

Number of teeth of the driven gear of second stage $= 18 \times 3.098 = 55.77$ say 56 teeth

Stage 1 Since the speed is high in first stage, helical gears are chosen.

$$T = \frac{60,000P}{2\pi N} = \frac{60,000 \times 7.5}{6.28 \times 960} = 74.642 \, \text{Nm}$$

Diameter of pinion $d_p = m \, Z_p = 18 \, m$

Tangential load on pinion $F_t = \dfrac{T}{0.5d_p} = \dfrac{74.642 \times 10^3}{0.5 \times 18m} = \dfrac{74,642}{0.5 \times 18m} = \dfrac{8,293.5}{m}$

Assume pressure angle $20°$ full depth teeth and face width $b = 10 \, m$.

Equivalent formulative number of teeth $Z' = \dfrac{Z}{\cos^3 \alpha} = \dfrac{18}{\cos^3 30°} = 27.7$

For $20°$, full depth involute teeth $y' = 0.154 - \dfrac{0.912}{Z'} = 0.154 - \dfrac{0.912}{27.7} = 0.121$

Assuming carefully cut gear, for which velocity factor $K_v = \dfrac{6}{6+5} = \dfrac{6}{6+5} = 0.545$

Calculate module from the Lewis equation:

$F_t = \sigma_b \, K_v \, b \, \pi \, m \, y'$ Substituting the values:

$$\frac{8,293.5}{m} = \frac{200}{2} \times 0.545 \times 10m \times 3.14m \times 0.121$$

Or, $8,293.5 = 207 \, m^3$ Or, $m = 3.4$, select 4 as a standard module.

Pitch circle diameter (PCD) of pinion $d_p = m \, Z_p = 18 \, m = 18 \times 4 = 72 \, \text{mm}$

Outside diameter of pinion $D_p = d_p + 2 \, m = 72 + 8 = 80 \, \text{mm}$

Check for the velocity assumed:

$$v = \frac{\pi d N}{60,000} = \frac{3.14 \times 72 \times 960}{60,000} = 3.6 \, \text{m/s}$$

It is lesser than assumed hence safe.

PCD of the gear $d_g = m \, Z_g = 56 \times 4 = 224 \, \text{mm}$

Outside diameter of pinion $D_g = d_g + 2 \, m = 224 + 8 = 232 \, \text{mm}$

Face width $b = 10 \, m = 10 \times 4 = 40 \, \text{mm}$

Stage 2   Since the speed is low in this stage, spur gears are chosen.

Power to this pinion will be $7.5 \times 0.95 = 7.125$

$$\text{Speed of intermediate shaft} = 960 \times \frac{18}{56} = 309 \text{ rpm}$$

$$T = \frac{60,000\,P}{2\pi N} = \frac{60,000 \times 7.125}{6.28 \times 309} = 220.3 \text{ Nm}$$

$$\text{Tangential load on pinion } F_t = \frac{T}{0.5 d_p} = \frac{220.3 \times 10^3}{0.5 \times 18 m} = \frac{24,478}{m}$$

For 20°, full depth involute teeth:

$$y = 0.154 - \frac{0.912}{Z} = 0.154 - \frac{0.912}{18} = 0.103$$

Calculate module from the Lewis equation:

$$F_t = \sigma_b\, K_v\, b\, \pi m\, y \qquad \text{Substituting the values:}$$

$$\frac{24,478}{m} = \frac{200}{2} \times 0.545 \times 10 m \times 3.14 m \times 0.103$$

Or, $24,478 = 172.6\ m^3$ Or, $m = 5.2$, select 6 as standard module.

PCD of pinion $d_p = m\,Z_p = 18\,m = 18 \times 6 = 108$ mm

Outside diameter of pinion $D_p = d_p + 2\,m = 108 + 12 = 120$ mm

$$\text{Check for the velocity assumed } v = \frac{\pi d N}{60,000} = \frac{3.14 \times 108 \times 280}{60,000} = 1.58 \text{ m/s}$$

It is lesser than assumed hence safe.

PCD of the gear $d_g = m\,Z_g = 56 \times 6 = 336$ mm

Outside diameter of pinion $D_g = d_g + 2\,m = 348 + 12 = 348$ mm

Face width $b = 10\,m = 10 \times 6 = 60$ mm

The gear train with the dimensions of PCD and face width is shown in Figure 9.S1.

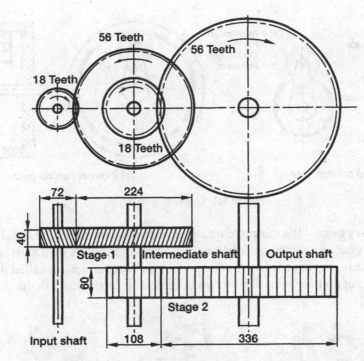

Figure 9.S1 Compound gear train two-stage gear box

Note Gears are generally provided with a hub. The outside hub diameter in larger gears should be 1.8 times the bore. Its length should be at least 1.25 times the bore or equal to the width of the gear.

9.6 Epicyclic Gear Trains

An epicyclic gear train consists of one or more rotating gears, revolving around a central gear. Epicyclic trains also called as planetary gear trains. These are suitable, where large speed reduction is required in a compact space. It has four elements, which are interconnected to each other as shown in Figure 9.3.

a. **Central gear** It rotates about its own central axis at a speed of ω_c. This gear can be external gear or internal gear. If it is external gear, it is also called as sun gear. Figure 9.3(a) shows sun gear as external gear and Figure 9.3(b) shows central gear as a ring or annulus gear as an internal gear.

b. **Arm / Spider** It also rotates about the same axis as that of central gear at angular velocity ω_a. If there is only one planetary gear, it is called arm. If there are three or four planetary gears, it is called planet carrier or spider (Figure 9.4).

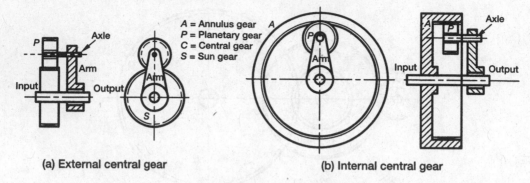

Figure 9.3 Epicyclic gear train

c. **Planetary gear** It is mounted on an axle fitted on one end of the arm. It meshes with the sun gear and rotates at rotational speed of ω_p about its own axis. Since it rotates in the same way as the planets rotate about the sun, this gear is called planetary gear. Number of plenary gears can be two or three or four as shown in the pictures below.

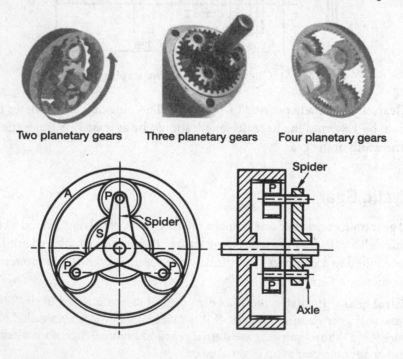

Figure 9.4 Epicyclic gear train with three planetary gears on a spider

Generally in gear boxes, three or four planetary gears are used to keep the rotating masses balanced. The arm is replaced by a spider having three or four arms as shown in Figure 9.4. Following are the unique features of this drive:

- If power is given to any one of the gears, no power flows out, since all gears are free to rotate; hence one of the elements has to be braked or locked for the power to flow out.
- If all gears are locked to each other, gear set moves as one solid mass and there is no change in speed, that is, speed ratio or gear ratio $R = 1$.
- For the power to flow, one of the elements has to be braked. Various combinations of braking, input and output are given in Table 9.2 with effect on speed:

Table 9.2 Combinations of Braking, Input, and Output Speed of an Epicyclic Gear

Brake	Input	Output	Effect on Speed
Annulus gear	Sun gear	Spider	Decreases
Annulus gear	Spider	Sun gear	Increases
Sun gear	Annulus gear	Spider	Decreases
Sun gear	Spider	Annulus gear	Increases
Spider	Sun gear	Annulus gear	Decreases
Spider	Annulus gear	Sun gear	Increases

9.7 Speed Ratio of Epicyclic Gear Trains

The speed ratio of these gear boxes can be found by any one of the following methods:

- Translation method (Section 9.7.1)
- Formula methods (Section 9.7.2 and 9.7.3)

9.7.1 Translation method

In this method, a table is prepared and the steps given below are followed:

Step 1 Lock all gear. All gears rotate as one block. Rotate any gear at some speed say +1 rpm. Counter clockwise (CCW) direction is taken positive. This condition is shown in first row of Table 9.3.

Step 2 Fix one of the elements, say arm. Give one rotation in CW direction to the driver (−1). Planetary gear rotates in CCW direction (+ sign) and the speed is in the ratio of their number of teeth, that is, $\dfrac{Z_S}{Z_P}$, where Z_S is number of teeth of sun gear and Z_P number of teeth of planet gear. Second row of the Table shows this condition.

Step 3 Add number of revolutions made by each element. The results are in the last third row, that is, when arm is rotated by +1 revolution the planet gears will rotate $1 + \dfrac{Z_S}{Z_P}$ revolution.

Table 9.3 Translation Method–External Central Gear as Sun Gear

Action Taken	Sun Gear N_s	Planet Gear N_p	Spider / Arm N_A
Lock all gears with each other	+1	+1	+1
Fix arm–Sun drives planet	−1	$+\dfrac{Z_S}{Z_P}$	0
Resultant rotation	0	$1+\dfrac{Z_S}{Z_P}$	+1

Example 9.2

Speed of sun gear from rotating arm with fixed ring gear

In an epicyclic gear train arm rotates at 15 rpm, while the ring gear having 70 teeth is fixed. If the number of teeth of sun gear and planet gear are 40 and 15, respectively, find the speed of the sun gear.

Solution

Given $N_A = 15$ rpm $Z_R = 70$ $N_R = 0$ $Z_S = 40$ $Z_P = 15$

Make a translation table with central gear as internal ring gear as under:

Action Taken	Ring Gear N_R	Sun Gear N_S	Planet Gear N_P	Arm N_A
Lock all gears with each other	+1	+1	+1	+1
Fix arm–Give one revolution CCW to ring gear to drive planet	−1	$+\dfrac{Z_R}{Z_S}$	$-\dfrac{Z_R}{Z_P}$	0
Resultant rotation	0	$1+\dfrac{Z_R}{Z_S}$	$1-\dfrac{Z_R}{Z_P}$	+1

From the above table, speed of sun gear $= 1+\dfrac{Z_R}{Z_S} = 1+\dfrac{70}{40} = 2.75$ for 1 rpm of the arm.

Hence for 15 rpm of the arm, sun gear will rotate at $N_S = 15 \times 2.75 = 41.25$ rpm.

Example 9.3

Speed of arm from rotating sun gear with fixed ring gear

In an epicyclic train, ring gear with 100 teeth is braked. The sun gear is rotated at 50 rpm in CCW direction. The number of teeth of planet gear is 40 and of sun gear 20. Using translation method, find output speed of the arm.

Solution

Given $N_R = 0$ $Z_R = 100$ $N_S = 50$ rpm (CCW) $Z_P = 40$ $Z_S = 20$

The translation table is given below:

Action Taken	Sun Gear N_S	Planet Gear N_P	Spider / Arm N_A	Ring Gear N_R
Fix the arm	–	–	0	–
Give 1 rpm to sun gear	+1	$-\dfrac{Z_S}{Z_P}$	–	$-\dfrac{Z_S}{Z_R}$
Give +x rpm to sun gear	+x	$-\dfrac{xZ_S}{Z_P}$	–	$-\dfrac{xZ_S}{Z_R}$
Give +y rpm to arm	$x + y$	$-\dfrac{xZ_S}{Z_P} + y$	+y	$-\dfrac{xZ_S}{Z_R} + y$

From the last row:

Since $N_R = 0$, hence, $-\dfrac{xZ_S}{Z_R} + y = 0$

$$\text{or} \quad -\frac{x \times 20}{100} + y = 0 \quad \text{or} \quad y = 0.2x \tag{a}$$

Given $N_S = 50$ rpm, hence from second column and last row of the table:

$$x + y = 50 \tag{b}$$

Solving Equations (a) and (b)

$$x + 0.2x = 50 \quad \text{or} \quad x = 41.67 \text{ rpm}$$

Since the output is from the arm, whose speed is: $y = 50 - 41.67 = 8.33$ rpm (CCW).

9.7.2 Formula method 1

Consider an epicyclic train as shown in Figure 9.5 with sun gear S, arm A, planetary gear P, and ring gear R. Let speeds of sun gear, arm, and ring gear be N_S, N_A, and N_R, respectively. Angular velocities of different gears are as under:

Relative angular velocity of ring gear with respect to arm is $N_{RA} = N_R - N_A$ (a)

Relative angular velocity of sun gear with respect to arm is $N_{SA} = N_S - N_A$ (b)

Dividing Equation (b) with Equation (a):

$$\text{Speed ratio} \quad R = \frac{\text{Speed of driver}}{\text{Speed of driven}} = \frac{N_{SA}}{N_{RA}} = \frac{N_S - N_A}{N_R - N_A} \tag{9.2}$$

Equation (9.2) gives the ratio of angular velocities of sun gear and ring gear both with respect to arm. The speed ratio will remain unchanged, even if the arm rotates as both velocities are with respect to arm.

This formula gives a simple way to calculate speed ratios for the planetary gear train under different conditions. Let $R = \dfrac{N_S}{N_R} = -\dfrac{Z_R}{Z_S}$

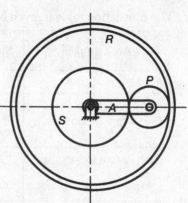

A. Condition 1: The carrier is fixed, that is, $N_A = 0$.

From Equation (9.2): $\dfrac{N_S - N_A}{N_R - N_A}$

Figure 9.5 An epicyclic train with sun gear and ring gear

Planet gear acts as an idler and speed ratio, $R = \dfrac{N_S - 0}{N_R - 0} = \dfrac{N_S}{N_R} = -\dfrac{Z_R}{Z_S}$ (9.3a)

Minus sign because the direction is reversed.

B. Condition 2: The ring gear is fixed, that is, $N_R = 0$. From Equation (9.2):

$$R = \frac{N_S - N_A}{0 - N_A} = -\frac{N_S}{N_A} + 1 \quad \text{or} \quad \frac{N_S}{N_A} = 1 - R \quad \text{Or,} \quad \frac{N_S}{N_A} = 1 + \frac{Z_R}{Z_S} \tag{9.3b}$$

C. Condition 3: The sun gear is fixed, that is, $N_S = 0$. From Equation (9.2):

$$R = \frac{0 - N_A}{N_R - N_A} \quad \text{or} \quad \frac{1}{R} = \frac{N_R - N_A}{-N_A} \quad \text{or} \quad \frac{1}{R} = -\frac{N_R}{N_A} + 1$$

$$\text{Or,} \quad \frac{N_R}{N_A} = 1 + \frac{Z_S}{Z_R} \tag{9.3c}$$

9.7.3 Formula method 2

The number of teeth of planetary gear does not affect the speed ratio. From simple geometry, it can be seen that diameter of ring gear is equal to diameter of sun gear plus twice the diameter of the planetary gear. Since the module of all the gears have to be same, number of teeth are module times the diameter. Hence

$$Z_R = Z_S + 2\,Z_P \tag{9.4}$$

To calculate speed ratio of an epicyclic gear, use equation given below:

$$(Z_R + Z_S)\,N_A = (Z_R \times N_R) + (Z_S \times N_S) \tag{9.5}$$

Where, Z_R Z_S and Z_P = Number of teeth of ring gear, sun gear and planetary gear respectively

N_R N_S and N_A = Speed of ring gear, sun gear and arm respectively

Speed of arm and other gears is related as:

$$N_A = N_S \times \frac{Z_S}{Z_R + Z_S} \tag{9.6}$$

Example 9.4

Speed ratio of an epicyclic gear train using different methods
An epicyclic train has ring gear of 60 teeth, which is braked to hold it stationary. The input power is given to the sun gear with 24 teeth, which rotates in CCW direction at 75 rpm. Using formula method, find:

a. Number of teeth of planetary gear.

b. Speed and direction of rotation of the arm if ring gear is held stationary.

c. Speed of planet gear.

d. Speeds of arm and planet gears using translation method.

Solution

Given $\quad N_R = 0 \quad\quad Z_R = 60 \quad\quad Z_S = 24 \quad\quad N_S = +75$ rpm (CCW)

a. From Equation (9.4), number of teeth of planetary gear $Z_P = \dfrac{Z_R - Z_S}{2} = \dfrac{60 - 24}{2} = 18$

b. For ring gear held stationary, condition 2 is applied for which, $\dfrac{N_S}{N_A} = 1 + \dfrac{Z_R}{Z_S}$
Substituting the values:

$$\frac{75}{N_A} = 1 + \frac{60}{24} \quad \text{or} \quad N_A = 21.43 \text{ rpm (CCW)}$$

Alternately using Equation (9.5), $(Z_R + Z_S)\,N_{Arm} = (Z_R \times N_R) + (Z_S \times N_S)$

$$\text{Or} \quad (60 + 24)\,N_{Arm} = (60 \times 0) + (24 \times 75) \quad \text{or} \quad N_{Arm} = \frac{24 \times 75}{84} = 21.43 \text{ rpm}$$

c. Using the same method described in Section 9.7.2, ratio of velocities for planet gear at speed N_P and N_A for arm: $R = \dfrac{N_{PA}}{N_{SA}} = \dfrac{N_P - N_A}{N_S - N_A}$

Velocity ratio between planetary gear P and sun gear S will be in the ratio of their number of teeth. That is, $\dfrac{N_{PA}}{N_{SA}} = \dfrac{Z_S}{Z_P} = \dfrac{24}{-18} = -1.333$ (minus as it will be CW)

Substituting this value in equation:

$$-1.333 = \frac{N_P - N_A}{N_S - N_A} = \frac{N_P - 21.43}{+75 - 21.43}$$

Or, $-1.333 \times 53.57 = N_P - 21.43$ or $N_P = -50$ rpm (CW)

d. Translation table is given below:

Action Taken	Sun Gear N_S	Planet Gear N_P	Spider / Arm N_A	Ring Gear N_R
Fix the arm	–	–	0	–
Give 1 rpm to sun gear	+1	$-\dfrac{Z_S}{Z_P}$	–	$-\dfrac{Z_S}{Z_R}$
Give +x rpm to sun gear	+x	$-\dfrac{xZ_S}{Z_P}$	–	$-\dfrac{xZ_S}{Z_R}$
Give + y rpm to arm	x + y	$-\dfrac{xZ_S}{Z_P}+y$	+ y	$-\dfrac{xZ_S}{Z_R}+y$

From the last row:

Since $N_R = 0$, $-\dfrac{xZ_S}{Z_R} + y = 0$

or $-\dfrac{x \times 24}{60} + y = 0$ or $y = 0.4x$ (a)

Given $N_S = +75$ rpm. Hence, from last row of the table

$x + y = 75$ (b)

Solving Equations (a) and (b), $x + 0.4x = 75$ or $x = 53.57$ rpm

Since output is from the arm, whose speed is:

$N_A = y = 75 - 53.57 = 21.43$ rpm (CCW)

From the third column and last row of the table:

Speed of planet gear $N_P = -\dfrac{xZ_S}{Z_P} + y = -\dfrac{53.57 \times 24}{18} + 21.43$

or $N_P = -71.43 + 21.43 = -50$ rpm

Note The speeds of arm and planet gears are same by both the methods.

9.7.4 Compound epicyclic gear trains

In compound epicyclic gear trains, output of one set is given to the other set of gear train. See a picture on the right side and Figure 9.S2. Input shaft in this figure is the spider arm, while the sun gear S1 is braked. Power output for the first gear set is through the planet gear P1. For the second train, input is through planet gear P2, which is rigidly connected to P1. Finally the power comes out from the sun gear S2. Method of solution for finding speed ratio is similar to simple epicyclic trains. It is demonstrated in Example 9.5.

Example 9.5

Compound gear train with sun gear braked

A compound gear train is shown in Figure 9.S2 below. Sun gear S1 with 50 teeth is fixed. Input shaft gives power to planet gears P1 having 25 teeth rotating around it. Planet gears P1 and P2 are rigidly connected and mounted on the same shaft to rotate at the same speed. Output is through the sun gear S2 with 25 teeth. Calculate the speed ratio between spider of stage 1 and sun gear 2.

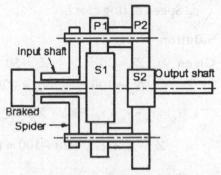

Figure 9.S2 Compound gear train with sun gear braked

Solution

Given $Z_{P1} = 25$ $Z_{S1} = 50$
            $Z_{P2} = 50$ $Z_{S2} = 25$

Gear S1 is braked; spider arm must rotate one revolution. It must be S1 to rotate – 1 revolution holding arm stationary.

Translation table is given below:

Action Taken	Sun Gear N_{S1}	Planet Gear N_{P1} or N_{P2}	Spider Arm N_A	Sun Gear N_{S2}
Revolve all by 1 rpm	1	1	1	1
Revolve S1 by –1 rpm	–1	2	0	– 4
Add all	0	3	1	–3

Hence, speed ratio of arm to sun gear 2 is 3 in the opposite direction.

Example 9.6

Compound gear train with rotating ring gear

In a compound epicyclic gear train shown in Figure 9.S3 has driver as sun gear A with 20 teeth rotates at 100 rpm. Planet gears B and C have 50 and 25 teeth, respectively, and are mounted on the same shaft of the arm. Ring gear rotates at -200 rpm and D meshes with small planetary gear C. Ring gear E meshes with large planetary gear C.

a. Assuming all gears having same module, calculate number of teeth for ring gears.

b. Speed of arm.

c. Speed of ring gear E.

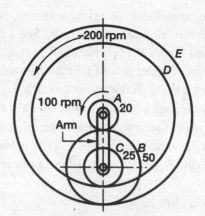

Figure 9.S3 Compound epicyclic gear train with rotating ring gear

Solution

Given $Z_A = 20$ $Z_B = 50$ $Z_C = 25$

$N_A = 100$ rpm $N_D = -200$ rpm

a. $r_D = r_A + r_B + r_C$. Hence $Z_D = Z_A + Z_B + Z_C = 20 + 50 + 25 = 95$

$Z_E = Z_A + 2Z_B = 20 + 100 = 120$

b. Speed ratio between gears A and D: $R1 = -\dfrac{Z_A}{Z_B} \times \dfrac{Z_C}{Z_D} = -\dfrac{20}{50} \times \dfrac{25}{95} = -\dfrac{2}{19}$ (a)

From Equation (9.2), $R1 = \dfrac{N_D - N_{Arm}}{N_A - N_{Arm}} = \dfrac{-200 - N_{Arm}}{100 - N_{Arm}}$ (b)

Equating Equations (a) and (b): $\dfrac{-200 - N_{Arm}}{100 - N_{Arm}} = -\dfrac{2}{19}$ or $N_{Arm} = 223.5$ rpm

c. Speed ratio between gears A and E: $R2 = -\dfrac{Z_A}{Z_E} = -\dfrac{20}{120} = -\dfrac{1}{6}$ (c)

From Equation (9.2), $R2 = \dfrac{N_E - N_{Arm}}{N_A - N_{Arm}} = \dfrac{N_E - 223.5}{100 - 223.5}$ (d)

Equating Equations (c) and (d): $\dfrac{N_E - 223.5}{100 - 223.5} = -\dfrac{1}{6}$ or $N_E = 244$ rpm

9.8 Torque Ratios of Epicyclic Gears

In epicyclic gear trains, two speeds should be known to find the third speed. At least one torque should be known to calculate other two torques. The equations, which relate different torque, are:

Torque on ring gear $T_R = T_S \times \dfrac{Z_R}{Z_S} = T_A \times \dfrac{Z_R}{Z_S + Z_R}$ (9.7a)

Where, T_S, T_A = Torque on sun gear and on arm respectively

Z_R, Z_S = Number of teeth of ring gear, and sun gear respectively

Torque on sun gear $T_S = T_R \times \dfrac{Z_S}{Z_R} = T_A \times \dfrac{Z_S}{Z_S + Z_R}$ (9.7b)

Torque on arm $\quad T_A = T_R \times \dfrac{Z_S + Z_R}{Z_R} = T_S \times \dfrac{Z_S + Z_R}{Z_S}$ (9.7c)

Holding torque T_H is the torque required to stop rotation of the braked gear. For equilibrium, the sum of all the torques has to be zero.

$$T_I + T_O + T_H = 0$$ (9.8)

Where, T_I = Input torque and T_O = Output torque

Example 9.7

Torque in an epicyclic gear train

A ring gear of an epicyclic gear train having 280 teeth is held stationary. A sun gear with 20 teeth drives planetary gears mounted on a spider to give output power. If the sun gear rotates at 300 rpm in CW direction with a torque of 20 Nm and efficiency 80 per cent, calculate:

a. Speed ratio.

b. Output speed and direction.

c. Power output.

d. Holding torque.

Solution

Given $\quad Z_R = 280 \quad\quad Z_S = 20 \quad\quad N_S = 300 \text{ rpm} \quad\quad T_A = 20 \text{ Nm}$
$\quad\quad\quad \eta = 80 \text{ per cent}$

a. From Equation (9.4), $Z_R = Z_S + 2Z_P$

Substituting the values:

$\quad 280 = 20 + 2Z_P \quad$ or $\quad Z_P = 130$

Using Equation (9.3b), $\dfrac{N_S}{N_A} = 1 + \dfrac{Z_R}{Z_S} = 1 + \dfrac{280}{20} = 15$

Speed ratio: $R = \dfrac{\text{Speed of driver}}{\text{Speed of driven}} = \dfrac{N_S}{N_R} = 15$

b. Output speed $= \dfrac{\text{Speed of driver}}{\text{Speed ratio}} = \dfrac{300}{15} = 20$ rpm

c. Power input $P_I = 2\,\pi n\, T = 2 \times 3.14 \times \dfrac{300}{60} \times 20 = 628$ W

Output power $P_O = P_I \times \eta = 628 \times 0.8 = 502.4$ W

d. Using Equation (9.7c), $T_A = T_S \times \dfrac{Z_S + Z_R}{Z_S} = 20 \times \dfrac{20 + 280}{20} = 300$ Nm

Using Equation (9.8), $T_I + T_O + T_H = 0$

Substituting the values:

$20 + 300 + T_H = 0 \qquad$ or $\qquad T_H = -320\,\text{Nm}$

9.9 Classification of Gear Boxes

Gear boxes can be classified in many ways:

A. **According to type of gear used:**
- **Spur gears**   Spur gears are used for heavy load but these are noisy.
- **Helical gears**   If the load is comparatively lesser, helical gears are preferred as these are silent in operation due to gradual engagement of teeth.
- **Bevel gears**   The axes of input and output shafts are at an angle.
- **Spiral gears**   Use is similar to bevel gears, but these are lesser noisy than bevel gears.
- **Hypoid gears**   These are used, if change of plane of rotation is required.
- **Planetary gears**   It offers high reduction in speed in a compact arrangement.

B. **According to number of output speeds:**
If the output speed is lesser than input shaft, the gear box is called a reducer unit. Generally, these are single fixed reduction ratio. As per AGMA (American Gear manufacturing Association) pinion speeds are less than 3,600 rpm or pitch line velocity less than 25 m/s. If speed is greater than this then the unit is called high-speed unit. Speed increaser gear boxes require special care in design and manufacturing. The output speed can be as under:
- **Single speed**   There are two shafts (driver and driven), with one set of gears as shown in Figure 9.6. These are generally helical to reduce noise. Speed reduction can

be in two stages as shown in Figure 9.7. It uses an additional shaft or more than one shaft called as intermediate shafts.

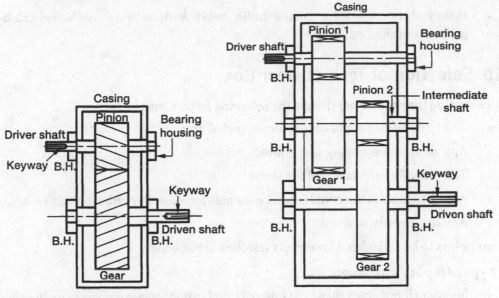

Figure 9.6 Single stage reduction gear box

Figure 9.7 Two-stage reduction gear box

- **Multiple speed** There can be many output speeds. There are more than two shafts (three or four) as shown in Figure 9.8. The gears shown joined by thick lines slide together on the splined shaft to have different gear ratio combinations. Movement of the coupled gears is shown by arrows.

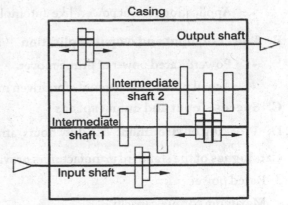

Figure 9.8 Three-stage reduction gear box

C. According to drive:

- **Mechanical drive** Using gears
- **Hydraulic drive** Using a torque converter (See Sections 9.26 and 9.27)

D. According to alignment of input and output shaft:

- **Inline** Both the axes of input and output shaft are in one straight line.
- **Angled** Axes between the two shafts are at angle but in the same plane.
- **Offset** Axes of the shafts are in different planes.

E. According to duty:

- **Light duty** Used for small gadgets, toys, etc. The gears may be either metal or plastic. This entire arrangement is enclosed in metallic or plastic housing.

- **Medium duty**   Used by most of the applications like machine tools, automobiles, escalator, and lift.

- **Heavy duty**   Used for rolling mills, heavy earth-moving machines, crushers, propeller of ship, etc.

9.10 Selection of Type of Gear Box

Before starting the preliminary design, the following factors must be known.

- The types of unit required: parallel or angled drive.

- Any abnormal operating conditions.

- The direction of rotation of the shafts.

- Any outside loads that could influence the unit, for example, overhung loads, brakes, etc.

- Any space restriction.

Salient points to be considered for correct selection are as under:

A.   Type of input / out power

- Source of input power: like I.C. (Internal Combustion) engines; single or multi-cylinder, electric motors.

- Application for out power: like automobile, machine tools, escalators, cranes, etc.

B.   Power of input and output application

- Power: Rated power of prime mover.

- Actual power requirements for driven machine.

C.   Speeds: Input and output speeds

D.   Working environment   like duty hours, atmospheric temperature.

Catalogues of the standard manufacturers provide the following information:

Rated power

Maximum torque capacity

Reduction ratio

Input / output speeds

Overall dimensions

Follow the procedure given below to select a gear box from a catalogue:

1.   Calculate the actual power required for the application.

2.   Get information on type of load: uniform or mild shock or heavy shock.

3.   Get value of service factor K_s from Table 9.4 given below:

Table 9.4 Values of Service Factor K_s

Input Source	Duty Hours	Type of Load			
		Uniform	Mild Shock	Heavy Shock Spur / Helical	Heavy Shock Worm Gear
I.C. engine: Single cylinder	8	1.35	1.65	2.15	1.95
	12	1.50	1.75	2.25	2.05
	16	1.65	1.85	2.45	2.30
	24	1.75	2.0	2.50	2.50
I.C. engine: Multicylinder	8	1.1	1.35	1.90	1.60
	12	1.25	1.50	2.00	1.75
	16	1.4	1.65	2.15	1.85
	24	1.5	1.75	2.25	2.0
Electric motor	8	0.9	1.1	1.65	1.45
	12	1.0	1.25	1.75	1.55
	16	1.15	1.4	1.85	1.65
	24	1.25	1.5	2.0	1.75

Note   For duty hours less than 8 hours, service factor for 8 hours can be reduced from 0.8 to 0.9 times suitably.

4. Gear boxes are designed for momentary over load of 100 per cent for 15 seconds. If peak load remains for a longer time, then increase the power to peak power.

5. Calculate equivalent power P_e = Actual power $P \times K_s$

6. Select a gear box from the catalogue, with rated power equal or more than equivalent power. Bearings selection is based on 90 per cent reliability for the following life:

   8 hours operation per day with life equal to 20,000 to 30,000 hours.

7. The output shaft of a gear box may have a pulley or sprocket or a gear, which is over hung from the gear box bearing. An overhung load factor K_L is to be considered for the drives as under:

   Flat belt: $K_L = 3$, V belt pulley $K_L = 1.5$, Spur gear $K_L = 1.25$, Sprocket $K_L = 1.0$

9.11 Speed Ratios in Geometric Progression

Following points are to be kept in mind, while designing for a specific transmission ratio and center distance between the two shafts. Transmission ratio or speed ratio may not be a whole number, whereas number of teeth of the gears has to be a whole number. Slight alterations are to be made in center distance or gear diameters to keep the number of teeth as an integer number.

Ideally a step less speed variation may be desired for machine tools, but most of the gear boxes provide a fixed number of speed sets. Increment of speeds can be arranged in arithmetical progression (AP) or (GP) geometrical progression. GP has proved to give minimum speed loss.

Peripheral speed of a gear v at PCD d for N rpm is: $v = \dfrac{\pi d N}{60,000}$

Or, $\dfrac{v}{d} = \dfrac{\pi N}{60,000} = k = \tan \Phi$

Where, Φ is called progression ratio.

If there are z number of desired speeds such as: $N_1, N_2, N_3, \ldots N_{z-1}, N_z$; these are in GP. Then these speeds can be written as:

$$N, N\Phi, N\Phi^2, N\Phi^3 \ldots \ldots N\Phi^{z-1}, N\Phi^z$$

$$\text{Or, } N\Phi, N\Phi^2, N\Phi^3 \ldots \ldots N\Phi^z, N\Phi^{z+1}$$

Standard values of Φ are 1.06, 1.12, 1.25, 1.41, 1.57, and 1.87.

Let N_{max}, N_{min} be the maximum and minimum desired speeds in rpm. Then the maximum transmission ratio R_n is given as:

$$R_n = \frac{N_{max}}{N_{min}} = \Phi^{z-1} \tag{9.9}$$

Or, $$\Phi = (R_n)^{\frac{1}{z-1}} \tag{9.10}$$

Or, $$z = \frac{\log(R_n) \times \Phi}{\log \Phi} \tag{9.11}$$

For a two-stage gear box, number of shafts is 3; hence the ratios of first stage R_1 and second stage R_2 are calculated from Equation (9.1) as: $R_n = R_1 \times R_2$ or $R_1 = R_2 = \sqrt{R_n}$

For S number of stages, there will be $S - 1$ intermediate shafts and $S + 1$ total shafts.

Ratio of each stage $R_1 = R_2 = R_3 \ldots\ldots\ldots R_S = \sqrt[S]{R_n}$ $\tag{9.12}$

Example 9.8

Selection of gear box and speed ratios

A reducer unit is to have maximum speed 1,440 rpm and output speed approximately 100 rpm with co-planer inline shafts.

   a. Select a suitable gear box with reasons.

   b. Output speeds, if 9 speeds are required.

Solution

Given $N_{max} = 1,440$ rpm $N_{min} = 100$ rpm $z = 9$

   a. Speed ratio $R_n = \dfrac{N_{max}}{N_{min}} = \dfrac{1,440}{100} = 14.4$

Although the gear ratio is high, even then worm gear cannot be used, as it is multi-speed.

Since the shafts are co-planer in line, bevel gear also will not be used.

Input speed is high; hence helical gears will be used instead of spur gears.

Single set of gear cannot provide such a high gear ratio (14.4); hence multi-stage gear box will be used. Initially assuming two-stage, that is, $S = 2$ and using Equation (9.12):

$$R_1 = R_2 = \sqrt[2]{R_n} = \sqrt{14.4} = 3.79$$

This ratio can be obtained by one set of gears; hence two-stage gear box is acceptable.

b. From Equation (9.10), progression ratio $\Phi = (R_n)^{\frac{1}{z-1}} = (14.4)^{\frac{1}{9-1}} = 1.397$

Selecting a standard progression ratio $\Phi = 1.41$

Hence speeds in GP are: 1,440, 1,021, 724, 514, 364, 258, 183, 130, and 101 rpm

9.12 Kinematic Diagram

Kinematic diagrams are useful for studying the inside mechanism. It is a graphic representation, which shows transmission sequence of a prime mover through an intermediate mechanism to the operative elements. It also shows how they are inter-related. It shows only those elements of an assembly, which participate in the transmission of motion. The elements can be idler gears, shafts, actuator pulleys, couplings, etc. Design of an assembly structure is drawn on the diagram in the form of solid lines. Each element is marked with numbers, which have a corresponding designation. Figure 9.9 shows a kinematic diagram of a two-stage gear box.

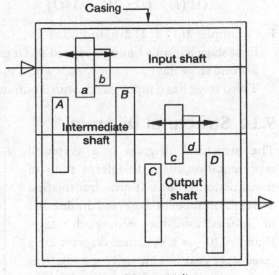

Figure 9.9 Kinematic diagram

There are three shafts; input shaft has two linked gears marked 'a' and 'b'. When they slide in left direction pinion 'a' gives power to gear 'A' on an intermediate shaft. When slided in the right side, pinion 'b' gives power to gear 'B' but with different speed ratio depending on their number of teeth.

Intermediate shaft rotates at the speed decided by gears pair a–A or gears b–B. On the same shaft, linked pinions 'c' and 'd' are also mounted. When pinion 'c' is slided to the left to engage the gear 'C' on the output shaft, power flows to the output shaft. When slided on the right side, speed of the output shaft is decided by the gear ratio of pinion 'd' and gear 'D'.

9.13 Structural Formula

Structural formula is used for gear trains to provide information about number of stages and available number of speed ratios. The transmission is divided in different stages of gear train. The first stage has one input speed N_o and some number of output speeds N_1, N_2, N_3, etc. Each stage is represented as $O(I)$

Where, O = Speed sets at output side and

I = Speeds available at input side

For example 3(2) means in this stage, there are 2 speeds available at input side and each speed gives 3 output speeds.

For a multi-stage gear train, the formula is written as $O1(I1)\ O2(I2)\ O3(I3)$ numbers 1, 2, 3, etc., represent the number of stage.

Number of output speeds of the first stage becomes the number of speeds at input side of the second stage. Similarly, number of input speeds for the third stage will be multiplication of output speeds of first and second stages. Thus formula for a three-stage gear train is written as under:

$$O1(I1)\ O2(O1)\ O3(O1 \times O2)$$

For example: 2(1) 3(2) 2(6) indicates

First stage has only one input speed and it gives 2 speeds at output.

Second stage has two input speeds with 3 output speed for each input.

Third stage has 6 input speeds with 2 output speed for each input. Thus total 12 speeds.

9.14 Structural Diagram

The structural diagram is a graphical representation, used to inform ratio of transmission groups. It gives information about the number of stages and the number of speeds available from each stage. Figure 9.10 is a structural diagram for a three-stage gear box giving 12 speeds. S1 is the input shaft, S2 and S3 are intermediate shafts and S4 is output shaft. It gives the desired spindle speeds, transmission range and characteristics of each group. **Stage one** (shown on the left side) has one input speed and 2 output speeds shown by inclined lines. Formula for this stage will be 2 (1).

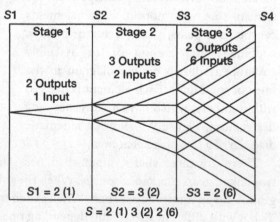

Figure 9.10 Structural diagram

Stage two (shown in the middle) has 2 input speeds received from stage 1 and each gives 3 output speeds. Formula for this stage will be 3 (2).

Finally, **stage three** has 3 input speeds from stage two and 2 output speeds from each input speed. Formula for this stage will be 2 (6)

Thus the structural formula for the whole transmission is: $S = 2\,(1)\,3\,(2)\,2\,(6)$

9.15 Number of Speeds and Stages

The number of output speeds depends on an application. It usually varies from 6 to 18, and also the number of stages generally varies from 2 to 3. Table 9.5 displays the various structural formulas, which are possible for a given number of output speeds.

Table 9.5 The Number of Output Speeds and Possible Structural Formulas

No. of Output Speeds	Factors of the Output Speeds	No. of Stages	Structural Formula with Number of Input Speeds in Parenthesis
4	2 × 2	2	2 (1) 2 (1)
6	3 × 2	2	3 (1) 2 (3)
	2 × 3	2	2 (1) 3 (2)
8	2 × 2 × 2	3	2(1) 2(2) 2(4)
	4 × 2	22	4(1) 2(4)
9	3 × 3	2	3(1) 3(3)
12	3 × 2 × 2	3	3 (1) 2(3) 2(6)
	2 × 3 × 2	3	2(1) 2(2) 2(6)
	2 × 2 × 3	3	2(1) 2(2) 3(4)
16	4 × 2 × 2	3	4 (1) 2 (4) 2 (8)
	2 × 4 × 2	3	2 (1) 4 (2) 2 (8)
	2 × 2 × 4	3	2 (1) 2 (2) 4 (4)
18	3 × 3 × 2	3	3 (1) 3 (3) 2 (9)
	3 × 2 × 3	3	3 (1) 2 (3) 3 (6)
	2 × 3 × 3	3	2 (1) 3 (2) 3 (6)

9.16 Alternate Structural Formulas

As seen in Table 9.5 that for the same number of output speed sets, different combinations are possible. In a structural formula, output speed sets $O1$, $O2$, $O3$, etc. remain as such, while the input speeds can take any position at stage 1, 2, or 3. Thus there can be six combinations for input speed sets for a three-stage gear box as under:

I1 I2 I3	I2 I1 I3	I3 I1 I2
I1 I3 I2	I2 I3 I1	I3 I2 I1

Based on these input speed sets, six structural formulas keeping $O1$, $O2$, and $O3$ at the same stage can be written as given below:

$$O1\ (I1)\ O2\ (I2)\ O3\ (I3) \qquad O1\ (I2)\ O2\ (I1)\ O3\ (I3) \qquad O1\ (I3)\ O2\ (I1)\ O3\ (I2)$$
$$O1\ (I1)\ O2\ (I3)\ O3\ (I2) \qquad O1\ (I2)\ O2\ (I3)\ O3\ (I1) \qquad O1\ (I3)\ O2\ (I2)\ O3\ (I1)$$

An example of 12 output speed sets is taken in three stages. Taking factors of 12, these are $2 \times 3 \times 2$, that is, $O1 = 2$, $O2 = 3$, and $O3 = 2$. Thus the general structural formula is:

$$2(I1)\ 3\ (I2)\ 2(I3)$$

Substituting the values of six input speed sets from above, six structural formulas are obtained as under:

(a) $2(I1)\ 3\ (I2)\ 2(I3)$ (b) $2(I2)\ 3(I1)\ 2(I3)$ (c) $2(I3)\ 3(I1)\ 2(I2)$

(d) $2(I1)\ 3\ (I3)\ 2(I2)$ (e) $2(I2)\ 3(I3)\ 2(I1)$ (f) $2(I3)\ 3(I2)\ 2(I1)$

Now input speed sets could be 1, 2, 3, 4, or 6 (factors of 12). Choosing $I1$, $I2$, and $I3$, any three at a time, the possible formulas for twelve output speed sets are as under:

Let there be three stages $O1$, $O2$, and $O3$ with $O1 = 2$, $O2 = 3$, and $O3 = 2$. Since $I1$ is the characteristics of $O1$, which will always be 1, as the input to the first stage is 1.

(a) $S = O1\ (I1)\ O2\ (I2)\ O3\ (I3)$

$I1 = 1$ $I2 = $ Characteristics of $I1 = O1 = 2$ and

$I3 = $ Characteristics of $I1 \times I2 = O1 \times O2 = 2 \times 3 = 6$

Hence $S = 2(1)\ 3(2)\ 2(6)$

(b) $S = O1\ (I2)\ O2\ (I1)\ O3\ (I3)$

$I1 = 1$ $I2 = $ Characteristics of $I1 = O2 = 3$ and

$I3 = $ Characteristics of $I1 \times I2 = O2 \times O1 = 3 \times I2 = 6$

Hence $S = 2(3)\ 3(1)\ 2(6)$

(c) $S = O1\ (I3)\ O2\ (I1)\ O3\ (I2)$

$I1 = 1$ $I2 = $ Characteristics of $I1 = O2 = 3$ and

$I3 = $ Characteristics of $I1 \times I2 = O2 \times O3 = 3 \times 2 = 6$

Hence $S = 2(6)\ 3(1)\ 2(3)$

(d) $S = O1\ (I1)\ O2\ (I3)\ O3\ (I2)$

$I1 = 1$ $I2 = $ Characteristics of $I1 = O1 = 2$ and

$I3 = $ Characteristics of $I1 \times I2 = O1 \times O3 = 2 \times 2 = 4$

Hence $S = 2(1)\ 3(4)\ 2(2)$

(e) $S = O1\ (I2)\ O2\ (I3)\ O3\ (I1)$

$I1 = 1$ $I2 = $ Characteristics of $I1 = O3 = 2$ and

$I3 = $ Characteristics of $I1 \times I2 = O3 \times O1 = 2 \times I2 = 4$

Hence $S = 2(2)\ 3(4)\ 2(1)$

(f) $S = O1\,(I3)\,O2\,(I2)\,O3\,(I1)$

$\qquad I1 = 1 \qquad\qquad I2 = \text{Characteristics of } I1 = O3 = 2 \text{ and}$

$\qquad I3 = \text{Characteristics of } I1 \times I2 = O3 \times O2 = 2 \times I3 = 6$

$\qquad$ Hence $S = 2(6)\,3(2)\,2(1)$

Thus the six structural formulas are:

$\qquad$ a. $2(1)\,3(2)\,2(6)$ $\qquad\qquad$ b. $2(3)\,3(1)\,2(6)$ $\qquad\qquad$ c. $2(6)\,3(1)\,2(3)$

$\qquad$ d. $2(1)\,3(4)\,2(2)$ $\qquad\qquad$ e. $2(2)\,3(4)\,2(1)$ $\qquad\qquad$ f. $2(6)\,3(2)\,2(1)$

Out of these structural formulas, which one will give optimum performance is described in Section 9.17.

9.17 Transmission Ratio of a Stage

If there are m stages, transmission ratio for each stage $R_m = \dfrac{I_{max}}{I_{min}}$ $\qquad\qquad$ (9.13)

Also transmission ratio of mth stage $R_m = \Phi^{(o_m - 1) \times I_m}$ $\qquad\qquad$ (9.14)

$\qquad$ Where, $O_m = $ Number of output speed sets of stage m

$\qquad\qquad I_m = $ Number of input speed sets of stage m

$\qquad\qquad \Phi = $ Progression ratio

Applying Equation (9.14) for formula (a), that is, $2(1)\,3(2)\,2(6)$

$\qquad$ For stage 1, $O1 = 2$ and $I1 = 1$; hence $R_1 = \Phi^{(2-1)\times 1} = \Phi^1 = \Phi$

$\qquad$ For stage 2, $O2 = 3$ and $I2 = 2$; hence $R_2 = \Phi^{(3-1)\times 2} = \Phi^4$

$\qquad$ For stage 3, $O3 = 2$ and $I3 = 6$; hence $R_3 = \Phi^{(2-1)\times 6} = \Phi^6$

Transmission ratios for all the formulas are calculated and tabulated below for each stage. It can be noted that sum of powers of Φ for all three stages is $(12 - 1)$, that is, 11.

Table 9.6 Transmission Ratios with Different Structural Formulas

No.	Structural Formula	Transmission Ratios			Structural Diagram
		R_1	R_2	R_3	
a	$2(I1)\,3(I2)\,2(I3)$ $2(1)\,3(2)\,2(6)$	Φ^1	Φ^4	Φ^6	
b	$2(I2)\,3(I1)\,2(I3)$ $2(3)\,3(1)\,2(6)$	Φ^3	Φ^2	Φ^6	

c	2(I3) 3(I1) 2(I2) 2(6) 3(1) 2(3)	Φ^6	Φ^2	Φ^3
d	2(I1) 3(I3) 2(I2) 2(1) 3(4) 2(2)	Φ^1	Φ^8	Φ^2
e	2(I2) 3(I3) 2(I1) 2(2) 3(4) 2(1)	Φ^2	Φ^8	Φ^1
f	2(I3) 3(I2) 2(I1) 2(6) 3(2) 2(1)	Φ^4	Φ^6	Φ^1

9.18 Optimum Structural Formula

For a good structure formula or diagram, the following points should be kept in mind:

1. Number of stages should not be more than four. In this example, there are only three stages; hence all six formulas satisfy this condition.

2. Transmission ratio should not be more than eight. For a value of $\Phi = 1.41$, $\Phi^8 = 15.6$ and $\Phi^6 = 7.8$. Formulas (d) and (e) do not satisfy this condition as value 15.6 is more than 8, hence rejected. Now options are (a), (b), (c), and (f).

3. Input speed sets: $I1$ should be less than $I2$ and $I2$ less than $I3$, that is, $I1 < I2 < I3$. Only formula (a) satisfies this condition, whereas (c), (e), and (f) do not satisfy.

4. Range of transmission: The output speeds provided by a stage must lie between $I_{max} = 2$ and $I_{min} = 0.25$ times the input speed.

5. Number of gears on the last shaft should be minimum possible.

6. The transmission ratio between spindle and the shaft preceding it should be the maximum possible, that is, speed reduction should be the maximum possible.

7. The structure diagram should be narrow towards the starting point, that is, at input shaft.

8. It should have concave shape (See Figure 9.13).

9. The sum of nodal points should be minimum.

Hence structural formula (a) is selected.

9.19 Ray Diagram

The structural diagram only informs the range of ratios, whereas ray diagram gives the transmission ratio of all transmissions and the rpm values of gear box shafts. So it is necessary to plot the speed chart to determine the transmission ratio. Vertical lines represent different shafts and the speed is plotted on Y axis (Figure 9.11). $N0$ is input speed and $N1, N2, \ldots N6$ are output speeds The line joining points of adjacent shafts in a speed chart is called a ray. It is shown by thick lines and gives the gear ratios. Meaning of these lines can be taken as given below:

- A horizontal line corresponds to transmission ratio $R = 1$, that is, no speed change.
- An upward inclined line is for transmission ratio $R > 1$, that is, speed increases.
- A downward inclined line is for transmission ratio $R < 1$, that is, speed reduces.

A ray diagram can be open type or crossed type as shown in Figure 9.11. An open type is one [Figure 9.11(a)], in which the rays do not intersect each other, whereas in crossed type they cross [Figure 9.11(b)].

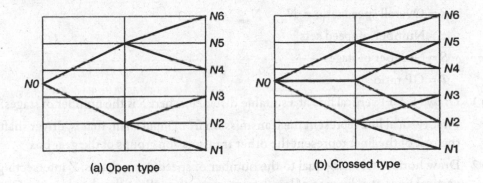

(a) Open type   (b) Crossed type

Figure 9.11 Open and crossed ray diagrams

Figure 9.12 is a ray diagram for six output speeds $N1$ to $N6$; $N1$ is minimum speed and $N6$ is the maximum speed. $N0$ is input shaft speed. The transmission is divided in two stages; stage 1 and stage 2. The first stage has one input speed $N0$ and two output speeds $N2$ and $N5$. It is represented as 2(1). The first number represents number of speed sets at output side and the number within parentheses speeds available at input side.

While plotting the speed chart, it is desirable to have the minimum transmission ratios. To illustrate the design procedure for optimal gear pair, consider the following example. Here, the gear box having 12 speed steps, that is, $z = 12$, has been selected for three stages, that is, $S = 3$. The maximum input speed N_{max} has been taken as 1,440 rpm and GP ratio Φ has been taken as 1.41. The 12 speed steps may be distributed in three stages.

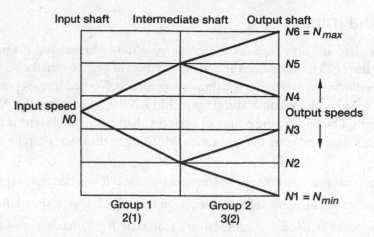

Figure 9.12 Ray diagram

Let N_{max} = Maximum speed

N_{min} = Minimum speed

R = Overall speed ratio = N_{max}/N_{min}

z = Number of speed sets

S = Number of stages

Φ = GP ratio

Step 1 Draw $(S+1)$ vertical line at a suitable distance, where S is the number of stages.

First vertical line represent the transmission from input shaft, that is, driver shaft and the rest of the lines represent the other transmission groups of the gear box.

Step 2 Draw horizontal lines equal to the number of speed steps, that is, Z intersecting the vertical lines at a distance of $\log \Phi$ from each other. Write the speeds in the GP with progression ratio 1.41 such as 1,440, 1,021, ... and 33 rpm. That is maximum speed at top and minimum speed at bottom as shown in Figure 9.13.

Step 3 Select a structure formula say 2(1) 3(2) 2(6) and then draw the rays showing input and output speeds in different stages. Figure 9.13(a) is with input speed 724 rpm and Figure 9.13(b) with 1,021 rpm. Figure 9.13(b) also has same structure formula, that is, 2(1) 3(2) 2(6). The inclination of lines will depend on the number of teeth on the driver and driven.

Step 4 Draw rays starting from the lowest rpm of last shaft; such that the transmission ratio is $I_{max} \leq 2$ and $I_{min} \geq 0.25$. In this problem, the transmission range of the last group is Φ^6 for $\Phi = 1.41$ on dividing only in one possible manner.

$$I_{max} = 2 = 1.41^2 = 1.9881 \text{ and } I_{min} = 0.25 = 1.41^4 = 3.952 \ (\Phi^6 = \Phi^2 \cdot \Phi^4)$$

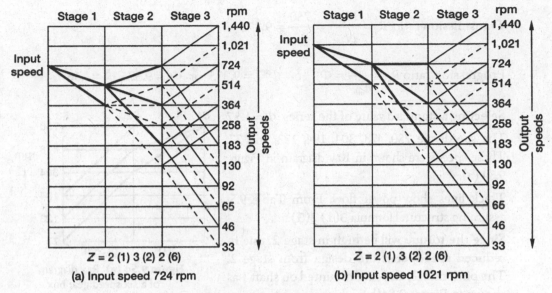

Figure 9.13 Ray diagrams

There can be many combinations to get the same number of speeds. For example, the options for the above example could be:

(a) 2(1) 3(2) 2(6)   (b) 2(1) 3(4) 2(2)   (c) 2(2) 3(4) 2(1)

(d) 2(3) 3(1) 2(6)   (e) 2(6) 3(1) 2(3)   (f) 2(6) 3(2) 2(1)

Generally, speed is reduced, but if input speed is less, sometimes it is required to increase speed also as shown in Figure 9.13(a) for stages 2 and 3.

Example 9.9

Design of a six speed gear box

Design a six speed gear box to transmit 3 kW of power at 1,500 rpm with minimum speed of about 80 rpm. The power is given to the gear box through a pulley system with speed ratio 2. Service factor is 1.25.

  a. Draw its kinematic and structural diagrams.

  b. Module and size of gears.

  c. Size of shafts.

Solution

Given $P = 3\,\text{kW}$ $N_{motor} = 1,500$ $S.R. = 2$ $N_{min} = 80$ $K_s = 1.25$

  a. Number of teeth on gears

$$\text{Speed of input shaft of gear box} = \frac{1,500}{2} = 750 \text{ rpm, that is, for gear box } N_{max} = 750\,\text{rpm}$$

Transmission ratio $R_n = \dfrac{N_{max}}{N_{min}} = \dfrac{750}{80} = 9.375$

Progression ratio $R_n = \dfrac{n_{max}}{n_{min}} = \Phi^{z-1}$ or $\Phi^{6-1} = 9.375$ or $\Phi^5 = 9.375$ $\Phi = 1.56$

Selecting a standard value of the series $\Phi = 1.57$

The speeds are 750, 478, 304, 194, 123, and 77. These speeds are shown in Ray diagram in Figure 9.S4 (a).

Thick lines show power flow. From Table 9.5, assuming structural formula 3(1) 2(3)

Since the torque will be high in stage 2, due to reduced speed, start the design from stage 2. The pinions a, b, and c are mounted on shaft 1 as shown in Figure 9.S4(b), which drives the gears

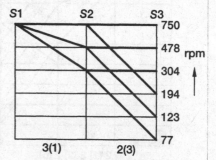

Figure 9.S4 (a) Ray diagram of a six speed gear box

A, B, and C, respectively, on the shaft 2. On the same shaft, pinions d and e are mounted which drive the gears D and E mounted on the output shaft 3.

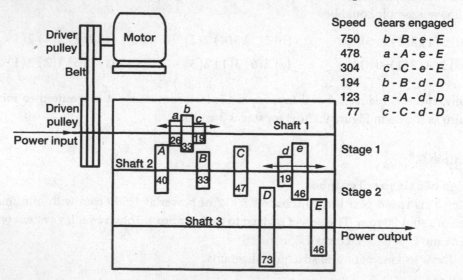

Speed	Gears engaged
750	b - B - e - E
478	a - A - e - E
304	c - C - e - E
194	b - B - d - D
123	a - A - d - D
77	c - C - d - D

Figure 9.S4 (b) Structural diagram

Stage 2   In stage 2, there are two outputs from one input speed. One output is at same speed as of shaft 2 and other reduced by the ratio Φ^3 as shown in Figure 9.S4(a) $[\Phi^3 = 3.87]$.

Gear set d and D   The driver on shaft 2 is the smallest pinion is d, whose number of teeth should be more than 17 to avoid interference. Assuming it as $Z_d = 19$ teeth.

$$\frac{\text{Number of teeth of } D}{\text{Number of teeth of } d} \frac{Z_D}{Z_d} = \frac{\text{Speed of } d}{\text{Speed of } D} = \frac{750}{194} = 3.87$$

Hence $Z_D = 73.45 \cong 73$

Gear set e and E   The second output is at the same speed as input; hence the number of teeth on pinion Z_e and on gear Z_E will be the same. The center distance for both the gear sets has to be same hence:

$$Z_d + Z_D = Z_e + Z_E = 2\,Z_e \quad \text{or} \quad 19 + 73 = 2\,Z_e \quad \text{or} \quad Z_e = Z_E = 46$$

Note   By choosing $Z_D = 73$ instead of 74, sum of $Z_d + Z_D$ is even, which helps for selecting Z_e.

Stage 1   In this stage, there are three outputs from one input. One output speed is same without reduction by gears b and B, second is reduced by Φ by gears a and A and the third reduced by Φ^2 by gears c and C as shown in Figure 9.S4(a) and structural diagram 9.S4(b).

Gear set c and C   Highest reduction. Speed of $C = \dfrac{750}{\Phi^2} = \dfrac{750}{2.465} = 304$

Smallest pinion is c, hence assuming its number of teeth $Z_c = 19$ teeth.

$$\frac{\text{Number of teeth of gear } Z_C}{\text{Number of teeth of pinion } Z_c} = \frac{\text{Speed of pinion } c}{\text{Speed of gear } C} = \frac{750}{304} = 2.467$$

Hence $Z_c = 19 \times 2.467 = 46.875$ say 47

Gear set b and B   No reduction—Equal number of teeth of pinion and gear $Z_b = Z_B$
To have same center distance as of first set of gears $Z_c + Z_C = Z_b + Z_B = 2Z_B$
Substituting the values : $19 + 47 = 2Z_B \quad$ or $\quad Z_b = Z_B = 33$

In sliding gear box, number of teeth on adjacent gears must differ by at least 4 to avoid interference of gears of one shaft with the gear of the other shaft while shifting.

$Z_b - Z_c = 33 - 19 = 14$. It is more than 4 hence OK.

Gear set a and A   Output speed in this step is $\dfrac{750}{\Phi} = \dfrac{750}{1.57} = 478$ rpm

$$\frac{\text{Number of teeth of gear } Z_A}{\text{Number of teeth of pinion } Z_a} = \frac{\text{Speed of pinion } a}{\text{Speed of gear } A} = \frac{750}{478} = 1.57$$

To have same center distance $Z_a + Z_A = Z_b + Z_B = 33 + 33 = 66$

$Z_a + 1.57 Z_a = 66 \quad$ or $\quad Z_a = 25.7$ say 26

Hence number of teeth of gear A, $Z_A = 66 - 26 = 40$

Difference in number of adjacent teeth $Z_a - Z_b = 26 - 19 = 7$. It is more than 4 hence accepted.

Arrangement of the gears is shown in Figure 9.S4(b).

Selection of material   Plain carbon steel 40C8 with allowable bending strength $\sigma_b = 200$ MPa.

b. Module calculation for stage 1 (Helical gears):

By belt and pulley, speed is reduced 0.5 times, that is, 750 rpm. Since the speed is high at stage 1, helical gears are selected with pressure angle 20° and helix angle 30°.

Torque at gear box input shaft $T = \dfrac{60,000\,P}{2\pi N} = \dfrac{60,000 \times 3}{6.28 \times 750} = 38.22$ Nm

Number of teeth of smallest pinion $Z_p = \dfrac{d_p}{m}$

PCD of pinion $d_p = 19\,m$

$$F_t = \frac{T}{0.5 d_p} = \frac{38.22 \times 10^3}{9.5 m} = \frac{4,023}{m}\,\text{N}$$

And, $Z_p' = \dfrac{Z_p}{\cos^3 \alpha} = \dfrac{19}{\cos^3 30} = 29$

Assuming material for gear and pinion same, pinion is always weaker than gear. So calculate form factor y_p' for the pinion.

Form factor for 20° is: $y_p' = 0.154 - \dfrac{0.912}{Z_p'}$

Or, $y_p' = 0.154 - \dfrac{0.912}{29} = 0.122$

Pitch line velocity $v = \dfrac{\pi d N}{60} = \dfrac{3.14 \times 19 m \times 10^{-3} \times 750}{60} = 0.746 m$ m/s

Assuming initially velocity 5 m/s

$$k_v = \frac{6}{6+5} = \frac{6}{11} = 0.545$$

Effective load $F_e = F_t \left(\dfrac{K_s}{K_v} \right) = \dfrac{4,023}{m} \left(\dfrac{1.25}{0.545} \right) = \dfrac{9,227}{m}$

From Lewis equation: $F_e = \sigma_b\, b\, \pi\, m_n\, y'$ (Assume face width as $10\,m$)

Substituting the values:

$$\frac{9,227}{m} = 200 \times (10m) \times 3.14 \times (m\cos 30°) \times 0.122$$

Or, $9,227 = 766.2 \, m^3$ or $m = 2.29$. Select a standard module **2.5**

PCD of pinion c, $d_c = Z_c \times m = 19 \times 2.5 = 47.5$ mm

Outside diameter of pinion c, $D_c = d_c + 2\,m = 47.5 + 5 = 52.5$ mm

Check for pitch line velocity: $v = \dfrac{\pi d\, N}{60} = \dfrac{3.14 \times 47.5 \times 10^{-3} \times 750}{60} = 1.86$ m/s

It is lesser than assumed 5 m/s, hence safe.

PCD of gear C, $d_c = Z_C \times m = 47 \times 2.5 = 117.5$ mm

Outside diameter of gear C, $D_C = d_C + 2\,m = 117.5 + 5 = 122.5$ mm

Since diameters of pinions a and b are more than pinion c, same module will be safe.

PCD of pinion b: $d_b = Z_b \times m = 33 \times 2.5 = 82.5$ mm, O.D., $D_b = 87.5$ mm

PCD of gear B: $d_B = Z_B \times m = 33 \times 2.5 = 82.5$ mm, O.D., $D_B = 87.5$ mm

PCD of pinion a: $d_a = Z_a \times m = 26 \times 2.5 = 65$ mm, O.D., $D_a = 70$ mm

PCD of gear A: $d_A = Z_A \times m = 40 \times 2.5 = 100$ mm, O.D., $D_A = 105$ mm

Width of all gears a, A, b, B, c, and C: $b1 = 10\,m = 10 \times 2.5 = 25$ mm

Module calculation for stage 2:

In this stage, maximum torque at shaft 2 will be at its slowest speed, that is, 304 rpm; hence torque will be calculated at this speed for the smallest pinion d having 19 teeth. Speed is reduced; hence spur gears with 20° pressure angle are selected.

Torque at gear box shaft 2, $T = \dfrac{60,000\,P}{2\pi N} = \dfrac{60,000 \times 3}{6.28 \times 304} = 94.28$ Nm

PCD of pinion $d_p = 19\,m$

$$F_t = \frac{T}{0.5 d_p} = \frac{94.28 \times 10^3}{9.5 m} = \frac{9,924}{m} \, \text{N}$$

Pitch line velocity $v = \dfrac{\pi d\, N}{60} = \dfrac{3.14 \times 19m \times 10^{-3} \times 304}{60} = 0.3m$ m/s

Velocity will be less than 1.86 m/s as calculated for stage 1, hence

$$k_v = \frac{3}{3 + 1.5} = \frac{3}{4.5} = 0.667$$

Dynamic load $F_d = F_t \left(\dfrac{K_s}{K_v}\right) = \dfrac{9,924}{m}\left(\dfrac{1.25}{0.667}\right) = \dfrac{18,598}{m}$

Assuming material for gear and pinion same, pinion is always weaker than gear. So calculate form factor y for the pinion.

$$y = 0.154 - \frac{0.912}{19} = 0.146$$

Lewis equation is: $F_t = \sigma b y \pi m$ Substituting the values:

$$\frac{18,598}{m} = 200 \times (10m) \times 0.146 \times 3.14 \times m$$

Or, $18,598 = 917 \, m^3$ Or, $m = 2.73$

Select a standard module **3**.

PCD of pinion d: $d_d = Z_d \times m = 19 \times 3 = 57$ mm, O.D., $D_d = 57 + (2 \times 3) \, 63$ mm

PCD of gear D: $d_D = Z_D \times m = 73 \times 3 = 219$ mm, O.D., $D_D = 219 + (2 \times 3) \, 225$ mm

Since diameters of pinions a and b are more than pinion e, same module will be safe.

PCD of pinion e: $d_e = Z_e \times m = 46 \times 3 = 138$ mm, O.D., $D_e = 138 + (2 \times 3) \, 144$ mm

PCD of gear E: $d_E = Z_E \times m = 46 \times 3 = 138$ mm, O.D., $D_E = 138 + (2 \times 3) \, 144$ mm

Width of all gears d, D, e, and E: $b2 = 10 \, m = 10 \times 3 = 30$ mm

c. Center distance between shafts:

Between shaft 1 and shaft 2

$$C1 = \frac{d_a + d_A}{2} = \frac{100 + 65}{2} = 82.5 \text{ mm}$$

Alternately, $\dfrac{d_b + d_B}{2} = \dfrac{82.5 + 82.5}{2} = 82.5$ mm

Between shaft 2 and shaft 3

$$C2 = \frac{d_d + d_D}{2} = \frac{57 + 219}{2} = 138 \text{ mm}$$

Alternately, $\dfrac{d_e + d_E}{2} = \dfrac{138 + 138}{2} = 138$ mm

d. Lengths of shafts

Assume the following lengths [See Figure 9.S4(c)]:

Length for bearing 30 mm on each side of the casing = 30 mm

Gap between gear and casing $x = 10$ mm

Gap between the gears of both the shafts: $2x = 20$ mm

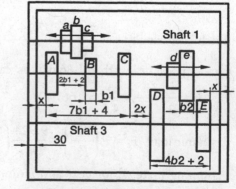

Figure 9.S4(c) Arrangement of gears and shafts

Length for a set of gears to slide:

Three gears, each of width $b1$ and a gap of 1 mm on both sides $2(2b1 + 2) + 3b1 = 7b1 + 4$ mm

Two gears, each of width $b2$ and a gap of 1 mm on both sides $(2b2 + 2) + 2b2 = 4b2 + 2$ mm

Width of gears calculated above are $b1 = 25$ mm and $b2 = 30$ mm.

Hence, length of shafts is: $L = 30 + 10 + (7b1 + 4) + 20 + (4b2 + 2) + 10 + 30$
$$= 40 + 179 + 20 + 122 + 40 = 301 \text{ mm}$$

Length of all the three shafts will be same as 301 mm.

e. Shaft design

Shaft 1

Worst loading causing bending moment will be gears in center. Actually these are not in center; hence even assuming in the center will be a conservative design.

Shaft 1 has helical gears which causes axial loads also.

Torque at shaft 1, $T = 38.22$ Nm [Calculated in part (b)]

$$F_t = \frac{T}{0.5 d_c} = \frac{38.22 \times 10^3}{23.75} = 1{,}609 \text{ N}$$

Normal load on tooth $F_n = \dfrac{F_t}{\cos \Phi} = \dfrac{1{,}609}{\cos 20°} = 1{,}712 \text{ N}$

$F_a = F_t \tan \alpha = 1{,}609 \tan 30° = 929$ N

For a simply supported beam,

Bending Moment: $M = \dfrac{F_n \times L}{4} = \dfrac{1{,}712 \times 301}{4} = 128{,}828 \text{ Nmm} = 128.828 \text{ Nm}$

Equivalent torque $T_e = \sqrt{M^2 + T^2} = \sqrt{(128.828)^2 + (38.22)^2} = 134.379 \text{ Nm}$

Take safe shear stress for the shaft as 60 MPa for the equation: $T_e = \dfrac{\pi}{16} d^3 \times \tau$

Substituting the values: $134.379 = \dfrac{\pi}{16} d^3 \times 60$

Or, $d = 22.5$. Selecting a standard diameter **25 mm**

Splined shaft: Sliding gears slide on a splined shaft; hence standard parallel side spline is chosen from Table 19.7 of volume 1 book. For 25 mm diameter, the number of splines is six; depth of spline is 2 mm; and width is 5 mm.

Shaft 2

Gears on the right side of this shaft are spur gears. Torque at this shaft will be maximum when the speed is minimum. It will increase by the transmission ratio (47 / 19).

Torque on shaft 2, $T = \dfrac{38,220 \times 47}{19} = 94,544$ Nmm

$F_t = \dfrac{T}{0.5 d_d} = \dfrac{94,544}{28.5} = 3,317$ N

Normal load on tooth $F_n = \dfrac{F_t}{\cos \Phi} = \dfrac{3,317}{\cos 20°} = 3,532$ N

For a simply supported beam, B.M.,

$$M = \frac{F_n \times L}{4} = \frac{3,532 \times 301}{4} = 265,762 \text{ Nmm} = 265.762 \text{ Nm}$$

Equivalent torque $T_e = \sqrt{M^2 + T^2} = \sqrt{(265.762)^2 + (94.544)^2} = 282.1$ Nm

$$T_e = \frac{\pi}{16} d^3 \times \tau \text{ Taking safe shear stress 60 MPa}$$

Substituting the values: $282,100 = \dfrac{\pi}{16} d^3 \times 60$

Or, $d = 28.9$ mm Selecting a standard diameter 30 mm

For 30 mm diameter, the number of splines is six; depth of spline is 2.5 mm; and width is 6 mm.

Shaft 3

Torque at this shaft will be maximum when the speed is minimum. It will increase by the transmission ratio (73 / 19).

Torque on shaft 3, $T = \dfrac{94,544 \times 73}{19} = 363,248$ Nmm

Smallest gear on this shaft is gear E, hence $F_t = \dfrac{T}{0.5 d_e} = \dfrac{363,248}{0.5 \times 138} = 5,264$ N

Normal load on tooth $F_n = \dfrac{F_t}{\cos \Phi} = \dfrac{5,264}{\cos 20°} = 5,605$ N

For a simply supported beam, B.M., $M = \dfrac{F_n \times L}{4} = \dfrac{5,605 \times 301}{4} = 421,751$ Nmm

Equivalent torque $T_e = \sqrt{M^2 + T^2} = \sqrt{(421,751)^2 + (363,248)^2} = 556,617$ Nmm

Also, $T_e = \dfrac{\pi}{16} d^3 \times \tau$ Taking safe shear stress 60 MPa

Substituting the values: $556{,}617 = \dfrac{\pi}{16}d^3 \times 60$ or $d = 36.2$ mm

Selecting a standard diameter 40 mm. Gears are fixed; hence no splines are required.

Example 9.10

Ray diagram from structural formula

A gear box is to be designed for 12 speeds with a structural formula 2(1) 3(2) 2(6). The input speed is 800 rpm and the GP ratio is to be kept 1.25. For maximum speed of 1,000 rpm, draw the following:

a. Kinematic diagram

b. Ray diagram

Solution

Given $Z = 12$ $\Phi = 1.25$ Structural formula 2(1) 3(2) 2(6) $N_{max} = 1{,}000$ rpm

From Equation (9.9), Transmission Ratio, $R_n = \dfrac{N_{max}}{N_{min}} = \Phi^{z-1} = (1.25)^{12-1} = 11.64$

$$N_{min} = \frac{1{,}000}{11.64} = 86 \text{ rpm}$$

Hence the speeds are: 1,000, 800 (1,000 / 1.25), 640 (800 / 1.25), 512, 410, 328, 262, 210, 168, 134, 108, and 86 rpm.

a. Structural formula 2(1) 3(2) 2(6). (Use Equation 9.14 to calculate R_m for each stage)

Stage 1 ($m = 1$) has one input ($I_m = 1$) and two outputs ($o_m = 2$) with transmission ratio $\Phi^{(2-1)\times 1}$, that is, $\Phi = 1.25$

Stage 2 ($m = 2$) has two input ($I_m = 2$) and three outputs ($o_m = 3$) with transmission ratio $\Phi^{(3-1)\times 2}$, that is, $1.25^4 = 2.44$

Stage 3 ($m = 3$) has six input ($I_m = 6$) and two outputs ($o_m = 2$) with transmission ratio $\Phi^{(2-1)\times 6}$, that is, $1.25^6 = 3.815$

Input speed at shaft 1 is 800 rpm. Output speed will be 800 rpm.

Draw the rays starting from the lowest rpm of last shaft, i.e. 86 rpm. The second output will be 6 steps above i.e. 328 rpm. Input speeds are set with the following restrictions.

Transmission ratio $R_1 = \dfrac{I_{max}}{I_{min}}$ $I_{max} < 4$ and $I_{min} > 0.25$

In this problem, the transmission range of the last group is Φ^6 for $\Phi = 1.25$ on dividing in one possible manner. $I_{max} = 2 = 1.25^2 = 1.56$; $I_{min} = 0.25 = 1.25^4 = 2.44$ ($\Phi^6 = \Phi^2 \cdot \Phi^4$).

b. Ray diagram for the above structural formula is shown in Figure 9.S5.

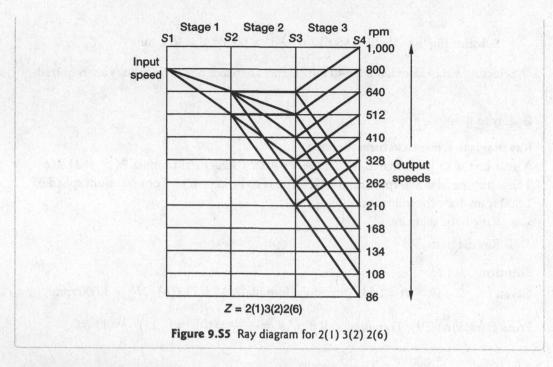

Figure 9.S5 Ray diagram for 2(1) 3(2) 2(6)

9.20 Two-stage Gear Box with Fixed Ratio

A two-stage reducer unit is shown in Figure 9.7 in Section 9.6. The gear ratios are fixed and cannot be varied. If speeds are to be varied, as in the case of gear box of an automobile, different types are available and are described in subsequent sections.

9.21 Sliding Mesh Gear Box

It is called sliding mesh, because a gear ratio is selected by sliding a gear on a splined shaft to mesh with another gear on a lay shaft (See Figure 9.14). There are three shafts; input shaft, lay shaft or intermediate shaft, and output shaft or gear box shaft. Input shaft has a pinion G1 called top pinion as the power goes through this in the top gear. Gear G2 mounted on lay shaft is always meshing and hence lay shaft always rotates. It also has three or gears G3, G4, and G5. For five speed gear box, there can be one more gear on this shaft. The gear box shaft is splined with gears G6, G7, and G8. When these are slided left or right by sleeves S1 or S2, a particular gear ratio is obtained. Gears G5 and G8 give the highest gear ratio. Gears G4–G7 have little less ratio and G3–G6 still lesser. When gear G6 meshes with gear G1 through a dog clutch, there is no reduction and output shaft rotates at the speed of input shaft.

Disadvantage of this gear box is that when the gears are engaged, there are chances of gear tooth breaking, when the peripheral speed of the gears to be meshed is different. An improvement to this gear box is constant mesh gear box described in the next section.

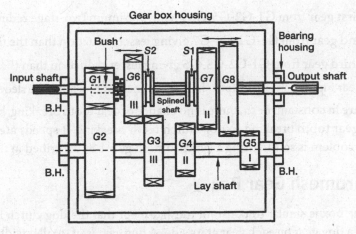

Figure 9.14 Sliding mesh gear box

9.22 Constant Mesh Gear Box

Its construction is similar to sliding mesh type gear box, except that helical gears are used in place of spur gears. This offers silent operation than spur gears. Further, the gears are not engaged but as its name suggests that the gears are not meshed by sliding but these are always meshing with each other. The gears on the gear box shaft are on sleeve bearings and all gears rotate freely even in neutral gear. These gears have a tooth on side similar to dog clutch as shown in Figure 9.15. Two sleeves S1 and S2 slide on the splined portion of the gear box shaft and have matching teeth to that of dog clutch on both sides. When these sleeves are slided on side, the teeth engage with the gear and hence power is transmitted from the gear to the shaft through the teeth of the sleeve.

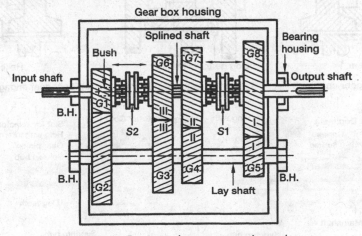

Figure 9.15 Four speed constant mesh gear box

Gear G1 on input shaft and G2 on lay shaft are always meshed. Power transmission is as under:

- In the first gear from G1-G2-G5-G8 giving maximum two-stage reduction.

- In second gear from G1-G2-G4-G7 giving lesser reduction than the first gear.

- In the third gear from G1-G2-G3-G6 giving lesser reduction than the second gear.

- In top gear input shaft and output shaft are connected by sliding the sleeve S2 to the left.

Since the gears are in constant mesh, there is no chance of gear tooth breaking, but the problem is shifted from gear tooth breaking to dog clutch teeth breaking, if speeds are unequal while meshing. This problem is solved using a synchromesh gear box described in the next section.

9.23 Synchromesh Gear Box

This type of gear box is similar to constant mesh, except that the dog clutch is replaced by a cone clutch and a dog gear on each gear at its side. A dog gear is of smaller width and diameter than main gear and is integral part of the helical gear (Figure 9.16). A synchromesh gear box uses friction cones to match the speed of gears before meshing. Sleeve is a sliding element, which has internal teeth. It slides over a mender having external teeth. The mender is fixed on the output shaft. Figure 19.16(b) and picture 19.16(c) shows the position of sleeve in neutral position.

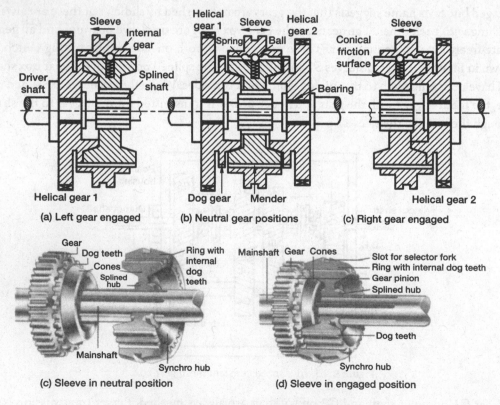

Figure 9.16 Working of synchromesh gear box

When the sleeve slides to engage a particular gear, before the sleeve reaches the dog gear, internal conical surface of the mender touches the cone on the side of the dog gear. See Figures 19.16 (a and c) and picture 19.16 (d). It acts as a cone clutch to bring the speeds equal. When the speed of the mender and dog gear become equal, the sleeve is further slided to engage with the teeth of dog gear. There are three spring-loaded balls placed in the radial holes in the mender. The ball keeps the sleeve in its position by fitting in a notch provided in the sleeve at neutral or engaged position.

9.24 Gear Box Housing

Gear box housing is a rectangular box in two halves. A flange is provided in middle to join both the halves using bolts and nuts. Ribs are provided at the vertical walls to provide strength at the bearing supports. The base is wider than the size and has holes for fixing.

Dimensions of the casing are fixed based on thumb rule given in Table 9.7. It gives wall thickness in mm of the gear boxes.

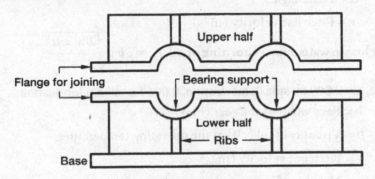

Figure 9.17 Gear box housing

Table 9.7 Wall Thickness of Gear Box Housing

Material	Wall Thickness (mm)	
	Without Case-Hardened Gears	With Case-Hardened Gears
Cast iron casting	0.007 L + 6	0.010 L + 6
Steel casting	0.005 L + 4	0.007 L + 4
Welded construction	0.004 L + 4	0.005 L + 4

If L = Largest dimension of the housing in mm and $S = 0.005\,L + 4$ mm

Top cover thickness $t_c = 0.8\,S$
Flange thickness $t_f = 2\,S$
Flange cover bolt diameter $d_{cb} = 1.5\,S$
Bolt spacing $= 6\,d_{cb}$
Foundation bolt diameter $d_{fb} = 2T^{1/3} \geq 12$ mm
Thickness of the foundation flange $t_{ff} \geq 1.5\,d_{fb}$

Width of the flange at the base $W_b = 2.5\, d_{fb}$

Width of the flanges at the two halves of the housing $W_f = 2.5\, d_{fb}$

   With welding bead of 5mm, $W_f = 45$ mm is taken.

Bearing housing 1.2–1.5 times the outside diameter of the bearing.

9.25 Power Losses in Gear Box

Every gear box has some transmission efficiency due to some power loss. The various factors adding to power losses are:

a.   Power loss in watts due to **tooth engagement** $P_t = F_n \left(\dfrac{0.1}{Z_p \cos \alpha} + \dfrac{0.3}{v+2} \right)$ (9.15)

   Where, F_n = Normal force coming on the tooth (N)

   Z_p = Number of pinion teeth

   α = Helix angle

   v = Pitch line velocity (m/s)

b.   Power loss in watts due to **churning of oil** $P_{ch} = c\, b\, v \sqrt{\dfrac{200\, v\, \mu}{Z_p + Z_g}}$ (9.16)

   Where, $c = 0.009$ for splash lubrication, 0.006 for stream lubrication.

   b = Face width of the gear (mm)

   μ = Viscosity of oil (cP) at the operating temperature

   v = Pitch line velocity (m/s)

Z_p, Z_g = Number of teeth of pinion and gear, respectively

c.   Power loss in watts **in bearings**: $P_b = F_r f d N$ (9.17)

   Where, F_r = Radial load on the bearing (N)

      f = Coefficient of friction at the bearing 0.005 to 0.01 for sleeve bearings

   For roller bearings from the catalog $f = 0.002$ and for ball bearings $f = 0.003$

   d = Shaft diameter (m)

   N = Shaft speed (rpm)

d.   Power loss in watts **in seals** $P_s = T_s \times \omega$ (9.18)

   Where, T_s = Seal friction torque

      ω = Angular velocity of the shaft

Total power loss $P_f = P_t + P_{ch} + P_b + P_s$ (9.19)

Example 9.11

Power loss in a gear box for given power

A single-stage gear box with spur pinion having 19 teeth of module 5 mm, width 50 mm and pressure angle 20° is transmitting power of 7.5 kW at 1,500 rpm with a gear ratio of 3. Assuming oil with viscosity 50 cP at operating temperature and friction coefficient for ball bearings 0.003, calculate the power loss in the gear box with splash lubrication, if its shaft diameter is 40 mm. Neglect power loss in seals.

Solution

Given $Z_p = 19$ $m = 5$ $b = 50$ mm $\Phi = 20°$ $P = 7.5$ kW
  $N = 1,500$ rpm $R = 3$ $\mu = 50$ cP $f = 0.003$ $d = 40$ mm

Torque at gear box input shaft: $T = \dfrac{60,000\,P}{2\pi N} = \dfrac{60,000 \times 7.5}{6.28 \times 1500} = 47.77$ Nm

PCD of pinion $d_p = Z_p\,m = 19 \times 5 = 95$ mm

Tangential force: $F_t = \dfrac{T}{0.5\,d_p} = \dfrac{47.77 \times 10^3}{0.5 \times 95} = 1{,}005.7$ N

Normal load on tooth: $F_n = \dfrac{F_t}{\cos \Phi} = \dfrac{1,005.7}{\cos 20°} = 1{,}070$ N

Radial load on bearing $F_r = F_t \tan \Phi = 1,005.7 \times \tan 20° = 366$ N

Pitch line velocity $v = \dfrac{\pi d N}{60} = \dfrac{3.14 \times 95 \times 10^{-3} \times 1,500}{60} = 7.46$ m/s

a.   From Equation (9.15) power loss due to tooth engagement

$$P_t = F_n \left(\dfrac{0.1}{Z_p \cos \alpha} + \dfrac{0.3}{v+2} \right)$$

Substituting the values: $P_t = 1,070 \left(\dfrac{0.1}{19 \times \cos 20°} + \dfrac{0.3}{7.46+2} \right) = 39.9$ W

b.   From Equation (9.16) power loss due to churning $P_{ch} = cbv\sqrt{\dfrac{200\,v\mu}{Z_p + Z_g}}$

Substituting the values: $P_{ch} = 0.009 \times 50 \times 7.46 \sqrt{\dfrac{200 \times 7.46 \times 50}{19 + (3 \times 19)}} = 105.2$ W

c.   From Equation (9.17) power loss in bearings: $P_b = F_r f d N$

$\qquad\qquad = 366 \times 0.003 \times 0.040 \times 1,500 = 65.9$ W

Total power loss $P_f = P_t + P_{ch} + P_b + P_s = 39.9 + 105.2 + 65.9 + 0 = 211$ W

9.26 Fluid Couplings

Fluid coupling came into use in 1920s to replace a conventional clutch. However, its patent is as old as 1,905. It is a device to transmit torque from one shaft to another using a hydraulic fluid such as oil. Fluid couplings can also be used as hydrodynamic brakes, dissipating rotational energy as heat through frictional forces.

Applications Fluid couplings are used in:

1. Applications, where heavy starting torque is required.
2. In automobiles as semi-automatic fully automatic transmission system.
3. Aeronautical applications.
4. Conveying systems.
5. Crane travel drives.
6. Processing equipment.
7. Filling and packaging equipment.

Construction

It consists of mainly two toroid (like donuts) parts; a pump also called as impeller and other a turbine contained in a sealed housing containing oil. Pump is attached to the driving shaft and turbine to the driven shaft. Both pump and turbine have semicircular annular groove with radial vanes as shown in Figure 9.18(a) and (c). Pump and turbine are placed side by side facing each other [Figure 9.18(b)]. Power from the prime mover is transferred through oil.

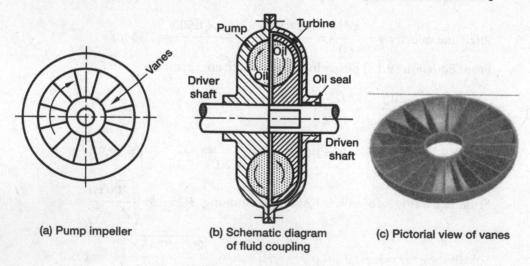

|(a) Pump impeller|(b) Schematic diagram of fluid coupling|(c) Pictorial view of vanes|

Figure 9.18 Fluid coupling

Working Imagine two fans facing each other and one of the fans is connected to the A.C. mains to rotate. It will throw air over the blades of the second fan, causing it to rotate without any physical connection. The other fan is forced to rotate by the air striking its blades like a wind mill. Thus both the fans are coupled together with the air in between them. But, if the second fan is jammed the first fan will keep on rotating. A fluid coupling works just on this principle, except that a fluid, generally oil, is used instead of air.

When the pump is rotated by the driver shaft, the fluid contained between the vanes of the pump is moved radially outwards due to centrifugal force. The curved shape of the pump changes the direction of oil from radial to axial. When it comes out of the pump vanes, it has some tangential velocity and some axial velocity depending on the rotational speed of the pump. The resultant direction is at some angle as shown in Figure 9.19(a). The spinning toroid moves the hydraulic fluid from the impeller and impinges on the blades of the turbine and forces it to rotate. The fluid gives up its energy and leaves the runner at low velocity. Difference between the angular velocity of the pump and turbine results in a net force on turbine causing it to rotate in same direction as pump.

At slow speed of pump, circumferential velocity is larger than the axial velocity and hence the fluid strikes at an angle [Figure 9.19(a)]. Oil coming from one compartment of vanes enters into the other compartment of turbine. As the speed of the pump increases, circumferential velocity is proportional to angular velocity, while centrifugal force increases as square of angular velocity. This increases radial velocity to greater extent, which becomes the axial velocity of the fluid due to curved shape of the pump impeller. The velocity triangle is shown in Figure 9.19(b) for high speed. The fluid now enters the compartment of the turbine just opposite to it. Thus a cylinder of whirling fluid is formed between the compartments of the pump and turbine. Both the parts are now coupled hydraulically to transfer power.

The slip characteristics of the fluid coupling allow to start nearly free of load and to accelerate the load to maximum torque.

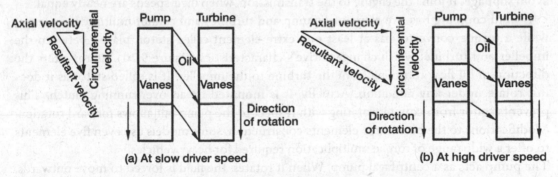

Figure 9.19 Direction of oil from pump entering turbine

Power transmitting capability of a fluid coupling is primarily related to pump speed to great extent. The torque transmitting capacity T of a fluid coupling is related as under:

$$T = 0.005\, \rho N^2 D^5 \tag{9.20}$$

Where, ρ = Mass density of the fluid (kg/m³)

N = Rotational speed of the pump (rad/s)

D = Diameter of the pump (m)

An important characteristic of a fluid coupling is **stall speed**. It is defined as the highest speed at maximum input power at which a pump can rotate, when the turbine is locked. In this condition, power is transferred as heat, which can damage the coupling.

Advantages   Fluid couplings can provide many advantages as given below:

1. Simple, robust, and reliable drive.
2. No wear of transmitting elements.
3. It accelerates fast as input shaft starts without any load.
4. Rate of acceleration can be controlled by varying the quantity of oil.
5. Oil absorbs shocks, hence shock free, and smooth load take up.
6. Protects overloads from jamming of machines. A fusible plug releases oil to isolate the motor from the load.
7. Fluid braking can be used by changing the direction of rotation of the motor.

9.27 Torque Converters

A torque converter is also a fluid coupling used to transfer rotating power from an internal combustion engine or electric motor, to a rotating driven load. The torque converter replaced a mechanical clutch in automatic transmission. It is located between the engine and the rest of transmission. The main characteristic of a torque converter is to multiply torque, when there is a substantial difference between input and output rotational speed. Thus it acts as a speed-reduction gear. Some of torque converters use a temporary locking mechanism to avoid slippage. It joins the engine to the transmission, when their speeds are nearly equal.

A fluid coupling has two elements (pump and turbine) and cannot multiply the torque, while a torque converter has at least one extra element called stator, placed between the impeller and turbine, which changes drive's characteristics (Figure 9.20). Stator alters the direction of oil flow returning from the turbine to the impeller. It is called stator as it does not rotate under any condition. Actually, it is mounted on an over-running clutch. This prevents stator from counter-rotating with respect to the pump but allows forward rotation. Modifications to the basic three elements construction, some models use even five elements to offer a wide range of torque multiplication required for heavy vehicles.

The pump acts as a centrifugal pump. When it rotates, the fluid is forced to move outwards. This causes a vacuum, at inner radius, which draws more fluid. The fluid coming out from the pump enters the blades of the turbine, which is connected to the transmission. The blades of the turbine are curved so that the fluid which enters the turbine from the outside has to change the direction, before it exits the center of the turbine. This directional change causes the turbine to rotate.

The fluid exits the turbine in a direction opposite to that of the pump. If this fluid hits the pump, it will slow the pump, wasting power; hence a stator is put in between the pump and turbine. It changes the direction of fluid again, which will be suitable for a smooth entry in the pump. The stator is mounted on a fixed shaft but has an internal one-way clutch, since it is required to freewheel at certain operating speeds. Torque multiplication increases the torque at the turbine shaft, and it also increases slippage inside the converter, raising the fluid temperature and reduce overall efficiency; hence internal parts and characteristics of a converter must be matched with the vehicle's specifications.

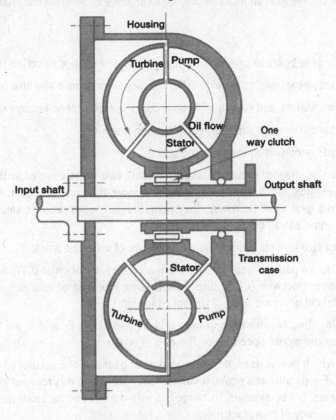

Figure 9.20 A torque converter

A torque converter has three stages of operation:

- **Stall** It is a condition when the driver shaft is developing full power but the driven shaft is prevented from rotating. Such a condition occurs in automobiles, when the engine is developing power but the vehicle is stationary, just at the time of start. This condition lasts for a small period till the vehicle starts moving.

- **Acceleration** There is a large difference between speeds of the driver and driven. In this condition, torque converter multiplies the torque to great extent but lesser than the stall condition. The amount of multiplication depends on the difference in speed.

- **Coupling** Speed of the turbine reaches about 90 per cent of the pump speed. The torque converter more or less works just like any fluid coupling. In this condition, some torque converters apply lock up between pump and turbine and eliminate any wasted power to improve fuel efficiency.

Summary

Function of a gear box It is used to change the speed (generally decrease), change torque and direction of rotation between input and output shafts. The point of contact of the gear teeth is well lubricated with gear oil. The gear oil must be very clean and free of abrasive materials to avoid wearing of the gears.

Applications:

- Automobile gear boxes to provide different speeds and change direction for reverse.
- Machine tools gear boxes to provide different speeds for spindle and also for automatic feeds.
- Elevators, escalators, and rolling mills gear boxes use as a speed reducer only at one speed.
- In many applications like turbines, earth moving machines.
- In multiblade windmills to increase the speed.

Construction In a single-speed gear box, there are only two gears coupled with one another put inside a casing. A gear box with many speeds may have more than one gear pair. If it has more than two pairs, it is called a gear train. Generally, they permit higher speed ratios in smaller packages than are feasible with a single pair of gears.

Gear train comprises two or more pairs of gears. These are of different types:

Simple gear train Uses a pair of a gear set, if the driver rotates in clockwise (CW) direction, the driven gear rotates in counter clockwise (CCW) direction. If same direction of rotation is required than an idler can be used. Helical gears are used, if the speed is high.

Compound gear train Two or more pairs of gear sets are used. If R_1, R_2, and R_3 are the speed ratios of the second pair, then the overall speed ratio $R = R_1 \times R_2 \times R_3$.

Pitch line velocity which can be used, depends on type of gear and the accuracy by which they have been manufactured. For parallel axis gears, it varies from 7 m/s to 10 m/s for spur gears and 10 m/s to 25 m/s for helical gears. For bevel gears, its range is 5 m/s–10 m/s and for spiral gears 10 m/s–25 m/s depending on the accuracy of manufacturing.

Epicyclic gear train consists of a sun gear, an arm, or a spider carrying 3 or 4 planetary gears and an annulus (also called as ring gear) meshing with the planetary gears. Many combinations of driver and driven inputs and outputs are possible as given below:

- No power flows unless one of the gears is braked.
- Speed ratio is one, if more than one gear is braked.
- Speed either increases or decreases, depending on which element is driver and which element is braked.
- **Speed ratio** of these gear boxes can be found by two methods:

a. *In translation method:* CCW direction is taken positive and a table is prepared:

Step 1 Lock all gear, all gears rotate at some speed say +1 rpm.

Step 2 Fix one of the elements. Give one rotation in CW direction to the driver (−1). Planetary gear rotates in CCW direction and the speed is in the ratio of their number of teeth, that is, Z_s / Z_p.

Step 3 Add number of revolutions made by each element. When arm rotates by +1 revolution the planet gears will rotate $1 + (Z_s / Z_p)$ revolution. Z_s and Z_p are number of teeth of sun gear and planet gear respectively.

b. *In formula method:* Number of teeth of ring gear, $Z_R = Z_S + 2Z_p$

Speed ratio $R = \dfrac{N_S - N_{Arm}}{N_R - N_{Arm}}$

Condition 1 The carrier is fixed, that is, $N_A = 0$, $R = \dfrac{N_S}{N_R} = -\dfrac{Z_R}{Z_S}$

Condition 2 The ring gear is fixed, that is, $N_R = 0$, $R = \dfrac{N_S}{N_A} = 1 + \dfrac{Z_R}{Z_S}$

Condition 3 The sun gear is fixed, that is, $N_S = 0$, $R = \dfrac{N_R}{N_A} = 1 + \dfrac{Z_S}{Z_R}$

Alternately, $R_1 = R_2 = R_3 = \sqrt[3]{R}$

Torque on ring gear $T_R = T_S \times \dfrac{Z_R}{Z_S} = T_A \times \dfrac{Z_R}{Z_S + Z_R}$

Where, T_S = Torque on sun gear and T_A = Torque on arm

Torque on sun gear $T_S = T_R \times \dfrac{Z_S}{Z_R} = T_A \times \dfrac{Z_S}{Z_S + Z_R}$

Torque on arm $\quad T_A = T_R \times \dfrac{Z_S + Z_R}{Z_R} = T_S \times \dfrac{Z_S + Z_R}{Z_S}$

Holding torque is the torque required to stop rotation of the braked gear. For equilibrium, sum of all the torques has to be zero.

$$T_I + T_O + T_H = 0$$

Where, T_I = Input torque and T_O = Output torque

Classification of gear boxes

A. *According to type of gear used:*

- Spur gears   Spur gears are used for heavy load but are noisy.

- Helical gears   For silent operation due to gradual engagement.

- Bevel gears   The axes of input and shafts are at an angle.

- Spiral gears   Use is similar to bevel gears, but these are lesser noisy than bevel gears.

- Hypoid gears   These are used, if change of plane of rotation is required.

- Planetary gears   It offers a compact arrangement.

B. *According to output speeds*   Single speed / Multiple speed

C. *According to drive*   Mechanical drive / Hydraulic drive

D. *According to alignment of input and output shaft*   In-line / Angled / Offset

E. *According to duty*   Light duty / Medium duty / Heavy duty

Selection of type of gear box   Following factors must be known for selection.

- The types of unit required—parallel or angled drive.
- Any abnormal operating conditions.
- The direction of rotation of the shafts.
- Any outside loads that could influence the unit, for example, overhung loads, brakes, etc.
- Any space restriction.

Other salient points to be considered for correct selection are as under:

a. *Source of input power*   like I.C. engines; single or multi-cylinder, electric motors.

b. *Application for out power*   like automobile, machine tools, escalators, cranes, etc.

c. *Power*   Rated power of prime mover

d. *Actual power requirements* for driven machine

e. *Speeds*   Input and output speeds

f. *Working environment*   Duty hours, atmospheric temperature

g. *Catalogues* providing information   Rated power; Torque capacity; Reduction ratio; Input / output speeds; Dimensions

Follow the procedure to select a gear box from a catalogue:

1. Calculate the actual power required for the application.
2. Get value of service factor K_s from Table 9.4 given below:
3. Calculate equivalent power P_e = Actual power $P \times K_s$
4. Select a gear from the catalogue, with rated power more than equivalent power.
5. If output shaft of a gear box has an overhung pulley or a gear, a load factor K_L is to be used: Flat belt: $K_L = 3$, V belt pulley $K_L = 1.5$, Spur gear $K_L = 1.25$.

Speed ratios for designing a gear box are kept in Geometrical Progression (GP). If there are z number of desired speeds, these are: $N, N\Phi, N\Phi^2, N\Phi^3 \ldots \ldots N\Phi^{z-1}$

Standard values of Φ are 1.06, 1.12, 1.25, 1.41, 1.57, and 1.87.

If N_{max}, N_{min} be the maximum and minimum desired speeds in rpm. Then the maximum transmission ratio R_n is given as: $R_n = \dfrac{N_{max}}{N_{min}} = \Phi^{z-1}$ or $\Phi = \left(R_n\right)^{\frac{1}{z-1}}$

Kinematic diagram   It is a graphic representation, which shows transmission sequence of an input shaft through an intermediate mechanism to the structure's operative elements. It also shows how they are interrelated. It shows only those elements of an assembly, which participate in the transmission of motion.

Structural diagram   It is a graphical representation, used to inform the ratio of transmission groups. It gives information about the number of shafts and the number of gears on each shaft.

Structural formula is used for gear trains to provide information about number of stages and available number of speed ratios. For a multi-stage gear train, the formula is written as $O1\ (I1)\ O2\ (I2)\ O3\ (I3)$ numbers 1, 2, and 3 represent the number of stage.

Where, O = Speed sets at output side and I = Speeds available at input side

The number of output speeds of the first stage becomes the number of speeds at input side of the second stage. Similarly, the number of input speeds for the third stage will be multiplication of output speeds of first and second stages. Thus formula for a three-stage gear train is written as under:

$O1\ (I1)\ O2\ (O1)\ O3\ (O1 \times O2)$

The number of output speeds depends on an application. It usually varies from 6 to 18. There can be many alternate structural formulas as given in Table 9.5 for same number of output speed sets. There can be six combinations for input speed sets for a three-stage gear box as under:

(a) $2(I1)\ 3\ (I2)\ 2(I3)$ (b) $2(I2)\ 3(I1)\ 2(I3)$ (c) $2(I3)\ 3(I1)\ 2(I2)$

(d) $2(I1)\ 3\ (I3)\ 2(I2)$ (e) $2(I2)\ 3(I3)\ 2(I1)$ (f) $2(I3)\ 3(I2)\ 2(I1)$

Transmission ratio of mth stage $R_m = \Phi^{(Om-1) \times Im}$

If there are m stages, transmission ratio for each stage $R_m = I_{max} / I_{min}$

Where, O_m and I_m = Number of output and in put speed sets of stage m respectively

Φ = Progression ratio

Optimum structural formula is selected based on the following points:

1. *Number of stages* should not be more than 4.
2. *Transmission ratio* should not be more than 8.
3. *Input speed sets:* $I1$ should be less than $I2$ and $I2$ less than $I3$, that is, $I1 < I2 < I3$.
4. *Range of transmission:* The output speeds provided by a stage must lie between $I_{max} = 2$ and $I_{min} = 0.25$ times the input speed.
5. *Number of gears* on the last shaft should be minimum possible.
6. The transmission ratio between spindle and the shaft preceding it should be the maximum possible, that is, speed reduction should be the maximum possible.
7. Structure diagram should be narrow towards the starting point of concave shape.
8. Nodal points should be minimum.

Ray diagram It gives the transmission ratio of all transmissions and the rpm values of gear box shafts. So it is necessary to plot the speed chart to determine the transmission ratio. Vertical lines represent different shafts and the speed is plotted on Y axis. The line joining points of adjacent shafts in a speed chart is called a ray gives the gear ratios. Meaning of these lines can be taken as given below:

- A horizontal line corresponds to transmission ratio $R = 1$, that is, no speed change.
- An upward inclined line is for transmission ratio $R > 1$, that is, speed increases.
- A downward inclined line is for transmission ratio $R < 1$, that is, speed reduces.

A ray diagram can be open type or crossed. An open type is the one in which the rays do not intersect each other, while in crossed type they cross.

Sliding mesh gear box has three shafts; input shaft, lay shaft or intermediate shaft, and output shaft. It is so called because a gear ratio is selected by sliding a gear on a splined shaft. Gear ratios are changed by selecting gear sets of different teeth on intermediate shaft and gear box shaft. For reverse gear, an idler is placed in between the output drive.

Disadvantage of this gear box is that when the gears are engaged, there are chances of gear tooth breaking, when the peripheral speed of the gears to be meshed is different.

Constant mesh gear box is similar to sliding mesh type gear box, except that helical gears are used in place of spur gears. This offers silent operation than spur gears. Since the gears are in constant mesh, there is no chance of gear tooth breaking while engaging, but the problem is shifted from gear tooth breaking to dog clutch teeth breaking, if speeds are unequal.

Synchromesh gear box is similar to constant mesh, except that the dog clutch is replaced by a cone clutch and a dog gear on each gear at its side.

Fluid coupling is a device to transmit torque from one shaft to another using a hydraulic fluid such as oil. Fluid couplings can also be used as hydrodynamic brakes.

Applications   Used where heavy starting torque is required, automobiles, conveying systems, crane travel drives, processing equipment, and filling and packaging equipment.

Construction   It consists of mainly a pump also called as impeller and a turbine contained in a sealed housing containing oil. Pump is attached to the driving shaft and turbine to the driven shaft. Both pump and turbine have semicircular annular groove with radial vanes. Pump and turbine are placed side by side facing each other. Power from the prime mover is transferred through oil. The torque transmitting capacity T of a fluid coupling is related as $T = 0.005 \, \rho N^2 D^5$.

Advantages   Simple, robust, and reliable drive; no wear of elements; input shaft starts without any load; accelerates fast, shock free, smooth load take up; protects overloads from jamming of machines.

Torque converter is a hydrodynamic fluid coupling. The main characteristic of a torque converter is to multiply torque, when there is a substantial difference between input and output rotational speed.

Construction   A torque converter has at least one extra element called stator, placed between the impeller and turbine. It is mounted on an over-running clutch. Some models use even five elements to offer a wide range of torque multiplication.

Working   The pump acts as a centrifugal pump. When it rotates, the fluid is forced to move outwards. This causes a vacuum, at inner radius, which draws more fluid. The fluid coming out from the pump enters the blades of the turbine, which is connected to the output shaft. The blades of the turbine are curved so that the fluid which enters the turbine from the outside, has to change direction before it exits the center of the turbine. This directional change causes the turbine to rotate.

The fluid exits the turbine in a direction opposite to that of the pump. If this fluid hits the pump, it will slow the pump, wasting power; hence a stator is put in between the pump and turbine. It changes the direction of fluid again, which will be suitable for a smooth entry in the pump. A torque converter has three stages of operation:

Stall   It is a condition when the driver shaft is developing full power but the driven shaft is prevented from rotating.

Acceleration   There is a large difference between speeds of the driver and driven. In this condition, torque converter multiplies the torque to great extent but lesser than the stall condition.

Coupling Speed of the turbine reaches about 90 per cent of the pump speed. The torque converter works more or less just like any fluid coupling.

Theory Questions

1. What are the functions of a gear box? How can these be classified?
2. Describe the construction of a gear box with a neat sketch and the various applications where these are used.
3. Differentiate between simple and compound gear trains with the help of sketches.
4. Write a short note on maximum pitch line velocity of different types of gears.
5. Describe an epicyclic train with a sketch. How does it work?
6. What are the different methods to find the speed ratio of an epicyclic gear train? Explain by taking an example.
7. What do you mean by holding torque? How is it calculated?
8. What points are to be considered in selecting a gear box?
9. How are the speeds decided for a gear box for given minimum speed, maximum speeds, and number of speed sets?
10. Differentiate between a kinematic diagram and structural diagram.
11. Discuss a structural formula for a three-stage gear box by taking any example.
12. How do you select transmission ratio of a stage?
13. What parameters are considered in selecting an optimum structural formula?
14. Differentiate between open and crossed ray diagrams with a sketch.
15. Describe a ray diagram with the help of a sketch for any structural formula.
16. Discuss the various power losses in a gear box.
17. Differentiate between a fluid coupling and torque converter.
18. Write the construction, working, and applications of fluid flywheel.
19. Describe the construction of a torque converter with a neat sketch.
20. How does torque multiplication takes place in a torque converter?
21. Describe the construction of a gear box and its size.

Multiple Choice Questions

1. What type of gear box could be used for right angle applications?
   (a) Spur gear and helical gear
   (b) Helical gear and bevel gear
   (c) Bevel gear and worm gear
   (d) Spiral gear and planetary gear
2. A compound gear train has pinions of 25 and 40 teeth and gears of 75 teeth and 80 teeth. The overall gear ratio is:
   (a) 6
   (b) 3.2
   (c) 1.75
   (d) 1.5

3. A three-stage gear box, has pairs of gear sets?
   (a) 3
   (b) 4
   (c) 6
   (d) Depends on number of speed sets

4. Speed ratio of an epicyclic gear train depends on:
   (a) Number of teeth of sun gear
   (b) Number of teeth of planetary gear
   (c) Number of teeth of ring gear
   (d) All given in (a), (b), (c), and which element is braked.

5. Which of the following speeds are good for a six-speed gear box having minimum speed 100 rpm?
   (a) 100, 125, 150, 175, 200, and 225
   (b) 100, 125, 156, 195, 244, and 305
   (c) 100, 150, 200, 250, 300, and 350
   (d) 100, 200, 300, 400, 500, and 600

6. A kinematic diagram informs about:
   (a) Input and output speeds
   (b) Speed ratios
   (c) Layout of different shafts and gears
   (d) None of the mentioned above

7. Structural formula is used for gear trains to provide information about:
   (a) Number of stages
   (b) Number of speed ratios
   (c) Both number of stages and speed ratios
   (d) Structure of the gear box

8. A structural diagram is a:
   (a) Table giving actual speeds of each gear set
   (b) A graphic display of input and out speeds of one stage
   (c) A graphic display of input and out speeds of all stages
   (d) Graph between input and output speed

9. A ray diagram displays:
   (a) Number of transmission groups
   (b) Number of speeds available
   (c) Speed of each gear selection
   (d) All given above

10. In a sliding mesh gear box, only the type of gears, which can be used is:
    (a) Spur gear
    (b) Helical gear
    (c) Bevel gear
    (d) Spiral gear

11. Helical gears are used for:
    (a) Slow speed sliding mesh gear box
    (b) High speed in constant mesh gear box
    (c) Gear box with high reduction
    (d) Input and output shafts are not parallel

12. Power losses in a gear box are due to:
    (a) Oil churning
    (b) Bearings
    (c) Seals
    (d) All given in (a), (b), and (c)

13. A fluid coupling is used:
    (a) To replace a clutch
    (b) To replace a gear box
    (c) Where large power is to be transmitted
    (d) Where speed is high

14. A torque converter is used as:

(a) Fluid coupling as clutch (b) Changing direction of rotation

(c) Clutch and also for torque multiplication (d) None of given in (a), (b), and (c)

Answers to multiple choice questions

1. (c) 2. (a) 3. (d) 4. (d) 5. (b) 6. (c) 7. (c) 8. (c) 9. (d) 10. (a)

11. (b) 12. (d) 13. (a) 14. (c)

Design Problems

Gear Trains

1. A simple gear train has three gears A, B, and C. Gear A having 60 teeth rotates at 1,500 rpm CW drives the gear having 180 teeth though gear B. If input torque to gear A is 10 Nm, calculate torque and power of output shaft, if efficiency is 80 per cent. $[T = 24\ \text{Nm}, P = 1.256\ \text{kW}]$

2. A compound gear train has four shafts S1, S2, S3, and S4. Shaft S1 has gear A, with 20 teeth drives gear B with 80 teeth on Shaft S2. Gear C with 30 teeth on shaft S2 drives D with 90 teeth on Shaft 3 and gear E on shaft S3 with 15 teeth drives gear F with 75 teeth on shaft S4. If shaft S1 rotates at 600 rpm in CW direction, calculate speed and direction of rotation of shaft S4. If input torque is 20 Nm and efficiency of the gear train is 75 per cent, calculate the output power and torque.

$[N = 10\ \text{rpm}, \text{CCW}, P = 0.942\ \text{kW}, T = 900\ \text{Nm}]$

3. An epicyclic gear train has fixed ring gear having 210 teeth and planetary gear 30 teeth. Power is given to the arm and taken out from the sun gear. Calculate the gear ratio. $[R = 1.4]$

4. Sun gear with 150 teeth is fixed in an epicyclic train, which meshes with a planetary gear of 30 teeth. Power is taken out from ring gear. If power is given to the arm carrying planetary gears. Calculate the number of teeth of ring gear and gear ratio. $[Z_A = 210, R = 1.714]$

5. An epicyclic gear train shown in Figure 9.P1 has sun gear with 60 teeth rotating at 150 rpm CW. Planetary gear has 30 teeth. Casing is allowed to rotate at 50 rpm in the same direction. Calculate the speed of the output shaft.
$[N = 25\ \text{rpm}, \text{CCW}]$

6. Annulus gear of an epicyclic gear train is braked. A sun gear with 30 teeth drives planetary gears with 120 teeth mounted on a spider to give output power. If the sun gear rotates at 250 rpm in CW direction with a torque of 30 Nm and efficiency 85 per cent, calculate:

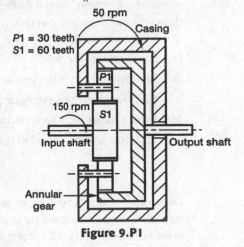

Figure 9.P1

    (a) Gear ratio.   (b) Output speed and direction.

    (c) Power output.   (d) Holding torque.

$$[(a)\ G = 10{:}1;\ (b)\ N = 25\ rpm\ (CW);\ (c)\ P = 667.2\ W;\ (d)\ 330\ Nm]$$

Gear Boxes

7.   Draw speed ray diagram and layout for a six speed gear box. The output speeds are 200 rpm minimum and 1,200 rpm maximum. The motor speed is 1,440 rpm.

$$[\text{Output speeds in rpm are: } 200, 286, 410, 586, 839, \text{ and } 1,200]$$

8.   A belt conveyor has to move a belt at maximum speed of 1.6 m/s. Driving pulley diameter is 280 mm. Torque transmitted to the driving pulley of the conveyor is 26 Nm. Speed of input shaft is 1,440 rpm. Design a two-stage reduction gear box for the drive, first stage helical and second spur gear. Assume allowable bending strength 200 MPa and helix angle for first stage 30°, minimum number of teeth 19. Also give a neat sketch.

$$[R_1 = R_2 = 3.63, m1 = 2, d_{p1} = 38\,mm, d_{g1} = 138\,mm, b1 = 20\,mm,$$
$$m2 = 3,\ d_{p2} = 57\,mm,\ d_{g2} = 207\,mm, b2 = 30\,mm]$$

9.   An electric motor running at 1,440 rpm is connected to a gear box through a belt drive and transmits of 5 kW. Design the gear box with nine speeds based on R 10 series. Minimum spindle speed is 150 rpm. Each stage has to have three speed sets. Determine:

    (a) Ratio of the belt pulley diameters.

    (b) Write the structural formula.

    (c) Number of teeth on each gear of with minimum number of teeth 17.

    (d) Draw the gear box layout.

    (e) Draw the structure and speed diagram for the arrangement.

$$[(a) = 1.61;\ (b)\ 3(1)\ 3(3);\ (c)\ \text{Stage 1: } Z_a = 17, Z_A = 65, Z_b = 28, Z_B = 54, Z_c = 41, Z_C = 41,$$
$$\text{Stage 2: } Z_d = 17, Z_D = 27, Z_e = 20, Z_E = 44, Z_f = 22, Z_F = 22]$$

10.   Design a multi-speed two-stage gear box for a machine tool for speed variation from 900 rpm to minimum 90 rpm using R5 series for spindle speeds. Power comes from a motor at 960 rpm through a pair of pulleys. Speed ratio in no set should be more than 6. Assuming minimum number of teeth 19, calculate:

    (a) Speed ratio of the pulleys.

    (b) Speeds in rpm from the gear box.

    (c) Number of teeth on each gear using structural formula 2 (1) 3(2).

    (d) Draw a speed and structural diagram.

$$[(a)\ 3.36 \text{ input speed to gear box 285 rpm; } (b) \text{ Output speeds 900, 586, 358, 226, 143, and 90;}$$
$$(c)\ \text{Stage 1: } Z_a = 19, Z_A = 38, Z_b = 38, Z_B = 19,$$
$$\text{Stage 2: } Z_c = 32, Z_C = 27, Z_d = 27, Z_D = 27, Z_e = 27, Z_E = 32]$$

11.   Design a suitable gear box for a lathe that has a variation of speed from 150 rpm to 900 rpm in nine steps. The power is supplied by an electric motor of 5 KW running at 1,440 rpm through a V-belt drive with speed ratio of 2:1. Using structural formula 3 (1) 3 (3) and allowable bending strength 180 MPa, calculate:

(a) Output speeds.   (b) Number of teeth.

(c) Module, face width and PCD of each gear.   (d) Draw the structural diagram and speed chart.

[(a) 150, 188, 235, 294, 67, 460, 575, 719, 900;

(b) Input 720 rpm **Stage 1:** $Z_a = 47, Z_A = 38, Z_b = 33, Z_B = 52, Z_c = 21, Z_C = 64$

Stage 2: $Z_d = 27, Z_D = 27, Z_e = 24, Z_E = 30, Z_f = 21, Z_F = 33;$

(c) Stage 1: $m1 = 3, b = 30$ mm, $d_{pa} = 141$ mm, $d_{gA} = 114$ mm, $d_{pb} = 99$ mm, $d_{gB} = 156$ mm, $d_{pc} = 63$ mm, $d_{gC} = 192$ mm,

Stage 2: $m2 = 4, b = 40$ mm, $d_{pd} = 108$ mm, $d_{gD} = 108$ mm, $d_{pe} = 96$ mm, $d_{gE} = 120$ mm, $d_{pf} = 84$ mm, $d_{gF} = 132$ mm]

Previous Competition Examination Questions

IES

1. The figure shows an epicyclic gear train with gears A and B having 40 and 50 teeth, respectively. If the arm C rotates at 200 rpm in anticlockwise direction about the center of gear A which is fixed, then speed of revolution of gear B in rpm would be:

    (a) 160

    (b) 250

    (c) 360

    (d) 450

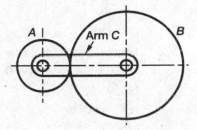

[Answer (c)] [IES 2013]

GATE

2. A gear train is made up of five spur gears as shown in Figure on right side. Gear 2 is driver and gear 6 is driven member. N_2, N_3, N_4, N_5 and N_6 represent number of teeth on gears 2, 3, 4, 5, and 6, respectively. The gear(s), which act(s) as idler(s) is / are:

    (a) Only 3   (b) Only 4

    (c) Only 5   (d) Both 3 and 5

    [Answer: c] [GATE 2015]

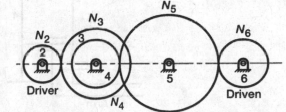

3. The number of degree of freedom of the planetary gear train shown in Figure on right side is:

    (a) 0

    (b) 1

    (c) 2

    (d) 3

    [Answer (c)] [GATE 2015]

4. Gear 2 rotates at 1,200 rpm is CCW direction and engages with gears 3 and 4 mounted on the same shaft (shown in figure below on right side). Gear 5 engages with gear 4. The number of teeth

on gears 2, 3, 4, and 5 are 20, 40, 15, and 30, respectively. The annular speed of gear 5 is:

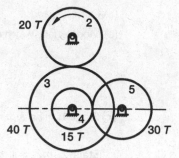
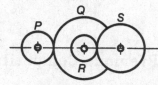

(a) 300 rpm CCW

(b) 300 rpm CW

(c) 4,800 rpm CCW

(d) 4,800 rpm CW

[Answer: (b)] [GATE 2014]

5. A compound gear train with gears P, Q, R, and S has number of teeth 20, 40, 15, and 20, respectively. Gears Q and R are mounted on the same shaft as shown in the figure on right side. The diameter of the gear Q is twice that of the gear R. If the module of the gear R is 2 mm, the center distance in mm between gears P and S is:

(a) 40   (b) 80

(c) 120   (d) 160   [Answer: (b)] [GATE 2013]

6. For the epicyclic gear arrangement shown in figure below, $\omega_2 = 100$ rad/s CW and $\omega_{arm} = 80$ rad/s CCW. The angular velocity ω_5 (in rad/s) is:

(a) 0   (b) 70 CW

(c) 140 CCW   (d) 140 CW

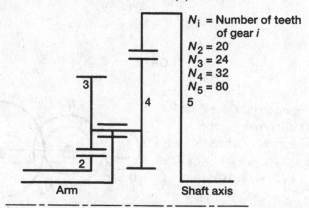

N_i = Number of teeth of gear i

$N_2 = 20$
$N_3 = 24$
$N_4 = 32$
$N_5 = 80$

Arm   Shaft axis

[Answer: (c)] [GATE 2010]

7. An epicyclic gear train is shown schematically in the figure on right side. The sun gear 2 on the input shaft is 20 teeth external gear. The plenary gear 3 is 40 teeth external gear. The ring gear 5 is 100 teeth internal gear. The ring gear is fixed and the gear 2 is rotating at 60 rpm CCW.

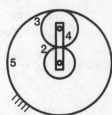

The arm 4 attached to the output shaft will rotate at:

(a) 10 rpm CCW   (b) 10 rpm CW

(c) 12 rpm CW   (d) 12 rpm CCW

[Answer: (a)] [GATE 2009]

Common Data Questions 8 and 9

A planetary gear train shown on right side has four gears and a carrier. Angular velocities of the gears are ω_1, ω_2, ω_3, and ω_4 respectively. The carrier rotates at angular velocity ω_5.

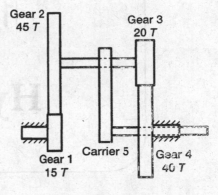

Gear 2 — 45 T
Gear 3 — 20 T
Gear 1 — 15 T
Carrier 5
Gear 4 — 40 T

8. What is the relation between the angular velocities of gear 1 and gear 4?

(a) $\dfrac{\omega_1 - \omega_5}{\omega_4 - \omega_5} = 6$

(b) $\dfrac{\omega_4 - \omega_5}{\omega_1 - \omega_5} = 6$

(c) $\dfrac{\omega_1 - \omega_2}{\omega_4 - \omega_5} = -\left(\dfrac{2}{3}\right)$

(d) $\dfrac{\omega_1 - \omega_5}{\omega_4 - \omega_5} = \left(\dfrac{8}{9}\right)$

[Answer: (a)] [GATE 2006]

9. For $\omega_1 = 60$ rpm CW when viewed from the left. What is the angular velocity of the carrier and its direction so that gear 4 rotates in CCW direction at twice the angular velocity of gear 1 when viewed from the left?

(a) 130 rpm CW

(b) 223 rpm CCW

(c) 256 rpm CW

(d) 156 rpm CCW

[Answer: (d)] [GATE 2006]

□□□

Hydrodynamic Bearings

Outcomes

➤ Learn about function and construction of sleeve bearings

➤ Classification of sleeve bearings

➤ Bearing materials and their desirable properties

➤ Hydrodynamic lubrication, Reynolds equation, short and long bearings

➤ Sommerfeld number, eccentricity ratio, critical pressure, unit load, and maximum pressure in finite bearings

➤ Oil flow through bearings

➤ Energy loss due to friction

➤ Heat generated and dissipated in these bearings

➤ Selection of parameters and design procedure of sleeve bearings

➤ Hydrodynamic thrust bearing

➤ Foot-step and collar thrust bearings

➤ Squeeze film journal bearings

10.1 Introduction

A rotating shaft has to be supported, so that it could rotate at its position and transmit power with minimum friction. Portion of the shaft inside the bearing is called a journal. Bearings permit low friction movement between the bearing and journal surfaces. Supports are used to provide support to the bearings, which carry the load. Friction is reduced either by providing a thin film of oil or by providing rolling elements such as balls or rollers, which are described in Chapter 11.

10.2 Construction of Bearings

Plain bearings are also called as bush or sleeve bearings, because their shape is like a sleeve. The bearing sleeve is also called bearing shell. Small bearings are in one piece as complete circular shown in Figure 10.1(a) and also in picture 10.1(d). Large bearings are made in two halves shown in Figure 10.1(b) and in picture 10.1(e) each of semicircular shape. Sometimes, it is provided with a collar to check its axial movement in its support. An oil hole on the top of the shell is provided to put oil [Figure 10.1(c)]. A circumferential groove is also sometimes provided for better supply of oil.

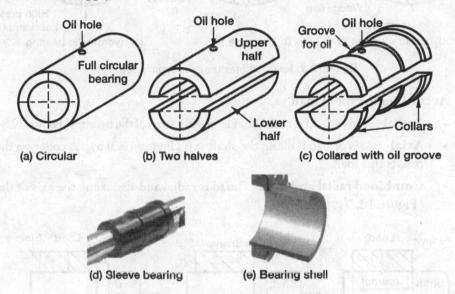

(a) Circular   (b) Two halves   (c) Collared with oil groove

(d) Sleeve bearing   (e) Bearing shell

Figure 10.1 Sleeve and bush-bearing shells

Small bearings use sintered metal, that contains a lubricant in its pores and does not need lubrication frequently. A heavily loaded bearing is provided with some arrangement to introduce oil under pressure using a pump. Such bearings are called hydrostatic bearings.

10.3 Classification of Bearings

Bearings can be classified in many ways as given below:

A. According to lubricant pressure

- **Hydrodynamic** Working fluid (generally oil) is supplied at atmospheric pressure. Pressure is generated due to wedge film formed between shaft and bearing space [Figure 10.2(a)].
- **Squeeze film** Pressure is generated by radial movement of the journal [Figure 10.2(b)].
- **Hydrostatic** Working fluid oil or air is supplied at high pressure [Figure 10.2(c)].

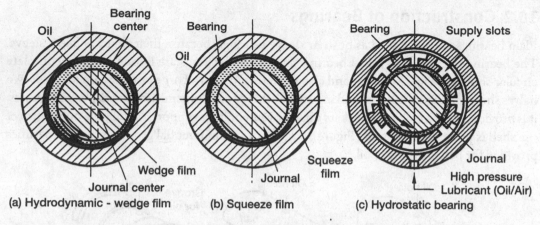

(a) Hydrodynamic - wedge film   (b) Squeeze film   (c) Hydrostatic bearing

Figure 10.2 Lubricant pressure generation in bearings

B. According to type of load

- **Radial**   Axis of load is radial, that is, at 90° to axis of the bearing [Figure 10.3(a)].

- **Axial**   Axis of load is along the shaft axis [Figure 10.3(b)]. A collar on the shaft resists this load.

- **Combined radial and axial**   Load is radial and also along the axis of the shaft [Figure 10.3(c)].

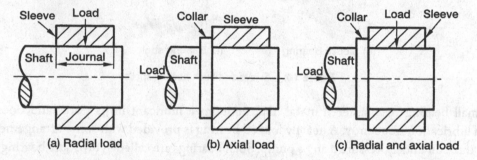

(a) Radial load   (b) Axial load   (c) Radial and axial load

Figure 10.3 Radial and axial loads on bearings

C. According to thickness of oil film

- **Thick film**   Bearing and journal surfaces are separated by lubricant and parts never touch each other under running conditions.

- **Thin film**   Lubricant is used but still for some time there may be metal to metal contact.

- **Zero film**   No lubricant is used between the surfaces.

D. According to relative movement between bearing and shaft

- Rotating: Shaft rotates in the bearing. Majority of the shafts rotate in the bearings [Figure 10.4(a)].

- **Sliding** Movement is linear along an axis, for example, carriage of type writer, Computer Numerical Control (CNC) machines slides [Figure 10.4(b)].
- **Oscillating** A loaded shaft oscillates in the bearing, for example, small end of connecting rod of an engine [Figure 10.4(c)].

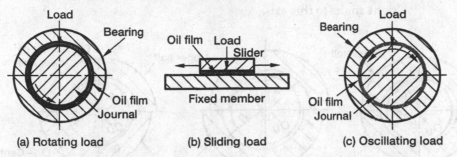

(a) Rotating load   (b) Sliding load   (c) Oscillating load

Figure 10.4 Relative movement between bearing and shaft

E. According to contact angle

- Full 360° Bearing Wrap angle of bearing over shaft is for 360° [Figure 10.5(a)].
- Split bearing Bearing is in two halves each of 180° [Figure 10.5(b)].
- Partial bearing Bearing supports even less than 180°, called pad or lobe, generally 120° [Figure 10.5(c)].

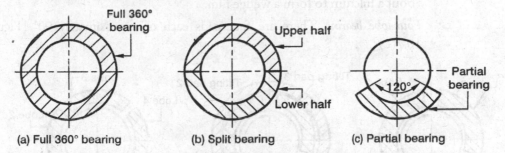

(a) Full 360° bearing   (b) Split bearing   (c) Partial bearing

Figure 10.5 Contact angle of bearings

F. According to number of pads

Multi-pad or multi-lobe bearings have more than one pad. These have either fixed pads or pads are allowed to tilt along a pivot point. Such bearings are called tilted pad bearings.

- *Two lobe bearing* These bearings have fixed pads and can be of the following types. It is same as split bearing but arrangement of the pads may differ as under:
  - *Elliptical bearing* Centers of the bearing halves are shifted to form an ellipse [Figure 10.6(a)]. Eccentricity due to shift is very less about 25 per cent–50 per cent of radial clearance between shaft and bearing. O_L and O_U are the centers of lower and upper-half pads, respectively.

- *Offset bearing* Centers of the bearing halves are shifted about 25 per cent–50 per cent of radial clearance along the split axis [Figure 10.6(b)].
- *Orthogonally displaced bearing* It is a combination of elliptical and offset bearings [Figure 10.6(c)]. Centers are shifted along split axis and also at right angles to this axis.

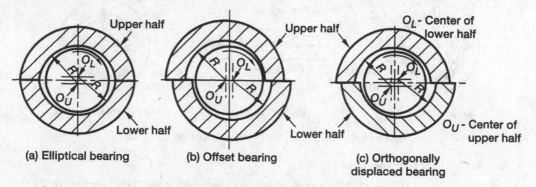

(a) Elliptical bearing   (b) Offset bearing   (c) Orthogonally displaced bearing

Figure 10.6 Types of two-lobe bearings

- *Three-lobe bearing* There are three pads each of about 120°. These can be fixed pad [Figure 10.7(a)] or tilting pad [Figure 10.7(b)]. Each pad can tilt about a fulcrum to form a wedge film.
- *Four-lobe bearing* There are four pads each of approximately 90° [Figure 10.7(c)].

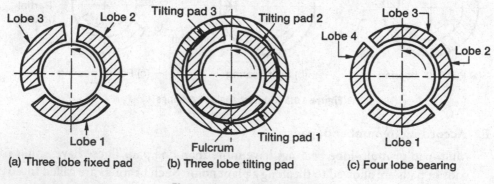

(a) Three lobe fixed pad   (b) Three lobe tilting pad   (c) Four lobe bearing

Figure 10.7 Multilobe bearings

G. According to freedom of pads

Multi-pad or multi-lobe bearings have more than one pad. These are:

- Fixed pads   Pads are fixed and not allowed to tilt [Figure 10.7(a) and (c)].
- Tilted pad bearings   Pads are allowed to tilt along a pivot point [Figure 10.7(b)].

10.4 Properties of Bearing Material

The material selected for a bearing should possess the following properties:

High compressive strength   The load coming on the shaft is conveyed to the bearing through the oil. Average pressure on the bearing is calculated by dividing the load W by projected area $(L \times D)$. The maximum bearing pressure p_{max} is much more than the average pressure p_{av}. The bearing material should be strong enough to withstand the maximum pressure without any permanent deformation or damage.

High fatigue strength   In case of varying loads, the bearing is subjected to fatigue loading. Endurance strength in fatigue is much lesser than the strength under constant loads. This becomes an important property, especially for automotive engines, where the loads vary even with shocks also.

Corrosion resistant   A bearing working under moist environment can get corrosion. The material should be able to resist corrosion.

High thermal conductivity   Heat generated in bearing due to friction is to be dissipated out. A material with high thermal conductivity is able to do this job satisfactorily.

Low thermal expansion   Bearing performance is very much dependent on radial clearance. A material with high coefficient of thermal expansion expands, resulting more radial clearance and thus affecting the bearing performance.

Embeddability   Small particles of dirt, etc., enter the oil, which can cause scratching on the bearing / journal surface. If the material is soft enough, the tiny particles of dirt embed in its surface and do not score the bearing surface.

Comformability   It is the ability of a material to accommodate shaft deflections or bearing inaccuracies by plastic deformation without excessive wear or temperature rise.

Bonding characteristics   Bearings are made by bonding one or more thin layers of a soft metal with another material of steel with high strength. The thin layer must bond easily with the bearing shell.

Machinability   The material should be easily machinable to give the desired accuracy with good surface finish.

10.5 Bearing Materials

Commonly used bearing materials are described below:

A. Babbitts

Sleeve bearings use tin or lead base alloys called babbitt. They are popular due to their ability to embed dirt, excellent properties even when oil film is very less. These can be used up to bearing pressures 14 MPa. Their compositions are as under:

Tin base babbitts   Tin 90 per cent, antimony 5 per cent, copper 4.5 per cent, and lead 0.5 per cent.

Lead base babbitts   Lead 84 per cent, antimony 9.5 per cent, tin 6 per cent, and copper 0.5 per cent.

B. Bronzes

Bronze is an alloy of copper, zinc, and tin. Many copper alloys such as aluminum bronze, tin bronze, phosphorous bronze, and lead bronze are used. These are generally used as a bush material, pressed into a steel shell. Most commonly used bronzes are:

Gun metal   It has copper 88 per cent, tin 10 per cent, and zinc 2 per cent. It is suitable for high journal speed and bearing pressures up to 10 MPa.

Phosphorous bronze   It has copper 80 per cent, tin 10 per cent, lead 9 per cent, and phosphorous 1 per cent.

It is also suitable for high speeds and high-bearing pressures up to 14 MPa.

C. Aluminum alloys

Alloys 750, A750, and B750 can be cast in sand or permanent molds, but not as die castings. Alloy X385 is preferred for die cast bearings. It is considered to have good machining and bearing properties. They offer high load-bearing capacity, thermal conductivity, fatigue, and wear resistance. Their application is for main and big-end bearings of connecting rod, steel mills, and reciprocating compressors.

D. Cast iron

A cast iron bearing can be used with a hardened steel shaft, because the coefficient of friction is relatively low. The cast iron glazes, therefore wear becomes negligible.

E. Plastics

Solid plastic plain bearings are now increasingly popular due to dry-running and lubrication-free behavior. Solid polymer plain bearings are corrosion resistant, have low weight and maintenance free. These materials soften at elevated temperatures, if used in applications at high bearing pressures than recommended values. Some materials such as Teflon and nylon offer excellent low-friction property. Common plastics are nylon, nylon, polyacetal, and poly-tetra-fluoro-ethylene (PTFE).

10.6 Properties of Lubricants

Lubricants should possess the following properties:
- Boiling point should be high.
- Freezing point should be low.
- Viscosity should not vary much with change in temperature.
- Thermal stability.
- Corrosion resistant.
- High resistance to oxidation.

10.6.1 Viscosity

Viscosity is the most important property of a lubricant. It is a measure of degree of fluidity. Viscosity is the internal frictional resistance offered by a fluid when it is subjected to change in shape either by shear force or tensile force. In any flow, liquid moves in layers at different velocities, which causes shear stress between the layers that ultimately opposes any applied force.

The relationship between the shear stress and the velocity gradient is obtained by considering two plates spaced at a small distance y, and separated by a homogenous substance. If plates are assumed large, with area A, the edge effects may be ignored. Let the lower plate be fixed, and a force F be applied to the upper plate. This force causes the substance between the plates to undergo shear flow with a velocity gradient du / dy (Figure 10.8).

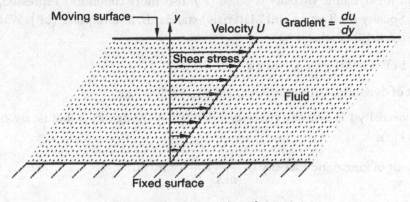

Figure 10.8 Newton's law of viscosity

The applied force is proportional to the area and velocity gradient in the fluid. Mathematically, $F \propto A\left(\dfrac{U}{y}\right)$. When sign of proportionality is converted to equality, the equation becomes:

$$F = \mu A\left(\frac{U}{y}\right) \tag{10.1}$$

Where, μ is the proportionality factor called dynamic viscosity or absolute viscosity. This equation in terms of shear stresses can be written as:

$$\tau = F / A \tag{10.2}$$

Shear stress between layers is proportional to the velocity gradient in the direction perpendicular to the layers. Thus:

$$\tau = \mu \frac{\partial u}{\partial y} \tag{10.3}$$

10.6.2 Units of viscosity

S.I. unit for dynamic viscosity μ is pascal-second (Pa.s), that is, N.s/m^2. As $N = \dfrac{kg.m}{s^2}$,

$$Pa.s = \frac{N.s}{m^2} = \left(\frac{kg.m}{s^2} \right) \times \left(\frac{s}{m^2} \right) = \frac{kg}{m.s}$$

Thus in SI units are: Pa.s = N.s/m^2 = kg/m·s

If a fluid with a viscosity of 1 Pa.s is placed between two plates of area 1 square meter, and one plate is moved sideways with a shear stress of one pascal, it moves a distance equal to the thickness of the layer between the plates in one second. Water at 20°C has a viscosity of 0.001 Pa.s.

CGS unit for dynamic viscosity is poise (P). It is more commonly expressed, in ASTM (American Society for Testing and Materials) standards, as centipoise (cP). Water at 20°C has a viscosity of 1.0 cP, that is,

$$1\ cP = 0.001\ Pa.s. = 1 mPa.s$$

British unit of dynamic viscosity is Reyn. 1 Reyn = 6,890 Pa.s

Kinematic viscosity v is dynamic viscosity μ divided by the density ρ that is, $\mu\,/\,\rho$. Unit of ρ is kg/m^3.

Hence, SI unit of kinematic viscosity is: $\dfrac{kg}{m.s} \times \dfrac{m^3}{Kg} = m^2/s$.

CGS unit for kinematic viscosity is stokes: $St = \dfrac{cm^2}{s} = 10^{-4}\ m^2/s$

It is also expressed in terms of centistokes: $cSt = mm^2/s = 10^{-6}m^2/s$

Water at 20°C has a kinematic viscosity of about 1 cSt.

10.6.3 SAE designation of oils

The Society of Automotive Engineers (SAE) has established a numerical code system for grading motor oils used in crank cases according to their viscosities from low to high: 0W, 5W, 10W, 15W, 20W, 25W, 20, 30, 40, 50, or 60. The numbers 0, 5, 10, 15, and 25 are suffixed with the letter W, meaning 'winter' or cold-start viscosity, which are suitable for lower temperatures. The number 20 comes with or without a W, depending on whether it is being used to denote a cold or hot viscosity grade.

The SAE has a separate viscosity rating system for gear oils used in gear boxes, axle, and manual transmission oils, which should not be confused with engine oil viscosity. The higher number of a gear oil (e.g., 75W-140) does not mean that it has higher viscosity than an engine oil.

10.6.4 Viscosity index

Viscosity index in short called as *VI* is a measure for the change of kinematic viscosity with temperature. It is used to characterize lubricating oil in the automotive industry. As the temperature increases, oil expands and inter-molecular force decreases. Thus viscosity decreases with increase in temperature as shown in Figure 10.9.

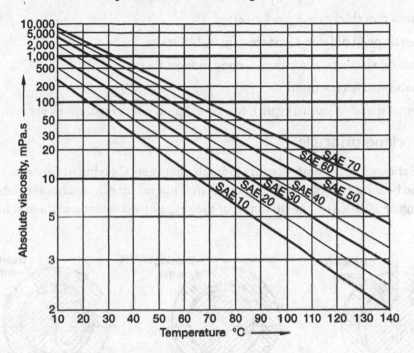

Figure 10.9 Viscosity variation with temperature

10.7 Hydrodynamic Lubrication

10.7.1 Terminology

Terms, notations, and their definitions used for hydrodynamic bearings are:

Bearing diameter, D = Inside diameter of the sleeve

Journal diameter, d = Diameter of portion of the shaft inside the bearing

Journal radius, $r = 0.5 \times d$

Diametric clearance, $c' = D - d$

Radial clearance, $c = 0.5 \times (D - d) = 0.5\,c'$

Eccentricity, e = Distance between centers of bearing and journal

Eccentricity ratio, $\varepsilon = \dfrac{\text{Eccentricity}}{\text{Radial clearance}} = \dfrac{e}{c}$

Attitude angle, Φ = Angle made by line of centers of bearing and journal with load line

Sommerfeld number, S = Dimensionless number $\left(\dfrac{r}{c}\right)^2 \left(\dfrac{\mu n}{p}\right)$.

Bearing modulus, $B = \left(\dfrac{\mu n}{p}\right)$. It is also known bearing characteristic number.

Minimum film thickness, $h_{min} = (c - e)$

Peripheral speed of journal surface in m/s, $U = \pi d n$

Absolute oil viscosity, μ in pascal-second in short Pa.s

Kinematic viscosity, v (mm²/s)

Rotational speed of journal (rpm), $N = 60 n$, Speed n is in revolutions per second (rps)

10.7.2 Working principle

Figure 10.8 shows a sleeve bearing with a journal in between. Oil is put in the clearance volume between the bearing and journal. When the shaft is not rotating, it touches the bottom of the bearing [Figure 10.10(a)]. The oil film is of the shape of a crescent and there is no pressure in oil.

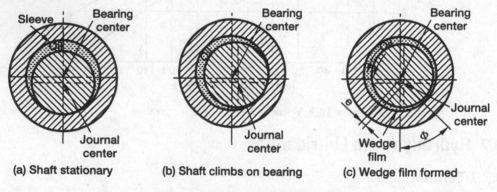

Figure 10.10 Formation of wedge film

As the shaft starts rotating, due to contact between bearing surface and shaft, it starts climbing the bearing surface in the direction of rotation [Figure 10.10(b)]. The oil film still is of the shape of crescent but it gets rotated by some angle. Now due to rotation of shaft, oil also starts rotating due to viscous forces. This rotational velocity of oil, which is half of peripheral speed of journal equal to $0.5U$ (average of journal speed U and bearing speed, which is zero), enters in wedge type clearance between journal and bearing and lifts it up, due to pressure caused by rotating oil [Figure 10.10(c)]. The journal center shifts upwards. The angle made by line of centers of bearing and journal with the load line is called attitude angle, which depends on load, speed of journal, and viscosity of oil.

10.7.3 Reynolds equation

Oil film between bearing and journal is analyzed mathematically using Reynolds equation. It is based on Newton's law of viscosity neglecting curvature of the bearing. It is the second order partial differential equation given as under:

$$\frac{\partial}{\partial x}\left[h^3 + \frac{\partial p}{\partial x}\right] + \frac{\partial}{\partial z}\left[h^3 + \frac{\partial p}{\partial z}\right] = 6\mu U\left(\frac{\partial h}{\partial x}\right) \tag{10.4}$$

Where, h = Radial film thickness at any radial location
x = Length along circumference of the bearing
z = Length along axis of the bearing
p = Oil pressure in radial direction
U = Peripheral velocity of journal

Exact solution of this equation is not available till date. Hence some assumptions like long bearing or short bearings are taken and approximate solutions are given. Finite bearing solution is done using numerical methods and these are described in Section 10.8.

10.7.4 Long bearings

It is assumed that the bearing is long, that is, L/D ratio is more than 1, that is, its length in Z direction (along the axis) is bigger than journal diameter. Pressure distribution of such a bearing is shown in Figure 10.11(a). This assumption means the pressure variation in Z direction can be neglected and hence Reynolds equation is simplified as:

$$\frac{\partial}{\partial x}\left[h^3 + \frac{\partial p}{\partial x}\right] = 6\mu U\left(\frac{\partial h}{\partial x}\right) \tag{10.5}$$

Solution to this equation was given by Sommerfeld for pressure p along circumference as:

$$p = \frac{\mu U r}{c^2}\left[\frac{6\varepsilon\,(\sin\theta)\,(2+\varepsilon\cos\theta)}{(2+\varepsilon^2)(1+\varepsilon\cos\theta)^2}\right] \tag{10.6}$$

Where, θ = Angle measured from load line

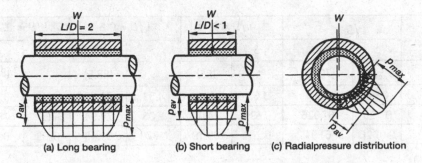

(a) Long bearing   (b) Short bearing   (c) Radial pressure distribution

Figure 10.11 Pressure distribution in hydrodynamic bearings

Pressure in a bearing increases from atmospheric pressure at ends to maximum pressure in middle. As the length of bearing is increased, pressure remains to maximum pressure for a greater length [Figure 10.10(a)] than a short bearing [Figure 10.10(b)]; hence long bearing has more load carrying capacity.

This bearing pressure distribution assumes that the supply of oil pressure is atmospheric. If it is more than atmospheric, the bearing oil pressure is also increased by the value of supply pressure.

Bearings with $L/d = 1$ are called as square bearings.

10.7.5 Short bearings

It is assumed that the length of bearing is short in comparison to diameter, that is, L/D ratio is less than 1. Flow of lubricant is less in circumferential direction in comparison to axial direction. See Figure 10.10(b) showing pressure distribution more predominant in z direction; hence the first term of the Reynolds equation is ignored to simplify the solution. Equation (10.4) for short bearing assumption reduces to:

$$\frac{\partial}{\partial z}\left[h^3 + \frac{\partial p}{\partial z}\right] = 6\mu U\left(\frac{\partial h}{\partial x}\right) \tag{10.7}$$

Ocvirk and DuBois gave the solution to this equation for oil pressure p along length z as:

$$p = \frac{\mu R}{rc^2}\left[\frac{L^2}{4} - z^2\right]\frac{3\varepsilon\sin\theta}{\left(1+\varepsilon\cos\theta\right)^2} \tag{10.8}$$

10.8 Finite Bearings

10.8.1 Eccentricity ratio and Sommerfeld number

Eccentricity ratio ε is the ratio of eccentricity 'e' to the radial clearance 'c'. Sommerfeld number 'S' for different eccentricity ratios and L/D ratios is tabulated in Table 10.1 For a very long bearing $L/D = \infty$ can be taken.

Table 10.1 Sommerfeld Number and Attitude Angle at Different Eccentricities Ratios

ε	$L/D = \infty$		$L/D = 1$		$L/D = 0.5$		$L/D = 0.25$	
	S	Φ	S	Φ	S	Φ	S	Φ
0	∞	(70.92)	∞	(85)	∞	(88.5)	∞	(89.5)
0.1	0.24	6,901	1.33	79.5	4.31	81.62	16.2	82.31
0.2	0.123	67,026	0.631	74.02	2.03	74.94	7.57	75.18
0.4	0.0626	61.94	0.264	63.10	0.779	61.45	2.83	60.86

0.6	0.0389	54.31	0.121	58.58	0.319	48.14	1.07	46.72
0.8	0.021	42.22	0.046	36.24	0.0923	33.31	0.261	31.04
0.9	0.115	31.62	0.0188	26.45	0.0313	23.66	0.074	21.86
0.97	–	–	0.0074	15.47	0.0061	13.75	0.010	12.22
1.0	0	0	0	0	0	0	0	0

Sommerfeld number as given in section 10.7.1, it is a dimensionless number relating viscosity μ in Pa.s, speed n in rps, pressure p in Pa i.e. N/m^2, journal radius r and radial clearance c in mm. It is very important number and all bearing parameters are related to this number.

To visualize the trend of variation between Sommerfeld number and eccentricity ratio at different L/D ratios, the results are plotted in Figure 10.12. Values given on left side of the plot are of minimum film thickness to radial clearance ratio, which is nothing but $(1 - \varepsilon)$. The plot also shows two conditions shown by dashed lines; one for maximum load and other for minimum friction. So the designer can choose a suitable value of Sommerfeld number according to selected L/D ratio.

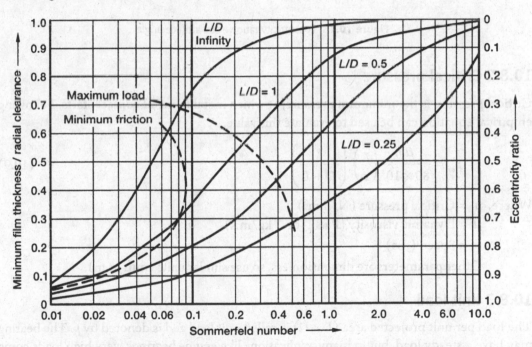

Figure 10.12 Eccentricity ratio versus Sommerfeld number for different L/D ratios

Variation of attitude angle with different eccentricity is also plotted in Figure 10.13. Arcs shown are for eccentricity ratios given on left side of plot.

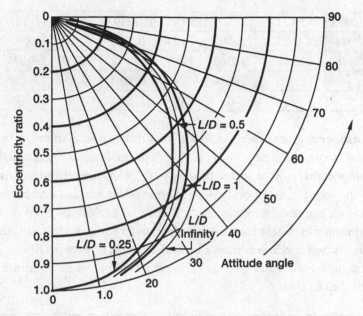

Figure 10.13 Eccentricity ratio versus attitude angle

10.8.2 Critical pressure

Critical pressure is the minimum pressure, at which metal to metal contact starts. Following empirical equation can be used to evaluate this value.

$$p_{cr} = \frac{\mu\, n}{80 \times 10^3} \left(\frac{r}{c}\right)^2 \left(\frac{L}{D+L}\right) \tag{10.9}$$

Where, p_{cr} = Critical pressure (N/mm²)

μ = Dynamic viscosity (Pa.s or kg/m.s)

n = Speed (rps)

Other parameters are dimensionless, so use similar units (mm or m).

10.8.3 Unit load

The load per unit projected area $(L \times D)$ is called unit load and is denoted by p. The bearing may have a steady load, but in many applications like engine bearings, very high loads come momentarily, which may be even 10 times the steady state value. Hydrodynamic bearings can handle such peaks without any problem. Table 10.2 can be used as a guideline to select a unit load $\left(\dfrac{p_{max}}{L \times D}\right)$ for a given application. p_{max} is the maximum pressure.

Table 10.2 Unit Loads (N/mm²) for Various Applications

Steady Loads			Fast Fluctuating Loads		
Application	Location	Unit Load (p)	Application	Location	Unit Load (p)
Compressors	Main bearing	1–1.5	Diesel engines	Main bearing	5–10
Compressors	Crank pin	1.5–3.0	Diesel engines	Crank pin	7–14
Centrifugal pumps	Main bearing	0.5–1.0	Petrol engines	Main bearing	4–5
Electric motors	Main bearing	0.7–1.5	Petrol engines	Crank pin	10–14
Machine tools	Main bearing	1.8–2.0	-	-	-
Steam turbines	Main bearing	1.0–2.0	-	-	-
Reduction gear boxes	Main bearing	0.8–1.5	-	-	-

10.8.4 Maximum pressure

The maximum oil pressure created in the bearing can be calculated from the ratio (p/p_{max}) given in Table 10.3. It is assumed that the oil supply pressure is atmospheric. If there is pressurized supply, that supply pressure may be added to this p_{max} value.

Table 10.3 Values of (p/p_{max}) for Different Values of L/D and Eccentricity Ratios

L/D	Eccentricity Ratio ε						
	0.1	0.2	0.4	0.6	0.8	0.9	0.97
∞	0.826	0.814	0.764	0.667	0.495	0.358	-
1	0.540	0.529	0.484	0.415	0.313	0.247	00152
0.5	0.523	0.506	0.441	0.365	0.267	0.206	0.126
0.25	0.515	0.489	0.415	0.334	0.240	0.180	0.108

Variation of the pressure ratio can be seen in Figure 10.14 for different L/D ratios.

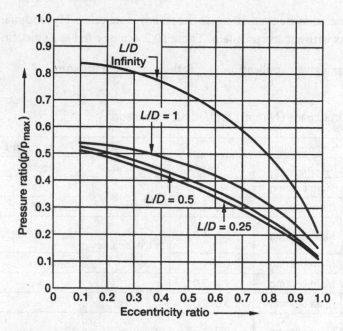

Figure 10.14 Plot of eccentricity ratio and pressure ratios

Example 10.1

Bearing parameters for a given bearing

A sleeve bearing of inside diameter 40.05 mm supports a journal of diameter 40 mm rotating at 1,440 rpm. If bearing length is 20 mm and oil used has a viscosity of 25 centipoise at the average operating temperature. Calculate:

    a. Radial distance between journal and bearing center for a load of 3 kN.

    b. Minimum oil film thickness.

    c. Angle made by line of centers with load line (attitude angle).

    d. Critical pressure.

    e. Maximum pressure in the bearing.

Solution

Given $D = 40$ mm $N = 1{,}440$ rpm $c = 40.05 - 40 = 0.05$ mm $L = 20$ mm

         $\mu = 25$ cP $W = 3$ kN

  a. Average pressure, $p = \dfrac{W}{LD} = \dfrac{3{,}000}{20 \times 40} = 3.75$ MPa $\left(\text{N/mm}^2\right) = 3.75 \times 10^6$ Pa

       Speed $n = \dfrac{1{,}440}{60} = 24$ rps

CentiPoise is CGS unit converting to SI units 1 cP = 1 mPa.s = 10^{-3} Pa.s

Sommerfeld number $S = \left(\dfrac{\mu n}{p}\right)\left(\dfrac{r}{c}\right)^2 = \left(\dfrac{25 \times 10^{-3} \times 24}{3.75 \times 10^6}\right)\left(\dfrac{40}{0.05}\right)^2 = 0.102$

From Figure 10.12, for $S = 0.102$ and $L/D = 0.5$,

Eccentricity ratio $\varepsilon = 0.79$.

Radial eccentricity, $e = \varepsilon \times c = 0.79 \times 0.05 = 0.0395$ mm

b. From Figure 10.12, minimum oil film thickness to clearance ratio

$\dfrac{h_{min}}{c} = 0.21.$ [It can also be found from $(1 - \varepsilon)$]

Hence, minimum oil film thickness $h_{min} = 0.21 \times 0.05 = 0.0105$ mm

c. From Figure 10.13, for eccentricity ratio = 0.79, and $L/D = 0.5$

Attitude angle $\Phi = 35°$

d. Critical pressure from Equation (10.9): $P_{cr} = \dfrac{\mu n}{80,000}\left(\dfrac{r}{c}\right)^2\left(\dfrac{L}{D+L}\right)$

Substituting the values: $P_{cr} = \dfrac{25 \times 10^{-3} \times 24}{80 \times 10^3}\left(\dfrac{40}{0.05}\right)^2\left(\dfrac{20}{40+20}\right) = 1.6$ MPa

e. From Figure 10.14, for $\varepsilon = 0.79$ and $L/D = 0.5$,

$\left(\dfrac{p}{p_{max}}\right) = 0.26$

Hence, maximum pressure $p_{max} = \dfrac{p}{0.26} = \dfrac{3.75}{0.26} = 14.4$ MPa

10.9 Oil Flow through Bearings

Let Q be the oil flow through bearing, which circulates in peripheral direction and Q_s, which leaks from the sides of the bearing. Flow Q can be converted to a dimensionless flow coefficient $\left(\dfrac{Q}{rcnL}\right)$. Its value at different eccentricity ratio is given in Table 10.4 for different L/D ratios and plotted graphically in Figure 10.15.

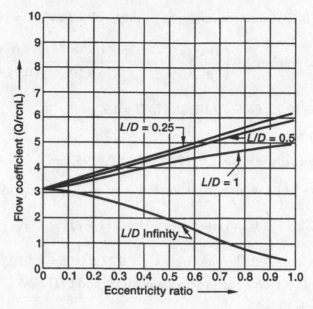

Figure 10.15 Eccentricity ratio versus flow coefficient

Owing to pressure in the bearing and being open from the sides, some oil (Q_s) leak. For an infinite long bearing, this can be neglected and assumed as zero because most of the oil remains in the bearing. The dimensionless side leakage parameter (Q_s/Q) is also tabulated in Table 10.4. It can be seen that as the pressure increases, the side leakage also increases (Figure 10.16).

Table 10.4 Values of Flow Coefficients for Different Values of Eccentricity Ratios and L/D Ratios

	$L/D = \infty$		$L/D = 1$		$L/D = 0.5$		$L/D = 0.25$	
ε	$\left(\dfrac{Q}{rcnL}\right)$	$\left(\dfrac{Q_s}{Q}\right)$	$\left(\dfrac{Q}{rcnL}\right)$	$\left(\dfrac{Q_s}{Q}\right)$	$\left(\dfrac{Q}{rcnL}\right)$	$\left(\dfrac{Q_s}{Q}\right)$	$\left(\dfrac{Q}{rcnL}\right)$	$\left(\dfrac{Q_s}{Q}\right)$
0	π	0	π	0	π	0	π	0
0.1	3.03	0	3.37	0.150	3.43	0.173	3.45	0.180
0.2	2.83	0	3.59	0.280	3.72	0.318	3.76	0.330
0.4	2.26	0	3.99	0.497	4.29	0.552	4.37	0.567
0.6	1.56	0	4.33	0.680	4.85	0.730	4.99	0.746
0.8	0.76	0	4.62	0.842	5.41	0.874	5.60	0.884
0.9	0.411	0	4.74	0.919	5.69	0.939	5.91	0.945
0.97	-	0	4.82	0.973	5.88	0.980	6.12	0.984
1.0	0	0	0	1.0	0	1.0	0	1.0

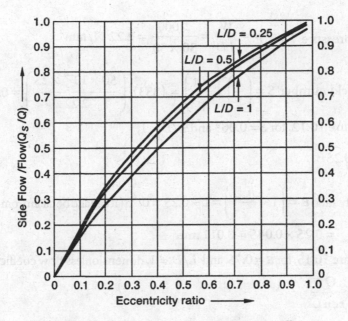

Figure 10.16 Eccentricity ratio versus side flow coefficient

Example 10.2

Oil flow and maximum pressure in a given bearing

A bearing of bore 30 mm and L/D ratio equal to 1 supports a vertical load of 2,000 N at 1,500 rpm. Radial clearance is 0.045 mm. Lubricant oil SAE 30 with viscosity 50 mPa.s, enters the bearing at 35°C. Calculate the following:

a. Minimum oil film thickness.

b. Total oil flow.

c. Maximum oil pressure in the bearing.

Solution

Given $D = 30$ mm $L/D = 1$ $W = 2,000$ N $N = 1,500$ rpm

$c = 0.045$ mm $T_i = 35°C$ $\mu = 50$ mPa.s

$$50 \text{ mPa.s} = 50 \times 10^{-3} \text{ Pa.s} = 50 \times 10^{-3} \text{ N.s/m}^2 = 50 \times 10^{-9} \text{ N.s/mm}^2$$

a. r/c ratio $= \dfrac{15}{0.045} = 333$

Speed $n = \dfrac{1,500}{60} = 25$ rps

Average pressure $p_{av} = p = \dfrac{W}{LD} = \dfrac{2,000}{30 \times 30} = 2.22 \text{ N/mm}^2$

Sommerfeld number $S = \left(\dfrac{r}{c}\right)^2 \left(\dfrac{\mu n}{p}\right) = (333)^2 \left(\dfrac{50 \times 10^{-9} \times 25}{2.22}\right) = 0.063$

From Figure 10.12, for $S = 0.063$ and $L/D = 1$,

$\dfrac{h_{min}}{c} = 0.25$

Eccentricity ratio $\varepsilon = \left(1 - \dfrac{h_{min}}{c}\right) = 1 - 0.25 = 0.75$ (It can also be taken from RHS of figure)

Hence, $h_{min} = 0.25 \times 0.045 = 0.011 \text{ mm}$

b. From Figure 10.15, for $\varepsilon = 0.75$ and $L/D = 1$, dimensionless flow coefficient $Q' = 4.6$.

$Q' = \dfrac{Q}{r c n L}$

Hence flow, $Q = r c n L Q'$

$Q = 15 \times 0.045 \times 25 \times 30 \times 4.6 = 2,329 \text{ mm}^3/\text{s} = 2.329 \times 10^{-6} \text{ m}^3/\text{s}$

c. From Figure 10.14, for $\varepsilon = 0.75$ and $L/D = 1$, dimensionless pressure ratio

$\dfrac{p}{p_{max}} = 0.35$

Hence, maximum pressure, $p_{max} = \dfrac{p}{0.35} = \dfrac{2.2}{0.35} = 6.28 \text{ N/mm}^2$

10.10 Energy Loss due to Friction

Owing to shearing between the layers of oils, there exists some friction. The dimensionless friction coefficient f' is friction coefficient f multiplied by (r/c) ratio. This value is tabulated in Table 10.5.

Table 10.5 Values of Dimensionless Friction Coefficients $f' = \left(\dfrac{r}{c}\right)f$ For Different L/D Ratios

L/D	Eccentricity Ratio ε						
	0.1	0.2	0.4	0.6	0.8	0.9	0.97
∞	4.80	2.57	1.52	1.20	0.961	0.756	-
1	26.4	12.8	5.79	3.22	1.70	1.05	0.514
0.5	85.6	40.9	17.0	8.10	3.26	1.60	0.610
0.25	322	153	61.1	26.7	8.80	3.50	0.922

Friction coefficient is also plotted in Figure 10.17. It can be seen that for short bearings the friction coefficient is high due to higher pressures inside.

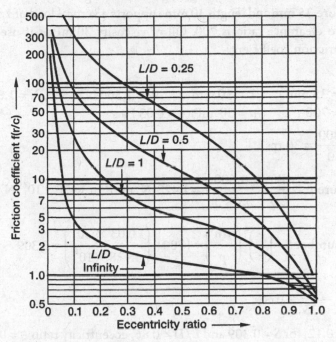

Figure 10.17 Friction coefficient variation with eccentricity ratio

If bearings are fully lubricated, coefficient of friction f can also be calculated from the empirical relation given by McKee:

$$f = \frac{0.33}{10^6} \times \left(\frac{\mu N}{p}\right)\left(\frac{r}{c}\right) + \Delta f \qquad (10.10)$$

Where, μ = Viscosity of oil in Pa.s (Note the units carefully)

$\quad N$ = Speed (rpm)

$\quad p$ = Unit load in N/m²

$\quad \Delta f$ = Correction factor for end leakage

Value of Δf is dependent on L/D ratio. Generally, the L/D ratio is near 1, Δf can be taken as 0.002.

Power lost in friction is equal to heat generated $H_g = f W v$ $\qquad (10.11)$

Where, v = Peripheral velocity in m/s

$\quad W$ = Load in N

Example 10.3

Friction factor due to oil flow through a given bearing

A bearing of bore 35 mm and length 30 mm supports a vertical load of 1,500 N at 1,800 rpm. Radius to clearance ratio is 700. Oil of viscosity 30 mPa.s is used as lubricant. Calculate the friction coefficient.

Solution

Given $D = 35$ mm $L = 30$ mm $W = 1,500$ N $N = 1,800$ rpm

    $r/c = 700$ $\mu = 30$ mPa.s $= 0.03$ Pa.s

Speed $n = \dfrac{1,800}{60} = 30$ rps

Average pressure $p = \dfrac{W}{LD} = \dfrac{1,500}{35 \times 30} = 1.428$ N/mm^2 $= 1.428 \times 10^6$ N/m^2

Sommerfeld number $S = \left(\dfrac{r}{c}\right)^2 \left(\dfrac{\mu n}{p}\right) = (700)^2 \left(\dfrac{0.03 \times 30}{1.428 \times 10^6}\right) = 0.309$

$L/D = \dfrac{30}{35} = 0.86$

From Figure 10.12, for $S = 0.309$ and $L/D = 0.86$, eccentricity ratio $\varepsilon = 0.4$.
From Figure 10.17, for $\varepsilon = 0.4$ and $L/D = 0.86$, dimensionless friction factor

$$f' = \left(\frac{r}{c}\right) f = 7$$

Hence friction coefficient $f = f'\left(\dfrac{c}{r}\right) = 7 \times \left(\dfrac{1}{700}\right) = 0.01$

Alternately:

From Equation (10.10):

$$f = \frac{0.33}{10^6} \times \left(\frac{\mu N}{p}\right)\left(\frac{r}{c}\right) + 0.002$$

or

$$f = \frac{0.33}{10^6} \times \left(\frac{0.03 \times 1,800}{1.428}\right) \times 700 + 0.002 = 0.0087 + 0.002 = 0.0107$$

10.11 Heat Generated and Temperature Rise

Owing to friction between the viscous layers of the oil, some power is consumed, which is converted to heat. This increases the temperature of oil and bearing. It can be assumed that the convection losses from the hot bearing are neglected and all the heat developed is carried by the oil circulating in the bearing.

If W is the radial load on bearing, frictional force is fW, then frictional torque T_f for a shaft of radius r is given as:

$$T_f = fW\,r \tag{a}$$

Power consumed in friction $p_f = 2\,\pi\,n \times T_f$ $\tag{b}$

This frictional power generates heat H_g which is equal to p_f

Substituting the value of T_f from Equation (a):

$$H_g = p_f = 2\,\pi\,n \times T_f = 2\,\pi\,n \times fW\,r \tag{c}$$

If p is pressure in oil, which is load per unit of projected area, then

$$W = p \times (2r\,L) = 2\,p\,r\,L \tag{d}$$

The dimensionless coefficient $f' = \left(\dfrac{r}{c}\right)f$

Or, $\qquad f = \left(\dfrac{c}{r}\right)f'$ $\tag{e}$

Substituting the values of f and W from equations (e) and (d), respectively, in Equation (c):

$$H_g = 2\pi n \left(\frac{c}{r}\right) f'(2\,prL) \times r = 4\pi\,n\,pr\,c\,L\,f' \tag{f}$$

Dimensionless oil flow coefficient: $Q' = \dfrac{Q}{rcnL}$

Or, $\quad Q = rcnLQ'$ $\tag{g}$

If Δt is temperature rise of oil due to this heat, then:

$$H_g = C_p\,m\,\Delta t = C_p\,(Q\rho)\,\Delta t \tag{h}$$

Where, C_p = Specific heat of oil (Approximately 1,700 J/kg/°C). See Table 10.6 for values.

ρ = Specific gravity of oil (Approximately 0.86)

m = Mass of oil being circulated in bearing = Volume flow × Specific gravity = $Q \times \rho$

Table 10.6 Specific Heat and Density of Engine Oils

Temperature (°C)	Specific Heat (C_p) (J/kg/°C)	Density(ρ) (kg/m³)	Temperature (°C)	Specific Heat (C_p) (J/kg/°C)	Density(ρ) (kg/m³)
0	1,796	899.1	80	2,131	852.2
20	1,880	888.2	100	2,219	840.0

| 40 | 1,964 | 876.1 | 120 | 2,307 | 829.0 |
| 60 | 2,047 | 864.0 | 140 | 2,395 | 816.9 |

Substituting the value of Q from Equation (g) in Equation (h):

$$H_g = C_p (r c n L Q' \rho) \Delta t \tag{10.12}$$

Substituting the value of H_g from Equation (f) in Equation (10.12):

$$4 \pi n p r c L f' = C_p r c n L Q' \rho \Delta t$$

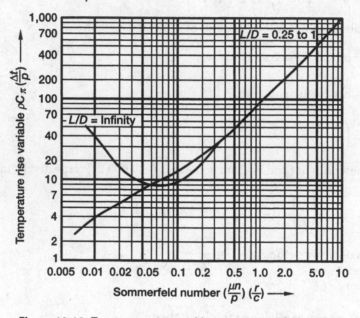

Figure 10.18 Temperature rise variable versus Sommerfeld number

Or, $\quad \Delta t = \left(\dfrac{4 \pi p}{\rho C_p} \right) \left(\dfrac{f'}{Q'} \right) \tag{10.13}$

If T_i is the inlet and T_{av} is the average oil temperature, then:

$$t_{av} = t_i + \left(\frac{\Delta t}{2} \right) \tag{10.14}$$

Temperature rise can also be found using the charts given by Raimondi for the temperature variable t_{var} in Figure 10.18 for different L/D ratios.

$$t_{var} = \rho C_p \left(\frac{\Delta t}{p} \right) \tag{10.15}$$

10.12 Heat Dissipated

Heat dissipating from a bearing depends on coefficient of heat transfer, difference in temperature, and projected area.

$$H_d = KA(t_b - t_a) \tag{10.16}$$

And, $(t_b - t_a) = \dfrac{(t_o - t_a)}{2}$

Where, $K = 140$ to 400 W/m²/°C for light construction bearings in still air

$\quad K = 500$ to $1,500$ W/m²/°C for well ventilated bearings

$\quad A = $ Projected area of bearing $(L \times D)$ (m²)

$\quad t_b = $ Bearing temperature (°C)

$\quad t_a = $ Atmospheric temperature (°C)

$\quad t_o = $ Oil outlet temperature (°C)

If heat dissipated is less than heat generated, it needs a pump for oil circulation to carry away heat by oil. Using the following equation, mass of oil can be calculated.

$$\text{Heat carried by oil } H_c = m\,C_p\,(t_o - t_a) \tag{10.17}$$

Where, $m = $ Mass of oil (kg/s)

$\quad C_p = $ Specific heat (J/kg/°C)

Example 10.4

Heat generated in a sleeve bearing of an application

Crank shaft of a four-stroke diesel engine of diameter 40 mm is supported on sleeve bearings of length 50 mm. It rotates at 1,200 rpm and the load on each bearing is 5 kN. Viscosity of oil used at operating temperature is 25 mPa.s. Radial clearance is 0.03 mm. Calculate the heat generated.

Solution

Given $\quad D = 40$ mm $\qquad L = 50$ mm $\qquad N = 1,200$ rpm $\qquad W = 5$ kN

$\qquad \mu = 25$ mPa.s $\qquad c = 0.03$ mm

Average pressure $p = \dfrac{W}{LD} = \dfrac{5,000}{50 \times 40} = 2.5$ MPa (N/mm²)

$\qquad \mu = 25$ mPa.s $= 25 \times 10^{-3}$ Pa.s

$\qquad L/D = 40/30 = 1.33$

From McKee equation (10.10): $f = \dfrac{0.33}{10^6} \times \left(\dfrac{\mu N}{p} \right) \left(\dfrac{r}{c} \right) + 0.002$

Substituting the values:

$$f = \frac{0.33}{10^6} \times \left(\frac{25 \times 10^{-3} \times 1,200}{2.5} \right) \left(\frac{15}{0.03} \right) + 0.002 = 0.00398$$

Velocity $v = \pi D n = \pi \times 0.04 \times 20 = 2.512$ m/s

Heat generated $H_g = fWv = 0.00398 \times 5,000 \times 2.512 = 50$ W

Example 10.5

Equilibrium temperature of oil for a sleeve bearing

A shaft of diameter 100 mm rotating at 1,800 rpm supports a load of 10 kN using a bearing having radial clearance of 0.08 mm and L/D ratio equal to 0.5. The oil used is SAE 40. Oil inlet temperature is 40°C. Find average oil temperature under steady state condition, neglecting heat dissipated from the bearing surface.

Solution

Given $D = 100$ mm $N = 1,800$ rpm $c = 0.08$ mm $L/D = 0.5$
                 $W = 10$ kN $t_i = 40°C$ Oil = SAE 40

Length of bearing, $L = 0.5 \times 100 = 50$ mm and radius $r = 0.5D = 50$ mm

Average pressure $= \dfrac{W}{LD} = \dfrac{10,000}{50 \times 100} = 2$ MPa $= 2 \times 10^6$ Pa

Speed $n = 1,800 / 60 = 30$ rps

Iteration 1   Assuming oil average temperature of 60°C.

From Figure 10.9, for SAE 40 oil, viscosity $\mu = 45$ mPa.s

Sommerfeld number $S = \left(\dfrac{\mu n}{p} \right) \left(\dfrac{r}{c} \right)^2 = \left(\dfrac{45 \times 10^{-3} \times 30}{2 \times 10^6} \right) \left(\dfrac{50}{0.08} \right)^2 = 0.264$

From Figure 10.18 at $S = 0.264$, $L/D = 0.5$, $t_{var} = 23°C$

From Equation (10.15) $t_{var} = \rho C_p \left(\dfrac{\Delta t}{p} \right)$ or $\Delta t = \dfrac{t_{var}\, p}{\rho C_p}$

From Table 10.6 for 60°C specific heat $C_p = 2,047$ J/kg/°C and density $\rho = 864$ kg/m³.

Substituting the values: $\Delta t = \dfrac{23 \times 2 \times 10^6}{864 \times 2,047} = 26°$ C

Average temperature $t_{av} = t_i + 0.5\Delta t = 40 + (0.5 \times 26) = 53°C$ (less than assumed).

Iteration 2   So assuming average temperature 55°C,

From Figure 10.9, viscosity of SAE 40 oil at this temperature is $\mu = 51$ mPa.s

Sommerfeld number $S = \left(\dfrac{\mu n}{p}\right)\left(\dfrac{r}{c}\right)^2 = \left(\dfrac{51 \times 10^{-3} \times 30}{2 \times 10^6}\right)\left(\dfrac{50}{0.08}\right)^2 = 0.299$

From Figure 10.18 at $S = 0.299$, $L/D = 0.5$, $t_{var} = 31°C$

From Table 10.6, for 55°C values are interpolated for specific heat $C_p = 2,026$ J/kg/°C and density $\rho = 867$ kg/m³.

$$\Delta t = \frac{35 \times 2 \times 10^6}{867 \times 2,026} = 35.7\,°C$$

Average temperature $t_{av} = t_i + 0.5\Delta t = 40 + (0.5 \times 35.7) = 57.8°C$

Iteration 3   Assuming 57°C as average temperature, the result will be very close to 57°C.

10.13 Selection of Parameters for Design of Bearings

In bearing design, the data that are known are:

- Journal diameter.
- Intensity of load (smooth or shocks) and type of load (radial or thrust) coming on bearing.
- Speed of rotation.

The designer has to select some parameters. These values depend on the application. Table 10.7 gives the range of the values for maximum pressure, L/D ratio, oil viscosity, and bearing modulus $\left(\dfrac{\mu n}{p}\right)$, which can be selected.

Table 10.7   Range of Design Parameters for Various Applications

Application	Location	MaximumPressure (MPa)	L/D Ratio	Viscosity (mPa.s)	Bearing Modulus
Automobile	Main	5.5–11.7	0.8–1.8	7–8	36.3
	Crank pin	10.4–24.4	0.7–1.4	7–8	24.2
	Wristpin	15–34	1.5–2.2	7–8	19.3
I.C. engines (four stroke)	Main	4.8–8.3	0.6–2.0	20–65	48.4
	Crank pin	4.8–12.4	0.6–1.5	20–65	24.2
	Wristpin	12.4–15.2	1.5–2.0	20–65	12.1
Reciprocating pumps compressors	Main	1.66	1.0–2.2	30	72.5
	Crank pin	4.1	1.5–2.0	80	24.2
	Wristpin	6.8	1.5–2.0	80	24.2

Generator, motor, centrifugal pump	Rotor	0.7–1.4	1.0–2.0	2.5	48.3
Steam turbine	Main	0.7–1.9	1.0–2.0	2.2	24.2
Machine tools	Main	2.1	1.0–1.4	40	97

The first group of selection parameters is:

- Lubricant viscosity μ. In the absence of specific information, it may be assumed that mineral lubricating oil with a density of about 870 kg/m^3 and a specific heat of about 2,000 J/kg/°C.
- Length/Diameter ratio (L/D).
- Load per unit projected area p.
- Rotational speed n.
- Radial clearance c.

Some parameters are under control of the designer, but there are some, which are dependent on some other parameters and define operational limits for the bearing. For example, it is assumed that the viscosity is constant throughout the film. It is not true, as the oil temperature rises, when it passes through the bearing. The viscosity reduces, as it is very much dependent on the temperature. Thus the bearing design involves some iterations.

The following parameters are dependent on the first group:

- Coefficient of friction f
- Temperature rise of oil Δt
- Oil flow rate Q
- Minimum oil film thickness h_{min}

10.13.1 Length to diameter ratio (L/D)

L/D ratios in most of the applications vary from 0.5 to 2. In old machines near to unity was used but in modern machines 0.25–0.75 are commonly used.

10.13.2 Radial clearance

It is very important parameter in the design of sleeve bearings. Higher clearance does not allow to build up hydrodynamic pressure to take up the load coming on the shaft, whereas less clearance may cause metal to metal contact, which can ruin the bearing. Generally, radial clearance is selected on the basis of surface roughness and material used for the bearing as given below:

Babbitt $\quad c = \dfrac{r}{600}$ to $\dfrac{r}{1,000}$

Copper–lead $c = \dfrac{r}{100}$ to $\dfrac{r}{1,000}$

10.13.3 Minimum oil film thickness

Minimum oil film thickness (h_{min}) depends on the surface roughness of the mating surfaces. It should allow expected size particles to pass through without causing damage. It can be assumed at least greater than five times the sum of roughness of journal and bearing. A thumb rule can be taken as: $h_{min} \geq 0.005 + 0.00004\ D$.

10.13.4 Clearance ratio

Clearance ratio $(c\,/\,r)$ for different applications is given as under for journal diameters 25 mm–150 mm:

Precision bearings $c\,/\,r = 0.001$.

Less precision machines $c\,/\,r = 0.002$.

Rough service machines $c\,/\,r = 0.004$.

10.13.5 Minimum oil thickness to clearance ratio

Selection of minimum oil film thickness to clearance ratio can be decided upon weather the bearing is to be designed for maximum load or minimum friction loss. Values can be selected from Figure 10.12.

10.13.6 Bearing pressure

Pressure p acting on the bearing surface is load per unit of projected area $(L \times D)$. Maximum allowable pressure p_{max} depends on bearing material and application. It varies from 1 to 14 N/mm² depending on application as given in Table 10.3.

10.13.7 Temperature rise

Maximum oil temperatures should not be excessive, as oxidation and degradation become rapid. Above 100°C, the rate of oxidation increases rapidly. Following guidelines can help to select maximum allowable temperature.

- For general purpose machinery, an oil-operating temperature of 60°C should give a good long life.
- In industrial equipment, temperatures should not be more than 120°C.

10.13.8 Oil viscosity

A parameter called as DN factor (multiplication of diameter in mm and speed in rpm) can be used to find out a suitable viscosity, assuming some average temperature of some SAE oil from Figure 10.9 and from Figure 10.19.

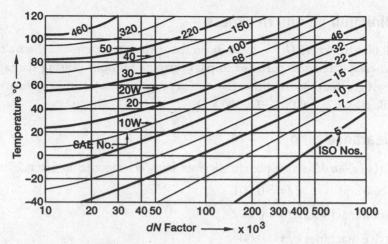

Figure 10.19 Chart for selection of viscosity

10.13.9 Bearing modulus

Bearing modulus is given as $B = \left(\dfrac{\mu n}{p}\right)$. In hydrodynamic lubrication, friction coefficient depends on this number. At start, its value is very high as shown in by point A in Figure 10.20. It decreases linearly up to point B. Performance of bearing in zone A to B is called thin film lubrication and is unstable. As the speed increases, friction coefficient value decreases nonlinearly to a minimum value C giving minimum friction as f_{min}. Hydrodynamic action starts after this critical point C. It again starts increasing but at a slow rate non-linearly up to point D and then linearly up to point E. The zone B to D offers partial lubrication. The zone C to D is called thick film lubrication and it is a stable zone. The bearings are design for this stable zone.

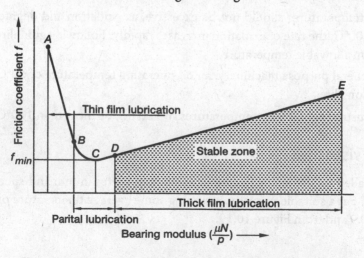

Figure 10.20 Variation of friction coefficient with bearing modulus

The minimum value of bearing modulus after which hydrodynamic lubrication starts depends on the shaft and bearing material. This value can be taken from Table 10.7. Value of bearing modulus selected should be at least three times the value at C.

Table 10.8 Minimum Value of Bearing Modulus (B_{min}) for different material combinations

Shaft Material	Bearing Material	
	Babbitt	Bronze
Hardened and ground steel	8.6	17.2
Machined soft steel	10.8	21.0

10.14 Design Procedure

Design procedure is explained by taking an example given below:

Example 10.6

Design of a sleeve bearing for given application

A pump running at 1,440 rpm has to support a load of 10 kN. Journal diameter is 130 mm. Design a suitable hydrodynamic journal bearing.

Solution

Given $N = 1,440$ rpm   or $n = 24$ rps $W = 10$ kN $= 10,000$ N $D = 130$ mm

Follow the steps given below to design a sleeve bearing:

Step 1 Select L/D ratio depending on application. If application is not known, $L/D = 1$ can be taken as a starting point. L/D ratio varies from 0.75 to 1, a value of $L/D = 0.75$ is chosen.

$L = 0.75\,D = 0.75 \times 130 = 97.5$ say 100 mm

Step 2 Calculate unit load from the given load and an appropriate nominal pressure. For pump recommended unit load from Table 10.2 is 0.5–1 MPa. Check with the selected length.

Unit load, $p = \dfrac{W}{LD} = \dfrac{10,000}{100 \times 130} = 0.77$ MPa $\left(\text{within the range}\right) = 0.77 \times 10^6$ Pa.

Step 3 Calculate minimum film thickness from: $h_{min} \geq 0.005 + 0.00004\,D$

$h_{min} = 0.005 + (0.00004 \times 130) = 0.01$ mm

Step 4 Select a suitable radial clearance, c based on fit (H8 / f 7) or (H9 / d9). (Chapter 3 Volume 1) Check r/c ratio. A value between 500 and 1,000 is average clearance ratio.

Clearance $c = \left(\dfrac{65}{1,000}\right) = 0.065$ mm (more than minimum thickness, hence safe).

Step 5 Check that radial clearance should be more than h_{min}. It is more, so OK.

Step 6 Calculate DN value $= 130 \times 1,440 = 187 \times 10^3$.

Step 7 Select an oil. Assuming average temperature of oil as 60°C, from Figure 10.20 oil for the value of DN from Figure 10.20 is SAE 20.

Step 8 Find viscosity. Using Figure 10.9 for the oil SAE 20, its viscosity at 60°C is 19 cP.

1 cP $= 10^{-3}$ Pa.s. Hence, $\mu = 19 \times 10^{-3}$ Pa.s

Because viscosity varies considerably with temperature, it is normally necessary to carry out calculations after finding average temperature in the first iteration.

Step 9 Calculate Sommerfeld number

$$S = \left(\dfrac{\mu n}{p}\right)\left(\dfrac{r}{c}\right)^2 = \left(\dfrac{19 \times 10^{-3} \times 24}{0.77 \times 10^6}\right)\left(\dfrac{65}{0.065}\right)^2 = 0.59$$

Step 10 Find eccentricity ratio from Figure 10.12 for the calculated values of S and L/D. For the value of $S = 0.59$ and $L/D = 0.75$ eccentricity ratio $\varepsilon = 0.35$.

Step 11 Calculate friction coefficient from the value of f' from Figure 10.17.

For $L/D = 0.75$, $f' = 13$, hence $f = f'\left(\dfrac{c}{r}\right) = 13 \times \left(\dfrac{0.065}{65}\right) = 0.013$

Step 12 Calculate circumferential velocity: $v = \pi\, d\, n = \pi \times 0.13 \times 24 = 9.79$ m/s

Step 13 Calculate flow coefficient Q' for the eccentricity ratio and L/D ratio from Figure 10.15 and side leakage coefficient Q'_s (Figure 10.16) or from Table 10.4.

After interpolations for $\varepsilon = 0.35$ and $L/D = 0.75$: $Q' = 4$ and $\left(\dfrac{Q_s}{Q}\right) = 0.47$.

Step 14 Find temperature rise variable t_{var} from Figure 10.18. For $S = 0.59$, values is 50.

Step 15 Calculate temperature rise Δt. Take value of C_p and ρ from Table 10.6.

$C_p = 2,047$ J/kg/°C for 60°C and $\rho = 864$ kg/m³

$$\Delta t = \dfrac{p\, t_{var}}{\rho C_p} = \dfrac{0.77 \times 10^6 \times 50}{864 \times 2,047} = 22°\,C$$

Step 16 Calculate heat generated from Equation (10.12), $H_g = C_p (r\, c\, n\, L\, Q'\, \rho)\, \Delta t$

$H_g = 2,047 \times (0.065 \times 0.065 \times 10^{-3} \times 24 \times 0.1 \times 4 \times 864) \times 22 = 1,578$ W.

Step 17 Calculate average temperature t_{av}, assuming inlet temperature t_i.

Assuming $t_i = 45°C$, $t_{av} = t_i + 0.5\ \Delta t = 45 + (0.5 \times 22) = 56°C$.

Step 18 Check average temperature with the assumed temperature. If different, then iterate for the new assumed oil temperature.

Temperature 56 °C is near to the assumed oil temperature, that is, 60 °C; hence OK.

Step 19 Calculate heat dissipated from Equation (10.16). Assuming atmospheric temperature t_a calculate oil outlet temperature t_o. Assume a suitable value of cooling constant from Section 10.12. Since heat generated is high, so selecting $K = 1,500$. Assuming t_a as 30 °C.

Heat dissipated,

$$H_d = K(L \times D)\frac{(t_o - t_a)}{2} = 1,500 \times (0.1 \times 0.13) \times \frac{60 - 30}{2} = 292 \text{ W}$$

Heat dissipated H_d is less than heat generated H_g. It needs a pump for oil circulation to carry away heat H_c by oil and cool externally. $H_c = H_g - H_d = 1,578 - 292 = 1,286$ W.

Step 20 Calculate heat carried by oil pump from Equation (10.17):

$$H_c = mC_p(t_o - t_a) \quad \text{Substituting the values:}$$

$1,286 = m \times 2,047 \times (60 - 30)$ or $m = 0.021$ kg/s

Hence the results of the bearing design are:

Bearing diameter $D = d + 2c = 130.13$ mm

Length of bearing $L = 100$ mm

Oil circulation pump capacity $Q = 0.021$ kg/s.

Example 10.7

Load capacity and power loss in friction for a given bearing
A shaft of diameter 60 mm rotating at 1,200 rpm is supported by a bearing having inside diameter 60.05 mm and L/D ratio equal to 1.2. The oil used is SAE 40 having absolute viscosity of 13 mPa.s at 80°C. Assume heat transfer coefficient 500 W/m²°C, room temperature 24°C and operating temperature 120°C Calculate the following:

a. Load, which the bearing can take safely.

b. Power loss due to friction.

Solution

Given $D = 60$ mm $= 0.06$ m $L/D = 1.2$ $N = 1,200$ rpm
$c = 60.05 - 60.0 = 0.05$ mm $\mu = 13$ mPa.s $= 13 \times 10^{-3}$ Pa.s
$K = 500$ W/m²/°C $t_o = 120$°C $t_a = 24$°C

a. Load capacity:

Speed: $n = \dfrac{1,200}{60} = 20$ rps

Average pressure

$$p = \frac{W}{LD} = \frac{W}{0.06 \times 1.2 \times 0.06} = 231.5\,W \; \left(\text{N/m}^2\right) = 231.5 \times 10^{-6}\,W \; \left(\text{N/mm}^2\right)$$

As per McKee Equation (10.10): $f = \dfrac{0.33}{10^6} \times \left(\dfrac{\mu N}{p}\right)\left(\dfrac{r}{c}\right) + 0.002$

Substituting the values: $f = \dfrac{0.33}{10^6} \times \left(\dfrac{13 \times 10^{-3} \times 1,200}{231.5 \times 10^{-6}\,W}\right)\left(\dfrac{30}{0.05}\right) + 0.002$

Or, $f = \dfrac{13.34}{W} + 0.002$ (a)

Peripheral velocity $v = \pi D n = 3.14 \times 0.06 \times 20 = 3.768$ m/s

Heat generated from Equation (10.11):

$$H_g = fWv = \left(\frac{13.34}{W} + 0.002\right) \times W \times 3.768$$

Bearing temperature: $t_b = \dfrac{t_o + t_a}{2} = \dfrac{120 + 24}{2} = 72\,°\text{C}$

From Equation (10.16) heat dissipated : $H_d = KLD(t_b - t_a)$

Substituting the values: $H_d = 500 \times 0.06 \times (72 - 24) = 103.5$ Watts

Because, $H_g = H_d$ or $\left(\dfrac{13.34}{W} + 0.002\right) \times W \times 3.768 = 103.5$

Or, $W = 7,064 \; N = 7.064$ kN

b. Power lost in friction

Friction coefficient as calculated above in Equation (a):

$$f = \frac{13.34}{W} + 0.002 \quad \text{Or,} \quad f = \frac{13.34}{7,064} + 0.002 = 0.0039$$

Power lost in friction P_f is equal to heat generated $H_g = fWv$

$$P_f = 0.0039 \times 7,064 \times 3.768 = 103.8 \; \text{W}$$

10.15 Thrust Bearings

Thrust bearings are used to take axial load acting along the axis of the shaft. These are of the following types:

a. Thick film lubrication
- Foot-step bearing
- Collar bearing: It could be single collar or multicollar

b. Hydrodynamic bearing
- Fixed pad
- Tilting pad

c. Hydrostatic bearing Uses externally pressurized fluid. These could be:
- Step bearing
- Conical bearing

10.15.1 Foot step bearing

Foot-step bearing is used for vertical shafts. Bottom end of the shaft can have a step, which rests over a bronze disc [Figure 10.21(a)]. If the shaft is not of steel, a steel disk is placed over the bronze disc [Figure 10.21(b)]. It is lubricated using an oil cup forming an annular area around the shaft. Pads used as disks are shown in picture 10.21(c).

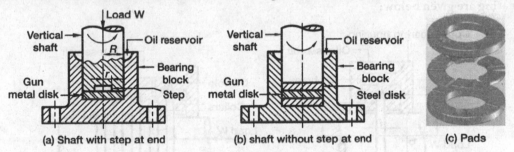

(a) Shaft with step at end   (b) shaft without step at end   (c) Pads

Figure 10.21 Foot-step bearing

Coefficient of friction varies from 0.015 to 0.02 for vertical and collar bearings. It can be calculated from the following relation:

$$f = 84 \sqrt{v} \, p^{-2/3} \tag{10.18}$$

Where, v = Average peripheral velocity (m/s)

p = Pressure per unit area of bearing (N/m²)

It is assumed that uniform pressure p exists over the circular area of the disk. If W is the axial load coming on the shaft, then the pressure is given as:

$$p = \frac{W}{\pi R^2}$$

Where, R = Radius of pad.

The allowable maximum pressure depends on peripheral velocity v.

For velocities v = 0.25 m/s – 1 m/s, $pv \leq 0.7$ where p is pressure in MPa

For velocities >1 m/s, pressure is limited to 0.7 MPa.

For intermittent service, p can be up to 10 MPa

For slow-speed operation, p can be up to 14 MPa

$$T_f = \frac{2}{3}\mu W R \tag{10.19}$$

The frictional torque

Wear of the disk is proportional to peripheral velocity; hence maximum wear is at the outermost edge of the disk. Sometimes, the disk is made as annular ring having outer radius R and inner radius r. In that case, the frictional torque is given by:

$$T_f = \frac{2}{3}fW\left(\frac{R^3 - r^3}{R^2 - r^2}\right) \tag{10.20}$$

10.15.2 Collar bearing

Collar bearing is suitable for horizontal shaft. The number of collar may be one [Figure 10.22(a)] or more than one for heavy axial loads [Figure 10.22(b)]. The axial force over the collars may be assumed uniformly distributed in single or multi-collar bearings. Proportions of the bearing are given below:

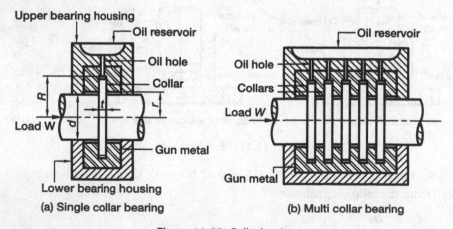

Figure 10.22 Collar bearings

Collar outside diameter d_2 is kept between 1.3 d and 1.8 d,

where, d = Shaft diameter

Thickness of collar $t = d / 6$.

Clearance between collars $c = d / 3$

Outer radius of collar $R = d_2 / 2$

Inside radius of collar $r = d / 2$

Number of collars $= n$

Pressure p on the collar is given by relation:

$$p = \frac{W}{\pi n (R^2 - r^2)} \tag{10.21}$$

Frictional torque as per Equation (10.20) is:

$$T_f = \frac{2}{3} f W \left(\frac{R^3 - r^3}{R^2 - r^2} \right)$$

Power lost in friction $P_f = 2 \pi n \times T_f W$ $\tag{10.22}$

Mean diameter d_m of the collar can be calculated as $d_m = \dfrac{d_2 + d}{2}$

Wear of collar bearing depends on $(p \times v)$ value and it should not be more than 60×10^4. Velocity v generally ranges from 0.25 m/s to 1 m/s.

Coefficient of friction $f = 84 \sqrt{v} p^{-2/3}$

Where, v = Velocity (m/s)

$\quad\quad\quad p$ = Fluid pressure (N/m²)

10.15.3 Hydrodynamic thrust bearing

Hydrodynamic action can be achieved in thrust bearings also. The annular surface right angle to the axis is segmented and each pad is slightly inclined to collar as shown in Figure 10.23. Owing to taper film between the pad and collar, wedge film is formed, which supports axial loads. The pad can be fixed or made tilted. These are used in high-speed turbine, generator, compressor, and gear drive applications.

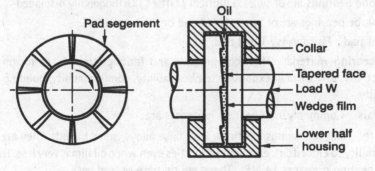

Figure 10.23 Hydrodynamic thrust bearing

10.16 Squeeze Film Journal Bearings

Flow of lubricant in a converging passage is called wedge film (bearings described in the sections above). The bearings, in which resistance to a viscous lubricant in squeezing out from the clearance between bearing and journal due to load is called squeeze film bearing [Figure 10.2(b)].

Under conditions of slow speed or oscillations, the wedge film is not formed to provide sufficient pressure to support load. If the load is uniform or varying in one direction only, the oil film thickness may possibly be zero, but if the load reverses in direction, squeeze film will develop to bear the dynamic loads, without any metal to metal contact. Such bearings are known as squeeze film bearings.

Summary

Journal bearings   Portion of the shaft inside the bearing is called a journal. Bearings permit low-friction movement between the bearing and shaft surfaces. Supports are used to provide support to the bearings, which carry the load.

Construction   These bearings are also called as bush or sleeve bearings. The bearing sleeve is also called bearing shell. Small bearings are in one piece as completely circular, whereas larger bearings are made in two halves each of semicircular shape.

Small bearings use sintered metal that contains a lubricant in its pores and does not need lubrication frequently. A heavily loaded bearing is provided with some arrangement to introduce oil under pressure using a pump. Such bearings are called hydrostatic bearings.

Classification of bearings   Bearings can be classified according to:

A.   Lubricant pressure: Hydrodynamic / Squeeze film / Hydrostatic

B.   Type of load: Radial / Axial / Combined radial and axial

C.   Thickness of oil film: Thick film / Thin film / Zero film

D.   Relative movement between bearing and shaft: Rotating / Sliding / Oscillating

E.   Contact angle: Full 360° bearing / Split bearing / Partial bearing

F.   Number of pads: Two-lobe bearing / Three-lobe bearing / Four-lobe bearing

    Two-lobe bearings are of types: Elliptical / Offset / Orthogonally displaced

    Three-lobe bearings are of types: Fixed pad or tilting pad.

G.   Type of pads: Fixed pads / Tilted pad.

Properties of bearing material   High compressive and fatigue strength, corrosion resistant, high thermal conductivity, low thermal expansion, embeddability, comformability, bonding characteristics, and machinability.

Bearing materials   Commonly used bearing materials are:

A.   *Babbitts*   Sleeve bearings use tin or lead base alloys called babbitt. They are popular due to their ability to embed dirt, excellent properties even when oil film is very less. These can be used up to bearing pressures 14 MPa. These are tin base or lead base.

B.   *Bronzes*   Bronze is an alloy of copper, zinc, and tin. Many copper alloys such as aluminum bronze, tin bronze, phosphorous bronze, and lead bronze are used. These are generally used as a bush material pressed into a steel shell. Most common bronzes are:

Gun metal It has copper 88%, tin 10%, and zinc 2%. It is suitable for high journal speed and bearing pressures up to 10 MPa.

Phosphorous bronze It has copper 80%, tin 10%, lead 9%, and phosphorous 1%. It is also suitable for high speeds and high bearing pressures up to 14 MPa.

C. *Aluminum alloys* Alloys 750, A750, and B750 can be cast in sand or permanent molds, but not as die castings. Alloy X385 is preferred for die cast bearings. They offer high load–bearing capacity, thermal conductivity, fatigue, and wear resistance.

D. *Cast iron* Coefficient of friction is relatively low. The cast iron glazes over; hence wear is negligible.

E. *Plastics* Solid plastic plain bearings are now increasingly popular due to dry-running and lubrication-free behavior. Some materials such as Teflon and nylon offer excellent low-friction property. Common plastics are nylon, and PTFE (Poly-tetra-fluoro-ethylene).

Properties of lubricants Boiling point should be high / Freezing point should be low / Viscosity should not vary much with change in temperature / Thermal stability / Corrosion resistant / High resistance to oxidation.

Viscosity S.I. unit for dynamic viscosity μ is pascal-second (Pa.s) = N·s/m² = kg/m·s

CGS unit for dynamic viscosity is Poise (P). 1 centiPoise = 0.001 Pa.s = 1 mPa.s.

SAE designation of oils: Their viscosities from low to high: 0W, 5W, 10W, 15W, 20W, 25W, 20, 30, 40, 50, or 60. The letter W, meaning 'winter' or cold-start viscosity, which are suitable for lower temperatures.

Viscosity index in short called as VI is a measure for the change of kinematic viscosity with temperature. Viscosity decreases with increase in temperature.

Terminology: Minimum film thickness h_{min} = (Radial clearance – Eccentricity) = $(c - e)$

Attitude angle Φ = Angle made by line of centers of bearing and journal with load line.

Sommerfeld Number $S = \left(\dfrac{r}{c}\right)^2 \left(\dfrac{\mu n}{p}\right)$

Where, viscosity μ is in Pa.s, speed n in rps, pressure p in Pa i.e. N/m², journal radius r and radial clearance c in mm Bearing characteristic number $B = \left(\dfrac{\mu n}{p}\right)$

Working Principle of hydrodynamic bearings

When a shaft starts rotating, it starts climbing the bearing surface in the direction of rotation. Oil also starts rotating at half of peripheral speed. Oil entering wedge type clearance lifts it up due to pressure caused by rotating oil.

Reynolds Equation: $\dfrac{\partial}{\partial x}\left[h^3 + \dfrac{\partial p}{\partial x}\right] + \dfrac{\partial}{\partial z}\left[h^3 + \dfrac{\partial p}{\partial z}\right] = 6\mu U\left(\dfrac{\partial h}{\partial x}\right)$

Where, z = Length along axis of bearing

h = Radial film thickness at any radial location

x = Length along circumference

p = Oil pressure in radial direction and

U = Peripheral velocity of journal.

Exact solution of this equation is not available and solved numerically for finite bearing. Approximate solutions are long bearing or short bearings.

Long bearings Assumed that L/D ratio >1. Pressure variation in Z direction can be neglected and hence Reynolds equation is:

$$\frac{\partial}{\partial x}\left[h^3 + \frac{\partial p}{\partial x}\right] = 6\mu U\left(\frac{\partial h}{\partial x}\right)$$

Its solution for pressure is: $P = \dfrac{\mu U R}{c^2}\left[\dfrac{6\varepsilon\,(\sin\theta)\,(2 + \varepsilon\cos\theta)}{(2 + \varepsilon^2)(1 + \varepsilon\cos\theta)^2}\right]$

Where, θ = Angle measured from load line.

Bearings with $L/D = 1$ is square bearing.

Short bearings Assumed that $L/D < 1$ Equation reduces to: $\dfrac{\partial}{\partial z}\left[h^3 + \dfrac{\partial p}{\partial z}\right] = 6\mu U\left(\dfrac{\partial h}{\partial x}\right)$

Ocvirk and DuBois gave pressure $p = \dfrac{\mu R}{r c^2}\left[\dfrac{L^2}{4} - z^2\right]\dfrac{3\varepsilon\sin\theta}{(1 + \varepsilon\cos\theta)^2}$

Finite bearings Sommerfeld number for different eccentricity ratios is tabulated in Table 10.1 and in Figure 10.12. Variation of attitude angle with different eccentricity is also plotted in Figure 10.13.

Critical pressure p_{cr} It is the minimum pressure, at which metal to metal contact starts.

$$p_{cr} = \frac{\mu n}{80 \times 10^3}\left(\frac{r}{c}\right)^2\left(\frac{L}{D + L}\right)$$

Unit load p Load per unit projected area ($L \times D$) is called unit load. Table 10.2 can be used as a guideline to select a unit load $\left(\dfrac{p_{max}}{L \times D}\right)$, where p_{max} is the maximum pressure.

Maximum pressure The oil pressure created in the bearing can be calculated from the ratio (p / p_{max}) given in Table 10.3. Variations of pressure ratio are shown in Figure 10.14.

Oil flow Oil flow $Q = r c n L Q'$ Where, Q' = Dimensionless oil flow coefficient.

Its value at different eccentricity ratio is given in Table 10.4 and plotted in Figure 10.15.

The dimensionless side leakage parameter (Q / Q_s) is also tabulated in Table 10.4.

Energy loss due to friction If f is friction coefficient then dimensionless friction coefficient $f' = f(r / c)$. This value is tabulated in Table 10.5 and plotted in Figure 10.17.

Coefficient of friction f can also be calculated from: $f = \dfrac{0.33}{10^6} \times \left(\dfrac{\mu N}{p}\right)\left(\dfrac{r}{c}\right) + 0.002$

Power lost in friction = Heat generated $H_g = f W v$ Where, W = Load and v = Velocity.

Heat generated Owing to friction, some power is converted to heat, which increases the temperature of oil and bearing. Frictional torque $T_f = f W r$

This frictional power generates heat $H_g = P_f = 2\pi n \times T_f = 2\pi n \times fWr$

The friction coefficient $f = \left(\dfrac{c}{r}\right) f'$; hence, $H_g = 2\pi n \left(\dfrac{c}{r}\right) f' (2prL) = 4\pi n prcLf'$

Temperature rise   If Δt is temperature rise of oil due to this heat, then:

Heat generated $H_g = C_p (rcnLQ'\rho)\Delta t$

Where, C_p = Specific heat of oil

ρ = Density of oil

$$4\pi n\,prcLf' = C_p\,rcnLQ'\rho\,\Delta t \quad\text{or}\quad \Delta t = \left(\dfrac{4\pi p}{\rho C_p}\right)\left(\dfrac{f'}{Q}\right)$$

If t_i is the inlet and t_{av} is the average oil temperatures, then: $t_{av} = t_i + \left(\dfrac{\Delta t}{2}\right)$

Raimondi gave charts for temperature variable $t_{var} = \rho C_p \left(\dfrac{\Delta t}{p}\right)$ in Figure 10.14.

Heat dissipated   Heat dissipated $H_d = KA(t_b - t_a)$ and $(t_b - t_a) = \dfrac{(t_o - t_a)}{2}$

Where, K = 140 W/m²/°C–400 W/m²/°C in still air and 500–1,500 for well ventilated bearings

A = Projected area of bearing ($L \times D$)

t_b = Bearing temperature (°C)

t_a = Atmospheric temperature (°C) and

t_o = Oil outlet temperature (°C)

If $H_d < H_g$ then it needs a pump for oil circulation to carry away heat by oil.

Mass of oil can be calculated from heat carried by oil $H_c = m C_p (t_o - t_a)$

Where, m = Mass of oil (kg/s) and

C_p = Specific heat (J/kg/°C)

Selection of parameters for design of bearings   The data that are known are: journal diameter, intensity and type of load on bearing, and speed of rotation. The designer has to select some parameters depending on the application. Table 10.7 gives range of maximum pressure, L/D ratio, viscosity, and bearing modulus for different applications.

a.   *L/D ratio*   It varies from 0.5 to 2. In new machines, 0.25–0.75 are commonly used.

b.   *Radial clearance (c)*   Generally, radial clearance is selected on the basis of surface roughness and material used for the bearing

Babbitt c = 0.017 r to 0.001 r

Copper–lead c = 0.01 r to 0.001 r

c.   *Minimum oil film thickness* (h_{min}) It can be assumed at least five times greater than the sum of roughness of journal and bearing.

d.   *Clearance ratio* (c/r) For journal diameters 25 mm–150 mm, it is as under:

Precision bearings c/r = 0.001, Less precision machines = 0.002, and Rough service = 0.004.

e. *Minimum oil thickness to clearance ratio*   The bearing is to be designed either for maximum load or minimum friction loss. Values can be selected from Figure 10.5.

f. *Bearing pressure* (p) Maximum allowable pressure depends on bearing material. It varies from 1 to 14 N/mm^2 depending on application as given in Table 10.3.

g. *Temperature rise*   For general purpose machinery, an oil operating temperature of 60°C gives a good long life. In industrial equipment, temperatures < 120°C.

h. *Oil viscosity*   DN factor (Diameter in mm × speed in rpm) can be used to find out a suitable viscosity assuming some average temperature from Figure 10.9 and SAE No. from Figure 10.19.

Design procedure

1. Select L/D ratio depending on application.
2. Calculate unit load from the given load and an appropriate nominal pressure. Check with the selected length.
3. Calculate minimum film thickness from $h_{min} \geq 0.005 + 0.00004\,D$
4. Select a suitable radial clearance, c based on fit (H8 / f7) or (H9 / d9). Check r/c ratio. A value between 500 and 1,000 is average clearance ratio.
5. Check radial clearance. It should be more than h_{min}.
6. Calculate $D\,N$ value (Shaft diameter D in mm and N speed in rpm).
7. Select an oil for assumed average temperature of oil, from Figure 10.20.
8. Calculate Sommerfeld number S using Equation (10.4).
9. Find eccentricity ratio from Figure 10.12 for the calculated values of S and L/D.
10. Calculate friction coefficient from the value of f' from Figure 10.17.
11. Calculate circumferential velocity; $v = \pi\,d\,n$.
12. Calculate flow coefficient for the eccentricity ratio and L/D ratio from Figure (10.15) and side leakage coefficient (10.16) or from Table 10.4.
13. Find temperature rise variable t_{var} from Figure 10.18.
14. Calculate temperature rise Δt. Take value of C_p and ρ from Table 10.6.
15. Calculate heat generated from Equation (10.12) $H_g = C_p (r\,c\,n\,L\,Q'\rho)\,\Delta t$
16. Calculate average temperature t_{av}, assuming inlet temperature t_r.
17. Check average temperature with the assumed temperature. If different, then iterate for the new assumed oil temperature.
18. Calculate heat dissipated from Equation (10.16); $H_d = 0.5K(L \times D)(t_o - t_a)$.
19. Calculate heat carried by oil pump from Equation (10.17); $H_c = m\,C_p(t_o - t_a)$.

Thrust bearings   These bearings are used to take axial load and are of following types:

a. *Thick film lubrication*   Foot-step bearing or collar bearing: Single collar or multi-collar.

b. *Hydrodynamic bearing*   Fixed pad or tilting pad

c. *Hydrostatic bearing*   Uses externally pressurized fluid. These could be: Step bearing or conical bearing.

Foot-step bearing   It is used for vertical shafts. Coefficient of friction $f = 84\sqrt{v}p^{-2/3}$. It varies from 0.015 to 0.02. It is assumed that uniform pressure p exists over the circular disk of radius R. Pressure is given as: $p = \dfrac{W}{\pi R^2}$.; where W is the axial load.

For velocities 0.25 m/s–1 m/s, $pv \le 0.7$; where p = pressure and v = velocity (m/s).

For velocities >1 m/s, pressure < 0.7 MPa; for intermittent service, up to 10 MPa; and for slow-speed operation up to 14 MPa. Frictional torque $T_f = \dfrac{2}{3}\mu W R$

Maximum wear is at the outermost edge of the disk. For disk as annular ring having outer radius R and inner radius r, Frictional torque $T_f = \dfrac{2}{3}fW\left(\dfrac{R^3 - r^3}{R^2 - r^2}\right)$

Collar bearing   It is suitable for horizontal shaft. The number of collar n may be one or more than one for heavy axial loads. Proportions of the bearing are:

Collar outside diameter d_2 is kept between 1.3d and 1.8d; where d = Shaft diameter.

Thickness of collar $t = d/6$ and clearance between collars $c = d/3$.

Pressure on the collar $p = \dfrac{W}{\pi n (R^2 - r^2)}$

Mean diameter of the collar $d_m = \dfrac{d_2 + d}{2}$

Wear of collar bearing depends on ($p \times v$) value and it should not be > 60 × 10⁴. Velocity generally ranges from 0.25 m/s to 1 m/s. Coefficient of friction $f = 84\sqrt{v}p^{-2/3}$.

Frictional torque $T_f = \dfrac{fW d_m}{2}$

Power lost in friction $P_f = 2\pi n \times T_f$

Hydrodynamic thrust bearing

Hydrodynamic action can be achieved in thrust bearings also. The annular surface right angle to the axis is segmented and each pad is slightly inclined to collar. Owing to taper film between the pad and collar, wedge film is formed, which supports axial loads. The pad can be fixed or made tilted. These are used in high-speed turbine, generator, compressor, and gear drive applications.

Squeeze film journal bearings   The bearings in which resistance to a viscous lubricant in squeezing out from the clearance between bearing and journal due to load is called squeeze film bearing. Under conditions of slow speed or oscillations, if the load is uniform or varying in one direction only, the oil film thickness may possibly be zero, but if the load reverses in direction, squeeze film will develop to bear the dynamic loads.

Theory Questions

1. Describe the construction of sleeve bearings with the help of a neat sketch.
2. How do you classify the bearings?

3. Write at least 10 properties of the bearing materials, which they should possess.

4. Write a note on the bearing materials, which are used for sleeve bearings.

5. What is meant by hydrodynamic lubrication? How is the oil pressure created in these bearings?

6. Write Reynolds equation and explain what approximations are done to solve it.

7. Describe long and short bearing theories and show pressure distribution.

8. Explain the terms Sommerfeld number, eccentricity ratio, and attitude angle.

9. Differentiate between critical pressure and unit load.

10. Describe the method to calculate oil flow through sleeve bearings. What is meant by side leakage and how is it calculated?

11. How do you calculate the energy loss due to friction?

12. Write the expression for the heat generated. What are the parameters affecting it?

13. How is the heat dissipated from the bearings and how do you calculate the equilibrium temperature?

14. Write a note on the selection of parameters for a hydrodynamic bearing design.

15. On what parameters, minimum oil film thickness depends?

16. Write the design procedure to design a sleeve bearing.

17. What is a thrust bearing and what are the different types of thrust bearings?

18. Describe the construction of a foot-step bearing with the help of a neat sketch.

19. When do you prefer to use multi-collar thrust bearing? Describe its construction with a neat sketch.

20. Describe a squeeze film bearing and explain its working.

Multiple Choice Questions

1. In a hydrodynamic bearing, pressure is developed:
   (a) Due to viscosity of oil
   (b) Due to rotation at any speed
   (c) Because wedge film created between journal and bearing
   (d) It depends on arc of contact

2. In a hydrostatic bearing, pressure is due to:
   (a) Its special geometry   (b) Pockets in bearing surface
   (c) Radial movement of journal   (d) High pressure fluid supplied by pump.

3. Babbitt material for bearings is composed of:
   (a) Tin, lead, copper, and antimony   (b) Aluminum, tin, and lead
   (c) Brass and zinc   (d) Copper, tin, and phosphorous

4. Teflon as a bearing material is used because it:

   (a) Is easily available   (b) Costs less

   (c) Has low coefficient of friction   (d) Is good conductor of heat

5. Sommerfeld number consists of parameters:

   (a) Viscosity, speed, load, and clearance

   (b) Viscosity, speed, journal radius, oil pressure, and clearance

   (c) Viscosity, speed, bearing length, and oil temperature

   (d) Journal radius, oil pressure, surface roughness, and clearance

6. Reynolds equation is applicable for:

   (a) Long bearings   (b) Short bearings

   (c) Any L/D ratio   (d) Bearings using high viscous oil

7. With increase of load, attitude angle:

   (a) Increases   (b) Decreases

   (c) Remains constant   (d) First increases and then decreases

8. If μ is absolute viscosity, N speed, and p bearing pressure, then bearing characteristic number is:

   (a) $\dfrac{\mu p}{N}$ (b) $\dfrac{\mu N}{p}$ (c) $\dfrac{\mu}{pN}$ (d) $\dfrac{pN}{\mu}$

9. Dimensionless friction coefficient depends on:

   (a) Eccentricity ratio   (b) L/D ratio

   (c) Both on eccentricity ratio and L/D (d) Any parameter

10. A hydrodynamic thrust bearing has two collars placed side by side and:

    (a) Both are plain collars   (b) One plain and other with radial slots

    (c) Both with radial slots   (d) One plain and other with tapered annular pads

11. The type of bearing used in railway axles is:

    (a) Full journal bearing   (b) Partial bearing

    (c) Hydrostatic bearing   (d) Fitted bearing

12. In a hydrodynamic bearing, the journal center under rotating condition is:

    (a) Just below the bearing center

    (b) Shifts from bearing center, depending on load

    (c) Shifts from bearing center, depending on speed

    (d) Shifts from bearing center, depending on load and speed

13. Attitude angle in a hydrodynamic bearing is the angle:

    (a) Of load line with vertical

    (b) Load line and the line joining bearing and journal centers

    (c) By which the journal climbs up in a bearing

    (d) None of the above

14. Bearing characteristic number contains the terms of:
    (a) Load, width, and length   (b) Speed, load, and clearance
    (c) Load, speed, and viscosity   (d) Speed, pressure, and viscosity

Answers to multiple choice questions

1. (c)   2. (d)   3. (a)   4. (c)   5. (b)   6. (c)   7. (d)   8. (b)   9. (c)   10. (d)
11. (b)   12. (d)   13. (b)   14. (d)

Design Problems

1. A 360° hydrodynamic bearing of length 50 mm has a radial load of 3,200 N and r / c ratio 1,000. Journal diameter is 50 mm, which rotates at 1,480 rpm. Oil used has viscosity of 25 mPa.s (cP). Calculate the following:
    (a) Coefficient of friction.   (b) Power loss in friction.
    (c) Minimum oil film thickness.   (d) Oil flow.
    (e) Temperature rise assuming that total heat generated is taken by oil in bearing.

    Following data may be used for solving the problem:

L/D	ε	h_{min}/c	S	Φ	$f(r/c)$	$Q/rcnL$	Q_s/Q	p/p_{axn}
1	0.4	0.6	0.264	63.1	5.79	3.99	0.497	0.484
	0.6	0.4	0.121	50.58	3.22	4.33	0.680	0.415

    $[f = 0.00644, P_f = 80.4 \text{ W}, Q = 6{,}720 \text{ mm}^3/\text{s}, h_{min} = 0.02 \text{ mm}, \Delta t = 7.8\,°C]$

2. Design a journal bearing for a centrifugal pump from the following data:

    Load on the journal = 20,000 N   Speed of journal = 900 rpm

    Type of oil SAE 10, for which the absolute viscosity at 55°C is 0.017 kg/m·s

    Ambient temperature of oil = 15.5°C

    Maximum bearing pressure for the pump = 1.5 N/mm²

    Calculate also the mass of lubricating oil required for artificial cooling, if rise of oil temperature is limited to 10°C. Heat dissipation coefficient = 1,932 W/m²/°C.   [L = 160 mm, m = 0.29 kg/min]

3. A sleeve bearing of length 100 mm and diameter 60 mm supports a journal with a load of 2,500 N rotating at 600 rpm. The diametric clearance is 0.06 mm. Energy dissipated coefficient based on projected area = 210W/m²/°C. Find the viscosity of oil so that bearing surface temperature does not exceed 60°C assuming room temperature is 20°C.   [18.67 mPa.s]

4. A 150 mm diameter shaft supporting a load of 10 kN has a speed of 1,500 rpm. The shaft runs in a bearing, where the length is 1.5 times the shaft diameter. If the diametrical clearance of the bearing is 0.15 mm and the absolute viscosity of the oil at the operating temperature is 0.11 kg/m·s, find the power wasted in friction. $[Q_g = 2.35 \text{ kW}]$

5. Design a journal bearing for supporting a generator shaft of 75 mm diameter with a load of 12 kN running at 1,440 rpm. Suitable data may be picked up from the design data book. [Q_g = 678 w]

6. A full journal bearing of 60 mm diameter and 110 mm length has a bearing pressure of 1.6 N/mm². The speed of the journal is 1,000 rpm and the ratio of journal diameter to diametrical clearance is 1,000. The bearing is lubricated with oil whose absolute viscosity at the operating temperature of 85°C may be taken as 0.012 kg/m·s. The room temperature is 40°C. Find:

   (a) The amount of artificial cooling required and

   (b) The mass of lubricating oil required per minute to circulate and difference between the outlet and inlet temperature of oil is 20°C. Take specific heat of the oil as 950 J/kg/°C.

   [Artificial cooling = 102 J/s, Oil mass m = 3.22 kg/min]

7. A hydrodynamic sleeve bearing is lubricated by SAE 20 oil having viscosity 0.01 kg/m·s. Oil temperature is 80°C and room temperature 25°C. A load of 4 kN acts on its shaft rotating at 960 rpm. The bearing modulus can be taken 10, value of L/D = 1.2, heat dissipation coefficient K = 400 W/m²/°C, and clearance ratio 0.001. Calculate:

   (a) Oil pressure   (b) Heat generated

   (c) Heat dissipated   [(a) p = 0.96 MPa; (b) H_g = 64 W; (c) H_d = 47.5 W]

8. Journal of a pump is of 100 mm diameter with a load of 20 kN rotating at 1,000 rpm. Design its bearing with length to diameter ration 1.5 and radius to radial clearance 1,000. Oil outlet temperature of SAE 10 oil is 65°C (μ = 0.0125 Pa.s). Find:

   (a) Bearing modulus   (b) Friction coefficient

   (c) Is natural cooling permissible?

   [(a) B = 9.4; (b) f = 0.05; (c) H_g = 523 W, H_d = 160 W, hence forced cooling required]

9. A sleeve bearing of length 80 mm supports a journal of 60 mm diameter runs at 1,000 rpm. Bearing temperature should not be more than 80°C. Diametric clearance ratio is 0.001. Calculate:

   (a) Permissible load, if bearing operates in still air at 30°C.

   (b) Power loss in friction, if oil viscosity is 0.013 kg/m·s and K = 400 W/m²/°C.

   [(a) W = 4431 N; (b) P_f = 96 W]

10. A centrifugal pump of rotor diameter 80 mm rotates at 1,000 rpm. Load on each bearing is 5 kN and atmospheric temperature is 25°C. Design the bearing.

   [L = 80 mm, D = 80.008 mm, Artificial cooling with SAE 50 oil]

11. Design main bearing for maximum load of a diesel engine assuming unit pressure is not to exceed 2 MPa for a load of 7,200 N for a maximum speed of crank shaft as 3,000 rpm for a journal diameter of 60 mm.

   [L = 60 mm, c = 0.1 mm, ε = 0.63, Oil = SAE 20, H_g = 1,162 W,
   H_d = 48 W, Coolant pump mass flow m = 0.029 kg/s]

12. A sleeve bearing of length 80 mm supports a journal of 80 mm diameter rotating at 1,000 rpm. Viscosity of oil used is 0.013 kg/Mrs. Bearing temperature should not increase 80°C in an environment of 30°C. Clearance ratio is 1,000. Calculate:

(a) Permissible load, if the bearing operates in still air with K = 400.

(b) Power lost in friction.   [(a) $W = 1,548$ N; (b) $p_f = 128$ W]

Previous Competition Examination Questions

IES

1. Consider that modern machines mostly use short bearings due to the following reasons:

    (1) L/D ratio of most of the bearings is 0.25 – 2.

    (2) No end leakage of oil from the bearing.

    (3) Shaft deflection and misalignment do not affect the operation.

    (4) Can be applied to both hydrodynamic and hydrostatic cases.

    Which of the above is correct?

    (a) 1 and 4   (b) 2 and 3   (c) 1 and 3   (d) 2 and 4

                                                        [Answer (c)] [IES 2015]

2. A 360° hydrodynamic bearing has a journal diameter and length 60 mm. It is running at a speed of 1,200 rpm. The radial clearance is 0.04 mm and minimum oil thickness is 0.008 mm. Sommerfeld number is 0.0446. Find the viscosity of the oil suitable for the bearing.

                                                    [$\mu = 8$ mPa.s] [IES 2014]

3. A full journal bearing has a journal diameter D of 25 mm, with a unilateral tolerance of –0.038 mm. The bushing bore has a diameter B of 25.038 mm and a unilateral tolerance of 0.075 mm. The L/D ratio is unity. The load is 1.1 kN and the journal runs at 18.33 rps. The average viscosity is 55.2 mPa.s. Minimum film thickness variable is 0.58 and coefficient of friction variable is 4.0. Find:

    (a) Sommerfeld number   (b) Minimum film thickness

    (c) Frictional torque

              [(a) $S = 0.25$; (b) $h_{min} = 0.011$ mm; (c) $T_f = 0.08$ Nm] [IES 2013]

4. In case of hydrodynamic lubrication in a journal bearing, the attitude is the ratio of:

    (a) Minimum film thickness and diametric clearance.

    (b) Eccentricity and minimum film thickness.

    (c) Eccentricity and diametric clearance.

    (d) Eccentricity and radial clearance.   [Answer (d)] [IES 2013]

5. The bearing characteristic number depends on:

    (a) Length and diameter of the bearing and the radial load.

    (b) Length and diameter of the bearing, bearing load, and speed of rotation.

    (c) Length and diameter of the bearing, bearing load, and viscosity of the lubricant.

    (d) Length and diameter of the bearing, bearing load, speed of rotation, and viscosity of the lubricant.

                                                        [Answer (d)] [IES 2013]

GATE

6. A hydrodynamic journal bearing is subjected to 2,000 N load at a rotational speed of 2,000 rpm. Both bearing bore diameter and length are 40 mm. If radial clearance is 20 am and bearing is lubricated with an oil having viscosity of 0.03 Pa.s, the Sommerfeld number of the bearing is:

    [Answer: S = 0.8] [GATE 2014]

7. Starting friction is low in:
    (a) Hydrostatic lubrication
    (b) Hydrodynamic lubrication
    (c) Mixed (or semifluid) lubrication
    (d) Boundary lubrication

    [Answer (a)] [GATE 2013]

8. In thick film hydrodynamic journal bearings, the coefficient of friction:
    (a) Increases with increase of load
    (b) Is independent of load
    (c) Decreases with increase of load
    (d) May increase or decrease with increase in load

    [Answer (c)] [GATE 2013]

Linked Answers Questions 9 and 10

A full journal bearing, 200 mm in diameter and 200 mm long is to support a radial load of 45 kN at an operating speed of 960 rpm. It is to operate at a Sommerfeld number of 0.08 when radial clearance is 0.2 mm.

9. The bearing pressure will be:
    (a) 0.125 MN/mm²
    (b) 1.125 MN/mm²
    (c) 0.0325 MN/mm²
    (d) 0.0425 MN/mm²

    [Answer (b)] [Gate 2013]

10. Power loss in the bearing if the resulting coefficient of friction value is 0.0035 will be:
    (a) 1,425 W
    (b) 1,225 W
    (c) 1,025 W
    (d) 825 W

    [Answer (a)] [GATE 2013]

Common Data Questions 11 and 12

A full journal bearing with a journal of 75 mm diameter and a bearing of 75mm length is subjected to a radial load of 2,500 N at 400 rpm. The lubricant is SAE 30 at 75°C having a viscosity of 16.5×10^{-3} kg/ms. Radial clearance is 0.03 mm. Eccentricity of the bearing is 0.27.

11. The Sommerfeld number will be:
    (a) 21.4×10^6
    (b) 23.4×10^6
    (c) 25.4×10^6
    (d) None of these.

    [Answer (b)] [GATE 2013]

12. Minimum film thickness will be:
    (a) 0.01 mm
    (b) 0.015 mm
    (c) 0.03 mm
    (d) 0.025 mm

    [Answer (b)] [GATE 2013]

13. A lightly loaded full journal bearing has journal diameter of 50 mm, bush bore 50.05 mm, and bush length of 20 mm. If rotational speed of journal is 1,200 rpm and average viscosity of liquid lubricant is 0.03 Pa.s, the power loss in W will be:
    (a) 37
    (b) 74
    (c) 118
    (d) 237

    [Answer (a)] [GATE 2010]

14. A journal bearing has shaft diameter of 40 mm and length 40 mm. The shaft is rotating at 20 red/s and the viscosity of the lubricant is 20 mPa.s. The clearance is 0.020 mm. The loss of torque due to viscosity of the lubricant is approximately:

(a)   0.040 Nm   (b)   0.252 Nm   (c)   0.400 Nm   (d)   0.652 Nm

[Answer (a)] [GATE 2008]

15. A natural feed journal bearing of diameter 50 mm and length 50 mm operating at 20 rps carries a load of 2.0 kN. The lubricant used has a viscosity of 20 mPa.s. The radial clearance is 50 μ m. The Sommerfeld number for the bearing is:

(a)   0.062   (b)   0.125   (c)   0.250   (d)0.785

[Answer (b)] [GATE 2007]

■■■

Rolling Bearings

11.1 Introduction

Bearings are used to support a rotating or stationary shaft. Bearings can be broadly classified as:

a. **Slide bearings** These are also called as sleeve bearings or hydrodynamic bearings, because the hydrodynamic pressure of the oil is used to support the shaft. These are described in detail in Chapter 10.

b. **Rolling bearings** In rolling bearings (shown in picture on right), friction is reduced by using a rolling element (ball or roller)

to roll between the bearing surfaces rather than sliding, as in slide bearings. Friction is less, as rolling friction is lesser than sliding friction. In these bearings, starting friction is about two times more than running friction but still lesser than the starting friction of sleeve bearings. Owing to low friction of rolling bearings, these are also called antifriction bearings. Friction variation with shaft speed is shown in Figure 11.1 for slide bearings and roller element bearings.

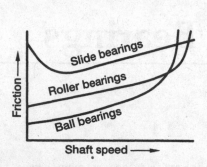

Figure 11.1 Friction versus Speed   **Figure 11.2** Bearing selection

Selection of rolling element ball or roller depends on load and speed. Figure 11.2 can be used to select type of rolling element for known load and speed.

Rolling bearings are manufactured in a large variety. Selection depends on the application. These bearings are manufactured for following types of loads:

- Pure radial load
- Pure axial load
- Combined radial and axial load

Advantages

- Low starting and running friction
- No continuous lubrication is required, hence clean
- Shaft center remains at one position, while in slide bearings it changes its position in bearing as the speed increases
- Reliable
- Low maintenance
- Easy to mount and replace

Disadvantages

- Bearing housings are complicated in construction and need close tolerances
- Need oil seals to prevent leakage of grease, which is used as lubricant
- Noisy at very high speeds

- Costlier than slide bearings
- Cannot tolerate misalignment of bearings
- Shocks can damage the bearing

11.2 Construction and Nomenclature

Rolling bearings are made of five elements described below and shown in Figure 11.3.

a. **Outer race** is a ring having a groove in the inner side periphery. Shoulder is the area on the side of groove. Outside diameter D of the ring is the maximum diameter. Axial width W of the outer race is the width of bearing.

b. **Inner race** also has a circumferential groove at its outer periphery to accommodate rolling elements. It is placed concentrically with the outer race. Bore is the size of shaft in the inner race. Corner radius r is the radius at the inside diameter of inner race. Generally, it is in one piece as shown in picture 11.3(c) but it can be in two pieces also as shown in picture 11.3(d).

c. **Rolling element** can be in the shape of ball, cylinder, or needle, of diameter d. There can be two rows of these elements for increasing load capacity. Figure 11.3(c) shows balls as the rolling element. Balls are inserted in the space formed by the grooves of the races as shown in Figure 11.3(b).

d. **Cage** keeps the rolling elements equally spaced in position. It is made in one or two parts and then joined with rivets.

e. **Rivet** joins two parts of the cage after putting the rolling elements [Figure 11.3(a)].

Ball bearings are discussed in Section 11.3.1 and roller bearings in Section 11.3.2.

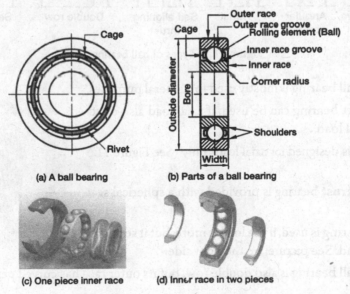

(a) A ball bearing

(b) Parts of a ball bearing

(c) One piece inner race

(d) Inner race in two pieces

Figure 11.3 A rolling bearing

11.3 Classification of Rolling Bearings

Rolling bearings can be classified according to the type of rolling element as under:

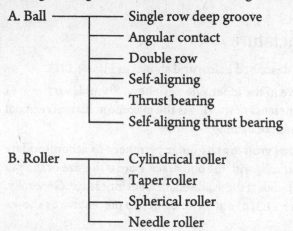

A. Ball
- Single row deep groove
- Angular contact
- Double row
- Self-aligning
- Thrust bearing
- Self-aligning thrust bearing

B. Roller
- Cylindrical roller
- Taper roller
- Spherical roller
- Needle roller

11.3.1 Ball bearings

Figure 11.4 shows different types of ball bearings. Only upper half, without cage is shown.

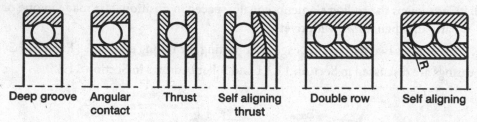

Deep groove Angular contact Thrust Self aligning thrust Double row Self aligning

Figure 11.4 Types of ball bearings

Deep groove ball bearing is mostly used for general purposes.

Angular contact bearing can be used, if axial load also exists along with radial load.

Thrust bearing is designed for axial loads only (See Figure 11.6 also).

Self-aligning thrust bearing is provided with a spherical seat for self-aligning.

Double row bearing is used, if the load is more, but it supports mainly radial load. See picture on the right side.

Self-aligning ball bearing is also double row, but its outer race has curved periphery.

11.3.2 Roller bearings

Ball bearing has a point contact with the races. In case of roller bearings, it is a line contact and hence these bearings can carry more load than ball bearings. Various types of roller bearings are shown in Figure 11.5.

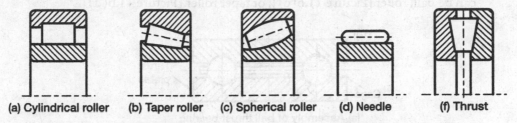

(a) Cylindrical roller   **(b) Taper roller**   **(c) Spherical roller**   **(d) Needle**   **(f) Thrust**

Pictures below show cylindrical, taper roller, double spherical, and needle bearings.

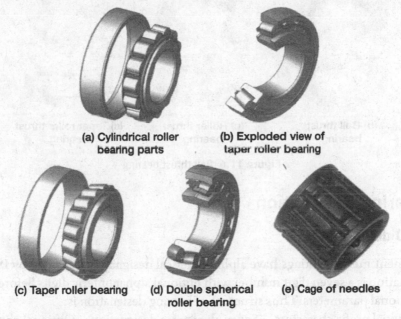

(a) Cylindrical roller
bearing parts

(b) Exploded view of
taper roller bearing

(c) Taper roller bearing

(d) Double spherical
roller bearing

(e) Cage of needles

Figure 11.5 Types of roller bearings (cage not shown in figures)

a. **Cylindrical roller**   bearing uses a cylindrical roller as rolling element and is used for heavy radial loads. These bearings cannot take any axial loads. See Figure 11.5(a).

b. **Taper roller**   has a roller is in the form of a frustum of a cone. Outer race is called cup and has a taper hole. Inner race is called cone and is conical with a bore to mount on shaft. It is shown in Figure 11.5(b). All lines extended from tapers meet at a common point. It can take heavy radial and axial loads.

c. **Spherical roller**   bearing is useful when along with heavy load; there could be misalignment also [Figure 11.5(c)]. Double spherical roller bearings can also be used.

d. **Needle roller** bearing uses roller of small diameter called needle. Roller diameters vary from 1.5 mm to 4 mm [See Figure11.5(e)].

e. **Thrust bearing** is meant for heavy axial loads only. Balls rotate in annular grooves in two thrust plates as shown in Figure 11.6(a) and Picture 11.6(b). Rolling element can be ball, roller [Picture 11.6(c)], or taper roller [Picture 11.6(d)].

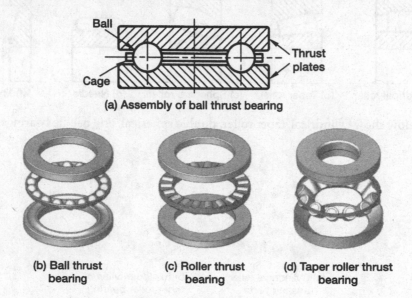

(a) Assembly of ball thrust bearing

(b) Ball thrust bearing   (c) Roller thrust bearing   (d) Taper roller thrust bearing

Figure 11.6 Ball thrust bearing

11.4 Bearing Designation

11.4.1 ISO designation

Rolling-element metric bearings have alphanumerical designations, defined by ISO 15. The main designation is a seven-digit number with optional alphanumeric digits before or after, to define additional parameters. Thus structure of bearing designation is:

Four optional prefix characters—Seven-digit main designation—Optional character.

a. Seven-digit main designation

Here the digits are defined as: 7654321. Zeros to the left of the last defined digit are not printed on bearing; for example, a designation of 0006206 is printed as 6,206.

First two digits on right extreme, that is, (2) and (1) give bore diameter.

00 means bore diameter 10 mm,   01 means bore diameter 12 mm,

02 means bore diameter 15 mm,   03 means bore diameter 17 mm,

From 04 onwards, bore diameter is found by multiplying with 5, for example, 05 means 25.

The third digit(3) from right indicates diameter series, which defines the outer diameter (Figure 11.7). The diameter series is defined in ascending order as:1, 2, 3, 4, 5, and 6.

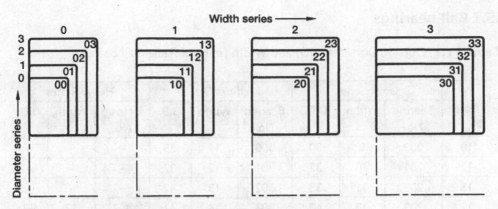

Figure 11.7 Width and diameter series of bearings

Extra light (100)   Light (200)   Medium (300)   Heavy (400)

For example a bearing number 301 means a medium series with bore 12 mm. A number 205 means, it is of light series with bore (05×5) 25 mm.

The fourth digit (4) defines the type of bearing:

0 = Ball single row,   1 = Ball two row,   2 = Short cylindrical roller,

3 = Spherical roller 2 rows,   4 = Needle roller,   5 = Spiral roller,

6 = Ball type thrust,   7 = Taper roller.

For example, 0400 is Ball single row bearing, and 7,204 = Taper roller bearing.

Fifth and sixth digits give structural modifications to the bearing.

The seventh digit gives the width series as shown in Figure 11.7

11.4.2 AFBMA designation

Anti-Friction Bearing Manufacturing Association (AFBMA) has recommended a five-digit code to specify bearing size. First number of code on right is width series (1, 2, 3, 4, 5, and 6) in increasing order and second number is outside diameter series (8, 9,..., 1, 2, 3, 4). Figure 11.7 shows sizes of the different series of bearings.

Some manufacturers designate a bearing with three or four or five-digit number.

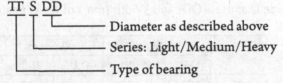

For example, bearing no. 6310 is ball bearing of diameter series 3 with bore (10×5) = 50 mm.

11.5 Size of Bearings

Series 1, 2, and 3 are most commonly used. Important dimensions of light, medium, and heavy series bearings are tabulated in Tables 11.1 for ball bearings and Table 11.2 for roller bearings.

11.5.1 Ball bearings

Table 11.1 Outside Diameter (OD) and Width (mm) of Radial Ball Bearings

Bore	Light Series			Medium Series			Heavy Series		
	Bearing No.	Width	O.D.	Bearing No.	Width	O.D.	Bearing No.	Width	O.D.
10	200	9	30	300	11	35	–	–	–
12	201	10	32	301	12	37	–	–	–
15	202	11	35	302	13	42	–	–	–
17	203	12	40	303	14	47	403	17	62
20	204	14	47	304	14	52	404	19	72
25	205	15	52	305	17	62	405	21	80
30	206	16	62	306	19	72	406	23	90
35	207	17	72	307	21	80	407	25	100
40	208	18	80	308	23	90	408	27	110
45	209	19	85	309	25	100	409	29	120
50	210	20	90	310	27	110	410	31	130
55	211	21	100	311	29	120	411	33	140
60	212	22	110	312	31	130	412	35	150
65	213	23	120	313	33	140	413	37	160
70	214	24	125	314	35	150	414	42	180
75	215	25	130	315	37	160	415	45	190
80	216	26	140	316	39	170	416	48	200
85	217	28	150	317	41	180	417	52	210
90	218	30	160	318	43	190	418	54	225

11.5.2 Roller bearings

Table 11.2 Outside Diameter (OD) and Width (mm) of Roller Bearings for Series 02. 03

Bore	02 Series		03 Series	
	O.D.	Width	O.D.	Width
25	52	15	62	17
30	62	16	72	19
35	72	17	80	21
40	80	18	90	23
45	85	19	100	24
50	90	20	110	27

55	100	21	120	29
60	110	22	130	31
65	120	23	140	33
70	124	24	150	35
75	130	24	160	37
80	140	26	170	39
85	150	28	180	41
90	160	30	190	43
95	170	32	200	45
100	180	36	215	47

11.6 Static Load Capacity

When a shaft does not rotate, the load on the bearing is called static load. It causes permanent deformation in races and balls. Static load capacity depends on acceptable permanent deformation, which is taken as 0.0001 times of rolling element diameter. If noise and smoothness of operation is not important, this capacity can be increased up to four times. As per IS 3823 – 1988, the static load capacity C_o is given as under:

Radial ball bearing $C_o = f_o\, i\, Z\, d \cos \alpha$ (11.1a)

Radial roller bearing $C_o = f_o\, i\, Z\, l_e\, d \cos \alpha$ (11.1b)

Thrust ball bearing $C_o = f_o\, i\, Z\, d^2 \sin \alpha$ (11.1c)

Thrust roller bearing $C_o = f_o\, i\, Z\, l_e\, d \sin \alpha$ (11.1d)

Where, C_o = Basic static load capacity of radial / axial in newtons

f_o = A constant depending on the type of bearing from Table 11.3

i = Number of rows of balls / rollers

Z = Number of balls / rollers in a row

d = Ball / roller diameter in mm

l_e = Effective contact length of roller and race (Total length – Chamfer)

Ratio of length l_e to diameter d of roller should lie between 1 and 1.4.

α = Nominal contact angle between line of action of resultant load and a plane perpendicular to the axis of the bearing.

Table 11.3 Values of f_o for Different Types of Bearings

Type of Bearing	Radial Ball		Radial Roller	Thrust Bearing	
	Self-Aligning	Radial / Angular		Ball	Roller
f_o	3.3	12.3	21.6	49	98

Angle Φ, in which rolling elements can be accommodated when placed touching to each other. It varies from 185° to 210°. A cage keeps them apart at some distance.

For ball bearings $\quad \Phi = 2(Z-1)\sin^{-1}\left(\dfrac{d}{D_m}\right)$ $\qquad$ (11.2a)

For roller bearings $\quad \Phi = 2Z\sin^{-1}\left(\dfrac{d}{D_m}\right)$ $\qquad$ (11.2b)

Where, d = Diameter of the rolling element

$\quad D_m$ = Mean pitch circle diameter of the set of rolling elements (Cage)

Example 11.1

Design of a roller bearing for given static load

Data of a lifting bridge is given below:

Load of bridge and counter weight on each bearing = 1,830 kN

Factor of safety = 2

Speed = 1 revolution per hour

Shaft diameter = 400 mm

Find:

    a. Number of rollers in a bearing

    b. Roller diameter $\qquad\qquad$ c. Length of roller

Solution

Given $\quad W = 1,830$ kN $\qquad FOS = 2 \qquad N = 1$ revolutions / hour $\qquad D = 400$ mm

Since the speed is very less, bearing will be designed for static loading.

There is no axial load and radial load is very high; hence a roller bearing with two rows $(i = 2)$ is selected.

For radial roller bearings, static load capacity should be equal to external load; hence using Equation (11.1b):

$$C_o = (W \times FOS) = f_o\, i\, Z\, l_e\, d \cos \alpha \quad \text{[Take value of } f_o \text{ from Table 11.3]}$$

Substituting the values: $1,830,000 \times 2 = 21.6 \times 2 \times (Z\, l_e d) \qquad [f_o = 21.6 \text{ and } \alpha = 0]$

Or, $\quad Z\, l_e d = 84,722$ N/mm² $\qquad\qquad\qquad\qquad\qquad\qquad$ (a)

First iteration

For bearings bigger than 100 mm diameter, special order is to be made; hence speed and life factors are ignored and can be specified to the manufacturer.

Assuming radial thickness of inner race as 20 mm and roller diameter 25 mm, mean pitch circle diameter of rollers is:

$$D_m = 400 + (2 \times 20) + 25 = 465 \text{ mm}$$

Circumference at mean circle $= \pi \times 465 = 1{,}461$ mm

Maximum number of rollers, $Z = \dfrac{1{,}461}{25} = 58$

Angle in which rollers can be accommodated given by Equation (11.2b) is:

$$\Phi = 2 \times 58 \times \sin^{-1}\left(\frac{25}{465}\right) = 397° \quad \text{It is more than } 360°$$

Hence, number of rollers that can be accommodated in 360°:

$$Z = \frac{58 \times 360}{397} = 53$$

Substituting value of Z and d in Equation (a) above: $l_e = \dfrac{84{,}722}{53 \times 25} = 63.9$ mm

$$\frac{l_e}{d} = \frac{63.9}{25} = 2.56$$

Ratio of length to diameter of roller should lie between 1 and 1.4, hence choosing roller diameter 50 mm.

Second iteration

Mean diameter of rollers in cage is: $D_m = 400 + (2 \times 20) + 50 = 490$ mm

Circumference at pitch circle $= \pi \times 490 = 1{,}554$ mm

Maximum number of rollers, $Z = \dfrac{1{,}554}{50} = 31$

Angle in which rollers can be accommodated is:

$$\Phi = 2 \times 31 \times \sin^{-1}\left(\frac{50}{490}\right) = 403.4° \quad \text{It is more than } 360°$$

Hence, number of rollers that can be accommodated in 360°:

$$Z = \frac{31 \times 365}{403.4} = 28$$

Substituting value of Z and d in Equation (a):

$$l_e = \frac{84{,}722}{28 \times 50} = 60.5 \text{ say } 61 \text{ mm}$$

$$\frac{l_c}{d} = \frac{61}{50} = 1.21 \quad \text{It is in tolerable range 1 to 1.4.}$$

Hence, Number of two rows, each of 28 rollers

Roller diameter = 50 mm

Length of roller = 61 mm

11.7 Static Equivalent Load

When there is an axial load in addition to radial load, a static equivalent load capacity C_o can be defined, which causes same permanent deformation as a mostly loaded ball / roller bearing.

$$C_o = X_o W_r + Y_o W_a \tag{11.3}$$

Where, X_o = Radial load factor $\qquad$ Y_o = Axial load factor

$\quad$ W_r = Radial load $\qquad\qquad$ W_a = Axial load

Values of X_o and Y_o are given in Table 11.4.

Notes on angular contact bearings

- Values of Y_o for angular contact bearing vary according to contact angle α.
- For two or more similar angular contact bearings in tandem use X_o and Y_o as for single row bearing.
- For two or more similar angular contact bearings used as back to back or face to face use X_o and Y_o as for double row bearing.

Table 11.4 Values of X_o and Y_o for Different Types of Bearings

Type of Bearing	Single Row		Double Row	
	X_o	Y_o	X_o	Y_o
Radial ball	0.6	0.5	0.6	0.5
Cylindrical roller	1	0	1	0
Spherical roller	1	2	1	2
Self-aligning ball / roller / taper roller	0.5	0.22 cot α	1	0.44 cot α
Angular contact bearing $\alpha = 15°$	0.5	0.46	1	0.92
$\alpha = 25°$	0.5	0.38	1	0.76
$\alpha = 35°$	0.5	0.29	1	0.66
$\alpha = 45°$	0.5	0.22	1	0.44

Notes on thrust bearings

With combined radial and axial loads, static equivalent axial load W_{oa} for thrust ball or roller bearings with angle of contact not 90° is given by the following relation:

$$W_{oa} = 2.3W_r \tan \alpha + W_a \tag{11.4}$$

Thrust bearings with contact angle 90° cannot take any radial load; hence equivalent axial load is equal to axial load, that is, $W_{oa} = W_a$.

11.8 Basic Dynamic Load Capacity

Basic dynamic load capacity C is a constant stationary load over a group of bearings, with inner race rotating and outer race stationary, for which a bearing can work satisfactorily for a rated life of one million revolutions. The load should be pure radial for radial bearings and pure axial for thrust bearings.

The load capacity depends on a factor f_c which depends on bearing geometry, hardness, smoothness, and accuracy in manufacturing. It also depends on number of rolling elements Z, number of rows of rolling elements i, contact angle α, ball diameter d for ball bearings / roller diameter d and equivalent length l_e for roller bearings. Relation of all these parameters with radial load capacity or axial load capacity for thrust bearings C is tabulated in Table 11.5 for different types of bearings. Value of f_c can be taken from Table 11.6.

It can be seen from Table 11.6 that value of f_c is between 67 and 78 for all the values of $[(d \cos \alpha) / D_m]$ more than 0.06.

Table 11.5 Relations for Dynamic Load Capacities of Bearings

Rolling Element	Type of Bearing	Dynamic Load Capacity Relation (c)
Balls < 25 mm	Radial / angular	$C = f_c Z^{2/3} d^{1.8} (i \cos\alpha)^{0.7}$
Balls > 25 mm	Radial / angular	$C = 3.65 f_c Z^{2/3} d^{1.4} (i \cos\alpha)^{0.7}$
Rollers any size	Radial / angular	$C = f_c Z^{3/4} d^{29/27} (i l_e \cos\alpha)^{7/9}$
Balls < 25 mm	Thrust $\alpha = 90°$	$C = f_c Z^{2/3} d^{1.8}$
Balls < 25 mm	Thrust $\alpha \neq 90°$	$C = f_c Z^{2/3} d^{1.8} (\cos\alpha)^{0.7} \tan\alpha$
Balls > 25 mm	Thrust $\alpha = 90°$	$C = 3.65 f_c Z^{2/3} d^{1.4}$
Balls > 25 mm	Thrust $\alpha \neq 90°$	$C = 3.65 f_c Z^{2/3} d^{1.4} (\cos\alpha)^{0.7} \tan\alpha$

Rollers any size	Thrust $\alpha = 90°$	$C = f_c Z^{3/4} d^{29/27} \left(l_e \right)^{7/9}$
Rollers any size	Thrust $\alpha \neq 90°$	$C = f_c Z^{3/4} d^{29/27} \left(l_e \cos\alpha \right)^{7/9}$

Table 11.6 Values of f_c for Various Bearings Geometries
(D_m is mean pitch circle diameter of rolling elements each of diameter d)

$\dfrac{d\cos\alpha}{D_m}$	0.01	0.02	0.03	0.04	0.05	0.06	0.07	0.08	0.09	0.1
f_c	45.7	53.45	56.45	62.27	65.02	66.78	69.43	71.69	72.77	73.84
$\dfrac{d\cos\alpha}{D_m}$	0.12	0.14	0.16	0.18	0.20	0.22	0.24	0.26	0.28	0.30
f_c	76.1	77.18	77.67	78.26	78.26	77.67	77.18	76.10	74.92	73.84

Since the data on number of rolling elements may not be available for all types of available bearings, Table 11.7 directly gives dynamic load capacity (C) for deep groove ball bearings. Consult *Design Data Book* for data on other type of bearings.

Table 11.7 Static Load (C_o) and Dynamic Load Capacities (C) in kN for Deep Groove Ball Bearings

Bore (mm)	Light Series			Medium Series			Heavy Series		
	Bearing No.	C_o	C	Bearing No.	C_o	C	Bearing No.	C_o	C
10	6200	2.24	4.0	6300	3.6	6.3	–	–	–
12	6201	3.1	6.89	6301	4.65	9.75	–	–	–
15	6202	2.5	5.59	6302	5.4	11.4	–	–	–
17	6203	4.5	9.56	6303	6.55	13.5	6403	11.8	22.9
20	6204	6.2	12.7	6304	7.8	15.9	6404	16.6	30.7
25	6205	6.95	14.0	6305	11.4	22.5	6405	19.6	35.8
30	6206	10.0	19.5	6306	14.6	28.1	6406	24.0	43.6
35	6207	13.7	25.50	6307	18.0	33.2	6407	31.0	55.3
40	6208	16.6	30.7	6308	22.4	41.0	6408	36.5	63.7

45	6209	18.6	33.2	6309	30.3	52.7	6409	45.5	76.1
50	6210	19.6	35.1	6310	36.0	61.8	6410	52.0	87.1
55	6211	25.0	43.6	6311	41.5	71.5	6411	63.0	99.5
60	6212	28.0	47.5	6312	48.0	81.9	6412	69.5	108.0
65	6213	34.0	55.9	6313	56.0	92.3	6413	78.0	119.0
70	6214	37.5	61.8	6314	63.0	104.0	6414	104.0	143.0
75	6215	40.5	66.3	6315	72.0	112.0	6415	114.0	153.0
80	6216	45.5	57.0	6316	80.0	96.5	6416	120.0	127.0
85	6217	55	65.5	6317	88.0	104.0	6417	132.0	134.0
90	6218	63.0	75.0	6318	98.0	112.0	6418	146.0	146.0
95	6219	72.0	85.0	6319	112.0.	120.0	6419	–	–
100	6220	81.5	96.5	6320	132.0	137.0	6420	–	–

11.9 Dynamic Equivalent Load

Equivalent dynamic load is defined as constant stationary radial / axial load (for thrust bearings), applied to rotating inner race and stationary outer race, which will give the same life under actual conditions of load and speed. Equivalent dynamic load W_e with combined radial load W_r and axial load W_a is given as under:

$$W_e = X\,VW_r + YW_a \qquad (11.5)$$

Where, X = Radial load factor for dynamically loaded bearing

Y = Axial load factor for dynamically loaded bearing

V = Rotation factor = 1.0 for outer race stationary and inner race rotating for all series of bearings

= 1.4 inner race stationary and outer race rotating for series 100 of bearings

= 1.5 inner race stationary and outer race rotating for series 200 of bearings

= 1.6 inner race stationary and outer race rotating for series 300 of bearings

= 1.7 inner race stationary and outer race rotating for series 400 of bearings

= 2.0 both inner and outer race rotating in opposite directions for all series

Values of X and Y are given in Table 11.8. The values are dependent on the axial load factor e. First assume value of e and then find values of X and Y from Table 11.8.

Table 11.8 Values of X and Y for Dynamically Loaded Bearings

Type of Bearing	Specifications	$\dfrac{W_a}{W_r} \le e$		$\dfrac{W_a}{W_r} > e$		e
		X	Y	X	Y	
Deep groove ball bearing	$\dfrac{W_a}{C_o} = 0.025$	1	0	0.56	2.0	0.22
	$= 0.04$				1.8	0.24
	$= 0.07$				1.6	0.27
	$= 0.13$				1.4	0.31
	$= 0.25$				1.2	0.37
	$= 0.50$				1.0	0.44
Angular contact ball bearing	Single row	1	0	0.35	0.57	1.14
	Two rows		0.73	0.62	1.17	0.86
	Two rows in tandem		0	0.35	0.57	1.14
	Two rows back to back		0.55	0.57	0.93	1.14
Self-aligning ball bearing	**Light series**					
	Bores 10–20 mm	1	1.3	0.65	2.0	0.5
	Bores 25–35 mm		1.7		2.6	0.37
	Bores 40–45 mm		2.0		3.1	0.31
	Bores 50–65 mm		2.3		3.5	0.28
	Bores 70–100 mm		2.4		3.8	0.26
	Medium series					
	Bore 12 mm	1	1.0	0.65	1.6	0.63
	Bores 15–20 mm		1.2		1.9	0.52
	Bores 25–50 mm		1.5		2.3	0.43
	Bores 55–90 mm		1.6		2.5	0.39
Spherical roller bearing	Bores 25–35 mm	1	2.1	0.67	3.1	0.32
	Bores 40–45 mm		2.5		3.7	0.27
	Bores 50–100 mm		2.9		4.4	0.23
Taper roller bearing	Bores 30–40 mm	1	0	0.4	1.6	0.37
	Bores 45–110 mm				1.45	0.44
	Bores 120–150 mm				1.35	0.41

11.10 Rated Life of a Bearing

Surfaces of the bearing races and rolling elements are subjected to repeated compressive stresses while rotating, even under normal working conditions, which causes formation of flakes due to fatigue loading. This results in high friction and finally the bearing fails. Life of a bearing is defined in number of revolutions without failure. Rated life of a bearing is specified in two ways:

- Life in million (10^6) of revolutions L_{10n}
- Life in hours L_{10h}

For a group of identical bearings, life of individual bearing differs from each other; hence life of a bearing is taken in number of revolutions L_{10n} or in hours L_{10h} that 90 per cent of the group run satisfactorily and will not show any flake.

Two standards are used for rated life:

a. **AFBMA** recommends rated capacity of bearing for a life of one million revolutions (or hours at a given speed) that 90 per cent of bearings will run safely, without any first evidence of failure. These parameters are related with each other as under:

$$L_n = \left(\frac{C}{W_e}\right)^k \tag{11.6}$$

Where, L_n = Life of bearing in number of million revolutions

C = Basic dynamic load rating

W_e = Equivalent dynamic load

$k = 3$ (A constant for ball bearings)

$k = \dfrac{10}{3}$ (A constant for roller bearings)

Life of a rolling bearing in hours L_h is related with life in revolutions L_n and speed N in rpm as:

$$L_h = \frac{L_n}{60N} = \frac{10^6}{60N}\left(\frac{C}{W}\right)^k \tag{11.7}$$

Rated life in hours L_{10h} is for 90 per cent of the bearings (10 per cent failure), which run satisfactorily for 500 hours at 33.3 rpm $[10^6 / (500 \times 60)]$

b. **Timken Co.** has standardized load ratings for 3,000 hours of L_h life at 500 rpm

Life of a rolling bearing depends on its basic dynamic load rating C, equivalent dynamic load and rated life of bearing.

Example 11.2

Equivalent dynamic load and bearing selection for given loads

A deep groove ball bearing is subjected to radial load of 5 kN and axial load of 2 kN. It rotates at 760 rpm. Shaft diameter should not be less than 50 mm and life has to be minimum 22,000 hours. Select a suitable bearing.

Solution

Given	$W_r = 5$ kN	$W_a = 2$ kN	$N = 760$ rpm	$L_H = 22,000$ h
	$D \geq 50$ mm			

Initially assuming $\dfrac{W_a}{W_r} > e$, values of X and Y from Table 11.8 are $X = 0.56$ and Y varies between 1 and 2. Assuming an intermediate value of $Y = 1.5$.

From Equation (11.5) : $W_e = X V W_r + Y W_a$

Substituting the values: $W_e = (0.56 \times 1 \times 5) + (1.5 \times 2) = 5.8$ kN

Bearings with 50 mm bore have following values from Table 11.7:

Bearing no. 6210	$C = 35.1$ kN	$C_0 = 19.6$ kN
Bearing no. 6310	$C = 61.8$ kN	$C_0 = 36$ kN
Bearing no. 6410	$C = 87.1$ kN	$C_0 = 52.0$ kN

Life required in revolutions $L_n = 60 \times 760 \times 22{,}000 = 1003.2 \times 10^6$ revolutions

From Equation (11.6): $L_n = \left(\dfrac{C}{W_e}\right)^k \times 10^6 \qquad$ (For balls $k = 3$)

$$1003.2 \times 10^6 = \left(\dfrac{C}{5.8}\right)^3 \times 10^6 \quad \text{or} \quad C = 58 \text{ kN}$$

Initially selecting bearing no. 6310 with $C = 61.8$ kN and $C_0 = 36$ kN

$\dfrac{W_a}{W_r} = \dfrac{2}{5} = 0.4$ and $\dfrac{W_a}{C_0} = \dfrac{2}{36} = 0.0555$. For these value from Table 11.8, $e \approx 0.25$

$\dfrac{W_a}{W_r} > 0.25$, hence by interpolation $Y = \dfrac{(1.8 - 1.6) \times (0.0555 - 0.04)}{(0.07 - 0.0555)} = 1.58$

Substituting the values in Equation (11.5): $W_e = (0.56 \times 1 \times 5) + (1.58 \times 2) = 5.96$ kN

From Equation (11.6): $1003.2 \times 10^6 = \left(\dfrac{C}{5.96}\right)^3 \times 10^6 \quad$ or $\quad C = 59.65$ kN

It is less than rated capacity, that is, 61.8 kN, hence safe. Hence bearing no. 6,310 is selected.

11.11 Reliability of Bearings

Reliability R of a bearing is defined as the ratio of number of bearings, which run successfully without failure for L million revolutions to the total number of bearings under test.

$$R = \dfrac{\text{Number of bearings sucessfully completed } L \text{ revolutions}}{\text{Total number of bearings tested}}$$

Weibull gave a relation between reliability R and bearing life L as:

$$L = 6.84 \left[\log_e \left(\frac{1}{R} \right) \right]^{0.855} \tag{a}$$

Let L_{10} be the life of bearing with 10 per cent failure, that is, 90 per cent reliability R_{90} then:

$$L_{10} = 6.84 \left[\log_e \left(\frac{1}{R_{90}} \right) \right]^{0.855} \tag{b}$$

Dividing Equation (a) with Equation (b):

$$\frac{L}{L_{10}} = \left[\frac{\log_e \left(\dfrac{1}{R} \right)}{\log_e \left(\dfrac{1}{R_{90}} \right)} \right]^{0.855} \tag{11.8}$$

Generally, bearing life is given with 90 per cent reliability. Equation (11.8) can be used to calculate the life for any reliability. Plot of these parameters is shown in Figure 11.8.

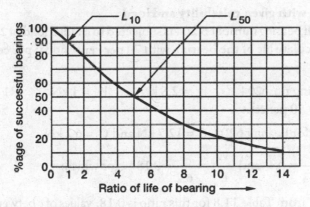

Figure 11.8 Reliability versus life of a bearing

Example 11.3

Variation of life with reliability
A ball bearing has a life of 2 million revolutions with 90 per cent reliability. Calculate its life with 95 per cent reliability. How much is reduction in life with this increased reliability.

Solution

Given $\qquad L_{10} = 2 \times 10^6 \qquad\qquad R = 95$ per cent

From Equation (11.8)

$$\frac{L}{L_{10}} = \left[\frac{\log_e \left(\dfrac{1}{R} \right)}{\log_e \left(\dfrac{1}{R_{90}} \right)} \right]^{0.855}$$

Substituting the values:

Life in revolutions, $L = 2 \times 10^6 \left[\dfrac{\log_e \left(\dfrac{1}{0.95} \right)}{\log_e \left(\dfrac{1}{0.9} \right)} \right]^{0.855} = 2 \times 10^6 \left[\dfrac{0.05129}{0.10536} \right]^{0.855} = 1.08 \times 10^6$

Reduction in life $= \dfrac{(2 - 1.08)}{2} \times 100 = 46$ per cent

Example 11.4

Life of a bearing with given reliability and load

A deep groove ball bearing number 6,204 is subjected to a radial load of 2,500 N and axial load of 1 kN. Calculate life of the bearing with 70 per cent and 50 per cent reliabilities.

Solution

Given   Bearing no. 6204 $W_r = 2.5$ kN $W_a = 1$ kN   R1 = 70 per cent
R2 = 50 per cent

From Table 11.7 for bearing 6204, $C = 12.7$ kN and $C_o = 6.2$ kN

$$\frac{W_a}{W_r} = \frac{1}{2.5} = 0.4 \quad \text{and} \quad \frac{W_a}{C_o} = \frac{1}{6.2} = 0.18 \quad \text{for which:}$$

Values of X and Y from Table 11.8 for this ratio is 0.18; values of e between 0.31 and 0.37.

$X = 0.56$ and Y varies between 1.2 and 1.4.

$$\frac{W_a}{W_r} > 0.18, \text{ hence by interpolation } Y = \frac{(1.4 - 1.2) \times (0.18 - 0.13)}{(0.25 - 0.13)} = 1.31$$

From Equation (11.5): $W_e = X \ VW_r + YW_a \quad (V = 1)$

Substituting the values: $W_e = (0.56 \times 1 \times 2.5) + (1.31 \times 1) = 2.71$ kN

From Equation (11.6): $L_{10n} = \left(\dfrac{C}{W_e} \right)^k \times 10^6 = \left(\dfrac{12.7}{2.71} \right)^3 \times 10^6 = 102.9$ million revolutions

From Equation (11.8) Life ratio for desired reliability R:

$$\frac{L}{L_{10}} = \left[\frac{\log_e \left(\dfrac{1}{R} \right)}{\log_e \left(\dfrac{1}{R_{90}} \right)} \right]^{0.855}$$

For reliability 0.7, Life: $L = L_{10n} \left[\dfrac{\log_e \left(\dfrac{1}{0.7} \right)}{\log_e \left(\dfrac{1}{0.9} \right)} \right]^{0.855} = 102.9 \times 2.83 = 291.5$ mil. rev.

For reliability 0.5, life: $L = L_{10n} \left[\dfrac{\log_e \left(\dfrac{1}{0.5} \right)}{\log_e \left(\dfrac{1}{0.9} \right)} \right]^{0.855} = 102.9 \times 5 = 514.5$ mil. rev.

11.12 Life with Varying Loads

If the load varies, then the life for each load is calculated separately and then combined. Let W_1, W_2, W_3, etc., be the loads for n_1, n_2, n_3, etc., number of revolutions, respectively, then L_1, L_2, L_3, etc., are the life of a bearing. From Equation (11.6) life L_1 is given as:

$$L_1 = \left(\frac{C}{W_1} \right)^k \times 10^6$$

Fraction of life with load W_1 for n_1 revolutions is:

$$\frac{n_1}{L_1} = n_1 \left(\frac{W_1}{C} \right)^k \times \frac{1}{10^6} \tag{a}$$

Similarly, for other loads fractional life is given as:

$$\frac{n_2}{L_2} = n_2 \left(\frac{W_2}{C} \right)^k \times \frac{1}{10^6} \tag{b}$$

$$\frac{n_3}{L_3} = n_3 \left(\frac{W_3}{C} \right)^k \times \frac{1}{10^6} \tag{c}$$

Sum of all the fractional lives is equal to one. That is,

$$\frac{n_1}{L_1} + \frac{n_2}{L_2} + \frac{n_3}{L_3} = 1$$

Substituting the values from Equations (a), (b), and (c) in the above equation:

$$n_1\left(\frac{W_1}{C}\right)^k \times \frac{1}{10^6} + n_2\left(\frac{W_2}{C}\right)^k \times \frac{1}{10^6} + n_3\left(\frac{W_3}{C}\right)^k \times \frac{1}{10^6} = 1 \qquad \text{(d)}$$

Or, $\quad n_1\left(W_1\right)^k + n_2\left(W_2\right)^k + n_3\left(W_3\right)^k = 10^6 C^k$

Introducing a constant equivalent load W_e for life of n revolutions then:

$$n = \left(\frac{C}{W_e}\right)^k \times 10^6 \qquad \text{(e)}$$

Where, $n = n_1 + n_2 + n_3$
From Equations (d) and (e):

$$n_1\left(W_1\right)^k + n_2\left(W_2\right)^k + n_3\left(W_3\right)^k = nW_e^k$$

Or, $\quad W_e = \left[\dfrac{n_1\left(W_1\right)^k + n_2\left(W_2\right)^k + n_3\left(W_3\right)^k}{n_1 + n_2 + n_3}\right]^{\frac{1}{k}} \qquad \text{(11.9a)}$

The above equation in terms of life can be written

$$W_e = \left[\frac{L_1\left(W_1\right)^k + L_2\left(W_2\right)^k + L_3\left(W_3\right)^k}{L_1 + L_2 + L_3}\right]^{\frac{1}{k}} \qquad \text{(11.9b)}$$

Example 11.5

Life with variable load for given speeds
A deep groove ball bearing is subjected to the following radial load cycle:
4,000 N for 20 per cent of time at 600 rpm, 8,000 N for 40 per cent of time at 720 rpm,
6,000 N for 30 per cent of time at 900 rpm, 1,000 N for 10 per cent of time at 900 rpm
If the bearing used is 6,306, with dynamic load capacity of 28,100 N, calculate:

  a. Expected life
  b. Select a bearing, if life desired is 5,000 hours and bore should not be less than 40 mm.

Solution

Given $\quad k = 3 \quad$ Bearing no. 6,306 $\quad\quad C = 28.1$ kN $\quad\quad L_D = 5,000$ h
$\quad\quad\quad\quad\quad D \geq 40$ mm

Cycle	Load W (kN)	Fraction of Time	n (rpm)	Revolutions/ Cycle
1	4	0.2	600	120
2	8	0.4	720	288
3	6	0.3	900	270
4	1	0.1	900	90
Total	–	1.0	–	768

a. From Equation (11.9a): $W_e = \left[\dfrac{n_1 (W_1)^k + n_2 (W_2)^k + n_3 (W_3)^k \ldots}{n_1 + n_2 + n_3 \ldots} \right]^{\frac{1}{k}}$

Or, $(W_e)^3 = \dfrac{120(4)^3 + 288(8)^3 + 270(6)^3 + 90(1)^3}{768} = 278$

Or, $W_e = 6.5$ kN

From Equation (11.6) Life in million revolutions: $L_n = \left(\dfrac{C}{W_e} \right)^k = \left(\dfrac{28.1}{6.5} \right)^3 = 80.79$ mil.rev.

Life in hours from Equation (11.7) Life in hours: $L_h = \dfrac{L_n}{60\,N} = \dfrac{80.79}{60 \times 768} = 1{,}753$ hours

b. Life in millions of revolutions $L_n = 60\,NL_h = 60 \times 768 \times 5{,}000 = 230.4$ mil. rev.

From Equation (11.6): $L_n = \left(\dfrac{C}{W_e} \right)^3$

Substituting the values: $230.4 = \left(\dfrac{C}{W_e} \right)^3$ or $C = W_e \sqrt[3]{230.4} = 6.5 \times 6.13 = 39.8$ kN

From Table 11.8, bearings which have dynamic load capacity close to 39.8 kN are:

No. 6406 has dynamic load capacity 43.6 with bore 30 mm

No. 6308 has dynamic load capacity 41.0 with bore 40 mm

No. 6211 has dynamic load capacity 43.6 with bore 55 mm

Owing to limitation of bore bearing no. 6308 is selected.

11.13 Cyclic Loads

In certain applications, the direction of radial load remains same but the magnitude of load varies sinusoidal in a revolution as shown in Figure 11.9.

Load W at any position θ can be found from the maximum load W_{max} using following relation:

$$W = \frac{W_{max}}{2} (1 - \cos \theta) \tag{11.10}$$

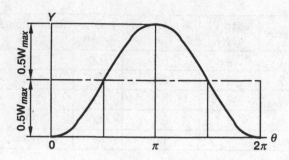

Figure 11.9 Cyclic load

Equivalent load W_e for one complete revolution 2π of can be found by integrating the above equation from the limit 0 to 2π.

$$W_e = \frac{1}{2\pi}\int_0^{2\pi} W^3\, d\theta = \left[\frac{1}{2\pi}\int_0^{2\pi} \frac{(W_{max})^3}{8}\left(1-\cos\theta\right)^3 d\theta\right]^{\frac{1}{3}}$$

Or, $\quad W_e = \frac{W_{max}}{2}\left[\frac{1}{2\pi}\int_0^{2\pi}\left(1-\cos\theta\right)^3 d\theta\right]^{\frac{1}{3}}$

Integrating and simplifying:

$$W_e = \frac{W_{max}\,(2.5)^{\frac{1}{3}}}{2} = 0.678\,W_{max} \tag{11.11}$$

Example 11.6

Bearing selection for a cyclic load

A single row ball bearing is subjected to a sinusoidal radial load, which varies between 0 and 2 kN at a speed of 960 rpm.

a. Calculate dynamic load capacity for a life of 6,000 hours

b. Select a suitable bearing from the catalog, if diameter should not be less than 20 mm.

Solution

Given $\quad W_{max} = 2$ kN $\qquad W_{min} = 0$ kN $\qquad N = 960$ rpm $\qquad L_h = 6,000$ hours

$\qquad\qquad D \geq 20$ mm

a. From Equation (11.11) equivalent load: $W_e = 0.678\,W_{max} = 0.678 \times 2 = 1.357$ kN

From Equation (11.7) life in hours: $L_h = \dfrac{L_n}{60\,N}$

Or, Life in revolutions: $L_n = 60\,N \times L_h = 60 \times 960 \times 6,000 = 345.6 \times 10^6$ revolutions

From Equation (11.6) Life in revolutions: $L_n = \left(\dfrac{C}{W_e}\right)^3$ or $C = W_e \left(L_n\right)^{\frac{1}{3}}$

Substituting the values, load capacity: $C = 1.357\left(345.6\right)^{\frac{1}{3}} = 9.523$ kN

b. From Table 11.7, to suit this dynamic load and bore 20 mm, bearing is: 6,204

11.14 Load Factor

The forces coming on a bearing may be more than calculated by equilibrium of forces, in certain applications. For example:

- In gear drive due to inaccuracies of the gear tooth profiles.
- In chain drive due to change in radius of the articulating links.
- In pulley drive due to vibrations of the belt on slack side.

To account for increase of forces due to these reasons, forces are multiplied by a load factor as given in Table 11.9 as these cannot be calculated exactly. The value depends on type of drive.

Table 11.9 Values of Load Factor

Type of Drive	Rotating Impact Free	Reciprocating	Impact Drive
Gear drive	1.2–1.4	1.4–1.7	2.5–3.5
Chain drive	1.5	1.5	1.5
Type of Belt	**V–Belt**	**Leather Single Ply**	**Leather Double Ply**
Belt drive	2	3	3.5

11.15 Design Procedure for Rolling Bearings

Following data are required for the design of rolling bearings, under normal working conditions without any constraints:

1. Load on bearing; radial and axial if it exists
2. Rotational speed
3. Application such as pump, compressor, and motor
4. Service hours per day
5. Reliability desired

Some additional data may be needed depending on the conditions as under:

1. Any shock coming
2. High temperature, dust, etc.
3. Any space constraints

4. Any bore constraints

5. Chances of misalignment

6. Load direction; fixed or oscillating

7. Fixed race; inner or outer

Step 1 Select desired life

Table 11.10 can be used to select life in thousands of hours L_h for 90 per cent reliability.

Table 11.10 Desired Life of Bearing for an Application

Application	Life in Thousand Hours	Application	Life in Thousand Hours
Automotive - Car, bus	2–5	Blowers: 8 hours / day	20–30
– Truck	1.5–2.5	Blowers: 24 hours / day	40–60
Compressors	40–60	Domestic applications	1–2
Machines for 8 h service	14–20	Electric motors: Commercial	20–30
Machines for 24 hours	50–60	Pumps	40–60

Step 2 Select a rolling element

Select rolling element; ball or roller from the given load and speed from Figure 11.2 in section 11.1 initially and finalize after various checks.

Step 3 Select a type of bearing

Selection of type of bearing also depends on rotational speed and mean diameter of rolling elements set as shown in Figure 11.10.

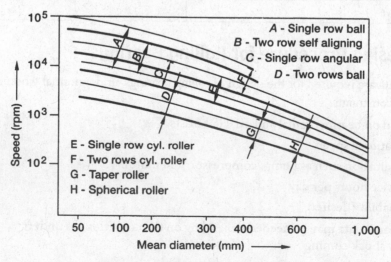

A - Single row ball
B - Two row self aligning
C - Single row angular
D - Two rows ball

E - Single row cyl. roller
F - Two rows cyl. roller
G - Taper roller
H - Spherical roller

Figure 11.10 Type of rolling element

Step 4 Calculate equivalent radial load

Calculate equivalent radial load W_e using Equation (11.5). Values of X and Y can be taken from Table 11.8. V is a factor depending on which race is rotating.

$$W_e = X\,VW_r + YW_a$$

Step 5 Calculate static equivalent load

Calculate equivalent static load W_o using Equation (11.3).

$$W_o = X_o W_r + Y_o W_a$$

If this $W_o > C_o$, permanent deformation will take place. When a bearing rotates, this load is distributed and the load capacity increases 2 to 4 times.

Oscillating loads are defined as, where rolling element rotates less than half revolution and thus approaches static load condition. This load induces rapid false Brinnelling and requires special lubrication techniques.

Step 6 Calculate required bearing capacity

Required load capacity, $C_r = W_e \times K_L \times K_r \times K_N \times K_h$

Where, K_L = Load factor (Section 11.14)

$\quad\quad K_r$ = Reliability factor (Section 11.11)

$\quad\quad K_N$ = Speed factor (Speed in rpm/33.3)$^{1/k}$

$\quad\quad K_h$ = Life factor (Desired life in hours/500)$^{1/k}$

Select a bearing whose C is greater than C_r

Step 7 Check for maximum permissible speed

Figure 11.10 can be used to check the maximum permissible speed of the selected bearing.

It can be calculated precisely from the mean diameter of the bearing:

$$d_m = \frac{\text{Outside diameter} + \text{Bore diameter}}{2}$$

Example 11.7

Bearing design for a mine shaft

A gear box for mine shaft lifts a load of 167 kN using a pulley of diameter 720 mm with a rope speed of 14.5 m/s. Pulley shaft diameter is 450 mm. Life expected for this bearing 60,000 hours. Take reliability 90% and shock factor = 1.1. Design a suitable bearing.

Solution

Given $W = 167$ kN $D' = 720$ mm $v = 14.5$ m/s $D = 450$ mm

               $L_h = 60{,}000$ hours $K_L = 1.1$ $K_r = 1$ (90% reliability)

$$\text{Pulley speed } N = \frac{60v}{\pi D'} = \frac{60 \times 14.5}{3.14 \times 720} = 38 \text{ rpm}$$

Load is high; hence roller bearing is selected. For this application, shock factor is taken as 1.1 and all other factors for dynamic loads are taken as 1.

For bearing desired life of 60,000 hours, life factor K_h is calculated from the following equation:

$$K_h = \left(\frac{L_D}{L_h}\right)^{\frac{1}{k}} = \left(\frac{60,000}{500}\right)^{\frac{3}{10}} = 4.20 \quad (k = 10 / 3 \text{ for roller bearings})$$

Speed factor for 38 rpm, $K_N = \left(\dfrac{N_D}{N_R}\right)^{\frac{1}{k}}$ [Standard speed $N_R = 33.3$ rpm for life of 500 hours]

Or, $\qquad K_N = \left(\dfrac{38}{33.3}\right)^{\frac{3}{10}} = 1.04$

Required capacity of the bearing $C_r = W \times K_L \times K_r \times K_N \times K_h$

Or, $\quad C_r = 167 \times 1.1 \times 1 \times 1.04 \times 4.20 = 802.4 \text{ kN}$

First iteration

Assuming race thickness 25 mm and roller diameter 25 mm.

Mean diameter of rolling elements $D_m = 450 + (2 \times 25) + 25 = 525$ mm

Circumference at pitch circle $= \pi \times 525 = 1,649$ mm

Maximum number of rollers $= \dfrac{1,649}{25} = 66$

Angle in which rollers can be accommodated

$$\Phi = 2 \times 66 \times \sin^{-1}\left(\frac{25}{525}\right) = 400°$$

Hence, the number of rollers that can be accommodated in 360°:

$$Z = \frac{66 \times 360}{400} = 60$$

$$\frac{d \cos \alpha}{D_m} = \frac{25 \cos 360}{525} = 0.0476$$

Assuming effective roller length l_e as 1.4d, that is, $l_e = 1.4 \times 25 = 35$ mm

From Table 11.6, value of $f_c = 65$. Dynamic load capacity of a roller bearing is given as:

$$C = f_c Z^{3/4} d^{29/27} \left(i l_e \cos \alpha\right)^{7/9} \quad [\alpha = 0 \text{ for radial bearing}]$$

Substituting the values: $C = 65 \times 60^{0.75} \times 25^{1.074} \times \left(1 \times 35 \times \cos 0\right)^{0.777}$

Or, $C = 65 \times 21.56 \times 31.72 \times 15.88 = 706,030 \text{ N} = 706.03 \text{ kN}$

This capacity is less than required capacity, that is, 802.4 kN; hence increase roller diameter to 30 mm, for second iteration.

Second iteration

Mean diameter of rolling elements $D_m = 450 + (2 \times 30) + 25 = 535 \text{ mm}$

Circumference at pitch circle $= \pi \times 535 = 1,680 \text{ mm}$

Maximum number of rollers $= \dfrac{1,680}{30} = 56$

Angle in which rollers can be accommodated:

$$\Phi = 2 \times 56 \times \sin^{-1}\left(\frac{30}{535}\right) = 404°$$

Hence, the number of rollers that can be accommodated in 360°:

$$Z = \frac{56 \times 360}{404} = 50$$

$$\frac{d \cos \alpha}{D_m} = \frac{30 \cos 0}{535} = 0.056$$

From Table 11.6, value of $f_c = 66$. Dynamic load capacity of a roller bearing is given as:

$$C = f_c \, Z^{3/4} d^{29/27} \left(il_e \cos \alpha\right)^{7/9} \, [i = 1 \text{ for single row bearing}]$$

Substituting the values: $C = 66 \times 50^{0.75} \times 30^{1.074} \times \left(1.4 \times 30 \times \cos 0\right)^{0.777}$

Or, $C = 66 \times 18.8 \times 38.58 \times 18.25 = 873,632 \text{ N} = 873.632 \text{ kN}$

This capacity is more than required capacity, that is, 802.4 kN; hence either decrease roller diameter or length.

Third iteration

Changing length of roller from 42 mm to 38 mm:

$$C = 66 \times 50^{0.75} \times 30^{1.074} \times \left(1 \times 38 \times \cos 0\right)^{0.777} = 822,000 \text{ N}$$

This capacity is close to desired capacity; hence final design is:

Roller diameter $d = 30 \text{ mm}$

Length of roller $l_e = 38 \text{ mm}$

Number of rollers $Z = 50$

11.16 Bearing Lubrication

Roller bearings are either lubricated with grease or they come prepacked with grease and enclosed from sides to prevent dirt and leakage of grease. Lubrication is done due to following reasons:

- It forms a thin film between sliding or rolling surfaces and thus avoids metal to metal contact.
- Reduces friction arising due to elastic deformation of rolling elements under load.
- It dissipates heat generated during running from the bearing and distributes throughout the bearing.
- It prevents bearing from rust and corrosion.
- It reduces foreign particles to enter.
- Reduces friction power losses.
- Helps bearing to attain the anticipated life.

Selection of lubricant

The lubricant can be either grease or oil. Grease is used when:

- Product of bore in mm × speed in rpm is less than 20,000.
- Temperature is less than 90 °C.
- Rotational speed is less.
- Long operation periods without attention.
- Simple bearing enclosures are to be used.
- Cheapest mode of lubrication.
- Lubricant should maintain desired viscosity at the operating temperature.
- It must be able to form a load sustainable film on the surface.
- Should be able to absorb water to some extent without affecting lubrication properties.

Lubrication properties of grease depend mainly on the properties of base oils used for grease. Therefore, base oils used are, of low viscosity for low temperature and high rotating speeds. High-viscosity base oils are used for heavy-loaded bearings. Base oils used are mainly silicon diester oil and flurocarbon oil. It is costlier than grease lubrication.

Oil is used when:

- Product of bore in mm × speed in rpm is more than 20,000.
- Temperature and rotational speeds are high.
- Bearings are attended regularly.
- Bearing enclosures are fitted with good oil seals.
- For small-size bearings, which operate at high speed, low viscosity oil is used.
- For large-sized heavily loaded bearings, high viscosity oil with some additives is used.
- A central lubrication is to be used to supply oil to all the bearings.

11.17 Mounting of Rolling Bearings

There are many methods of mounting antifriction bearings, but only a few are commonly used which are described below. Housing bore and shaft diameter must be manufactured in close limits for a good mounting. While mounting ball bearings, the revolving race is kept press fit on the shaft or its mating surface and stationary race is kept in transitional fit. Theoretically, these bearings do not need any lubrication, but to protect them from corrosion and ensure smooth running, these are lubricated, generally with grease.

Generally, there are two bearings; one at each end of the shaft. Inner race is backed up against the step of the shaft (Figure 11.11). Outer race is backed up against the housing shoulder and is kept pressed from both sides by end cover plates.

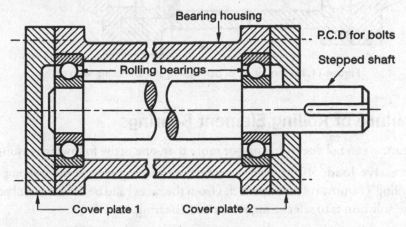

Figure 11.11 Mounting of ball bearings in a housing with fixed bearings

If there are any chances of expansion of shaft, the other side for outer race for the second bearing may be kept floating (Figure 11.12). The inner race is kept in position by nuts on threaded shaft.

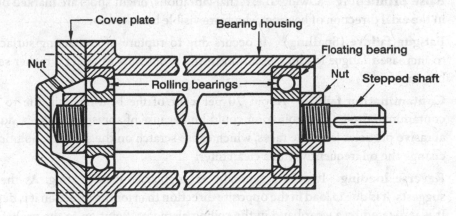

Figure 11.12 Mounting of ball bearings in a housing with floating bearing

While mounting bearings, it should be ensured that the lubricant does not leak. Its leakage is checked by providing lip seals (Figure 11.13).

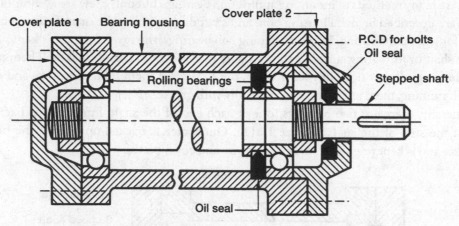

Figure 11.13 Mounting of ball bearings in a housing with seals

11.18 Failure of Rolling Element Bearings

A rolling bearing can fail due to any one or more than one of the following reasons:

1. **Excessive load** If the load is more than the capacity of the bearing, it leads to spalling (removal of metal particles from the races) and bearing may fail prematurely. The solution is to select a higher series of bearing.

2. **Overheating** At operating temperatures more than 200°C, annealing (reduction of hardness) of the bearing starts, which softens the races and the rolling elements. Reasons for this failure are insufficient lubrication, debris in oil, and insufficient cooling of the bearing.

3. **False Brinnelling** Owing to external vibrations, bright spots are marked on races in the axial direction of bearing, which are visible by naked eye.

4. **Fatigue failure (Spalling)** It occurs due to rupture of contacting surfaces due to increased fatigue loading. Solution to the problem is to select a higher series of bearing.

5. **Contamination failure** About 70 per cent of the bearings fail due to use of contaminated oil. Contamination could be because of abrasive particles, dust, and abrasive particles from the races, which cause scratch on the surfaces. Solution is to change the oil frequently with a clean filter.

6. **Reverse loading** It occurs generally in angular contact bearing. As the name suggests, it is due to load in the opposite direction than for the direction it is designed. It is indicated by a wear band in the rolling elements. Solution to the problem is to put in a direction, for which it is designed.

7. **Misalignment** Causes of misalignments in a rotor system are due to bending of shafts, misaligned shoulders on the shaft, nonconcentric bores on which bearing is mounted, and lock nut surfaces that may not be perpendicular. It is indicated by non-parallel running mark of ball on outer race or extra wide ball pathway on the inner race.

8. **Improper lubrication** It is the main item to keep the bearing fit. Lack of lubrication or use of incorrect viscosity of oil leads to premature bearing failure. Discolored races (blue black) indicate this type of failure. It happens due to metal to metal contact of races and rolling elements, which increases temperature.

9. **Improper fit** It could be due to improper fit for shaft that is, shaft diameter is lesser than 0.03 mm than the inside diameter of inner race or bore of bearing housing is more than 0.02 mm than the outside diameter of the outer race.

10. **Fretting** It is generation of metal particles, which oxidize, leaving brown color marks on the surfaces. These are abrasive particles and ruin the surfaces.

Summary

Rolling bearings In these bearings, friction is reduced using a rolling element (ball or a roller) to roll between the bearing surfaces. Friction is less, as rolling friction is lesser than sliding friction. Owing to low friction of rolling bearings, these are also called antifriction bearings. Selection of rolling element ball or roller depends on load and speed.

Rolling bearings are manufactured in a large variety. Selection depends on the application and types of loads: pure radial load, pure axial load, or combined radial and axial load.

Advantages Low starting and running friction, no continuous lubrication is required, shaft center remains at one position, reliable, low maintenance, and easy to mount and replace.

Disadvantages Bearing housings are complicated in construction and need close tolerances, need oil seals to prevent leakage of grease, noisy at very high speeds, costlier than slide bearings, cannot tolerate misalignment, and shocks can damage the bearing.

Construction and nomenclature Rolling bearings are made of five elements:

    a. Outer race   b. Inner race   c. Rolling element   d. Cage   e. Rivet

Classification These are classified according to type of rolling element: Ball or roller.

Ball bearings Different types of ball bearings are:

Deep groove ball bearing is mostly used for general purposes.

Angular contact bearing can be used, if axial load also exists along with radial load.

Thrust bearing is designed for axial loads only. See Figure 11.6 also.

Self-aligning thrust bearing is provided with a spherical seat for self - aligning.

Double row bearing is used, if the load is more, but it supports mainly radial load.

Self-aligning ball bearing is also double row, but its outer race has curved periphery.

Roller bearings The different types of rollers used are:

Cylindrical roller | Taper roller | Spherical roller / Needle roller | Thrust bearing

Bearing designations   ISO or AFBMA (Anti Friction Bearing Manufacturing Association)

ISO designation

The main designation is a seven-digit number. The structure of bearing designation is:

Four optional prefix characters: *Seven-digit main designation—Optional character.*

Seven—digit main designation

Here the digits are defined as: 7 6 5 4 3 2 1. Zeros to the left of the last defined digit are not printed on bearing; for example, a designation of 0006206 is printed as 6,206.

First two digits on right extreme, that is, (2) and (1) give *bore diameter.*

> 00 means bore diameter 10 mm   01 means bore diameter 12 mm
>
> 02 means bore diameter 15 mm   03 means bore diameter 17 mm

From 04 onwards, bore diameter is found by multiplying with 5, for example, 05 means 25.

The *third digit* (3) from right indicates *diameter series*, which defines the outer diameter (Figure 11.7). The diameter series is defined in ascending order as: 1, 2, 3, 4, 5, and 6.

Extra light (100)   Light (200)   Medium (300)   Heavy (400)

The *fourth digit* (4) defines the type of bearing:

> 0 = Ball radial singlerow, for example, 0,400 or 400, 7,204 = Taper roller bearing

Anti-Friction Bearing Manufacturing Association (AFBMA) has recommended two-digit code to specify bearing size. First number of code on right is width series (1, 2, 3, 4, 5, and 6) in increasing order, second number is outside diameter series (8, 9, ...1, 2, 3, 4).

For example, bearing no. 6310 is (6) ball bearing of diameter series 3 with bore 50 mm.

Size of bearings: Series 1, 2, and 3 are most commonly used. Important dimensions of light, medium, and heavy series bearings are tabulated in Tables 11.1 for ball bearings and Table 11.2 for roller bearings.

Static load capacity: When a shaft does not rotate, the load on the bearing is called static load. It causes permanent deformation in races and balls. Static load capacity depends on acceptable permanent deformation, which is taken as 0.0001 times of rolling element diameter. If noise and smoothness of operation is not important, this capacity can be increased up to four times. Static load capacity C_o is given as under:

> Radial ball bearing $C_o = f_o i Z d^2 \cos \alpha$ Radial roller bearing $C_o = f_o i Z l_e d \cos \alpha$
>
> Thrust ball bearing $C_o = f_o i Z d^2 \sin \alpha$ Thrust roller bearing $C_o = f_o i Z l_e d \sin \alpha$

Where, C_o = Basic static load capacity of radial / axial in newtons

f_o = A constant depending on the type of bearing from Table 11.3

i = Number of rows of balls

Z = Number of balls in a row

d = Ball / roller diameter (mm)

l_e = Effective contact length of roller and race (Total length – Chamfer)

α = Nominal angle of contact.

Angle Φ in which rolling elements can be accommodated varies between 185° and 210°:

For ball bearings $\quad \Phi = 2(Z-1)\sin^{-1}\left(\dfrac{d}{D_m}\right)$

For roller bearings $\Phi = 2Z\sin^{-1}\left(\dfrac{d}{D_m}\right)$

Where, D_m = Mean diameter of rolling elements

Static equivalent load: When there is an axial load in addition to radial load, a static *equivalent load capacity* C_o can be defined, which causes same permanent deformation as a mostly loaded ball / roller bearing.

$$C_o = X_o W_r + Y_o W_a$$

Where, X_o = Radial load factor $\qquad\qquad Y_o$ = Axial load factor

$\quad W_r$ = Radial load $\qquad\qquad\qquad W_a$ = Axial load.

Rated life of a bearing Life of a bearing is defined in number of revolutions without failure. Rated life of a bearing is specified in two ways:

Life in million of revolutions L_{10n} $\qquad$ and $\qquad$ Life in hours L_{10h}

Hence, life of a bearing is taken in *number of revolutions* L_{10n} or *in hours* L_{10h} that 90 per cent of the group run satisfactorily without showing any flake. Two standards are:

a. AFBMA recommends rated capacity of bearing for a life of one million revolutions (or hours at a given speed) that 90 per cent of bearings will run safely. These parameters are related with each other as under: $L_n = \left(\dfrac{C}{W_e}\right)^k$

Where, L_n = Life of bearing in number of million revolutions

$\quad C$ = Basic dynamic load rating

$\quad W_e$ = Equivalent dynamic load

$\quad k$ = 3 for ball bearings and k = (10 / 3) for roller bearings.

Life of a rolling bearing in hours L_h is related with life in revolutions L_n and speed N in rpm as:

$$L_h = \frac{L_n}{60\,N} = \frac{10^6}{60\,N}\left(\frac{C}{W}\right)^k$$

Rated life in hours L_{10h} is for 90 per cent of the bearings (10 per cent failure), which run satisfactorily for 500 h at 33.3 rpm $[10^6 / (500 \times 60)]$

b. **Timken Co.** has standardized load ratings for 3,000 h of L_h life at 500 rpm. Life of a rolling bearing depends on its basic dynamic load rating C, equivalent dynamic load, and rated life of bearing.

Reliability of bearing Reliability R of a bearing is defined as the ratio of number of bearings, which run successfully without failure for L million revolutions to the total number of bearings under test. If L_{10} is life of bearing with 10 per cent failure (R = 90 per cent).

$$R = \frac{\text{Number of bearings sucessfully completed } L \text{ revolutions}}{\text{Total number of bearings tested}}$$

Weibull gave a relation between reliability R and bearing life $L = 6.84\left[\log_e\left(\frac{1}{R}\right)\right]^{0.855}$

Generally, it is given with 90 per cent reliability $\dfrac{L}{L_{10}} = \left[\dfrac{\log_e\left(\dfrac{1}{R}\right)}{\log_e\left(\dfrac{1}{R_{90}}\right)}\right]^{0.855}$

Life with varying loads If the load varies then the life for each load is calculated separately and then combined. Let W_1, W_2, W_3, etc., be the loads for n_1, n_2, n_3, etc., then L_1, L_2, L_3, etc., are the lives of a bearing. For the life of n revolutions $n = n_1 + n_2 + \ldots$ revolutions

Introducing equivalent load $W_e = \left[\dfrac{L_1\left(W_1\right)^k + L_2\left(W_2\right)^k + L_3\left(W_3\right)^k}{L_1 + L_2 + L_3}\right]^{\frac{1}{k}}$

Cyclic loads In certain applications, direction of radial load remains the same but the magnitude of load varies sinusoidal in a revolution.

Equivalent load for cyclic loads: $W_e = \dfrac{W_{max}\left(2.5\right)^{\frac{1}{3}}}{2} = 0.678\, W_{max}$

Load factor The forces coming on a bearing may be more than calculated by equilibrium of forces. It happens in gear drive due to inaccuracies of the gear tooth profiles, in chain drive due to change in radius of the articulating links, and in pulley drive due to vibrations of the belt on slack side.

To account for increase of forces due to these reasons, forces are multiplied by a load factor as given in Table 11.9. The value depends on type of drive.

Design procedure for rolling bearings

Following data are required for the design of rolling bearings without any constraints:

1. Load on bearing; Radial and axial, if it exists
2. Rotational speed
3. Application such as pump, compressor, and motor.
4. Service hours per day
5. Reliability desired

Some additional data may be needed depending on the conditions as under:

1. Any shock coming
2. High temperature, dust, etc.
3. Any space constraints
4. Any bore constraints
5. Chances of misalignment
6. Fixed race; inner or outer
7. Load direction; fixed or oscillating

Follow the steps given below to design a bearing:

1. Select desired life from Table 11.10.

2. Select a rolling element: Ball or roller from given load and speed from Figure 11.2.

3. Select a type of bearing; Selection of type of bearing also depends on rotational speed and mean diameter of rolling elements set. Select from Figure 11.10.

4. Calculate equivalent load W_e from given data Values of X and Y can be taken from Table 11.8. V is a factor depending on which race is rotating. $W_e = X V W_r + Y W_a$

5. Calculate static equivalent load $W_o = X_o W_r + Y_o W_a$

6. Calculate required bearing capacity $C_r = W_e \times K_L \times K_r \times K_N \times K_h$

   Where, K_L = Load factor $\qquad\qquad$ K_r = Reliability factor

   $\qquad$ K_N = Speed factor and $\qquad\qquad$ K_h = Life factor.

   Select a bearing whose C is greater than C_r.

7. Check for maximum permissible speed; Use Figure 11.10 to check the maximum permissible speed of the selected bearing for the mean diameter of rolling elements d_m.

Bearing lubrication Roller bearings are either lubricated with grease or they come pre-packed with grease and enclosed from sides to prevent dirt and leakage of grease. Lubrication is done due to following reasons:

- It forms a thin film between sliding or rolling surfaces avoiding metal to metal contact.

- Reduces friction arising due to elastic deformation of rolling elements under load.

- It dissipates heat generated and distributes throughout the bearing.

- It prevents bearing from rust and corrosion.

- It reduces foreign particles to enter.

- Reduces friction power losses.

- Helps bearing to attain the anticipated life.

Selection of lubricant The lubricant can be either grease or oil.

Grease is the cheapest mode of lubrication and used when:

- Product of bore in millimeters × speed in revolutions per meter is less than 20,000.

- Rotational speed and temperature is less than 90°C.

- Long operation periods without attention.

- Simple bearing enclosures are to be used.

- Lubricant should maintain desired viscosity at the operating temperature.

- It must be able to form a load sustainable film on the surface.

- Be able to absorb water to some extent without affecting lubrication properties.

Base oils used for making grease are with low viscosity for low temperature and high rotating speeds. High-viscosity base oils are used for heavy-loaded bearings. Base oils used are mainly silicon diester oil and flurocarbon oil.

Oil is costlier than grease lubrication and used when:

- Product of bore in mm × speed in rpm is more than 20,000.

- Temperature and rotational speed is high.

- Bearings are attended regularly and enclosed with good oil seals.

- A central lubrication can be used to supply oil to all the bearings.

- For small-size bearings, which operate at high speed, low-viscosity oil is used.

- For large-sized heavily loaded bearings, high-viscosity oil with some additives is used.

Mounting of rolling bearings   While mounting ball bearings, the revolving race is kept press fit on the shaft and stationary race is kept in transitional fit. Inner race is backed up against the step of the shaft. Outer race is backed up against the housing shoulder and is kept pressed from both sides by end cover plates.

If there are any chances of expansion of shaft, the other side for outer race for the second bearing may be kept floating. The inner race is kept in position by nuts on threaded shaft. While mounting bearings, it should be ensured that the lubricant does not leak. Its leakage is checked by providing lip seals.

Failure of bearing is due to any one or more of the reasons: Excessive load, overheating, false Brinnelling, contamination, reverse loading, misalignment, improper lubrication, improper fit, and fretting.

Theory Questions

1. Differentiate between sleeve and rolling bearing.

2. What are the advantages and disadvantages of rolling bearings?

3. Describe the construction of a ball bearing with the help of a neat sketch.

4. What are the various types of ball bearings? Give a simple sketch of each type.

5. Sketch different types of roller bearings. Discuss the typical use of each type.

6. Define static load capacity of a bearing. What are the parameters that affect it? Give the relation for a ball and a roller bearing.

7. What is meant by static equivalent load? How is it calculated?

8. Define dynamic load capacity of a bearing. Write the governing equation.

9. What is meant by rated life of a bearing? Describe two methods to specify rated life.

10. Define reliability of a bearing. How does it affect bearing life?

11. Derive an equation to calculate equivalent load for varying loads.

12. What is the dynamic equivalent load? Describe the method to calculate it.

13. Write the procedure to design a rolling bearing.

14. How a bearing is designated? Explain it by giving an example.

15. What is the purpose of lubrication?

16. What are the different methods of mounting bearings? Explain by neat sketches.

17. What could be the possible reasons for a failure of a rolling element bearing?

Multiple Choice Questions

1. Friction coefficient of rolling bearing in comparison to sleeve bearing is:
   (a) Lesser at slow speed
   (b) Greater at slow speeds
   (c) Lesser at high speed
   (d) Almost same

2. A cage in a rolling bearings keeps the:
   (a) Outer and inner race together
   (b) Balls and outer race together
   (c) Balls and inner race together
   (d) Balls together at required distance

3. A self-aligning ball bearing, whose inner surface is cylindrical, has outer race:
   (a) Plain
   (b) Convex.
   (c) Concave
   (d) Half-convex half-concave

4. Spherical roller bearing is used where:
   (a) Load is less and speed high
   (b) Load is high for any speed
   (c) Load and speed both are medium
   (d) Environment is dirty

5. As per ISO designation of rolling bearings, diameter series as 200 means:
   (a) Extra light
   (b) Light
   (c) Medium
   (d) Heavy

6. Taper roller bearing is used where load is:
   (a) High radial
   (b) Axial
   (c) Radial and axial
   (d) With shocks

7. Life of a bearing is taken as one million revolutions for:
   (a) 100 per cent of bearings
   (b) 98 per cent of bearings
   (c) 95 per cent of bearings
   (d) 90 per cent of bearings

8. Timken Co. has standardized load ratings for rolling bearings as:
   (a) 3,000 h at 500 rpm
   (b) 2,000 h at 1,000 rpm
   (c) 1,000 h at 2,000 rpm
   (d) 500 h at 1,500 rpm

9. Grease is used as lubricant for roller bearings when:
   (a) Temperature is less than 90°C
   (b) Rotational speed is less
   (c) Long operation periods without attention
   (d) All given in (a), (b), and (c)

10. If the load remains constant for certain number of revolutions, and then changes for another number of revolutions, life is calculated as:
    (a) Life for average speed
    (b) Life for average load
    (c) Life for each load and speed separately and then combined
    (d) None of the above

11. Which of the following gives least starting friction:
    (a) Rolling
    (b) Hydrodynamic
    (c) Hydrostatic
    (d) Both (a) and (c)

12. A self-aligning bearing has:
    (a) Spherical rollers
    (b) Double row with concave outer race
    (c) Taper roller
    (d) Both (a) and (b)

13. The bearing, which can take heavy radial and axial load, is:
    (a) Cylindrical roller
    (b) Spherical roller
    (c) Taper roller
    (d) None of the above

14. Needle bearing is used where:
    (a) Space for outer race is small
    (b) Loads fluctuate
    (c) Both (a) and (b)
    (d) None of the above

15. A suitable bearing for heavy load and slow speed is:
    (a) Ball bearing
    (b) Roller bearing
    (c) Hydrodynamic bearing
    (d) Hydrostatic bearing

16. For radial loads at high speed along with axial load, the suitable bearing is:
    (a) Single row ball bearing
    (b) Double row ball bearing
    (c) Deep groove ball bearing
    (d) Cylindrical roller bearing

17. Taper roller bearings take:
    (a) Only axial loads
    (b) Only radial loads
    (c) Both with ratio of axial to radial more than one
    (d) Both with ratio of axial to radial less than one

Answers to multiple choice questions

1. (a) 2. (d) 3. (c) 4. (b) 5. (b) 6. (c) 7. (d) 8. (a) 9. (d) 10. (c)
11. (c) 12. (d) 13. (c) 14. (c) 15. (d) 16. (c) 17. (d)

Design Problems

1. A load of 5 kN acts on a ball bearing. Calculate the dynamic load for the ball bearing for a life of 2 million revolutions with 90 reliability. [6.3 kN]

2. A single row deep groove ball bearing no. 6,202 is subjected to an axial thrust of 1,000 N and a radial load of 2,200 N. Find the expected life that 90 per cent of the bearings will complete under these conditions.

[L = 14.13 million revolutions]

3. A radial ball bearing is subjected to radial load of 10 kN. If reliability required is 90 per cent and life of 10,000 h at 1,500 rpm calculate dynamic load capacity. [96,527 N]

4. Dynamic load capacity of a roller bearing is 20 kN. If it rotates at 600 rpm and desired life is 6,000 h with 90 per cent reliability, calculate equivalent load. [3,334 N]

5. Select a suitable deep groove ball bearing for the following data:

   Radial load = 1,500 N   Speed = 1,000 rpm
   Reliability = 90 per cent   Expected life = 9,000 h

   Load factor = 2   [Bearing no. 6404]

6. A shaft rotates at 1,000 rpm and is subjected to a load of 800 N at its ball bearing. The bearing has to have a life of 20,000 h with 98 per cent reliability. Find the dynamic load rating of the bearing at 90 per cent reliability. [6.31 kN]

7. A horizontal shaft of diameter 50 mm supports a load of 10 kN centrally between two radial bearings. Shaft rotates for 8 h per day. Select a ball bearing for a life of 10 years of service with speed of 1,500 rpm. [Bearing no. 6409]

8. A shaft rotates at 1,440 rpm with a radial load of 3 kN. Life of 5 years with 8 h/day working with 90 per cent reliability is desired. Suggest suitable ball bearings for shaft diameters 25 mm, 35 mm, and 45 mm. [Bearing no. 6209, Bearing no. 6307, Bearing no. 6,405]

9. Select a suitable light series antifriction bearing for radial load of 2,000 N with operation at 1,200 rpm for 2,200 h. Assume axial load, if any to be negligible. [Bearing no. 6202]

10. The rolling contact ball bearings are to be selected to support the overhung crank shaft. The shaft rotates at 720 rpm. The bearings are to have 90 per cent reliability corresponding to a life of 24,000 h. The bearing is subjected to an equivalent radial load of 1 kN. Consider life adjustment factor for operating conditions and material as 0.9 and 0.85, respectively. Find the basic dynamic load rating of the bearing from manufacturer's catalogue specified at 90 per cent reliability. [C = 21.62 kN]

11. The radial load on a pump shaft end is 2,250 N. There is a dead axial load of 750 N. The shaft rotates at 900 rpm. Suggest a suitable ball bearing. Use of pump is 8 h/day for a life of 6 years is expected. The shaft end diameter from strength considerations is 20 mm. [Bearing no. 6,204]

12. A ball bearing is operating on a work cycle consisting of three parts a radial load of:

    3,000 N at 1,440 rpm for one quarter of a cycle,

    5,000 N at 720 rpm for one half of a cycle, and

    2,500 N at 1,440 rpm for remaining cycle.

    The expected life of the bearing is 10,000 h. Calculate the dynamic load carrying capacity of the bearing. [33 kN]

13. A ball bearing is subjected to the following radial load cycle:

    5 kN for 5 per cent of time   4 kN for 15 per cent of time

    3 kN for 25 per cent of time   2 kN for 35 per cent of time

    0 kN for rest of time   Calculate equivalent radial load.

    [2.94 kN]

Previous Competition Examination Questions

IES

1. If the equivalent load in case of a radial ball bearing is 500 N and the basic dynamic load rating is 62,500 N, then L_{10} life of this bearing is:

   (a) 1.958 million of revolutions.
   (b) 3.765 million of revolutions.
   (c) 6.953 million of revolutions.
   (d) 9.765 million of revolutions.

   [Answer (a)] [IES 2015]

2. Consider the following statements in connection with thrust bearings:

   1. Cylindrical thrust bearings have higher coefficient of friction than ball thrust bearings.
   2. Taper rollers cannot be employed for thrust bearings.
   3. Double row thrust bearing is not possible.
   4. Lower race, outer race, and retainer are readily separable in thrust bearings.

   Which one of the above statements is correct?

   (a) 1 and 2
   (b) 2 and 3
   (c) 3 and 4
   (d) 1 and 4

   [Answer (d)] [IES 2015]

3. The shoulder provided on the shaft using antifriction bearing is to prevent:

   (a) Axial movement of the shaft.
   (b) Tangential movement of the shaft.
   (c) Rotation of the outer sleeve of the bearing.
   (d) Axial movement of the outer sleeve.

   [Answer (a)] [IES 2013]

4. If the load on a ball bearing is reduced to one third, then its life would increase by:

   (a) 3 times
   (b) 9 times
   (c) 27 times
   (d) 81 times

   [Answer (b)] [IES 2013]

5. Removal of metal particles from the raceway of a rolling contact bearing is a kind of failure known as:

   (a) Pitting
   (b) Wearing
   (c) Spalling
   (d) Scuffing

   [Answer (c)] [IES 2013]

GATE

6. For ball bearings, the fatigue life measured in number of revolutions and the radial load F are related by $FL^{1/3} = K$, where K is a constant. It withstands a radial load of 2 kN for a life of 450 million revolutions. The load (in kN) for a life of one million revolutions is_____

   [Answer: F = 16.286 kN] [GATE 2015]

7. Ball bearings are rated by a manufacturer for a life of 10^6 revolutions. The catalogue rating of a particular bearing is 16 kN. If the design load is 2 kN, the life of the bearing will be $p \times 10^6$ revolutions, where p is equal to_____

[Answer: $p = 512$] [GATE 2014].

8. Spherical roller bearings are normally used:

(a) For increased radial load

(b) For increased thrust load

(c) When there is less radial space

(d) To compensate for angular misalignment

[Answer: (a, d)] [GATE 2013]

9. The life of a ball bearing is inversely proportional to:

(a) $(Load)^{1/3}$

(b) $(Load)^3$

(c) $(Load)^{3.3}$

(d) $(Load)^2$

[Answer: (b)] [GATE 2013]

10. The basic load rating of a ball bearing is:

(a) The maximum static radial load that can be applied without causing any plastic deformation of bearing components.

(b) The radial load at which 90 per cent of the group of apparently identical bearings run for one million revolutions before the first evidence of failure.

(c) The maximum radial load that can be applied during operation without any plastic deformation of bearing components.

(d) A combination of radial and axial loads that can be applied without any plastic deformation.

[Answer: (b)] [GATE 2013]

11. The life of a ball bearing at a load of 10 kN is 8,000 hours. Its life in hours, if its load is increased to 20 kN, keeping all other conditions the same is:

(a) 4,000

(b) 2,000

(c) 1,000

(d) 500

[Answer: (c)] [GATE 2013]

12. Two identical ball bearings P and Q are operating at loads of 30 kN and 45 kN, respectively. The ratio of the life of bearing P to the life of bearing Q is:

(a) 81 / 16

(b) 27 / 8

(c) 9 / 4

(d) 3 / 2

[Answer: (b)] [GATE 2011]

13. A ball bearing operating at a load F has 8,000 hours of life. The life of a bearing in hours when the load is doubled to $2F$ is:

(a) 8,000

(b) 6,000

(c) 4,000

(d) 1,000

[Answer: (d)] [GATE 2005]

□□□

Cylinder of an I.C. Engine

12.1 Introduction to I.C. Engines

An engine is a device that converts chemical energy of fuel into mechanical energy. If the burning of the fuel is inside the cylinder of an engine, it is called internal combustion engine or in short I.C. engine (See Picture on right). If the fuel is burnt outside the cylinder, it is called external combustion engine, for example, a steam engine, in which the fuel is burnt outside the cylinder in a boiler to make steam, which is used in the cylinder to develop power.

I.C. engine uses vaporized fuel in a cylinder, which increases pressure of the gas, when burnt in a cylinder. This pressure pushes a piston in a cylinder. The force on the piston, due to expanding gases in the cylinder is transferred through a connecting rod to a crank shaft, which converts reciprocating motion to rotary motion (Figure 12.1). I.C. engines are of two types; two-stroke and four-stroke as shown in picture below. Small engines are generally two-stroke type. The description given in this chapter is for four-stroke engines only.

Main parts of an engine are cylinder, piston, connecting rod, crank shaft, valves, etc. Function and design of these parts is described in the subsequent chapters / sections.

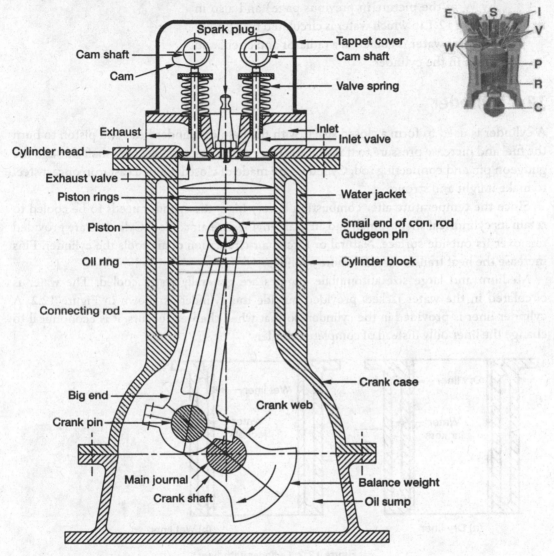

Figure 12.1 An I.C. engine

12.2 Types of Cylinders

An engine needs cooling to save the materials from high temperature in the cylinder. There are two types of cylinders:

- **Air cooled**   Small engines are air cooled. Cylinder is provided with fins to increase surface area for more heat transfer as shown in Picture on the right side.

- **Water cooled**   Water jackets are provided outside the cylinder and cylinder head (marked by W in the picture in previous page) and also in Figure 12.1 in which water is circulated by a pump. The hot water is cooled in a radiator and circulated again in the cylinder.

12.3 Cylinder

A cylinder is used to form a closed space, with the help of cylinder head and piston to burn the fuel and increase pressure to transmit the force of the gases to the crank shaft through the gudgeon pin and connecting rod. Cylinders are made of closed grained cast iron or cast steel to make it light and strong.

Since the temperature after combustion is very high, the cylinder needs to be cooled to retain its cylindrical shape. Small and air craft engines are air cooled. Cylinders are provided fins over its outside surface. Natural or forced air circulation of air cools the cylinder. Fins increase the heat transfer area for more cooling.

Medium and large size automobile engines are generally water cooled. The water is circulated in the water jackets provided outside the cylinder as shown in Figure 12.2. A cylinder liner is provided in the cylinder, so that when the wear occurs, it is economical to change the liner only instead of complete cylinder.

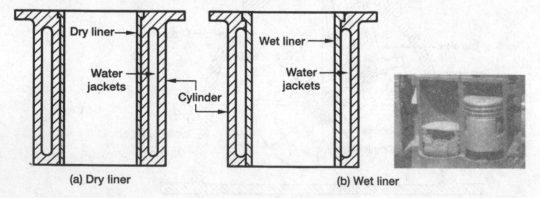

Figure 12.2 Cylinder with liner

There are two types of liners, dry liner shown in Figure 12.2(a) and wet liner in Figure 12.2(b) and picture. In dry liner, water does not touch the surface of the liner. It is fitted tightly in the cylinder. In a wet liner, water touches the outer surface and it forms one side of the water jacket, hence has to be thicker than the dry liner to account for corrosion.

Liners are made of closed grained cast iron, nickel cast iron, and nickel–chromium cast iron. The inside surface is machined, ground, and lapped for a high-grade surface finish to reduce friction and heat treated to reduce wear.

A piston is a component in reciprocating engines / pumps, which slides in a cylinder and is made gas tight with piston rings. Gas sealing is achieved with two or more piston rings (See Chapter 13).

12.4 Mean Effective Pressure

Pressure in the cylinder is not constant, but varies in each stroke. In suction stroke, it is less than atmosphere, and in compression stroke it increases, as the piston moves upwards. Maximum pressure occurs in power stroke, when the fuel mixture is burnt. This pressure is very high and may be even eight to ten times the average pressure. The burnt gases are expelled from the cylinder during exhaust stroke at a pressure slightly above atmospheric. The mean pressure p_m is the average pressure of all the four strokes in a cycle. It is measured from the indicator diagram. Area of the diagram divided by the length gives indicated mean power. This value can be used to compare the performance of engines. Actually, it indicates the power developed per unit volume. Higher value of this pressure indicates more power per unit swept volume (area × stroke) of the cylinder. Following values of the mean effective pressure can be taken to calculate cylinder volume described in the next section.

Petrol engines	0.85 MPa – 1.05 MPa
Turbo charged four-stroke engines	1.25 MPa – 1.7 MPa
Large two-stroke diesel engines	1.9 MPa

12.5 Size of Cylinder and Power Developed

The size of the cylinder depends on the power, which an engine develops in one cylinder, speed of operation, and mechanical efficiency. The power developed in the cylinder is called indicated power P_i, which depends on the mean pressure, cylinder diameter, length of stroke of piston l, and engine speed N. It can be calculated using the relation given by Equation (12.1).

$$P_i = \frac{p_m l A n}{60,000} \tag{12.1}$$

Where, P_i = Indicated power (kW)

p_m = Mean effective pressure (MPa)

l = Stroke of piston (mm)

A = Area of cylinder $(0.785D^2)$, D is diameter of cylinder (mm^2)

N = Speed of engine (rpm)

n = Number of working strokes, $n = N$ for two-stroke engines; and

$n = 0.5\ N$ for four-stroke engines.

Out of this indicated power, some power is wasted in friction called as frictional power P_f and hence the net power available at the crank shaft is brake power P, that is, $P = P_i - P_f$.

Mechanical efficiency η_{mech} of an engine is the ratio of brake power P and indicated power P_i. It can be assumed as 80 per cent, if not given. Thus P and P_i are related as:

$$\eta_{mech} = \frac{P}{P_i} \tag{12.2}$$

From Equations (12.1) and (12.2)

$$\text{Brake power}\quad P = \frac{p_m\, l A n}{60,000} \times \eta_{mech} \tag{12.3}$$

Length of stroke 1 and diameter ratio varies from 1 to 1.5. If not known, it can be assumed as 1.25, that is, $l = 1.25\ D$. Assuming a suitable value of mean effective pressure as given in Section 12.4, cylinder diameter can be calculated from the given power. Length of cylinder L is kept 1.15 times of stroke.

Hence, length of cylinder, $L = 1.15l$

$$\tag{12.4}$$

Example 12.1

Cylinder size for given power and speed
A four-stroke petrol engine develops 10 kW at 1,000 rpm. Assuming mean effective pressure of 0.9 MPa, L/D ratio 1.2, and mechanical efficiency 80 per cent calculate diameter and length of cylinder.

Solution

Given $P = 10\ \text{kW}$ $N = 1,000\ \text{rpm}$ $p_m = 0.9\ \text{MPa}$ $L/D = 1.2$
$\eta_{mech} = 80$ per cent

From Equation (12.3) brake power: $P = \dfrac{p_m\, l A n}{60,000} \times \eta_{mech}$

Substituting the values in the above equation:

$$10 = \frac{0.9 \times \left(1.2D\right) \times \left(0.785 D^2\right) \times \left(0.5 \times 1,000\right) \times 0.8}{60,000} = 0.00565 D^3$$

Or, Cylinder diameter $D = 56\ \text{mm}$

Length of cylinder $L = 1.15\ l = 1.15 \times 1.2\ D = 1.15 \times 1.2 \times 56 = 77.3\ \text{mm}$

12.6 Design of Cylinder

Design of the cylinder involves calculation of its wall thickness, its flange, and studs.

12.6.1 Stresses in cylinder

There are three types of stresses acting in a cylinder (Figure 12.3):

- Longitudinal stress along the axis of cylinder, due to gas pressure on piston and head [Figure 12.3(a)].
- Circumferential stress due to radial pressure of the gases. Symbol σ_h, that is, hoop stress is used to avoid confusion of σ_c symbol used earlier for compressive stress [Figure 12.3(b)].
- Bending stress, due to side thrust of the piston. It is quite less in comparison to above-mentioned stresses, hence generally neglected.

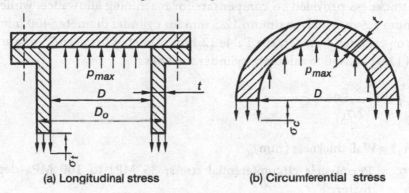

(a) Longitudinal stress   (b) Circumferential stress

Figure 12.3 Stresses in a cylinder

Longitudinal apparent stress $\sigma_{la} = \dfrac{\text{Force along axis of cylinder}}{\text{Area of cylinder}} = \dfrac{0.785 D^2 \times p_{max}}{0.785\left(D_o^2 - D^2\right)}$

Or, $\quad \sigma_{la} = \dfrac{D^2 \times p_{max}}{\left(D_o^2 - D^2\right)}$ (12.5)

Where, D_o = Outside diameter of cylinder $(D + 2t)$
$\quad\quad D$ = Inside diameter of cylinder
$\quad\quad p_{max}$ = Maximum pressure in cylinder. It varies from 8 to 10 times the mean pressure.

Apparent hoop stress $\sigma_{ha} = \dfrac{\text{Force at right angle to axis}}{\text{Wall area, ressisting force}} = \dfrac{(D \times L) \times p_{max}}{2t \times L}$

Length of cylinder $L = 1.15l$, where l is stroke length of piston

Or, $\quad \sigma_{ha} = \dfrac{D \times p_{max}}{2t}$ (12.6)

Where, L = Length of cylinder; $(D \times L)$ is the projected area for the radial pressures

t = Wall thickness of cylinder

Since these two stresses are at right angle to each other, the net stresses are related with Poisson's ratio μ given as under:

Net longitudinal stress $\sigma_l = \sigma_{la} - \mu \sigma_{ha}$ (12.7)

Net circumferential stress $\sigma_h = \sigma_{ha} - \mu \sigma_{la}$ (12.8)

12.6.2 Wall thickness of cylinder

An engine cylinder is considered as thin cylinder to calculate its wall thickness. There exists wear of cylinder; hence the thickness is slightly increased to allow for re-boring. It is extra thickness provided to compensate for machining allowance, while re-boring. It is minimum 1.5 mm and maximum 12.5 mm for cylinder diameter 400 mm and more. It depends on cylinder diameter; see Table 12.1 for diameters between 75 mm and 400 mm. Equation (12.9) is used to calculate cylinder thickness.

$$t = \frac{D \times p_{max}}{2\sigma_h} + C$$ (12.9)

Where, t = Wall thickness (mm)

σ_h = Permissible circumferential stress; 35 MPa to 100 MPa depending on material

p_{max} = Maximum gas pressure in MPa after combustion. If not given, assume 10 times p_m

D = Cylinder diameter (mm)

C = Re-boring allowance (mm)

Table 12.1 Reboring Allowance C

D (mm)	75	100	150	200	250	300	350	400
C (mm)	1.5	2.4	4.0	6.3	8.0	9.5	11	12.5

Value of cylinder wall thickness depends on its diameter and varies from 5 mm to 25 mm. Following empirical relations can be used to calculate its thickness:

$$t = 0.045\,D + 1.5 \text{ mm}$$ (12.10a)

Thickness of dry liner $t = 0.03\,D$ to $0.035\,D$ (12.10b)

Thickness of water jacket wall $t_j = 0.032\,D + 1.5 \text{ mm}$ (12.11a)

Or, $t_j = 0.33\,t$ to $0.75\,t$ (12.11b)

Space between outside of cylinder and inside surface of jacket wall t_w is:

$$t_w = 0.08D + 6.5 \text{ mm} \qquad (12.12)$$

Example 12.2

Wall thickness and stresses in a cylinder

A four-stroke petrol engine has cylinder diameter 100 mm and peak pressure 5 MPa. Cylinder is made from cast iron having safe tensile strength 42 MPa and Poisson's ratio 0.25. Calculate:

a. Wall thickness
b. Net longitudinal stress
c. Net circumferential stress

Solution

Given $D = 100$ mm $\qquad p_{max} = 5\,\text{MPa}$ $\qquad \sigma_t = 42$ MPa $\qquad \mu = 0.25$

a. From Equation (12.9): $t = \dfrac{D \times p_{max}}{2\sigma_t} + C$

From Table 12.1 for $D = 100$ mm, $C = 2.4$ mm.

Substituting the values in the above equation:

$$t = \frac{100 \times 5}{2 \times 42} + 2.4 = 8.35 \text{ say 9 mm}$$

Cylinder outside diameter $D_o = D + 2\,t = 100 + (2 \times 9) = 118$ mm

From Equation (12.5), $\sigma_{la} = \dfrac{D^2 \times p_{max}}{\left(D_o^2 - D^2\right)} = \dfrac{100^2 \times 5}{\left(118^2 - 100^2\right)} = 12.74$ MPa

From Equation (12.6), $\sigma_{ha} = \dfrac{D \times p_{max}}{2t} = \dfrac{100 \times 5}{2 \times 9} = 27.7$ MPa

From Equation (12.7), net longitudinal stress:

$$\sigma_l = \sigma_{la} - \mu\sigma_{ha} = 12.74 - \left(0.25 \times 27.7\right) = 5.8 \text{ MPa}$$

From Equation (12.8), net circumferential stress:

$$\sigma_h = \sigma_{ha} - \mu\sigma_{la} = 22.7 - \left(0.25 \times 12.74\right) = 24.5 \text{ MPa}$$

12.6.3 Flange size

Flanges are casted as integral part of the cylinder. A cylinder may have flange only on top to fix the cylinder head (Figure 12.1), and is casted integral with the crankcase or a flange at bottom also to fix on the flange of the crank case, if casted separately. The cylinder head is fixed on this flange using bolts / studs and nuts. The flange is drilled with holes equal to the number of bolts at a pitch circle diameter (PCD)D_p. It can be calculated as:

$$D_p = D + 3\,d \tag{12.13}$$

Where, D = Cylinder diameter
$\quad\quad d$ = Bolt diameter

Outside diameter of the flange D_f can be taken as:

$$D_f = D_p + 3\,d = D + 6\,d \tag{12.14}$$

Thickness of the flange t_f is kept 20 per cent to 40 per cent more than the wall thickness of the cylinder, that is,

$$t_f = 1.2\,t \text{ to } 1.4\,t \tag{12.15}$$

12.6.4 Studs / Bolts

Number of bolts / studs depends on cylinder diameter and the gas pressure inside. The force exerted on the cylinder head due to gas pressure is:

$$F = \frac{\pi}{4}D^2\,p_{max}$$

Where, D = Cylinder diameter
$\quad\quad p_{max}$ = Maximum gas pressure in the cylinder

This force is to be resisted by all the bolts together. Equating the gas force with the resisting force:

$$\frac{\pi}{4}D^2\,p_{max} = z \times \left(\frac{\pi}{4}d_c^2\right)\sigma_t \tag{12.16}$$

Where, z = Number of bolts / studs
$\quad\quad d_c$ = Core diameter of threaded portion of bolt. It can be assumed as $0.84\,d$.
$\quad\quad \sigma_t$ = Safe tensile stress for the bolt / stud (It varies from 35 MPa to 70 MPa).

The number of bolts depends on cylinder diameter D. The number z can be checked using the following empirical relations:

Minimum number of bolts $z_{min} = 0.01\,D + 4$ $\hspace{3cm}$ (12.17a)

Maximum number of bolts $z_{max} = 0.02\,D + 4$ $\hspace{3cm}$ (12.17b)

The actual distance between two bolts is called pitch and can be calculated as:

$$\text{Pitch} = \frac{\pi D_p}{z} \tag{12.18}$$

Pitch can vary between the following minimum and maximum values:

$$\text{Minimum pitch} = 19\sqrt{d} \tag{12.19a}$$

$$\text{Maximum pitch} = 28\sqrt{d} \tag{12.19b}$$

These bolts are subjected to variable loads due to varying gas pressure. A gasket is also used between the cylinder flange and cylinder head to make it leak proof. Hence the treatment of such bolts is different than normal bolts in tension. This type of loading is described in Chapter 16 of volume 1.

12.7 Cylinder Head

It is used to cover the cylinder at top and is fixed on the flange of cylinder using studs / bolts. It is of considerable depth because it has to accommodate ports and valves for inlet and outlet of gases. It also accommodates a spark plug for a petrol engines or an injector for a diesel engine. A cylinder head with overhead double cam shaft is shown in Picture on the right side.

Thickness of cylinder head t_H (See Figure 12.1) is calculated using the following relation:

$$t_H = D\sqrt{\frac{p_{max}}{10 \times \sigma_t}} \tag{12.20}$$

Where, D = Cylinder diameter (mm)

p_{max} = Maximum pressure in the cylinder (MPa)

σ_t = Safe tensile strength (MPa) (If not given, assume between 35 MPa and 50 MPa)

PCD of the bolts $D_p = D + 3\,d$

Diameter of head $D_H = D_F = D + 6\,d$

Where, D = Cylinder diameter (mm)

d = Nominal diameter of the bolt

PCD and number of bolts for cylinder head have necessarily be the same, as of flange of cylinder.

Example 12.3

Design of a cylinder head

Design a cylinder head for an engine having cylinder diameter 80 mm and maximum gas pressure 4.5 MPa. Also calculate bolt diameter, number of bolts, and pitch. Assume safe tensile strength for the head 40 MPa and for bolts 60 MPa.

Solution

Given $D = 80$ mm $p_{max} = 4.5$ MPa

For head $\sigma_t = 40$ MPa   For bolts $\sigma_t = 60$ MPa

Using Equation (12.20) $t_H = D\sqrt{\dfrac{p_{max}}{10 \times \sigma_t}} = 80\sqrt{\dfrac{4.5}{10 \times 40}} = 8.5$ mm

Maximum gas force $F = \dfrac{\pi}{4}D^2 \times p_{max} = 0.785 \times 80^2 \times 4.5 = 22{,}608$ N

From Equations (12.17a): $z_{min} = 0.01\,D + 4 = 0.01 \times 80 + 4 = 4.8$

From Equations (12.17b): $z_{max} = 0.02\,D + 4 = 0.02 \times 80 + 4 = 5.6$

So take number of bolts $z = 5$.

Equating resisting force of core diameter of bolts $(0.84d)$ with maximum gas pressure:

$$z \times \frac{\pi}{4}\left(0.84d\right)^2 = F \text{ Substituting the values:}$$

$5 \times 0.785 \times (0.84\,d)^2 = 22{,}608$ Or, $d = 11.6$ say 12 mm

From Equation (12.13) PCD: $D_p = D + 3d = 80 + (3 \times 12) = 116$ mm

Outside diameter of the flange D_f can be taken from Equation (12.14):

$$D_f = D + 6d = 80 + (6 \times 12) = 152 \text{ mm}$$

From Equation (12.15), thickness of the flange t_f is kept 20 per cent to 40 per cent more than the wall thickness of the cylinder. Assuming 30 per cent more:
$t_f = 1.3\,t = 1.3 \times 8.5 = 11$ mm

From Equations (12.18):   Pitch $= \dfrac{\pi D_p}{z} = \dfrac{3.14 \times 116}{5} = 72.5$ mm

From Equation (12.19a) Minimum pitch $= 19\sqrt{d} = 19\sqrt{12} = 65.8$ mm

From Equation (12.19b) Maximum pitch $= 28\sqrt{d} = 28\sqrt{12} = 97$ mm

The calculated value 72.5 mm is within the range, hence accepted.

Example 12.4

Design of cylinder and cylinder head

A diesel engine working on four-stroke cycle develops 10 kW brake power at 1,500 rpm. Mean effective pressure is 0.6 MPa and mechanical efficiency 78 per cent. Maximum pressure may be taken eight times the mean effective pressure. Cylinder is made of cast iron having ultimate tensile strength 280 MPa and Poisson's ratio 0.25. Ultimate tensile strength of bolt material is 350 MPa. Assuming factor of safety as 5, calculate:

  a. Cylinder diameter and length assuming l/D ratio 1.2

  b. Thickness of liner

  c. Net stresses

  d. Diameter and number of bolts, flange thickness and diameter

  e. Thickness of cylinder head

Solution

Given $P = 10\,\text{kW}$ $N = 1,500\,\text{rpm}$ $p_m = 0.6\,\text{MPa}$ $l/D = 1.2$

       $\eta_{mech} = 78$ per cent $\mu = 0.25$ $p_{max} = 8\,pm$ $FOS = 5$

       For cylinder $S_{ut} = 280\,\text{MPa}$ For bolts $S_{ut} = 350\,\text{MPa}$

  a. From Equations (12.3), brake power: $P = \dfrac{p_m l A n}{60,000\,\eta_{mech}}$

Substituting the values (l and D in meters) (For four-stroke, $n = 0.5\,N$)

$$10 = \frac{0.6 \times \left(1.2 \times 10^{-3} D\right) \times \left(0.785 D^2\right) \times \left(0.5 \times 1,500\right)}{60,000 \times 0.78} = 9.06 D^3$$

Cylinder diameter $D = 103$ mm

Length of stroke $l = 1.2\,D = 1.2 \times 103 = 126$ mm

Length of cylinder $L = 1.15\,D = 1.15 \times 126 = 145$ mm

  b. From Table 12.1, by interpolation re-boring factor $C = 2.5$ mm

Safe circumferential stress $\sigma_h = \dfrac{S_{ut}}{FOS} = \dfrac{280}{5} = 56$ MPa

From Equation (12.10), thickness of liner is, $t = \dfrac{D \times p_{max}}{2\sigma_h} + C$

Substituting the values:

Wall thickness $t = \dfrac{103 \times 8 \times 0.6}{2 \times 56} + 2.5 = 6.5$ mm

  c. Outside diameter $D_o = D + 2\,t = 103 + (2 \times 6.5) = 116$ mm

From Equation (12.5): $\qquad \sigma_{la} = \dfrac{D^2 \times p_{max}}{\left(D_o^2 - D^2\right)}$

Substituting the values:

Apparent longitudinal stress $\sigma_{la} = \dfrac{103^2 \times 8 \times 0.6}{\left(116^2 - 103^2\right)} = 17.9 \text{ MPa}$

From Equation (12.6): $\sigma_{ha} = \dfrac{D \times p_{max}}{2t}$

Substituting the value, apparent circumferential stress is:

$$\sigma_{ha} = \dfrac{103 \times 8 \times 0.6}{2 \times 6.5} = 38 \text{ MPa} \left(\text{less than 126 MPa, hence safe}\right)$$

From Equation (12.7) net longitudinal stress:

$$\sigma_t = \sigma_{la} - \mu\sigma_{ha} = 17.9 - 0.25 \times 38.6 = 8.4 \text{ MPa}$$

From Equation (12.8) net circumferential stress:

$$\sigma_c = \sigma_{ha} - \mu\sigma_{la} = 38.6 - 0.25 \times 17.9 = 33.5 \text{ MPa}$$

d. From Equation (12.17a): $Z_{min} = 0.01 D + 4 = (0.01 \times 103) + 4 = 5$

From Equation (12.17b): $Z_{max} = 0.02 D + 4 = (0.02 \times 103) + 4 = 6$

So taking number of bolts $z = 6$

Safe tensile strength $\sigma_t = \dfrac{S_{ut}}{FOS} = \dfrac{350}{5} = 70 \text{ MPa}$

Maximum gas force $F = \dfrac{\pi}{4}D^2 \times p_{max} = 0.785 \times 103^2 \times 0.8 \times 6 = 39{,}975 \text{ N}$

Resisting force by all bolts $= z \times \left(\dfrac{\pi}{4}d_c^2\right)\sigma_t = 6 \times 0.785 \times d_c^2 \times 70 = 330 d_c^2$

Equating this force with gas force:

$330 d_c^2 = 39{,}975$, or, $d_c = 11 \text{ mm}$

Hence, nominal bolt diameter, $d = \dfrac{\text{Core diameter}}{0.84} = \dfrac{11}{0.84} = 14 \text{ mm}$

e. Pitch circle diameter, $D_p = D + 3d = 103 + (3 \times 14) = 145 \text{ mm}$

$$\text{Pitch} = \dfrac{\pi D_p}{z} = \dfrac{3.14 \times 145}{6} = 75.9 \text{ mm}$$

From Equation (12.19a): Minimum pitch $= 19\sqrt{d} = 19\sqrt{14} = 71.1$ mm

From Equation (12.19b): Maximum pitch $= 28\sqrt{d} = 28\sqrt{14} = 104.7$ mm

Pitch 75.9 mm is in the desired range, hence selected.

f. Thickness of cylinder head is given by Equation (12.20) as:

$$t_H = D\sqrt{\frac{p_{max}}{10 \times \sigma_t}} = 103\sqrt{\frac{0.8 \times 6}{10 \times 56}} = 9.5 \text{ say } 10 \text{ mm}$$

Summary

Internal combustion engine Fuel is burnt inside the cylinder to get mechanical energy. Cooling of cylinders of I.C. engines is of two types; air cooled and water cooled. I.C. engines are of two types; two-stroke and four-stroke. Main parts of an engine are cylinder, piston, connecting rod, crank shaft, valves, etc.

Cylinder A cylinder is used to form a closed space with the help of cylinder head. Cylinders are made of closed grained cast iron or cast steel to make it light and strong. Natural or forced air circulation of air cools the cylinder. Fins increase the heat transfer area for more cooling.

Medium and large-size automobile engines are generally water cooled. The water is circulated in the water jackets provided outside the cylinder. A cylinder liner is provided in the cylinder so that when the wear occurs, it can be easily changed.

Liners are of two types *Dry liner* and *wet liner*. In dry liner, water does not touch the surface of the liner. It is fitted tightly in the cylinder. In a wet liner, water touches the outer surface of the liner and it forms one side of the water jacket, hence has to be thicker than the dry liner to account for corrosion. Liners are made of closed grained cast iron, nickel cast iron, and nickel–chromium cast iron. The inside surface is machined, ground, and lapped for a high-grade surface finish to reduce friction and heat treated to reduce wear.

Mean effective pressure Pressure in the cylinder varies in each stroke. Maximum pressure occurs in power stroke, when the fuel mixture is burnt. This pressure is very high and may be even eight to times the average pressure. The mean pressure p_m is the average pressure of all the four strokes in a cycle. It can be taken as: for petrol engines 0.85 MPa – 1.05 MPa, turbo charged four-stroke engines 1.25 MPa –1.7 MPa.

Size of cylinder and power Size of the cylinder depends on the power, developed in cylinder, speed of operation, and mechanical efficiency. The power developed in the cylinder called as *Indicated power* $P_i = \dfrac{p_m l A n}{60,000}$

Where, l = Stroke of piston

A = Area of cylinder $(0.785 D^2)$

D = Diameter of cylinder

Number of working strokes, $n = N$, Speed of two-stroke engines and $n = 0.5 N$ for four-stroke engines

Power wasted in friction is called *frictional power* P_f and hence the net power available at the crank shaft is *brake power* P, that is, $P = P_i - P_f$.

Mechanical efficiency η_{mech} of an engine is the ratio of brake power and indicated. It can be assumed as 80 per cent, if not given. Thus P and P_i are related as: $\eta_{mech} = \dfrac{P}{P_i}$

From the above Equations:

$$\text{Brake power} = \frac{p_m \, l \, A \, n}{60,000} \times \eta_{mech}$$

Length of stroke and diameter ratio varies from 1 to 1.5. If not known, it can be assumed as 1.25, that is, $l = 1.25 \, D$. Length of cylinder $L = 1.15 l$.

Stresses in cylinder   There are three types of stresses acting in a cylinder.

- *Longitudinal stress* along the axis of cylinder due to gas pressure on piston and head.

- *Circumferential stress* due to radial pressure of the gases.

- *Bending stress* due to side thrust of the piston. It is quite less, hence neglected.

Longitudinal apparent stress $\sigma_{la} = \dfrac{0.785 D^2 \times p_{max}}{0.785 \left(D_o^2 - D^2 \right)} = \dfrac{D^2 \times p_{max}}{\left(D_o^2 - D^2 \right)}$

D = Inside diameter of cylinder

D_o = Outside diameter, $(D + 2t)$

t = Wall thickness

p_{max} = Maximum pressure in cylinder. It varies from 8 to 10 times the peak pressure.

Apparent hoop stress $\sigma_{ha} = \dfrac{(D \times L) \times p_{max}}{2t \times L} = \dfrac{D \times p_{max}}{2t}$

These two stresses are related with Poisson's ratio μ given as under:

Net longitudinal stress $\sigma_l = \sigma_{la} - \mu \, \sigma_{ha}$

Net circumferential stress $\sigma_h = \sigma_{la} - \mu \, \sigma_{ha}$

Wall thickness of cylinder   An engine cylinder is considered as thin cylinder to calculate its wall thickness. Extra thickness is provided to compensate for machining allowance while reboring. It is minimum 1.5 mm and maximum 12.5 mm for cylinder diameter 400 mm and more. It depends on cylinder diameter; see Table 5.1 for diameters between 75 mm and 400 mm.

Calculate cylinder thickness, $t = \dfrac{D \times p_{max}}{2 \sigma_h} + C$ (C = Reboring allowance in mm)

$\sigma_h = 35 - 100$ MPa (depending on material)

p_{max} = Maximum gas pressure in Mpa after combustion, if not given, assume 10 times p_m.

Value of cylinder wall thickness depends on its diameter and varies from 5 mm to 25 mm.

Following empirical relations can be used to calculate its thickness:

$t = 0.045 \, D + 1.5$ mm

Thickness of dry liner, $t = 0.03\,D$ to $0.035\,D$

Thickness of water jacket wall, $t_j = 0.032\,D + 1.5$ mm

Or, $t_j = 0.33t$ to $0.75t$

Space between outside of cylinder and inside surface of jacket wall $t_w = 0.08\,D + 6.5$ mm

Flange size Flanges are casted as integral part of the cylinder. The cylinder head is fixed on this flange using bolts / studs and nuts. The flange is drilled with holes equal to the number of bolts at a pitch circle diameter, $D_p = D + 3d$ where, $d =$ Bolt diameter

Outside diameter of the flange, $D_f = D_p + 3\,d = D + 6\,d$

Thickness of the flange t_f is kept 20 per cent to 40 per cent more than the wall thickness $t_f = 1.2\,t$ to $1.4\,t$

Studs / Bolts Force exerted on cylinder head due to gas pressure is: $F = \dfrac{\pi}{4}D^2\,p_{max}$

Equating the gas force with the resisting force: $\dfrac{\pi}{4}D^2\,p_{max} = z \times \left(\dfrac{\pi}{4}d_c^2\right)\sigma_t$

z = Number of bolts / studs

d_c = Core diameter of threaded portion of bolt. It can be assumed as $0.84\,d$

σ_t = Safe tensile stress for the bolt / stud (from 35 MPa to 70 MPa)

Minimum number of bolts = $0.01\,D + 4$ and maximum number of bolts = $0.02\,D + 4$

The actual distance between two bolts is called pitch $= \dfrac{\pi\,D_p}{z}$

Pitch can vary between minimum pitch $= 19\sqrt{d}$ and maximum pitch $= 28\sqrt{d}$

A *gasket* is used between the cylinder flange and cylinder head to make it leak proof.

Cylinder head It is used to cover the cylinder at top and is fixed on the flange of cylinder using studs / bolts.

Thickness of head $t_H = D\sqrt{\dfrac{p_{max}}{10 \times \sigma_t}}$

PCD of the bolts $D_p = D + 3\,d$

Diameter of head $D_H = D_F = D + 6\,d$

Theory Questions

1. Name the important parts of an I.C. engine and function of each.
2. What is the function of a liner? Describe the two types of liners and compare their performance.
3. What is meant by mean effective pressure? How can it be used to calculate cylinder volume?
4. Describe the following terms:
   a. Indicated power
   b. Brake power
   c. Frictional power
   d. Mechanical efficiency

5. Describe the various stresses coming in a cylinder.

6. What is re-boring allowance? How do you calculate the wall thickness of a cylinder?

7. Write a note on fixing the size of a cylinder flange.

8. How do you calculate the number of studs or bolts required to fix cylinder head with cylinder?

9. Write the method to design a cylinder head.

10. Write the method to calculate bolt size for cylinder head.

Multiple Choice Questions

1. Cylinder volume is calculated from:
   (a) Maximum pressure
   (b) Half of maximum pressure
   (c) Mean effective pressure
   (d) Average pressure

2. Cylinders are made of:
   (a) High carbon steel
   (b) Mild steel
   (c) Aluminium alloy
   (d) Cast iron

3. Pressure calculated from the indicator diagram is:
   (a) Brake mean effective pressure
   (b) Indicated mean effective pressure
   (c) Maximum pressure
   (d) Average pressure

4. Stresses in a cylinder are:
   (a) Hoop
   (b) Longitudinal
   (c) Both hoop and longitudinal
   (d) Shear

5. Wall thickness of a cylinder is calculated using:
   (a) Hertz stresses
   (b) Hoop stresses
   (c) Compressive stresses
   (d) Shear stresses

6. Flange diameter of a cylinder diameter D and bolt diameter d is:
   (a) $1.25\,D$
   (b) $1.5\,D$
   (c) $D + 3\,d$
   (d) $D + 6\,d$

Answers to multiple choice questions

1. (c)   2. (d)   3. (b)   4. (c)   5. (b)   6. (d)

Design Problems

1. A four-stroke engine produces 5 kW at 3,000 rpm with mean effective pressure of 1.1 MPa, L/D ratio 1.2 and mechanical efficiency 82 per cent calculate diameter and length of cylinder.

   [D = 62 mm]

2. A four-stroke petrol engine has cylinder diameter 80 mm and peak pressure 5.5 MPa. Cylinder is made from cast iron having safe tensile strength 48 MPa and Poisson's ratio 0.25. Calculate:

    a. Wall thickness with reboring allowance of 1.5 mm

    b. Net longitudinal stresses

    c. Net circumferential stresses

    $$[t = 6.2 \text{ mm}, \sigma_l = 7.63 \text{ MPa}, \sigma_h = 31.4 \text{ MPa}]$$

3. Design a cylinder head for an engine having cylinder diameter 100 mm and maximum pressure 4.0 MPa. Also calculate bolt diameter, number of bolts and pitch. Assume safe tensile strength for the head 42 MPa and for bolts 55 MPa. Check if the joint is leak proof.

    $$[t_H = 9.8 \text{ mm}, D_H = 190 \text{ mm}, d = 15 \text{ mm}, z = 6, p = 75.9 \text{ mm, Yes, leak proof}]$$

4. A diesel engine cylinder is made of cast iron. It works on four-stroke cycle with the following data:

    Brake power = 7.5 kW at 2,000 rpm

    Mean effective pressure = 0.75 MPa

    Maximum pressure may be taken eight times the mean effective pressure

    Mechanical efficiency = 76 per cent

    Ultimate tensile strength for cylinder = 280 MPa for bolt = 350 MPa

    Poisson's ratio = 0.25

    Factor of safety = 5

    Calculate:

    a. Cylinder diameter and length assuming L / D ratio 1.2

    b. Thickness of cylinder

    c. Net longitudinal and hoop stresses

    d. Number of bolts and diameter

    e. Flange thickness and diameter

    f. Thickness of cylinder head

    $$[D = 93 \text{ mm}, t = 6.5 \text{ mm}, \sigma_l = 9.3 \text{ MPa}, \sigma_h = 38 \text{ MPa}, z = 6,$$
    $$d = 14 \text{ mm}, t_f = 9.6 \text{ mm}, D_f = 177 \text{ mm}, t_H = 9.6 \text{ mm}]$$

Pistons

13.1 Definition and Function

Piston is a reciprocating hollow cylinder closed at one end with a transverse hole in the middle for a pin and grooves on periphery on upper half to provide piston rings to stop leakage of gas from the clearance between piston and cylinder. It is shown in picture on right side.

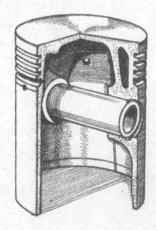

Main function of the piston is to transmit force created by burning fuel on the upper side through piston pin and connecting road. A crank shaft provided at the other end of the connecting rod converts reciprocating motion into rotary motion. Piston also takes heat through rings provided on it and transfers to the cylinder walls.

13.2 Desirable Characteristics of Piston

Pistons are designed with features to perform specific functions during engine operation. A piston should possess the following qualities:

- Strong enough to withstand heavy gas pressure.
- Should withstand high temperature and maintain its shape at high temperature.
- Should have good conductivity to transfer heat from piston to cylinder walls.
- It should be light to reduce inertia forces.

13.3 Piston Materials

Pistons are generally made of cast iron, steel, or aluminum alloys. Aluminum is light and hence can operate at higher speeds. To withstand high temperature and to reduce weight further, new aluminum alloys are developed.

Selection of material depends on piston speed, which is equal to $2LN / 60$, L is stroke of piston in meters and N speed in revolutions per minute (rpm). Cast iron pistons are used for engine having piston speed less than 6 m/s. Earlier, pistons were casted, but for better strength, forged steel pistons are used. For very high-speed engines, aluminium alloy called Y alloy is also used as piston material.

Aluminium pistons are used for high-speed engines, that is, piston speed more than 6 m/s. It has coefficient of thermal expansion 2.5 times that of cast iron; hence it requires greater clearance between cylinder and piston. Owing to this more clearance piston tilts and gives a slap (called piston slap) on the cylinder wall, till it heats up and the clearance reduces. If the clearance is reduced when cold, on heating its diameter may increase more than cylinder diameter causing piston seizure. Conductivity of aluminium is higher than cast iron; hence their cooling is better than cast iron pistons. Table 13.1 compares a cast iron piston with an aluminium piston.

Table 13.1 Comparison of Cast Iron and Aluminium Pistons

S.No.	Parameter	Cast Iron	Aluminium
1	Strength	Good	Less and reduces to 50 per cent at temperature 325°C
2	Expansion coefficient (a)	0.1×10^{-6}	0.24×10^{-6}
3	Conductivity (k)	Less: 46.6 W/m/°C	More: 175 W/m/°C
4	Weight and density (ρ)	Heavy: 7,200 kg/m³	Light: 2,700 kg/m³
5	Temperature at center	425°C–450°C	260°C–290°C
6	Temperature at edge	220°C–225°C	185°C–215°C

13.4 Types of Pistons

a. **Trunk pistons** Trunk pistons are used for internal combustion engines both for petrol and diesel engines. These are longer, relative to their diameter. As the connecting rod is at some angle to axis of the piston for part of its rotation, a side force acts along the side of the piston against the cylinder wall. A longer piston helps to support this force. These are described in detail in Section 13.5 and shape shown in Figure 13.2.

b. Deflector pistons   Deflector pistons are used for two-stroke engines with crankcase compression, where the charge (petrol + air mixture) flows across the cylinder. For cross scavenging, the transfer port (inlet to the cylinder) and exhaust ports are directly facing sides of the cylinder wall. To prevent the incoming mixture passing straight across from one port to the other, the piston has a raised surface called deflector on its crown [Figure 13.1(a)]. This is intended to deflect the incoming mixture upwards, around the combustion chamber.

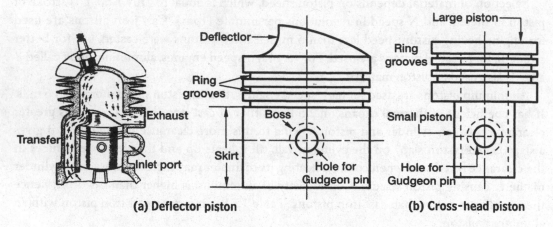

Figure 13.1 Types of pistons

c. Crosshead pistons   Large slow-speed diesel engines require additional support for the side forces on the piston and hence use crosshead pistons. The main piston has a large piston rod extending downwards from the piston, which is effectively a second smaller-diameter piston. The main piston is responsible for gas sealing and carries the piston rings [Figure 13.1(b)]. The smaller piston is purely a mechanical guide. It runs within a small cylinder as a trunk guide and also carries the gudgeon pin. Because of the additional weight of these pistons, they are not used for high-speed engines.

13.5 Construction of Pistons

Following are names of the parts of a piston doing specific functions:

Piston head   It receives the majority of the force caused by the combustion process. Top area of the piston is called the crown (Figure 13.2), above which the combustion takes place. The temperature is the highest at its surface; maximum in the center and decreases towards the edge. A piston head for a diesel engine may have a hemispherical cavity to form combustion chamber.

Top land   The circumferential area at top of the piston is called top land.

Ring section   Circumferential grooves are cut below the top land area. The piston compression rings fit in these grooves to make it gas tight.

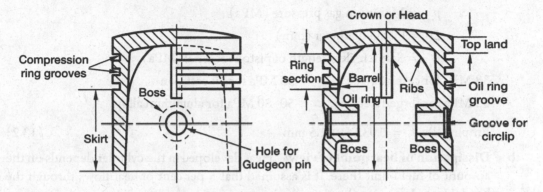

Figure 13.2   Names of areas of a piston

Oil ring grooves   Below the compression rings grooves, there are grooves for oil rings. The number may be one or more than one. This groove has radial holes for the oil to pass from outside of the piston to inside, from where it returns to the crank case. During the operation, a significant amount of oil is accumulated in the piston oil ring groove. This oil is returned to the reservoir through piston windows or through a machined channel near the piston pin. Piston windows are a series of small holes into the oil ring groove surface of the piston. The oil ring collects excess oil from the surface of cylinder.

Bosses   These are provided below the ring area with holes to fix piston pin.

Skirt   The bottom most portion of the piston, after the ring area is called a skirt. Bosses are provided in the top area of the skirt. The thickness of the skirt is reduced towards free end to reduce weight of the piston.

13.6 Piston Design

Design of a piston involves the design of piston head, ring grooves, barrel, skirt, and bosses. Each is described below:

13.6.1 Piston head

Head design is based on the following two parameters, strength, and heat transfer capacity.

a. **Strength point of view**   It should be strong enough to withstand high gas force. The crown is considered as a solid circular flat disc fixed at the outer edge. Gas pressure is assumed acting uniformly over the entire crowned circular surface. Relation to calculate head thickness is given as:

$$t_H = \sqrt{\frac{3\,p_{max}\,D^2}{16\sigma_t}} = 0.43D\sqrt{\frac{p_{max}}{\sigma_t}} \qquad (13.1)$$

Where, t_H = Thickness of piston head (mm)

p_{max} = Maximum gas pressure (MPa)

D = Piston diameter (mm)

σ_t = Safe tensile strength of piston material (MPa)

38 MPa for cast iron   55 MPa for Ni cast iron

82 MPa for forged steel   50–80 MPa for aluminum alloy

Empirically, $t_H = 0.032\,D + 1.5$ mm $\qquad (13.2)$

b **Dissipation of heat point of view**   Heat developed in the cylinder depends on the amount of fuel burnt there. It is assumed that 5 per cent of heat flows through the head. Heat H entering the piston is given as:

$$H = 0.05 \times m \times \text{HCV} \times P \qquad (13.3)$$

Where, H = Heat developed in cylinder (kJ/s)

m = Specific fuel consumption (kg/kW/s)

It depends on the efficiency of engine. However following average values can be assumed

For petrol engines $m = 9 \times 10^{-5}$ kg/kW/s

For diesel engines $m = 8 \times 10^{-5}$ kg/kW/s

HCV = Higher calorific value of the fuel (kJ/kg)

For petrol HCV = 47×10^3 kJ/kg

For diesel HCV = 45×10^3 kJ/kg

P = Power developed (kW)

Thickness of piston head t_H is calculated from this heat flow by the following relation:

$$t_H = \frac{H}{12.56k\left(T_C - T_E\right)} \qquad (13.4)$$

Where, H = Heat developed in cylinder (kJ/s)

k = Thermal conductivity of piston material(W/m/°C)

For cast iron $k = 46.6$. For steel $k = 51.2$ and for aluminium alloy $k = 175$

T_C = Temperature at center of piston (°C)

For cast iron $T_C = 444$ °C   For aluminium alloy $T_C = 225$ °C

T_E = Temperature at edge of piston (°C)

For cast iron $T_E = 222$ °C   For aluminium alloy $T_E = 111$°C

The thickness of piston head can also be calculated from piston diameter using the following relation:

$$t_H = \frac{D^2 q}{12.56 k (T_C - T_E)}$$ (13.5)

Where, D = Piston diameter (mm)

k = Thermal conductivity of piston material (W/m/°C)

For cast iron $k = 46.6$ W/m/°C For aluminium alloy $k = 175$ W/m/°C

q = Heat flow from gases (W/m²)

For aluminium $q = 64,000$ to $256,000$ For cast iron $q = 3,260$ to $13,050$

Take the higher value of t_H as calculated in section (a) and section (b).

Notes:
If head thickness is less than 6 mm, provide ribs (See Figure 13.2) to strengthen the head. Thickness of ribs can be taken $0.33 t_H$ to $0.5 t_H$.

13.6.2 Ring grooves

Below top land, there are circumferential grooves provided on the periphery of the piston for compression and oil rings. Top land length h_1 is taken empirically 20 per cent more than head thickness, that is, $h_1 = 1.2 t_H$.

The number of grooves equal to rings depends on the maximum pressure and the clearance between cylinder and piston. Radial thickness t_R of the ring in mm is taken as:

$$t_R = D \sqrt{\frac{3 p_w}{\sigma_t}}$$ (13.6)

Where, p_w = Radial pressure against cylinder wall (0.025 MPa to 0.042 MPa)
D = Piston diameter (mm)

σ_t = Safe tensile strength of ring material (85 MPa to 110 MPa for cast iron)

Depth of the groove is taken 0.4 mm more than the radial thickness of the ring, that is, $t_R + 0.4$.

Axial width of ring t_A along the axis of the cylinder can be found using the relation:

$$t_A = \frac{0.1 D}{n_g}$$ (13.7)

Where, D = Piston diameter (mm)
n_g = Number of rings

Axial width of top land, that is, gap between top and first groove $h_1 = t_A$ to $1.2\, t_A$.

Axial width of land, that is, gap between adjacent grooves is taken as $h_2 = 0.75 t_A$ to t_A.

13.6.3 Barrel

Radial thickness of barrel at top t_{B1} depends on piston diameter D, radial width of piston rings t_R and can be calculated using Equation (13.8). Radial thickness at bottom is taken as 25 per cent to 35 per cent of t_{B1} as given by Equation (13.9)

Radial thickness of barrel at top, $t_{B1} = 0.03\,D + t_R + 5$ mm $\qquad\qquad$ (13.8)

Radial thickness of barrel at bottom, $t_{B2} = 0.25t_{B1} + t_R + 5$ mm $\qquad\qquad$ (13.9)

13.6.4 Bosses

Two diametrically opposite bosses are inside the barrel of the piston below the rings area to support gudgeon pin. These are exposed to a heavy force due to rapid directional changes. These are also subjected to thermal expansion caused by heat transfer from the piston head to the body of the piston. The piston pin area is subjected to more thermal expansion than the other areas of the piston. A circumferential groove in this hole is used to put a circlip so that the pin does not come out. A hole of diameter d_o is drilled in the boss, which is same as piston pin. It is calculated as described in Section 13.8.

Mean boss diameter of cast iron piston = $1.4d_o$

Mean boss diameter of aluminium alloy piston = $1.5d_o$

13.6.5 Skirt

The portion of piston below the ring section is called skirt. When the gas pressure acts on piston, piston tries to tilt in cylinder, especially when the engine is cool and clearance between cylinder and piston is more. This tilt causes side thrust on the projected area of the piston (Diameter × Length of skirt). Some pistons are provided with holes in the barrel and skirt to make them light and reduce inertia effects as shown in the picture below. The tilt of piston striking the cylinder wall is called piston slap. It is more in power stroke, in comparison to other strokes.

$$\text{Side thrust, } R = D \times L_s \times p_b \qquad\qquad (13.10)$$

Where, D = Piston diameter (mm)

$\qquad L_s$ = Length of skirt (mm).

Empirically it can be taken as: $0.6D$ to $0.8D$

$\qquad p_b$ = Bearing pressure (MPa).

Side thrust R is taken as 10 per cent of the maximum gas force.

$$\text{Hence, side thrust } R = 0.1 \times \frac{\pi}{4}D^2 \times p_{max} \qquad (13.11)$$

Equating Equations (13.10) and (13.11) for side thrust, bearing pressure can be calculated.

$$D \times L_s \times p_b = 0.1 \times \frac{\pi}{4} D^2 \times p_{max}$$

Or, $p_b = \dfrac{\pi D\, p_{max}}{40\, L_s}$ (13.12)

Generally it should not be more than 0.25 MPa but for high-speed engines it can be up to 0.5 MPa.

13.7 Piston Rings

Piston rings are of two types:

a. **Compression rings** are used to seal the annular clearance between cylinder and piston by applying radial pressure against cylinder walls.

b. **Oil rings** are used to scrap the lubricating oil from cylinder walls back to oil sump. Their number varies from 1 to 3.

13.7.1 Compression rings

Cross section of these rings is rectangular with radial thickness t_R, and axial width t_A. These values can be calculated using Equations (13.6) and (13.7). They also transfer heat from piston to cylinder walls. Rings are made of cast iron, because of good wearing properties and also capacity of retaining springiness at high temperatures. Their numbers may be 2–6 depending on gas pressure.

These are made of diameter slightly bigger than cylinder diameter. A cut called free gap is given and then these are inserted in the grooves by expanding. The cut is inclined or stepped as shown in Figure 13.3.

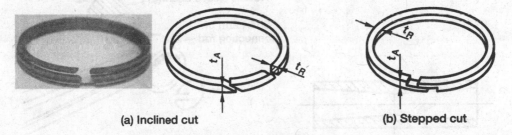

(a) Inclined cut (b) Stepped cut

Figure 13.3 Cut on a compression ring

Free gap When the ring is not inside the cylinder, the gap is kept between $3.5 t_R$ and $4\, t_R$.

The rings are put in the grooves such that the gaps due to cut do not fall in one line but at different angular positions to offer a path of maximum resistance to the leaking gases. These are then compressed circumferentially with a special tool and inserted in cylinder along with piston. The length of fee cut reduces, when the rings are inside the cylinder. This gap is very important for the satisfactory working of an engine. This gap is necessary for the thermal expansion of the rings. If it is more, gases will leak even when the piston rings are hot. If it is less, rings may expand more than cylinder and cause piston seizure; hence it is measured by a feeler gauge, by substituting only a ring in cylinder, when the engine is cool.

Radial thickness of these rings is calculated using Equation (13.6) $t_R = D \sqrt{\dfrac{3 p_w}{\sigma_t}}$

Axial width of ring t_A is found using Equation (13.7) $t_A = \dfrac{0.1 D}{n_g}$

Gap, when the ring is inside the cylinder is 0.002 D to 0.004 D

13.7.2 Oil rings

These rings scrap the oil up and down along with piston and thus spread oil all over the surface of cylinder. Their cross section is of U section; top face of U is towards cylinder wall. It has radial holes at the bottom of U, through which oil drips back to oil sump.

13.8 Gudgeon Pin

Gudgeon pin is a hollow cylindrical pin and is used to join piston with connecting rod. The hole tapers towards center to increase cross-sectional area to withstand maximum bending moment at center (Figure 13.4). It is also called wrist pin or piston pin.

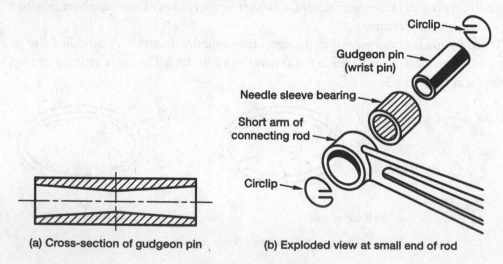

(a) Cross-section of gudgeon pin   (b) Exploded view at small end of rod

Figure 13.4 Gudgeon pin

It is fixed in the holes of the bosses of the piston Needle bearing is used to reduce friction between gudgeon pin and small end of the connecting rod. The holes for the gudgeon pin are located from 0.02D to 0.04D above the middle of the skirt to reduce piston slap.

13.8.1 Fixing of gudgeon pin

Gudgeon pin is fixed to the piston in different ways as given below:

a. **Fully floating**   The pin is free to turn in the piston. Axial movement of the gudgeon pin is restricted by putting a circlip in the annular groove in the boss, on each side. See Figure 13.5.

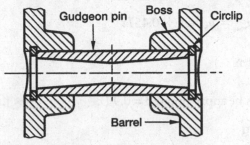

Figure 13.5 Fully floating

b. **Semi floating**   There are two ways to fix in this method:

The pin is fixed in the boss [Figure 13.6(a)] and floats in small end.

The pin is fixed in the connecting rod [Figure 13.6(b)] and floats in bosses.

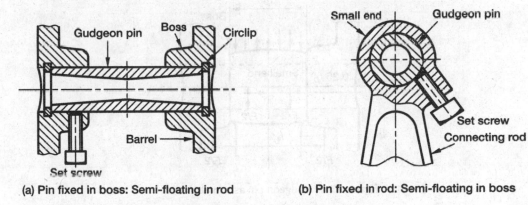

(a) Pin fixed in boss: Semi-floating in rod   (b) Pin fixed in rod: Semi-floating in boss

Figure 13.6 Fixing of gudgeon pin

13.8.2 Design of gudgeon pin

Design of gudgeon pin involves calculating its length, diameter, hole size, and check for bending and bearing pressures. It is made of carbon steel or alloy steel. It is hardened and ground to reduce friction. The pin is designed for maximum gas force or inertia force of piston, whichever is larger. The pin diameter is calculated by equating maximum load capacity to the bearing pressure capacity at small end, that is:

$$\frac{\pi}{4} D^2 \times p_{max} = p_b \times d_o \times l_p$$

Where, p_{max} = Maximum gas pressure (MPa)

$\quad\quad p_b$ = Safe-bearing pressure (MPa);can be taken as 25 MPa.

$\quad\quad l_p$ = Length of pin in the small end of the connecting rod = $0.45D$ mm

$\quad\quad d_o$ = Outside diameter of the gudgeon pin (mm)

Substituting the value of l_p the above equation:

$$0.785D^2 \times p_{max} = p_b \times d_o \times 0.45D$$

Or, $\quad l_p = \dfrac{1.75D\, p_{max}}{p_b}$ (13.13)

Overall length of pin can be approximated = 0.9 D.

Checking for bending

The pin is assumed supported at the center of the bosses (Figure 13.7); hence distance between centers of the bosses l_b is given as (D is piston diameter):

$$l_b = l_p + \frac{D - l_p}{2} = \frac{D + l_p}{2}$$ (13.14)

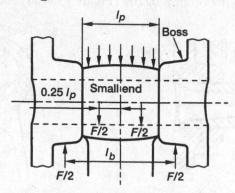

Figure 13.7 Gudgeon pin as a loaded beam

Maximum gas load $F = \dfrac{\pi}{4}D^2 \times p_{max}$ is considered uniformly distributed over the length of pin in the small end for length l_p. Reaction at each boss = 0.5 F. Maximum bending moment M will be in the center and is found by taking moments at the center of the pin.

$$M = \left(0.5F \times 0.5l_b\right) - \left(0.5F \times 0.25l_p\right) = 0.25Fl_b - 0.125F\, l_p$$

Substituting the value of l_b from Equation (13.14):

$$M = 0.25F\left(\frac{D + l_p}{2}\right) - 0.125Fl_p = 0.125FD$$ (13.15)

Modulus of section for a hollow cylinder $Z = \dfrac{\pi}{32}\left[\dfrac{d_o^4 - d_i^4}{d_o}\right]$ (13.16)

Hole diameter d_i is taken as $0.6d_o$. Substituting the value of d_i in Equation (13.16)

$$Z = 0.0854 d_o^3$$

Using the relation: $\sigma_b = \dfrac{M}{Z}$ bending stress σ_b can be calculated.

It varies between 85 MPa to 140 MPa depending on material.

Example 13.1

Design of gudgeon pin

Design a gudgeon pin for a piston of diameter 150 mm having maximum pressure 4 MPa. Safe bearing pressure is 12 MPa. Check the design in bending, if safe bending stress for the pin material is 115 MPa.

Solution

Given $D = 150$ mm $p_b = 25$ MPa $p_{max} = 4$ MPa $\sigma_b = 115$ MPa

Maximum gas force, $F = \dfrac{\pi}{4}D^2 \times p_{max} = 0.785 \times 150^2 \times 4 = 70{,}650$ N

Length of pin in the small end of the rod $l_p = 0.45\,D = 0.45 \times 150 = 67.5$ say 68 mm

Gas force should be equal to bearing capacity of the pin $F = p_b \times d_o \times l_p$

$$70{,}650 = 25 \times d_o \times 68 \quad \text{or} \quad d_o = 41.6 \text{ say } 42 \text{ mm}$$

Maximum bending moment $M = 0.125\,F\,D = 0.125 \times 70{,}650 \times 150 = 1{,}324{,}725$ Nmm

From Equation (13.16): Modulus of section, $Z = \dfrac{\pi}{32}\left[\dfrac{d_o^4 - d_i^4}{d_o}\right]$

Assuming hole diameter $d_i = 0.6d_o$

$$Z = 0.098\left[\dfrac{d_o^4 - 0.1296 d_o^4}{d_o}\right] = 0.0854 d_o^3 = 0.0854 \times 42^3 = 12{,}008 \text{ mm}^3$$

Bending stress $\sigma_b = \dfrac{M}{Z} = \dfrac{1{,}324{,}725}{12{,}008} = 110 \text{ MPa}$

It is less than safe stress 115 MPa, hence safe in bending.

Hence, outside diameter of piston pin $d_o = 42$ mm

Inside diameter $d_i = 42 \times 0.6 = 25.2$ say 26 mm

Length of pin in small end $l_p = 68$ mm.

Overall length of pin that can be approximated $= 0.9 D = 0.9 \times 150 = 135$ mm

Example 13.2

Design of piston

Design a piston for a four-stroke petrol engine having cylinder diameter 85 mm running at 3000 rpm with mean effective pressure 0.8 MPa and 84 per cent mechanical efficiency. Specific fuel consumption is 0.2 kg/kW/h and its higher calorific value is 47×10^3 kJ/kg. Assume the following:

Maximum gas pressure = 7 times the mean effective pressure
Length to diameter ratio = 1.2
Conductivity for the material = 50 W/m/°C
Piston material Ni–cast iron with safe tensile strength = 55 MPa
Tensile strength of ring material = 80 MPa and for gudgeon pin = 95 MPa
Radial pressure against cylinder wall = 0.03 MPa
Bearing pressure for small end bearing = 25 MPa
Temperatures at center and edge of piston are 440° C and 290° C respectively

Solution

Given $D = 85$ mm $\qquad N = 3{,}000$ rpm $\qquad p_m = 0.8$ MPa

$\eta_{mech} = 84$ per cent $\qquad p_{max} = 7p_m \qquad L/D = 1.2$

$k = 50$ W/m/°C $\qquad m = 0.2$ kg/kW/h $\qquad$ HCV $= 47 \times 10^3$ kJ/kg

$\sigma_t = 55$ MPa for piston $\qquad \sigma_t = 80$ MPa for rings $\qquad p_b = 15$ MPa

$\sigma_t = 95$ MPa for gudgeon pin $\quad p_w = 0.03$ MPa

$T_C = 440°C \qquad\qquad T_E = 290°C$

Piston crown

Head thickness from Equation(13.1)

$$t_H = 0.43 D \sqrt{\frac{p_{max}}{\sigma_t}} = 0.43 \times 85 \sqrt{\frac{0.8 \times 7}{55}} = 12 \text{ mm} \qquad\qquad \text{(a)}$$

Area of cylinder $A = \dfrac{\pi}{4} D^2 = 0.785 \times 85^2 = 5{,}672 \text{ mm}^2$

Brake power $P = \dfrac{p_m l A n}{60{,}000} \times \eta_{mech}$

$$P = \frac{0.8 \times (1.2 \times 85) \times 5{,}672 \times (0.5 \times 3{,}000)}{60{,}000} \times 0.84 = 9{,}720 \text{ W} = 9.72 \text{ kW}$$

Specific fuel consumption $m = \dfrac{0.2}{3,600} = 55 \times 10^{-6}\,\text{kg/kW/s}$

From Equation (13.3): Heat through head is $H = 0.05 \times m \times \text{HCV} \times P$

Or, $H = 0.05 \times (55 \times 10^{-6}) \times (47 \times 10^{3}) \times 9.72 = 1.256\,\text{kW} = 1,256\,\text{W}$

From Equation (13.4), Thickness of piston head $t_H = \dfrac{H}{12.56k\left(T_C - T_E\right)}$

Or, $t_H = \dfrac{1,256}{12.56 \times 50\left(440 - 290\right)} = 0.013\,\text{m} = 13\,\text{mm}$ \hfill (b)

Taking the larger value calculated in Equation (a) and (b) $t_H = 13\,\text{mm}$

Thickness of ribs can be taken $0.33\,t_H$ to $0.5\,t_H$, that is, 4 mm to 6 mm. So selecting intermediate value of 5 mm.

Since piston diameter is not large, taking number of ribs as 4.

Piston ring grooves:

Radial thickness of ring (Equation 13.6) $t_R = D\sqrt{\dfrac{3p_w}{\sigma_t}} = 85\sqrt{\dfrac{3 \times 0.03}{80}} = 2.85\,\text{mm}$

Depth of the groove $= t_R + 0.4 = 2.85 + 0.4 = 3.25\,\text{mm}$

Axial width of ring (Equation 13.7) $t_A = \dfrac{0.1D}{n_g} = \dfrac{0.1 \times 85}{3} = 2.83\,\text{mm}$

Axial width of land h_2 is $0.75\,t_A$ to t_A; hence choosing $h_2 = 2.5\,\text{mm}$

Gudgeon pin:

Maximum gas force $F = \dfrac{\pi}{4}D^2 \times p_{max} = 0.785 \times 85^2 \times (7 \times 0.8) = 31,761\,\text{N}$

Length of pin in the small end of the rod: $l_p = 0.45\,D = 0.45 \times 85 = 38\,\text{mm}$

Gas force should be equal to bearing capacity of the pin $F = p_b \times d_o \times l_p$

$31,761 = 25 \times d_o \times 38,$ or, $d_o = 33.5$ say 35 mm

Bending moment $M = 0.125\,FD = 0.125 \times 31,761 \times 85 = 337,460\,\text{Nmm}$

From Equation (13.16): Modulus of section $Z = \dfrac{\pi}{32}\left[\dfrac{d_o^4 - d_i^4}{d_o}\right]$

Assuming hole diameter d_i as $0.6\,d_o$ and substituting the values:

$$Z = 0.098\left[\dfrac{d_o^4 - 0.1296d_o^4}{d_o}\right] = 0.0854 \times d_o^3 = 0.0854 \times 35^3 = 3,657\,\text{mm}^3$$

Bending stress

$$\sigma_b = \frac{M}{Z} = \frac{337,460}{3,657} = 92\,\text{MPa}\;(\text{Less than given safe value 95 MPa, hence safe})$$

Bosses   Mean diameter of piston bosses $= 1.4\,d_o = 1.4 \times 36 = 50.4$ say 51 mm

Barrel   Radial thickness of barrel at top $t_{B1} = 0.03D + t_R + 5$ mm

$$= (0.03 \times 85) + 2.85 + 5 = 10.4 \text{ say } 11 \text{ mm}$$

Radial thickness of barrel at bottom $t_{B2} = 0.25\,t_{B1} + t_R + 5$ mm

$$= (0.25 \times 11) + 2.85 + 5 = 10.6 \text{ say } 11 \text{ mm}$$

So assume constant radial thickness 11 mm.

Skirt   Empirically, the length of skirt, $L_s = 0.6\,D$ to $0.8\,D = 51$ mm to 68 mm

So assuming the length of skirt $L_s = 65$ mm.

From Equation (13.12) bearing pressure:

$$p_b = \frac{\pi \cdot D\, p_{max}}{40\,L_s} = \frac{3.14 \times 85 \times (7 \times 0.8)}{40 \times 65} = 0.57\,\text{MPa}$$

It should not be more than 0.5 MPa

It is more than 0.5 MPa, hence increase skirt length = 75 mm, which gives $p_b = 0.49$ MPa.

Summary

Definition   Piston is a reciprocating hollow cylinder closed at one end with a transverse hole in the middle for a pin and grooves on periphery on upper half to provide piston rings to stop leakage of gas from the clearance between piston and cylinder.

Function   The main function of the piston is to transmit force through piston pin, and connecting rod. A crank shaft provided at the other end of the connecting rod converts its reciprocating motion into rotary motion.

Desirable qualities of a piston   A piston should possess the following qualities:

-   Strong enough to withstand heavy gas pressure.
-   Should withstand high temperature and maintain its shape at high temperature.
-   Should have good conductivity to transfer heat from piston to cylinder walls.
-   It should be light to reduce inertia forces.

Piston materials   Pistons are generally made of cast iron, steel, or aluminum alloys. Aluminum is light and hence can operate at higher speeds.

Selection of material depends on piston speed $= 2\,l\,N\,/\,60$, where N is rotational speed of engine in rpm and l is piston stroke in m.

Cast iron pistons are used for speeds less than 6 m/s. Earlier, pistons were casted, but for better strength *forged steel* pistons are used. For very high-speed engines, aluminium alloy called Y alloy is also used as piston material.

Aluminium pistons are used for speed > 6 m/s. It has coefficient of thermal expansion 2.5 times that of cast iron, which requires more clearance. Owing to this, piston tilts and gives a slap (called piston slap) to the cylinder wall. If the clearance is reduced when cold, on heating its diameter may increase more than cylinder diameter, causing piston seizure. Table 13.1 compares a cast iron piston with an aluminium piston.

Types of pistons

   a.   *Trunk pistons*   These are longer, relative to their diameter and help support side force.

   b.   *Deflector pistons*   Deflector pistons are used for two-stroke engines with crankcase compression, where petrol + air mixture flows across the cylinder. To prevent the incoming mixture passing straight across from one port to the other, the piston has a raised surface called deflector on its crown.

   c.   *Cross head pistons*   Large slow-speed diesel engines require additional support for the side forces on the piston and hence use cross head pistons. The main piston has a large piston rod extending downwards from the piston, which is effectively a second smaller-diameter piston.

Construction of pistons   Following are the names of parts of piston doing specific functions:

   a.   *Piston head*   It is also called crown. It receives the majority of the initial pressure and force caused by the combustion process. A piston head for a diesel engine may have a hemispherical cavity to form combustion chamber. Crown of a two-stroke engine can have a deflector at top.

   b.   *Top land*   The circumferential area at top of the piston is called top land.

   c.   *Ring section*   Circumferential grooves are cut below the top land area. The piston compression rings fit in these grooves to make it gas tight.

   d.   *Oil ring grooves*   Below the compression rings grooves there are oil grooves for oil rings.

   e.   *Bosses*   These are provided below the ring area with holes to fix piston pin.

   f.   *Skirt*   The bottom most portion of the piston, after the ring area is called a skirt. The thickness of the skirt is reduced towards free end to reduce weight of the piston.

Piston design   Design of a piston involves the design of piston head, ring grooves, barrel, skirt, and bosses. Each is described below:

Piston head   Head design is based on following two parameters; strength and heat-transfer capacity.

   a.   *Strength point of view*   It should be strong enough to withstand high gas force. To calculate head thickness $t_H = 0.43D \sqrt{\dfrac{p_{max}}{\sigma_t}}$

Empirically, $t_H = 0.032\,D + 1.5$ mm

Maximum gas pressure p_{max} = 38 MPa for cast iron;   55 MPa for Ni cast iron; 82 MPa for forged steel;   50 MPa to 80 MPa for aluminium alloy

b. *Dissipation of heat point of view*   It is assumed that 5 per cent of heat in kJ/s flows through the head. Heat entering the piston $H = 0.05 \times m \times HCV \times P$

Where, P = Power (kW)

m = Specific fuel consumption (kg/kW/s). For petrol engines $m = 9 \times 10^{-5}$ kg/kW/s

and for diesel engines $m = 8 \times 10^{-5}$ kg/kW/s

HCV, for petrol = 47×10^3 kJ/kg; for diesel= 45×10^3 kJ/kg

Thickness of piston head can be calculated from $t_H = \dfrac{H}{12.56\,k\,(T_C - T_E)}$

k = Thermal conductivity of piston material (W/m/°C)

For cast iron k = 46.6; for steel k = 51.2; for aluminium alloy k = 175

T_C = Temperature at center of piston (°C)

For cast iron T_C= 444 ° C; for aluminium alloy T_C= 225 °C

T_E = Temperature at edge of piston (°C)

For cast iron T_E= 222 °C; for aluminium alloy T_E= 111°C

Alternately, thickness of piston head: $t_H = \dfrac{D^2 q}{12.56\,k\,(T_C - T_E)}$

Where, q = 64,000 W/m² to 256,000 W/m²

For aluminium q = 3,260 to 13,050 for cast iron

Take the higher value of t_H as calculated in section (a) and section (b).

Notes   If head thickness is less than 6 mm, provide *ribs* to strengthen the head.

Thickness of *ribs* can be taken $0.33t_H$ to $0.5t_H$.

Ring grooves   Top land length h_1= $1.2t_H$.

The number of grooves depends on the radial wall pressure p_w and diameter D.

Radial thickness $t_R = D\sqrt{\dfrac{3p_w}{\sigma_t}}$

Take p_w = 0.025 MPa–0.042 MPa

σ_t = Safe tensile strength of ring material (85 MPa–110 MPa for cast iron)

Depth of the groove is taken 0.4 mm more than radial thickness of the ring, that is, t_R+ 0.4.

Axial width of ring t_A along the axis of the cylinder is: $t_A = \dfrac{0.1D}{n_g}$ Where, n_g is number of rings

Axial width of top land h_1, that is, gap between top and first groove = t_A to $1.2t_A$.

Axial width of land h_2, that is, gap between adjacent grooves is taken as $0.75t_A$ to t_A.

Barrel   Radial thickness of barrel at top, $t_{B1} = 0.03D + t_R + 5$ mm

Radial thickness of barrel at bottom, $t_{B2} = 0.25t_{B1} + t_R + 5$ mm

Bosses   Bosses have a hole of diameter d_o to support gudgeon pin. A circumferential groove in this hole is cut to put a circlip so that the pin does not come out.

Mean boss diameter of cast iron piston $= 1.4d_o$

Mean boss diameter of aluminium alloy piston $= 1.5d_o$

Skirt   Empirically, the length of skirt $L_s = 0.6\,D$ to $0.8\,D$

It takes the side thrust, $R = D \times L_s \times p_b$

Bearing pressure $p_b = \dfrac{\pi D p_{max}}{40 L_s}$

It should not be more than 0.25 MPa or 0.5 MPa for high-speed engines.

Piston rings   These rings are of two types: Compression rings and Oil rings

a   **Compression rings**   are used to seal the clearance gap between cylinder and piston. Rectangular section rings are made of cast iron because of good wearing properties. Their numbers may be 2–6 depending on gas pressure.

Radial thickness of these rings is calculated using relation $t_R = D\sqrt{\dfrac{3 p_w}{\sigma_t}}$

Axial width of ring t_A is found using the relation: $t_A = \dfrac{0.1D}{n_g}$

An inclined or stepped cut is given in ring called free gap of size $3.5\,t_R$ to $4\,t_R$

Gap when the ring is inside the cylinder is 0.002 D to 0.004 D.

b.   **Oil rings**   are used to scrap the lubricating oil from cylinder walls. Their cross section is of U section; top face of U is towards cylinder wall. It has radial holes at the bottom of U, through which oil drips back to oil sump.

Gudgeon pin   It is also called wrist pin or piston pin, which is a hollow and is fixed in the holes of the bosses to join piston with connecting rod. The holes are located 0.02 D to 0.04, D above the middle of the skirt to reduce piston slap.

Fixing of gudgeon pin   It is fixed to the piston in different ways as given below:

a.   *Fully floating*   The pin is free to turn in the piston. Axial movement of the pin is restricted by substituting a circlip in the annular groove in the boss, on each side.

b.   *Semifloating*   There are two ways to fix in this method:

- The pin is fixed in the boss and floats in small end.
- The pin is fixed in the connecting rod and floats in bosses.

Design of gudgeon pin   It is made of carbon steel or alloy steel and is hardened and ground to reduce friction. The pin diameter is calculated from:

$$\frac{\pi}{4}D^2 \times p_{max} = p_b \times d_o \times l_p$$

Where, p_b = Safe bearing pressure (MPa) (25 MPa)

l_p = Length of pin in the small end of the connecting rod = 0.45 D

d_o = Outside diameter of gudgeon pin

Substituting the value of l_p in the above equation:

$$0.785 D^2 \times p_{max} = p_b \times d_o \times 0.45\, D$$

From Equation (13.13) $l_p = \dfrac{1.75 D\, p_{max}}{p_b}$

Overall length of pin that can be approximated = 0.9 D

Checking for bending The pin is assumed supported at the center of the bosses.

Distance between centers of the bosses, $l_b = l_p + \dfrac{D - l_p}{2} = \dfrac{D + l_p}{2}$

Where, D is piston diameter

Maximum gas load $F = \dfrac{\pi}{4} D^2 \times p_{max}$ is considered uniformly distributed over the length of pin in the

small end for length l_p. Reaction at each boss = 0.5 F.

Maximum bending moment M at the center of the pin is:

$$M = (0.5 F \times 0.5\ l_b) - (0.5\ F \times 0.25\ l_p) = 0.25\ F l_b - 0.125\ F l_p$$

Substituting the value of l_b: $M = 0.25 F\left(\dfrac{D + l_p}{2}\right) - 0.125 F l_p = 0.125 F D$

Modulus of section for a hollow cylinder, $Z = \dfrac{\pi}{32}\left[\dfrac{d_o^4 - d_i^4}{d_o}\right]$

Hole diameter d_i is taken as 0.6 d_o. Substituting this value $Z = 0.0854\, d_o^3$.

Using the relation: $\sigma_b = \dfrac{M}{Z}$, bending stress σ_b can be calculated.

It varies between 85 MPa and 140 MPa depending on the material.

Theory Questions

1. What is a piston and what does it do in an internal combustion engines?
2. Write a note on piston materials. What are the desirable qualities of the material?
3. What are the different types of pistons? Describe them with the help of a neat sketch.
4. Describe the use of different types of pistons.
5. With the help of a neat sketch describe construction of a trunk piston.
6. Describe the method to design a piston head.

7. Describe the proportions of compression ring grooves in a piston.

8. What is the function and construction of a gudgeon pin? What are the different methods to fix it in a piston?

9. Write the method to design a gudgeon pin.

10. Describe the method to design skirt of a piston.

Multiple Choice Questions

1. Piston slap occurs due to:
   (a) High speed
   (b) Heavy load
   (c) More clearance between cylinder and piston
   (d) High engine temperature

2. Piston skirt helps the piston for:
   (a) Transfer of heat
   (b) Reducing gas leakage
   (c) Better lubrication
   (d) Taking side thrust due to obliquity of connecting rod

3. If p_{max} is maximum pressure in cylinder, thickness of piston head is proportional to:
   (a) p_{max}
   (b) $\sqrt{p_{max}}$
   (c) $\left(p_{max}\right)^{1.2}$
   (d) $\left(p_{max}\right)^{1.5}$

4. Ring pressure against cylinder wall is:
   (a) 0.025 MPa to 0.042 MPa
   (b) 0.05 MPa to 0.07 MPa
   (c) 0.07 MPa to 0.09 MPa
   (d) 0.09 MPa to 1.1 MPa

5. Bearing pressure for the skirt area is taken as:
   (a) 0.10 MPa to 0.15 MPa
   (b) 0.15 MPa to 0.25 MPa
   (c) 0.25 MPa to 0.5 MPa
   (d) 0.5 MPa to 0.6 MPa

6. Gudgeon pin in a piston is used for:
   (a) Strengthening the barrel
   (b) Providing a pin joint for connecting rod
   (c) Reducing weight of piston
   (d) None of the above

7. Forces coming on a gudgeon pin are:
   (a) Shear
   (b) Bending
   (c) Bearing
   (d) All given in (a), (b), and (c)

Answers to multiple choice questions

1. (c)   2. (d)   3. (b)   4. (a)   5. (c)   6. (b)   7. (d)

Design Problems Questions

1. A piston of diameter 90 mm is subjected to maximum gas pressure of 5.5 MPa. Design its gudgeon pin, if bearing pressure at small end is not to exceed 30 MPa and safe bending stress 130 MPa. Pin can be hollow with hole diameter half of its outside diameter. Assume bearing length as 40 per cent of piston diameter. Check it for the bending stress.

$$[d_o = 33 \text{ mm}, d_i = 16.5 \text{ mm}, \sigma_b = 119 \text{ MPa}]$$

2. Design a piston for a four-stroke petrol engine having cylinder of 100 mm running at 2,500 rpm with mean effective pressure 1 MPa and 85 per cent mechanical efficiency. Specific fuel consumption is 0.22 kg/kW/h and its higher calorific value is 45×10^3 kJ/kg. Assume the following:

   Maximum gas pressure 6 MPa

   Length to diameter ratio = 1.2

   Conductivity for the material = 45 W/m/°C

   Piston material cast iron with safe tensile strength = 35 MPa

   Tensile strength of ring material = 60 MPa

   Tensile strength of ring material = 90 MPa

   Radial pressure against cylinder wall = 0.03 MPa

   Bearing pressure for small end bearing = 25 MPa

$$[t_H = 18 \text{ mm}, t_R = 4 \text{ mm}, t_A = 3 \text{ mm}, d_o = 42 \text{ mm},$$
$$d_i = 25 \text{ mm}, t_{B1} = 10 \text{ mm}, t_{B1} = 10 \text{ mm}, L_s = 95 \text{ mm}]$$

CHAPTER 14

Connecting Rod

14.1 Introduction

Connecting rod shown in Plate 14.1 is used to join piston with a pin joint at its one end and on crank shaft at the other end. The end fixed on piston with the gudgeon pin is called small end and the end mounted on the crank pin is called big end. Function of the connecting rod is to transmit gas

Plate 14.1 A connecting rod

force from reciprocating motion of piston to the crank shaft for converting to rotary motion.

The length of rod is the distance between center of gudgeon pin to center of crank pin. Small rod length increases the angularity of connecting rod, which increases side thrust of piston and hence more wear. On the other hand, larger length increases its weight, which increases inertia force; hence, ratio of its length to radius of crank shaft (l / r) is kept between 4 and 5.

Connecting rod is supposed to have high strength, hence made by drop forging or die forging to provide strength and stiffness. It is made of following materials:

0.4 per cent carbon steel, with ultimate strength of 650 MPa.

0.5 per cent heat-treated carbon steel, with ultimate strength of 750 MPa.

Nickel–chromium alloy steel, having ultimate strength of 1,050 MPa.

14.2 Construction

Small end of the rod is made circular to accommodate a bush made of phosphorous bronze as shown in Figure 14.1.

Cross section of the rod can be circular or rectangular for small speed engines, but high-speed engines use I section to make it light to reduce inertia forces. If t is the thickness of web of I section, width of the connecting rod is kept $4t$. Its width remains constant from small end to big end, while height increases from $H1$ at small end to $H2$ at big end. Height in middle H is kept $5t$ (Figure 14.1). Thickness t is calculated on the basis of gas force coming on the rod causing compression and buckling.

Small end height $H1 = 0.75\ H$ to $0.95\ H$

Big end height $H2 = 1.1\ H$ to $1.25\ H$.

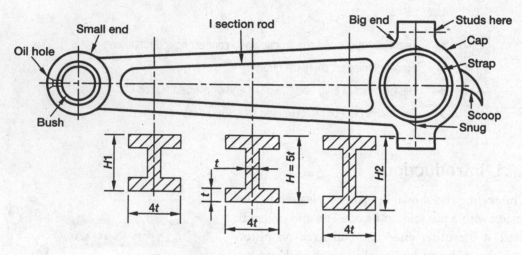

Figure 14.1 Construction of connecting rod

The big end is split into two halves; one half is integral with the rod and is flared to increase height. The other half is called cap, which is fixed to the big end of the rod using two studs and nuts (Figure 14.1). This end is provided with a bearing shell made of steel, brass, or bronze with a lining of soft metal like tin or babbitt.

The shells are provided with a snug at one edge, which fits into the cavity of the rod and cap. It is provided so that the shell does not rotate in the hole due to friction between pin and shell. While boring the big end, the two parts are assembled with a thin copper shim of about 0.05 mm thickness. When the bore wears it becomes oblong, the shim is removed between the two parts and again assembled.

14.3 Forces on Connecting Rod

Forces coming on a connecting rod are given below and described in sub-sections:

a. Axial force due to gas pressure and inertia of parts
b. Bending due to inertia of reciprocating parts
c. Bending due to inertia of connecting rod
d. Frictional forces due to friction between piston rings and cylinder
e. Frictional forces due to friction between gudgeon pin and crank pin

14.3.1 Axial force due to gas pressure

The gas force coming on the rod is:

$$F = \frac{\pi}{4} D^2 \, p_{max} \tag{14.1}$$

Where, D = Piston diameter (mm)

p_{max} = Maximum pressure in cylinder (MPa)

14.3.2 Axial force due to inertia of reciprocating parts

Mass of the reciprocating parts comprises of mass of piston and gudgeon pin. The connecting rod does not reciprocate but oscillates, where as its small end reciprocates. Hence one-third of mass of connecting rod m_C is added to the reciprocating mass of piston m_P. Thus reciprocating mass m_R is given as:

$$m_R = m_P + \frac{1}{3} m_C \tag{14.2}$$

Where, m_R = Mass of reciprocating parts

m_P = Mass of piston

m_C = Mass of connecting rod

Acceleration of parts is given as:

$$f = \omega^2 \, r \left(\cos\theta + \frac{\cos 2\theta}{n} \right)$$

Where, ω = Angular speed of crank shaft (rad/s)

θ = Angle made by crank arm from the line joining piston pin and center of crank shaft at top dead center (TDC) position.

n = Ratio of length of connecting rod l to crank radius r, that is, (l / r)

Inertia force F_i opposes the piston force, when it moves from top to bottom for vertical engines (inner dead center to outer dead center for horizontal engines) helps when moving from bottom to top, that is, it varies from $+ F_i$ to $- F_i$.

Inertia force F_i = Mass × Acceleration = $m_R \times \omega^2\, r \left(\cos\theta + \dfrac{\cos 2\theta}{n} \right)$ (14.3)

Hence, the axial force at piston is given as: $F_p = F \pm F_i$

Where, F = Gas force given by Equation (14.1)

F_i = Inertia forces given by equation (14.3)

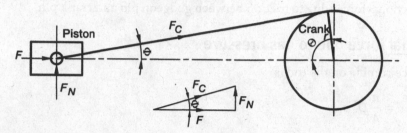

Figure 14.2 Forces at small end of the connecting rod though gudgeon pin

This force is maximum when the crank arm and connecting rod are at right angles (Figure 14.2). At this position, gas force reduces and hence can be neglected.

14.3.3 Bending force due to inertia of reciprocating parts

Inertia forces act at right angle to the connecting rod (Figure 14.3). Its value is zero at piston pin P and varies linearly from piston to maximum at crank pin C. Thus the loading is triangular from zero to maximum.

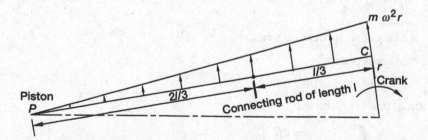

Figure 14.3 Acceleration forces on connecting rod

Inertia force at P = 0

Inertia force at $C = m_R\, \omega^2 r$

Where, m_R = Mass of connecting rod

Hence average inertia force $F_i = 0.5 m_R\, \omega^2 r$

This force acts at a distance $2l / 3$ from piston pin.

It is assumed that one-third mass of connecting rod acts at piston pin and two-thirds at crank pin. Therefore:

$$\text{Force at piston pin } F_P = \frac{F_i}{3} \text{ and force at crank pin } F_C = \frac{2F_i}{3}$$

Maximum bending moment M_{max} for triangular loading acts at $\left(l/\sqrt{3}\right)$ from piston pin. Maximum bending moment at this location is:

$$M_{max} = \frac{F_i l}{9\sqrt{3}} \tag{14.4}$$

$$\text{Maximum stress } \sigma_{max} = \frac{M_{max}}{Z}$$

This is also called as **whipping stress**.

It may be noted that inertia force increases with square of speed (ω^2) and maximum bending stress occurs when crank angle θ is between 65° and 70° from TDC.

Bending of the connecting rod can be in the direction of depth [Figure 14.4(a)] or in width [Figure 14.4(b)].

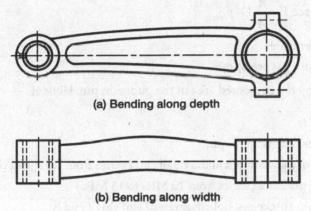

(a) Bending along depth

(b) Bending along width

Figure 14.4 Bending of connecting rod

14.3.4 Frictional forces due to friction between piston rings and cylinder

Frictional force F_f due to piston rings rubbing on cylinder wall is given by Equation (14.5):

$$F_f = \pi D t_A n p_r \mu \tag{14.5}$$

Where, D = Cylinder bore
t_A = Axial thickness of rings
n = Number of piston rings
p_r = Radial pressure of rings (varies from 0.025 MPa to 0.4 MPa)
μ = Coefficient of friction (~ 0.1)

This force is much less than the other forces and hence can be neglected.

14.3.5 Frictional forces due to friction between gudgeon pin and crank pin

The frictional force at the gudgeon pin and small end of connecting rod tries to bend the connecting rod. This can also be neglected.

14.4 Design of Connecting Rod

The design of connecting rod involves, calculation of dimensions for the following:
- Hole size at small end
- Cross section of connecting rod at small end, middle, and big end
- Thickness of big end cap
- Size of cap bolts.

14.4.1 Small end of the rod

Gudgeon pin diameter is calculated on the basis of gas forces as described in Chapter 13. A bush made of phosphorous, about 2 mm to 3 mm thick is put over it to reduce friction.

$$\text{Maximum gas force } F = \frac{\pi}{4} D^2 p_{max}$$

Where, D = Cylinder bore

p_{max} = Maximum gas pressure

This force is taken by the projected area of the gudgeon pin. Hence:

$$F = d_p l_p p_{bp} \tag{14.6}$$

Where, d_p = Diameter of gudgeon pin

l_p = Length of gudgeon pin in the small end (varies from 1.5 d_p to 2 d_p)

p_{bp} = Bearing pressure (varies from 10 MPa to 15 MPa)

Using a bush of 3 mm thickness, hole diameter D_p of small end is:

$$D_p = d_p + 6 \text{ mm} \tag{14.7}$$

The thickness of small end can be found by considering tearing at eye end. If D_s is the diameter at small end, it has to resist the gas force given by Equation (14.8)

$$F = \left(D_s - D_p\right) l_p \sigma_t \tag{14.8}$$

Where, D_s = Outside diameter of small end

D_p = Outside diameter of bush

l_p = Length of gudgeon pin in the small end

σ_t = Safe tensile stress of rod material

Example 14.1

Design of small end of a connecting rod
A diesel engine has cylinder diameter of 120 mm and maximum gas pressure 4.2 MPa. Length to diameter ratio for the gudgeon pin is 1.6 and safe bearing pressure is 15 MPa. Determine the size of the pin and outside diameter of small end, if safe tensile strength of rod is 80 MPa.

Solution

Given $D = 120$ mm $\quad p_{max} = 4.2$ MPa $\quad l_p/d_p = 1.6$ $\quad p_{bp} = 15$ MPa $\quad \sigma_t = 80$ MPa

Force due to gas pressure $F = \dfrac{\pi}{4} D^2 p_{max} = 0.785 \times 120^2 \times 4.2 = 48{,}400$ N

From Equation (14.6)

$$F = d_p l_p p_{bp} = d_p \times (1.6 d_p) \times 15 = 24 d_p^2$$

Using the value of F, $48{,}400 = 24 d_p^2$

Hence, diameter of gudgeon pin $d_p = 44.9$ say 45 mm

Using a bush of thickness 3 mm,

Hole diameter of the small end $D_p = 45 + (2 \times 3) = 51$ mm

Length of pin in connecting rod $l_p = 1.6 \times d_p = 1.6 \times 42 = 67$ mm

Using Equation (14.8): $F = (D_s - D_p) l_p \sigma_t$

$$48{,}400 = (D_s - 51) \times 67 \times 80 \quad \text{Or,} \quad D_s = 60 \text{ mm}$$

14.4.2 Cross section of connecting rod

The compressive force on the rod tries to buckle the rod, and hence Rankine formula is used to design. The cross section is generally taken I section for high-speed engines and circular for slow-speed engines. For I section, the proportions taken are given in Figure 14.5. A damaged connecting rod due to buckling is shown in Plate 14.2.

Buckling load about X axis (at right angle to the axis of the gudgeon pin) is:

$$F_x = \frac{\sigma_C A}{1 + a\left(\dfrac{l}{k_{xx}}\right)^2}$$

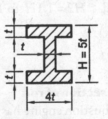

Figure 14.5 | cross section of connecting rod

Plate 14.2
A damaged bent rod

Where, a = Constant = $\dfrac{1}{7{,}500}$ for mild steel and $a = \dfrac{1}{1{,}600}$ for cast iron

Since the ends are constrained in Y direction, effective length is half the length of rod; hence the buckling load about Y axis is:

$$F_y = \dfrac{\sigma_c\, A}{1 + a\left(\dfrac{l}{2k_{yy}}\right)^2}$$

For equal buckling strengths in X and Y directions

$$\left(\dfrac{l}{k_{xx}}\right)^2 = \left(\dfrac{l}{2k_{yy}}\right)^2 \quad \text{or} \quad (k_{xx})^2 = 4(k_{yy})^2$$

Since $I = Ak^2$, hence $I_{xx} = 4I_{yy}$

Area of cross section A for the I section (Refer Figure 11.5) in terms of thickness t is:

$$A = (4t \times t) + (3t \times t) + (4t \times t) = 11\,t^2$$

$$I_{xx} = \dfrac{1}{12}\left[4t \times (5t)^3\right] - \left[3t \times (3t)^3\right] = \dfrac{419\,t^4}{12} \tag{14.9}$$

And, $$I_{yy} = \dfrac{1}{12}\left[2t \times (4t)^3\right] + \left[3t \times t^3\right] = \dfrac{131\,t^4}{12} \tag{14.10}$$

In actual practice, I_{xx} is kept between 3. to 3.5 times I_{yy}. These dimensions are at the middle of the connecting rod. The width is kept constant, while the height is tapered such that:

Height at small end $H1 = 0.75H$ to $0.9H$ (14.11)

Height at big end $H2 = 1.1H$ to $1.25H$ (14.12)

Example 14.2

Design of cross section of rod

An internal combustion engine has cylinder diameter 80 mm and stroke 100 mm. The length of rod / crank radius ratio $(l\,/\,r)$ is 4. Maximum pressure is 3.5 MPa. Assuming compressive strength of the rod 350 MPa, determine the cross section of I section of rod with factor of safety 5. Value of 'a' in Rankine's formula can be taken $\dfrac{1}{7{,}500}$.

Solution

Given $D = 80$ mm $L = 100$ mm $l\,/\,r = 4$ $p_{max} = 3.5$ MPa

$$\sigma_c = 350 \text{ MPa} \quad FOS = 5 \qquad\qquad a = \frac{1}{7,500}$$

Force due to gas pressure $F = \frac{\pi}{4} D^2 p_{max} = 0.785 \times 80^2 \times 4 = 17,584 \text{ N}$

Assuming width of I section of rod as $4t$ and height as $5t$ at the middle of the rod (Figure 14.5).

Area of the rod $A = (4t \times t) + (3t \times t) + (4t \times t) = 11\ t^2$

From Equation (14.9); $I_{xx} = \frac{419t^4}{12}$ and from Equation (14.10): $I_{yy} = \frac{131t^4}{12}$

$$\frac{I_{xx}}{I_{yy}} = \frac{419}{131} = 3.2;\text{ It is less than 4, hence safe.}$$

With FOS for rod, design gas force $= F \times FOS = 17,584 \times 5 = 87,920 \text{ N}$

$$K_{xx} = \sqrt{\frac{I_{xx}}{A}} = \sqrt{\frac{419t^4}{12} \times \frac{1}{11t^2}} = 1.78t$$

Radius of crank $r = 0.5 \times \text{stroke} = 0.5 \times 100 = 50 \text{ mm}$

Hence the length of connecting rod $L = 4 \times 50 = 200 \text{ mm}$ (as $l\,/\,r = 4$ is given)

Using Rankine's Equation $F_x = \dfrac{\sigma_C A}{1 + a\left(\dfrac{l}{k_{xx}}\right)^2}$

Substituting the values:

$$87,920 = \frac{350 \times 11t^2}{1 + \dfrac{1}{7,500}\left(\dfrac{200}{1.78t}\right)^2} \qquad \text{Or,} \qquad 22.83 = \frac{t^2}{1 + \dfrac{1.68}{t^2}}$$

Or, $t^4 - 22.83t^2 - 38.43 = 0$

Solving this quadratic equation: $t^2 = 21$; Hence $t = 4.6$ say 5 mm

Width of I section at middle of rod $B = 4\ t = 4 \times 5 = 20 \text{ mm}$

Height of I section at middle of rod $H = 4\ t = 5 \times 5 = 25 \text{ mm}$

Width of rod remains constant as 20 mm while height varies from small end to big end.

Height at small end $H1 = 0.85\ H = 0.85 \times 25 = 21.2 \text{ mm}$

Height at big end $H2 = 1.2\ H = 1.2 \times 25 = 30 \text{ mm}$

14.4.3 Big end of the rod

The big end of the connecting rod is in two parts (Plate 14.3). Part one is integral with the rod and has a semicircular hole of diameter to suit the crank pin. The other half of semicircular shape is called strap or bearing cap, which is bolted the rod using two bolts / studs.

Plate 14.3 Big end of rod and its cap

The big end is also designed such that the bearing pressure at the projected area of the bearing is supposed to take the gas force F.

$$F = d_c l_c p_{bc} \tag{14.13}$$

Where, l_c = Length of crank pin $(1.25d_c$ to $1.5d_c)$

p_{bc} = Bearing pressure at the crank pin (7 MPa to 12 MPa)

Removable bearing shells made of about 3 mm thickness are put in the semicircular cavity. Thus the hole diameter is: $D_c = d_c + 6$ mm.

Example 14.3

Design of big end of a connecting rod
For the example given in 14.1, if the length to diameter ratio for crank pin is 1.4 and safe bearing pressure is 9 MPa. Calculate the size of big end.

Solution
Given $D = 120$ mm $p_{max} = 4.2$ MPa $l_c / d_c = 1.4$ $p_{bc} = 9$ MPa

Force due to gas pressure $F = \dfrac{\pi}{4} D^2 p_{max} = 0.785 \times 120^2 \times 4.2 = 47{,}477$ N

From Equation (14.13): $F = d_c l_c p_{bc}$

Substituting the values:

$$47{,}477 = d_c \times (1.4\, d_c) \times 9 = 12.6\, d_c^2 \quad \text{Or,} \quad d_c = 61.4 \text{ say } 62 \text{ mm}$$

With shells of 3 mm, hole diameter $D_c = 62 + (2 \times 3) = 68$ mm

Length of pin in connecting rod $l_c = 1.4 \times d_c = 1.4 \times 62 = 86.7$ say 87 mm

---

14.4.4 Cap bolts

The big end of the connecting rod is in two parts. Part one is integral with the rod and the other is strap. Both are joined with bolts called cap bolts. Bolts used are of uniform strength to reduce stress concentration. See Plate 14.4 that the diameter of threaded portion is more than the shank of the bolt.

Plate 14.4 Cap bolt

Since 2 bolts are used to fix cap, hence $F_i = 2\dfrac{\pi}{4}d_c^2\sigma_{tb}$ $\qquad$ (14.14)

Where, F_i = Inertia force coming on the bolt

$\qquad d_c$ = Core diameter of bolts

$\qquad \sigma_{tb}$ = Safe tensile strength of bolts

14.4.5 Strap thickness

Center distance between bolts x = Core diameter of cap bolts, d_c + (2 × Bush thickness) + Crank pin diameter, d + 3 mm

Moment of inertia $I = \dfrac{1}{12}bt_s^3 = \dfrac{l_c}{12}t_s^3$ (As width $b = l_c$ and t_s is thickness of strap)

Distance of extreme fiber, $y = 0.5t_s$

Maximum bending moment $M = \dfrac{F_i x}{6}$ $\qquad$ (14.15)

$$\sigma_{bc} = \frac{My}{I} = \frac{F_i\,x\,(0.5t_s)}{\dfrac{1}{12}l_c t_s^3} = \frac{6F_i x}{l_c t_s^2} \qquad (14.16)$$

Example 14.4

Strap thickness and bolts for a connecting rod

A connecting rod of an internal combustion (I.C.) engine has the following data:

Speed of rotation = 2,000 rpm

Diameter of crank pin = 60 mm

Length of crank pin = 75 mm

Bush thickness = 3 mm

Stroke length = 150 mm

Length of connecting rod = 300 mm

Mass of reciprocating parts = 2 kg

Tensile strength of bolt material = 65 MPa

Safe bending strength of the strap material = 80 MPa

Calculate bolt diameter and strap thickness.

Solution

Given $N = 2{,}000$ rpm $\qquad d_c = 60$ mm $\qquad l_c = 75$ mm $\qquad$ Bush thickness = 3 mm

$\qquad\qquad 2r = 150$ mm $\qquad l = 300$ mm $\qquad m_R = 2$ kg $\qquad \sigma_{tb} = 65$ MPa

$\qquad\qquad \sigma_{bc} = 80$ MPa $\qquad l/r = 300/75 = 4$

Rotational speed $\omega = \dfrac{2\pi N}{60} = \dfrac{2\times 3.14\times 2{,}000}{60} = 209.3$ rad/s

From Equation (14.3): Inertia force for $\theta = 0$, $F_i = m_R \times \omega^2 r \left(1 + \dfrac{r}{l} \right)$

Substituting the values: $F_i = 2 \times (209.3)^2 (0.075) \left(1 + \dfrac{75}{300} \right) = 8{,}216 \text{ N}$

Bolt diameter: From Equation (14.14): $F_i = 2 \dfrac{\pi}{4} d_c^2 \sigma_{tb}$

Substituting the values: $8{,}216 = 2 \times 0.785 \times d_c^2 \times 65$ Or, $d_c^2 = 80.5$

Hence, core diameter of bolt $d_c = 8.97$ mm

Nominal diameter of bolt $d = \dfrac{d_c}{0.84} = \dfrac{8.97}{0.84} = 10.7$ mm

Choosing a standard bolt diameter $d = 12$ mm

Strap thickness:

Center distance between bolts $x = d_c + (2 \times \text{Bush thickness}) + d + 3$ mm

Or, $x = 60 + (2 \times 3) + 12 + 3 = 81$ mm

From Equation (14.15): Bending moment $M = \dfrac{F_i x}{6} = \dfrac{8{,}216 \times 81}{6} = 110{,}916 \text{ Nmm}$

From Equation (14.16) $\sigma_{bc} = \dfrac{M y}{I} = \dfrac{F_i x (0.5 t_s)}{\dfrac{1}{12} l_c t_s^3} = \dfrac{6 F_i x}{l_c t_s^2}$

$\sigma_{bc} = \dfrac{M y}{I}$ Putting values: $80 = \dfrac{110{,}916 \times 0.5 t_s}{(75/12) t_s^3}$ or $t_s^2 = 111$

Hence, thickness of strap $t_s = 10.5$ say 11 mm.

Example 14.5

Whipping stress in a connecting rod

A connecting rod of length 320 mm of I section has height in middle $5t$, width $4t$ and web thickness $t = 6$ mm. The engine rotates at 1,500 rpm with a stroke of 160 mm. Calculate whipping stress, if density of the rod material is 7,800 kg/m³.

Solution

Given $N = 1{,}500$ rpm $t = 6$ mm $2r = 160$ mm $L = 320$ mm

           $\rho = 7{,}800$ kg/m³ $H = 5t$ $W = 4t$

Rotational speed $\omega = \dfrac{2\pi N}{60} = \dfrac{2 \times 3.14 \times 1{,}500}{60} = 157 \text{ rad/s}$

Area of the rod (Refer Figure 14.5): $A = (4t \times t) + (3t \times t) + (4t \times t) = 11\,t^2$

Mass per unit length $m_i = A \times \rho = 11(0.006)^2 \times 7{,}800 = 3.09$ kg/m

Maximum bending moment $M_{max} = m_i\,\omega^2 r\dfrac{L^2}{9\sqrt{3}} = 3.09 \times (157)^2 \times 0.08 \times \dfrac{(0.32)^2}{9\sqrt{3}}$

Or, $\quad M_{max} = 40$ Nm $= 40{,}000$ Nmm

From Equation (14.9) Moment of inertia $I = \dfrac{419\,t^4}{12} = \dfrac{419 \times 6^4}{12} = 45{,}252$ mm^4

$\qquad y = 2.5\,t = 2.5 \times 6 = 15$ mm

Whipping bending stress, $\sigma_b = \dfrac{M\,y}{I} = \dfrac{40{,}000 \times 15}{45{,}252} = 13.25$ MPa

Example 14.6

Design of connecting rod

An internal combustion engine runs at 1,500 rpm and develops maximum pressure of 4 MPa. Design its connecting rod of I section with FOS = 5 and the following data:

Piston diameter = 110 mm

Stroke length of piston = 200 mm

Length of rod / crank radius ratio $(l/r) = 4$

Mass of reciprocating parts = 2.5 kg

Compressive strength = 300 MPa

Density of rod material = 8,000 kg/m^3

Allowable bearing pressure for small end = 15 MPa

Allowable bearing pressure for big end = 10 MPa

Length to pin diameter ratio for small end = 1.5

Length to pin diameter ratio for big end = 1.3

Safe bending stress for the rod material = 75 MPa

Safe bending stress for the bolts = 58 MPa

Constant for Rankine's formula $a = \dfrac{1}{7{,}500}$

Solution

Given $\quad N = 1{,}500$ rpm $\qquad p_{max} = 4$ MPa $\qquad D = 110$ mm $\quad 2r = 200$ mm

$\qquad\qquad l/r = 4 \qquad\qquad m_R = 2.5$ kg $\qquad \sigma_c = 300$ MPa $\quad \rho = 8{,}000$ kg/m^3

$\qquad\qquad p_{bp} = 15$ MPa $\qquad p_{bc} = 10$ MPa $\quad l_p/d_p = 1.5 \qquad l_c/d_c = 1.3$

$\qquad\qquad \sigma_{bb} = 58$ MPa $\qquad \sigma_{bc} = 75$ MPa $\quad FOS = 5 \qquad\qquad a = 1/7{,}500$

a. **Small end**

Force due to gas pressure $F = \dfrac{\pi}{4} D^2 \, p_{max} = 0.785 \times 110^2 \times 4 = 48{,}400$ N

From Equation (14.6): $F = d_p l_p p_{bp} = d_p \times (1.5 d_p) \times 15 = 22.5 d_p^2$

Using the value of F, $48{,}400 = 22.5 d_p^2$

Hence, diameter of gudgeon pin, $d_p = 46.5$ mm

Using a bush of thickness 3 mm,

Hole diameter of the small end $D_p = 46.5 + (2 \times 3) = 52.5$ mm

Length of pin in connecting rod $l_p = 1.5 \times d_p = 1.5 \times 46.5 = 70$ mm

b. **Cross section of the rod**

Assuming width of I section of rod as $4t$ and height as $5t$ at the middle of the rod.

Area of the rod $A = [(4t \times t) + (3t \times t) + (4t \times t)] = 11t^2$

From Equation (14.9); $I_{xx} = \dfrac{419 t^4}{12}$ and from Equation (14.10): $I_{yy} = \dfrac{131 t^4}{12}$

$\dfrac{I_{xx}}{I_{yy}} = \dfrac{419}{131} = 3.2$; It is less than 4 hence safe.

$K_{xx} = \sqrt{\dfrac{I_{xx}}{A}} = \sqrt{\dfrac{419 t^4}{12} \times \dfrac{1}{11 t^2}} = 1.78 t$

Radius of crank $r = 0.5 \times \text{stroke} = 0.5 \times 200 = 100$ mm

Hence length of connecting rod $l = 4 \times 100 = 400$ mm (As $l/r = 4$ is given)

Using Rankine Equation: $F = \dfrac{\sigma_c A}{1 + a\left(\dfrac{l}{k_{xx}}\right)^2}$

Design force is: $F \times FOS = 48{,}400 \times 5 = 242{,}000$ N

Substituting the values:

$$242{,}000 = \dfrac{300 \times 11 t^2}{1 + \dfrac{1}{7{,}500}\left(\dfrac{400}{1.78 t}\right)^2} \qquad \text{Or,} \qquad 73.3 = \dfrac{t^2}{1 + \dfrac{6.73}{t^2}}$$

Or, $t^4 - 73.3\, t^2 - 493.5 = 0$ Or, $t^2 = 73$; Hence $t = 8.5$ mm

Width of I section at middle of rod $B = 4\,t = 4 \times 8.5 = 34$ mm

Height of I section at middle of rod $H = 5\,t = 5 \times 8.5 = 42$ mm

The width of rod remains constant as 34 mm while height varies from small end to big end.

Height at small end $H1 = 0.85\,H = 0.85 \times 42 = 36$ mm

Height at big end $H2 = 1.2\,H = 1.2 \times 42 = 50.4$ say 51 mm

c. Big end

From Equation (14.13): $F = d_c\,l_c\,p_{bc}$

Substituting the values:

$$48,400 = d_c \times (1.3d_c) \times 10 = 13d_c^2 \qquad \text{Or,} \qquad d_c = 61 \text{ mm}$$

Length of pin in connecting rod, $l_c = 1.3 \times d_c = 1.3 \times 61 = 79.3$ say 80 mm

With shells of 3 mm, hole diameter $D_c = 61 + (2 \times 3) = 67$ mm

d. Bolt diameter

Rotational speed in radians per second $\omega = \dfrac{2\pi N}{60} = \dfrac{2 \times 3.14 \times 1,500}{60} = 157\,\text{rad/s}$

From Equation (14.3): Inertia force $F_i = m_R \times \omega^2 r\left(\cos\theta + \dfrac{\cos 2\theta}{n}\right)$

Substituting the values for $\theta = 0$: $F_i = 2.5 \times 157^2 \times 0.1\left(1 + \dfrac{1}{4}\right) = 7,703$ N

From Equation (14.14): Inertia force $F_i = 2\dfrac{\pi}{4}d_c^2\sigma_{tb}$

Substituting the values: $7,703 = 2 \times 0.785 \times d_c^2 \times 65 \qquad \text{Or,} \qquad d_c^2 = 75.5$

Hence, core diameter of bolt $d_c = 8.7$ mm (Note here d_c is core diameter of bolt and not crank pin diameter)

Nominal diameter of bolt $d = \dfrac{d_c}{0.84} = \dfrac{8.7}{0.84} = 10.4\,\text{mm}$

Choosing a standard bolt diameter $d = 12$ mm

e. Strap

Center distance between bolts $x = d_c + (2 \times \text{Bush thickness}) + d + 3$

$$= 61 + (2 \times 3) + 12 + 3 = 82 \text{ mm}$$

Moment of section, $Z = \dfrac{1}{6}bt_s^2 = \dfrac{l_c}{6}t_s^2 = \dfrac{80}{6}t_s^2 = 13.33t_s^2 \text{ mm}^3$

Maximum bending moment, $M = \dfrac{F_i x}{6} = \dfrac{7,703 \times 82}{6} = 105,274$ Nmm

$$\sigma_{bc} = \dfrac{M}{Z} \quad \text{or} \quad 75 = \dfrac{105,274}{13.33 t_s^2}$$

Or, $t_s^2 = 105$; Hence, thickness of strap $t_s = 10.3$ say 11 mm.

Summary

Connecting rod is used to join piston with a pin joint at its one end on crank shaft pin at the other end. The end on piston side is called *small end* and the end mounted on the crank pin is called *big end*. The function of the connecting rod is to transmit gas force from reciprocating motion of piston to the crank shaft, for converting to rotary motion. The length of rod is the distance between centers of gudgeon pin and center of crank pin. Ratio of its length to radius of crank (l/r) is kept between 4 and 5. Connecting rod is supposed to have high strength, hence made by drop forging or die forging to provide strength and stiffness. It is made of following materials:

0.4% carbon steel, 0.5% heat treated carbon steel, and nickel–chromium alloy steel.

Construction

Small end is made circular to accommodate a bush made of phosphorous bronze.

Body   Cross section of the rod can be circular or rectangular for small speed engines but high-speed engines use I section to make it light to reduce inertia forces. If t is the thickness of web of I section, width of the connecting rod is kept $4t$ and height $5t$ in middle. Its width remains constant, while height increases from *H1* at small end to *H2* at big end.

$H1 = 0.75\ H$ to $0.95\ H$ and $H2 = 1.1\ H$ to $1.25\ H.$

Big end is split into two halves; one half is integral with the rod. The other half is called cap or strap, which is fixed to the big end using two studs and nuts. This end is provided with a bearing shell made of steel, brass, or bronze with a lining of soft metal like tin or babbitt.

The shells are provided with a *snug* at one edge, which fits into the cavity of the rod and cap so that the shell does not rotate in the hole, due to friction between pin and shell.

Forces on connecting rod   Following forces act on a connecting rod:

a.   Axial force due to gas pressure p_{max} over a cylinder of diameter D is: $F = \dfrac{\pi}{4} D^2 p_{max}$

b.   Axial force due to inertia of reciprocating parts   Mass of the reciprocating parts m_R is mass of piston m_P and gudgeon pin. The connecting rod of mass m_C is taken one-third of mass of connecting rod m_C. Thus: $m_R = m_P + \dfrac{1}{3} m_C$

Acceleration of parts is given as: $f = \omega^2 r \left(\cos\theta + \dfrac{\cos 2\theta}{n} \right)$ Where, $n = l/r$ ratio

Inertia force $F_i = m_R \times \omega^2 r \left(\cos\theta + \dfrac{\cos 2\theta}{n} \right)$

Axial force at piston is $F_p = F \pm F_r$

c. **Bending force due to inertia of reciprocating parts** Inertia forces act at right angle to the connecting rod. Its value is zero at piston pin and varies linearly from piston to maximum at crank pin. Thus the loading is triangular from zero to maximum $m_R \omega^2 r$ Hence average inertia force $F_i = 0.5 m_R \omega^2 r$. It acts at a distance $2l/3$ from piston pin.

Maximum bending moment M_{max} for triangular loading acts at $(l/\sqrt{3})$ from piston pin.

$$M_{max} = \frac{F_i l}{9\sqrt{3}}; \text{Maximum stress } \sigma_{max} = \frac{M_{max}}{Z}.$$ This is called whipping stress.

d. **Frictional forces due to friction between piston rings and cylinder** Frictional force due to piston rings rubbing on cylinder wall is: $F_f = \pi D\, t_A n\, p_r\, \mu$

t_A = Axial thickness of rings

n = Number of piston rings

p_r = Radial pressure rings (from 0.025 MPa to 0.4 MPa) and

μ = Coefficient of friction

e. **Frictional forces due to friction between gudgeon pin and crank pin:** This force at the gudgeon pin and small end of connecting rod tries to bend the connecting rod.

Design of connecting rod

Small end of the rod Gudgeon pin diameter d_p is calculated on the basis of gas pressure p_{max} resisted by projected area of the gudgeon pin.

$$\text{Maximum gas force } F = \frac{\pi}{4} D^2 p_{max} = d_p\, l_p\, p_{bp}$$

Where, d_p = Diameter of gudgeon pin

l_p = Length of gudgeon pin

p_{bp} = Bearing pressure (varies from 10 MPa to 15 MPa).

Using a bush of 3 mm thickness hole diameter D_p of small end is: $D_p = d_p + 6$ mm

Thickness of small end can be found by considering tearing at eye end. If D_s is the diameter at small end, it has to resist the gas force $F = (D_s - D_p)\, l_p\, \sigma_t$

Where, D_s = Outside diameter of small end

D_p = Outside diameter of bush

Cross section of connecting rod Buckling load about X axis (at right angle to the axis of the gudgeon pin) is:

$$F_x = \frac{\sigma_c A}{1 + a\left(\dfrac{l}{k_{xx}}\right)^2}$$

Where, Constant $a = \dfrac{1}{7,500}$ for mild steel

For equal buckling strengths in X and Y directions: $\left(\dfrac{l}{k_{xx}}\right)^2 = \left(\dfrac{l}{2k_{yy}}\right)^2$

$\qquad (k_{xx})^2 = 4(k_{yy})^2 \qquad$ Since $I = AK^2$, $I_{xx} = 4I_{yy}$

Area of cross section for I section for width $4t$ and height $5t$ in terms of thickness t is: $A = 11t^2$

$\qquad I_{xx} = 419t^4/12$ and $I_{yy} = 131t^4/12$

In actual practice, I_{xx} is kept between 3 to 3.5 times I_{yy}. Height at small end $H1 = 0.75H$ to $0.9H$ and height at big end $H2 = 1.1H$ to $1.25H$.

Big end of the connecting rod   The big end is designed such that the bearing pressure at the projected area of the bearing is supposed to take the gas force:

$$F = d_c\, l_c\, p_{bc}$$

Where, l_c = Length of crank pin

$\qquad p_{bc}$ = Bearing pressure at the crank pin

Removable bearing shells made of about 3 mm thickness is put in the semicircular cavity. Thus the hole diameter is: $D_c = d_c + 6$ mm.

Cap bolts   Both parts of the big end are joined with two bolts called cap bolts.

Inertia force $F_i = 2\dfrac{\pi}{4}d_c^2\,\sigma_{tb}$,

$\qquad d_c$ = Core diameter of bolts

$\qquad \sigma_{tb}$ = Safe tensile strength.

Strap thickness   Center distance between bolts, $x = d_c(2 \times$ Bush thickness$) + d + 3$ mm

Moment of inertia, $I = \dfrac{1}{12}bt^3 = \dfrac{l_c}{12}t^3$ (As width $b = l_c$)

$\qquad y = 0.5\,t$

Maximum bending moment $M = \dfrac{F_i x}{6}$

Or, $\sigma_{bc} = \dfrac{My}{I} = \dfrac{6F_i x}{l_c t^2}$

Theory Questions

1. What is the function of connecting rod? Mention the materials used for it.
2. Describe its construction with the help of a neat sketch.
3. Discuss the various forces coming in it.
4. What are the inertia forces? How are these accounted in design?
5. Describe the procedure to calculate the cross section of the rod.
6. Describe the design procedure to design small and big ends of the rod.
7. How do you calculate the bolt diameter for the strap of connecting rod?

8. Explain the method to calculate strap thickness of connecting rod.

9. What is meant by whipping stress? Drive an equation to calculate it.

Multiple Choice Questions

1. Cross section of connecting rod of a high-speed I.C. engine is generally:
   (a) Circular
   (b) Rectangular
   (c) I section
   (d) T section

2. In design of connecting rod for bending, ends are considered as:
   (a) Both fixed
   (b) One fixed and other hinged
   (c) One fixed and other free
   (d) Both hinged

3. Inertia force of connecting rod:
   (a) Has no effect
   (b) Opposes the piston force while moving from inner to outer dead center
   (c) Helps the piston force, while moving from inner to outer dead center
   (d) Always opposes the piston force

4. Maximum bending stress in a connecting rod occurs when crank angle is:
   (a) Between 65° and 70° from TDC
   (b) Between 50° and 60° from TDC
   (c) Between 35° and 50° from TDC
   (d) 45°

5. While calculating mass of reciprocating parts, mass of connecting rod m_c is added to mass of piston as:
   (a) $0.45 \, m_c$
   (b) $0.40 \, m_c$
   (c) $0.37 \, m_c$
   (d) $0.33 \, m_c$

6. Bearing pressure at small end is taken as:
   (a) 4 MPa to 6MPa
   (b) 7 MPa to 9 MPa
   (c) 10 MPa to 15 MPa
   (d) 16 MPa to 20 MPa

7. If M_{max} is maximum bending moment and Z modulus of section of a connecting rod $\frac{M_{max}}{Z}$ is called as:
   (a) Bending stress
   (b) Whipping stress
   (c) Inertial stress
   (d) Residual stress

8. Strap of a connecting rod is:
   (a) Fitted on the forked big end to cover the big end bearing
   (b) An additional strap put on small end for increasing strength
   (c) An extra plate attached to rod for balancing purpose
   (d) None of the above

Answers to multiple choice questions

1. (c)   2. (d)   3. (b)   4. (a)   5. (d)   6. (c)   7. (b)   8. (a)

Design Examples

1. An I.C. engine having cylinder diameter 100 mm develops maximum gas pressure 3.8 MPa. Length to diameter ratio for the gudgeon pin is 1.7 and safe bearing pressure is 14 MPa. Determine the size of the pin.

$$[d_p = 35 \text{ mm}, l_p = 60 \text{ mm}]$$

2. An internal combustion engine has length of rod / crank radius ratio $l/r = 4$. Crank radius is 45 mm and length of stroke is 25 per cent more than the cylinder diameter. Maximum pressure is 4.0 MPa. Assuming compressive strength of the rod 330 MPa, determine the cross section of I section of rod with FOS = 6. Value of 'a' in Rankine's formula can be taken: $\dfrac{1}{7,500}$.

$$[t = 5.2 \text{ mm}, \ B = 21 \text{ mm}, \ H1 = 22 \text{ mm}, H2 = 31 \text{ mm}]$$

3. An engine with cylinder diameter 85 mm shows maximum load on connecting rod as 40 kN. If length to diameter ratio of the pin is 1.4 and safe bearing pressure is 9 MPa, calculate the size of crank pin.

$$[d_c = 57 \text{ mm}, l_c = 80 \text{ mm}]$$

4. An I.C. engine of stroke length 130 mm rotates at 2,400 rpm. Its diameter is 100 mm and l/r ratio for connecting rod is 3.8. Diameter of crank pin is 60 mm and length is 25 per cent more than its diameter. Mass of reciprocating parts is 1.8 kg. Take bush thickness 3 mm, tensile strength of bolt and strap material 70 MPa. Calculate bolt diameter and strap thickness.

$$[d = 12 \text{ mm}, t_s = 12 \text{ mm}]$$

5. An engine rotates at 2,000 rpm with a stroke of 140 mm and length of rod 280 mm. Its connecting rod has flange and web thickness of I section 5 mm. Calculate whipping stress, if density of the rod material is 8,000 kg/m³.

$$[\sigma_b = 19.43 \text{ MPa}]$$

6. A connecting rod of length 280 mm has I section of thickness 5 mm, width 20 mm, and depth 25 mm in the middle. If the safe compressive strength of the material is 300 MPa, calculate the maximum buckling load, which it can take.

$$[F = 72.88 \text{ kN}]$$

7. A four-stroke I.C. engine develops 3.5 MPa maximum pressure in a cylinder of length 100 mm and diameter 75 mm. Design I section connecting rod with the following data:

    Mass of reciprocating parts = 1.5 kg

    Density of rod material = 8,000 kg/m³

    l/r ratio = 4

    l/d ratio for small end 1.8 with safe bearing pressure = 14 MPa

    l/d ratio for big end 1.4 with safe bearing pressure = 10 MPa

    Safe stress for bolt material = 65 MPa and for strap = 80 MPa

Compressive yield stress = 300 MPa

FOS in buckling = 4

[t = 6 mm, B = 24 mm, H = 30 mm, $H1$ = 42 mm, $H2$ = 28 mm, d_p = 25 mm, l_p = 45 mm, d_c = 35 mm, l_c = 45 mm, d_b = 6 mm, t_s = 6 mm]

Previous Competition Examination Questions

IES

1. In the design of connecting rod small end bearing, the value of permissible bearing pressure to be used is:

   (a) Less than used for big end bearing

   (b) More than used for big end bearing

   (c) Equal to that used for big end bearing

   (d) None of the above

   [Answer (b)] [IES 2013]

□□□

Crank Shaft

Outcomes

➤ Learn about function and construction of a crank shaft

➤ Types of crank shafts and materials used for them

➤ Forces on crank shaft and its design

➤ Design of side crank

15.1 Introduction

Crank shaft shown in picture on right side for a multi-cylinder engine converts reciprocating motion of piston to rotary motion using connecting rods. It can be divided into four portions as shown in Figure 15.1.

a. Crank pin, on which the big end is fixed using strap and cap bolts.

b. Main bearing journal, where the crank shaft is supported on bearings (in two halves).

c. Crank web, which joins crank shaft and the crank pin.

d. Balance weight is a dead weight on webs opposite to the crank pin. It can be integral with the web or bolted separately on the web.

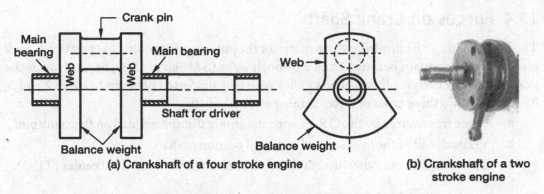

Figure 15.1 Parts of a center crank shaft

15.2 Types of Crank Shafts

Crank shafts can be mainly divided into two types:

a. Center crank shaft: Crank pin is in between two webs as shown in Figure 15.1(a). Crank shaft for a two stroke engine is shown in picture 15.1(b).

b. Side crank shaft or overhung crank shaft: Crank pin is overhung on one web called disk (Figure 15.2)

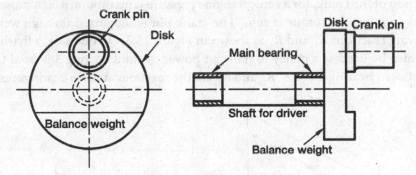

Figure 15.2 Side crank shaft

15.3 Crank Materials

Crank shaft is subjected to shocks and fatigue and hence has to be tough. These are usually forged to impart strength and good grain structure. The crank pin and journals are hardened by carburizing or nitriding and ground. The material used depends on the type of application as given below:

Industrial engines: Plain carbon steel— 40C8, 45C8, 50C4, 55C8 and 60C4

Automobile engines: Manganese steels— 20Mn2, 27Mn2, and 37Mn2

Aero engines: Ni–Cr alloy steels—16Ni3Cr2, 35Ni5Cr2 and 40Ni2Cr1Mo28

15.4 Forces on Crank Shaft

The crank shaft can be divided into two parts for the purpose of analysis. One part is the crank pin supported on main bearings 1 and 2 through webs (See Figure 15.3). Second part is the portion of shaft carrying flywheel or a pulley with belt supported between bearings 2 and 3. Mainly there are three types of forces coming on crank shaft:

a. Force transmitted by the C.R. causing uniformly distributed load on the crank pin.

b. Vertical load of the flywheel in the second portion of shaft.

c. Torsional stresses, when the crank web is at an angle from top dead center (TDC).

15.5 Design of Crank Shaft

A crank shaft is designed for two positions of the crank shaft. Stresses are calculated for both the positions and maximum stress is considered for design. These two positions are:

- Crank at dead center position. TDC for vertical engine and IDC for horizontal engine. position (Section 15.5.1)

- Crank at position of maximum torque (Section 15.5.2)

15.5.1 Design of crank at dead center position

At this position of the crank, for a vertical engine pressure is maximum, which causes bending of the crank pin, but the torque is zero. The crank pin is supported through webs on two bearings giving reactions R_{1v} and R_{2v} as shown in Figure 15.3. On one side, a flywheel is put, which can also be used as a pulley to transmit power. A third bearing 3 is used to support flywheel between bearing 2 and 3. R_{2v}' and R_{3v}' are the reactions at these bearings respectively.

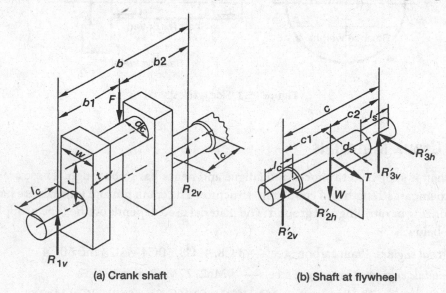

(a) Crank shaft (b) Shaft at flywheel

Figure 15.3 Crank shaft at TDC position

The following assumptions are made before starting design:

- Distance (b) between the centers of crank shaft main bearings 1 and 2, is not given, hence initially assumed as twice the cylinder diameter, that is, $b = 2D$.

- Distance between the centers of bearings 2 and 3 for the shaft at flywheel, (c) is assumed as three times the cylinder diameter, that is, $c = 3D$

- The crank pin and the flywheel are placed centrally between the bearings. Thus the distance of center of crank pin from the centers of bearings 1 and 2 is equal. That is, $b1 = b2 = 0.5b$ and $c1 = c2 = 0.5c$

The design is done in the following four steps:

Step 1 *Calculate reactions at the bearings*

Maximum gas force at this position $F = \dfrac{\pi}{4}D^2 p_{max}$ (15.1)

Where, D = Cylinder bore (mm)

p_{max} = Maximum gas pressure (MPa)

Taking moments about point 2, $F \times b2 = R_{1v} \times b$ or $R_{1v} = \dfrac{F \times b2}{b}$

Taking moments about point 1, $F \times b1 = R_{2v} \times b$ or $R_{2v} = \dfrac{F \times b1}{b}$

Similarly reactions due to horizontal force P: $R_{1h} = \dfrac{P \times b2}{b}$ and $R_{2h} = \dfrac{P \times b1}{b}$

For the flywheel of weight W at middle, reactions are calculated as given below. Superscript ($'$) indicates the reactions due to flywheel.

Taking moments about point 2: $W \times c1 = R'_{3v} \times c$ or $R'_{3v} = \dfrac{W \times c1}{c}$

Taking moments about point 3, $W \times c2 = R'_{2v} \times c$ or $R'_{2v} = \dfrac{W \times c2}{c}$

If T is the total pull of the belt tensions and assuming that the belt is horizontal:

Taking moments about point 2: $T \times c1 = R'_{3h} \times c$ or $R'_{3h} = \dfrac{T \times c1}{c}$

Taking moments about point 3, $T \times c2 = R'_{2h} \times c$ or $R'_{2h} = \dfrac{T \times c2}{c}$

Therefore, total reactions at section 1, in this position of crankshaft

$R_1 = R_{1v}$ as $R_{1h} = 0$

Resultant reaction at section 2, $R_2 = \sqrt{\left(R_{2v} + R'_{2v}\right)^2 + \left(R'_{2h}\right)^2}$

Resultant reaction at section 3, $R_3 = \sqrt{\left(R'_{3v}\right)^2 + \left(R'_{3h}\right)^2}$

Step 2 *Calculate crank-pin size*

Bending moment at center of crank pin: $M = R_{1v} \times b1$

Moment of inertia for crank pin of diameter d_c is:

$$I = \frac{\pi}{64}d_c^4, \quad y = \frac{d_c}{2}; \quad \text{Hence } Z = \frac{I}{y} = \frac{\pi}{32}d_c^3$$

Using the relation $M = \sigma_b Z$, calculate crank pin diameter for the safe bending stress σ_b.

For the design to be safe, maximum gas force is to be taken by the projected area of the crank pin of length l_c, without increasing the safe bearing pressure p_b. Thus:

$$F = d_c \times l_c \times p_b \tag{15.2}$$

Assuming a suitable bearing pressure, l_c/d_c ratio between 1 and 1.2, calculate bearing pressure. It should be less than allowable limit.

Step 3 *Calculate size of web*

Both the left and right side webs are made similar. The web is designed using following empirical relations and then checked for the stresses.

Width of web $w = 1.14d_c$ or 0.4 to $0.6d_s$ or $0.22D$ to $0.32D$

Where, d_s = Shaft diameter at flywheel and D is cylinder bore.

Thickness of web $t = 0.67\, d_c$

There are two types of stresses coming in web:

(i) Direct compressive stress $\sigma_c = \dfrac{\text{Load}}{\text{Web area}} = \dfrac{R_{1v}}{w \times t}$ $\qquad$ (15.3)

(ii) Compressive stress due to bending.

Bending moment at central plane of web $M = R_{1v}\left[b1 - \dfrac{t_c}{2} - \dfrac{l_c}{2}\right]$

Section modulus of the web $Z = \dfrac{wt^2}{6}$

Bending stress $\sigma_b = \dfrac{M}{Z}$ $\qquad\qquad\qquad\qquad\qquad$ (15.4)

Total compressive stress $\sigma_{ct} = \sigma_c + \sigma_b$ $\qquad\qquad\qquad$ (15.5)

This stress should be less than allowable compressive stress.

Step 4 *Calculate size of shaft at flywheel*

Bending moments are calculated at center of flywheel

Bending moment in vertical direction $M_v = R_{3v}' \times c2$

Bending moment in horizontal direction $M_h = R'_{3h} \times c2$

Resultant bending moment $M = \sqrt{M_v^2 + M_h^2}$ (15.6)

Knowing the value of safe bending stress σ_b, calculate the shaft diameter d_s at flywheel using the following relation:

$$M = \frac{\pi d_s^3}{32} \sigma_b$$ (15.7)

15.5.2 Design of crank at position of maximum torque

When crank rotates, it turns an angle θ from dead center piston of crank shaft. The connecting rod is inclined at an angle Φ with the line of centers joining gudgeon pin P and crank shaft O as shown in Figure 15.4.

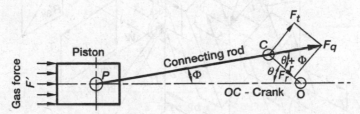

Figure 15.4 Forces on crank pin at an angular position of crank shaft

Step 5 *Calculation of component of forces*

In the inclined position of the crank shaft, component of force along the axis of the connecting rod is F_q as shown in Figure 15.4. This force can be resolved in radial component F_r along the crank web and tangential component F_t at right angle to the web. The torque is maximum, when tangential component of force is maximum. This condition occurs when the crank angle θ is between 25° and 35°. This range of angle is for petrol engines, while for diesel engine its range is 30° to 40°. At this position, the gas pressure is denoted by p'. Thus, gas force at this position F' is given by the following relation:

$$F' = \frac{\pi}{4} D^2 p'$$ (15.8)

The angle of crank shaft and C.R. are related by Equation (15.9) due to the geometry.

$$\sin \theta = \frac{L}{r} \sin \Phi \text{ or } \Phi = \sin^{-1} \left(\frac{r}{L} \sin \theta \right)$$ (15.9)

Where, r = Radius of crank

L = Length of connecting rod (C.R.)

Force along the connecting rod, $F_q = \dfrac{F'}{\cos \Phi}$ (15.10)

Tangential component of force $\quad F_t = F_q = \sin(\theta + \Phi)$ (15.11)

Radial component of force $\quad\quad F_r = F_q = \cos(\theta + \Phi)$ (15.12)

Step 6 *Calculation of reactions at bearings*

Figure 15.5 shows crank shaft at maximum torque position.

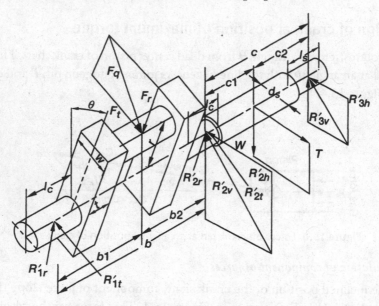

Figure 15.5 Crank shaft at maximum torque position

The analysis is done first considering crank pin supported on two bearings 1 and 2. This pin is subjected to radial force and tangential force as calculated above. The reactions will also be in the radial and tangential directions. Superscript ($'$) is used for this position of crank shaft.

Taking moments about bearing 2: $F_t \times b2 = R'_{1t} \times b \quad$ or $\quad R'_{1t} = \dfrac{F_t \times b2}{b}$

Taking moments about bearing 1: $F_t \times b1 = R'_{2t} \times b \quad$ or $\quad R'_{2t} = \dfrac{F_t \times b1}{b}$

Similarly, radial force will give reaction, $R'_{1r} = \dfrac{F_r \times b2}{b} \quad$ and $\quad R'_{2r} = \dfrac{F_r \times b1}{b}$

Step 7 *Calculate crank pin size*

Bending moment at center of crank pin in tangential direction $M_t = R'_{2t} \times b2$

Bending moment in radial direction $\quad M_r = R'_{2r} \times b2$

Resultant bending moment at crank pin $M_c = \sqrt{M_t^2 + M_r^2}$ \hfill (15.13)

Twisting moment on crank pin $\qquad\qquad T_c = R'_{1t} \times r$ \hfill (15.14)

Equivalent twisting moment at crank pin $T_e = \sqrt{\left(M_c\right)^2 + \left(T_c\right)^2}$ \hfill (15.15)

Knowing the value of safe shear stress τ, calculate the crank pin diameter d_c using the following relation:

$$T_e = \frac{\pi d_c^3}{16}\tau \hfill (15.16)$$

Step 8 *Calculate size of shaft at junction of right hand crank web and crank shaft*

Reactions of bearings 2 and due to fly wheel will be same as calculated for dead center position.

At the junction of shaft at flywheel and right crank web following forces act:

Bending moment due to radial forces, $M_{2r} = R'_{1r}(b1 + t + 0.5\ lc) - F_r(t + 0.5\ lc)$

Bending moment due to tangential forces, $M_{2t} = R'_{1t}(b1 + 0.5\ lc + t) - F_t(0.5\ lc + t)$

Resultant bending moment due to above moments $M_2 = \sqrt{\left(M_{2r}\right)^2 + \left(M_{2t}\right)^2}$

Torsion due to tangential force $T_2 = F_t \times r$

Equivalent torque due to B.M. and torsion $T_e = \sqrt{M_2^2 + T_2^2}$

Knowing the value of safe shear stress τ, calculate the shaft diameter d_s at flywheel using the following relation:

$$T_e = \frac{\pi d_s^3}{16}\tau$$

Step 9 *Calculate size of shaft at flywheel*

Forces acting at the center of the shaft at flywheel give rise to bending moments:

Bending moment in vertical direction $M_v = R'_{3v} \times c2$

Torsion due to tangential force $T = F_t \times r$

Equivalent torque $T_e = \sqrt{M_v^2 + T^2}$ \hfill (15.17)

Knowing the value of safe shear stress τ, calculate the shaft diameter d_s at flywheel using the following relation:

$$T_e = \frac{\pi d_s^3}{32}\tau \hfill (15.18)$$

Step 10 *Calculate size of right side web (Flywheel side)*

Right hand web is subjected to the following forces:

a. Bending moment at central plane of web due to radial force

$$M_{2r} = R'_{2r} \times \left[b2 - \frac{t}{2} - \frac{l_c}{2} \right]$$

Modulus of section $Z = \frac{1}{6}wt^2$

Using the relation $M_{2r} = Z\sigma_{br} = \frac{1}{6}wt^2\sigma_{br}$ or $\sigma_{br} = \frac{6M_{2r}}{wt^2}$ (a)

Bending moment due to tangential force $M_{2t} = F_t \times \left[r - \frac{d_{s1}}{2} \right]$

Where, d_{s1} = Shaft diameter at the junction of right web and r = Crank radius

Modulus of section in this plane $Z = \frac{1}{6}tw^2$

Using the relation $M_{2t} = Z\sigma_b = \frac{1}{6}tw^2\sigma_b$ or $\sigma_{bt} = \frac{6M_{2t}}{tw^2}$ (b)

b. Direct bending stress due to radial force F_r (0.5 F_r for one web)

$$\sigma_{bd} = \frac{F_r}{2wt}$$ (c)

Total bending stress $\sigma_{bc} = \sigma_{br} + \sigma_{bt} + \sigma_{bd}$ (15.19)

c. Torsion in the web due to arm length is:

$$T = R'_{1t} \times \left[b2 + \frac{l_c}{2} \right] - F_t \times \frac{l_c}{2} = R'_{2t} \times \left[b1 - \frac{l_c}{2} \right]$$

Polar moment of the section $Z_p = \frac{tw^2}{4.5}$

Shear stress on the arm $\tau = \frac{T}{Z_p} = \frac{4.5T}{tw^2}$ (15.20)

Maximum combined stress $\sigma_{bmax} = \frac{\sigma_{bc}}{2} + \frac{1}{2}\sqrt{\sigma_{bc}^2 + 4\tau^2}$ (15.21)

Stress calculated from Equation (15.21) should be less than safe stress.

Left hand web is not loaded as right web, hence its size it taken same as right side web to maintain symmetry and balance.

Step 11 *Calculate size of bearings*

Bearing 2 is usually most heavily loaded due to crank pin forces on one side and flywheel forces from the other side; hence bearing 2 is designed assuming a safe bearing pressure. Other bearings, that is, 1 and 3 are taken of same size to help interchangeability.

Reaction at bearing 2 is $R_2 = \sqrt{\left(R_{2r} + R'_{2r}\right)^2 + \left(R_{2t} + R'_{2t}\right)^2}$ (15.22)

Bearing pressure $\qquad p_b = \dfrac{R_2}{l_2 \times d_{s1}}$ or $l_2 = \dfrac{R_2}{p_b \times d_{s1}}$ (15.23)

Calculate bearing length assuming a suitable bearing pressure p_b.

Example 15.1

Design of center crank shaft for given size of cylinder

A single cylinder vertical diesel engine has the following data. Design the crank shaft and its bearings.

Cylinder bore (D) = 100 mm
Stroke length = 120 mm
Length of connecting rod = 240 mm
Weight of flywheel = 800 N
Horizontal total belt pull = 1.5 kN
Maximum torque position of crank = 30°
Gas pressure at TDC = 3 MPa
Gas pressure at 30° of crank = 2.2 MPa
Safe bending stress = 80 MPa
Safe bearing pressure = 10 MPa
Distance between crank pin bearings = 2 D
Distance between flywheel bearings = 3 D

Assume crank pin and flywheel are placed centrally.

Solution

Given

D = 100 mm	$2r$ = 120 mm	W = 800 N
P = 1.5 kN	θ = 30°	p_{max} = 3 MPa
p' = 2.2 MPa	σ_b = 80	p_b = 10 MPa
b = 2D	c = 3D	L = 240

$b = 2D = 2 \times 100 = 200$, hence $b1 = b2 = 0.5b = 100$ mm
$c = 3D = 3 \times 100 = 300$, hence $c1 = c2 = 0.5b = 150$ mm

Position I: Crank at top dead center position

Step 1 *Calculate reactions at the bearings*

Maximum gas force at this position $F = \frac{\pi}{4} D^2 p_{max} = \frac{\pi}{4} 100^2 \times 3 = 23,550$ N

Reactions due to vertical forces $R_{1v} = R_{2v} = \frac{F}{2} = \frac{23,550}{2} = 11,755$ N

Reactions due to flywheel $R'_{2v} = R'_{3v} = \frac{W}{2} = \frac{800}{2} = 400$ N

$$R'_{2h} = R'_{3h} = \frac{P}{2} = \frac{1,500}{2} = 750 \text{ N}$$

Step 2 *Calculate crank pin size*

Bending moment $M = R_{1v} \times b1 = 11,755 \times 100 = 1,175,500$

Modulus of section $Z = \frac{\pi}{32} d_c^3$

Bending moment at center of crank pin $M = \sigma_b Z$

Substituting values in relation

$$1,175,500 = 80 \times \frac{\pi}{32} d_c^3 \quad \text{or} \quad d_c = 53.1 \text{ say } 55 \text{ mm}$$

$F = d_c \times l_c \times p_b$

Assuming $l_c = d_c$, length of bearing $l_c = 55$ mm

$R_{1v} = d_c \times d_c \times p_b$

$$11,755 = 55 \times 55 \times p_b \quad \text{or} \quad p_b = \frac{11,755}{55 \times 55} = 3.9 \text{ MPa}$$

It is less than safe value, hence safe.

Step 3 *Calculate size of web and check its strength*

Assuming thickness of web, $t_c = 0.7 \, d_c = 0.7 \times 55 = 38.5$ say 40 mm

And, width $w = 1.14 \, d_c = 1.14 \times 55 = 63$ mm

(i) Direct compressive stress $\sigma_c = \frac{R_{1v}}{w \times t} = \frac{11,755}{63 \times 40} = 4.7$ MPa, which is safe.

(ii) Bending moment at central plane of crank pin $M = R_{1v} \left[b1 - \frac{t_c}{2} - \frac{l_c}{2} \right]$

Section modulus of the web $Z = \frac{wt^2}{6}$

Bending stress $\sigma_b = \frac{M}{Z} = \frac{6 R_{1v}}{wt^2} \left[b1 - \frac{t_c}{2} - \frac{l_c}{2} \right]$

Or, $\sigma_b = \dfrac{M}{Z} = \dfrac{6 \times 11,755}{63 \times 40^2}\left[100 - \dfrac{40}{2} - \dfrac{55}{2}\right] = 36.8 \text{ MPa}$

Total stress $\sigma_t = \sigma_c + \sigma_b = 4.67 + 36.8 = 41.5 \text{ MPa}$

It is less than 80 MPa, hence safe.

Step 4 *Calculate size of shaft at flywheel*

Bending moment in vertical direction

$M_v' = R_{3v}' \times c2 = 400 \times 150 = 60,000 \text{ Nmm}$

Bending moment in horizontal direction

$M_h' = R_{3h}' \times c2 = 750 \times 150 = 112,500 \text{ Nmm}$

Resultant bending moment $M' = \sqrt{M_v^2 + M_h^2}$

$= \sqrt{(60,000)^2 + (112,500)^2} = 127,500 \text{ Nmm}$

$M = \dfrac{\pi d_s^3}{32}\sigma_b \quad \text{or} \quad 127,500 = \dfrac{\pi d_s^3}{32} \times 80 \quad \text{or} \quad d_s = 25.3 \text{ mm}$

Position II: Design of crank at position of maximum torque

Step 5 *Calculation of component of forces*

Use Equations (15.8) to (15.12)

$F' = \dfrac{\pi}{4}D^2 p' - 0.785 \times 100^2 \times 2.2 = 17,270 \text{ N}$

$\Phi = \sin^{-1}\left(\dfrac{r}{L}\sin\theta\right) = \sin^{-1}\left(\dfrac{60}{240}\sin 30°\right) = 7.24°$

Force along the C.R. $F_q = \dfrac{F'}{\cos\Phi} = \dfrac{17,270}{\cos(7.24°)} = 17,444 \text{ N}$

Tangential component of force

$F_t = F_q \sin(\theta + \Phi) = 17,444 \times \sin(30° + 7.24°) = 10,556 \text{ N}$

Radial component of force

$F_r = F_q \cos(\theta + \Phi) = 17,444 \times \cos(30° + 7.24°) = 13,887 \text{ N}$

Step 6 *Calculation of reactions at bearings*

Reactions in tangential and radial directions at bearings 1 and 2:

$R_{1t}' = R_{2t}' = \dfrac{F_t \times b1}{b} = \dfrac{10,556 \times 100}{200} = 5,278 \text{ N}$

$$R'_{1r} = R'_{2r} = \frac{F_r \times b1}{b} = \frac{13{,}887 \times 100}{200} = 6{,}944 \text{ N}$$

Reactions in vertical and horizontal direction at points 2 and 3 are as under:

$$R'_{2v} = R'_{3v} = \frac{W}{2} = \frac{800}{2} = 400 \text{ N}$$

$$R'_{2h} = R'_{3h} = \frac{P}{2} = \frac{1{,}500}{2} = 750 \text{ N}$$

Step 7 *Calculate crank pin size*

Bending moment in tangential direction $M_t = R_{1t} \times b1 = 5{,}278 \times 100 = 527{,}800 \text{ Nmm}$

Bending moment in radial direction $M_r = R_{1r} \times b1 = 6{,}944 \times 100 = 694{,}400 \text{ Nmm}$

Resultant bending moment at crank pin $M_c = \sqrt{M_t^2 + M_r^2}$

Substituting the values $M_c = 10^4 \sqrt{(52.78)^2 + (69.44)^2} = 87.22 \times 10^4 \text{ Nmm}$

Twisting moment on crank pin $T_c = R_{1t} \times r = 5{,}278 \times 60 = 31.67 \times 10^4 \text{ Nmm}$

Equivalent twisting moment at crank pin $T_e = \sqrt{(M_c)^2 + (T_c)^2}$

Or, $T_e = 10^4 \sqrt{(87.22)^2 + (31.67)^2} = 91.88 \times 10^4 \text{ Nmm}$

Calculate the crank pin diameter d_c using the relation $T_e = \frac{\pi d_c^3}{16} \tau$

$$91.88 \times 10^4 = \frac{3.14 \times d_c^3}{16} \times 40 \left(\text{Shear stress is taken as half of bending stress} \right)$$

$$d_c = \sqrt[3]{\frac{918{,}800 \times 16}{3.14 \times 40}} = 48.8 \quad \text{say 50 mm}$$

In position I at dead center position, crank shaft diameter was calculated 55 mm, hence the **higher value 55 mm** is taken.

Step 8 *Calculate size of shaft at junction of right hand crank web and crank shaft*

Taking value of l_c and t from step 2: $0.5l_c + t = (0.5 \times 55) + 40 = 67.5 \text{ mm}$

Bending moment due to radial forces $M_{2r} = R'_{1r}(b1 + t + 0.5l_c) - F_r(t + 0.5l_c)$

$= 6{,}944 \times (100 + 67.5) - (13{,}877 \times 67.5) = 0.226 \times 10^6 \text{ Nmm}$

Bending moment due to tangential forces, $M_{2t} = R'_{1t}(b1 + 0.5l_c + t) - F_t(0.5l_c + t)$

$= 5{,}278 \times (100 + 67.5) - (10{,}556 \times 67.5) = 0.1715 \times 10^6 \text{ Nmm}$

Resultant B.M. $(M_2)^2 = (M_{2r})^2 + (M_{2t})^2 = 10^8 [22.6^2 + 17.15^2] = 804.9 \times 10^8$

Or $M_2 = 28.37 \times 10^4$ Nmm

Torsion due to tangential force:

$T_2 = F'_t \times r = 10,556 \times 60 = 63.34 \times 10^4$ Nmm

Equivalent torque $T_e = \sqrt{(M_2)^2 + (T_2)^2} = 69.4 \times 10^4$ Nmm

Using relation $(d_s)^3 = 16 \dfrac{T_e}{\pi \tau} = \dfrac{(16 \times 69.4 \times 10^4)}{(3.14 \times 40)} = 8.84 \times 10^4$

Or $d_s = 44.5$ say 45 mm

Step 9 *Calculate size of shaft at flywheel*

Reaction at point 3 $R'_3 = \sqrt{\left(R'_{3v}\right)^2 + \left(R'_{3h}\right)^2} = \sqrt{(400)^2 + (750)^2} = 850$ N

Bending moment at point 3

$M'_3 = R'_3 \times c2 = 850 \times 150 = 127,500 = 12.75 \times 10^4$ N

Torsion due to tangential force

$T'_2 = F'_t \times r = 10,556 \times 60 = 577,920 = 63.33 \times 10^4$ N

Equivalent torque $T_e = \sqrt{M_3^2 + T_2^2}$

$\qquad\qquad = 10^4 \sqrt{(12.75)^2 + (63.33)^2} = 64.6 \times 10^4$ N

$T_e = \dfrac{\pi d_s^3}{16} \times \tau$ or $d_s^3 = \dfrac{16 T_e}{\pi \times \tau}$

Substituting the values

$d_s^3 = \dfrac{16 \times 646,000}{3.14 \times 40} = 82,293$ or $d_s = 43.5$ say 45 mm

In position I of dead center position shaft diameter at flywheel was calculated 25.3 mm in step 4, hence the **higher value 45 mm** is taken.

Step 10 *Calculate size of right side web*

a. Bending moment at center plane of web due to radial force $M_{2r} = R_{2r} \times \left[b2 - \dfrac{t}{2} - \dfrac{l_c}{2} \right]$

$M_{2r} = 6,944 \times \left[100 - \dfrac{40}{2} - \dfrac{55}{2} \right] = 364,560$ Nmm

Modulus of section $Z = \dfrac{1}{6} wt^2 = \dfrac{1}{6} \times 63 \times 40^2 = 16,800$ mm³

Using the relation: $\sigma_{br} = \dfrac{M_{2r}}{Z} = \dfrac{364,560}{16,800} = 21.7$ MPa

b. Bending moment due to tangential force $M_{2t} = F_t \times \left[r - \dfrac{d_s}{2} \right]$

$$M_{2t} = 5,278 \times \left[60 - \dfrac{45}{2} \right] = 197,925 \text{ Nmm}$$

Modulus of section in this plane $Z = \dfrac{1}{6} t w^2 = \dfrac{1}{6} \times 40 \times 63^2 = 26,460 \text{ mm}^3$

$$\sigma_{bt} = \dfrac{M_{2t}}{Z} = \dfrac{197,925}{26,460} = 7.5 \text{ MPa}$$

c. Direct compressive stress in one web due to radial force F_r calculated in step 5.

$$\sigma_{bd} = \dfrac{F_r}{2wt} = \dfrac{13,877}{2 \times 63 \times 40} = 2.7 \text{ MPa}$$

Total stress $\sigma_{bc} = \sigma_{br} + \sigma_{bt} + \sigma_{bd} = 21.7 + 7.5 + 2.7 = 31.9$ MPa

Torsional stress in the web is given as:

$$T = R_{2t} \times \left[b1 - \dfrac{l_c}{2} \right] = 5,278 \times \left[100 - \dfrac{55}{2} \right] = 382,655 \text{ Nmm}$$

Polar moment of the section $Z_p = \dfrac{tw^2}{4.5} = \dfrac{40 \times 63^2}{4.5} = 35,280 \text{ mm}^3$

Shear stress on the arm $\tau = \dfrac{T}{Z_p} = \dfrac{382,655}{35,280} = 10.8$ MPa

Maximum combined stress $\sigma_{bmax} = \dfrac{\sigma_{bc}}{2} + \dfrac{1}{2} \sqrt{\sigma_{bc}^2 + 4\tau^2}$

Or, $\sigma_{bmax} = \dfrac{33.8}{2} + \dfrac{1}{2} \sqrt{(33.8)^2 + 4 \times (10.8)^2} = 35.5$ MPa

It is less than safe stress, that is, 80 MPa, hence safe.

Left-hand web is not loaded as right web, hence web size 40 mm × 63 mm as in step 3, is taken same as right side web to maintain symmetry and balance.

Step 11 *Calculate size of bearings*

Bearing 2 is usually most heavily loaded, hence bearing 2 is designed

Reaction at bearing 2 is:

$$R_2 = \sqrt{\left(R_{2v} + R'_{2v}\right)^2 + \left(R_{2h} + R'_{2h}\right)^2}$$

$$R_2 = \sqrt{\left(11{,}755 + 400\right)^2 + \left(0 + 750\right)^2} = 12{,}178 \text{ N}$$

Bearing pressure from Equation(15.23):

Length of bearing 2: $l_2 = \dfrac{R_2}{P_b \times d_s}$ or $l_2 = \dfrac{12{,}178}{10 \times 25.3} = 48$ mm

15.6 Design of Side Crank

A side crank is shown in Figure 15.2. It is supported between two bearings and a driver such as flywheel or pulley is normally placed in the middle between the bearings. Design of overhung or side crank is also similar to the design of center crank shaft. The forces are calculated at TDC position and at position of crank for maximum torque. Bigger of the forces are selected for calculating size of parts. Design procedure is explained by taking an example given below:

Example 15.2

Design of side crank for an engine of given size
A four-stroke horizontal single cylinder engine has cylinder diameter 150 mm and stroke 200 mm. The length of connecting rod is 2.2 times the stroke. Maximum gas pressure is 3.5 MPa. The power developed by engine is transmitted through a heavy pulley of width 130 mm in the center weighing 20 kN and a horizontal belt giving total pull of 6 kN. Gas pressure at maximum torque position when crank is at 35° is 1.8 MPa. Safe bending stress for the crank material is 75 MPa and bearing stress 10 MPa. Design a side crank shaft, assuming suitable proportions.

Solution

Given $D = 150$ mm $l = 200$ mm $L/l = 2.2$

$p_{max} = 3.5$ MPa $W = 20$ kN $P = 6$ kN

$\theta = 35°$ $p' = 1.8$ MPa $\sigma_b = 75$ MPa

$p_b = 10$ MPa

Position I - Crank at TDC position
The design procedure can be followed as given below:

1. **Crank pin design**

   - From the maximum gas pressure p_{max} calculate gas force $F = \dfrac{\pi}{4}D^2 p_{max}$

   $F = 0.785 \times 150^2 \times 3.5 = 61{,}820$ N

- Assume ratio of length l_c to diameter of crank pin d_c for the crank pin, which varies between 0.6 and 1.4 as 0.8. Bearing pressure p_b varies between 10 MPa and 12 MPa. Calculate crank pin diameter from: $F = d_c \times l_c \times p_b$

$$61{,}820 = d_c \times \left(0.8 d_c\right) \times p_b = 0.8\, d_c^2 \times 10$$

$$d_c^2 = \frac{61{,}820}{8} \quad \text{or} \quad d_c = 88 \text{ mm}$$

Length of crank pin $l_c = 0.8 \times 88 = 70$ mm

- Gas force is neither concentrated nor UDL. Assuming that it acts at $0.75 l_c$. Calculate bending moment M.

$$M = 0.75\, l_c\, F = 0.75 \times 70 \times 61{,}820 = 3.25 \times 10^6 \text{ Nmm}$$

- Knowing safe bending stress σ_b, calculate crank pin diameter using $M = \sigma_b Z$

Modulus of section $Z = \dfrac{\pi}{32} d_c^3 = 0.0981 \times 88^3 = 66{,}869 \text{ mm}^3$

Bending stress $\sigma_b = \dfrac{M}{Z} = \dfrac{3.25 \times 10^6}{66{,}869} = 48.6 \text{ MPa}$

It is lesser than safe value 75 MPa, hence safe.

2. **Size of bearings**

Refer Figure 15.6

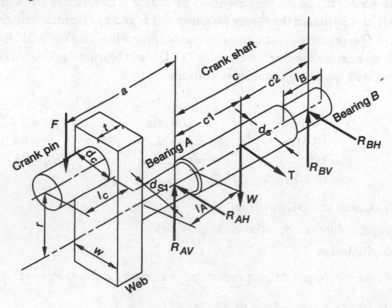

Figure 15.6 Bearing distances

- Assuming thickness of crank web $t = 0.6\, d_c$ and length of bearing A, $l_A = 1.75\, d_c$, calculate distance between center of bearing to axis of load $a = 0.75\, l_c + t + 0.5\, l_A$

  $t = 0.6\, d_c = 0.6 \times 88 = 53$ mm and $l_A = 1.75\, d_c = 1.75 \times 88 = 154$ mm

  $a = (0.75 \times 70) + 53 + (0.5 \times 154) = 183$ mm

- Calculate bending moment at center of bearing $M = F \times a$

  $M = 61,820 \times 183 = 11.31 \times 10^6$ Nmm

- Using relation for modulus of section $Z = \dfrac{M}{\sigma_b}$, calculate Z.

  $$Z = \frac{11.31 \times 10^6}{75} = 150,841 \text{ mm}^3$$

- Using relation $Z = \dfrac{\pi}{32} d_A^3$ calculate diameter d_A. $\left(\dfrac{\pi}{32} = 0.0981 \right)$

  $150,841 = 0.0981\, d_A^3$ Or,   Diameter of bearing A: $d_A = 115$ mm

- Check for bearing pressure.

  $$p_b = \frac{F}{d_A\, l_A} \quad \text{or} \quad p_b = \frac{61,820}{115 \times 154} = 3.5 \text{ MPa. Lesser than safe value 10 MPa.}$$

3. Distance between bearings

See Figure 15.6 and 15.S1 (a)

- Calculate distance (c) between bearing A and B:

  $c = 0.5\, l_A +$ Width of fly wheel $+ 0.5\, l_B +$ margin. Margin may be suitably assumed.

  Assuming margin as 16 mm, $c = (0.5 \times 154) + 130 + (0.5 \times 154) + 16 = 300$ mm

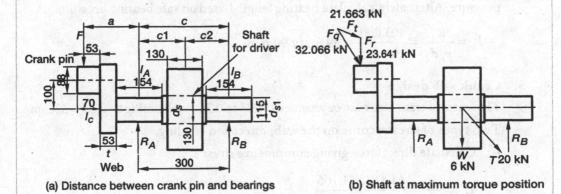

(a) Distance between crank pin and bearings   (b) Shaft at maximum torque position

Figure 15.S1 Side crank (154)

4. **Calculate bearing reactions**

- Calculate horizontal reactions R_{AH} and R_{BH} for bearing A and B, respectively, by taking moments.

  Since the engine is horizontal axis, gas force will be in horizontal direction. Belt is horizontal. Refer Figure 15.6. Distance $a + c = 183 + 300 = 483$ mm

  Taking moments about bearing center B:

  $$(61,820 \times 483) + (6,000 \times 150) = (R_{AH} \times 300) \quad \text{or} \quad R_{AH} = 102,530\,\text{N} = 102.53\,\text{kN}$$

  Taking moments about bearing center A:

  $$(61,820 \times 183) + (R_{BH} \times 300) = 6,000 \times 150 \quad \text{or} \quad R_{BH} = 34,710\,\text{N} = -34.71\,\text{kN}$$

- Calculate vertical bearing reactions R_{AV} and R_{BV} also.

  The vertical load is only the weight of flywheel, which is put centrally between the two bearings A and B; hence both the reactions will be equal and half of weight.

  $$R_{AV} = R_{BV} = \frac{20,000}{2} = 10,000 = 10\,\text{kN}$$

- Calculate resultant reactions R_A and R_B from the relations: $R_A = \sqrt{R_{AH}^2 + R_{AV}^2}$

  $$R_A = \sqrt{102.53^2 + 10^2} = 103\,\text{kN}$$

  Similarly, $R_B = \sqrt{R_{BH}^2 + R_{BV}^2} = \sqrt{(-34.71)^2 + 10^2} = 36.1\,\text{kN}$

  Check for bearing pressure $p_b = \dfrac{R_A}{l_A\,d_A}$ if it is safe.

  Since bearing A is heavily loaded, that will be considered for checking for bearing pressure. Alternately, calculate bearing length based on safe bearing pressure:

  $$l_A = \frac{R_A}{p_b\,d_A} = \frac{103,000}{10 \times 115} = 89.5 \text{ say } 90 \text{ mm}$$

5. **Crank web design**

   Thickness of crank web t can be assumed $0.45\,d_c$ to $0.75\,d_c$. Assuming $t = 0.6d_c = 53$ mm

   Two types of stresses come on the web; direct and bending.

   - Calculate direct force giving compressive stress

     $$\sigma_c = \frac{F}{w\,t} = \frac{61,820}{w \times 53} = \frac{1,166}{w}$$

   - Calculate bending stresses due to eccentricity of the load.

Bending moment at central plane of web $M = F \times (0.75\, l_c + 0.5\, t)$

$$M = 61,820 \times [(0.75 \times 70) + (0.5 \times 53)] = 4,884,000 \text{ N}$$

- Calculate modulus of section $Z = \dfrac{w\, t^2}{6} = \dfrac{w\, 53^2}{6} = 468.2w \text{ mm}^3$

- Calculate bending stress $\quad \sigma_b = \dfrac{M}{Z} = \dfrac{4,884,000}{468.2\, w} = \dfrac{11,598}{w}$

- Calculate total stress $\quad \sigma_t = \sigma_c + \sigma_b = \dfrac{1166}{w} + \dfrac{11,598}{w}$

- Total stress should be equal to safe bending stress, calculate width w.

$$75 = \frac{1,166}{w} + \frac{11,598}{w} \quad \text{or} \quad \text{Width of crank web, } w = 154.6 \text{ say 155 mm}$$

6. **Shaft diameter at flywheel**

    - From the maximum resultant bearing reactions, calculate bending moment M.
      B.M. at central plane of flywheel $M = R_A \times (0.5\, c) = 103,000 \times (0.5 \times 300)$

    - Calculate shaft diameter d_s from the relation:

$$M = \sigma_b \times \frac{\pi}{32} d_s^3 = \sigma_b\, 0.0981 d_s^3$$

    Substituting the values: $103,000 \times (0.5 \times 300) = 75 \times 0.0981\, d_S^3$

    Or, $d_S = 128 \text{ mm say 130 mm}$

Position II - Crank at maximum torque position

When crank is in this position (angle θ), angle of the C.R. with line of centers is Φ and C.R. force F_q acts along its axis. It can be resolved into forces in tangential and radial directions. Follow the similar procedure as for center crank. The method is demonstrated in the continuing Example 15.2.

7. **Radial and tangential forces**

    - At this position, calculate gas force $F' = \dfrac{\pi}{4} D^2 p' = 0.785 \times 150^2 \times 1.8 = 31,792 \text{ N}$

    - Calculate angle of C.R. for this position, $\Phi = \sin^{-1}\left(\dfrac{r}{L}\sin\theta\right)$

$$\Phi = \sin^{-1}\left(\frac{1}{4.4}\sin 35°\right) = 7.5°$$

    - Force along the connecting rod $F_q = \dfrac{F'}{\cos\Phi} = \dfrac{31,792}{\cos 7.5°} = 32,066 \text{ N}$

- Tangential component of force, $F_t = F_q \sin(\theta + \Phi)$

  Putting the values, $F_t = 32{,}066 \times \sin(35° + 7.5°) = 21{,}663$ N

- Radial component of force, $F_r = F_q \cos(\theta + \Phi)$

  Putting the values, $F_r = 32{,}066 \times \cos(35° + 7.5°) = 23{,}641$ N

8. **Bearing reactions**

   - Calculate bearing reactions due to tangential forces: Refer Figure 15.S1(b).
   - Taking moments about central axis of bearing B:

   $$R_{At} = \frac{F_t \times (a+c)}{c} \tag{15.24a}$$

   Putting the values, $R_{At} = \dfrac{21{,}602 \times (183 + 300)}{300} = 34{,}877$ N

   - Taking moments about central axis of bearing A:

   $$R_{Bt} = \frac{F_t \times a}{c} \tag{15.24b}$$

   Putting the values, $R_{Bt} = \dfrac{21{,}602 \times 183}{300} = 13{,}214$ N

   - Calculate bearing reactions due to radial forces:

   Taking moments about central axis of bearing B:

   $$R_{Ar} = \frac{F_r \times (a+c)}{c} \tag{15.25a}$$

   Putting the values, $R_{Ar} = \dfrac{23{,}641 \times (183 + 300)}{300} = 38{,}062$ N

   Taking moments about central axis of bearing A:

   $$R_{Br} = \frac{F_r \times a}{c} \tag{15.25b}$$

   Putting the values, $R_{Br} = \dfrac{23{,}641 \times 183}{300} = 14{,}421$ N

   - Calculate resultant bearing reactions:

   Resultant reactions at bearing A: $R_A = \sqrt{R_{At}^2 + R_{Ar}^2}$ \hfill (15.26a)

   $$R_A = \sqrt{(34{,}877)^2 + (38{,}662)^2} = 52{,}069 \text{ N}$$

Resultant reactions at bearing B: $R_B = \sqrt{R_{Bt}^2 + R_{Br}^2}$ (15.26b)

$$R_B = \sqrt{(13,214)^2 + (14,421)^2} = 19,559 \text{ N}$$

9. **Design of crank web**

Critical section for a side crank is where web and shaft join. Three types of stresses that occur at this place are direct compressive stress, bending stresses, and torsion.

a. *Direct compressive stress*

- Calculate direct compressive stress σ_c due to radial force F_r:

$$\sigma_c = \frac{F_r}{w\,t} \tag{15.27}$$

$$\sigma_c = \frac{F_r}{wt} = \frac{23,641}{155 \times 53} = 2.9 \text{ MPa}$$

b. *Bending stresses*

- Calculate bending moment due to radial force F_r:

$$M_{br} = F_r\,(0.75\,l_c + 0.5\,t) \tag{15.28}$$

$$M_{br} = 23,641 \times [(0.75 \times 70) + (0.5 \times 53)] = 1,867,560 \text{ Nmm}$$

- Calculate bending stress due to radial force:

$$\sigma_{br} = \frac{M_{br}}{Z} = \frac{6M_{br}}{wt^2} \tag{15.29}$$

$$\sigma_{br} = \frac{6 \times 1,867,560}{155 \times 53^2} = 25.7 \text{ MPa}$$

- Calculate bending moment due to tangential force F_t:

$$M_{bt} = F_t \times r \tag{15.30}$$

$$M_{bt} = 21,663 \times 100 = 2,166,300 \text{ Nmm}$$

- Calculate bending stress due to tangential force:

$$\sigma_{bt} = \frac{M_{bt}}{Z} = \frac{6M_{bt}}{w\,t^2} \tag{15.31}$$

$$\sigma_{bt} = \frac{6 \times 2,166,300}{155 \times 53^2} = 29.9 \text{ MPa}$$

- Calculate total bending stresses:

$$\sigma_{bc} = \sigma_c + \sigma_{br} + \sigma_{bt} \tag{15.32}$$

$$\sigma_{bc} = 2.9 + 25.7 + 29.9 = 58.5 \text{ MPa}$$

Shear stress due to torsion

Shear stress τ is due to torque given by tangential force F_t.

- Calculate torque

$$T = F_t \,(0.75 l_c + 0.5\, t) \tag{15.33}$$

$$T = 21{,}663 \times [(0.75 \times 70) + (0.5 \times 53)] = 1{,}711{,}377 \text{ Nmm}$$

- Calculate polar section modulus for rectangular section

$$J = \frac{wt^2}{4.5} \tag{15.34}$$

$$J = \frac{155 \times 53^2}{4.5} = 435{,}395 \text{ mm}^3$$

- Calculate shear stress $\tau = \dfrac{T}{J} = \dfrac{1{,}711{,}377}{435{,}395} = 3.9$ MPa

- Calculate combined stress σ_{max} due to bending and shear stress:

$$\sigma_{max} = 0.5 \sigma_{bc} + 0.5 \sqrt{(\sigma_{bc})^2 + 4 \tau^2} \tag{15.35}$$

$$= (0.5 \times 58.5) + 0.5 \sqrt{(58.5)^2 + 4(3.9)^2} = 58.7 \text{ MPa}$$

This stress should be less than allowable stress. It is less than 75 MPa, hence safe.

10. Shaft diameter at the junction of crank and web

Let d_{S1} be the diameter of the shaft at junction of crank.

- Calculate bending moment at this junction:

$$M = F_q \,(0.75 l_c + t) \tag{15.36}$$

$$M = 32{,}066 \times [(0.75 \times 70) + 53)] = 3{,}382{,}963 \text{ Nmm}$$

- Calculate torque at this junction:

$$T = F_t \times r = 21{,}663 \times 100 = 2{,}166{,}300 \text{ Nmm}$$

- Calculate equivalent twisting moment: $T_e = \sqrt{M^2 + T^2}$

$$T_e = 10^6 \sqrt{(3.382)^2 + (2.166)^2} = 4.016 \times 10^6 \text{ Nmm}$$

- Calculate diameter d_{S1} from relation:

$$T_e = \frac{\pi}{16} (d_{S1})^3 \times \tau \tag{15.37}$$

$$4.016 \times 10^6 = \frac{\pi}{16} (d_{S1})^3 \times 37.5 \quad \text{or} \quad d_{S1} = 81.7 \text{ say } 82 \text{ mm}$$

It is lesser than the diameter calculated in step 2 for bearing A that is, 115 mm, so higher value of shaft diameter at flywheel is taken as 115 mm.

11. Shaft diameter at flywheel

Let d_S be diameter of the shaft at flywheel. Its distance from crank pin force is $(a + c1)$.

- Calculate B.M. due to gas force

$$M_g = F(a + c1) - (R_A \times c2) \tag{15.38}$$

$$M_g = [61,820 \times (183 + 150)] - (103,000 \times 150) = 5,136,060 \text{ Nmm}$$

This force is horizontal as the engine has horizontal axis.

- Calculate horizontal B.M. due to belt tensions

$$M_b = \frac{(T_1 + T_2)c1}{c} \times c2 \tag{15.39}$$

$$M_b = \frac{(T_1 + T_2)c1}{c} \times c2 = \frac{6,000}{300} \times 150 = 3,000 \text{ Nmm}$$

- Calculate total horizontal bending moment

$$M_H = M_g + M_b \tag{15.40}$$

$$M_H = M_g + M_b = 5,136,060 + 3000 = 5,139,060 \text{ Nmm}$$

- Calculate vertical B.M. due to weight of flywheel

$$M_V = \frac{Wc1}{c} \times c2 \tag{15.41}$$

Substituting values, $M_V = \dfrac{20,000 \times 150}{300} \times 150 = 1,500,000 \text{ Nmm}$

- Calculate resultant bending moment $M_R = \sqrt{(M_H)^2 + (M_V)^2}$

$$M_R = 10^6 \sqrt{(5.139)^2 + (1.5)^2} = 5.353 \times 10^6 \text{ Nmm}$$

- Calculate twisting moment:

$$T = F_t \times r = 21,663 \times 100 = 2,166,300 \text{ Nmm}$$

- Calculate equivalent twisting moment

$$T_e = \sqrt{(M_R)^2 + T^2} \tag{15.42}$$

$$T_e = 10^6 \sqrt{(5.353)^2 + (2.166)^2} = 5.755 \times 10^6 \text{ Nmm}$$

- Calculate shaft diameter d_s from relation:

$$T_e = \frac{\pi}{16}(d_s)^3 \times \tau \tag{15.43}$$

$$5.775 \times 10^6 = \frac{\pi}{16}(d_s)^3 \times 37.5$$

Shaft diameter under fly wheel $d_s = 92.2$ mm say 93 mm.

It is lesser than 130 mm calculated in step 6, hence higher value of shaft diameter at flywheel is taken as 130 mm.

Summary

Crank shaft in an engine converts reciprocating motion of piston to rotary motion using a connecting rod. It can be divided into four portions:

- *Crank pin* on which the big end is fixed using strap and cap bolts.
- *Main bearing journal*, where the crank shaft is supported on bearings.
- *Crank web*, which joins crank shaft and the crank pin.
- *Balance weight* is a dead weight on webs opposite to the crank pin for balancing the eccentric weight. It can be integral with the web or bolted separately on the web.

Types of crank shafts   Crank shafts can be mainly divided into two types:

- *Center crank shaft:* Crank pin is in between two webs.
- *Side crank shaft* or *overhung crank shaft:* Crank pin is overhung on one web.

Crank materials   Crank shaft is subjected to shocks and fatigue and hence has to be tough. These are usually forged to impart strength. Crank pin and journals are hardened by carburizing or nitriding and ground. The materials used are:

Industrial engines   Plain carbon steel: 40C8, 45C8, 50C4, 55C8 and 60C4

Automobile engines   Manganese steels: 20Mn2, 27Mn2, and 37Mn2

Aero engines   Ni–Cr alloy steels:16Ni3Cr2, 35Ni5Cr2 and 40Ni2Cr1Mo28

Forces on crank shaft   The crank shaft can be divided into two parts for the purpose of analysis. *One part* is the crank pin supported on main bearings through webs. *Second part* is the portion of shaft carrying flywheel or a pulley supported between bearings.

Mainly there are three types of forces coming on crank shaft:

    a.   Force transmitted by the connecting rod causing uniformly distributed load on the crank pin.

    b.   Vertical load of the flywheel in the second portion of shaft.

    c.   Torsional stresses, when the crank web is at an angle from top dead center.

Design of crank shaft   A crank shaft is designed for two positions of the crank shaft. (a) Crank at dead center position and (b) Crank at position of maximum torque.

Position I: Design of crank at top dead center position   At this position of the crank, pressure is maximum, which causes bending of the crank pin, but the torque is zero. Two bearings reactions are R_{1v} and R_{2v}. A third bearing with reaction R_{3v} is used to support flywheel. Assumptions:

- Gap between crank shaft main bearings assumed $b = 2$ times cylinder diameter D.

- Gap between the bearings 2 and 3 for the shaft at flywheel (c) is assumed as three times the cylinder diameter, that is, $c = 3D$

- Crank pin and the flywheel are placed centrally between the bearings. That is,

$b1 = b2 = 0.5b$ and $c1 = c2 = 0.5\ c$

The design is done in the following four steps:

    1. *Calculate reactions at the bearings*   Maximum gas force when pressure p_{max} is maximum,

$$F = \frac{\pi}{4}D^2 p_{max}$$

Reactions due to horizontal forces $R_{1h} = \dfrac{F \times b2}{b}$ and $R_{2h} = \dfrac{F \times b1}{b}$

Flywheel of weight W is at the middle. Superscript ($'$) indicates reactions due to flywheel.

Taking moments $R_{3v}' = \dfrac{W \times c1}{c}$ and $R_{2v}' = \dfrac{W \times c2}{c}$

If T is the total pull of the belt tensions and assuming belt is horizontal.

Taking moments $R_{3h}' = \dfrac{T \times c1}{c}$ and $R_{2h}' = \dfrac{T \times c2}{c}$

Therefore, total reactions at section 1, $R_1 = R_{1v}$

Resultant reaction at section 2, $R_2 = \sqrt{\left(R_{2v} + R_{2v}'\right)^2 + \left(R_{2h}'\right)^2}$

Resultant reaction at section 3, $R_3 = \sqrt{\left(R_{3v}'\right)^2 + \left(R_{3h}'\right)^2}$

    2. *Calculate crank pin size*   Bending moment at center of crank pin $M = R_{1v} \times b1$

Moment of inertia of crank pin $I = \dfrac{\pi}{64}d_c^4$, $y = \dfrac{d_c}{2}$; hence $Z = \dfrac{I}{y} = \dfrac{\pi}{32}d_c^3$

Using $M = \sigma_b Z$, calculate crank pin diameter d_c for the safe bending stress σ_b.

For the design to be safe, bearing pressure p_b, $F = d_c \times l_c \times p_b$

Assuming a suitable bearing pressure, l_c / d_c ratio between 1 and 1.2 calculate bearing pressure. It should be less than allowable limit.

3. *Calculate size of web* Both the left and right side webs are made similar.

Width of web $w = 1.14\, d_c$ or $0.4d_s$ to $0.6d_s$ or $0.22D$ to $0.32D$

Where, d_s = Shaft diameter at flywheel and D = Cylinder bore

Thickness of web $t = 0.67\, d_c$

There are two types of stresses coming in web:

(i) Direct compressive stress $\sigma_c = \dfrac{\text{Load}}{\text{Web area}} = \dfrac{R_{1v}}{w \times t}$

(ii) Compressive stress due to bending.

Bending moment at central plane of crank pin $M = R_{1v}\left[b1 - \dfrac{t_c}{2} - \dfrac{l_c}{2}\right]$

Section modulus of the web $Z = \dfrac{w\,t^2}{6}$ Bending stress $\sigma_b = \dfrac{M}{Z}$

Total compressive stress $\sigma_{ct} = \sigma_c + \sigma_b$, σ_{ct} should be lesser than allowable compressive stress.

4. *Calculate size of shaft at flywheel* B.M. are calculated at center of flywheel

Bending moment in vertical direction $M_v = R'_{3v} \times c2$

Bending moment in horizontal direction $M_h = R'_{3h} \times c2$

Resultant bending moment $M = \sqrt{M_v^2 + M_h^2}$

Calculate the shaft diameter at flywheel d_s^3 using relation: $M = \dfrac{\pi d_s^3}{32}\sigma_b$

Position II: Design of crank at maximum torque position This occurs when crank angle θ for petrol engines is 25°–35° and for diesel engine its range is 30° – 40°.

5. *Calculation of component of forces* In inclined position of the crank shaft, component of force along the axis of the C.R. is F_q. This force is resolved in radial component F_r along crank web and tangential component F_t at right angle to the web. At this position, the gas pressure is denoted by

p' and gas force $F' = \dfrac{\pi}{4}D^2\, p'$

Angle of C.R. $\varPhi = \sin^{-1}\left(\dfrac{r}{L}\sin\theta\right)$

Where, r = Radius of crank   and L = Length of C.R.

Force along C.R. $F_q = \dfrac{F'}{\cos\varPhi}$

Tangential force $F_t = F_q \sin(\theta + \Phi)$ and Radial force $F_r = F_q \cos(\theta + \Phi)$

6. *Calculation of reactions at bearings* The center crank shaft has three bearings. The analysis is done first considering crank pin supported on two bearings 1 and 2

   Taking moments $R'_{1t} = \dfrac{F_t \times b1}{b}$ and $R'_{2t} = \dfrac{F_t \times b2}{b}$

   Similarly, radial force will give reaction, $R'_{1r} = \dfrac{F_r \times b1}{b}$ and $R'_{2r} = \dfrac{F_r \times b2}{b}$

   Reactions of bearings 2 and due to fly wheel

   Taking moments $R'_{3h} = \dfrac{T \times c1}{c}$ and $R'_{2h} = \dfrac{T \times c2}{c}$

7. *Calculate crank pin size*

   B.M. in tangential direction $M_t = R'_{2t} \times b2$ and in radial direction $M_r = R'_{2r} \times b2$

   Resultant B.M. at crank pin, $M_c = \sqrt{\left[(M_r)^2 + (M_t)^2 \right]}$

   Twisting moment $T_c = R'_{1t} \times r$

   Equivalent twisting moment at crank pin $T_e = \sqrt{(M_c)^2 + (T_c)^2}$

   Calculate the crank pin diameter, d_c using relation: $T_e = \dfrac{\pi d_c^3}{16} \tau$

8. *Calculate size of shaft at junction of right hand crank web and crank shaft* At the junction of shaft at flywheel and right crank web following forces act:

   B.M. due to radial forces, $M_{2r} = R'_{1r}(b1 + t + 0.5\, l_c) - F_r(t + 0.5\, l_c)$

   B.M. due to tangential forces, $M_{2t} = R'_{1t}(b1 + t + 0.5\, l_c) - F_t(t + 0.5\, l_c)$

   Resultant B.M. due to above moments $M_2 = \sqrt{M_{2r}^2 + M_{2t}^2}$

   Torsion due to tangential force $T_2 = F_t \times r$ and Equivalent torque $T_e = \sqrt{M_2^2 + T_2^2}$

   Calculate the shaft diameter d_s at flywheel using $T_e = \dfrac{\pi d_s^3}{16} \tau$

9. *Calculate size of shaft at flywheel* Forces acting at center of flywheel give rise to B.M.

   B.M. in vertical direction $M_v = R'_{3v} \times c2$

   Torsion due to tangential force $T_2 = F_t \times r$ and Equivalent torque $T_e = \sqrt{M_v^2 + T^2}$

   Calculate shaft diameter at flywheel using relation: $T_e = \dfrac{\pi d_s^3}{32} \tau$

10. *Calculate size of right side web* Right hand web is subjected to the following forces:

    (a) B.M. due to radial force $M_{2r} = R_{2r} \times \left[b2 - \dfrac{t}{2} - \dfrac{l_c}{2} \right]$

    Modulus of section $Z = \dfrac{1}{6} wt^2$ Hence, stress: $\sigma_{br} = \dfrac{6 M_{2r}}{wt^2}$

B.M. due to tangential force $M_{2t} = F_t \times \left[r - \dfrac{d_{s1}}{2} \right]$

where, d_{s1} = Shaft diameter at the junction of right web.

$Z = \dfrac{1}{6} t w^2$; hence $\sigma_{bt} = \dfrac{6 M_2 t}{t w^2}$

(b) Direct compressive stress due to radial force F_r, (0.5 F_r for one web) $\sigma_{bd} = \dfrac{F_r}{2wt}$

Total compressive stress $\sigma_{bc} = \sigma_{br} + \sigma_{bt} + \sigma_{bd}$

Torsion in the web is: $T = R_{1t} \times \left[b2 + \dfrac{l_c}{2} \right] - F_t \times \dfrac{l_c}{2} = R_{2t} \times \left[b1 - \dfrac{l_c}{2} \right]$

Polar moment of the section $Z_p = \dfrac{t w^2}{4.5}$

Shear stress on the arm $\tau = \dfrac{T}{Z_p} = \dfrac{4.5T}{t w^2}$

Therefore maximum combined stress $\sigma_{bmax} = \dfrac{\sigma_{bc}}{2} + \dfrac{1}{2} \sqrt{\sigma_{bc}^2 + 4\tau^2}$

Stress σ_{bmax} should be lesser than safe stress.

Left web is not loaded as right web; hence its size it taken same as right side.

11. *Calculate size of bearings*   Bearing 2 is usually most heavily loaded; hence bearing 2 is designed assuming a safe bearing pressure. Bearings 1 and 3 are taken of same size.

Reaction at bearing 2 is: $R_2 = \sqrt{\left(R_{2v} + R'_{2v} \right)^2 + \left(R_{2h} + R'_{2h} \right)^2}$

Bearing pressure $P_b = \dfrac{R_2}{l_2 \times d_{s1}}$ or $l_2 = \dfrac{R_2}{P_b \times d_{s1}}$

Calculate bearing length l_2 assuming a suitable bearing pressure P_b

Design of side crank   It is supported between two bearings and a driver such as flywheel or pulley normally placed in the middle between the bearings. It is also designed for TDC position and maximum torque position and maximum value is selected.

Position I: Crank at dead center position

1. *Crank pin design* Maximum gas force $F = \dfrac{\pi}{4} D^2 p_{max}$. Take (l_c / d_c) ratio between 0.6 to 1.4 and bearing pressure p_b 10 MPa to 12 MPa. Calculate crank pin diameter from: $F = d_c \times l_c \times p_b$.
   Bending moment $M = 0.75\, l_c\, F$. Calculate crank pin diameter using $M = \sigma_b Z$

2. *Bearings* Assuming thickness of crank web $t = 0.6\, d_c$ and length of bearing A, $l_A = 1.75\, d_c$.
   Calculate distance between center of bearing $a = 0.75\, l_c + t + 0.5\, l_A$

   B.M., $M = F \times a$. Using $Z = \dfrac{\pi}{32} d_A^3$ calculate diameter d_A. Distance from bearing A to B $c = 0.5\, l_A +$ Width of fly wheel $+ 0.5\, l_B +$ margin.

3. Bearing reactions Taking moments find horizontal reactions R_{AH} and R_{BH} and vertical reactions R_{AV} and R_{BV} for bearing A and B. Resultant reactions are:

$$R_A = \sqrt{R_{AH}^2 + R_{AV}^2} \quad \text{and} \quad R_B = \sqrt{R_{BH}^2 + R_{BV}^2}$$

Check for bearing pressure $p_b = \dfrac{R_A}{l_A d_A}$ if it is safe.

4. Crank web design Thickness of crank web t 0.45 d_c to 0.75 d_c. Length of bearings $l_1 = 1.5\, d_c$ and 2 d_c.

Two types of stresses come on the web; direct and bending.

Direct compressive stress $\sigma_c = \dfrac{P}{w\,t}$

B.M., $M = F \times (0.75\, l_c + 0.5t)$, $Z = \dfrac{wt^2}{6}$, and bending stress $\sigma_b = \dfrac{M}{Z}$

Total stress: $\sigma_t = \sigma_c + \sigma_b$. Equating this with safe bending stress, calculate width w.

5. Shaft diameter at flywheel B.M. at central plane of flywheel $M = R_A \times (0.5\ c)$

Calculate shaft diameter d_s from $M = \sigma_b \times \dfrac{\pi}{32} d_s^3 = \sigma_b\, 0.0981\, d_s^3$

Position II: Crank at maximum torque position

6. Radial and tangential forces At this position gas force $F' = \dfrac{\pi}{4} D^2 p'$

Angle for this position $\phi = \sin^{-1}\left[\dfrac{r}{L}\sin\theta\right]$

Force along the C.R. $F_q = \dfrac{F'}{\cos\phi}$

Tangential component $F_t = F_q \sin(\theta + \Phi)$ Radial component $F_r = F_q \cos(\theta + \Phi)$

7. Bearing reactions Due to tangential forces:

Reaction at bearing A: $R_{At} = \dfrac{F_t \times (a+c)}{c}$ and at bearing B $R_{Bt} = \dfrac{F_t \times a}{c}$

Bearing reactions due to radial forces: $R_{Ar} = \dfrac{F_r \times (a+c)}{c}$ and $R_{Br} = \dfrac{F_r \times a}{c}$

Resultant reactions at bearing A and B: $R_A = \sqrt{R_{At}^2 + R_{Ar}^2}$ and $R_B = \sqrt{R_{Bt}^2 + R_{Br}^2}$

8. Design of crank web Critical section for a side crank is, where web and shaft join. Three types of stresses which occur at this place are direct compressive stress, bending stresses, and torsion.

   (i) Direct compressive stress: $\sigma_c = \dfrac{F_r}{wt}$

   (ii) Bending stresses

   B.M. due to radial force F_r: $M_{br} = F_r\,(0.75\, l_c + 0.5\ t)$

Bending stress $\sigma_{br} = \dfrac{6M_{br}}{wt^2}$

B.M. due to tangential force F_t: $M_{bt} = F_t \times r$

Bending stress $\sigma_{bt} = \dfrac{M_{bt}}{Z} = \dfrac{6M_{bt}}{wt^2}$

Total bending stresses: $\sigma_{bc} = \sigma_c + \sigma_{br} + \sigma_{bt}$

(iii) *Shear stress due to torsion* It is due to torque given by tangential force F_t.

Torque $T = F_t (0.75\, l_c + 0.5\, t)$

Polar section modulus for rectangular section $J = \dfrac{wt^2}{4.5}$

Shear stress $\tau = \dfrac{T}{J}$

Combined stress $\sigma_{max} = 0.5\sigma_t + 0.5\sqrt{(\sigma_t)^2 + 4\tau^2}$. It should be less than allowable.

9. *Shaft diameter d_{s1} at the junction of crank and web*

B.M. at this junction: $M = F_q (0.75\, l_c + t)$ and Torque at this junction: $T = F_t \times r$

Equivalent twisting moment: $T_e = \sqrt{M^2 + T^2}$

Calculated diameter d_{s1} from relation: $T_e = \dfrac{\pi}{16}(d_{s1})^3 \times \tau$

10. *Shaft diameter d_s at flywheel*

Its distance from crank pin force is $(a + c1)$.

Horizontal B.M. due to gas force $M_g = F(a + c1) - (R_A \times c2)$

B.M. due to belt tensions $M_b = \dfrac{(T_1 + T_2)c1}{c} \times c2$

Total horizontal bending moment $M_H = M_g + M_b$

Vertical B.M. due to weight of flywheel $M_V = \dfrac{wc1}{c} \times c2$

Resultant B.M. $M_R = \sqrt{(M_H)^2 + (M_V)^2}$

Twisting moment: $T = F_t \times r$ Equivalent twisting moment $T_e = \sqrt{(M_R)^2 + T^2}$

Calculate shaft diameter d_s from relation: $T_e = \dfrac{\pi}{16}(d_s)^3 \times \tau$

Theory Questions

1. What is the function of a crank shaft? Sketch a crank shaft and name its parts.

2. What are the different types of crank shafts? Sketch a crank shaft for a single cylinder engine and describe its construction.

3. What are the different materials used for a crank shaft?
4. Discuss the various forces coming in a crank shaft.
5. Describe the design procedure to design a center crank shaft.
6. What is an overhung crank? Write its design procedure.

Multiple Choice Questions

1. Number of main bearings in an overhang crank are:
   (a) Two   (b) Three
   (c) One long bearing   (d) Depends upon type of engine

2. Crank shaft is used to:
   (a) Transfer force of piston to connecting rod
   (b) Convert rotary motion to reciprocating motion
   (c) Convert reciprocating motion to rotary motion
   (d) Open and close the valves

3. Torque on a crank shaft is maximum, when crank angle from TDC position is:
   (a) Between 20° and 24°   (b) Between 25° and 40°
   (c) 45°   (d) 65°

4. Stresses on the crank web are:
   (a) Compressive   (b) Bending
   (c) Shear   (d) All given in (a), (b), and (c)

5. A side crank is used for:
   (a) In line multi-cylinder engine   (b) Two cylinders placed side by side
   (c) Single cylinder engine   (d) None of the above

6. Gas pressure at the maximum torque position of the crank shaft is:
   (a) Same as maximum gas pressure
   (b) Mean effective pressure
   (c) Average of maximum and atmosphere pressure
   (d) None of the above

7. In an overhung crank, most heavily loaded bearing is:
   (a) Nearest to crank web   (b) Farthest from crank web
   (c) Both bearings are equally loaded   (d) Depends on type of engine

Answers to multiple choice questions

1. (a)   2. (c)   3. (b)   4. (d)   5. (c)   6. (d)   7. (a)

Design Problems

1. A crank shaft of a four-stroke petrol engine has the following data.

   Cylinder diameter = 80 mm   Stroke length = 110 mm

   l/r ratio = 4 l/d ratio of crank pin = 1

   Weight of flywheel = 1,000 N   Horizontal total belt pull = 1.4 kN

   Gas pressure at TDC = 3.3 MPa   Gas pressure at 35° of crank = 2.4 MPa

   Safe bending stress = 80 MPa   Safe bearing pressure = 10 MPa

   Distance between crank pin bearings = 2D   Distance between flywheel bearings = 3D

   Maximum torque position of crank = 35° from TDC

   Assume crank and flywheel are placed centrally

   Calculate bigger size from TDC and maximum torque position:

   (a) Diameter and length of crank pin.   (b) Width and thickness of crank web.

   (c) Shaft diameter under flywheel.   (d) Length of most heavily loaded bearing.

   [(a) d_c = 45 mm, l_c = 45 mm; (b) t = 32 mm, w = 52 mm;

   (c) d_s = 39 mm, (d) l_2 = 23 mm]

2. A four-stroke single cylinder vertical engine has cylinder diameter 100 mm and stroke 150 mm. The length of C.R. is four times the crank radius. Crank pin length to diameter ratio can be taken 0.85. Maximum gas pressure is 2.8 MPa. The power developed by engine is transmitted through a heavy pulley of width 200 m min the center weighing 30 kN and a horizontal belt giving total pull of 20 kN. Gas pressure at maximum torque position when crank is at 33° is 1.6 MPa. Safe bending stress 65 MPa and bearing pressure 12 MPa. Design a side crank shaft.

   [d_c = 48 mm, l_c = 41 mm, t= 35 mm, w = 91 mm, d_A = 75 mm, l_2 = 54 mm, d_s = 104 mm]

Previous Competition Examination Questions

GATE

1. In a certain slider crank mechanism, lengths of crank and C.R. are equal. If the crank rotates with a uniform angular speed of 14 rad/s and the crank length is 300 mm, the maximum acceleration of the slider (m/s) of the slider is _____

   [Answer 117.6 m/ s²] [GATE 2015]

2. A slider crank mechanism with crank radius 60 mm and C.R. 240 mm is shown in figure on right side. The crank rotates counterclockwise with uniform angular speed of 10 rad/s. For the given configuration, the speed (m/s) of the slider is _____

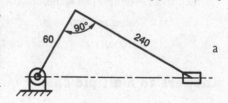

   [Answer 0.6 m/s] [GATE 2014]

□□□

CHAPTER 16

Valve Gears

Outcomes

➤ Learn design of all the components, which transfer power from cam shaft to valves of an internal combustion engine

➤ Calculations for finding out port sizes for inlet and exhaust valves

➤ Design for diameter and thickness of valves, stem size, lift, and materials used

➤ Various forces coming on inlet and exhaust valves

➤ Design of valve spring for strength and natural frequency

➤ Design of rocker arm, boss, eye end and forked end, fulcrum, and roller pin

➤ Design of pushrod and materials used

➤ Cam shaft and cams for different types of lift accelerations and decelerations

➤ Drawing lift diagram and cam profile for flat and roller followers

16.1 Valve Gear Mechanism

Valve gear mechanism in an engine is used to operate inlet and outlet valves. The mechanism depends on the type of internal combustion (I.C.) engine. Maximum number of parts are in I head engine, in which the cylinder head is fitted on the top of a vertical engine (See picture on right). In a four-stroke cycle engine, the cycle is completed in two revolutions, and hence the drive shaft for the valves called cam shaft rotates at half the speed of crank shaft. A set of helical gears with gear ratio two is used or a toothed belt is used to drive the cam shaft. The cams mounted on this shaft move the

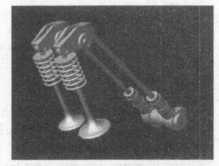

cam follower up and down. The motion is transferred through push rod and rocker arm to the valve [Figure 16.1(a)]. The rocker arm transfers upward movement of push rod to downward movement of the valve to open. The valves are spring loaded and close by themselves due to the action of spring. To keep the valve aligned, valve stem is guided in guide bush placed in the cylinder head. The valve fits tightly on a valve seat to stop any leakage.

Thus the main parts involved in a valve gear arrangement for I head engine are:

Cam shaft	Cam	Cam follower	Push rod	Adjusting screw
Rocker arm	Rocker shaft	Spring	Valve guide	Valve

In an L head engine, the valves are placed by the side of cylinder. The valves are inverted as shown in Figure 16.1(b). Rocker arm is placed below the cam shaft. The length of push rod is less than I head engine, thus mass of reciprocating parts gets reduced. This reduces the inertia effects; hence it can be used for speeds more than I head engines.

New high-speed engines further reduce the inertia masses by eliminating even the push rod and rocker arm. Cam shaft is directly placed over the valves as shown in Figure 16.1(c) that is why these engine are called overhead valve engines.

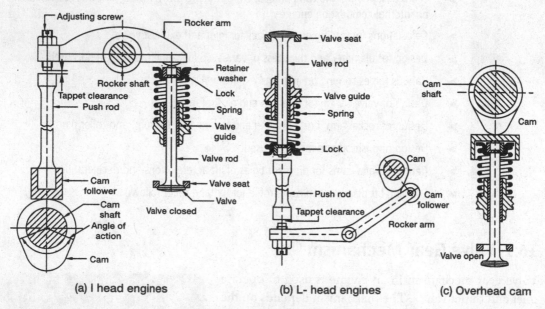

Figure 16.1 Valve gear mechanism

16.2 Ports

Port is a passage, which connects the inlet manifold up to the valve, as shown in Figure 16.2. Atmospheric air passes through the inlet port to inlet valve and the exhaust gases pass through exhaust valve to exhaust port.

Port diameter depends on allowable gas velocity given in Table 16.1 and mean piston velocity given by Equation (16.1).

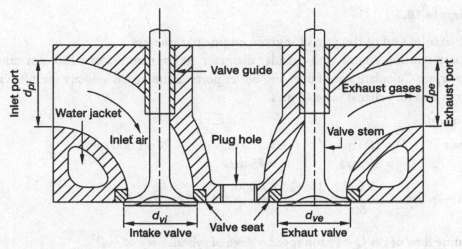

Figure 16.2 Inlet and outlet ports

Table 16.1 Maximum Allowable Mean Gas Velocities

Name of Valve	Speed of Engine	Mean Gas Velocity (m/s)
Inlet valve	Slow speed	25–35
	Medium speed	35–50
	High speed	50–80
Exhaust valve	Slow speed	40–50
	Medium speed	50–70
	High speed	70–100

Mean piston velocity $v = \dfrac{2\,L\,N}{60}$ (16.1)

Where, L = Stroke of piston

N = Speed of engine

Volume flow of gas Q = Piston speed × Area of cylinder $= v \times \dfrac{\pi}{4} D^2$ (a)

Also discharge Q = Port area × Gas velocity $= A_p \times v_g$ (b)

Equating equations (a) and (b)

$$v \times \frac{\pi}{4} D^2 = A_p \times v_g = \frac{\pi}{4} \left(d_p\right)^2 \times v_g$$

Port diameter $d_p = \sqrt{\dfrac{v}{v_g}}\, D$ (16.2)

Example 16.1

Size of inlet and outlet ports for given engine parameters

A spark ignition engine with cylinder diameter 100 mm and stroke 120 mm runs at 1,000 rpm. Calculate size of inlet and outlet ports assuming gas velocity for inlet valve 30 m/s and for exhaust valve 45 m/s.

Solution

Given $D = 100$ mm $L = 120$ mm $N = 1,000$ rpm

$v_{gi} = 30$ m/s $v_{ge} = 45$ m/s

Piston speed $v = \dfrac{2\,LN}{60} = \dfrac{2 \times 120 \times 1,000}{60} = 4,000$ mm/s $= 4$ m/s

Volume flow of gas $Q =$ Piston speed $\times$ Area of cylinder $= v \times \dfrac{\pi}{4}D^2$

Substituting the values: $Q = 4 \times \dfrac{\pi}{4}(0.1)^2 = 0.0314$ m^3/s

Also discharge $Q =$ Port area $\times$ Gas velocity $= A_p \times v_g$

Inlet valve port area $A_{pi} = \dfrac{Q}{v_{gi}} = \dfrac{0.034}{30} = 1.03 \times 10^{-3}$ m^2

Also, $\dfrac{\pi}{4}\left(d_{pi}\right)^2 = A_{pi} = 1.03 \times 10^{-3}$ or, Inlet port diameter, $d_{pi} = 0.0365$ m $= 36.5$ mm

Alternately   Equation (16.2) can also be directly used:

$$d_{pi} = \sqrt{\dfrac{v}{v_{gi}}}\,D = \sqrt{\dfrac{4}{30}} \times 100 = 36.5 \text{ mm}$$

Note   Additional suffices i and e are used for inlet and outlet valves.

Exhaust valve port area $A_{pe} = \dfrac{Q}{v_{ge}} = \dfrac{0.034}{45} = 0.755 \times 10^{-3}$ m^2

Also, $\dfrac{\pi}{4}\left(d_{pe}\right)^2 = A_{pe} = 0.755 \times 10^{-3}$ or, Exhaust port diameter, $d_{pe} = 0.031$ m $= 31$ mm

16.3 Valves

Various types of valves are used such as sleeve valves, rotary valves, but the valves used for I. C. engine are poppet valves as shown in Figure 16.3 and picture on its right. The valve is kept closed against its seat by a compression spring. A retainer and lock transfer the spring force to the valve. The valve is guided by a bush fixed in the cylinder head.

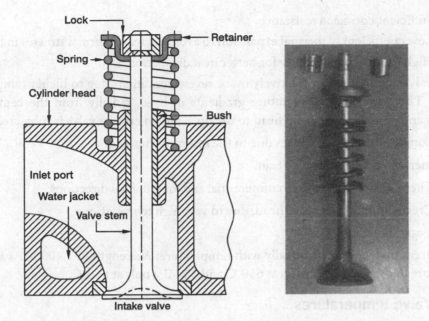

Figure 16.3 Valve arrangement

A poppet valve is shown in Figure 16.4. The valve head is like a circular disk with a margin of about 1 mm and then an inclined face, which fits tightly against the valve seat. The valve stem is long enough so that its tip at the top is above the spring. The other side of the stem has a groove, in which a lock in two halves is fixed. The head and stem are provided with a large fillet radius so that gases can flow easily without much resistance. The bottom portion of the valve called face, is exposed inside the cylinder to face the gases.

Advantages of the poppet valves over other types of valves are:

- Offer more flow area with respect to piston area.
- High coefficient of discharge.
- Easy to manufacture in comparison to other types of valves.
- Offer very less friction.
- Require very less lubrication.

Figure 16.4 Parts of a poppet valve

Due to adverse operating conditions of temperature and cyclic loading, material for exhaust valve should have the following requirements.

- High strength and hardness to resist cyclic loads and stem wear.
- High strength at elevated temperature.
- Hard to resist cupping of head and wear of seats.
- High fatigue and creep resistance.

- Sufficient corrosion resistance.
- Low coefficient of thermal expansion to avoid decrease thermal stresses in head.
- High thermal conductivity for better heat dissipation.

Exhaust valves operate under relatively more severe conditions due to higher temperatures involved. There are large temperature gradients in head, radially from the center to its periphery and axially in stem from head to the tip. Thus an exhaust valve is subjected to:

- Longitudinal cyclic stresses due to the return spring load.
- Inertia forces due to valve train.
- Thermal stresses in the circumferential and longitudinal directions.
- Creep conditions at valve head, due to very high temperatures.
- Corrosion conditions.

Tensile strength decreases drastically with temperature. A strength of 1,400 MPa at normal temperature decreases to 290 MPa at 650°C and 32 MPa only at 850°C.

16.3.1 Valve temperatures

Temperatures inside the intake valve are of the order of 550°C. The intake valves are cooled by incoming gases. Corresponding values inside the exhaust valve are 700°C–750°C. The exhaust valve temperature can go up to 900°C. Since the exhaust valves operate at high temperatures, they are exposed to thermal load and chemical corrosion. Some exhaust valves are made hollow and filled with sodium to improve heat removal from the valves. Valve guides are made longer to provide more contact area for heat transfer. Materials used for such high temperatures are given below.

16.3.2 Valve materials

a. **Nickel–base alloys**

They can operate at temperature up to 1,020°C but have low fatigue strength.

b. **Austenitic valve steels**

To face high temperatures, recently so-called austenitic, nonhardening steels have been introduced as a material for exhaust valves. It has good strength at high temperature, impact, and resistance to oxidation and corrosion. Its disadvantage is that it has higher coefficient of thermal expansion, hence needs more clearance at valve stem and guide.

It contains carbon 0.3–0.45, manganese 0.80–1.30, silicon 2.50–3.25, chromium 17.50–20.50, nickel 7.00–9.00, and phosphorus and sulphur 0.03 each.

Its tensile strength at 850°C is 120 MPa.

c. **Silcrome steel**

The latest addition to the list of exhaust-valve materials is a steel intermediate between the ferritic and the austenitic types. This steel, silcrome XCR contains carbon

0.40–0.50, manganese 1.00, chromium 23.25–24.25, nickel 4.50–5.00, molybdenum 2.50–3.00, and phosphorus and sulphur 0.035 each.

Its tensile strength at 850°C is 140 MPa.

16.3.3 Size of valves

The size of valves depends on cylinder diameter and number of valves per cylinder. For better volumetric efficiency, the number of valves per cylinder is increased. Figure 16.5 shows a single valve and two valves arrangements. Inlet valves are made bigger than exhaust valves. The ratio of exhaust valve and inlet valve diameter varies from 0.83 to 0.87.

The sizes of valve to cylinder diameter ratios for different types of engines are given in Table 16.2.

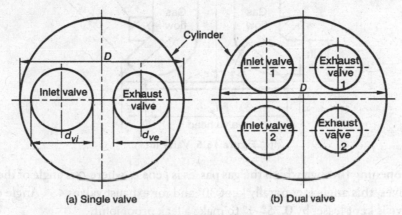

(a) Single valve   (b) Dual valve

Figure 16.5 Valve arrangements

Table 16.2 Range of Valve Sizes

Type of Engine	Inlet Valve Sizes d_{vi}/D		Exhaust Valve Sizes d_{ve}/D	
	One Valve	Two Valves	One Valve	Two Valves
Petrol engine	0.39–0.49	0.38	0.36–0.41	0.36
Diesel engine	0.45–0.49	0.34	0.31–0.39	0.34

Where, D = Cylinder diameter

d_{vi} = Diameter of inlet valve

d_{ve} = Diameter of exhaust valve

Valve seat is inclined at an angle α, which is either 30° or 45° as shown in Figure 16.6. The inclined length of the valve seat is kept between 0.07 and 0.1 times the port diameter d_p. Assuming a mean value 0.85 d_p the width of seat w along the face of valve:

For inlet valve $w = 0.085\, d_p \cos 30° = 0.075\, d_p$

For exhaust valve $w = 0.085\, d_p \cos 45° = 0.06\, d_p$

Thus inlet valve diameter $d_{vi} = d_p + 2w = d_p + 0.15w$ (16.3a)

Exhaust valve diameter $d_{vi} = d_p + 2w = d_p + 0.12w$ (16.3b)

16.3.4 Lift of valves

Lift of the valve is calculated assuming that the gas flow velocity remains same from port to valve exit. Thus the area at the port is same as across the valve. If l is the maximum lift of valve, then the gas passes through the area around the valve with circumference πd_p (Figure 16.6).

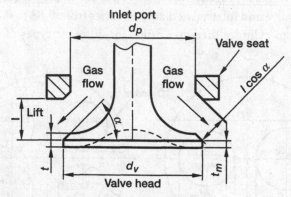

Figure 16.6 Valve lift

Inclined opening through which the gas passes is $l \cos \alpha$, where α is angle of the valve seat. For inlet valves, this angle is generally kept 30° and for exhaust valves 45°. Angle of the valve seat and valve is kept lesser by 0.75°–1° to make a leak proof joint.

Thus, $\dfrac{\pi}{4}(d_p)^2 = \pi d_p \times l \cos \alpha$ or $l = \dfrac{d_p}{4 \cos \alpha}$ (16.4)

Lift of the valve is kept about 25 per cent of the valve diameter.

16.3.5 Thickness of valve

Mass of the engine valves varies from 0.18 kg to 0.5 kg depending on the size of ports. Its mass can be calculated from the disk and stem volume and specific gravity. Thickness of the disk is calculated for safe bending stress by considering valve a circular disk simply supported at the valve seat periphery, loaded uniformly with the maximum pressure in the cylinder. Thickness t (mm) for the valve made of steel is determined using the following relation:

Total thickness of head disk $t = 0.45\, d_p \sqrt{\dfrac{p_{max}}{\sigma_b}}$ (16.5)

Where, p_{max} = Maximum gas pressure in cylinder (MPa)

     σ_b = Allowable bending stress (MPa)

For carbon steels 50 MPa to 60 MPa
For alloy steels 80 MPa to 120 MPa

d_p = Port diameter (mm)

Empirically, total thickness of head $t = 0.10 \, d_p$ to $0.14 \, d_p$

The edge of disk called margin $t_m = 0.025 \, d_p$ to $0.045 \, d_p$ (Minimum 0.7 mm)

Inclined seat length $c = 0.10 \, d_p$ to $0.12 \, d_p$

The disk head is initially provided with an angle of 10°–15° with horizontal followed with a small fillet radius to match to stem.

16.3.6 Size of valve stem

Valve stem is subjected to spring force when seated, which causes axial tensile force in the stem. Its size can be assumed empirically using the relations given below.

Stem diameter for inlet valve $d_s = 0.18 \, d_p$ to $0.24 \, d_p$ (16.6a)

Stem diameter for exhaust valve $d_s = 0.22 \, d_p$ to $0.28 \, d_p$ (16.6b)

To improve heat transfer, stem diameter of the exhaust valve is made 10 to 15 per cent greater than that of the inlet valve.

16.3.7 Valve Timings

Valves do not open and close at dead centers, but open before or close after dead center. Thus their opening time is more than 180° as given in Table 16.3. This duration helps in calculating inertia forces for calculating acceleration (Section 16.3.8).

Table 16.3 Duration of Valve Opening

Valve	Conventional	High Performance
Intake	230°	270°
Exhaust	235°	285°

Example 16.2

Valve head diameters and lifts with double valves per cylinder

An I.C. engine having cylinder diameter 120 mm and stroke 150 mm runs at 1,200 rpm. Calculate:

    a. Size of inlet and outlet valve heads assuming two valves per cylinder.

    b. Velocity of gases, assuming port diameter as 90 per cent of valve diameter.

    c. Lift of inlet and outlet valves, assuming angle of seat 30° for inlet and 45° for exhaust.

Solution

Given $D = 120$ mm $L = 150$ mm $N = 1,200$ rpm $d_p = 0.9d_v$
$\alpha_i = 30°$ $\alpha_e = 45°$

a. Size of valve heads

From Table 16.2,

Inlet valve size $d_{vi} = 0.38 \times D = 0.38 \times 120 = 45.6$ say 46 mm

Exhaust valve size $d_{ve} = 0.36 \times D = 0.36 \times 120 = 43.2$ say 43 mm

b. Velocity of gases

Piston speed $v = \dfrac{2LN}{60} = \dfrac{2 \times 150 \times 1,200}{60 \times 1,000} = 6$ m/s

Volume flow of gas Q = Piston speed × Area of cylinder $= v \times \dfrac{\pi}{4}D^2$

Substituting the values: $Q = 6 \times \dfrac{\pi}{4}(0.12)^2 = 0.0678$ m^3/s

Inlet port diameter $d_{pi} = 0.9 \times 46 = 41.4$ mm $= 0.0414$ mm

Exhaust port diameter $d_{pe} = 0.9 \times 43 = 38.7$ mm $= 0.0387$ mm

Since there are two valves, discharge through each port $= \dfrac{0.0678}{2} = 0.0339$ m^3/s

Also discharge Q = Port area × Gas velocity $= A_p \times v_g = \dfrac{\pi}{4}\left(d_{pi}\right)^2 \times v_g$

Gas velocity in inlet port:

  $0.0339 = 0.785 \times (0.0414)^2 \times v_g$ or $v_g = 25.2$ m/s

Gas velocity in exhaust port:

  $0.0339 = 0.785 \times (0.0387)^2 \times v_g$ or $v_g = 28.8$ m/s

c. Lift of valves

From Equation (16.4), Lift of inlet valve $l_i = \dfrac{d_{pi}}{4\cos\alpha_i} = \dfrac{41.4}{4 \times \cos 30°} = 11.9$ mm

Lift of exhaust valve $l_e = \dfrac{d_{pe}}{4\cos\alpha_e} = \dfrac{38.7}{4 \times \cos 45°} = 13.7$ mm

16.3.8 Forces on valves

A closed valve is loaded by spring force and pressure inside the cylinder, which varies periodically, peak value is about 15 MPa. Such high pressures inside the cylinder cause bending of the valve cone, which results in a sliding motion and improper contact between valve face and seat insert, thus eventually leading to wear failure.

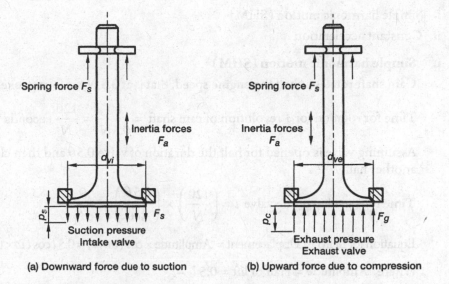

Figure 16.7 Forces on inlet and exhaust valves

There are three types of forces acting on a valve as explained below:

a. **Spring force** The valve is exposed inside the cylinder; hence the gas pressures act on the valve face area. During suction stroke, pressure inside the cylinder is less than atmospheric pressure, which tries to pull the valve inside the cylinder. If the spring force is not enough, even the exhaust valve may open during suction stroke as shown in Figure 16.7(a); hence the spring force has to be at least equal to force created due to suction pressure p_s. Weight of the valve (w = mass of valve $\times g$) adds to spring force and hence net initial spring force required for an I head engine is:

$$\text{Initial spring force } F_i = \frac{\pi}{4}\left(d_{vi}\right)^2 p_s + w \tag{16.7}$$

For an L head engine, where valves are on side of cylinder [Figure 16.1(b)], weight of valve acts opposite to spring force, and hence it has to be subtracted.

b. **Gas force** During other strokes like power and compression, the pressure is above atmospheric pressure and the valves are closed as shown in Figure 16.7(b). Exhaust valve opens before bottom dead center during power stroke hence some pressure p_c acts on exhaust valve giving a gas force F_g, which is given by the following relation. Weight of the valve w helps in opening, and hence it decreases the force on valve.

$$\text{Gas force } F_g = \frac{\pi}{4}\left(d_{ve}\right)^2 p_c - w \tag{16.8}$$

For L head engine, add weight of valve instead of subtracting.

c. **Inertia force** When the valves are pushed from rest position to some velocity, acceleration of the mass of the valve train causes inertia forces. For the analysis of acceleration, two types of assumptions for opening and closing of valves are used:

i. Simple harmonic motion (SHM)
ii. Constant acceleration

i. Simple harmonic motion (SHM)

Cam shaft rotates at half the engine speed, that is at 0.5 N for four stroke engine.

Time for rotation for 1 revolution of cam shaft $= \dfrac{60}{0.5N} = \dfrac{120}{N}$ seconds

Assuming valve is opened for half the duration of valve 0.5θ and then closes for another half.

Time for opening of the valve $t = \left(\dfrac{120}{N}\right) \times \left(\dfrac{0.5\,\theta}{360}\right) = \dfrac{\theta}{6N}$ \hfill (16.9)

Equation for SHM is: Displacement $=$ Amplitude $\times \cos(\omega \times t) = 0.5\,l \cos(\omega \times t)$ \hfill (a)

Where, Amplitude $=$ Half of lift $= 0.5\,l$

$\qquad\qquad t =$ Time and

$\qquad\qquad \omega =$ Angular velocity of the system (rad/s) $= \dfrac{2\pi}{t}$ \hfill (b)

Differentiating the expression (a): Velocity $= -\,0.5\,\omega l \sin(\omega \times t)$

Differentiating once again: Acceleration $\alpha = 0.5\,\omega^2 l \cos(\omega \times t)$

Maximum acceleration occurs when the term $\cos(\omega \times t)$ is maximum, that is, this occurs at the extremes of the motion.

Hence maximum acceleration $a = 0.5\,\omega^2 l$

Putting the value of ω, from equation (b) and value of t from Equation (16.9):

Acceleration, $\alpha = 0.5l\left(\dfrac{2\pi}{t}\right)^2 = 0.5\,l\left(\dfrac{12\pi N}{\theta}\right)^2 = 710\,l\left(\dfrac{N}{\theta}\right)^2$ \hfill (16.10)

Inertia force due to acceleration $F_a = m\,\alpha$ \hfill (16.11)

Or, $\quad F_a = 710\,ml\left(\dfrac{N}{\theta}\right)^2$ \hfill (16.12)

Where, $m =$ Mass of the valve (kg)

$\qquad\qquad \alpha =$ Acceleration (m/s^2)

Thus total force on exhaust valve is

$\qquad F_{ve} = F_g + F_a$ \hfill (16.13)

Total force on inlet valve is

$\qquad F_{vi} = F_i + F_a$ \hfill (16.14)

ii. Constant acceleration

Newton's second law of motion $s = ut + 0.5 at^2$ is used for calculating acceleration. When a valve is lifted, it starts from zero velocity ($u = 0$) for half of lift $s = 0.5 \, l$, and half of the lift it is accelerated at constant acceleration and the remaining half of the lift is with constant deceleration. Calculate total time t for the valve to remain open using Equation (16.9). Half of the time is rise period and half fall period; hence time for acceleration will be only one quarter of the total time, that is, $0.25t$. Then use the relation $s = ut + 0.5 \, at^2$ with initial velocity $u = 0$.

$$0.5 \, l = 0.5 \alpha \left(0.25 \, t\right)^2 \quad \text{or} \quad \alpha = \frac{16l}{t^2} \tag{16.15}$$

Where, t = Total time for the valve to remain open

l = Total lift

Calculate inertia force F_a using Equation (16.11). With this value of F_a in Equations (16.13) and (16.14) calculate total forces F_{ve} and F_{vi}.

The keeper groove area in stem is subjected to tensile stresses and becomes a critical section due to geometric stress concentrations.

Example 16.3

Sizes, lifts and forces on valves

Cylinder of an S.I. engine running at a speed of 1,400 rpm has a cylinder of 150 mm diameter and stroke 180 mm with a suction pressure of 0.15 MPa and pressure at the opening of exhaust valve 0.6 MPa. Take air velocity for inlet port 45 m/s and for exhaust gases 55 m/s. Inlet valve opens for 210° and exhaust valve for 240°. Mass of both the valves can be taken approximately 0.25 kg. Calculate:

a. Sizes of inlet and exhaust valves.

b. Lifts of both the valves assuming seat angle 45° for both the valves.

c. Forces on inlet and exhaust valves.

Solution

Given $D = 150$ mm $L = 180$ $N = 1,400$ rpm $p_s = 0.15$ MPa

$p_c = 0.6$ MPa $v_{gi} = 40$ m/s $v_{ge} = 55$ m/s $m = 0.25$ kg

$\theta_i = 210°$ $\theta_e = 240°$ $\alpha_i = \alpha_e = 45°$

a. Size of valves

Piston speed $v = \dfrac{2LN}{60} = \dfrac{2 \times 180 \times 1,400}{60 \times 1,000} = 8.4$ m/s

Inlet valve: From Equation (16.2)

Inlet port diameter, $d_{pi} = \sqrt{\dfrac{v}{v_{gi}}} D = \sqrt{\dfrac{8.4}{45}} \times 150 = 64.8$ mm

From Equation (16.3b) for 45° angle, inlet valve diameter, $d_{vi} = 1.12\, d_{pi} = 72$ mm

Exhaust valve: From Equation (16.2)

$$d_{pe} = \sqrt{\dfrac{v}{v_{ge}}} D = \sqrt{\dfrac{8.4}{55}} \times 150 = 58.6 \text{ mm}$$

Exhaust valve diameter $d_{ve} = 1.12\, d_{pe} = 66$ mm

b. Lifts

For inlet valve $l_i = \dfrac{d_{pi}}{4 \cos \alpha_i} = \dfrac{64.8}{4 \cos 45°} = 23$ mm

For exhaust valve $l_e = \dfrac{d_{pe}}{4 \cos \alpha_e} = \dfrac{58.6}{4 \cos 45°} = 21$ mm

c. Force on inlet valve

Spring force $F_i = \dfrac{\pi}{4}(d_{vi})^2 p_s = 0.785 \times (72)^2 \times 0.15 = 610.4$ N

Acceleration force $F_a = 710\, ml\left(\dfrac{N}{\theta}\right)^2 = 710 \times 0.25 \times 0.023 \left(\dfrac{1,400}{210}\right)^2 = 181.4$ N

From Equation (16.14)

Total force on inlet valve is $F_{vi} = F_i + F_a = 610.4 + 181.4 = 791.8$ N

Force on exhaust valve:

Spring force $F_i = \dfrac{\pi}{4}(d_{ve})^2 p_s = 0.785 \times (66)^2 \times 0.15 = 513$ N

From Equation (16.8),

Gas force $F_g = \dfrac{\pi}{4}(d_{ve})^2 p_c = 0.785 \times (66)^2 \times 0.6 = 2,052$ N

Acceleration force $F_a = 710\, ml\left(\dfrac{N}{\theta}\right)^2 = 710 \times 0.25 \times 0.021\left(\dfrac{1,400}{240}\right)^2 = 126.8$ N

From Equation (16.13)

Total force on exhaust valve is $F_{ve} = F_g + F_a$

$F_{ve} = 2,052 + 126.8 = 2,178.8$ N

16.4 Valve Spring

The purpose of the spring is just to keep the valve closed against its seat unless opened whenever desired. It is shown in picture on the right side. The forces acting on the valves described in Section 16.3.8 are summarized as under: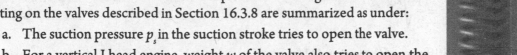

a. The suction pressure p_s in the suction stroke tries to open the valve.

b. For a vertical I head engine, weight w of the valve also tries to open the valve. For L head engine, valves are on side of cylinder and inverted, hence wight of valve is subtracted. Here the analysis is for I head engine, in which valves are on top of cylinder.

$$\text{Initial minimum force on spring } F_{min} = \frac{\pi}{4}(d_{vi})^2 \, p_s + w \qquad \text{(a)}$$

c. When a valve is opened, the gas force inside the cylinder opposes to opening of the valve. Thus the gas force also comes on the spring.

$$\text{Gas force } F_g = \frac{\pi}{4}(d_{ve})^2 \, p_c - w \qquad \text{(b)}$$

Thus the maximum force on the valve is $F_{max} = F_{min} + F_g$

Substituting the value of F_{min} from Equation (a), weight of valve gets cancelled and the maximum force is:

$$F_{max} = \frac{\pi}{4}\left[(d_{vi})^2 \, p_s + (d_{ve})^2 \, p_c\right] \qquad (16.16)$$

If spring stiffness is not given, then for slow-speed engines, total load on spring can be taken to give a load of $0.07 \times$ valve area in millimeters.

Spring of a valve is designed by the same method as described in Chapter 22 of Volume 1 for variable loads. Minimum and maximum forces are calculated as given in equations above.

One more parameter, which is considered for the design of valve spring, is its natural frequency. When a spring is compressed and released suddenly, it vibrates longitudinally till its energy is damped by air friction. This frequency is called natural frequency.

When there are quick repeated loads, a spring is compressed at one end and a sudden rapid compression at the other end. The end coil is pushed against its adjacent coil, before the remaining coils get time to share the displacement. After sufficiently rapid compressions, the excessive displacement moves along spring touching the first coil, then second and then third etc. until it reaches the last coil of the opposite end. From here, the disturbance is reflected back towards the displacement end and so on, until the energy is dissipated. This phenomenon is called **spring surge**. It decreases the ability to control the motion of the engine valve. This condition may result in very large deflections and it is possible that a spring may even fail.

If natural frequency of the spring is near to the external cyclic forces, the resonance may increase deflections so high, that a spring can break. For an engine running at 4,800 rpm, cam shaft runs at 2,400 cycles per minute, that is, 40 c/s. Natural frequency of the spring should be at least 13 times the disturbing cycle, that is, $40 \times 13 = 520$ Hz. Natural frequency f_n of a spring in hertz can be calculated using the following equation:

$$f_n = \frac{1}{2}\sqrt{\frac{kg}{m}} \tag{16.17}$$

Where, k = Stiffness of the spring (N/m)

$\quad m$ = Mass of active turns of spring (kg)

$\quad g$ = Acceleration due to gravity = 9.810 m/s²

Spring constant $k = \dfrac{Gd^4}{8D^3 n}$ (See Chapter 22 of Volume 1) $\tag{16.18}$

Mass of the spring m = Area of wire × Length of wire × Density = $\dfrac{\pi}{4}d^2 (\pi Dn)\rho$

Where, ρ = Mass density (kg/mm³)

$\quad n$ = Number of active turns

Substituting the values

$$f_n = \frac{d}{2\pi n D^2}\sqrt{\frac{G \times g}{2\rho}} \tag{16.19}$$

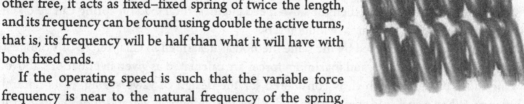

This relation is for fixed–fixed end conditions (that is, both ends fixed) of a helical spring. If one end is fixed and the other free, it acts as fixed–fixed spring of twice the length, and its frequency can be found using double the active turns, that is, its frequency will be half than what it will have with both fixed ends.

If the operating speed is such that the variable force frequency is near to the natural frequency of the spring, two springs (one inside the other) of different stiffness are used (See picture above). Thus stiffness of a spring is changed. For dual spring arrangement, outer spring is designed to take 60 per cent to 70 per cent of maximum spring force and the rest of the force is taken by the inner spring.

Design procedure

1. Calculate minimum force on the spring using equation $F_{min} = \dfrac{\pi}{4}(d_{vi})^2 p_s + w$

2. Force required to open this valve = $k\,\delta$

    Where, k = Stiffness of the spring (N/m)

    $\quad w = m \times g$, and m = Mass of spring (kg)

    Thus maximum force = $F_{max} = F_{min} + k\,\delta$ Where, δ = Deflection

3. Calculate wire diameter d using equation:

$$\tau = \frac{8 K F_{max} C}{\pi d^2} \quad \text{Or,} \quad d = \sqrt{\frac{8 K F_{max} C}{\pi \tau}} \tag{16.20}$$

Where, τ = Allowable shear stress

C = Spring index D / d

D = Mean coil diameter

K = Wahl factor = $\dfrac{4C - 1}{4C - 4} + \dfrac{0.615}{C}$

4. Calculate actual number of turns:

Active number of turns, $n = \dfrac{G d^4}{8 D^3 k}$ \qquad (16.21)

Total number of turns $n' = n + 2$ (2 is for square ends)

Where, G = Modulus of rigidity (MPa)

k = Spring stiffness (N/mm)

5. Gap between two adjacent coils varies from 1.25 mm to 1.5 mm

Free length of spring FL = SL + Gap = $(d \times n') + 1.5 (n' - 1)$

Where, SL = Solid length $(d \times n')$

n' = Total number of turns

6. Calculate natural frequency f_n of spring:

$$f_n = \frac{1}{2} \sqrt{\frac{kg}{m}}$$

(It is half when both ends are fixed)

7. If it is near to disturbing cycle frequency, use two concentric springs

---

Example 16.4

Design of valve spring for a diesel engine

A spring is to be designed for a four-stroke vertical diesel engine having suction valve diameter 42 mm and exhaust valve 36 mm. Suction pressure is 0.03 MPa below the atmospheric pressure and gas pressure when exhaust valve opens is 0.35 MPa. Weight of the valves can be taken 100 gm. Lift of the valves is 10 mm. Safe shear strength is 280 MPa and modulus of rigidity 82 GPa. Mean coil diameter to wire diameter ratio is 8. Gap between coils is 1.5 mm when the valve is fully open. Find:

a. Wire and coil diameter.

b. Number of total turns assuming square ground ends.

c. Free length.

d. Natural frequency assuming density of wire 7,800 kg/mm³.

Solution

Given

$$d_{vi} = 42\,\text{mm} \qquad d_{ve} = 36\,\text{mm} \qquad p_s = 0.03\,\text{MPa} \qquad p_c = 0.35\,\text{MPa}$$

$$l = 10\,\text{mm} \qquad \tau = 280\,\text{MPa} \qquad G = 82\,\text{GPa} \qquad D/d = 8$$

$$w = 0.1\,\text{kg} = 9.81\,\text{N} \qquad \text{Gap} = 1.5\,\text{mm}$$

a. Minimum force on spring

$$F_{min} = \frac{\pi}{4}\left(d_{vi}\right)^2 p_s + w$$

$$= 0.785 \times (42)^2 \times 0.03 + 9.81 = 51.3\,\text{N}$$

Gas force on exhaust valve: $F_g = \dfrac{\pi}{4}\left(d_{ve}\right)^2 p_c - w$

$$= 0.785 \times (36)^2 \times 0.35 - 9.81 = 346.3\,\text{N}$$

Maximum spring force: $F_{max} = F_{min} + F_g = 346.3 + 51.3 = 397.6\,\text{N}$

Spring index $k = \dfrac{F_{max} - F_{min}}{l} = \dfrac{397.6 - 51.3}{10} = 34.6\,\text{N/mm}$

Wahl factor $K = \dfrac{4C-1}{4C-4} + \dfrac{0.615}{C} = \dfrac{(4\times8)-1}{(4\times8)-4} + \dfrac{0.615}{8} = 1.184$

Wire diameter $d = \sqrt{\dfrac{8KF_{max}C}{\pi\tau}} = \sqrt{\dfrac{8 \times 1.184 \times 387.8 \times 8}{3.14 \times 280}} = 5.8$ say 6 mm

b. Number of turns:

Number active of turns, $n = \dfrac{G\,d^4}{8D^3k} = \dfrac{82 \times 10^3 \times 6^4}{8 \times \left(8\times6\right)^3 \times 34.6} = 3.5$ say 4

Total number of turns: $n' = n + 2 = 4 + 2 = 6$ turns

c. Free length of spring, FL = Solid length + Gap when valve is open + Initial compression to give minimum force + Lift

$FL = (d \times n') + 1.5\,(n' - 1) + \text{Initial compression} + 1 = (6 \times 6) + 1.5\,(6 - 1) +$
$(51.3 / 34.6) + 10 = 55\,\text{mm}$

d. Length of wire $= \pi D\,n' = 3.14 \times (6 \times 8) \times 6 = 904.3\,\text{mm}$

Volume of wire $= \dfrac{\pi}{4} d^2 \left(\pi D n' \right) = 0.785 \times 6^2 \times 904.3 = 25{,}556\ \text{mm}^3$

Mass of spring $= \text{Volume} \times \text{Density} = 25{,}556 \times 10^{-9} \times 7{,}800 = 0.199\,\text{kg}\ = 1.953\,\text{N}$

Natural frequency $f_n = \dfrac{1}{2}\sqrt{\dfrac{k}{m}} = \dfrac{1}{2}\sqrt{\dfrac{34.6 \times 1{,}000}{1.953}} = 66.7\ \text{c/s}$

16.5 Rocker Arm

A rocker arm is used to convert the direction of force from push rod to the axis of the valve stem at its tip. If axis of push rod and valve both are vertical, it simply acts as a straight lever with fulcrum in between the ends. If the valves are inclined it takes the shape of a bell crank lever. Included angle depends on the axis of the valve.

One end of the rocker arm is made flat with a tapped hole to fix tappet screw. As the valves heat up, their length of stem increases, and hence certain amount of clearance between the push rod and tappet screw is necessary to compensate this thermal expansion. The tappet clearance in cold conditions is so adjusted that at hot conditions it remains almost zero. That is why tappet clearance is kept more for exhaust valve and less for inlet valve.

Material used for rocker arm for car engines is generally steel stampings. It provides a reasonable balance of strength, weight and cost. Because the rocker arms are considered as a part of reciprocating mass, excessive mass especially at high speeds causes high inertia forces. Truck diesel engines use stronger and stiffer rocker arms made of ductile cast iron or forged carbon steel.

Although force on inlet valve is lesser than exhaust valve, but even then, rocker arms for both the valves are kept identical. Fulcrum is provided with a hole for a phosphorous bronze bush of thickness 2 mm to 3 mm. Rocker shaft (Section 16.6) fits in the hole of the bush (Figure 16.8).

It can be seen from Picture and Figure 16.8 that the arm lengths from the center of the boss are different on each side to provide some ratio called rocker arm ratio. It is the ratio of the distance from the boss center to the tip divided by the distance to the axis of push rod. New automotive rocker arm ratios are of about 1.5:1 to 1.8:1. The force exerted on the valve stem is determined by this ratio. In the past, smaller positive ratios (valve lift > cam lift) and even negative ratios (valve lift < cam lift) have been used. Many engines of 1950s used neutral ratio 1:1.

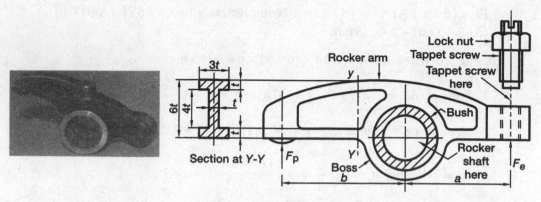

Figure 16.8 A rocker arm

Force on exhaust valve F_e is more than the suction valve F_i. Because both the rocker arms are identical, it is designed for the exhaust valve. Force on one side of the rocker arm is taken as F_e, force on the other side is same if leverage ratio is 1:1. But to reduce the force on cam, some leverage (b/a) is provided.

Thus force on push road side $F_p = \dfrac{a}{b} F_e$ (16.22)

For straight rocker arm, reaction at the fulcrum is $R = F_p + F_e$ (16.23)

If arms are inclined and make angle β as shown in Figure 16.9, then reaction is:

$$R = \sqrt{\left(F_p\right)^2 + \left(F_e\right)^2 - 2\,F_p\,F_e\cos\beta}$$ (16.24)

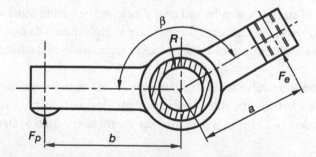

Figure 16.9 Rocker arm with arms at angle

Consider double shear at the rocker shaft of diameter d_r

$$R = 2 \times \frac{\pi}{4}(d_r)^2 \times \tau$$ (16.25)

Rocker shaft diameter d_r and length of bush l_r for safe bearing pressure are related as:

$$R = d_r\, l_r\, p_b$$ (16.26)

Length of fulcrum bush l_r is kept 1.2 d_r to 1.3 d_r. Assuming an average value $l_r = 1.25\ d_r$

Hence Equation (16.26) becomes:

$$R = d_r \times 1.25\ d_r \times p_b = 1.25\ (d_r)^2 p_b \tag{16.27}$$

Select the bigger diameter calculated from shear and bearing criteria.

A phosphorous bush of thickness 2 mm to 3 mm is placed in the hole of boss to reduce friction; hence hole diameter at fulcrum $d_h = d_r + 4$ mm.

Boss diameter is kept $D_b = 2d_h = 2\ d_r + 8$ $\qquad$ (16.28)

Bending moment (BM) at the center of the boss $M = F_p \times b$

Thickness of rocker arm is kept as length l_r and depth equal to D_h.

Mass moment of inertia at center of boss

$$I = \frac{1}{12}(l_r)(D_b)^3 - \frac{1}{12}(l_r)(d_h)^3 \tag{16.29}$$

Bending stress: $\sigma_b = \dfrac{M\,y}{I}$

Where, y is maximum distance from central axis $= 0.5\ D_b$.

Check that the bending stresses σ_b is less than safe bending stress.

Mass moment of area of I section shown in Figure 16.8 with width $3t$ and depth $6t$ is:

$$I = \frac{1}{12}(3t)(6t)^3 - \frac{1}{12}(2t)(4t)^3 = 54t^4 - 10.7t^4 = 43.7t^4 \cong 44t^4 \tag{16.30}$$

Bending moment M at the end of boss at section $y - y$.

$$M = F_p(a - 0.5\ D_b)$$

$$\sigma_b = \frac{M\,y}{I} = \frac{F_p(a - 0.5D_b) \times (0.5D_b)}{44\ t^4} \tag{16.31}$$

Calculate value of t from the value of safe bending stress.

Example 16.5

Design of rocker arm for an engine

A four-stroke I.C. engine uses a rocker arm for an exhaust valve having lengths 85 mm towards push rod and 70 mm towards valve as shown in Figure 16.S1. Engine runs at 1,200 rpm. Diameter of the valve is 40 mm, mass 0.35 kg, and lift of 10 mm. Maximum suction pressure is 0.05 MPa and cylinder pressure when the valve opens is 0.4 MPa. The valve is actuated by a cam giving SHM. The valve remains open for 220°. Take safe bending stress 65 MPa, bearing pressure 6 MPa, and safe shear stress 40 MPa. Design:

a. Rocker shaft diameter.
b. Cross section of rocker arm.

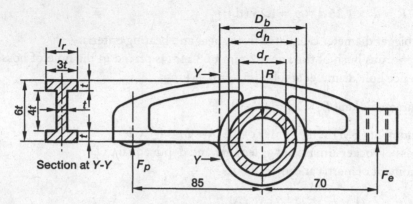

Figure 16.S1 A rocker arm

Solution

Given $a = 85$ mm $b = 75$ mm $N = 1,200$ rpm $d_{ve} = 40$ mm
              $p_s = 0.05$ MPa $p_c = 0.4$ MPa $m = 0.35$ kg $l = 10$ mm
              $\theta_e = 220°$ $\sigma_b = 65$ MPa $p_b = 6$ MPa $\tau = 40$ MPa

a. Suction force $F_s = \dfrac{\pi}{4}(d_{ve})^2 p_s - w = 0.785 \times (40)^2 \times 0.05 - 3.56 = 59.5$ N

Gas force $F_g = \dfrac{\pi}{4}(d_{ve})^2 p_c + w = 0.785 \times (40)^2 \times 0.4 + 3.5 = 505.9$ N

Acceleration force $F_a = 710ml\left(\dfrac{N}{\theta}\right)^2 = 710 \times 0.35 \times 0.01\left(\dfrac{1,200}{220}\right)^2 = 73.9$ N

Total force on exhaust valve is $F_{ve} = F_s + F_g + F_a = 59.5 + 505.9 + 73.9 = 639.3$ N

Force on other side of the rocker arm $F_p = \dfrac{a}{b}F_e = \dfrac{70}{85} \times 639.3 = 526.3$ N

Reaction at the fulcrum $R = F_{ve} + F_p = 639.3 + 526.3 = 1,165.6$ N

Considering double shear at the rocker shaft $R = 2 \times \dfrac{\pi}{4}(d_r)^2 \times \tau$

Substituting the values

$1,165.6 = 2 \times 0.785\,(d_r)^2 \times 40$ or $d_r = 4.3$ mm

Assuming an average value $l_r = 1.25\,d_r, R = d_r \times 1.25\,d_r \times p_b$

Substituting the values

$1165.6 = 1.25\,(d_r)^2 \times 6$ or $d_r = 12.5$ mm

Select the bigger diameter calculated from shear and bearing criteria, that is, 12.5 mm.

Hence the length of boss $l_r = 1.25\, d_r = 1.25 \times 12.5 = 15.6$ mm

Hole diameter $d_h = d_r + 4$ mm $= 12.5 + 4 = 16.5$ mm

Boss diameter $D_b = 2 d_h = 2 \times 16.5 = 33$ mm

b. B. M. at the center of the boss $M = F_p \times a = 526.3 \times 85 = 44.735 \times 10^3$ Nmm

Thickness of rocker arm is kept as length l_r and depth equal to D_b.

At center of boss $I = \dfrac{1}{12}(l_r)(D_b)^3 - \dfrac{1}{12}(l_r)(d_h)^3$

$$= \frac{15.6}{12} \times (33)^3 - \frac{15.6}{12} \times (16.5)^3 = 40.9 \times 10^3 \text{ mm}^4$$

$$\sigma_{b1} = \frac{My}{I} = \frac{44.735 \times 10^3 \times 16.5}{40.9 \times 10^3} = 10.94 \text{ MPa}$$

Bending moment, M at section $Y - Y$:

$$M = F_p(a - 0.5\,D_b) = 526.3 \times (85 - 16.5) = 36.05 \times 10^3 \text{ Nmm}$$

From Equation (16.31): $\sigma_{b2} = \dfrac{M \times 0.5\,D_b}{44\,t^4}$

Substituting the values: $65 = \dfrac{36.05 \times 10^3 \times 16.5}{44\,t^4}$ or $t = 3.8$ say 4 mm

16.6 Rocker Shaft

Rocker shaft is a shaft, on which the rocker arms are placed to suit the location of valve stem. Each cylinder has two valves and hence two rocker arms. Picture on right side shows a cylinder head with two rocker shafts, one for inlet valves and other for exhaust valves.

For the design of rocker shaft, reactions are to be calculated at each position and then the shaft diameter is calculated on the basis of bending moment due to these forces, considering it as a semi floating beam at the supports.

It may be noted, that in no case rocker shaft diameter should be more than rocker arm bush diameter d_r calculated by Equation (16.26), otherwise the bush will not be inserted.

16.7 Push Rod

Push rod is the weakest link in the valve gear train. Push rods transfer and redirect the upward motion of the cam in one direction, to the valves through rocker arms, in another direction. Consequently, they are subjected to bending forces as the load and engine speed increases.

Compression and deflection are the two fundamental forces acting on push rods. They deflect under high loads, because of differences in angularity between the cams and the cups or adjustment screws in the rocker arms. Oblique angles contribute to side loading, leading to deflection, if the push rod is not rigid.

If the push rod deflects, it reduces net lift of the valve, which hurts breathing and the engine's power potential. It also shortens valve duration and retards valve timing. Most of the deflection occurs at the lower end of the push rod, where the side loads are typically the greatest.

Push rod deflection can also be reduced using single taper or offset dual taper push rods. The thickest part of the push rod is positioned closest to the cam to take advantage of the larger cross section. Taper push rod represents a difference of about 2.5 per cent of effective weight. Taper provides a huge increase in valve gear train stability. The valve side of the valve train is the critical side, where any weight saving makes marked improvements. Something else to consider when choosing push rods is, whether the engine has guide for the push rod or not.

16.7.1 Materials

Materials used for push rod are:

- High carbon steel is used for slow-speed engines. For slow-speed engines, even mild steel push rods are adequate. But for performance applications, with stiffer valve springs and higher loads, stronger and stiffer push rods are needed to withstand the forces.

- Chromium and molybdenum alloy steel is used for stronger rods and for higher speeds. Tensile strength varies from 950 MPa up to 1,650 MPa depending on the heat treatment.

- Aluminum is lighter but softer than steel, or its alloys. If used for push rods, diameters are more. These push rods are quiet in an engine, due to good damping characteristics, but flex under extreme loads. These are not for high speeds, because of their deformation.

- Titanium is a less common material for push rods because of its higher cost and more limited availability. It offers lower weight than steel and better strength. These push rods resist flexing and deformation.

- Aluminum Matrix Composite called AMC in short combines Nextel Ceramic Fibres with aluminum, which is more than 40 per cent lighter, stiffer and offers better damping characteristics than chrome molybdenum steel.

Push rod is designed as a long column between rocker arm and cam. Since its weight is added to the reciprocating masses causing inertia forces, it is made hollow to reduce its weight. Ratio of inside to outside diameter ratio varies between 0.6 to 0.8.

Buckling weight is a way to define rigidity and resistance to bending in a push rod. It affects engine power, displacement and speed of engine in the calculations.

16.7.2 Design of push rod

For a hollow rod, let D be the outside diameter and d as inside diameter.

Cross-sectional area of hollow rod: $A = \dfrac{\pi}{4}\left(D^2 - d^2\right)$

Moment of inertia $I = \dfrac{\pi}{64}\left(D^4 - d^4\right)$ (16.32)

Radius of gyration $k = \sqrt{\dfrac{I}{A}} = \sqrt{\dfrac{\left(D^2 + d^2\right)}{16}}$ (16.33)

Design load F_{des} = Force on push rod F × Factor of safety, that is, $F_{des} = F \times FOS$

For the design of push rod, length is known and diameter is to be calculated considering it solid or hollow. Since L/k ratio for the push rod is large, it has to be designed considering buckling (Chapter 5, Section 5.11 in Volume 1) using Rankine's or Euler's formula given below:

Crippling force on push rod (Rankine's)

$$F_{ur} = \frac{\sigma_c A}{1 + a\left(L/k\right)^2}$$ (16.34)

Crippling force on push rod (Euler's)

$$F_{cre} = \frac{c\pi^2 EA}{\left(L/k\right)^2}$$ (16.35)

Where, σ_c = Safe compressive stress (N/mm²)

A = Area of rod (mm²)

L = Length of rod (mm)

a = A constant depending on end conditions and material. For mild steel $a = \dfrac{1}{7,500}$

End conditions depend on the shape of push rod at the ends. Guided push rods may have flat end or spherical in a hemispherical seat. F_{des} should be less than crippling load.

Euler's formula is valid only if L/k is more than 80. Because L/k ratio is not known, solution is done by the following method:

1. First assume a safe compressive stress according to type of rod from Table 16.4.
2. Calculate the diameter using Rankine's formula.

Table 16.4 Safe Compressive Stresses σ_c (MPa) for Solid / Hollow Push Rods with Different L/k Ratios

Type of Rod	L/k Ratios									
	10	20	30	40	50	60	70	80	90	100
Solid	140	98	68	47	33	24	19	15	13	10
Hollow	130	77	49	30	21	16	12	9	8	7

3. Calculate L/k ratio.
4. Check, if this ratio is less or more than 80. Then use appropriate formula.
5. Check the new L/k ratio gives safe stress less than the value given in Table 16.4. Repeat the above process with this new safe stress.

Example 16.6

Design of push rod for given load and length
Design a hollow push rod for a load of 1,500 N and length 350 mm. Assume ratio of inner to outer diameter 0.75.

Solution

Given $\quad F = 1,500\,\text{N} \qquad L = 350\,\text{mm} \qquad d/D = 0.75$

Step 1 Assuming safe stress initially $\sigma_c = 65$ MPa.

Step 2 Cross-sectional area of hollow rod $A = \dfrac{\pi}{4}\left(D^2 - (0.75D)^2\right) = 0.343D^2$

Radius of gyration $k^2 = \dfrac{D^2 + d^2}{16} = \dfrac{D^2 + (0.75D)^2}{16} = 0.0977D^2$

Using Rankine's equation: Crippling load, $F_{crr} = \dfrac{\sigma_c A}{1 + a\left(L/k\right)^2}$

$$1,500 = \dfrac{65 \times 0.343D^2}{1 + \dfrac{1}{7,500}\left(\dfrac{350^2}{0.0977D^2}\right)} = \dfrac{22.3D^2}{1 + \dfrac{167.2}{D^2}}$$

Or, $0.0149D^4 - D^2 - 167.2 = 0$ or $D^2 = 144$ or $D = 12$ mm

Step 3 $L/k = \dfrac{350}{12} = 29$. It is less than 80; hence Rankine formula is OK.

Step 4 Safe stress for this L/k ratio in Table 16.4 is 49 MPa, so take little less say 45 MPa.

Step 5 Using Rankine equation again for safe compressive stress 45 MPa:

$$1,500 = \frac{45 \times 0.343 D^2}{1 + \dfrac{1}{7,500}\left(\dfrac{350^2}{0.0977 D^2}\right)} = \frac{15.43 D^2}{1 + \dfrac{167.2}{D^2}}$$

$$0.00914 D^4 - D^2 - 167.2 = 0 \quad \text{or} \quad D^2 = 186 \quad \text{or} \quad D = 13.7 \text{ say } 14 \text{ mm}$$

$$L/k = \frac{350}{14} = 25. \text{ Safe stress for this ratio is more than 45 MPa, hence safe.}$$

16.8 Cam Shaft

A cam shaft contains many cams along its axis (Figure 16.10). Number of cams depends on the number of cylinders. If single cam shaft is used, two cams are provided for each cylinder; one for inlet valve and other for exhaust valve. An extra cam is for a fuel pump.

Cams are forged integral with the shaft. Their lobed portion is at predetermined orientation to actuate the valves at the desired timings. See the orientations of various cams in the side view. I1 is for inlet valve and E1 for exhaust valve for cylinder number 1. One end of the shaft is mounted with a timing gear for getting power from crank shaft. Cam shaft of a petrol engine is provided a bearing B1, B2, etc., after two cams, while for a diesel engine, a bearing is provided after every cam.

Diameter of cam shaft = (0.16 × Cylinder diameter) + 12.5 mm

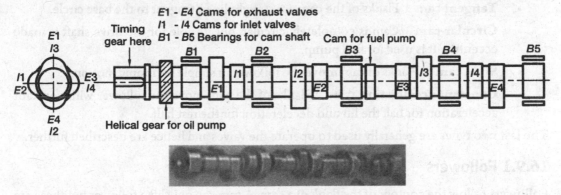

Figure 16.10 Cam shaft

16.9 Cams

Cam is a lobed machine element, which when rotated, moves its follower in a desired manner. In an I.C. engine, radial cams are used to operate inlet and exhaust valves. For a radial cam, follower moves perpendicular to the axis of rotation. Dwell period is the angle, when the follower remains stationary at its position. Its minimum radius is called base circle and its radius changes with change in angle, except at dwell period. Peripheral face is used to drive the follower. When the cam rotates, its lobed portion moves the follower radially outward during lift period. It then descends during the fall period back to the base circle. The tip at its top is called nose. A generalized cam is shown in Figure 16.11(a) and a cam used for an I. C. engine in Figure 16.11(b).

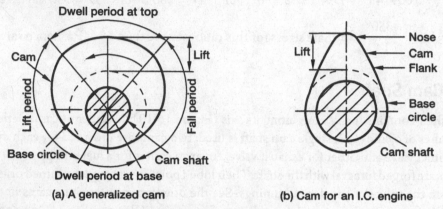

Figure 16.11 A radial cam

There are many types of cams. These are classified according to the shape and type of motion given by the cam. Velocity and acceleration during lift and fall period depend on the shape of the flank.

- **Tangent cam**   Flanks of the cam are straight lines, tangent to the base circle.
- **Circular cam**   Cam is completely circular, but its hole for the drive shaft is made eccentric. It is used for fuel pump.
- **SHM cam**   Flanks of cam move the follower in simple harmonic motion.
- **Constant acceleration cam**   Flanks of the cam move the follower with constant acceleration for half the lift and deceleration for the rest half.

The last two types are generally used to operate the valves and hence are described further.

16.9.1 Followers

Followers follow the contour of the flank of a cam. A cam can only lift a follower but these are brought back during fall period by the help of external force, such as spring or dead weight. Two important types of followers are:

Flat follower, which is just a circular disk as shown in Figure 16.12(a). If its diameter is large, then the radial lift given by the cam and distance moved by the follower may be different, because the edge of the follower may touch the cam not at the center line, but somewhere else. Flat follower is mounted slightly eccentric so that the friction between cam and follower gives a slight rotation to the follower. This makes the follower to wear uniformly. This type of follower is widely used in I. C. engines.

Roller follower is supported on a forked end of the push rod with a pin as shown in Figure 16.12(b). Its rolling action offers low friction. A roller follower modifies the lift of the cam, but to lesser extent than flat follower.

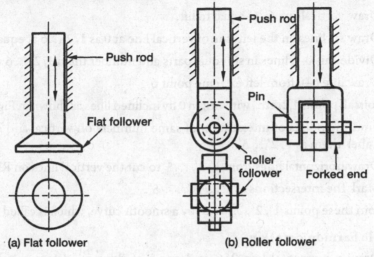

Figure 16.12 Types of followers

16.9.2 Lift diagrams

A lift diagram for the rise and fall period is necessary to draw a cam profile. It is just a plot, showing rotation angle on X axis and movement of the follower on Y axis. These diagrams are made full size so that the lift is transferred to the drawing of the cam profile directly. Construction for two types of motions, that is, constant acceleration is shown in Figure 16.13(a) and SHM in Figure 16.13(b). The left side of the vertical line at 0 shows the construction required.

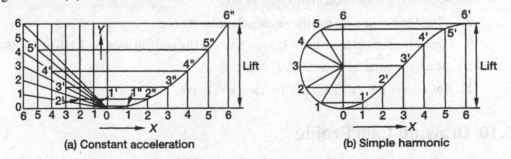

Figure 16.13 Construction for different lift diagrams

To draw a lift diagram, draw a horizontal line 0–6 and a vertical line at point at 0 equal to the lift (Figure 16.13). The stepwise procedure is given below:

a. **Constant acceleration**

1. Refer Figure 16.13(a). Draw a horizontal line 0–6 as the rise period or fall in degrees to a suitable scale.

2. Divide the base line 0–6 in six equal parts and number them as 1, 2, …6. Number of points can be increased, if base diameter is large.

3. Draw vertical line 6–6" equal to lift.

4. Draw six lines on the left side of vertical line at 0 as 1, 2,…6 at equal interval.

5. Divide this 0–6 lines in six equal parts and number them 1, 2, …6 as shown.

6. Draw line 6–0 from left extreme point 6.

7. Join all points 1,2, etc., with origin 0 by inclined lines as shown in Figure 16.13(a).

8. Find the points of intersection of same numbers on vertical and inclined lines. Label them as 1', 2',…5'.

9. Draw horizontal lines from 1', 2', …5' to cut the vertical lines on RHS of line 0.

10. Mark the intersections as 1", 2", ….. 5".

11. Join these points 1", 2", ….. 6" by a smooth curve, which is called a lift diagram.

b. **Simple harmonic motion**

1. Draw a horizontal line 0–6 as degrees of rise period to a suitable scale. See Figure 16.13(b).

2. Divide the base line 0–6 in six equal parts and number them as 1, 2, … 6.

3. Draw vertical line 6–6' equal to lift.

4. Draw lines on right side of vertical line 0–6 at points 1, 2,…5.

5. Draw line 0–6 at left point 0 of the base line.

6. Draw a semicircle with diameter 0–6 as shown in the figure.

7. Divide the semicircle at intervals of 30°.

8. Draw radial lines to cut the semicircle from its center at points 1, 2, 3, … 5.

9. Draw horizontal lines from 1, 2, … 5 to cut the vertical lines on RHS of line 0–6.

10. Mark the intersections as 1', 2', … 5'.

11. Join these points 1', 2', … 5' by a smooth curve.

16.10 Drawing Cam Profile

Shape of cam depends on the lift diagram and the types of follower. Figure 16.14(a) shows the lift diagram with SHM and Figure 16.14(b) shows cam profile with roller follower. Stepwise method to draw a cam profile is given in subsequent sections.

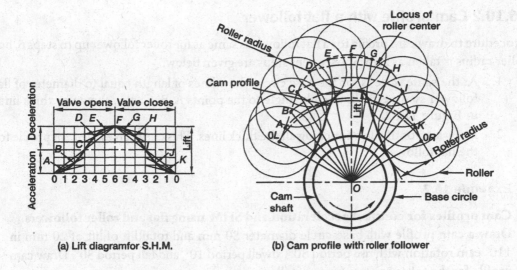

(a) Lift diagram for S.H.M.

(b) Cam profile with roller follower

Figure 16.14 Drawing cam profile for roller follower

16.10.1 Cam profile with a roller follower

Follow the steps given below:

1. Draw the lift diagram according to the motion of the follower required as described in Section 16.9.2. Refer Figure 16.14(a) for SHM. Lift duration is from 0 to 6 and then fall period from 6 to 0 on X axis. Each period is divided into six parts. The lifts are marked by letters A, B, ... K.

2. Draw center lines for X and Y axes and then base circle to a suitable scale.

3. Lift period in degrees is half the period of valve opening because the cam shaft rotates at half the engine speed. Draw radial lines O–0L as shown in Figure 16.14(b) at this angle on left and O–0R on right side of the vertical axis.

4. Divide the lift and fall period in equal number of radial lines as in lift diagram. Draw radial lines in six divisions in the figure as shown.

5. Draw an arc 0L–0R with center O and radius base circle + roller radius shown by center line.

6. Draw arcs at a distance equal to the lifts from the lift diagram parallel to arc 0L–0R. These arcs are also shown by centerlines. Mark points A, B, C, etc., where these arcs intersect the radial lines. The radial lines are O–A, O–B,... O–F for lift period and O–G, O–H, ... O–K for fall period.

7. Draw circles of follower radius with centers A, B, C, etc., shown by thin lines.

8. Draw a smooth curve shown by thick line touching all the circles. This curve gives the cam profile for the given roller radius and the type of motion desired.

16.10.2 Cam profile with a flat follower

Procedure to draw cam profile for a flat follower is same as for roller follower up to steps 6, but roller radius is taken as zero. Subsequent steps are given below:

1. At the same scale as for the base circle, draw lines of length equal to diameter of flat follower symmetrically at right angle to the points A, B, C, etc., shown by thick lines in Figure 16.S2.

2. Draw a smooth curve touching all the thick lines. This curve gives the cam profile for the flat follower.

Example 16.7

Cam profiles for constant acceleration and SHM using flat and roller followers
Draw a cam profile with base circle diameter 50 mm and total lift of lift of 20 mm in 110° cam rotation with rise period 50°, dwell period 10°, and fall period 50°. Draw cam profile for the following two types of followers:

a. Constant acceleration and deceleration with flat follower of diameter 15 mm.

b. SHM with roller follower of diameter 30 mm.

Solution

Given $D = 50$ mm $\quad l = 20$ mm $\quad \theta = 110°$ $\quad \theta_r = 50°$

$\theta_d = 10°$ $\quad \theta_f = 50°$ $\quad d_r = 30$ mm

a. Lift diagram with constant acceleration for a lift of 20 mm is shown in Figure 16.S2(a). The cam profile drawn by the method described in Section 16.10.2 is shown in Figure 16.S2(b).

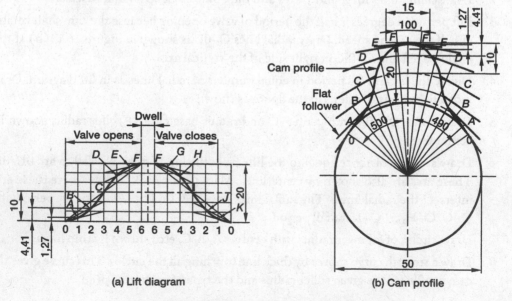

(a) Lift diagram   (b) Cam profile

Figure 16.S2 Drawing cam profile with flat follower

b. Lift diagram with SHM for a lift of 20 mm is shown in Figure 16.S3(a). The cam profile drawn by the method described in Section 16.10.1 is shown in Figure 16.S3(b).

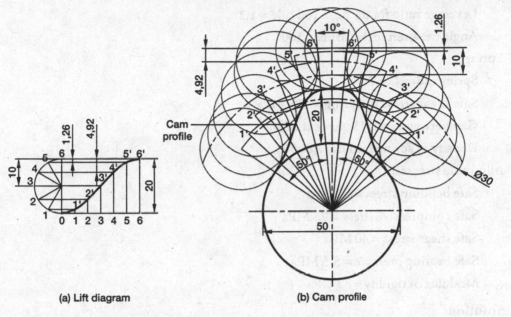

(a) Lift diagram   (b) Cam profile

Figure 16.S3 Drawing cam profile with roller follower

Example 16.8

Design of valve gear components for an engine of given size

Design valve gear components for an exhaust valve for an I head vertical four-stroke I.C. engine with the following data: Speed = 600 rpm, cylinder diameter = 180 mm, and stroke = 250 mm. Cam is directly mounted over the roller of rocker arm, that is, no push rod.

Valve data:

> Maximum gas pressure = 3.8 MPa
>
> Weight of valve = 0.35 kg
>
> Pressure when exhaust valve opens = 0.35 MPa
>
> Valve opens 40° before BDC and closes at 5° after TDC
>
> Seat angle of valve = 45°
>
> Suction pressure = 0.02 MPa below atmosphere
>
> Valve opens and closes with SHM
>
> Valve seat pressure = 0.07 MPa
>
> Gas velocity at port = 40 m/s

Rocker arm:

Rocker arm length for tappet screw from fulcrum pin = 110 mm

Leverage ratio for the push rod side = 1.2

Angle between rocker arms = 160°

Spring:

Spring index $C = 7$

Safe shear stress for spring = 385 MPa

Gap between coils 15 per cent of lift

Density of spring material 7,800 kg/m³

Allowable stresses:

Safe bending stress = 65 MPa

Safe compressive stress = 75 MPa

Safe shear stress = 40 MPa

Safe bearing pressure = 5.5 MPa

Modulus of rigidity = 82 GPa

Solution

Given

$N = 600$ rpm	$D = 0.18$ m	$L = 0.25$ m
$P_{max} = 3.8$ MPa	$P_s = 0.02$ MPa	$P_e = 0.35$ MPa
$m = 0.35$ kg	$\theta = 40° + 180° + 5° = 225°$	
$\alpha = 45°$	$v_{ge} = 40$ m/s	$p_s = 0.07$ MPa
$a = 0.11$ m	$b = 1.2a$	$\beta = 160°$
$C = 7$	$\sigma_b = 65$ MPa	$\sigma_c = 75$ MPa
$p_b = 5.5$ MPa	$\tau = 40$ MPa	For spring $\tau = 385$ MPa
$G = 82$ GPa	Gap = 15 % of lift	$\rho = 7,800$ kg/m³

a. Ports

Piston speed $v = \dfrac{2LN}{60} = \dfrac{2 \times 0.25 \times 600}{60} = 5$ m/s

Volume flow of gas $Q = v \times \dfrac{\pi}{4}D^2 = 5 \times 0.785 \times 0.18^2 = 0.1272$ m³/s

Exhaust valve port area $A_{pe} = \dfrac{Q}{v_{ge}} = \dfrac{0.1272}{40} = 3.18 \times 10^{-3}$ m²

Or, $\dfrac{\pi}{4}\left(d_{pe}\right)^2 = A_{pe} = 3.18 \times 10^{-3}$ or Exhaust port diameter, $d_{pe} = 0.064$ m = 64 mm

b. Valve size

Projected width of seat $w = 0.06\, d_{pe} = 0.06 \times 64 = 3.84$ mm

Valve diameter $d_{ve} = d_{pe} + 2w = 64 + (2 \times 3.84) = 71.7$ say 72 mm

Valve thickness $t = 0.4 d_p \sqrt{\dfrac{p_{max}}{\sigma_b}} = 0.4 \times 64 \sqrt{\dfrac{3.8}{65}} = 6.2$ mm

Valve stem diameter $d_s = 0.25\, d_p = 0.25 \times 64 = 16$ mm

Valve lift $l = \dfrac{d_p}{4 \cos \alpha} = \dfrac{64}{4 \times \cos 45°} = 22.6$ say 23 mm

c. Forces on valve

Initial force $F_i = \dfrac{\pi}{4} (d_v)^2\, p_s + w = 0.785 \times 72^2 \times 0.02 + 3.5 = 84.9$ N

Gas force $F_g = \dfrac{\pi}{4} (d_v)^2\, p_c - w = 0.785 \times 72^2 \times 0.35 - 3.5 = 1{,}420.8$ N

From Equation (16.12): Acceleration for SHM $\alpha = 710\, l \left(\dfrac{N}{\theta}\right)^2$

Substituting the values: $= 710 \times 0.023 \left(\dfrac{600}{225}\right)^2 = 116$ m/s^2

Inertia force $F_a = m\alpha = 0.35 \times 116 = 40.6$ N

Total force on exhaust valve is

$$F_{ve} = F_i + F_g + F_a = 84.9 + 1420.8 + 40.6 = 1{,}546.3 \text{ N}$$

d. Fulcrum pin

Force on the other side of rocker arm $F_p = \dfrac{F_{ve} \times a}{1.2a} = \dfrac{1{,}546.3}{1.2} = 1{,}288.6$ N

Reaction on fulcrum pin $R = \sqrt{(F_p)^2 + (F_e)^2 - 2 F_p F_e \cos \beta}$

Or, $R = \sqrt{(1{,}288.6)^2 + (1{,}546.3)^2 - (2 \times 1{,}288.6 \times 546.3 \times \cos 160°)} = 2{,}791$ N

Consider double shear at the pin and find diameter from shear criterion.

$$R = 2 \times \frac{\pi}{4} (d_r)^2 \times \tau$$

Substituting the values: $2{,}791 = 2 \times 0.785\,(d_r)^2 \times 40$ or $d_r = 6.3$ mm

Assuming value of $l_r = 1.2\, d_r$, Reaction on pin, $R = d_r \times 1.2\, d_r \times p_b$

Substituting the values find diameter from bearing criterion:

$$2{,}791 = 1.2 \, (d_r)^2 \times 5.5 \qquad \text{or} \qquad d_r = 20.6 \text{ say } 21 \text{ mm}$$

Select the bigger diameter from shear and bearing criteria, that is, 21 mm.

Hence, length of boss $l_r = 1.2 \, d_r = 1.2 \times 21 = 25$ mm

Hole diameter $d_h = d_r + 6$ mm $= 21 + 6 = 27$ mm

Boss diameter $D_b = 2 \, d_h = 2 \times 27 = 54$ mm

e. Rocker arm

Width cannot be more than length of boss; hence width is taken as 25 mm.
Refer Figure 16.S4.

Moment of inertia at boss $I = \dfrac{1}{12}\left(l_r\right)\left(D_b\right)^3 - \dfrac{1}{12}\left(l_r\right)\left(d_h\right)^3$

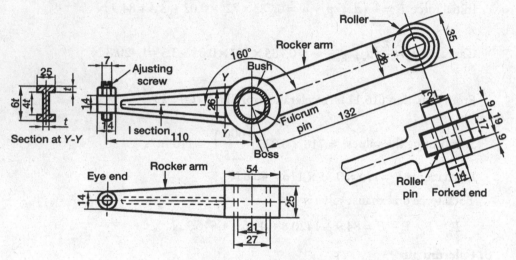

Figure 16.S4 Rocker arm

Substituting the values:

$$I = \frac{1}{12} \times 25 \times (54)^3 - \frac{1}{12} \times 25 + (27)^3 = 287.1 \times 10^3 \text{ mm}^4$$

Modulus of section $Z = \dfrac{I}{y} = \dfrac{287.1 \times 10^3}{27} = 10.63 \times 10^3 \text{ mm}^3$

Bending moment $M = a \times F_{ve} = 110 \times 1{,}546.3 = 170 \times 10^3 \text{ Nmm}$

Bending stress $\sigma_b = \dfrac{M}{Z} = \dfrac{170 \times 10^3}{10.63 \times 10^3} = 16 \text{ MPa (Less than allowable, so safe)}$

For section of rocker arm, I section of height 6 t and width 25 mm is used. Refer Figure 16.8.

Moment of inertia at $Y - Y$,

$$I = \frac{1}{12} \times 25 \times (6t)^3 - \frac{1}{12} \times (25 - t) \times (4t)^3 = (316t^3 + 5.3t^4) \text{ mm}^4$$

Modulus of section $Z = \dfrac{I}{y} = \dfrac{(316t^3 + 5.3t^4)}{3t} = 1.7t^3 + 105.2t^2 \text{ mm}^3$

Bending moment $M = (a - 0.5 D_b) \times F_v = (110 - 27) \times 1{,}546.3 = 128.35 \times 10^3 \text{ Nmm}$

Bending stress

$$\sigma_b = \frac{M}{Z} = \frac{128.35 \times 10^3}{1.7t^3 + 105.2\, t^2} \quad \text{or} \quad 65 = \frac{128.35 \times 10^3}{1.7t^3 + 105.2t^2} \quad \text{By trial and error:} \quad t = 4.2 \text{ mm}$$

Hence depth of section $= 6\, t = 6 \times 4.2 = 26$ mm

f. Eye end of rocker arm

Core diameter d_c of tappet screw: $F_v = \dfrac{\pi}{4}(d_c)^2 \times \sigma_c$

Substituting the values:

$1{,}546.3 = 0.785\,(d_c)^2 \times 75 \quad \text{or} \quad d_c = 5.1$ mm

Screw diameter $= \dfrac{d_c}{0.8} = \dfrac{5.1}{0.8} = 6.4$ say 7 mm

Diameter of eye end $= 2 \times$ Screw diameter $= 2 \times 7 = 14$ mm

Depth of eye end $=$ diameter of eye end $= 14$ mm

g. Forked end of rocker arm

Load on roller pin F_p calculated in step (d) $= 1{,}288.6$ and bearing pressure is 5.5 MPa.

Assuming length of pin 1.2 times diameter $1{,}288.6 = d_p \times 1.2\, d_p \times 5.5$

Or, $d_p = 14$ mm

A bush of phosphorous bronze of thickness 3 mm is inserted over the pin.

Diameter of the eye $= 2\, d_p + 6 = (2 \times 14) + 6 = 34$ mm

Hence, the hole in the roller $= d_p + 6 = 14 + 6 = 20$ mm

Roller thickness $=$ Length of pin $l_p = 1.2 \times d_p = 1.2 \times 14 = 16.8$ say 17 mm

Assuming a clearance of 1 mm on each side of roller [See Figure (16.S4)].

Gap between forked eyes $= 17 + 2 = 19$ mm

Thickness of each eye $= 0.5\, l_p = 0.5 \times 17 = 8.5$ say 9 mm

Total length of pin = 9 + 1 + 17 + 1 + 9 = 37 mm

Head diameter of roller pin = 1.5 d_p = 1.5 × 14 = 21 mm

Thickness of head of roller pin = 0.5 d_p = 0.5 × 14 = 7 mm

Assuming uniform load from roller to pin and triangular variation of reactions in forked eyes,

Bending moment $M = \dfrac{5}{24} F_p l_p$ = 0.208 × 1,288.6 × 19 = 5,101 Nmm

Modulus of section $Z = \dfrac{\pi}{32} \left(d_p \right)^3$ = 0.098 × 14³ = 269 mm³

Bending stress $\sigma_b = \dfrac{M}{Z} = \dfrac{5,101}{269}$ or σ_b = 18.9 MPa (Less than allowable, hence safe)

h. Spring

Load on spring $F_{max} = \dfrac{\pi}{4} \left(d_v \right)^2 \times 0.07 = 0.785 \times 72^2 \times 0.07 = 284.8$ N

Stiffness of spring $k = \dfrac{F_{max}}{l} = \dfrac{284.8}{23} = 12.38$ N/mm

Wahl factor $K = \dfrac{4C - 1}{4C - 4} + \dfrac{0.615}{C} = \dfrac{(4 \times 7) - 1}{(4 \times 7) - 4} + \dfrac{0.615}{7} = 1.213$

Wire diameter $d = \sqrt{\dfrac{8 K F_{max} C}{\pi \tau}} = \sqrt{\dfrac{8 \times 1.213 \times 284.8 \times 7}{3.14 \times 385}} = 4$ mm

Actual number of turns: $n = \dfrac{G d^4}{8 D^3 k} = \dfrac{82 \times 10^3 \times 4^4}{8 \times \left(7 \times 4 \right)^3 \times 12.38} = 9.6$ say 10

Total number of turns: $n' = n + 2 = 10 + 2 = 12$ turns

Solid length of spring $SL = (d \times n') = 4 \times 12 = 48$ mm

Free length of spring $FL = SL + 1.15\, l = 48 + (1.15 \times 23) = 74.5$ mm

To allow for adjustment free length = 80 mm

Mass of the spring m = Area of wire × Length of wire × Density $= \dfrac{\pi}{4} d^2 \left(\pi D n \right) \rho$

Or, $m = 0.785 \times (4)^2 \{3.14 \times (4 \times 7) \times 10\} \times (7,800 \times 10^{-9}) = 0.0861$ kg

Natural frequency $f_n = \dfrac{1}{2\pi} \sqrt{\dfrac{k \times g}{m}} = \dfrac{1}{2\pi} \sqrt{\dfrac{34.6 \times 9,810}{0.0861}} = 316$ c/s

Natural frequency of spring should be at least 13 times more than cam shaft speed 300 rpm or 5 c/s in this example, hence it is safe.

i. **Cam:** Diameter of cam shaft is taken $= 0.16\,D + 12.5 = (0.16 \times 180) + 12.5 = 41.5$ mm

Base circle diameter of cam should be 3 mm to 5 mm more than cam shaft diameter; hence it is taken as 45 mm. Draw the lift diagram for a lift of 23 mm as calculated above. It is shown in Figure 16.S5(a).

Draw the cam profile for roller follower of diameter 35 mm as described in section 16.10.1. The lift diagram and cam profile are shown in Figure 16.S5.

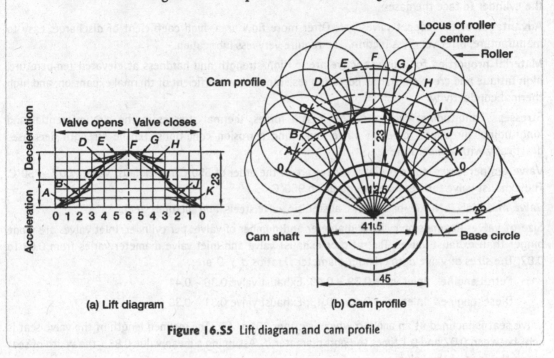

(a) Lift diagram   (b) Cam profile

Figure 16.S5 Lift diagram and cam profile

Summary

Valve gear mechanism in an engine is used to operate inlet and outlet valves. In a four-stroke cycle engine, the cycle is completed in two revolutions, and hence the drive shaft for the valves called cam shaft rotates at half the speed of crank shaft. A set of helical gears with gear ratio two is used or a toothed belt is used to drive the cam shaft. The cams mounted on this shaft move the cam follower up and down. The motion is transferred through push rod and rocker arm to the valve. The rocker arm transfers upward movement of push rod to downward movement of the valve to open. The valves are spring loaded and close by themselves due to the action of spring. In an L head engine, the valves are placed by the side of cylinder. The valves are inverted. This reduces the inertia effects. In new high-speed engines, cam shaft is directly placed over the valves, which gives minimum inertia effects.

Port is a passage, which connects the inlet manifold up to the valve. Port diameter depends on allowable gas velocity given in Table 16.1. For a stroke L, mean piston velocity $v = \dfrac{2\,L\,N}{60}$

Volume flow of gas = Piston speed × Area of cylinder of diameter D. Hence, flow $Q = v \times \dfrac{\pi}{4} D^2$

Also, discharge $Q =$ Port area × Gas velocity $= A_p \times v_g = \dfrac{\pi}{4}(d_p)^2 \times v_g$

Port diameter $d_p = \sqrt{\dfrac{v}{v_g}}\, D$

Types of valves used for I.C. engine are poppet valves. The valve is guided by a bush fixed in the cylinder head. In poppet valve, head is a circular disk with a margin of about 1 mm and then an inclined face, which fits tightly against the valve seat. Bottom portion of the valve called face, is inside the cylinder to face the gases.

Advantages of the poppet valves are: Offer more flow area, high coefficient of discharge, easy to manufacture, offer very less friction, and require very less lubrication.

Material properties for exhaust valve are: High strength and hardness at elevated temperature, high fatigue and creep resistance, corrosion resistance, low coefficient of thermal expansion, and high thermal conductivity.

Stresses Longitudinal cyclic stresses, inertia forces, thermal stresses in the circumferential and longitudinal directions, creep at valve head, and corrosion conditions. Tensile strength decreases drastically with temperature.

Valve temperatures inside the intake valve are of the order of 550°C and exhaust valve 700°C–750°C. The exhaust valve temperature can go up to 900°C.

Valve materials Nickel–base alloys, austenitic valve steels, and silcrome steel.

Size of valves depends on cylinder diameter and number of valves per cylinder. Inlet valves are made bigger than exhaust valves. The ratio of exhaust valve and inlet valve diameter varies from 0.83 to 0.87. The sizes of valve d_v to cylinder diameter D ratios d_v / D are:

    Petrol engine   Inlet valve 0.39 – 0.49; Exhaust valve 0.36 – 0.41

    Diesel engine   Inlet valve 0.45 – 0.49; Exhaust valve 0.31 – 0.39

Valve seat is inclined at an angle α which is either 30° or 45°. The inclined length of the valve seat is kept between 0.07 and 0.1 times the port diameter d_p. Assuming a mean value $0.85 d_p$ the width of seat along the face of the valve:

    $w = 0.06 d_p$ to $0.075 d_p$

Thus for inlet valve diameter $d_v = 1.15 d_p$ for exhaust valve $d_v = 1.12 d_p$

Lift of valves is calculated assuming that gas flow velocity remains same from port to valve exit.

    Maximum lift of valve $l = \dfrac{d_p}{4 \cos \alpha}$

    Lift of the valve is about 25 per cent of valve diameter

Thickness of valve disk is calculated for safe bending stress $t = 0.45 d_p \sqrt{\dfrac{p_{max}}{\sigma_b}}$

Where, $p_{max} =$ Maximum gas pressure cylinder

    $d_p =$ Port diameter

    $\sigma_b =$ Safe bending stress

Empirically total thickness of head $t = 0.10 d_p$ to $0.14 d_p$

The edge of disk called margin $t_m = 0.025d_p$ to $0.045d_p$ (Minimum 0.7 mm)

Inclined seat length $c = 0.10d_p$ to $0.12d_p$

Size of valve stem for inlet valve $d_s = 0.18d_p$ to $0.24d_p$. For exhaust valve $d_s = 0.22d_p$ to $0.28d_p$.

Valve timings   Valve opening: Intake valve 230°– 270°, Exhaust value 235° – 285°.

Forces on valve   There are three types of forces acting on a valve:

a.   *Spring force*   Initial spring force $F_i = \dfrac{\pi}{4}(d_{vi})^2 p_s + w$, Where, w = weight of piston

b.   *Gas force* $F_g = \dfrac{\pi}{4}(d_{ve})^2 p_c - w$

c.   *Inertia force*   Assumptions for opening / closing of valves are: SHM or constant acceleration.

SHM   Inertia force due to acceleration $F_a = 710\, ml\left(\dfrac{N}{\theta}\right)^2$

Where, m = Mass of the valve in Kg

   N = Speed in rpm

   θ = Valve opening period in degrees

   l = Total lift in mm

Constant acceleration   Acceleration $\alpha = \dfrac{16l}{t^2}$ where, t = Total time for valve to remain open.

   Inertia force due to acceleration $F_a = m\alpha$

   Thus total force on exhaust valve is $F_{ve} = F_i + F_g + F_a$

   Total force on inlet valve is $F_{vi} = F_i + F_a$

Valve spring   Initial minimum force on spring $F_{min} = \dfrac{\pi}{4}(d_{vi})^2 p_s + w$

Where, p_s = Suction pressure, and w = weight of the valve.

$$\text{Gas force } F_g = \frac{\pi}{4}(d_{ve})^2 p_c + w$$

Hence, maximum force on the valve is: $F_{max} = F_{min} + F_g = \dfrac{\pi}{4}\left[(d_{vi})^2 p_s + (d_{ve})^2 p_c\right]$

When a spring is compressed and released suddenly it vibrates longitudinally till its energy is damped by air friction. This frequency is called natural frequency f_n

$$f_n = \frac{d}{2\pi n D^2}\sqrt{\frac{G \times g}{2\rho}} \quad \text{Where, } p \text{ is the density of material}$$

Design procedure for spring

1.   Calculate minimum force on the spring using equation $F_{min} = \dfrac{\pi}{4}(d_{vi})^2 p_s + w$

2.   Force required to open this valve $F_{max} = F_{min} + k\delta$

3.   Calculate wire diameter d using equation: $\tau = \sqrt{\dfrac{8K F_{max} C}{\pi \tau}}$ where, τ = Allowable shear stress

   C = Spring index D/d, D = Mean coil diameter, and K = Wahl factor $= \dfrac{4C - 1}{4C - 4} + \dfrac{0.615}{C}$

4. Calculate actual number of turns: $n = \dfrac{G\,d^4}{8\,D^3\,k}$

    Total number of turns $n' = n + 2$

5. Free length of spring $FL = SL + \text{Gap} = (d \times n') + 1.5\,(n' - 1)$; where $SL = $ Solid length $(d \times n')$

6. Calculate natural frequency $\omega_n\ (= 2\,\pi\,f_n)$ of spring: $\omega_n = \dfrac{1}{2\pi}\sqrt{\dfrac{k \times g}{m}}$

7. If it is near to disturbing cycle frequency, use two concentric springs.

Rocker arm   It is used to convert the direction of force from push rod to the axis of the valve stem at its tip. Fulcrum is provided with a hole for a phosphorous bronze bush of thickness 2 mm–3 mm. Rocker shaft fits in the hole of the bush. Arm lengths from the center of the boss are different on each side to provide some ratio called rocker arm ratio. New automotive rocker arm ratios are of about 1.5:1 to 1.8:1

Material used for rocker arm are steel stampings, ductile cast iron, or forged steel.

To reduce force on cam, some leverage (b/a) is provided. Thus force on push road side $F_p = \dfrac{a}{b}F_e$

For straight rocker arm, reaction at the fulcrum is $R = F_p + F_e$

If arms are inclined and make angle β then reaction is:

$$R = \sqrt{\left(F_p\right)^2 + \left(F_e\right)^2 - 2 F_p\,F_e \cos\beta}$$

Consider double shear at the rocker shaft $R = 2 \times \dfrac{\pi}{4}\left(d_r\right)^2 \times \tau$

For safe bearing pressure $R = d_r\,l_r\,p_b$. Assuming $l_r = 1.25\,d_r$, $R = 1.25\,(d_r)^2\,p_b$

Phosphorus bronze bush of thickness 2 mm–3 mm is placed in hole. So hole diameter $d_h = d_r + 4$ mm.

   Boss diameter is kept $D_b = 2d_h$

   BM at the center of the boss $M = F_p \times b$

Thickness of rocker arm is kept as length l_r and depth equal to D_b.

Mass moment of inertia for I section with width $3t$ and depth $6t$ is: $I = 44\,t^4$

BM at the end of boss at section $(y - y)$ will be, $M = F_p(a - 0.5 D_b)$

Bending stress $\sigma_b = \dfrac{My}{I} = \dfrac{F_p(a - 0.5 D_b) \times (0.5 D_b)}{44 t^4}$

Calculate t from safe bending stress value.

Push rods are subjected to bending forces as the load and engine speed increases. It is designed as a long column between rocker arm and cam. To reduce its weight, it is made hollow. Ratio of inside to outside diameter ratio varies between 0.6 and 0.8.

Materials used for push rod are: High carbon steel, chromium and molybdenum alloy steel, and aluminum matrix composite.

Design of push rod   For hollow rod, of outside diameter D and inside diameter d.

Radius of gyration $k = \sqrt{\dfrac{I}{A}} = \sqrt{\dfrac{\left(D^2 + d^2\right)}{16}}$

Design load F_{des} = Force on push rod $F \times FOS$

Since L/k ratio for the push rod is large, it has to be designed considering buckling using Rankine's or Euler's formula given below:

Crippling force (Rankine's) $F_{crr} = \dfrac{\sigma_c A}{1 + a(L/k)^2}$

Crippling force (Euler's) $F_{cre} = \dfrac{c\pi^2 EA}{(L/k)^2}$

a = A constant depending on end conditions and material. For M.S., $a = \dfrac{1}{7,500}$

Euler's formula is valid only if L/k is more than 80. Because L/k ratio is not known, solution is done by the following method:

1.   First assume a safe compressive stress according to type of rod from Table 16.4.
2.   Calculate the diameter using Rankine's formula.
3.   Calculate L/k ratio.
4.   Check, if this ratio is less or more than 80. Then use appropriate formula.
5.   Check if the new L/k ratio gives safe stress less than the value. Repeat the above process with this new safe stress.

Cam shaft   It contains many cams along its axis. Number of cams depends on the number of cylinders. Cams are forged integral with the shaft. Their lobed portion is at predetermined orientation to actuate the valves at the desired timings.

Cam shaft diameter = (0.16 × Cylinder diameter) + 12.5 mm

Cam   It is a lobed machine element, which when rotated, moves its follower in a desired manner. In an I.C. engine, radial cams are used to operate inlet valve, exhaust valves. For a radial cam, follower moves perpendicular to the axis of rotation. Dwell period is the angle, when the follower remains at its position. Its minimum radius is called base circle and its radius changes with change in angle except at dwell period. When the cam rotates, its lobed portion, moves the follower radially outward during lift period. The tip at its top is called nose.

Types of cams

SHM cam   Flanks of cam move the follower in simple harmonic motion.

Constant acceleration cam: Flanks of cam move the follower with constant acceleration.

Followers   They follow the contour of the flank of a cam. A cam can only lift a follower but these are brought back by the help of external force, such as spring force or dead weight.

Two important types of followers are: Flat follower and roller follower.

Lift diagrams   It is used for the rise and fall period to draw a cam profile. It is just a plot, showing rotation angle on X axis and movement of the follower on Y axis.

Drawing cam profile   Following stepwise method describes to draw a cam profile.

1.   Draw the lift diagram according to the motion of the follower.
2.   Draw center lines for X and Y axis and then draw base circle to a suitable scale.

3. Mark lift period in degrees equal to period of valve opening symmetrically about vertical axis. Draw radial lines at suitable number of equal intervals.

4. Draw arcs at a distance equal to the roller radius + lifts (from the lift diagram) parallel to base circle. For flat follower, this radius is taken as zero.

5. At the intersections of radial lines and corresponding arcs, draw circles of follower radius or flat lines of diameter of flat follower.

6. Draw a smooth curve touching all the circles or flat follower lines.

Theory Questions

1. Name the components involved in a valve gear mechanism of an I head system with a simplified sketch.

2. How does a valve system of an I head engine differs from L head engine? Explain with the help of a diagram.

3. Describe the method to calculate the size of ports.

4. What are the materials used for the valves? What are the temperatures involved for an exhaust valve?

5. Sketch a multi valve arrangement and write the range of sizes of the inlet and exhaust valves in relation to cylinder diameter.

6. How do you calculate valve diameter, thickness, and stem diameter?

7. Sketch a valve and port to demonstrate the gas flow area and lift of valve.

8. What do you mean by valve timing diagram? Sketch one for a normal speed engine.

9. What are various forces acting on inlet and exhaust valve?

10. How do you calculate time in seconds for opening of a valve from the valve timing diagram and speed of engine?

11. Discuss acceleration involved for the valve movement and the assumptions made to calculate it. How do you calculate inertia force?

12. Write the design method to design a spring for a valve.

13. Sketch a rocker arm with some included angle between the arms. Write the method to design section of arm, fulcrum pin, and both the ends of rocker arm.

14. What are the materials used for push rod? Mention the forces coming on it and its design method considering buckling.

15. Sketch a cam shaft for a four-cylinder engine. How many cams are used on it?

16. Describe a generalized shape of a cam. How does it differ from the cams used for engine valves?

17. Sketch and describe the different types of followers used for cams of an engine.

18. Write the procedure with a diagram to draw a lift diagram for SHM of valve.

19. Write the stepwise method to draw a lift diagram with constant acceleration and deceleration for a given lift and valve opening angle.

20. Describe the method to draw cam profile from the given lift, base circle, and roller diameter. How does the method differ for a flat follower?

Multiple Choice Questions

1. For a high-speed engine the most efficient engine is:
   - (a) I head
   - (b) L head
   - (c) Overhead cam shaft
   - (d) Horizontal engine

2. Port is a passage:
   - (a) From air filter to engine manifold
   - (b) In the inlet manifold
   - (c) In the cylinder head
   - (d) None of the above

3. A suitable material for an exhaust valve is:
   - (a) Aluminium
   - (b) Copper
   - (c) Mild steel
   - (d) Silcrome steel

4. Diameter of valve is calculated from port diameter d_p as:
   - (a) $0.85\, d_p$
   - (b) $0.9\, d_p$
   - (c) Equal to d_p
   - (d) $1.1\, d_p$

5. Thickness of the valve depends on:
   - (a) Maximum gas pressure in cylinder
   - (b) Mean effective pressure
   - (c) Back pressure during exhaust stroke
   - (d) Suction pressure

6. Lift of a valve depends on:
   - (a) Port diameter
   - (b) Port diameter and gas velocity
   - (c) Valve diameter
   - (d) Port diameter, gas velocity, and seat angle

7. Force on exhaust valve is:
   - (a) Spring force
   - (b) Spring force + gas force
   - (c) Spring force + gas force + inertia force
   - (d) Spring force + gas force − inertia force

8. For a high-speed engine cross section of rocker arm is kept:
   - (a) T section
   - (b) I section
   - (c) Rectangular
   - (d) Circular

9. A push rod is made:
   - (a) Square
   - (b) Rectangular
   - (c) Solid circular
   - (d) Hollow circular

10. Cam profile can be based on assumption:
   - (a) SHM only
   - (b) Constant acceleration only
   - (c) Constant deceleration only
   - (d) Either given in (a) or given in (b) and (c)

Answers to multiple choice questions

1. (c)   2. (c)   3. (d)   4. (d)   5. (a)   6. (d)   7. (c)   8. (b)   9. (d)   10. (d)

Design Problems

1. An engine having cylinder diameter 75 mm and stroke 80 mm runs at 3,000 rpm. Assuming maximum gas pressure 4.5 MPa, gas velocity 22 m/s, and allowable stress 40 MPa, calculate:

    a. Diameter of port.

    b. Size of inlet valve, if seat angle is 30°.

    c. Valve lift

    $$[\text{(a)} \, d_p = 45 \text{ mm} \quad \text{(b)} \, d_v = 52 \text{ mm}, t = 7.5 \text{ mm} \quad \text{(c)} \, l = 13 \text{ mm}]$$

2. Determine the lift, head, and diameter of a valve having maximum gas pressure 5.5 MPa, cylinder diameter 85 mm, and mean piston speed 25 m/s. Assume allowable strength of valve 40 MPa and seat angle 30° $[l = 11 \text{ mm}, t = 7 \text{ mm}, d_v = 54 \text{ mm}]$

3. A hollow push rod of 300 mm length with ratio of inside to outside diameter 0.75 has to transfer an axial load of 2 kN. Assuming an *FOS* 2.5 and Young's modulus of elasticity 210 kN / mm², calculate the outside diameter of the rod. $[D = 9 \text{ mm}]$

4. A four-stroke I.C. engine running at 1,600 rpm has suction pressure of 0.12 MPa and cylinder pressure at the time of opening exhaust valve 0.55 MPa has the following data for the valves:

	Inlet valve diameters	Exhaust valve
Valve diameter	75 mm	67 mm
Lift	20 mm	18 mm
Mass of the valve	175 gm	150 gm
Valve opening duration	205°	235°

    Calculate the forces on the valves. $[F_{vi} = 684 \text{ N}, F_{ve} = 2{,}027 \text{ N}]$

5. Design a helical compression spring for a four-stroke petrol engine running at 1,800 rpm. Inlet and outlet valve diameters are 35 mm and 39 mm, respectively. Suction pressure is 0.1 MPa and gas pressure at opening of exhaust valve is 0.45 MPa. Valve opening duration for inlet valve is 210° and for exhaust valve 235° and mass is 85 gm and 75 gm, respectively. Lift of both the valves is 10 mm. Take safe value of shear stress 250 MPa and spring index 7. Gap between the coils in open condition is 1.4 mm. Density of the spring wire material is 7,800 kg/mm³. Calculate its natural frequency also.

    $$[d = 6 \text{ mm}, D = 42 \text{ mm}, k = 21 \text{ N/mm}, n' = 10, FL = 72.6 \text{ mm}, f_n = 43 \text{ c/s}]$$

6. An I.C. engine uses valve spring such that load at open and closed conditions are 300 N/mm and 180 N/mm, respectively. Spring length in closed condition of the valve is 45 mm and lift is 9 mm. Allowable safe shear stress is 320 MPa and modulus of rigidity 82 GPa. Design the spring if due to space limitations outside diameter of spring should not be more than 30 mm and gap between the coils is 15 per cent of deflection. $[d = 4.5 \text{ mm}, k = 13.3 \text{ N/mm}, n' = 14, FL = 88.9 \text{ mm}]$

7. Design an I section rocker arm for the exhaust valve of a four-stroke I.C. engine with the following data:

    Speed of engine 1,000 rpm

    Length of arm on valve and push rod sides are 65 mm and 75 mm, respectively.

    Angle between the arms = 150°

Valve diameter = 60 mm

Valve lift = 18 mm

Mass of valve and parts moving with it = 0.3 kg

Suction pressure = 0.022 MPa

Gas pressure at exhaust valve opening = 0.45 MPa

Cam lifts the valve with constant acceleration and deceleration for 230° of crank without any dwell period.

Allowable stress = 70 MPa

Safe bearing pressure = 5 MPa

$[t = 4$ mm, $w = 25$ mm, $H = 24$ mm, $d_h = 27$ mm, $d_b = 54$ mm$]$

8. A hollow push rod has inner to outside diameter ratio 0.65 and transfers a load of 1,000 N and length 300 mm. Design the rod, if bending stress is not to exceed 50 MPa. Use the Table given below for the safe stress for different L / K ratios.

L / K	10	20	30	40	50	60
σ_c	130	77	49	30	21	16

$[D = 10$ mm, $d = 6.5$ mm$]$

10. A hollow push rod with d / D ratio 0.7 has outside diameter 15 mm and length 180 mm. It pushes a rocker arm of leverage 1:1.2 . Safe compressive stress is 40 MPa. If the gas pressure in the cylinder is 0.6 MPa, what size of valve it can push?

$[D_{ve} = 75$ mm$]$

11. Draw cam profile for a flat follower of diameter 30 mm. Base circle diameter of cam is 60 mm and lift of 25 mm with rise period 50° at constant acceleration and deceleration, dwell period 10° and fall period 50° with similar accelerations as for rise period.

12. A cam drives a roller follower of diameter 25 mm with SHM. Cam shaft diameter is 25 mm and base circle is bigger than camshaft diameter by 10 mm. Draw the cam profile for lift period of 115°.

13. A four-stroke I.C. engine has cams with constant acceleration and deceleration. Its inlet valve opens 10° before TDC and closes after 45° of BDC with a lift of 16 mm. Draw the cam profile and find the acceleration of a roller follower of diameter 20 mm if the engine runs at 1,200 rpm. Base circle diameter can be taken 50 mm.

$[\alpha = 240$ m/s$^2]$

Previous Competition Examination Questions

IES

1. Draw the profile of a cam operating a knife edge follower having a lift of 30 mm. The cam raises the follower with SHM for 150° of its rotation, followed by a period of dwell for 60°. The follower descends for the next 100° rotation of the cam with uniform velocity, again followed by a dwell period. The cam rotates at a uniform speed of 120 rpm.

What are the maximum velocity and acceleration of the follower during lift and return?

[Answer: Lift period $v = 0.113$ m/s, $\alpha = 0.85$ m/s^2,

Return period $v = 0.055$ m/s, $\alpha = 0$] [IES 2014]

GATE

2. In a cam follower mechanism, the follower needs to rise through 20 mm during 60° of cam rotation, the first 30° with constant acceleration and then the deceleration of the same magnitude. The initial and final speeds of the follower are zero. The cam rotates at uniform speed of 300 rpm. The maximum speed of the follower is:

(a) 0.6 m/s   (b) 1.2 m/s

(c) 1.68 m/s   (d) 2.4 m/s

[Answer (b)] [GATE 2005]

☐☐☐

Fly Wheels

➤ Learn about function and construction of a flywheel

➤ Design of shaft, hub, and key

➤ Fluctuation in energy and speed

➤ Rim velocity and stresses in a flywheel

➤ Mass and energy stored in a flywheel

➤ Types of flywheels; solid, webbed, with arms and split

➤ Size calculation for flywheel for engines and punches

17.1 Function

A flywheel is used as a reservoir to store energy and give it back when required. For example,

in an internal combustion (I.C.) engine, power is developed only in power stroke, which is partially stored in a flywheel shown in picture on right side. It gives back in suction, compression, and exhaust stroke, when no power is developed in engine. Its use can also be seen in a power press, where heavy energy is required in a shear or blanking process. Flywheel stores energy between two consecutive punching processes, when there is no press work and delivers energy, when the punching or blanking process is carried out. A flywheel can be used for any one of the following functions:

- Store energy, when idle and release energy when required.
- If the load is fluctuating, capacity of the prime mover is reduced with its use.
- Its use reduces variations in speed.

17.2 Construction of a Flywheel

Construction of a flywheel depends on its size. Four types of flywheels used from small to big are shown in Figure 17.1. Smallest type, which is just a sold disk is shown in Figure 17.1(a). Generally, a flywheel has three parts; a heavy rim to store energy, a hub to mount on a shaft, and a disk or arms to join rim with hub. Since the energy stored is maximum in the outer periphery of the flywheel, the rim is made heavy. It is joined with hub either with a web [Figure 17.1(b)] or arms [Figure 17.1(c)] to reduce its weight. Sometimes, holes are made in the web to facilitate handling. A big flywheel is made in two halves and then joined together with bolts and nuts as shown in Figure 17.1(d).

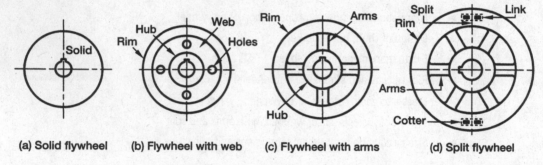

(a) Solid flywheel   (b) Flywheel with web   (c) Flywheel with arms   (d) Split flywheel

Figure 17.1 Construction of flywheels

17.3 Design of Shaft, Hub, and Key

Shaft of a flywheel is designed for the maximum torque T_{max} in the torque diagram. Shaft diameter d is calculated for the safe shear stress τ of the material using the relation given below again.

$$T_{max} = \frac{\pi}{16} d^3 \times \tau$$

Hub is the central part of the flywheel, which fits over the shaft. Its hole diameter is equal to shaft diameter d as calculated above, and outside diameter d_h is calculated assuming it as a hollow shaft, using the relation as given below:

$$T_{max} = \frac{\pi}{16} \left(\frac{(d_h)^4 - d^4}{d} \right) \times \tau$$

Empirically, hub outside diameter is taken about 2 to 2.5 times the shaft diameter. The length of hub L is taken same as the axial width of the rim.

Key is designed on the basis of its shear strength. A sunk key of cross section $\frac{d}{4} \times \frac{d}{4}$ is used. The length of the key (same as hub length L) is calculated using the relation:

$$T_{max} = \text{(Sheared area of key)} \times \text{(Shaft radius)} \times \text{(Safe shear stress)} = \left(L \times \frac{d}{4}\right) \times \frac{d}{2} \times \tau$$

Example 17.1

Shaft hub and key design

75 kW of power is transmitted at 350 rpm with a cast iron flywheel. The shaft on which it is mounted and the key have safe shear strength 35 MPa. Calculate shaft diameter, hub, and key size, assuming maximum torque to be 1.3 times the mean torque.

Solution

Given $P = 75$ kW $N = 350$ rpm $T_{max} = 1.3 T_m$

For shaft and key $\tau = 35$ MPa

$$P = \frac{2\pi N T_m}{60,000} \quad \text{or} \quad T_m = \frac{60,000\,P}{2\pi N} = \frac{60,000 \times 75}{2 \times 3.14 \times 350} = 2,047 \text{ Nm}$$

$$T_{max} = 1.3 T_m = 2,047 \times 1.3 = 2,661 \text{ Nm} = 2,661 \times 10^3 \text{ Nmm}$$

$$T_{max} = \frac{\pi}{16} d^3 \times \tau \quad \text{or} \quad 2,661 \times 10^3 = \frac{3.14}{16} d^3 \times 35 \quad \text{or} \quad d = 73 \text{ mm}$$

Hub diameter is taken as: $2d = 2 \times 73 = 146$ mm

Length of hub (L) is taken = Length of key

Cross-section of key is taken as $\frac{d}{4} \times \frac{d}{4}$, that is, 18 × 18 mm

If L is length of key, $T_{max} = \left(L \times \frac{d}{4}\right) \times \frac{d}{2} \times \tau$

Or, $2,661 \times 10^3 = L \times 18 \times 36.5 \times 35$ or $L = 116$ mm Also, hub length = 116 mm

17.4 Fluctuation in Energy and Speed

A flywheel mounted on a shaft is shown in Figure 17.2. T_i is the input torque to the shaft and T_o is the output torque from the flywheel. If T_i is more than T_o, the shaft accelerates and if less it decelerates.

Energy stored in such a flywheel $E = \frac{1}{2} I \omega^2$ 　　　　　　(17.1)

Where, I = Mass moment of in inertia

ω = Mean rotational speed of flywheel in radians per second = $\dfrac{(\omega_{max} + \omega_{min})}{2}$

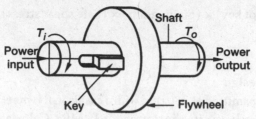

Figure 17.2 A solid flywheel on a shaft

A schematic torque diagram is shown in Figure 17.3(a) for an I.C. engine, where torque is negative in suction, compression, exhaust stroke, and positive in power stroke.

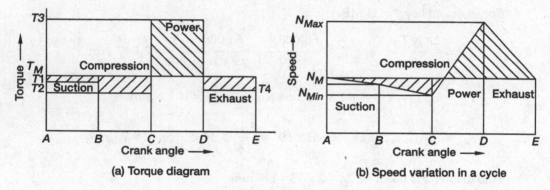

(a) Torque diagram   (b) Speed variation in a cycle

Figure 17.3 Variation of torque and speed

In suction stroke AB, torque is negative; hence speed decreases Figure 17.3(b).

In compression stroke BC also, torque is negative; hence speed decreases further.

In power stroke CD, torque is positive; hence speed increases.

In exhaust stroke DE, again the torque is negative; hence speed decreases.

The mean torque T_m is calculated such that negative area below the mean torque line up to E is equal to positive area above this line. Work done W can be found from the mean torque T_m for one cycle using relation:

$$\text{Work done } W = 2\,\pi\,s\,T_m \tag{17.2}$$

Where, T_m = Mean torque(Nm)

     s = 1 for two-stroke engines

     s = 2 for four-stroke engines

The energy supplied by the flywheel E_i during strokes AB, BC, and DE is:

$$E_i = \int_A^B \left(T_m - T_1\right)d\theta + \int_B^C \left(T_m - T_2\right)d\theta + \int_D^E \left(T_m - T_4\right)d\theta$$

The energy supplied to the flywheel E_o during power stroke CD is:

$$E_o = \int_C^D \left(T_3 - T_m\right)d\theta$$

Change in kinetic energy from point C to D due to increase in speed is:

$$\Delta E = E_i - E_o \tag{17.3}$$

Also $\quad \Delta E = \dfrac{1}{2}I\left(\omega_{max}^2 - \omega_{min}^2\right) = I\dfrac{\left(\omega_{max} + \omega_{min}\right)}{2}\left(\omega_{max} - \omega_{min}\right)$

$$\Delta E = I\,\omega\left(\omega_{max} - \omega_{min}\right) = I\,\omega\,\Delta\omega \tag{17.4}$$

Where, ω_{max} = Maximum speed of flywheel

$\quad\quad \omega_{min}$ = Minimum speed of flywheel

$\quad\quad \Delta\omega$ = Variation in speed = $\left(\omega_{max} - \omega_{min}\right)$

$\quad\quad \omega$ = Mean rotational speed in radians per second $= \dfrac{\left(\omega_{max} + \omega_{min}\right)}{2}$

Fluctuation of energy can be found from the torque diagram for one cycle. Coefficient of fluctuation of energy C_E is defined as the ratio of energy variation in a cycle and the work done in one cycle, that is,

$$C_E = \frac{E_i - E_o}{\text{Work done per cycle}} = \frac{\Delta E}{W} \tag{17.5}$$

Where, E_i = Energy supplied by the flywheel

$\quad\quad E_o$ = Energy supplied to the flywheel

Coefficient of fluctuation of energy for different types of engines is given in Table 17.1.

Coefficient of fluctuation in speed is defined as:

$$C_S = \frac{\left(\omega_{max} - \omega_{min}\right)}{\omega}$$

Or, $\quad C_S = \dfrac{\Delta\omega}{\omega} \tag{17.6a}$

Table 17.1 Coefficient of Fluctuation of Energy C_E for Engines

Type of Engine	No. of Cylinders	C_E		Type of Engine	No. of Cylinders	C_E
		Two Stroke	Four Stroke			
I.C. engines	1	0.95–0.1	2.35–2.4	Steam engine	1	0.27–0.35
	2	0.2–0.25	0.15–1.6		2	0.07–0.8
	4	0.075–0.1	0.15–0.2		3	0.03–0.04
	6	0.016–0.2	0.1–0.12		Compound	0.96

The term coefficient of fluctuation of speed can also be written as:

$$C_S = \frac{2\left(\omega_{max} - \omega_{min}\right)}{\omega_{max} + \omega_{min}} = \frac{2\left(N_{max} - N_{min}\right)}{N_{max} + N_{min}} = \frac{N_{max} - N_{min}}{N} \qquad (17.6b)$$

Where, $N = \dfrac{N_{max} + N_{min}}{2}$ Mean speed (rpm)

N_{max} = Maximum speed (rpm)

N_{min} = Minimum speed (rpm)

Values of coefficient of speed for different applications are given in Table 17.2.

Table 17.2 Value of Coefficient of Speed C_S for Various Applications

Application	C_S	Application	C_S
Compressor: belt driven	0.12	Automobiles	0.1–0.2
Compressor: gear driven	0.02	Machine tools	0.03–0.2
Floor mills	0.02	Pumps	0.03–0.05
Rolling mills	0.025	Punch press	0.07–0.2
Weaving machines	0.025	Spinning mills	0.017–0.02
I.C. engines	0.03	Generator: AC	0.003–0.005
Crushers, power hammers	0.2	Generator: DC	0.007–0.01

Substituting the values of $\Delta\omega$ from Equation (17.6a) in Equation (17.4)

$$\Delta E = I\, \omega^2 C_s \qquad (17.7)$$

Substituting value of $(I\, \omega^2)$ from Equation (17.1) variation in energy

$$\Delta E = 2\, E\, C_S \qquad (17.8)$$

Example 17.2

Coefficient of fluctuation of speed and mass of flywheel

An engine is connected to a load, which varies the speed of engine from 1,550 rpm to 1,450 rpm. Difference between input and output energies is 35 kNm. Find:

    a. Coefficient of fluctuation of speed.

    b. Mass moment of inertia of the flywheel.

    c. Mass of a solid disk flywheel, if its radius is 500 mm

Solution

Given $N_{max} = 1{,}550 \text{ rpm}$ $N_{min} = 1{,}450 \text{ rpm}$ $\Delta E = 35 \text{ kNm}$

         $R = 500 \text{ mm}$

    a. Coefficient of fluctuation of speed.

$$\text{Mean speed, } C_s = \frac{N_{max} + N_{min}}{2} = \frac{1{,}550 + 1{,}450}{2} = 1{,}500 \text{ rpm}$$

$$\text{Coefficient of fluctuation of speed } C_s = \frac{N_{max} - N_{min}}{N} = \frac{(1{,}550 - 1{,}450)}{1{,}550} = 0.067$$

    b. Angular speed: $\omega = \dfrac{2\pi N}{60} = \dfrac{2 \times 3.14 \times 1{,}500}{60} = 157 \text{ rad/s}$

    From Equation (17.7), $\Delta E = I \omega^2 C_s$

    Substituting the values:

$$35{,}000 = I \times (157)^2 \times 0.067 \quad \text{or} \quad I = 21.32 \text{ kg m}^2$$

    c. For a solid disk mass moment of inertia $I = 0.5 \, m \, R^2$

    Or, $21.32 = 0.5 \times m \times (0.5)^2$ or $m = 170.8 \text{ kg}$

Example 17.3

Energy variation and mass of flywheel for a steam engine

A single cylinder double acting steam engine, develops 60 kW at 200 rpm. Speed variation is to be within ± 1.5 per cent and energy fluctuation 10 per cent per revolution. Circumferential velocity of the rim is limited to 10 m/s. Find:

    a. Maximum and minimum speeds and coefficient of speed variation.

    b. Variation in energy.

    c. Radius and mass of flywheel.

Solution

Given $P = 60 \text{ kW}$ $N = 200 \text{ rpm}$ $\Delta E = 0.1 \, W$ $v = 10 \text{ m/s}$

           $100 \, \Delta N / N = \pm 1.5$ per cent

a. 1.5 per cent of $\Delta N = \dfrac{200 \times 1.5}{100} = 3$ rpm

$N_{max} = 200 + 3 = 203$ rpm and, $N_{min} = 200 - 3 = 197$ rpm

$$C_S = \frac{N_{max} - N_{min}}{N} = \frac{203 - 197}{200} = 0.03$$

b. Mean torque $T_m = \dfrac{60,000\,P}{2\pi N} = \dfrac{60,000 \times 60}{2 \times 3.14 \times 200} = 2{,}866 \ \text{Nm}$

Work done per revolution $W = 2{,}866 \times 2\pi = 18{,}000$ Nm

Variation in energy $\Delta E = 0.1\,W = 0.1 \times 18{,}000 = 1{,}800$ Nm

c. Angular speed: $\omega = \dfrac{2\pi N}{60} = \dfrac{2 \times 3.14 \times 200}{60} = 20.93$ rad/s

Circumferential velocity $v = \omega R$

Or, $\quad 10 = 20.93\,R \qquad\qquad$ Or, $\qquad R = 0.477$ m

Also $\Delta E = mv^2 C_s$

Or, $\quad 1{,}800 = m \times 10^2 \times 0.03 \quad$ Or, $\qquad m = 600$ kg

17.5 Rim Velocity

Rim velocity is important in the design of flywheel. It is the velocity of the outermost surface of the flywheel. Shaft speed N is fixed by the application; hence rim velocity depends on outside diameter D_o, which a designer can select suitably.

$$\text{Rim velocity } v = \frac{\pi D_o N}{60} \tag{17.9}$$

It may be noted that bigger the outside diameter of a flywheel, lesser is the weight required for the given fluctuation of speed. But the size is limited due to following reasons:

- Available space.
- Larger diameter causes more centrifugal force, causing higher tensile stresses, which may burst the rim. So stress has to be within the allowable stress.
- Larger the peripheral velocity, more is the stress. Table 17.3 gives maximum allowable rim velocity for different materials.

Table 17.3 Maximum Allowable Rim Velocity

Material	Rim Velocity (m/s)
Cast iron up to 75 kW power	25
Cast iron more than 75 kW power	30
Good quality casting	35
Fully machined C.I. for automobiles	50
Cast steel	60
Forged rim with welded arms	75

17.6 Stresses in Flywheel

Flywheels are manufactured generally by sand casting process, which causes local stress concentration near arms joining the rim, while cooling. It is not possible to calculate but can be neglected, if the manufacturing is done carefully to avoid those stresses. A flywheel rim has to withstand the following stresses:

- Tensile stresses due to centrifugal force.
- Bending stresses due to constrained arms.

17.6.1 Tensile stresses due to centrifugal force

Mass of flywheel M for an element of the rim subtending a small angle $\delta\theta$ at the center (Figure 17.4) is given as:

$$M = \rho A R \, \delta\theta \tag{17.10}$$

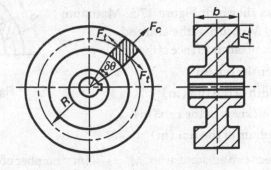

Figure 17.4 Centrifugal force on the rim

Where, ρ = Mass density (kg/m³)

$A = (b \times h)$ cross-sectional area of the rim (m²)

b = Axial width of the rim (m)

h = Radial thickness of the rim (m)

R = Mean radius of the rim (m)

Centrifugal force is given as: $F_c = \dfrac{Mv^2}{R}$

Substituting the value of M from Equation (17.10)

$$F_c = \frac{\rho A R \delta\theta \, v^2}{R} \qquad (17.11)$$

Where, F_c = Centrifugal force (N)

v = Circumferential velocity at mean radius (m/s)

This centrifugal force F_c is resisted by the tensile force F_t in circumferential direction. Both these forces are related as:

$$F_c = 2 F_t \sin\left(\frac{\delta\theta}{2}\right) = F_t \sin(\delta\theta) \quad \left(\text{Because } \delta\theta \text{ is small}\right)$$

Or, $\qquad F_t = \dfrac{F_c}{\delta\theta} = \dfrac{\rho A R \delta\theta \, v^2}{R \delta\theta}$ (Substituting the value of F_c from Equation 17.11)

$$\text{Tensile stress } \sigma_t = \frac{F_t}{A} = \frac{F_c}{A\delta\theta} = \frac{\rho A R \delta\theta \, v^2}{R A \, \delta\theta} = \rho v^2 \qquad (17.12)$$

17.6.2 Bending stresses due to constrained arms

The portion of the rim between the two adjacent arms can be considered as a curved beam, which is loaded due to the centrifugal forces as shown in Figure 17.5. Maximum bending moment (BM) M occurs at the arm end.

Mass of the rim per mm of circumference of the rim $w = \rho b h$

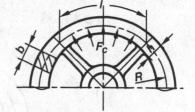

Figure 17.5 Bending stresses on the rim

Where, ρ = Mass density (kg/m³)

b = Axial width of the rim (m)

h = Radial thickness of the rim (m)

R = Mean radium of the rim (m)

Mass of the rim between two adjacent arms, $M' = \dfrac{M}{m}$ (m = number of arms)

Length of beam between two adjacent arms, $l = \dfrac{2\pi R}{m}$

Maximum bending moment is: $M_b = \dfrac{wl^2}{12} = \dfrac{\rho h b \omega^2 R \left(\dfrac{2\pi R}{m}\right)^2}{12}$

Modulus of section $Z = \dfrac{bh^2}{6}$

Also $\qquad M_b = \sigma_b Z \qquad$ Or, $\qquad \sigma_b = \dfrac{M_b}{Z}$

Using $v = \omega R$ and simplifying after substituting the values:

$$\sigma_{b'} = \frac{2\rho(\pi v)^2 R}{m^2 h} \tag{17.13}$$

G- Lanza gave that the resultant stress in the rim is 75 per cent of tensile and 25 per cent bending stress, that is, $\sigma_{total} = 0.75\sigma_t + 0.25\sigma_b$

$$\sigma_{total} = \rho v^2 \left(0.75 + \frac{\pi^2 R}{2m^2 h}\right) \tag{17.14}$$

17.7 Mass and Energy Stored in Flywheel

As described in Section 17.1, flywheel can be solid as shown in picture on right side or with web or armed. Which type of flywheel is to be used depends on the outside diameter D_o. Following guide lines can be used to select a type of flywheel:

a. Solid flywheel, if $D_o < 200$ mm
b. Flywheel with disk, if $D_o < (7.5\ d + 80$ mm$)$; where, d is shaft diameter in mm.
c. Flywheel with arms, if $D_o > (7.5\ d + 80$ mm$)$
d. Split flywheel with arms, if $D_o > 2{,}500$ mm

17.7.1 Solid flywheel

Figure 17.1(a) and picture above shows a solid disk flywheel. This type is used for small diameters up to 200 mm. It is just a circular disk of width b and outside diameter D_o. Mass moment of inertia I of such a disk is given as:

$$I = \frac{1}{2}m(R_1)^2 \quad (R_1 = \text{Outside radius} = 0.5D_o) \tag{a}$$

Mass of flywheel $m = \text{Area} \times \text{Thickness} \times \text{Density}$

$$= \pi\,(R_1)^2 = t\rho \tag{b}$$

Where, $t = $ Axial thickness of disk (m)

$\rho = $ Density of material (kg/mm³)

For C.I. $\rho = 7,200$ and for steel $\rho = 7,820$

From Equations (a) and (b):

$$I = \frac{1}{2}\pi\left(R_1\right)^2 t\rho \times \left(R_1\right)^2 = \frac{\pi}{2}\left(R_1\right)^4 t\rho \qquad (17.15)$$

Energy stored in a flywheel

$$E = \frac{1}{2}I\omega^2 = \frac{\pi}{4}\left(R_1\right)^4 t\rho\omega^2 \qquad (17.16)$$

From Equations (17.7 and 17.8):

$$\Delta E = I\omega^2 C_S = 2EC_S = \frac{\pi}{2}\left(R_1\right)^4 t\rho\omega^2 C_S \qquad (17.17)$$

Example 17.4

Mass of solid flywheel for given variation in torque

An electric motor supplying constant torque to a machine, whose turning moment diagram is shown in Figure 17.S1(a). Mechanical efficiency of the drive from motor to machine shaft is 85 per cent. Torque first increases from 300 Nm to 1,200 Nm for half revolution and then decreases to 300 Nm in quarter revolution at 400 rpm. No load for the rest quarter revolution. Coefficient of fluctuation of speed is limited to 0.15. Density of flywheel material is 7,200 kg/m³ Calculate:

a. Power required for the motor.

b. Mass and size of disk flywheel, if radius to thickness ratio is limited to 15.

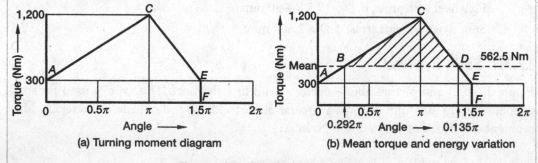

Figure 17.S1 Turning moment diagrams

Solution

Given $N = 400$ rpm $T_{max} = 1,200$ Nm $C_s = 0.15$

$\eta_{mech} = 0.85$ $\rho = 7,200$ kg/m³ $R/t = 15$

a. Mean torque $T_m = \dfrac{\text{Area OACEF}}{2\pi}$

Or, $T_m = \dfrac{\left[0.5 \times (300 + 1{,}200) \times \pi\right] + \left[0.5 \times (300 + 1{,}200) \times 0.5\,\pi\right]}{2\pi} = 562.5 \text{ Nm}$

Power required $P = \dfrac{2\pi N T}{60{,}000} = \dfrac{2 \times 3.14 \times 400 \times 562.5}{60{,}000} = 2.355 \text{ kW}$

Power of motor $= \dfrac{2.355}{0.85} = 2.77 \text{ kW}$

b. Maximum variation of energy is the area BCD. Calculate position of points B and D.

By linear interpolation point B is at: $\dfrac{\pi \times (562.5 - 300)}{1{,}200 - 300} = 0.292\,\pi$

By linear interpolation point D is at: $1.5\,\pi - \dfrac{0.5\pi \times (562.5 - 300)}{1{,}200 - 300} = 1.35\,\pi$

Area BCD $= \Delta E = \dfrac{1}{2}(1{,}200 - 562.5) \times (1.35\pi - 0.292\pi) = 1{,}051 \text{ Nm}$

Rotational speed $\omega = \dfrac{2\pi N}{60} = \dfrac{2 \times 3.14 \times 400}{60} = 41.9 \text{ rad/s}$

Mass moment of inertia $I = \dfrac{\Delta E}{\omega^2 C_S} = \dfrac{1051}{(41.9)^2 \times 0.15} = 4 \text{ kg m}^2$

Also from Equation (17.15) if R is the outside radius:

$$I = \dfrac{\pi}{2} R^4 t\rho = \dfrac{3.14}{2} \times 7{,}200 \times \dfrac{R}{15} \times R^4$$

Or, $\quad 4 = 753.6\,R^5 \quad$ Or, $\quad R = 0.35 \text{ m} = 350 \text{ mm}$

Thickness $t = \dfrac{R}{15} = \dfrac{350}{15} = 23.3 \text{ mm say 24 mm}$

Mass $= \pi R^2 t\rho = 3.14 \times 0.35^2 \times 0.24 \times 7{,}200 = 66.5 \text{ kg}$

17.7.2 Flywheel with web

A flywheel with web joining rim and hub is shown in Figure 17.1(b) and picture on right. Generally, the energy stored by the hub is neglected, as its diameter is much smaller than rim diameter.

Thickness of web $t = 0.05\, b$

Where, b is width of rim in mm

Mass of the flywheel m = Mass of rim + Mass of disk

Calculate mass of the web, as given in Section 17.7.1 and the rim as given in Section 17.7.3.

Radial thickness of rim h is taken equal or more than axial width b. It can vary from 0.7 to 2. Sometimes flywheel is used as a pulley also. In that case, width of flywheel is kept about 20 mm–30 mm more than width of belt.

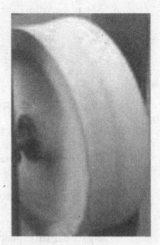

Example 17.5

Size of a flywheel with web for an engine of given power
A flywheel with web is to be designed for a four-stroke diesel engine developing 50 kW power at 500 rpm. Assume 90 per cent of energy is to be absorbed by rim and rest by web, coefficient of fluctuation of speed is 0.03, and coefficient of fluctuation of energy 2.3. Density of material is 7,200 kg/m³. Neglect the mass of hub. If rim velocity is limited to 25 m/s, calculate:

  a. Thickness, radius and mass of web

  b. Mass and size of rim, assuming radial thickness 25 per cent of inside radius

Solution

Given $P = 50\,\text{kW}$ $s = 2$ $N = 500\,\text{rpm}$ $\Delta E_R = 0.9\Delta E$

         $C_S = 0.03$ $C_E = 2.3$ $\rho = 7{,}200\,\text{kg/m}^3$ $v = 25\,\text{m/s}$

      $R_w / t = 10$ $h = 0.25\, R_w$

  a. Web size:

Mean torque $T_m = \dfrac{60{,}000\,P}{2\pi N} = \dfrac{60{,}000 \times 50}{2 \times 3.14 \times 500} = 955\,\text{Nm}$

From Equation (17.2):

Work done: $W = 2\,\pi s\, T_m = 2 \times 3.14 \times 2 \times 955 = 12{,}000\,\text{Nm}$

From Equation (17.5):

$C_E = \dfrac{\Delta E}{W}$ or $\Delta E = W\,C_E$ or $\Delta E = 12{,}000 \times 2.3 = 27{,}600\,\text{Nm}$

Energy to be absorbed by web $\Delta E_w = 0.1\ \Delta E = 0.1 \times 27{,}600 = 2{,}760$ Nm

Rotational speed $\omega = \dfrac{2\pi N}{60} = \dfrac{2 \times 3.14 \times 500}{60} = 52.3$ rad/s

Circumferential velocity $v = \omega R$ or, $25 = 52.3\,R_1$ or $R_1 = 0.478$ m $= 478$ mm

Radius of rim $R_1 = $ Radius of web $R_w + $ radial height of rim $h = R_w + 0.25\,R_w = 1.25\,R_w$

or Web radius, $R_w = 0.478/1.25 = \mathbf{0.382\ m}$

Mass moment of inertia $I = \dfrac{\pi}{2}\left(R_w\right)^4 t\rho = \dfrac{3.14}{2}(0.382)^4 \times t \times 7{,}200 = 241.7\,t$

From Equation (17.7): $\Delta E_w = I\,\omega^2\,C_s$

$2{,}760 = 241.7\,t\,(52.3)^2 \times 0.03$ Or, $t = 0.139$ mm say 140 mm

Mass of web $m_w = \pi\,(R_w)^2 t\rho = 3.14\,(0.382)^2 \times 0.14 \times 7{,}200 = 463$ kg

b. Energy to be absorbed by rim:

$\Delta E_r = 0.9 \Delta E = 0.9 \times 27{,}600 = 24{,}840$ Nm

Inside radius of rim = Outside radius of web = 0.382 m

Mean diameter of rim $R = 0.5 \times (478 + 382) = 430$ mm

Radius of gyration for rim can be taken as the mean radius of the rim.

Mass moment of inertia $I_r = mk^2 \cong mR^2 \cong m_r\,(0.430)^2 = 0.185\,m_r\ \text{kg m}^2$

From Equation (17.7) : $\Delta E_r = I_r\,\omega^2\,C_s$

Substituting the values: $24{,}840 = 0.185\,m_r \times (52.3)^2 \times 0.03$

Or, Mass of rim $m_r = 1{,}636$ kg

Also, Mass of rim $m_r = 2\,\pi R\,(h \times b)\,\rho$ $\qquad h = 438 - 382 = \mathbf{96\ mm}$

$1{,}415 = 2 \times 3.14 \times 0.4626 \times (96 \times b) \times 7{,}200$ or Width of rim $b = 0.88\,m = \mathbf{880\ mm}$

Radial height of rim $h = 478 - 382 = \mathbf{96\ mm}$

Outside radius $R_1 = R + 0.5\,h = 430 + (0.5 \times 96) = \mathbf{478\ mm}$

Inside radius $R_2 = R - 0.5\,h = 430 - (0.5 \times 96) = \mathbf{382\ mm}$

17.7.3 Flywheel with arms

Fly wheels with diameter more than 500 mm are provided with arms. Number of arms can vary from 4 to 12 depending on the size of flywheel. Generally, six arms are used. Arms are subjected to variable loads due to variation in incoming torque; hence minimum factor of safety is taken as 8. If there are heavy shocks such as for punching machines, *FOS* can be still higher such as 12.

Stress on arms of flywheel is more in the circumferential direction; hence its width in the plane of rotation a is kept more than along the axis of flywheel. A circular section will give equal strength in both the directions and rectangular section offers more air resistance to the rotating arms; hence elliptical section of the arm is most suitable.

Number of arms depends on outside diameter:

Diameter < 400 mm	Number of arms $m = 4$
Diameter 400 mm to 3,000 mm	Number of arms $m = 6$
Diameter > 3,000 mm	Number of arms $m = 8$ to 12

Analysis of flywheel with arms is done assuming that only 10 per cent is contributed by arms and the rest 90 per cent by the rim. If I is moment of inertia required for flywheel, then moment of inertia of rim $I_r = C_r I$. Value of constants for rim C_r is taken as 0.9.

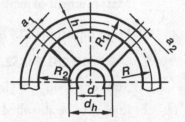

Radial thickness of rim $h = R_1 - R_2$ (Refer Figure 17.6)

Where, R_1 = Outer diameter of rim

R_2 = Inner diameter of rim

Figure 17.6 Hub, arms, and rim of a flywheel

Since h is very small in comparison to rim radii, radius of gyration of a rectangular rim can be taken at mean rim radius of rim $R = \dfrac{R_1 + R_2}{2}$

Mass of rim m_r = Mean circumference × Area × Density = $2\pi R (h \times b) \rho$

Radius of gyration of rim $K_r = R$ (As the thickness of rim is small)

Moment of inertia of rim $I_r = m_r (K_r)^2 = 2\pi R.(h \times b) \rho R^2$

From Equation (17.16):

Energy stored in rim $E_r = \dfrac{1}{2} I_r \omega^2 = \pi R h b \rho v^2$ $\quad (v = \omega R)$

From Equation (17.17):

$$\Delta E = I_r \omega^2 C_s = 2\pi R h b \rho v^2 C_s = m_r v^2 C_s$$

Design of arms

Bending on the arms is more near the hub, and hence these are made tapered with more thickness at the hub than at the rim. Taper per meter length of arm for major axis is about 20 mm and 10 mm for the minor axis. $a_1 \times b_1$ is the size of arm near the hub and $a_2 \times b_2$ is the size of arm near the rim as shown in Figure 17.7. Arms can be straight or curved as shown in picture below.

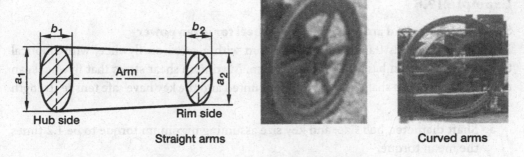

Figure 17.7 Cross section of flywheel arms

Section modulus of elliptical cross section is:

$$Z = \frac{\pi}{32}(a_1)^2 b_1 \qquad (17.18)$$

Where, a_1 = Major axis of arm at hub

b_1 = Minor axis of arm at hub. It is taken as 0.5 a_1.

Let, D_r = Diameter of rim

D = Diameter of hub

T = Torque

m = Number of arms

l = Length of arms

Bending moment M for each arm is given as:

$$M = \frac{T}{m \times 0.5 D_r} \times l = \frac{T}{m \times 0.5 D_r} \times \frac{(D_r - D)}{2} = \frac{T \times (D_r - D)}{m \times D_r}$$

Bending stress $\sigma_b = \dfrac{M}{Z} = \dfrac{T \times (D_r - D)}{m \times D_r} \times \dfrac{32}{\pi (a_1)^2 b_1}$

Assuming the value $b_1 = 0.5\, a_1$

$$\sigma_b = \frac{20.4\, T \times (D_r - D)}{m \times (a_1)^3 \times D_r} \qquad (17.19)$$

This stress should be less than allowable bending stress.

Example 17.6

Shaft, hub, rim, and arm design of a flywheel for given power

150 kW of power is transmitted at 300 rpm with a cast iron flywheel with elliptical tapered arms, which has to absorb 500 kNm. Maximum shear stress that flywheel can take is 10 MPa. The shaft, on which it is mounted and the key have safe tensile strength 40 MPa. Calculate:

a. Shaft diameter, hub size, and key size assuming maximum torque to be 1.2 times the mean torque.

b. Weight of flywheel and rim size, if density of the material is 7,200 kg/m³.

c. Size or arms, if six arms are used.

Solution

Given $P = 150$ kW $N = 300$ rpm $E = 500$ kNm $\rho = 7,200$ kg/m³ $m = 6$

For flywheel $\sigma_t = 10$ MPa   For shaft and key $\tau = 40$ MPa

a. $P = \dfrac{2\pi N T_m}{60,000}$ or $T_m = \dfrac{60,000\,P}{2\pi N} = \dfrac{60,000 \times 150}{2 \times 3.14 \times 300} = 4,777$ Nm

$T_{max} = 1.2\,T_m = 4,777 \times 1.2 = 5,732$ Nm $= 5,732 \times 10^3$ Nmm

$T_{max} = \dfrac{\pi}{16}d^3 \times \tau$ or $5732 \times 10^3 = \dfrac{3.14}{16}d^3 \times 40$ or $d = 90$ mm

Hub diameter is taken as $2d = 2 \times 90 = 180$ mm

Length of hub is taken as $2.5\,d = 2.5 \times 90 = 225$ mm

Key size is taken as $\dfrac{d}{4} \times \dfrac{d}{4}$, that is, 22.5 mm × 22.5 mm

If L is length of key $T_{max} = \left(L \times \dfrac{d}{4}\right)\dfrac{d}{2} \times \tau$

Or, $5,732 \times 10^3 = L \times 22.5 \times 45 \times 40$ or $L = 142$ mm

Length of key should not be less than length of hub, hence $L = 225$ mm

b. Rotational speed $\omega = \dfrac{2\pi N}{60} = \dfrac{2 \times 3.14 \times 300}{60} = 31.4$ rad/s

Circumferential velocity $v = \omega R = 31.4\,R$

Maximum stress in rim $\rho v^2 = \sigma_t$

Or, $7,200 \times (31.4 \times R)^2 = 10 \times 10^6$ or $R = 1.18$ m

Energy $E = \dfrac{1}{2}mv^2$ or $500 \times 10^3 = \dfrac{1}{2}m(31.4 \times 1.18)^2$ or $m = 728.4$ kg

Ratio of length of hub and width of flywheel is kept = 1.2.

Hence width of flywheel $b = \dfrac{L}{1.2} = \dfrac{225}{1.2} = 187.5$ say 188 mm

Also mass of flywheel $2\pi R (b \times h) \rho = m$

Or, $2 \times 3.14 \times 1.18 \times 188 \times h \times 7{,}200 = 728.4$ or $h = 72.6$ say 73 mm

Outside radius of rim $R_1 = 1.18$ m = 1,180 mm

Inside radius of rim $R_2 = R_1 - h = 1{,}180 - 73 = 1107$ mm

c. Bending moment for one arm $M = \dfrac{T_{max}}{6} = \dfrac{5{,}732}{6} = 955.4$ Nm $= 955.4 \times 10^3$ Nmm

Modulus of section $Z = \dfrac{\pi}{32}(a_1)^2 b_1$

Assuming $b_1 = 0.5 a_1$, $Z = \dfrac{\pi}{64}(a_1)^3$

Also $M = Z\sigma_t$ or $955.4 \times 10^3 = \dfrac{\pi}{64}(a_1)^3 \times 10$ Or, $a_1 = 125$ mm

Hence, $b_1 = 0.5 \times 125 = 62.5$ mm

Taper given to arms for major axis is 20 mm per meter.

Length of arm $= R_2 -$ Hub radius $= 1{,}107 - \dfrac{180}{2} = 1{,}017$ mm say 1 m

Hence $a_2 = a_1 - 20 = 125 - 20 = 105$ mm

Taper given to arms for minor axis is 10 mm per meter.

Hence $b_2 = b_1 - 10 = 62.5 - 10 = 52.5$ mm.

17.7.4 Split flywheel

Very large-sized flywheels are made in two halves, due to size limitations of castings. When a flywheel is casted, during cooling when the rim contracts its radius decreases, which causes compression in the arms. This split gives an additional advantage that during cooling arms and hub are free to move and hence work stresses do not appear. The two halves are then joined using bolts and nut for hub and the rim as shown in Figure 17.8 (*a* and *b*) and picture on the right side.

Cotters are also used for the rim joint for slow-speed flywheels. A rim insert link is inserted in a rectangular hole of the rim and then cotters are used to make a cotter joint as shown in Figure 17.8 (c). The mass and energy calculations remain the same as given in Section 17.7.3.

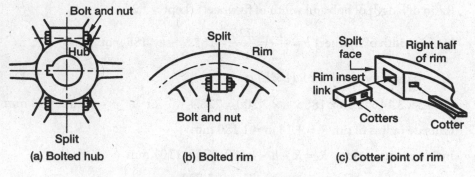

Figure 17.8 Split flywheel

Tensile stress due to centrifugal force $\sigma_t = \rho v^2$

Cross-sectional area of rim $A = h \times b$

Force F = Tensile stress × Area = $\sigma_t A$

From Equation (17.13): Bending stress $\sigma_b = \dfrac{2\rho\left(\pi v^2\right)R_m}{m^2 h}$

Total stress in rim $\sigma_{total} = \dfrac{3}{4}\sigma_t + \dfrac{1}{4}\sigma_b$

Total tensile force on the rim $F = F = \sigma_{total} \times$ Area

Force on each bolt $= \dfrac{F}{Z} = \dfrac{\pi}{4}\left(d_c\right)^2 \sigma_t$ $\left(Z = \text{Number of bolts}\right)$

Calculate core diameter and then choose a standard bolt.

Example 17.7

Design of split flywheel for given power

A 250 kW steam engine running at 180 rpm has a split flywheel of outside radius 1.5 m with six tapered radial arms of elliptical cross section. The two halves are joined using 8 bolts. Ratio of major to minor axis for arms is 2. Radial thickness of rim is 20 per cent of its outside radius. Ratio of width of rim and radial thickness is 1.2. Take fluctuation of speed 0.15, safe tensile stress 60 MPa and shear stress 20 MPa and density = 7,200 kg/m³. Calculate:

  a. Total stress in rim.

  b. Size of bolts.

Solution

Given $P = 250\,\text{kW}$ $N = 180\,\text{rpm}$ $R_1 = 1.5\,\text{m}$ $h = 0.2R_1$
$\quad\quad\quad\quad C_S = 0.15$ $a/b = 2$ $m = 6$ $b/h = 1.2$
$\quad\quad\quad\quad \sigma_t = 60\,\text{MPa}$ $\tau = 20\,\text{MPa}$ $\rho = 7{,}200\,\text{kg/m}^3$ $Z = 8$

a. Radial thickness of rim $h = 0.2\,R_1 = 0.2 \times 1.5 = 0.3$ m

Inside radius of rim $R_2 = R_1 - h = 1.5 - 0.3 = 1.2$ m

Mean radius of rim $R_m = \dfrac{R_1 + R_2}{2} = \dfrac{1.5 + 1.2}{2} = 1.35$ m

Peripheral velocity at mean radius $v = \dfrac{2\pi R_m N}{60} = \dfrac{2 \times 3.14 \times 1.35 \times 180}{60} = 25.43$ m/s

Tensile stress due to centrifugal force $\sigma_t = \rho v^2 = \dfrac{7,200 \times (25.43)^2}{10^6} = 4.65$ MPa

Cross-sectional area of rim $A = h \times b = h \times 1.2\,h = 0.3 \times 1.2 \times 0.3 = 0.108$ m²

Area of any section across the flywheel resisting centrifugal force $= 2A = 0.216$ m²

Force = Tensile stress × Area $= 4.65 \times 0.216 \times 10^6 = 1 \times 10^6$ N $= 1,000$ kN

From Equation (17.13)   Bending stress $\sigma_b = \dfrac{2\rho(\pi v)^2 R_m}{m^2 h}$

$$\sigma_b = \dfrac{2 \times 7,200 \times (3.14 \times 25.43)^2 \times 1.35}{6^2 \times 0.3 \times 10^6} = 11.47 \text{ MPa}$$

Total stress in rim $\sigma_{total} = \dfrac{3}{4}\sigma_t + \dfrac{1}{4}\sigma_b = (0.75 \times 4.65) + (0.25 \times 11.47) = 6.36$ MPa

b. Total tensile force on the rim $F = \sigma_{total} \times$ Area $= 6.36 \times 10^6 \times 0.108 = 0.687 \times 10^6$ N

Force on each bolt $= \dfrac{F}{Z} = \dfrac{0.687 \times 10^6}{8} = 0.086 \times 10^6$ N

For bolts $F = \dfrac{\pi}{4}(d_c)^2 \sigma_t$ or $86,000 = 0.785 \times (d_c)^2 \times 60 = d_c = 42.7$ mm

Standard bolt size for core diameter near to 42.7 mm is M 48.

17.8 Flywheels for Engines

I.C. engines are either two-stroke or four-stroke. In four-stroke engines, the power stroke is only once in two revolutions and hence lot of fluctuation of energy. Turning moment diagram of a four-stroke engine is shown in Figure 17.3. In suction, compression and exhaust stroke, the power is given to engine by the flywheel and hence negative, that is, below the atmospheric pressure line. But due to inertial forces of the reciprocating and rotating masses, shown in Figure 17.9(a) the turning moment diagram gets modified as shown in Figure 17.9(b). The mean torque of all the four strokes is shown by chain line.

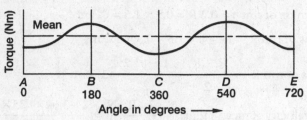

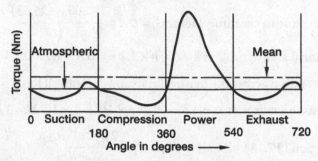

(a) Torque variation with crank angle due to inertial effects

(b) Torque variation with crank angle due to
gas pressure and inertial effects

Figure 17.9 Turning moment diagrams

Average energy delivered by an engine is given by following relation:

$$\text{For four-stroke engine } E = \frac{60,000 \, P}{0.5 \, N}$$

Where, P = Power (kW)

N = Speed of engine (rpm)

Fluctuation of energy $\Delta E = C_E E$

Value of C_E for I.C. engines and steam engines is given in Table 17.2.

$$C_E E = m_r v^2 C_S \quad \text{or} \quad m_r = \frac{E C_E}{v^2 C_S} \tag{17.20}$$

Example 17.8

Mass of a solid disk flywheel for an I.C. engine

Turning moment diagram of a four-stroke petrol engine running at 1,500 rpm is shown in Figure 17.S2(a). Y axis shows turning moment at a scale of 150 Nm per mm and X axis crank shaft rotation at a scale of 5° per mm. Areas measured are given in mm² as $A = -28$, $B = 40$, $C = -74$, $D = 350$, $E = -36$, and $F = 32$. If coefficient of fluctuation of speed is 0.02 and diameter of flywheel is not to exceed 400 mm. Calculate mass of a solid disk flywheel assuming radius to thickness ratio 20 and density as 7,200 kg/m³.

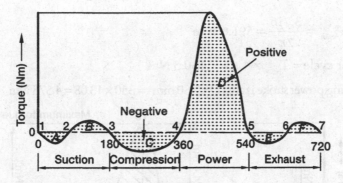

Figure 17.S2(a) Turning moment diagram of a petrol engine

Solution

Given
$N = 1,500$ rpm Y scale $= 150$ Nm/mm X scale $= 5°/\text{mm}$
$C_S = 0.02$ $D = 400$ mm $\rho = 7,200$ kg/m³
$R/t = 20$

Let energy at point 1 be E. Energy at point $2 = E - 28$

Energy at point $3 = E - 28 + 40 = E + 12$

Energy at point $4 = E + 12 - 74 = E - 62$

Energy at point $5 = E - 62 + 350 = E + 288$

Energy at point $6 = E + 288 - 36 = E + 252$

Energy at point $7 = E + 252 + 32 = E + 284$

Y scale $= 150$ Nm/mm and X scale $= 5°/\text{mm}$

Hence 1 mm² of turning moment diagram $= \dfrac{150 \times 5 \times \pi}{180} = 13.08$ Nm

Maximum energy $= E + 288$ and Minimum energy $= E - 62$

Hence variation in energy $\Delta E = (E + 288) - (E - 62) = 350$ mm²

Or, $\Delta E = 350 \times 13.08 = 4,578$ Nm

$$\omega = \frac{2\pi N}{60} = \frac{2 \times 3.14 \times 1,500}{60} = 157 \text{ rad/s}$$

From Equation (17.20): $\Delta E = m_r\, \omega^2 R^2 C_S$

Substituting the values:

$4,578 = m \times (157)^2 \times (0.2)^2 \times 0.02$ or $m = 232$ kg

In one cycle, from point 1 to 7, extra energy available is 284 has to be same. Extra 284 is the net energy available at crank shaft.

284 mm² $= 284 \times 13.08 = 3,716$ Nm

Mean torque $T_{mean} = \dfrac{3,716}{2\pi} = 591.6$ Nm

Work done per cycle $= T_{mean} \times 4\pi = 7,430.5$ Nm

Work done during power stroke is area $D = 350$ mm$^2 = 350 \times 13.08 = 4,578$ Nm

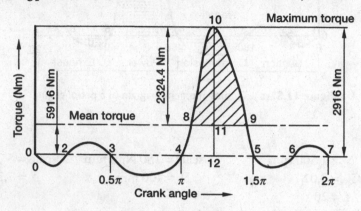

Figure 17.S2(b) Maximum and mean torque

Work done during power stroke (Area 4 – 10 – 5) $= 0.5\,\pi\,T_{max} = 4,578$

Hence $T_{max} = \dfrac{4,578}{0.5\pi} = 2,916$ Nm

Maximum fluctuation of energy $\Delta E = $ Shaded area 8 – 9 – 10 [Figure 17.S2(b)]

From similar triangles: $\dfrac{\text{Area 8-9-10}}{\text{Area 4-10-5}} = \left(\dfrac{\text{Length 10-11}}{\text{Length 10-12}}\right)^2 = \left(\dfrac{T_{max} - T_{mean}}{T_{max}}\right)^2$

$$\Delta E = \left(\text{Area } 4-10-5\right) \times \left(\dfrac{T_{max} - T_{mean}}{T_{max}}\right)^2 = 4,578 \times \left(\dfrac{2,324.4}{2,916}\right)^2 = 2,908 \text{ Nm}$$

$$\Delta E = I\,\omega^2\,C_{,_S} \quad \text{or} \quad 2,908 = I \times (157)2 \times 0.02 \quad \text{or} \quad I = 5.9 \text{ kg m}^2$$

From Equation (17.15): $I = \dfrac{\pi}{2}\left(R_1\right)^4 t\rho$

Substituting the values: $5.9 = 1.57\left(R_1\right)^4 \left(\dfrac{R_1}{20}\right) \times 7,200$ or $\left(R_1\right)^5 = 0.01$

Outer radius of flywheel $R_1 = \sqrt[5]{0.01} = 0.4\,\text{m} = 400$ mm

Thickness $t = \dfrac{R_1}{20} = \dfrac{0.4}{20} = 0.02\,\text{m} = 20$ mm

Mass of flywheel $m = 2\pi R t\rho = 2 \times 3.14 \times 0.4 \times 0.02 \times 7,200 = 362$ kg

Example 17.9

Size of rim of a flywheel for a steam engine with given torque

Turning moment diagram of a double acting steam engine running at 300 rpm is shown in Figure 17.S3(a). For one side turning moment is 20 kNm and for other side 16 kNm both for 50 per cent cut off. Safe tensile stress for the flywheel is 5 MPa and width to thickness ratio of flywheel rim is 2. Density of material is 7,200 kg/m³. Calculate:

a. Power developed.

b. Diameter of fly wheel, if speed variation is limited to ± 1.5 per cent.

c. Cross section of rim.

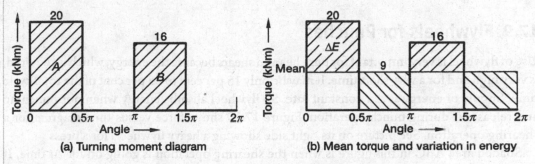

(a) Turning moment diagram

(b) Mean torque and variation in energy

Figure 17.S3 Torque diagram of a double acting steam engine

Solution

Given $N = 300$ rpm $T_1 = 2,000$ kNm $T_2 = 1,600$ kNm
            $C_s = \pm 1.5$ per cent $\sigma_t = 5$ MPa $b/h = 2$
            $\rho = 7,200$ kg/m³

Area of moment diagram $= (20 \times 0.5\,\pi) + (16 \times 0.5\,\pi) = 18\,\pi$

Mean torque $= \dfrac{18\pi}{2\pi} = 9$ kNm $= 9,000$ Nm

Maximum torque $= 20,000$ Nm

a. Power developed $P = \dfrac{2\pi NT}{60,000} = \dfrac{2 \times 3.14 \times 300 \times 9,000}{60,000} = 283$ kW

b. From Equation (17.12):

$\sigma_t = \rho v^2$ or $5 \times 10^6 = 7,200\ v^2$ or $v = 26.3$ m/s

$v = \dfrac{\pi DN}{60}$ or $26.3 = \dfrac{3.14 \times D \times 300}{60}$ or $D = 1.678$ m

c. Area of rim $A = b \times h = 2h \times h = 2h^2$

Maximum variation of energy is shaded area: [Refer Figure 17.S3(b)]

$\Delta E = (20,000 - 9,000) \times 0.5\,\pi = 5,500\,\pi$

From Equation (17.20):

$\Delta E = m v^2 C_s$ or $5,500\,\pi = m \times (26.3)^2 \times 0.03$ or $m = 832\ \text{kg}$

Also, mass $m = \pi D A \rho = 832 = 3.14 \times 1.678 \times 2\,h^2 \times 7,200$

Or, $h = 0.105\ \text{m} = 105\ \text{mm}$

Width $b = 2h = 2 \times 105 = 210\ \text{mm}$

17.9 Flywheels for Punches

Use of flywheel is very important for punches and shears because the energy, which is required, is very high and for a very short time. It may be only 15 per cent to 25 per cent of the total cycle time. Supply of energy is at a constant rate. So flywheel absorbs energy when not punching and releases it during punch operation. Figure 17.10 shows force versus time diagram for a shearing operation. See picture on its right side showing a heavy flywheels for a press

Shaded area ABC in this figure is when the shearing operation is going on for Δt time. It can be seen, it is quite less than the cycle time T shown on the X axis. Line DE shows constant force supplied by the motor continuously. During shearing time, part of the force is supplied by the motor and rest mostly by flywheel.

Force required to shear F = Sheared area × Ultimate shear stress

$F = A \times S_{su}$

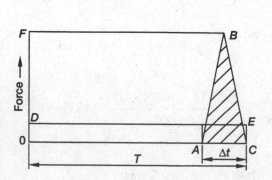

A press

Figure 17.10 Force-time diagram

Force required to punch a hole of diameter d' and thickness t', $F = \pi d' t' S_{su}$

Force required to shear a plate of length L' and thickness t', $F = L' t' S_{su}$

This force F varies from zero to F linearly; hence average force can be taken as $0.5\,F$. Work done during shearing operation: $W = $ Average force $\times$ Distance $= 0.5\,F \times t'$ Energy required for shearing operation is shaded area, that is,

$$E = W = 0.5\,F \times t' \tag{17.21}$$

During this time, energy supplied by the motor is:

$$E_m = \frac{\Delta t \times E}{T} \tag{17.22}$$

Energy supplied by flywheel $E_f = E - E_m$ $\tag{17.23}$

Example 17.10

Mass and size of a flywheel to punch a hole of given size

A punching machine is designed to punch a hole of 20 mm diameter in 10 mm thick steel plate having ultimate shear strength 320 MPa. Design force should be taken 30 per cent more than the force required to punch and mechanical efficiency can be taken 90 per cent. Speed of flywheel is 150 rpm and velocity of flywheel rim should not exceed 10 m/s. Fluctuation in speed is 15 per cent. Take the density of material 7,200 kg/m³. Assume width to height of rim 1.5 and neglect mass of arms and hub. Calculate mass and size of rim.

Solution

Given
$d' = 20$ mm $\qquad t' = 10$ mm $\qquad S_{su} = 320$ MPa
$N = 150$ rpm $\qquad \eta_{mech} = 90$ per cent $\qquad F_{design} = 1.3\,F$
$v_1 = 10$ m/s $\qquad v_2 = 85 \times v_1 \qquad \rho = 7{,}200$ kg/m³
$b/h = 1.5$

Shearing force $F = \pi d' t' S_{su} = 3.14 \times 20 \times 10 \times 320 = 201$ kN

Design force $F_{design} = 1.3\,F = 1.3 \times 201 = 261.3$ kN

Work done, $W = $ Average force $\times$ Distance $= 0.5\,F \times t' = 0.5 \times 261.3 \times 10 = 1{,}306$ Nm

Energy required at flywheel $E = \dfrac{W}{\eta_{mech}} = \dfrac{1{,}306}{0.9} = 1{,}451.4$ Nm $\qquad\qquad$ (a)

Rim velocity before punch $v_1 = \dfrac{\pi D_o N}{60}$ $\quad$ or $\quad$ $D_o = \dfrac{60 v_1}{\pi N} = \dfrac{60 \times 10}{3.14 \times 150} = 1.274$ m

Velocity after punch $v_2 = 0.85 \times 10 = 8.5$ m/s

Energy in flywheel $E = \dfrac{1}{2} m \left(v_1^2 - v_2^2 \right) = \dfrac{1}{2} m \left(10^2 - 8.5^2 \right) = 13.875\,m$

Substituting value of this energy in Equation (a):

$$13.875\,m = 1{,}451.4 \qquad \text{or} \qquad m = 104.6 \text{ kg}$$

Mass of rim = Circumference at mean radius × Area of rim × Density

Or, $m = \pi(D_o - 2h) \times (h \times b)\rho = 3.14 \times (1.274 - h) \times 1.5h \times 7{,}200$

Or, $104.6 = (1.274 - h) \times 33{,}912\,h^2$

Solving by trial and error: $h = 0.05$ m = 50 mm

Hint Find approximately value, by first substituting $(1.274 - h)$ as 1.274 because h is small.

Width of rim $b = 1.5\,h = 1.5 \times 50 = 75$ mm

Inside diameter of rim = Outside diameter $- 2\,h = 1{,}274 - 2 \times 50 = 1{,}174$ mm

Example 17.11

Force by a punching machine for given energy

A punching machine has a flywheel, which can store 15 kNm at 200 rpm. During punching stroke, its speed decreases to 150 rpm.

   a. What percentage of energy is given by the flywheel in this operation?

   b. If only 90 per cent of the energy is delivered to the punching machine with a stroke of 10 mm, how much force can it exert?

Solution

Given $E_1 = 15$ kNm $N_1 = 200$ rpm $N_2 = 150$ rpm $E = 0.9(E_1 - E_2)$ $d = 10$ mm

   a. Energy before punching $E_1 = 0.5 I \omega_1^2$ and energy after punching $E_2 = 0.5 I \omega_2^2$

$$\frac{E_2}{E_1} = \frac{0.5 I \omega_2^2}{0.5 I \omega_1^2} = \frac{\omega_2^2}{\omega_1^2} = \frac{N_2^2}{N_1^2} = \frac{150^2}{200^2} = 0.5625$$

Energy given by the flywheel $= E_1 - E_2 = E_1 - (0.5625\,E_1) = 0.4375\,E_1$

Percentage of energy given by the flywheel $= \dfrac{E_1 - E_2}{E_1} \times 100$

$$= \frac{0.4375\,E_1}{E_1} \times 100 = 43.75 \text{ per cent}$$

   b. Energy given to punching machine $= 0.9 \times (E_1 - E_2) = 0.9 \times 0.4375 E_1$

$$= 0.9 \times 0.4375 \times 15{,}000 = 5{,}906 \text{ Nm}$$

Energy given by the flywheel is converted to the work done = Force × distance

$$5{,}906 = F \times 0.01 \quad \text{or} \quad F = 590{,}600 \text{ N} = 590.6 \text{ kN}$$

Example 17.12

Size of a flywheel rim to shear a plate of given size

A sharing machine shown in Figure 17.S4(a) is designed to shear a mild steel 3 mm thick plate of width 1 m having ultimate shear strength 350 MPa. Shearing operation takes 60° of crank rotation, which runs at 200 rpm. There are 10 cuts per minute and mechanical efficiency is 90 per cent. Assume that 90 per cent mass is of rim, width to height ratio 1.5, and coefficient of fluctuation of speed is 0.15. Maximum allowed velocity is 9 m/s. Calculate mass and size of rim.

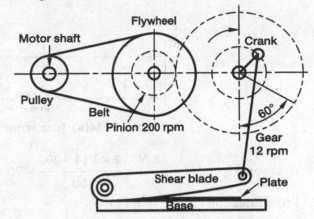

Figure 17.S4(a) Shearing machine

Solution

Given
$L' = 1,000$ mm $\qquad t' = 3$ mm $\qquad\qquad S_{su} = 350$ MPa
$\theta = 60° \qquad\qquad N = 200$ rpm $\qquad$ Cuts per minute $= 10$
$\eta_{mech} = 90$ per cent $\qquad \rho = 7,200$ kg/m³ $\qquad b/h = 1.5$
$C_S = 0.15 \qquad\qquad v = 9$ m/s

Shearing force $F = \theta' t' S_{su} = 1,000 \times 3 \times 350 = 1,050,000$ N $= 1,050$ kN

Work done for one shear= Average force × Distance $= 0.5\, F \times t'$

Or, $\quad W = 0.5 \times 1,050 \times 10^3 \times 3 \times 10^{-3} = 1,575$ Nm

Work done per second $= \dfrac{W \times 10}{60} = \dfrac{1,575 \times 10}{60} = 2{,}62.5$ Nm

Power required $= \dfrac{W}{\eta_{mech}} = \dfrac{262.5}{0.9} = 292$ say 300 W

There are 10 cuts per minute hence cycle time $T = \dfrac{60}{10} = 6$ s

Shearing time $\Delta t = \dfrac{60 \times 6}{360} = 1$ s [See Figure 17.S4(b)]

Energy required to shear is: $E = W = 1,575$ Nm

Energy supplied by the motor is $E_m = \dfrac{\Delta t \times E}{T} = \dfrac{1 \times 1,575}{6} = 262.5$ Nm

Energy supplied by flywheel $E_f = E - E_m = 1,575 - 262.5 = 1,312.5$ Nm

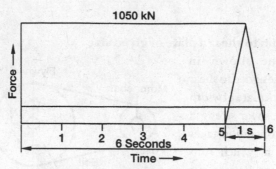

Figure 17.S4(b) Force versus time diagram

Rotational speed $\omega = \dfrac{2\pi N}{60} = \dfrac{2 \times 3.14 \times 200}{60} = 20.93$ rad/s

From Equation (17.7): $E_r = I_r \omega^2 C_s$

Rim has 90 per cent energy; hence substituting the values:

$0.9 \times 1{,}312.5 = I_r \times (20.93)^2 \times 0.15$ or $I_r = 18$ kg m²

Velocity $v = \omega R = 20.93\, R$

Hence, $R = \dfrac{9}{20.93} = 0.43$ m

Mass of rim $m_r = \dfrac{I_r}{R^2} = \dfrac{18}{(0.43)^2} = 97.3$ kg

Also, $m_r = 2\pi R\,(b \times h)\,\rho$

Substituting the values: $97.3 = 2 \times 3.14 \times 0.43 \times \left(\dfrac{1.5h}{10^3} \times \dfrac{h}{10^3} \right) \times 7{,}200$

Or, $h = 58$ mm, and width of rim $b = 1.5\,h = 1.5 \times 58 = 87$ mm

Example 17.13

Size of solid disk flywheel of a press for given torque variation

An electric motor running at 1,500 rpm delivering a constant torque is to be used for a dual crank press with 180° angle in between. The torque increases linearly from 0 Nm to 100 Nm and then decreases linearly to 0 Nm in half revolution In other half revolution torque varies from 0 Nm to 80 Nm and again to 0 Nm as shown in Figure 17.S5(a). If coefficient of fluctuation of speed is limited to 0.1, calculate the size of a solid disk flywheel, if it is driven by a gear set of ratio 7.5. Take density 7,800 kg/m³. Calculate:

    a. Power of motor required.

    b. Size of flywheel, assuming radius 20 times the thickness.

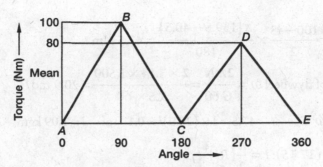

Figure 17.S5(a) Torque diagram of a press

Solution

Given $N_1 = 1,500$ rpm $\quad C_s = 0.1 \quad G = 7.5 \quad \rho = 7,800$ kg/m³ $\quad R/t = 20$

Area of triangle ABC $= \dfrac{100 \times \pi}{2} = 157$ Nm

Area of triangle CDE $= \dfrac{80 \times \pi}{2} = 125.6$ Nm

Total area of ABC and CDE triangles $W = 282.6$ Nm

Mean torque $T_{mean} = \dfrac{282.6}{2\pi} = 45$ Nm

Power by motor $P = \dfrac{2\pi NT}{60,000} = \dfrac{2 \times 3.14 \times 1,500 \times 45}{60,000} = 7.065$ kW

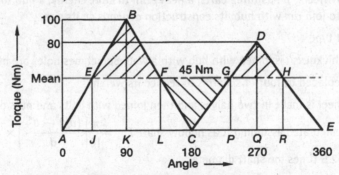

Figure 17.S5(b) Fluctuation of energy from mean torque

Maximum variation in energy is area EBF. To find this area, calculate angle for points E and F using linear interpolation or similar triangles.

Angle of point $J = \dfrac{45 \times 90}{100} = 40.5°$

Angle of point $L = 180 - 40.5 = 139.5°$

Area EBF, $\Delta E = \dfrac{100 - 45}{2} \times \dfrac{\pi\left(139.5 - 40.5\right)}{180} = 47.5 \text{ Nm}$

Angular speed of flywheel $\omega = \dfrac{2\pi N}{G\,60} = \dfrac{2 \times 3.14 \times 1,500}{7.5 \times 60} = 20.9 \text{ rad/s}$

$\Delta E = I\,\omega^2 C_s \quad \text{or} \quad 47.5 = I \times (20.9)^2 \times 0.1 \quad \text{or} \quad I = 1.09 \text{ kgm}^2$

From Equation (17.15): $I = \dfrac{\pi}{2}\left(R_1\right)^4 t\rho$

Substituting the values: $1.09 = 1.57\left(R_1\right)^4\left(\dfrac{R_1}{20}\right) \times 7,800 \quad \text{or} \quad \left(R_1\right)^5 = 0.0018$

Outer radius of flywheel $R_1 = \sqrt[5]{0.0018} = 0.280 \text{ m} = 280 \text{ mm}$

Thickness $t = \dfrac{R_1}{20} = \dfrac{280}{20} = 0.014 \text{ m} = 14 \text{ mm}$

Mass of flywheel $m = 2\pi R\,t\rho = 2 \times 3.14 \times 0.28 \times 0.014 \times 7,800 = 192 \text{ kg}$

Summary

Flywheel acts as a reservoir to store energy and gives it back when required. It is used in I.C. engine, power press, etc., where heavy energy is required in a shear or blanking process. Its main function is to store energy when idle and release energy when required.

Construction of flywheel It has three parts; a heavy rim to store energy, a hub to mount on a shaft and a disk or arms to join rim with hub. Its construction depends on its size.

Sold disk Smallest type

Webbed Rim of thickness t is joined with hub with a web. Sometimes holes are made in the web.

Armed Web is replaced by four or six arms to reduce its weight.

Split Big flywheel is made in two halves and then joined with bolts and nuts or cotters.

Hub diameter d_h is calculated assuming it as hollow shaft, $T_{max} = \dfrac{\pi}{16}\left(\dfrac{\left(d_h\right)^4 - d^4}{d}\right) \times \tau$

Empirically, $d_h = 2\text{--}2.5$ times the shaft diameter.

Length of hub L is taken same as the axial width of the rim Or empirically $2.5d$.

Sunk key of cross section $\dfrac{d}{4} \times \dfrac{d}{4}$ is used. Length of key = Hub length L

L is calculated using relation: $T_{max} = \left(L \times \dfrac{d}{4}\right) \times \dfrac{d}{2} \times \tau$

Fluctuation in energy and speed

If input torque T_i is more than output torque T_o, the shaft accelerates and if less, it decelerates.

Energy stored in a flywheel $E = \dfrac{1}{2} I \omega^2$

Where, I = Mass moment of in inertia

ω = Mean rotational speed of flywheel in radians per second = $\dfrac{\omega_{max} + \omega_{min}}{2}$

Mean torque T_m is calculated such that negative area below the mean torque line up to base line is equal to positive area above this line.

Work done $W = 2\pi s\, T_m$

$s = 1$ for two-stroke engines; $s = 2$ for four-stroke engines

Change in energy $\Delta E = \dfrac{1}{2}I\left(\omega_{max}^{2} - \omega_{min}^{2}\right) = I\,\dfrac{\omega_{max} + \omega_{min}}{2}\left(\omega_{max} - \omega_{min}\right) = I\,\omega\,\Delta\omega$

Coefficient of fluctuation of energy $C_E = \dfrac{E_i - E_o}{\text{Work done per cycle}} = \dfrac{\Delta E}{W}$

Coefficient of fluctuation of energy for engines is given in Table 17.1.

Coefficient of fluctuation in speed is defined as $C_s = \dfrac{\omega_{max} - \omega_{min}}{\omega} = \dfrac{\Delta\omega}{\omega}$

Coefficient of fluctuation $C_s = \dfrac{2\left(\omega_{max} - \omega_{min}\right)}{\omega_{max} + \omega_{min}} = \dfrac{2\left(N_{max} - N_{min}\right)}{N_{max} + N_{min}} = \dfrac{N_{max} - N_{min}}{N}$

$$\Delta E = I\,\omega^{2}\,C_s = 2\,E\,C_s$$

Stresses in flywheel A flywheel rim has following stresses: Tensile and Bending

a. **Tensile stresses due to centrifugal force**

Mass of flywheel $M = \rho\, A\, R\, \delta\theta$

ρ = Mass density $\qquad\qquad A = (b \times h)$ cross-sectional area of rim

b = Axial width of rim $\qquad\qquad h$ = Radial thickness of the rim

R = Mean radius of the rim

Centrifugal force $F_c = \dfrac{Mv^{2}}{R}$

Substituting the value of M: $F_c = \dfrac{\rho\, A\, R\, \delta\theta\, v^{2}}{R}$

Tensile stress $\sigma_t = \dfrac{F_c}{A\,\delta\theta} = \dfrac{\rho\, A\, R\, \delta\theta\, v^{2}}{R\, A\, \delta\theta} = \rho v^{2}$

b. **Bending stresses due to constrained arms**

The portion of the rim between the two adjacent arms can be considered as a curved beam, loaded due to the centrifugal forces. Maximum BM occurs at arm end.

Mass of the rim per mm of circumference of the rim $w = \rho\, b\, h$

Mass of the rim between two adjacent arms $M' = \dfrac{M}{m}$ (m = number of arms)

Length of beam between two adjacent arms $l = \dfrac{2\pi R}{m}$

Maximum BM. is: $M_b = \dfrac{w l^2}{12} = \dfrac{\rho h b \omega^2 R \left(\dfrac{2\pi R}{m}\right)^2}{12}$

Modulus of section $Z = \dfrac{b' h^2}{6}$

$\sigma_b = \dfrac{M_b}{Z}$ and $v = \omega R$, on simplification $\sigma_b = \dfrac{2\rho (\pi v)^2 R}{m^2 h}$

G–Lanza gave that total resultant stress $\sigma_{total} = 0.75\sigma_t + 0.25\sigma_b = \rho v^2 \left(0.75 + \dfrac{\pi^2 R}{2m^2 h}\right)$

Guide lines to select a type of flywheel depending on outside diameter D_o.

- Solid flywheel, if $D_o < 200$ mm
- Flywheel with disk, if $D_o < (7.5d + 80$ mm), where d is shaft diameter in mm
- Flywheel with arms, if $D_o > (7.5d + 80$ mm)
- Split flywheel with arms, if $D_o > 2{,}500$ mm

Solid flywheel is just a circular disk of radius outer R_1 of width b

Moment of inertia I of such a disk is: $I = \dfrac{1}{2} m (R_1)^2$

If t = Axial thickness of disk (m), then mass of flywheel $m = \pi (R_1)^2 t \rho$

For C.I, $\rho = 7{,}200$ kg/m³, for steel, $\rho = 7{,}820$ kg/m³

$$I = \dfrac{1}{2}\pi (R_1)^2 t \rho \times (R_1)^2 = \dfrac{\pi}{2}(R_1)^4 t \rho$$

Energy stored in a flywheel $E = \dfrac{1}{2} I \omega^2 = \dfrac{\pi}{4}(R_1)^4 t \rho \omega^2$

$$\Delta E = I \omega^2 C_S = 2 E C_S = \dfrac{\pi}{2}(R_1)^4 t \rho \omega^2 C_S$$

Webbed flywheels   Energy stored by the hub is neglected as its diameter is much smaller than rim diameter. Thickness of web $t = 0.05b$, where, b is width of rim in mm.

Mass of the flywheel m = Mass of rim + Mass of disk

Radial thickness of rim h is taken greater than or equal to axial width b. Value of h varies from $0.7b$ to $2b$.

Flywheels with arms   Number of arms m depends on outside diameter:

Diameter < 400 mm, $m = 4$, Diameter 400 mm to 3,000 mm, $m = 6$, Diameter > 3,000 mm, $m = 8$ to 12

Minimum FOS is taken as 8. For heavy shocks like punching FOS can be 12.

Width of arm in the plane of rotation a is kept more than along the axis of flywheel. It is assumed that only 10 per cent of torque is contributed by arms and the rest 90 per cent by the rim

Moment of inertia of rim $I_r = C_r I$. Value of constant for rim C_r is taken as 0.9.

Radial thickness of rim $h = R_1 - R_2$, where, R_1, R_2 = Outer and inner diameter of rim respectively.

Since h is very small in comparison to rim radii, radius of gyration $R = \dfrac{R_1 + R_2}{2}$

Mass of rim $m_r = 2 \pi R (h \times b) \rho$

Radius of gyration of rim $K_r = R$

Moment of inertia of rim $I_r = m_r (K_r)^2 = 2 \pi R (h \times b) \rho R^2$

Energy stored in rim $E_r = \dfrac{1}{2} I_r \omega^2 = \pi R h b \rho v^2$ $\quad (v = \omega R)$

$$\Delta E = I_r \omega^2 C_s = 2 \pi R h b v^2 C_s = m_r v^2 C_s$$

Design of arms   Section modulus of elliptical cross section is: $Z = \dfrac{\pi}{32} (a_1)^2 b_1$.

a_1 = Major axis of arm at hub. Minor axis of arm at hub $b_1 = 0.5 a_1$.

a_2 and b_2 are sizes of arm near the rim.

Taper per meter length of arm for major axis is about 20 mm and 10 mm for minor axis.

If D_r = Diameter of rim, D = Diameter of hub, l = Length of arms

BM for each arm for a torque T is:

$$M = \frac{T \times l}{m \times 0.5 D_r} = \frac{T}{m \times 0.5 D_r} \times \frac{(D_r - D)}{2} = \frac{T \times (D_r - D)}{m \times D_r}$$

Bending stress $\sigma_b = \dfrac{M}{Z} = \dfrac{T \times (D_r - D)}{m \times D_r} \times \dfrac{32}{\pi (a_1)^2 b_1}$ (Assuming the value $b_1 = 0.5\, a_1$)

$$\sigma_b = \frac{20.4 T \times (D_r - D)}{m \times (a_1)^3 \times D_r}$$

This stress should be less than allowable bending stress.

Split flywheels   Very large-sized flywheels are made in two halves, due to size limitations of castings.

Tensile stress due to centrifugal force $\sigma_t = \rho v^2$

Cross-sectional area of rim $A = h \times b$

Force F = Tensile stress × Area = $\sigma_t A$

Bending stress $\sigma_b = \dfrac{2 \rho (\pi v)^2 R_m}{m^2 h}$

Total stress in rim $\sigma_{total} = 0.75\, \sigma_t + 0.25\, \sigma_b$

Total tensile force on the rim $F = \sigma_{total} \times$ Area

Force on each bolt $= \dfrac{F}{Z} = \dfrac{\pi}{4} (d_c)^2 \sigma_t$ (Z = Number of bolts)

Calculate core diameter and then choose a standard bolt.

Flywheel for I.C. engines For four-stroke engines, the power stroke is only once in two revolutions causing a lot of fluctuation of energy. If P power in kW and N speed in rpm:

Average energy delivered by an engine for four-stroke engine $E = \dfrac{60,000\,P}{0.5\,N}$

Fluctuation of energy $\Delta E = C_E E$

Value of C_E for engines is given in Table 17.2.

$$C_E E = m_r\, v^2\, C_S \qquad \text{or} \qquad m_r = \frac{E\,C_E}{v^2\,C_S}$$

Flywheel for punches Flywheel is very important for punches and shears because the energy required is very high and for a very short time. It may be only 15 per cent to 25 per cent of total cycle time.

Force required to shear a plate of length L' and thickness t', $F = L'\,t'\,S_{su}$

This force F varies from zero to F linearly, hence average force can be taken as 0.5 F.

Work done during shearing operation: $W = $ Average force × Distance = 0.5 $F t'$

Energy required for shearing operation is $E = W = 0.5\,F t'$

During this time, energy supplied by the motor is $E_m = \dfrac{\Delta t \times E}{T}$

Energy supplied by flywheel $E_f = E - E_m$

Theory Questions

1. Write the function of a flywheel. Give some typical applications where it is used.

2. Describe the construction of various types of flywheels with a neat simple sketch.

3. How do you design shaft, hub, and key for a flywheel?

4. Define coefficient of fluctuation of speed. Derive an expression for it.

5. What do you mean by fluctuation of energy? Give its values for some typical applications.

6. Draw a turning moment for a four-stroke I.C. engine and with the help of this sketch explain mean torque and fluctuation of energy.

7. Give reasons as to why rim velocity is important in the design for flywheel. Give safe values of rim velocity for different materials.

8. Derive an expression for the tensile stress in a flywheel, due to centrifugal force.

9. Discuss the bending stresses developed in the rim of a flywheel. Derive an expression to calculate total tensile stresses in the rim of a flywheel.

10. Write a note on a solid disk flywheel, mentioning its moment of inertia, relation of energy, and coefficient of fluctuation of speed and its mass.

11. Write the method to design a flywheel with a web and rim.

12. Describe the design procedure for the design of a flywheel with rim and arms.

13. What do you mean by a split flywheel? When it is used and what are its advantages? Describe with sketches the methods to join two halves of hub and rim.

14. How do you design a flywheel for an I.C. engine and steam engine?

15. Differentiate the difference in design procedure for a flywheel for engine and a punching machine.

16. How do you calculate the work done in punching a hole of given diameter and thickness of a plate.

Multiple Choice Questions

1. A flywheel is used to:

    (a) Control the maximum speed of an engine

    (b) Control the minimum speed of an engine

    (c) Stores energy in power stroke and give back when required in other strokes

    (d) None of the above

2. If N is mean rotational speed, and ΔN difference of maximum and minimum speed, then coefficient of fluctuation of speed is:

    (a) $\dfrac{\Delta N}{N}$
    (b) $\dfrac{2\Delta N}{N}$
    (c) $\dfrac{\Delta N}{0.5N}$
    (d) $\dfrac{N}{\Delta N}$

3. In a turning moment diagram, fluctuation of energy is:

    (a) Variation in energy above and below the mean torque line

    (b) Difference between maximum energy and mean torque line

    (c) Difference between mean torque line and minimum energy

    (d) Difference between maximum energy and minimum energy

4. If W is work done per cycle, ΔE maximum fluctuation of energy and C_s coefficient of fluctuation of speed, then coefficient of fluctuation of energy is:

    (a) $\dfrac{WC_s}{\Delta E}$
    (b) $\dfrac{2\,\Delta E}{C_s W}$
    (c) $\dfrac{\Delta E}{2W}$
    (d) $\dfrac{\Delta E}{W}$

5. Centrifugal force of the rim in a big flywheel causes stress in arms which is:

    (a) Shear
    (b) Tensile
    (c) Compressive
    (d) Bending

6. If σ_t is tensile stress in a rim of a flywheel and σ_b is bending stress due to centrifugal force then the total stress is taken as:

    (a) $\sigma_t + \sigma_b$
    (b) $0.25\,\sigma_t + 0.75\,\sigma_b$

    (c) $0.5\sigma_t + 0.5\,\sigma_b$
    (d) $0.75\sigma_t + 0.25\,\sigma_b$

7. Cross section of flywheel arms is generally kept as:

    (a) Circular
    (b) Rectangular

    (c) Elliptical
    (d) I section

8. FOS for arms of flywheel with heavy shocks is taken as:

    (a) 3 to 5
    (b) 6 to 10
    (c) 10 to 15
    (d) More than 15

9. Length of hub is taken as:

    (a) Equal to shaft diameter
    (b) Twice the shaft diameter

    (c) Equal to hub diameter
    (d) Equal to width of rim

10. If ΔE is variation in energy, I moment of inertia, ω rotational speed, and C_s coefficient of speed variation. These are related as:

    (a) $\Delta E = I\,\omega^2 C_s$ (b) $\Delta E = \dfrac{1}{2}\omega^2 C_s$ (c) $\Delta E = \dfrac{1}{2}I\omega^2 C_s$ (d) $2\,I\,\omega^2 C_s$

11. If F is the maximum force to punch a hole of diameter d in a plate of thickness t, the work done is:

    (a) $F\,d\,t$ (b) $0.5\,F\,t$ (c) $0.9\,F\,t$ (d) F/dt

12. Bending stress in a flywheel depends on:

    (a) Rim velocity   (b) Radius of flywheel

    (c) Number of arms   (d) All given in (a), (b), and (c)

Answers to multiple choice questions

1. (c)   2. (a)   3. (b)   4. (d)   5. (b)   6. (b)   7. (c)   8. (c)   9. (d)   10. (a)

11. (b)   12. (d)

Design Problems

1. A four-stroke I.C. engine is developing 5 kW at 1,800 rpm. Coefficient of fluctuation of energy is 2.2. A solid cast iron flywheel of density 7,200 kg/m³ is to be designed such that speed variation is within 1 per cent. Maximum allowable rim velocity is 30 m/s. Find outside diameter, width and mass for a solid flywheel.   [D = 320 mm, t = 142 mm, m = 81 kg]

2. A single cylinder four-stroke diesel engine develops 25 kW at 400 rpm. Work done in power stroke is 2.5 times the work done in compression stroke. Neglect the work done in suction and exhaust strokes. If speed is to be maintained within ± 1.5 per cent, find mass of a solid disk flywheel if its radius is to be kept 0.6 m.   [560 kg]

3. A crank rotating at 300 rpm is driven by a gear set with gear ratio 5:1. Flywheel is mounted on the pinion shaft and gets power from belt. An overhang crank on the flywheel shaft drives a punch to make 20 holes per minute of diameter 15 mm in a plate of thickness 8 mm with ultimate shear strength 320 MPa. Cutting operation takes for quarter of a revolution. Overall transmission efficiency of the drive is 80 per cent and coefficient of speed variation is 0.15. Mass of flywheel is 200 kg. Calculate:

    (a) Power of motor   (b) Radius of flywheel.

                                                [(a) P = 0.2 kW; (b) R = 110 mm]

4. A punching machine is designed to punch 15 holes per minute of diameter 25 mm in a plate of thickness 10 mm. Machine is driven from a pulley working as flywheel at 270 rpm with 200 mm overhung. Punching time is 10 per cent of cycle time. Diameter of flywheel should not be more than 1 m with maximum four arms, height of rim to width ratio 0.6, and coefficient of fluctuation of speed 0.1. Mechanical efficiency of the drive can be assumed 85 per cent, FOS = 4, ultimate shear stress 240 MPa, and tensile strength 360 MPa Density of material = 7,200 kg/m³ Calculate:

(a) Size of rim and its mass, assuming 90% of flywheel effect is by rim.

(b) Tensile stress in rim.

$$[(a)\ D_o = 0.94\ m,\ b = 62\ mm,\ h = 37\ mm\ and\ m = 47\ kg \qquad (b)\ \sigma_t = 1.2\ MPa]$$

5. A punching machine is driven by an electric motor running at 1,500 rpm through a pulley acting as flywheel with speed ratio of 4. Stroke length of punch is 200 mm. Punching load is 30 kN, which acts during the last is 18 per cent of stroke. Coefficient of fluctuation of speed is 0.02. Diameter of flywheel should not exceed 500 mm. Calculate power of motor, assuming transmission efficiency 80% and mass of flywheel. $\qquad [P = 4.22\ kW,\ m = 241\ kg]$

Previous Competition Examination Questions

IES

1. The flywheel of a machine having weight of 4,500 N and radius of gyration 2 m has cyclic fluctuation of speed from 125 rpm to 120 rpm. Assuming $g = 10\ m/s^2$, the maximum fluctuation of energy is:

    (a) 12,822 N-m   (b) 24,200 N-m   (c) 14,822 N-m   (d) 12,100 N-m

    [Answer: (d)] [IES 2015]

2. If the speed of an engine varies between 390 rpm and 410 rpm in a cycle of operation, the coefficient of fluctuations of speed would be:

    (a) 0.01   (b) 0.02   (c) 0.04   (d) 0.05

    [Answer: (d)] [IES 2013]

GATE

3. The torque in newton meters exerted on the crank shaft of a two-stroke engine can be described as $T = 10,000 + 1,000 \sin 2\theta - 1,200 \cos 2\theta$, where θ is the crank angle as measured from the inner dead center position. Assuming the resisting torque to be constant, the power (kW) developed by the engine at 100 rpm is _____ [Answer: $P = 104.7$ kW] [GATE 2015]

4. Torque and angular speed data for one cycle for a shaft carrying a flywheel are shown in the figure below. Moment of inertia (kgm^2) of the flywheel is _____

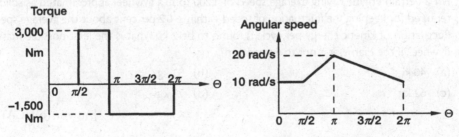

[Answer: $I = 32$ kg m²] [GATE 2014]

5. Maximum fluctuation of kinetic energy in an engine is calculated to be 2,600 J. Assuming that the engine runs at an average speed of 200 rpm, the polar mass moment of inertia in (kgm²) of a flywheel to keep the speed fluctuation within ± 0.5 per cent of the average speed is_____

   [Answer: I = 593 kg m²] [GATE 2014]

6. An annular disk has a mass m, inner radius R, and outer radius $2R$. The disc rolls on a flat surface without slipping. If the velocity of the center of mass is v, the kinetic energy of the disk is:

   (a) $\dfrac{9}{16}mv^2$ (b) $\dfrac{11}{16}mv^2$ (c) $\dfrac{13}{16}mv^2$ (d) $\dfrac{15}{16}mv^2$

   [Answer: (c)] [GATE 2014]

7. A flywheel connected to a punching machine has to supply energy of 400 Nm while running at a mean angular speed of 20 rad/s. If total fluctuation of speed is not to exceed within ± 0.2 cent, the mass moment of inertia of the flywheel in kg m² is:

   (a) 25   (b) 50   (c) 100   (d) 120

   [Answer: (a)] [GATE 2013]

8. A circular solid disc of uniform thickness 20 mm, radius 200 mm, and mass 200 kg is used as a flywheel. If it rotates at 600 rpm, the kinetic energy of the flywheel in joules is:

   (a) 395   (b) 790   (c) 1,580   (d) 3,160

   [Answer: (a)] [GATE 2012]

9. The speed of an engine varies from 210 rad/s to 190 rad/s. During cycle the change in kinetic energy is found to be 400 Nm. The inertia of flywheel in kg m² is:

   (a) 0.1   (b) 0.2   (c) 0.3   (d) 0.4

   [Answer: (a)] [GATE 2007]

10. If C_f is the coefficient of speed fluctuation of a flywheel then the ratio of $\omega_{max}/\omega_{min}$ will be:

    (a) $\dfrac{1-2C_f}{1+2C_f}$ (b) $\dfrac{2-C_f}{2+C_f}$ (c) $\dfrac{1+2C_f}{1-2C_f}$ (d) $\dfrac{2+C_f}{2-2C_f}$

    [Answer: (d)] [GATE 2006]

11. For a certain engine having average speed of 1,200 rpm, a flywheel approximated as solid disc, is required for keeping the fluctuation of speed within ± 0.2 per cent about the average speed. The fluctuation of kinetic energy per cycle is found to be 2 kJ. What is the least possible mass of the flywheel, if the diameter is not to exceed 1 m?

    (a) 40 kg   (b) 51 kg

    (c) 62 kg   (d) 73 kg

    [Answer: (b)] [GATE 2003]

□□□

Clutches

18.1 Definition and Function

A clutch is a device, which can be used to connect or disconnect the incoming power from a driver shaft to a driven shaft at same speed or less than driver speed, whenever desired without stopping the driver shaft.

It is used invariably in an automobile, shown in picture on right side, where engine rotates continuously, while the wheels rotate at reduced speeds from zero to maximum. The power comes from

engine to clutch at speed N. Speed after the clutch can have any value between zero to N and then to gear box to allow different speed to get high torque at starting as shown pin Figure 18.1. In top gear, there is no reduction. Universal couplings help in transferring power at any angle. A permanent reduction of about 4–5 is provided in differential, to further reduce the speed of the propeller shaft and transfer power at right angles to the wheels.

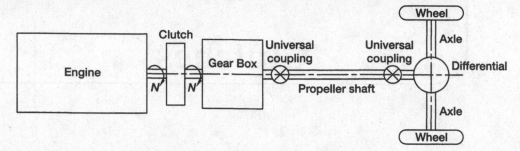

Figure 18.1 Location of a clutch in an automobile

Main functions of a clutch are:

- Connect or disconnect the power from a driver shaft to the driven shaft. The shafts should be collinear, that is, their axes should be in one straight line, otherwise its performance deteriorates.
- Provide any speed between zero to maximum of the driver shaft speed.
- Absorb torsional shocks.

18.2 Types of Clutches

There are following four types of clutches. Selection of a clutch depends on the application and requirements.

a. **Positive drive clutch** The driver and driven shafts are mounted with a flange having jaws or teeth on one side. When these jaws mesh with each other, driven shaft starts rotating at the driver speed. There cannot be any slip and hence these are called positive drive clutches. See Section 18. 3 for its types and design.

b. **Friction clutch** Most of the automobile clutches are of this type. The power from driver shaft is transmitted to the driven shaft through a friction plate called clutch plate. It has friction lining on both the faces, which when contacted with the driver surface start rotating and transfer power to the driven shaft. See Section 18.4 for its construction in detail.

c. **Hydraulic clutch** It is also called fluid coupling as a fluid is used to connect driver and driven rotor. The rotor on driver side is called pump, which sends a fluid at high velocity to a member in front of it called turbine. When this oil strikes on the blades of the turbine it starts rotating. It offers very smooth start and absorbs torsional shocks also. See Chapter 9 for their construction and working.

d. Electromagnet clutch In this type of clutch, power is transmitted by a magnetic field. These clutches use magnetic particles or eddy current to have magnetic coupling. There is no friction and hence no wear. These clutches respond quickly, offer smooth start, and easy to control. These are not described in this book.

18.3 Positive Drive Clutch

Figure 18.2 shows two different types of positive drive clutches; square jaw and spiral jaw. The driver shaft has a flange with either square jaws on sides [Figure 18.2(a)] or spiral jaws [Figure 18.2(b)]. The driven shaft also has similar jaws and the driven member slides over a splined shaft. When the projections of the jaws fit into the recesses, the power is transferred from the driver to the driven shaft. The driven shaft slides with a grooved sleeve to engage or disengage the clutch.

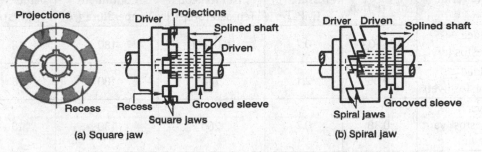

Figure 18.2 Types of positive drive clutches

18.4 Friction Clutch

A friction clutch can be of a single friction plate (Figure 18.3), or multiple plates (Figure 18.5 in Section 18.8) or cone clutch (Section 18.9). Figure 18.3 shows multiple springs to apply force on pressure plate.

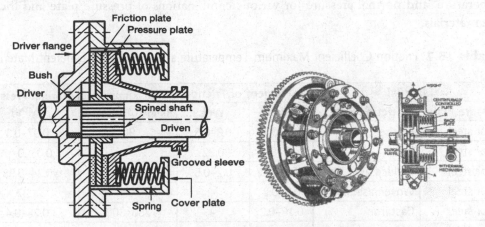

Figure 18.3 Assembly of a single plate friction clutches

Picture of the same type of clutch is shown on its right side with additional axial force caused by centrifugal force of dead weights. Some clutches operate under dry conditions, but if the heat is more due to friction, wet clutches are also used to carry out the heat generated due to sliding friction. Small vehicles such as scooters use wet clutches, in which the plates are dipped in oil to carry out heat.

18.4.1 Friction materials

There are many friction materials for dry and wet clutches. Coefficient of friction, maximum pressure, and temperature, which these materials can bear, are tabulated in Table 18.1.

Table 18.1 Properties of Friction Materials

Material	Coefficient of Friction	Maximum Pressure (p) (MPa)	Maximum Instantaneous Temperature(°C)	Maximum Continuous Temperature (°C)	Maximum Velocity(v) (m/s)
Molded asbestos (dry)	0.35–0.41	0.7	350–400	180	18
Molded Asbestos (wet)	0.06	2.1	180	180	18
Woven Asbestos yarn and wire	0.38	0.7	260	130	18
Woven cotton	0.47	0.7	110	75	18
Resilient paper	0.09–0.15	2.8	150	–	$pv < 18$

18.4.2 Coefficient of friction

Friction coefficient depends not only on one material but also on the mating surface, and its condition (dry or wet). Table 18.2 gives coefficient of friction, maximum allowable temperature, and normal pressure for various combinations of pressure plate and friction plate materials.

Table 18.2 Friction Coefficient Maximum Temperature and Pressure for Different Materials

Material		Coefficient of Friction		Maximum Temperature(°C)	Normal Pressure (MPa)
Pressure Plate	Friction Plate	Dry	Wet		
C.I. or steel	Leather	0.12–0.15	0.3–0.5	90	0.07–0.3
C.I. or steel	Felt	0.18	0.22	90	0.03–0.07
C.I. or steel	Molded asbestos	0.08–0.12	0.2–0.5	260	0.34–0.98
C.I. or steel	Woven asbestos	0.1–0.2	–	–	1.37
C.I. or steel	Cast iron	0.15–0.2	–	250–300	0.25–0.4
C.I. or steel	Cast iron	–	0.06	250–300	0.6–0.8

| Hardened steel | Hardened steel | – | 0.08 | 250 | 0.8 |
| C.I. or steel | Bronze | – | 0.06 | 150 | 0.4 |

18.4.3 Variation of bearing pressure

A friction plate with friction lining on both the sides is shown in Figure 18.4(a). Its outer radius is r_1 and inner radius is r_2.

When the friction plate is new, mating surfaces are parallel and hence pressure is uniform over the entire surface of the friction material [Figure 18.4(b)]. Wear is proportional to rubbing velocity, which is maximum at its outer periphery, and hence the outer annular area wears faster. The wear at this edge reduces the pressure as the pressure plate is rigid, and hence the wear increases now at the inner edge. This cycle of wear at outer edge and then at inner edge continues till the wear is adjusted in such a manner that product of pressure and radius $(p \times r)$ is constants.

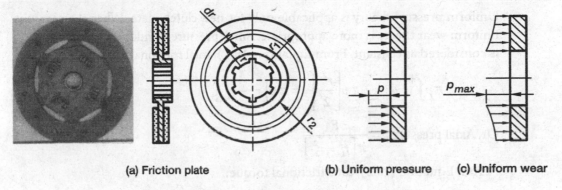

(a) Friction plate   **(b) Uniform pressure**   **(c) Uniform wear**

Figure 18.4 Pressure distribution on friction plate

From the above, it is concluded that a for new friction plate uniform pressure can be assumed, while for a worn or used friction plate, uniform wear is to be assumed [Figure 18.4(c)]. The pressure profile for both the cases is shown in Figure 18.4.

18.4.4 Torque transmitting capacity

Torque that a friction clutch can transmit depends on:

- Number of friction surfaces
- Surface area of the friction disc
- Pressure over the disc

An annular ring of radial length dr [Figure 18.4(a)] is considered at any radius r. Considering this ring:

Annular area of the ring $A = 2 \pi r\, dr$

If axial pressure at this radius is p MPa, axial force $F = p\,(2 \pi r\, dr)$

With coefficient of friction μ, frictional force $= \mu\, p\,(2 \pi r\, dr)$ (a)

Frictional torque on the ring $T_r = \mu p (2 \pi r dr) r = 2 \pi \mu p r^2 dr$ (b)

To get the total axial force integrate Equation (a) within the limits from r_2 to r_1.

From Equation (a): Total frictional force $F = \int_{r_2}^{r_1} p 2 \pi r \, dr = 2 \pi \int_{r_2}^{r_1} p r \, dr$ (18.1)

To get the total frictional torque, integrate Equation (b) within the limits from r_2 to r_1.
From Equation (b),

Total frictional torque $T = \int_{r_2}^{r_1} 2 \pi \mu p r^2 \, dr = 2 \pi \mu \int_{r_2}^{r_1} p r^2 \, dr$ (18.2)

The torque capacity depends on which theory is used; uniform pressure or uniform wear. Both the conditions are analyzed separately as given below.

a. **Uniform pressure**

Uniform pressure theory is applicable only for new clutch plate. When it wears, out uniform wear theory is more appropriate. Since pressure is independent of radius, it is considered as constant. From Equation (18.1), total frictional force:

$$F = 2 \pi p \int_{r_2}^{r_1} r \, dr = 2 \pi p \left[\frac{r^2}{2} \right]_{r_2}^{r_1} = \pi p \left[r_1^2 - r_2^2 \right]$$ (18.3a)

Or, Axial pressure $p = \dfrac{F}{\pi \left[r_1^2 - r_2^2 \right]}$ (18.3b)

From Equation (18.2), total frictional torque:

$$T = \int_{r_2}^{r_1} 2 \pi \mu p r^2 \, dr = 2 \pi \mu p \left[\frac{r^3}{3} \right]_{r_2}^{r_1} = 2 \pi \mu p \left[\frac{r_1^3 - r_2^3}{3} \right]$$ (18.4a)

Substituting the value of p from Equation (18.3b):

$$T = 2 \pi \mu \left[\frac{F}{\pi \left[r_1^2 - r_2^2 \right]} \right] \left[\frac{r_1^3 - r_2^3}{3} \right] = \frac{2}{3} \mu F \left[\frac{r_1^3 - r_2^3}{r_1^2 - r_2^2} \right]$$ (18.4b)

If a term mean radius r_{mp} for uniform pressure theory is defined as:

$$r_{mp} = \frac{2}{3} \left[\frac{r_1^3 - r_2^3}{r_1^2 - r_2^2} \right]$$ (18.5)

Single clutch plate has two surfaces; one on each side. Hence for two surfaces ($z = 2$)

$T = z \mu F r_{mp}$ (18.6)

b. **Uniform wear**

This theory is applicable for worn clutch plates. Wear is proportional to the product of p and r. Let p_{max} be axial pressure at minimum radius r_2. Thus the term $(p_{max} r_2)$ is considered constant.

From Equation (18.1) total frictional force:

$$F = 2\pi p_{max} r_2 \int_{r_2}^{r_1} dr = 2\pi p_{max} r_2 [r]_{r_2}^{r_1} = 2\pi p_{max} r_2 (r_1 - r_2)$$

Or, axial Axial pressure $p_{max} = \dfrac{F}{2\pi r_2 [r_1 - r_2]}$ (18.7)

From Equation (18.2) total frictional torque:

$$T = \int_{r_2}^{r_1} 2\pi \mu p_{max} r^2 \, dr = 2\pi \mu p_{max} r_2 \int_{r_2}^{r_1} r \, dr$$

Or, $T = 2\pi \mu p_{max} r_2 \left[\dfrac{r^2}{2}\right]_{r_2}^{r_1} = 2\pi \mu p_{max} r_2 \left[\dfrac{r_1^2 - r_2^2}{2}\right]$ (18.8)

A single plate clutch, has friction lining on both sides and hence, number of friction surfaces $z = 2$ Substituting the value of p_{max} from Equation (18.7) for z surfaces:

$$T = \dfrac{zF\mu}{2[r_1 - r_2]}\left[r_1^2 - r_2^2\right] = zF\mu \dfrac{r_1 + r_2}{2} = zF\mu r_m$$ (18.9)

Where, Mean radius $r_m = \dfrac{r_1 + r_2}{2}$

Example 18.1

Clutch plate size for given power with uniform pressure

A single plate clutch transmits 50 kW at 960 rpm. Its outer diameter is 250 mm, coefficient of friction 0.3, and axial pressure 0.2 MPa. Using uniform pressure theory, calculate:

a. Inner diameter

b. Axial force

c. Mean radius

Solution

Given $z = 2$ $P = 50$ kW $N = 960$ rpm $2r_1 = 250$ mm
$\mu = 0.3$ $p = 0.2$ MPa

a. Torque $T = \dfrac{60,000 P}{2\pi N} = \dfrac{60,000 \times 50}{2 \times 3.14 \times 960} = 723.8$ Nm

Using Equation (18.3a)

Axial force $F = $ Pressure $\times$ Area $= p\,\pi\left[r_1^2 - r_2^2\right]$

Or, $F = 0.2 \times 3.14\left[125^2 - r_2^2\right]$

Mean radius with uniform pressure $r_{mp} = \dfrac{2}{3}\left[\dfrac{r_1^3 - r_2^3}{r_1^2 - r_2^2}\right] = \dfrac{2}{3}\left[\dfrac{125^3 - r_2^3}{125^2 - r_2^2}\right]$

Using Equation (18.6) $T = z\,\mu\,F\,r_m$

Substituting the values of F and r_{mp}:

$$723.8 \times 10^3 = 2 \times 0.3 \times \left[0.2 \times 3.14\left(125^2 - r_2^2\right)\right] \times \dfrac{2}{3}\left[\dfrac{125^3 - r_2^3}{125^2 - r_2^2}\right]$$

Or, $723{,}800 = 0.251\left(125^3 - r_2^3\right)$ or $r_2 = 97.5$ mm

Hence inside diameter $2r_2 = 195$ mm

b. Axial force $F = p\pi\left[r_1^2 - r_2^2\right] = 0.2 \times 3.14\left[125^2 - 97.5^2\right] = 8{,}160$ N

c. Mean radius $r_m = \dfrac{2}{3}\left[\dfrac{r_1^3 - r_2^3}{r_1^2 - r_2^2}\right] = \dfrac{2}{3}\left[\dfrac{125^3 - 97.5^3}{125^2 - 97.5^2}\right] = 111.8$ mm

Example 18.2

Clutch plate size for given power with uniform wear

A single plate clutch rotating at 600 rpm transmits 80 kW. Assuming coefficient of friction 0.3 and maximum axial pressure 0.3 MPa, find the outside and inside diameters using uniform wear theory. Ratio of inside to outside diameters can be assumed 0.5.

Solution

Given $N = 600$ rpm $P = 80$ kW $\mu = 0.3$ $p_{max} = 0.3$ MPa
$r_2 / r_1 = 0.5$

Torque $T = \dfrac{60{,}000\,P}{2\pi N} = \dfrac{60{,}000 \times 80}{2 \times 3.14 \times 600} = 1{,}274$ Nm $= 1{,}274 \times 10^3$ Nmm

Mean radius for uniform wear $r_m = \dfrac{r_1 + r_2}{2} = \dfrac{r_1 + 0.5\,r_1}{2} = 0.75 r_1$

Using Equation (18.7) Axial force $F = 2\,\pi\,p_{max}\,r_2(r_1 - r_2)$

Substituting the values: $F = 2 \times 3.14 \times 0.3 \times \left(0.5\,r_1\right)\left[r_1 - 0.5\,r_1\right]$

Or, $F = 0.942\,r_1\left[r_1 - 0.5\,r_1\right] = 0.942\,r_1 \times 0.5\,r_1 = 0.471\,r_1^2$

Using Equation (18.9): $T = z\,F\,\mu\,r_m$ and Substituting the values:

$1,274 \times 10^3 = 2 \times 0.471\,r_1^2 \times 0.3 \times 0.75\,r_1 = 0.212\,r_1^3$ or $r_1 = 182$ mm

Hence outside diameter $d_1 = 2 \times r_1 = 364$ mm

Inside diameter $d_2 = 0.5 \times d_1 = 182$ mm

Example 18.3

Pressure variation with uniform wear theory

In a single plate friction clutch, axial force is 5 kN, outer and inner radii are 140 mm and 80 mm, respectively. Assuming uniform wear theory, find:

 a. Pressure at mean radius

 b. Maximum pressure

 c. Minimum pressure

Solution

Given $F = 5$ kN $r_1 = 140$ mm $r_2 = 80$ mm

 a. Average pressure $p_{av} = \dfrac{\text{Force}}{\text{Area}} = \dfrac{F}{\pi\left(r_1^2 - r_2^2\right)}$

Substituting the values: $p_{av} = \dfrac{5{,}000}{3.14 \times \left(140^2 - 80^2\right)} = 0.12$ N/mm^2

 b. Mean radius $r_m = \dfrac{r_1 + r_2}{2} = \dfrac{140 + 80}{2} = 110$ mm

     $p_{av} \times r_m = 0.12 \times 110 = 13.2$ N/mm

Maximum pressure is at inner radius (80 mm) and minimum at outer radius (140 mm). Using uniform wear theory:

     $p_{av} \times r_m = p_{max} \times r_2 = p_{min} \times r_1$

Maximum pressure, $p_{max} = \dfrac{p_{av} \times r_m}{r_2} = \dfrac{13.2}{80} = 0.165$ N/mm^2

 c. Minimum pressure, $p_{min} = \dfrac{p_{av} \times r_m}{r_1} = \dfrac{13.2}{140} = 0.094$ N/mm^2

18.4.5 Maximum torque transmitting capacity

Annular area of friction surface depends on inner diameter $2r_2$ and for a given outside diameter $2r_1$. For an axial pressure p_a, equation (18.8) for uniform wear theory can be written as:

$$T = 2\pi\,\mu\,p_a\,r_2\left[\frac{r_1^2 - r_2^2}{2}\right]$$

If x is taken as ratio $\dfrac{r_2}{r_1}$, then $r_2 = xr_1$. Substituting this value in equation:

$$T = 2\pi\,\mu\,p_a\,r_1\,x\left[\frac{r_1^2 - x^2 r_1^2}{2}\right] = \mu\,p_a\,r_1^3\left[x\left(1 - x^2\right)\right]$$

For finding maximum torque, differentiate the above equation w.r.t. x and equate it to zero. All the terms on left side of parentheses do not depend on x, hence

$$\frac{d\left(x - x^3\right)}{dx} = 0 \quad \text{or} \quad 3\,x^2 = 1 \quad \text{or} \quad x = 0.577$$

Thus, torque transmission capacity for uniform wear condition is maximum when inside diameter is 0.577 times the outside diameter.

Similarly for uniform pressure this optimum ratio of inside to outside diameters is: $x = 0.5$.

18.5 Design of a Single Plate Clutch

In the design of a clutch, the power P and speed N of rotation are given. In some cases, space limitations are also given, which decide the maximum outside diameter. If this is not given, the outside diameter can be decided on the basis of maximum allowable axial pressure p_a. Following steps can be followed to design a single plate friction clutch:

Step 1 *Calculate torque:* Torque $T = \dfrac{60,000\,P}{2\,\pi\,N}$

Increase this value by 50 per cent to account for stress concentration and key way.

Step 2 *Select material:* Use Table 18.1 and find coefficient of friction for dry surface.

Step 3 *Calculate inside diameter r_2:*

If outside space limitations are given, select outside diameter $(2r_1)$ about 80 per cent of that space, to account for the cover plate of the friction discs. Inside diameter $(2r_2)$ can be taken from 0.5 to 0.7 times the outside diameter.

Step 4 *Calculate friction area of one side:* Area $A = \pi\left[r_1^2 - r_2^2\right]$

Step 5 *Check for the axial force:* Assume axial pressure p_a between 0.5 MPa and 1.2 MPa and calculate axial force.

$$F = p_a \left[r_1^2 - r_2^2 \right]$$

Step 6 *Calculate mean radius:* Assuming uniform wear $r_m = \dfrac{r_1 + r_2}{2}$

Step 7 *Calculate torque T:* Use equation $T = z F \mu r_m$

This value of torque should be close to the design torque calculated in step 1. If not, alter values of axial pressure or inner radius suitably. If design torque is much more than the torque capacity of a single plate clutch, then a multi-plate clutch is to be used as described in Section 18.8.

Example 18.4

Clutch plate size with specified spring stiffness
An engine of an automotive vehicle develops 80 kW at 2,000 rpm. Maximum torque for design is 25 per cent more than the average torque developed. Ratio of outer to inner radii is 1.3. Coefficient of friction is 0.3 and the pressure is not to exceed 0.09 MPa. The pressure plate is pressed using six helical compression springs of stiffness 35 N/mm. Calculate:

  a. Size of friction plate

  b. Initial compression of the spring

Solution

Given $P = 80$ kW $N = 2{,}000$ rpm $T_{design} = 1.25\,T$ $r_1/r_2 = 1.3$
            $\mu = 0.3$ $p = 0.09$ MPa $m = 6$ $K = 35$ N/m

a. Torque $T = \dfrac{60{,}000\,P}{2\pi N} = \dfrac{60{,}000 \times 80}{2 \times 3.14 \times 2{,}000} = 382$ Nm

$$T_{design} = 1.25\,T = 1.25 \times 382 = 477 \text{ Nm}$$

Constant $c = p\,r_2 = 0.09\,r_2$ N/mm

Using Equation (18.7), axial force $F = 2\pi p\,r_2\,(r_1 - r_2)$ Substituting the values:

$$F = 2 \times 3.14 \times 0.09\,r_2(1.3\,r_2 - r_2) = 0.17\,r_2^2$$

Mean radius $r_m = \dfrac{1.3\,r_2 + r_2}{2} = 1.15\,r_2$

$$T_{design} = z\mu F r_m \qquad \text{Substituting the values:}$$

$$477 = 2 \times 0.3 \times 0.17\,r_2^2 \times 1.15\,r_2 = 0.117\,r_2^3$$

Hence $r_2 = 160$ mm

b. Axial force $F = 0.17\, r_2^2 = 0.17 \times 160^2 = 4,352$ N

Stiffness of 6 springs $S = m \times k = 6 \times 35 = 210$ N/mm

Initial compression to give that force $= \dfrac{F}{S} = \dfrac{4,352}{210} = 20.7$ mm

18.6 Time for Clutch Engagement

Time for engagement of clutch is the time taken, when the driven shaft speed becomes equal to the driver speed. It depends on the relative velocities of the two shafts, mass, and inertia of the rotating shafts. Driver shaft decelerates (hence minus sign), while driven shaft accelerates.

$$\text{Torque } T = -I_1 \alpha_1 = I_2 \alpha_2 \tag{18.10}$$

Where, I_1 and I_2 = Moment of inertia of the driver shaft and driven shafts, respectively.

α_1 and α_2 = Angular deceleration of the driver shaft and acceleration of driven shaft, respectively.

$$\text{Angular deceleration } \quad \alpha_1 = \frac{d^2\theta_1}{dt^2} = -\frac{T}{I_1} \tag{18.11a}$$

$$\text{And, angular acceleration } \alpha_2 = \frac{d^2\theta_2}{dt^2} = \frac{T}{I_2} \tag{18.11b}$$

Where, θ_1, θ_2 = Angular movement of driver and driven shafts, respectively.

Integrating angular accelerations with time:

$$\frac{d\theta_1}{dt} = -\frac{Tt}{I_1} + C1 \text{ and } \frac{d\theta_2}{dt} = \frac{Tt}{I_2} + C2$$

Where, t = Time taken in clutching operation

C1 and C2 are constants of integration. Conditions to evaluate these are:

$$\text{At } t = 0 \quad \frac{d\theta_1}{dt} = \omega_1 \quad \text{and} \quad \frac{d\theta_2}{dt} = \omega_2$$

Substituting these values in equations above: $C1 = \omega_1$ and $C2 = \omega_2$

Using these values, angular velocities of the driver and driven shafts are:

$$\frac{d\theta_1}{dt} = -\frac{Tt}{I_1} + \omega_1 \text{ and } \frac{d\theta_2}{dt} = \frac{Tt}{I_2} + \omega_2$$

Relative angular velocity ω_r between the two shafts is the difference in their speeds, that is,

$$\omega_r = \frac{d\theta_1}{dt} - \frac{d\theta_2}{dt} = \left[-\frac{Tt}{I_1} + \omega_1 \right] - \left[\frac{Tt}{I_2} + \omega_2 \right]$$

Or, $\qquad \omega_r = (\omega_1 - \omega_2) - Tt\left[\dfrac{I_1 + I_2}{I_1 I_2} \right]$

Or, $\qquad \omega_r = (\omega_1 - \omega_2) - T t I' \qquad\qquad\qquad\qquad$ (18.12)

Where, $\qquad I' = \dfrac{I_1 + I_2}{I_1 I_2}$

After time t, when the clutch operates, relative velocity of both the shafts will be zero.

$$\text{Hence } 0 = (\omega_1 - \omega_2) - TtI' \quad \text{or} \quad t = \frac{\omega_1 - \omega_2}{T I'} \qquad\qquad (18.13)$$

Thus the time of engagement depends on relative velocity, inertia of the rotating members and inversely proportional to the torque.

During engagement, input shaft decelerates, while output shaft accelerates. At the end of engagement after time t, speeds of both the shaft are equal, that is,

$$\omega_1 - t\alpha_1 = \omega_2 + t\alpha_2$$

Thus, time for engagement $t = \dfrac{\omega_1 - \omega_2}{\alpha_2 - \alpha_1} \qquad\qquad\qquad\qquad$ (18.14)

Example 18.5

Time of engagement of a clutch for given power

A single plate clutch rotating at 1,500 rpm transmits 10 kW. Moment of inertia of the driver and driven shafts are 0.8 kg m² and 0.6 kg m², respectively. Find the time of engagement, if:

   a. Driven shaft is stationary

   b. Driven shaft rotates at 500 rpm

Solution

Given $\qquad N_1 = 1{,}500 \text{ rpm} \qquad P = 10 \text{ kW} \qquad I_1 = 0.8 \text{ kg m}^2$

$\qquad\qquad\quad I_2 = 0.6 \text{ kg m}^2 \quad$ (a) $N_2 = 0 \text{ rpm} \quad$ (b) $N_2 = 500 \text{ rpm}$

   a. Torque $T = \dfrac{60{,}000\, P}{2\pi N} = \dfrac{60{,}000 \times 10}{2 \times 3.14 \times 1{,}500} = 63.7 \text{ Nm}$

Angular velocity of the driver shaft $\omega_1 = \dfrac{2\pi N_1}{60} = \dfrac{2 \times 3.14 \times 1{,}500}{60} = 157$ rad/s

Angular velocity of the driven shaft $\omega_2 = 0$ rad/s

Moment inertia of the system $I' = \dfrac{I_1 + I_2}{I_1 I_2} = \dfrac{0.8 + 0.6}{0.8 \times 0.6} = 2.917$

Using Equation (18.13): Time for engagement $t = \dfrac{\omega_1 - \omega_2}{T I'}$

Substituting the values: $t = \dfrac{157 - 0}{63.7 \times 2.917} = 0.845$ s

b. Angular velocity of the driver shaft is same, that is, $\omega_1 = 157$ rad/s

Angular velocity of the driver shaft $\omega_2 = \dfrac{2\pi N_2}{60} = \dfrac{2 \times 3.14 \times 500}{60} = 53.3$ rad/s

Using Equation (18.13): Time for engagement $t = \dfrac{\omega_1 - \omega_2}{T I'}$

Substituting the values: $t = \dfrac{157 - 53.3}{63.7 \times 2.917} = 0.56$ s

18.7 Heat Generated during Clutching

When a clutch is engaged, heat is generated due to friction between the slipping surfaces of the friction plate and pressure plate.

Rate of energy generation $U = T\dfrac{d\theta}{dt}$

Where, T = Torque being transmitted

$\dfrac{d\theta}{dt}$ = Relative angular velocity ω, given by Equation (18.12). Substituting the values:

$$U = T\left[\omega_r - T t I'\right] \tag{18.15}$$

Total energy generated E during clutch operation time t is given as:

$$E = T \int_0^t U\,dt = T \int_0^t \left[\omega_r - T t I'\right] dt$$

Or, $\qquad E = T\omega_r - T^2 I' \int_0^t t\,dt = T\omega_r t - T^2 I' \dfrac{t^2}{2}$

Substituting the value of t from Equation (18.13):

$$E = T\,\omega_r \times \left(\frac{\omega_r}{T\,I'}\right) - 0.5 \times T^2 I' \left(\frac{\omega_r}{T\,I'}\right)^2$$

Or, $\qquad E = \dfrac{\omega_r^2}{2I'}$ $\hfill$ (18.16)

From the above equation, it can be seen that the energy generated is independent of the torque and is proportional to the square of relative velocity between the driver and driven shafts.

Example 18.6

Heat generated during clutching for given drive system
A single plate friction plate clutch is transmitting a torque of 25 Nm at speed of 1,800 rpm. Data for driver and driven shafts are as under:
Mass of driver and driven shafts are 10 kg and 25 kg, respectively.
Radii of gyration driver and driven shafts are 50 mm and 85 mm, respectively.
Calculate:

a. Time for bringing the driven shaft from stationary condition to driver speed.

b. Heat generated during clutch operation

Solution

Given $\qquad T = 25$ Nm $\qquad N_1 = 1,800$ rpm $\qquad m_1 = 10$ kg $\qquad m_2 = 25$ kg
$\qquad\qquad\quad k_1 = 50$ mm $\qquad k_2 = 85$ mm $\qquad N_2 = 0$ rpm

a. Time for engagement:

Mass moment of inertia of the driver shaft $I_1 = m_1 k_1^2 = 10 \times (0.05)^2 = 0.025$ kg m^2

Mass moment of inertia of the driven shaft $I_2 = m_2 k_2^2 = 25 \times (0.085)^2 = 0.18$ kg m^2

$$I' = \frac{I_1 + I_2}{I_1 I_2} = \frac{0.025 + 0.18}{0.025 \times 0.18} = 45.55$$

Angular velocity of the driver shaft $\omega_1 = \dfrac{2\pi N_1}{60} = \dfrac{2 \times 3.14 \times 1,800}{60} = 188.4$ rad/s.

Angular velocity of the driven shaft $\omega_2 = 0$ rad/s

Using Equation (18.13): Time for engagement $t = \dfrac{\omega_1 - \omega_2}{T\,I'}$

Substituting the values: $t = \dfrac{188.4 - 0}{25 \times 45.55} = 0.16$ s

b. Heat generated during clutch operation

Using Equation(18.16), Heat generated $E = \dfrac{\left(\omega_1 - \omega_2\right)^2}{2I'}$

Substituting the values: $E = \dfrac{\left(188.4 - 0\right)^2}{2 \times 45.5} = 390 \text{ J}$

Example 18.7

Clutch of an automobile for an automobile

A diesel engine of a vehicle transmits 50 kW at 1,500 rpm through a single plate dry friction clutch with coefficient of friction 0.3 (Figure 18.S1). Axial pressure on the pressure plate is not to exceed 0.2 MPa. Space for the largest friction plate is 250 mm. Inner radius can be taken as half the outside radius. Clutch is engaged when the vehicle speed is 40 km/h in top gear. Engine speed at start of engagement is 1,200 rpm. Other data for the vehicle is as under:

Torque, when the clutch is engaged is 35 per cent of full torque.

Weight of loaded vehicle 13 kN.

Wheel diameter 650 mm.

Inertia of flywheel = 1 kg m².

Speed reduction in differential = 4.5.

Torque at rear wheels = 170 Nm.

Calculate:

a. Inside radius of the clutch plate.

b. Angular acceleration of the input and output shaft.

c. Time of engagement.   d.   Slip angle in revolutions.

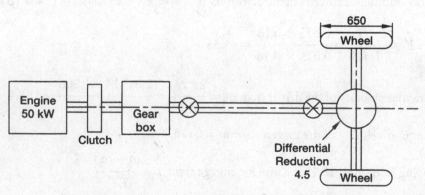

Figure 18.S1 Power transmission system for an automobile

Solution

Given
$P = 50\,kW$ $N_1 = 1,500\,rpm$ $\mu = 0.3$ $p_{max} = 0.2\,MPa$

$2r_1 = 250\,mm$ $r_2 = 0.5r_1$ $V = 40\,km/h$ $N_2 = 1,200\,rpm$

$I_1 = 1\,kg\,m^2$ $T_c = 0.35\,T$ $W = 13\,kN$ $d_w = 650\,mm$

$G = 4.5$ $T_w = 170\,Nm$

a. Torque $T = \dfrac{60,000\,P}{2\pi N} = \dfrac{60,000 \times 50}{2 \times 3.14 \times 1,500} = 318\,Nm$

Assuming uniform wear theory, mean radius $r_m = \dfrac{r_1 + r_2}{2} = 62.5 + 0.5r_2$

From Equation (18.3a): Axial force $F = p\pi\left[r_1^2 - r_2^2\right]$

$$F = 0.2 \times 3.14 \times \left[125^2 - r_2^2\right] = 0.628 \times \left[15,625 - r_2^2\right]$$

From Equation (18.9): Torque $T = \mu z F r_m$

Where, z = Number of friction surfaces. For single friction plate, $z = 2$

Substituting the values: $318,000 = 0.3 \times 2 \times F \times r_m = 0.6\,F\,r_m\,Nm$

Or, $318,000 = 0.6 \times 0.628\left(15,625 - r_2^2\right)\left(62.5 + 0.5\,r_2\right)$

It is a cubical equation, solving by trial and error method. Taking a starting value of inner radius r_2 about 55 per cent of outside radius r_1, that is ($r_2 = 0.55 \times 125 = 70\,mm$):

r_2	r_2^2	r_m $(62.5 + 0.5r_2)$	$F =$ $0.628(15,625 - r_2^2)$	T $0.6\,F \times r_m$	Action taken
70	4,900	97.5	6,735	394,015	$> T$, increase r_2
80	6,400	102.5	5,793	356,288	$> T$, increase r_2
90	8,100	107.5	4,725	304,808	$< T$, decrease r_2
85	7,225	105	5,275	332,338	Close to 318,000, accepted

Value of torque 332,338 Nm is close to the engine torque 318 Nm. Allowing some margin for wear, a value of inner radius $r_2 = 85\,mm$ is selected.

b. Acceleration of input shaft

Angular speed of the input shaft $\omega_1 = \dfrac{2\pi \times 1,500}{60} = 157\,rad/s$

Torque, when the clutch is engaged is 35 per cent of full torque, that is, $T_c = 0.35\,T$

Torque for acceleration for engine = Torque at engagement – Torque of engine

$$T_a = 0.35\,T - T = -0.65\,T = -0.65 \times 318 = -206.7\,Nm$$

From Equation (18.11a)

$$\alpha_1 = -\frac{T_a}{I_1} = -\frac{206.7}{1} = -206.7 \text{ rad/s}^2$$

Acceleration of output shaft

Circumference of tyre $= \pi\, d_w = 3.14 \times 650 = 2{,}041 \text{ mm} = 2.041 \text{ m}$

Rotational speed of tyre in rps at 40 km/h $= \dfrac{40{,}000}{3{,}600 \times 2.041} = 5.45 \text{ rps}$

Since there is speed reduction of 4.5 in differential, speed of propeller shaft or output shaft of gear N_2. (Gear ratio in gear box in top gear is 1). Hence:

$N_2 = 5.45 \times 4.5 = 24.51 \text{ rps}$

Angular speed of the output shaft $\omega_2 = 2\,\pi \times 24.51 = 154 \text{ rad/s}$

Accelerating force of wheel $F_w = \dfrac{T_w}{0.5\, d_w} = \dfrac{170}{0.5 \times 0.65} = 523 \text{ N}$

Acceleration of wheel $= \dfrac{F_w}{m} = \dfrac{523 \times 9.81}{13{,}000} = 0.394 \text{ m/s}^2$

Angular acceleration of clutch output shaft:

$$\alpha_2 = \frac{\text{Acceleration} \times \text{Gear ratio}}{\text{Radius of wheel}} = \frac{0.394 \times 4.5}{0.5 \times 0.65} = 5.45 \text{ rad/s}^2$$

c. **Time for engagement**

Using Equation (18.14): Time for engagement $t = \dfrac{\omega_1 - \omega_2}{\alpha_2 - \alpha_1}$

Substituting the values: $t = \dfrac{157 - 154}{5.45 - (-206.7)} = 0.014 \text{ s}$

d. **Angle through which input shaft rotates during engagement:**

$$\theta_1 = \omega_1 t + 0.5\,\alpha_1 t^2 = 157 \times 0.014 + 0.5(-206.7)(0.014)^2$$

Or, $\theta_1 = 2.2 - 0.02 = 2.18 \text{ rad}$

Revolutions of input shaft $= \dfrac{2.18}{2\,\pi} = 0.35 \text{ rev.}$

Angle through which output shaft rotates during engagement:

$$\theta_2 = \omega_2 t + 0.5\,\alpha_2 t^2 = 154 \times 0.014 + 0.5(5.45)(0.014)^2$$

Or, $\theta_2 = 2.16 + 0.0005 = 2.16 \text{ rad}$

Revolutions of output shaft $= \dfrac{2.16}{2\pi} = 0.34$ rev.

Slip angle in revolutions $= \dfrac{\theta_1 - \theta_2}{2\pi} = 0.35 - 0.34 = 0.01$ rev.

18.8 Multi-plate Clutch

The size of a clutch depends on the torque, friction coefficient, surface area and axial pressure. Friction coefficient is the property of the material and can hardly be changed to more than 0.4. The axial pressure cannot be increased beyond a certain limit, as it increases the force required to disengage the clutch. Increasing area increases the size of clutch. Hence, where there is space limitation, multi-plate clutch is used. It is shown in Figure 18.5 and in picture on its right side. It can be seen that friction plate and pressure plates are arranged alternately. Thus the effective friction area is the annular area of one friction surface multiplied by the number of pairs of the surfaces z.

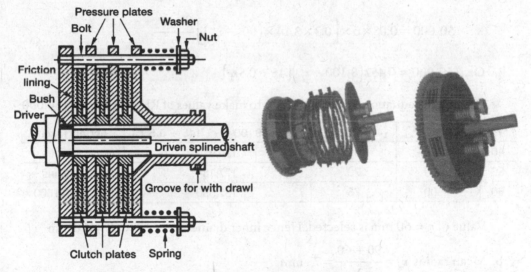

Figure 18.5 Multiplate clutch

If there are z_1 pressure plates and z_2 friction plates, the number of pair of friction surfaces is given as:

$$z = z_1 + z_2 - 1 \qquad (18.17)$$

Design of multi-plate clutch is similar to the single plate friction clutch except that the number of friction surfaces is more than 2.

Example 18.8

Size of discs and axial pressure for a multi-plate clutch

A multiple plate clutch uses 4 steel plates and 3 friction plates. Outside radius is limited to 90 mm. Torque to be transmitted is 150 Nm, axial pressure is not to exceed 0.3 MPa, and $\mu = 0.08$. Calculate:

    a. Inner diameter   b.   Mean radius

    c. Axial force   d.   Axial pressure

Solution

Given $T = 150$ Nm $z_1 = 4$ $z_2 = 3$ $\mu = 0.08$

                 $p_a = 0.3$ MPa $r_1 = 90$ mm

    a. Number of friction surfaces:

$$z = z_1 + z_2 - 1 = 4 + 3 - 1 = 6$$

Axial force $F =$ Pressure $\times$ Area $= 0.3 \times \pi\left(90^2 - r_2^2\right)$

Using Equation (18.9): Torque $T = \mu z F r_m$ Substituting the values:

$$150,000 = 0.08 \times 6 \times \left[0.3 \times 3.14 \times \left(90^2 - r_2^2\right)\right]\frac{90 + r_2}{2}$$

Or, $150,000 = 0.452\left(8,100 - r_2^2\right)\left(45 + 0.5 r_2\right)$

Solving this equation by trial and error to make values of RHS = LHS i.e. 150,000:

r_2	$(8100 - r^2_2)$	$(45 + 0.5\ r_2)$	$0.452\ (8100 - r^2_2)(45 + 0.5\ r_2)$	Action Taken
40	6,500	65	190,970	Increase radius
50	5,600	70	177,184	Increase radius
60	4,500	75	152,550	Close to 150,000

Value of $r_2 = 60$ mm is selected. Hence inner diameter $2r_2 = 2 \times 60 = 120$ mm

    b. Mean radius $r_m = \dfrac{90 + 60}{2} = 75$ mm

    c. With these dimensions F is calculated using:

From Equation (18.9): $T = \mu z F r_m$ Substituting the values:

$$150,000 = 0.08 \times 6 \times F \times 75 \quad \text{Or,} \quad F = 4,170 \text{ N}$$

    d. Area $A = \left(90^2 - 60^2\right)$

$$\text{Axial pressure} = \frac{\text{Axial force}}{\text{Area}} = \frac{4,170}{\left(90^2 - 60^2\right)} = 0.293 \text{ MPa}$$

18.9 Cone Clutch

In a cone clutch, the frictional surface is frustums of cone. The outer hollow cone is keyed to the driver shaft and the inner cone having friction lining on the outer conical surface is mounted on a splined driven shaft, so that it can slide axially for engagement. The inner cone is pressed axially with a helical compression spring. It also has a groove for withdrawal. A cone clutch is shown in Figure 18.6.

In case of a plate clutch, axial force is normal to the friction surface, but in cone clutch the normal force F_n to the friction surface is related with axial force through cone angle α as given below.

$$\text{Axial force } F_a = F_n \sin \alpha$$

$$\text{Frictional force } F_f = \mu \times F_n = \frac{\mu F_a}{\sin \alpha}$$

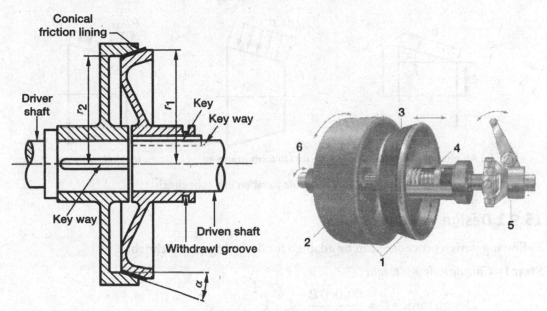

Figure 18.6 A cone clutch

$$\text{Mean radius of the cone } r_m = \frac{r_1 + r_2}{2}$$

Where, r_1 = Bigger radius of the cone and
 r_2 = Smaller radius of the cone

$$\text{Torque that can be transmitted } T = F_f \times r_m = \frac{\mu F_a (r_1 + r_2)}{2 \sin \alpha} \tag{18.18}$$

It can be seen from Equation (18.18), that the torque capacity of a cone clutch increases by a factor ($\operatorname{cosec} \alpha$) times than for single plate friction clutch given by Equation (18.9).

Main objection to the use of cone clutch is that there is difficulty while disengaging due to wedging action, dust, etc. Hence the cone angle 2α should not be less than 15°. Its value varies from 15° to 60° but generally kept between 20° and 30°. Face width b of the inclined surface of the cone can be found using the relation:

$$\text{Face width } b = \frac{F_n}{2\pi \, pr_m} \tag{18.19}$$

Dimensions of a cone clutch are decided empirically as under:

Mean radius of the cone r_m = 2.5 to 5 times shaft diameter

Face width b = 0.25 to 0.45 times the mean radius r_m

Peripheral velocity at outside radius of cone v = 10 – 25 m/s

Pressure variation for a cone clutch also depends on the theory used, that is, uniform pressure or uniform wear. Pressure distribution is shown in Figure 18.7.

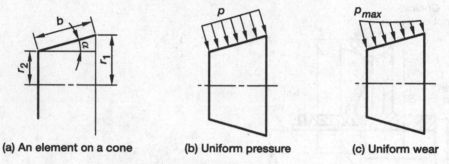

(a) An element on a cone   (b) Uniform pressure   (c) Uniform wear

Figure 18.7 Pressure variation in a cone clutch

18.9.1 Design of a cone clutch

Following design procedure can be adapted for designing a cone clutch:

Step 1 *Calculate design torque:*

$$\text{Design torque } T_d = \frac{60,000 \, P}{2\pi N} \times k_t \times k_s$$

k_t and k_s are stress concentration factor and service factor. If not known, design torque can be increased by 50 per cent to the torque calculated without these factors.

Step 2 *Select lining material:* Use Table 18.1 and select a material for dry surface.

Step 3 *Calculate shaft diameter:* $T_d = \dfrac{\pi}{16}d^3\tau$

Step 4 *Calculate mean radius and width of the cone:*

From Table 18.2, for the pair of materials selected, find its coefficient of friction μ and the pressure p. Initially, assume face width b = 0.3 times mean radius r_m.

Normal force $F_n = 2 \pi b p r_m$

Calculate torque $T = F_n \mu r_m = 2 \pi (0.3 r_m)(p r_m) p r_m (\mu r_m) = 0.6 \pi p r_m^3$

Calculate mean radius r_m from above equation. It should be between 2.5 d and 5d.

Calculate face width b. It should be between 0.25 and 0.45 the mean radius r_m.

Step 5 *Calculate cone diameters*:

Difference of big and small radius of the cone: $(r_1 - r_2) = b \sin \alpha$

Big diameter of the cone $2 r_1 = 2 r_m + b \sin \alpha$

Small diameter of the cone $2 r_2 = 2 r_m - b \sin \alpha$

Step 6 *Calculate axial force*:

Axial force $F_a = F_n \sin \alpha = 2 \pi b p r_m \sin \alpha$

Example 18.9

Size of cone and axial force for a cone clutch for given power

A cone clutch transmits 20 kW at 900 rpm from an electric motor to a centrifugal pump. Assuming coefficient of friction 0.25, inclined width 0.3 times the mean radius, normal pressure 0.15 MPa, service factor 1.2, stress concentration factor due to key way 1.5, and safe shear stress for the shaft 40 MPa. Calculate the main dimensions of the clutch, if cone angle is 25°.

   a. Shaft diameter.   b. Face width.

   c. Big and small end radii of the cone.   d. Axial force.

Solution

Given $P = 20$ kW $N = 900$ rpm $\mu = 0.25$ $b/r_m = 0.3$

                 $p = 0.15$ MPa $k_s = 1.2$ $k_t = 1.5$ $\tau = 40$ MPa

                 $2\alpha = 25°$

  a. Torque $T = \dfrac{60,000 P}{2 \pi N} = \dfrac{60,000 \times 20}{2 \times 3.14 \times 900} = 212.3$ Nm

Design torque $T_d = k_s k_t T = 1.2 \times 1.5 \times 212.3 = 382.2$ Nm $= 382,200$ Nmm

$T_d = \dfrac{\pi}{16} d^3 \tau$ or $382,200 = \dfrac{3.14}{16} d^3 \times 40$ or $d = 36.5$ mm

To account for weakening of the shaft due to key way shaft diameter $d = 40$ mm

  b. Normal force $F_n = 2 \pi b p r_m = 2 \times 3.14 (0.3 r_m) \times 0.15 r_m = 0.282 r_m^2$

Torque $T = F_n \mu r_m$ Substituting the values:

$$382{,}200 = \left(0.282\, r_m^2\right) \times 0.25 \times r_m = 0.07\, r_m^3 \quad \text{or} \quad r_m = 176 \text{ mm}$$

It should vary between $2.5d$ to $5d$, that is, 100 mm to 200 mm. Hence OK

As given, inclined length of cone $b = 0.3\, r_m = 0.3 \times 176 = 52.8$ mm say 55 mm

c. Cone angle $2\,\alpha = 25°$, hence semi-cone angle $\alpha = 12.5°$

Difference of big and small radius of the cone $(r_1 - r_2) = b \sin 12.5°$

$$r_1 - r_2 = 55 \times 0.216 = 11.9 \text{ mm}$$

Also, $r_1 = r_m + \dfrac{r_1 - r_2}{2} = 176 + \dfrac{11.9}{2} = 182$ mm

And, $r_2 = r_m - \dfrac{r_1 - r_2}{2} = 176 - \dfrac{11.9}{2} = 170$ mm

d. Normal force $F_n = 2\,\pi\, b\, p\, r_m = 2 \times 3.14 \times 55 \times 0.15 \times 176 = 9{,}118$ N

Axial force $F_a = F_n \sin \alpha = 9{,}118 \times \sin 12.5° = 1{,}973$ N

18.10 Centrifugal Clutch

This type of clutch can be considered as an automatic clutch as there is no lever or pedal to operate it. The output shaft automatically gets connected as the speed of the input shaft is increased. Some mopeds use this type of the clutch. This type of clutch has the following advantages:

- Permits startup under no load conditions
- As the input shaft speed increases, output shaft picks up gradually, smoothly, and without any shock.
- Prime mover will not stall due to overload as the clutch starts slipping after the speed drops beyond a certain limit.

18.10.1 Construction and working

It consists of a spider, which is mounted on the driver shaft. It can have two, three (Figure 18.8) or four radial arms. An alternative arrangement of mounting the springs to pull the shoes is shown in the picture on the right side. A hollow drum is mounted on driven shaft covering the spiders and aligns axially with the driver shaft. A clearance of about 1 mm is kept between the lining and the drum. Each arm of the spider contains a floating shoe, whose outer surface with friction lining has the same radius as the drum.

When the driver shaft rotates, the floating weights in the spider also start rotating. The centrifugal force of these shoes moves them radially outward to touch the drum. The drum starts rotating due to friction between friction lining and inner surface of the drum with some slip. This is called engagement speed, which is kept about 75 per cent of the full speed. As the speed increases, centrifugal force increases at the rate of square of the speed. At high speeds, force is enough that the slip is almost reduced to zero.

When the driver speed decreases, centrifugal force on the shoes decreases and the clutch again starts slipping. Thus it is automatic in operation.

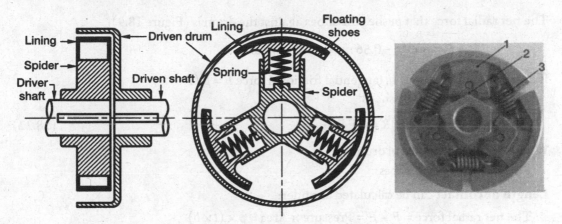

Figure 18.8 A centrifugal clutch

18.10.2 Design of a centrifugal clutch

The centrifugal force F_c on the shoe is given by the relation:

$$F_c = m \, \omega^2 \, r \tag{18.20}$$

Where, m = Mass of one shoe

ω = Angular speed of the driver in rad/s = $\dfrac{2\pi N}{60}$ (N is speed in rpm)

r = Distance between shaft axis and the center of gravity of the shoe (Figure 18.9).

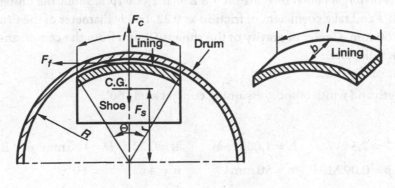

Figure 18.9 Shoe of a centrifugal clutch

Engagement speed is taken about 0.75 of driver's full speed, and hence the angular speed of the driver at the time of engagement can be found by the relation: $\omega_1 = 0.75\ \omega$.

The spring force $F_s = m\ \omega_1^2\ r = m\left(0.75\ \omega\right)^2 r = 0.56\ m\ \omega^2\ r$ (18.21)

The net radial force that pushes the shoes against the drum is (Figure 18.9):

$$F_c - F_s = m\omega^2 r - 0.56\ m\omega^2 r = 0.44 m\ \omega^2 r$$

If coefficient of friction is μ, tangential frictional force $F = \mu\ (F_c - F_s)$

Torque transmitted by shoes

$$T = n\ \mu(F_c - F_s)R = n\,F\,R \qquad (18.22)$$

Where, R = Inside radius of drum

n = Number of shoes

Length of contact can be calculated as under:

The net radial force = $F_c - F_s$ = Pressure $\times$ Area = $p \times (l \times b)$

Where, p = Allowable pressure on lining (about 0.1 MPa)

l = Length of contact of all the shoes

b = Width of the shoes

If θ is the angle sub tended by the lining of one shoe at the center of the shaft, total contact length by n shoes:

$$l = n\,R\,\theta \qquad (18.23)$$

Example 18.10

Mass and size of shoe for a centrifugal clutch for given power

Design a centrifugal clutch to transmit 7.5 kW at 1,000 rpm. Limit the lining pressure to 0.09 MPa and take coefficient of friction as 0.22. Inside diameter of the drum can be taken as 160 mm, Center of gravity of the shoe is 50 mm from the center, and number of shoes 4. Calculate:

a. Mass of the shoe.

b. Length and width of shoe, assuming contact angle 60°.

Solution

Given $P = 7.5\,\text{kW}$ $N = 1,000\,\text{rpm}$ $\mu = 0.22$ $D = 160\,\text{mm}$ or $R = 0.08\,\text{m}$

           $p = 0.09\,\text{MPa}$ $r = 50\,\text{mm}$ $n = 4$ $\theta = 60°$

a. Torque $T = \dfrac{60,000\,P}{2\,\pi\,N} = \dfrac{60,000 \times 7.5}{2 \times 3.14 \times 1,000} = 71.65\,\text{Nm}$

Angular speed of the driver $\omega = \dfrac{2\pi N}{60} = \dfrac{2 \times 3.14 \times 1,000}{60} = 104.7 \text{ rad/s}$

Angular speed at 75 per cent speed $\omega_1 = 0.75 \, \omega = 0.75 \times 104.7 = 78.52 \text{ rad/s}$

Centrifugal force $F_c = m \, \omega^2 \, r = m \, (104.7)^2 \times 0.05 = 547 \, m$

Spring force $F_s = m \, \omega_1^2 \, r = m \, (78.52)^2 \times 0.05 = 308 \, m$

Net radial force $F_c - F_s = 547 \, m - 308 \, m = 239 \, m$

Tangential frictional force $F = \mu(F_c - F_s) = 0.22 \times 239 \, m = 52.6 \, m$

Torque $T = n \, F \, R$ Substituting the values:

$\qquad 71.65 = 4 \times 52.6 \, m \times 0.08 \quad$ or $\quad m = 4.26 \text{ kg}$

b. Net radial force $F_c - F_s = 239 \, m = 239 \times 4.26 = 1,018 \text{ N}$

Length of contact $l = n \, R \, \theta = 4 \times 0.08 \times \dfrac{\pi}{3} = 0.335 \text{ m} = 335 \text{ mm}$

The net radial force = Pressure × Area

$\qquad 1,018 = 0.09 \times (335 \times b) \quad$ or, $\quad b = 33.7 \text{ say } 34 \text{ mm}$

Summary

Clutch is a device used to connect or disconnect incoming power from a driver shaft to the driven shaft at same speed or speed lesser than driver speed, whenever desired without stopping the driver shaft. It is used in automobiles invariably.

Functions of a clutch are:

- Connect or disconnect the power from a driver shaft to the driven shaft.

- Provide any speed between zero and maximum that of the driver shaft speed.

- Absorb torsional shocks.

Types of clutches There are following four types of clutches.

a. *Positive drive clutch* The driver and driven shafts are mounted with a flange having jaws or teeth on one side. When these jaws mesh with each other, the driven shaft starts rotating at the driver speed.

b. *Friction clutch* Power from driver shaft is transmitted to the driven shaft through a friction plate called clutch plate.

c. *Hydraulic clutch* It is also called fluid coupling, as a fluid is used to connect driver and driven.

d. *Electro-magnet clutch* Power is through a magnetic field. There is no friction and no wear.

Positive drive clutch   Driver shaft has a flange with either square jaws or spiral jaws. Driven shaft also has flange with similar jaws, which slides over a splined shaft. When projections of jaws fit into the recesses, power is transferred.

Friction clutch can be of single plate, multiple plate, or cone clutch. Some clutches operate under dry conditions, but if the heat is more, wet clutches are also used.

Friction materials   There are many friction materials for dry and wet clutches. Coefficient of friction, maximum pressure, and temperature are tabulated in Table 18.1.

Coefficient of friction   It depends on material and also upon condition of mating surface, dry or wet. Table 18.2 gives the values for various combinations.

Variation of bearing pressure   For new friction plate, uniform pressure can be assumed while for a worn or used friction plate, uniform wear is to be assumed.

Torque transmitting capacity depends on: number of friction surfaces, surface area of the disc, and pressure over the disc.

A friction disc has outside diameter $2r_1$ and inside diameter $2r_2$.

$$\text{Total frictional force}\quad F = 2\pi \int_{r_2}^{r_1} pr\, dr$$

$$\text{Total frictional torque}\quad T = 2\pi \mu \int_{r_2}^{r_1} pr^2\, dr$$

Torque capacity depends on which theory is used; uniform pressure or uniform wear.

a.   **Uniform pressure**   Axial pressure $p = \dfrac{F}{\pi\left[r_1^2 - r_2^2\right]}$
    Where, F is force on friction plate.

$$\text{Torque } T = 2\pi\mu p\left[\frac{r_1^3 - r_2^3}{3}\right] = 2\pi\mu\left[\frac{F}{\pi\left[r_1^2 - r_2^2\right]}\right]\left[\frac{r_1^3 - r_2^3}{3}\right] = \frac{2}{3}\mu F\left[\frac{r_1^3 - r_{12}^3}{r_1^2 - r_2^2}\right]$$

If mean radius $r_{mp} = \dfrac{2}{3}\left[\dfrac{r_1^3 - r_2^3}{r_1^2 - r_2^2}\right]$, Hence for two surfaces ($z = 2$) $T = z\mu F r_{mp}$

b.   **Uniform wear**   Wear is proportional to the product of p and r. Let p_{max} be axial pressure at minimum radius r_2. Thus the term $(p_{max}\, r_2)$ is considered constant.

Axial pressure $p_{max} = \dfrac{F}{2\pi r_2\left[r_1 - r_2\right]}$

$$\text{Torque } T = 2\pi\mu p_{max}\, r_2 \int_{r_2}^{r_1} r\, dr = 2\pi\mu p_{max} r_2\left[\frac{r^2}{2}\right]_{r_2}^{r_1} = 2\pi\mu p_{max}\, r_2\left[\frac{r_1^2 - r_2^2}{2}\right]$$

For z surfaces $T = \dfrac{zF\mu}{2\left[r_1 - r_2\right]}\left[r_1^2 - r_2^2\right] = zF\mu\dfrac{r_1 + r_2}{2} = zF\mu r_m$

Where, $r_m = \dfrac{r_1 + r_2}{2}$

Maximum torque transmitting capacity: If x is taken as ratio $\frac{r_2}{r_1}$, then $r_2 = xr_1$. Then

$$T = 2\pi \mu p_a r_1 \times \frac{r_1^2 - x^2 r_1^2}{2} = \mu p r_1^3 \left[x\left(1 - x^2\right) \right]$$

To find maximum torque, differentiate the above equation w.r.t. x and equate it to zero. $3x^2 = 1$ or $x = 0.577$

Thus, torque transmission capacity for uniform wear condition is maximum, when $x = 0.577$. Similarly, for uniform pressure this optimum ratio $x = 0.5$.

Design of a single plate clutch   In the design, power P and speed of rotation N are given.

1. *Calculate torque*   Torque $T = \dfrac{60,000\, P}{2\pi N}$

   Increase 50 per cent for stress concentration and weakening due to key way.

2. *Select material*   Select material from Table 18.1 and get coefficient of friction from Table 18.2.

3. *Calculate inside diameter r_2*   Select outside diameter $2r_1$ about 80 per cent of available space, to account for the cover plate. Inside diameter $2r_2$ from 0.5 to 0.7 times outside diameter.

4. *Calculate friction area of one side*   Area $A = \pi\left[r_1^2 - r_2^2 \right]$

5. *Calculate axial force*   Assume axial pressure p_a between 0.5 MPa and 1.2 MPa, and calculate axial force $F = p_a\left[r_1^2 - r_2^2 \right]$

6. *Calculate mean radius r_m*   Assuming uniform wear $r_m = \dfrac{r_1 + r_2}{2}$

7. *Calculate torque*   Use equation $T = z F \mu\, r_m$

This value of torque should be close to the design torque calculated in step 1. If not, alter values of axial pressure or inner radius suitably. If design torque is much more than the torque capacity of a single plate clutch, then use multi-plate.

Time for clutch engagement   It is the time taken, when the speed of driven shaft becomes equal to the speed of driver shaft. It depends on the relative velocities of the two shafts, mass and inertia.

Torque $T = -I_1 \alpha_1 = I_2 \alpha_2$

Where, I_1 and I_2 = Moment of inertia of the driver and driven shafts, respectively.

α_1 and α_2 = Angular deceleration of the driver and angular acceleration driven shafts, respectively.

At $t = 0$, $\dfrac{d\theta_1}{dt} = \omega_1$ and $\dfrac{d\theta_2}{dt} = \omega^2$

After time t, when the clutch operates, relative velocity of both the shafts will be zero.

Hence, $0 = \left(\omega_1 - \omega_2 \right) - T t I'$ or $t = \dfrac{\omega_1 - \omega_2}{T I'}$

Where, $I' = \dfrac{I_1 + I_2}{I_1 I_2}$

Time for engagement $t = \dfrac{\omega_1 - \omega_2}{\alpha_2 - \alpha_1}$

Heat generated during clutching: When a clutch is engaged, heat is generated due to friction between the slipping surfaces of the friction plate and pressure plate. Rate of energy generation

$U = T\dfrac{d\theta}{dt}$ Where, $\dfrac{d\theta}{dt}$ = Relative angular velocity ω_r

$U = T[\omega_r - T t l']$

Energy generated E during clutch operation time t:

$$E = T\int_0^t U\,dt = T\int_0^t [\omega_r - T t l']\,dt$$

Or, $E = T\omega_r - T^2 l' \int_0^t t\,dt = T\omega_r t - T^2 l'\dfrac{t^2}{2} = T\omega_r \times \left(\dfrac{\omega_r}{Tl'}\right) - 0.5\,T^2 l'\left(\dfrac{\omega_r}{Tl'}\right)^2 = \dfrac{\omega_r^2}{2\,l'}$

From the above equation, it can be seen that the energy generated is independent of the torque and is proportional to the square of relative velocity between the driver and driven shafts.

Multi–plate clutch   Where there is space limitation, multi-plate clutch is used. Effective friction area is the annular area of one friction surface multiplied by the number of pairs of the surfaces. If there are z_1 pressure plates, and z_2 friction plates, the number of pair of friction surfaces is given as: $z = z_1 + z_2 - 1$

Design of multi-plate clutch is similar to the single plate friction clutch except that the number of friction surfaces is more than 2.

Cone clutch   Frictional surface is frustum of cone. Force F_n, normal to the friction surface is related with axial force, through cone angle α. Axial force $F_a = F_n \sin \alpha$.

Frictional force $F_f = \mu \times F_n = \dfrac{\mu F_a}{\sin \alpha}$

Mean radius of the cone $r_m = \dfrac{r_1 + r_2}{2}$

Where, r_1 = Bigger radius of the cone and r_2 = Smaller radius of the cone

Torque $T = F_f \times r_m = \dfrac{\mu F_a (r_1 + r_2)}{2 \sin \alpha}$

Thus, torque capacity of a cone clutch increases by cosec α times than for single plate clutch. Cone angle 2α should not be less than 15°. Its value generally kept between 20° and 30°.

Face width b of the cone can be found using the relation: $b = \dfrac{F_n}{2\pi p r_m}$

Mean radius of the cone r_m = 2.5 to 5 times the shaft diameter

Face width b = 0.25 to 0.45 times the mean radius r_m

Peripheral velocity at outside radius of cone v = 10 m/s to 25 m/s

Design of a cone clutch: First 2 steps are same as for single plate clutch.

1. *Calculate design torque*

2. *Select lining material*   Select material from Table 18.1.

3. *Calculate shaft diameter*   Using equation $T_d = \dfrac{\pi}{16} d^3 \tau$

4. *Calculate mean radius and width of the cone*   From Table 18.2, find coefficient of friction μ and the pressure p. Initially assume face width b = 0.3 times mean radius r_m

   Normal force $F_n = 2 \pi b p r_m$

   Calculate torque $T = F_n \mu r_m = 2 \pi (0.3 r_m) p r_m (\mu r_m) = 0.6 \pi p r_m^3$

   Calculate mean radius r_m from the above equation. It should be 2.5d to 5d

   Calculate face width b. It should be between 0.25 and 0.45 times the mean radius.

5. *Calculate cone diameters:* Big diameter of cone $2r_1 = 2r_m + b \sin \alpha$

   Small diameter of cone $2r_2 = 2r_m - b \sin \alpha$

6. *Calculate axial force:* Axial force $F_a = F_n \sin \alpha = 2 \pi b p r_m \sin \alpha$

Centrifugal clutch   It is as an automatic clutch, as there is no lever or pedal to operate it. The output shaft automatically gets connected as the speed of the input shaft is increased. Its advantages are:

- Permits startup under no load conditions

- As input shaft speed increases, output shaft picks up speed smoothly without any shock.

- Prime mover will not stall due to overload as clutch starts slipping.

Construction and working   It consists of a spider, mounted on the drivers haft. It can have two, three, or four radial arms. A hollow drum mounted on the driven shaft aligns axially with the driver shaft. Each arm of the spider contains a floating shoe, whose outer surface with friction lining has the same radius as the drum.

When the driver shaft rotates, the centrifugal force on shoes moves them radially outward to touch the drum. Engagement speed is kept about 75 per cent of the full speed. As the speed increases, centrifugal force increases at the rate of square of the speed.

Design of a centrifugal clutch : The centrifugal force F_c on the shoe is $F_c = m \omega^2 r$

m = Mass of one shoe, ω = Angular speed of the driver, r = Distance between the shaft axis and the center of gravity of the shoe. Engagement speed $\omega_1 = 0.75 \omega$.

The spring force $F_s = m \omega_1^2 r = m (0.75 \omega)^2 r = 0.56 m \omega^2 r$

Radial force pushing shoes against drum $F_c - F_s = m \omega^2 r - 0.56 m \omega^2 r = 0.44 m \omega^2 r$

If coefficient of friction is μ, tangential frictional force $F = \mu (F_c - F_s)$

Torque transmitted by n shoes $T = n \mu (F_c - F_s) R = n F R$, where, R = Inside radius of drum.

The net radial force = $F_c - F_s$ = Pressure × Area = $p \times (l \times b)$

Where, p = Allowable pressure on lining (about 0.1 MPa)

l = Length of contact of all the shoes and

b = Width of the shoes.

If θ is the angle sub tended by the lining of one shoe at the center of the shaft, total contact length by n shoes: $l = n R \theta$

Theory Questions

1. What are the functions of a clutch? Give some examples of its use.

2. What are the various types of clutches? Describe a positive drive clutch with the help of a neat sketch.

3. How does a friction clutch work Sketch a single plate friction clutch and explain its construction and working.

4. List the various friction materials. What are the important properties to be seen for these materials?

5. What are the two theories to design a single plate friction clutch? Derive an expression to find the mean radius using uniform pressure theory.

6. Describe the pressure variation with a sketch using uniform wear theory and torque transmission capacity with this theory.

7. How do you calculate the time for engagement of a friction clutch? Derive the expression.

8. Derive an expression to calculate the heat generated during engagement time.

9. Write the steps in the design of a single plate clutch.

10. When a multi-plate clutch is used? Describe its construction and working with the help of a neat sketch.

11. Write the steps in the design of a multi-plate clutch.

12. How a cone clutch is different than a single plate friction clutch? Describe its construction with the help of a sketch.

13. What are the advantages and disadvantages of a cone clutch over single plate clutch?

14. Derive an expression for the torque transmission capacity of a cone clutch.

15. What steps are to be followed in the design of a cone clutch?

16. How does a centrifugal clutch work? Describe its construction and working.

17. Describe the steps to calculate spring force for a centrifugal clutch?

Multiple Choice Questions

1. A clutch is used to:

    (a) Connect or disconnect driver and driven shafts

    (b) Provide smooth start when a rotating shaft is connected to a stationary shaft.

    (c) Absorb torsional shocks

    (d) All given above

2. Centrifugal clutch transmits power by:

    (a) Friction between shoe and drum
    (b) Centrifugal force
    (c) Spring force and centrifugal force
    (d) None of the above

3. In uniform wear theory, axial pressure is assumed:

    (a) Constant all over the surface
    (b) Variable linearly with radius and maximum at outer radius
    (c) Variable linearly with radius maximum at inner radius
    (d) Variable at square of radius and constant all over the surface

4. Mean radius in uniform wear is:

    (a) $\dfrac{2}{3}\left[\dfrac{r_1^3 - r_2^3}{r_1^2 - r_2^2}\right]$
    (b) $\dfrac{r_1 + r_2}{2}$

    (c) $\dfrac{r_1^2 - r_2^2}{2}$
    (d) None of these

5. Torque transmission capacity is maximum, if the ratio of inside to outside radius for uniform wear theory is:

    (a) 0.5
    (b) 0.577
    (c) 0.65
    (d) 0.75

6. Time for engagement of a clutch depends on:

    (a) Driver shaft speed
    (b) Driven shaft speed
    (c) Both driver and driven shaft speeds
    (d) Does not depend on speed

7. Heat generated during clutching operation depends on:

    (a) Moment of inertia of both the shafts and rotors
    (b) Driver shaft speed
    (c) Driven shaft speed
    (d) All given above

8. Multiple plate clutch is used when:

    (a) Power transmitted is more
    (b) Space is less
    (c) Long life is desired
    (d) All given in a, b, and c

9. For the same outer radius, ratio of power transmitting capacity of a cone clutch and single plate friction clutch is:

    (a) More than one
    (b) Less or more depends of cone angle
    (c) Less than one
    (d) Depends on coefficient of friction

10. In a centrifugal clutch, the speed at which it engages depends on:

    (a) Mass of shoes
    (b) Stiffness of springs
    (c) Gap between drum and shoe
    (d) All given in a, b, and c

Answers to multiple choice questions

1. (d)   2 (a)   3. (c)   4. (b)   5. (b)   6. (c)   7. (d)   8. (d)   9. (a)   10. (d)

Design Problems

Single plate clutch

1. A plate clutch with maximum diameter 200 mm has maximum lining pressure of 0.35 MPa. The power to be transmitted is 13.5 kW at 400 rpm and coefficient of friction is 0.35. Using uniform pressure theory, find inside diameter and spring force required to engage the clutch.

$$[d = 144 \text{ mm}, F = 5{,}293 \text{ N}]$$

2. A single dry plate friction clutch transmits 15 kW at 1,200 rpm. Coefficient of friction is 0.2 and axial pressure 0.09 MPa. Ratio of inner to outer radii = 0.56. The pressure plate is pushed by six springs. Assuming uniform wear theory, calculate:

    (a) Outside and inside radii
    (b) Face width
    (c) Axial force
    (d) Load per spring

$$[(a)\ r_1 = 125 \text{ mm},\ r_2 = 70 \text{ mm}; (b)\ b = 55 \text{ mm}; (c)\ 3{,}031 \text{ N}; (d)\ 505 \text{ N}]$$

3. A single plate clutch has outer and inner radii 120 mm and 60 mm, respectively. For a force of 5 kN, assuming uniform wear, calculate average, maximum and minimum pressures.

$$[(a)\ p_{av} = 0.147 \text{ MPa}; (b)\ p_{max} = 0.221 \text{ MPa}; (c)\ p_{min} = 0.11 \text{ MPa}]$$

4. A single plate dry friction clutch has to transmit 80 Nm using maximum pressure 0.15 MPa on pressure plate, coefficient of friction is 0.25, and ratio of inner to outer diameters is 0.7. Assuming uniform wear, find:

    (a) Size of friction plate
    (b) Axial force

$$[(a)\ d_1 = 196 \text{ mm}, d_2 = 138 \text{ mm}; (b)\ 1{,}818 \text{ N}]$$

5. An engine of a vehicle using a single plate friction clutch develops 50 kW at 1,800 rpm. $\mu = 0.3$, $p = 0.15$ MPa, and outer radius of the friction plate = 120 mm. The other relevant data are as under:

    Torque at the engagement of clutch = 105 Nm

    Torque for acceleration = 150 Nm

    Moment of inertia of fly wheel = 0.85 kg m^2

    Weight of loaded vehicle = 14 kN

    Wheel diameter = 700 mm

    Inertia of driven shaft = 1 kgm^2

    Speed reduction in differential = 4

    Vehicle speed when the clutch is engaged = 45 km/h

    Calculate:

    (a) Time for engagement
    (b) Heat generated during this period

$$[(a)\ t = 0.18 \text{ s}; (b)\ H = 344 \text{ J}]$$

6. A single plate friction clutch transmits 7.5 kW at 1,500 rpm. Mass of driver and driven shafts are 5 kg and 9 kg, respectively. Radius of gyration of driver and driven rotors are 70 mm and 110 mm, respectively. Calculate:

(a) Time of engagement, if the driven shaft is stationary

(b) Heat generated

$$[(a)\ t = 0.066\ s;\ (b)\ H = 247\ J]$$

Multi-plate clutch

7. A multi-plate clutch transmits 5 kW at 720 rpm though steel and friction plates arranged alternately. Outer and inner radii of the friction plate are 80 mm and 45 mm, respectively. The plates are immersed in oil to give a coefficient of friction 0.08. If maximum pressure on the plate is not to exceed 0.25 N/mm², calculate:

(a) Number of discs (b) Axial force required

$$[(a)\ \text{Steel plate} = 3,\ \text{friction plate} = 2;\ (b)\ 3,434\ N]$$

8. A multi-plate clutch has seven steel discs and six friction discs of outer and inner radii 100 mm and 65 mm, respectively. Maximum pressure 0.1 MPa and coefficient of friction 0.28. Find the maximum torque which it can transfer with uniform wear. [396 Nm]

9. A multi-plate clutch transmits 20 kW at 1,500 rpm though three pairs of friction plates. Inner radius is 50 mm, μ = 0.3. If pressure on the plate is not to exceed 0.12 N/mm², calculate outer radius assuming uniform wear. $[r_1 = 79\ mm]$

Cone clutch

10. Cone clutch transmits 5 kW power at 960 rpm with the following data:

Cone angle 30°, μ = 0.2, maximum pressure = 0.2 MPa. Calculate size of cone assuming uniform wear.

$$[r_1 = 167\ mm,\ r_2 = 154\ mm,\ b = 48\ mm]$$

11. A cone clutch transmits 120 Nm torque. Bigger diameter of the cone is 250 mm. Cone angle is 20°, width b = 8 mm, μ = 0.2. Assuming uniform wear condition, find:

(a) Axial force (b) Average pressure

(c) Maximum pressure

$$[(a)\ 1,790\ N;\ (b)\ p_{av} = 0.766\ MPa;\ (c)\ p_{max} = 0.776\ MPa]$$

Centrifugal clutch

12. Four shoes of a centrifugal clutch transmit 10 kW at 1,200 rpm. Inside radius of the drum is 140 mm. Coefficient of friction is 0.3. C.G. of the shoes can be taken at 80 per cent of the maximum radius of the shoe. Shoe sub tends and angle of 60° at the center of spider. Clutch engagement begins at 75 per cent of running speed. Radial pressure on shoe should not exceed 0.04 MPa. Determine:

(a) Mass of the shoe (b) Width of shoe

$$[(a)\ m = 0.613\ kg;\ (b)\ b = 21.6\ mm]$$

13. A centrifugal clutch having three shoes each of 1 kg runs at 850 rpm, spring force is such that it exerts a force of 600 N at the time of engagement. Drum inside radius is 175 mm and coefficient of friction is 0.25. C.G of the shoe is at radius 140 mm, calculate:

   (a) Percentage of running speed at which engages

   (b) Power which it can transmit

[(a) 74 per cent; (b) $P = 7$ kW]

Previous Competition Examination Questions

IES

1. A plate clutch has a single surface with an outside diameter of 250 mm and inside diameter 100 mm with a coefficient of friction 0.2. Find the required axial force to develop a maximum pressure of 0.65 MPa. Under this pressure, find the torque capacity of the clutch.

   [Answer: $F = 268$ N, $T = 4.68$ Nm] [IES 2014]

2. Sketch and explain the working of centrifugal clutch. Give one application of it.

   [IES 2013]

3. Match list I with list II and select the correct answer using the codes given below:

List I	List II
A. Single friction clutch plate	1. Scooters
B. Multi-plate friction clutch plate	2. Rolling mills
C. Centrifugal clutch	3. Trucks
D. Single friction clutch plate	4. Mopeds

   Code:

	A	B	C	D
(a)	1	3	4	2
(b)	1	3	2	4
(c)	3	1	2	4
(d)	3	1	4	2

   [Answer: (d)] [IES 1998]

4. A multiple disc friction clutch is installed on a shaft whose angular velocity is 100 rad/s and which transmits a power of 5 kW. Check whether the number of friction surface pairs $z = 3$ is sufficient, if inside and outside radii of the friction surface are 50 mm and 75 mm, respectively. Take $\mu = 0.08$, $p = 0.49$ MPa. [Answer: Sufficient] [IES 1998]

5. In the multiple disc clutch, if there are six discs on the driving shaft and five discs on the driven shaft, then the number of pairs of contact surfaces will be:

   (a) 11   (b) 12

   (c) 10   (d) 22

   [Answer: (c)] [IES 1997]

6. In designing a clutch plate, assuming of uniform wear condition is made because:

    (a) It is closer to real life situation   (b) It leads to a safer design

    (c) It leads to cost effective design   (d) No other assumption is possible

    [Answer: (a)] [IES 1996]

GATE

7. A single plate clutch is designed to transmit 10 kW power at 2,000 rpm. The equivalent mass and radius of gyration of the input shaft are 20 kg and 75 mm, respectively. The equivalent mass and radius of gyration of the output shaft are 35 kg and 125 mm, respectively. Calculate the time required to bring the output shaft from rated speed to rest.

    [Answer: $t = 0.41$ s] [GATE 2015]

8. A disc clutch with a single friction surface has coefficient of friction equal to 0.3. The maximum pressure that can be imposed on the friction material is 1.5 MPa. The outer diameter of the clutch plate is 200 mm and the internal diameter is 100 mm. Assuming uniform wear theory for the clutch plate, the maximum torque (Nm) that can be transmitted is_____

    [Answer: $T = 530$ Nm] [GATE 2014]

Common data questions 9 and 10.

A multiple disc clutch is to transmit 4 kW at 750 rpm. Available steel and bronze discs of 40 mm inner radius and 70 mm outer radius are to be assembled alternately in appropriate numbers. The clutch is to operate in oil with an expected coefficient of friction of 0.1 and maximum allowable pressure not to exceed 350 kPa. Assume uniform wear condition to prevail.

9. The number of steel (driving) and bronze (driven) discs required respectively will be:

    (a) 3,2   (b) 2,3

    (c) 2,2   (d) 3,3

    [Answer: (c)] [GATE 2013]

10. Axial force to be applied to develop the full torque will be:

    (a) 2,160 N   (b) 2,260 N

    (c) 2,360 N   (d) 2,560 N

    [Answer: (c)] [GATE 2013]

Linked answers Q 11 and Q 12.

A clutch plate with outside radius 120 mm and inside radius half of outer radius. Axial force of 5 kN is applied on the plate. Assume uniform wear condition.

11. Maximum pressure on clutch plate:

    (a) 0.22 MPa   (b) 0.24 MPa

    (c) 0.27 MPa   (d) 0.29 MPa

    [Answer: (a)] [GATE 2013]

12. Minimum pressure on clutch plate:

    (a) 0.08 MPa   (b) 0.10 MPa   (c) 0.11 MPa   (d) 0.15 MPa

    [Answer: (c)] [GATE 2013]

13. A clutch has outer and inner dimensions 100 mm and 40 mm, respectively. Assuming a uniform pressure of 2 MPa and coefficient of friction of liner material 0.4, the torque carrying capacity of the clutch is:

    (a) 148 Nm   (b) 196 Nm   (c) 372 Nm   (d) 490 Nm

    [Answer: (b)] [GATE 2008]

14. A disc clutch is required to transmit 5 kW at 2,000 rpm. The disc has friction lining with coefficient of friction 0.25. Bore radius of friction lining is 25 mm. Assume uniform pressure of 1 MPa. The value of outside radius of friction lining is:

    (a) 39.4 mm   (b) 49.5 mm

    (c) 97.9 mm   (d) 142.9 mm

    [Answer: (a)] [GATE 2008]

15. A clutch has outer and inner diameters 100 mm and 40 mm, respectively. Assuming a uniform pressure of 2 MPa and coefficient of lining material 0.4, the torque carrying capacity of the clutch is:

    (a) 148 Nm   (b) 196 Nm

    (c) 372 Nm   (d) 490 Nm

    [Answer: (b)] [GATE 2008]

16. A multiple disc clutch is to transmit 4 kW at 750 rev/min. Available steel and bronze discs of 40 mm inner radius and 70 mm outer radius are to be assembled alternately in appropriate numbers. The clutch is to operate in oil with expected coefficient of friction of 0.1 and maximum allowable pressure not to exceed 350 kPa. Assume uniform wear conditions to prevail. Specify the number of steel (driving) and bronze (driven) discs required. Also determine what axial force is to be applied to develop the full torque.

    [Steel plates = 3, Bronze plates = 2] [GATE 1997]

17. A cone clutch is to be designed to transmit a torque of 1,000 Nm. Space restrictions limit the outside and inside radii to the values of 75 mm and 45 mm, respectively and the semi cone angle is 15°. A moulded friction lining with dry conditions is to be used for which average coefficient of friction is 0.35. Using uniform wear theory, find the required clamping force. The clamping force is supplied by one spring of a spring constant = 1,500 N/mm. If the friction lining wears out 0.4 mm, what is the reduction in torque capacity.

    [12.3 kN, 190 Nm] [GATE 1992]

□□□

Brakes

19.1 Definition and Functions

A brake is a mechanical device, used to stop or slow down a moving part, by absorbing its kinetic or potential energy by friction. See use of disc brakes in picture on right side. The energy absorbed due to friction results in heat, causing rise in temperature. This heat is dissipated to the atmosphere using generally natural or sometimes forced cooling.

A brake is used for any one of the following functions:

- To stop or slow down a moving part or system, for example, automobiles, trains, etc.
- To hold a part in position, for example, hand brake of a vehicle.
- To absorb potential energy released by moving parts, as in hoists.

19.2 Types of Brakes

Brakes are of many types. They can be classified as under:

a. On the basis of its operation:

- Mechanical brakes: Operated by lever / pedal.
- Hydraulic brakes: Operated by oil pressure in a master cylinder.
- Pneumatic: Operated by air pressure in a cylinder.
- Magnetic: Operated by magnetic forces, eddy currents, and magnetic particles.

b. On the basis of type of shoes:

- Fixed block brakes [Figure 19.1(a)] or pivoted shoe as in railways (See Section 19.7.5).
- Internally expanding shoe brakes [Figure 19.1(b)] as in old automobiles.

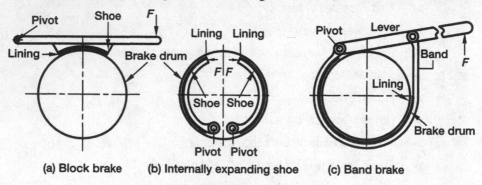

(a) Block brake   (b) Internally expanding shoe   (c) Band brake

Figure 19.1 Types of brakes

- Externally contracting shoe brakes. Shoes are outside the drum and contract while braking (See Section 19.11).
- Band brakes [Figure 19.1(c)] as in hoists. These are described in detail in Section 19.9.
- Disc brakes: Shoes are on the sides of a rotating disc. The force is applied axially (See Section 19.12).

19.3 Materials for Brake Lining

A material used as a brake lining should possess the following properties:

- High coefficient of friction.
- Value of coefficient of friction should not change with increase in temperature.

- Low wear rate.
- Withstand heat without losing properties.
- High mechanical strength.
- Not affected much by oil or moisture.

Materials commonly used for brake lining are tabulated in Table 19.1 with their useful properties.

Table 19.1 Properties of Brake Lining Materials

Material		Coefficient of Friction		Allowable Pressure (N / mm²)
Brake Drum	Lining	Dry and Clean	Wet	
Cast iron	Cast iron	0.15–0.2	0.05–0.1	1.0–1.75
Cast iron	Bronze	0.2–0.25	0.05–0.1	0.55–0.85
Cast iron	Steel	0.2–0.3	0.07–0.12	0.85–1.4
Cast iron	Wood	0.2–0.35	0.08–0.12	0.4–0.6
Steel	Aluminium	0.6	–	0.2–0.3
Steel	Brass	0.35	0.19	0.6–0.9
Metal	Leather	0.3–0.5	0.15–0.2	0.07–0.3
Metal	Asbestos	0.4–0.5	0.25–0.3	0.3–1.1
Metal	Wire asbestos	0.35–0.5	0.25–0.3	0.2–0.55

19.4 Energy Absorbed by Brakes

The energy to be absorbed depends on type of motion of moving body. The motion can be of any one type of the following:

- Pure rotation
- Pure translation
- Combined rotation and translation

19.4.1 Pure rotation

Braking of pure rotating part can be seen for hoists. Following notations are used for finding energy:

E_R = Energy absorbed while braking, due to pure rotation(J)

I = Mass moment of inertia(kg-m²)

ω_1 = Angular velocity before braking (rad/s)

ω_2 = Angular velocity after braking (rad/s)

Energy absorbed while braking, due to rotation:

$$E_R = \frac{1}{2}I(\omega_1^2 - \omega_2^2)$$

(19.1a)

If the moving body is completely stopped $\omega_2 = 0$, then the energy is:

$$E_R = \frac{1}{2}I\omega_1^2 \tag{19.1b}$$

19.4.2 Pure translation

This condition is applicable for objects of mass (N) moving with linear velocity, for example, a moving car. If initial velocity before braking is v_1 (m/s) and final velocities after braking v_2 (m/s) then:

Energy absorbed while braking due to pure translation:

$$E_T = 0.5\,I\left[\left(v_1\right)^2 - \left(v_2\right)^2\right] \tag{19.2a}$$

If the moving body is completely stopped $v_2 = 0$, the energy is:

$$E_T = \frac{1}{2}Iv_1^2 \tag{19.2b}$$

If the movement is in vertical direction, potential energy is to be considered. Such an energy is given as:

$$E_P = mg(h_1 - h_2) \tag{19.3}$$

Where, E_P = Energy absorbed while braking due to vertical movement

m = Mass of moving part (kg)

g = Acceleration due to gravity (9.81 m/s²)

h_1 = Initial height before braking (m)

h_2 = Final height after braking (m)

19.4.3 Combined rotation and translation

Such a condition exists when the motion is linear as well as rotation is also there. For example, in a moving train or vehicle, linear velocity is there and the wheels rotate also. The kinetic energy absorbed due to combined effect is:

$$E = E_R + E_T + E_P \tag{19.4}$$

When brakes are applied with some force, work done $W = E = F_f\,\pi D N t$

Or, $$F_f = \frac{E}{\pi D N t} \tag{19.5}$$

Where, F_f = Tangential force on the brake drum

$D = 2\,r$ = Brake drum diameter (mm)

N = Speed of rotation of the drum (rpm)

t = Time for which the brakes are applied (s)

The frictional torque T_f is given as:

$$T_f = 0.5 \, D \, F_f = rF_f \qquad (19.6)$$

19.5 Heat Dissipated

The energy generated due to friction produces heat and thus increases the temperature of the drum and lining. Heat dissipated depends on the following parameters:

- Coefficient of friction
- Braking force
- Mass of the brake drum
- Surface area of the drum
- Braking time
- Heat dissipating capacity of the drum

Some of the heat is dissipated to the surrounding, while the rest increases the temperature of drum and lining. The allowable temperatures for the different lining materials are as under:

Wood	60 °C
Fibre and leather	70 °C
Asbestos without oil	90 °C
Asbestos with oil	105 °C
Forodo(commercial name)	180 °C

Heat generated per unit time is given as:

$$H_g' = \mu N v = \mu p A v \qquad (19.7)$$

Where, H_g' = Heat generated per second (J/s)

μ = Coefficient of friction

N = Force on the brake shoe = Reaction on brake shoe (N)

A = Projected area of contact surface (m²)

p = Normal pressure (N/m²)

v = Peripheral velocity of drum (m/s)

Heat dissipated is given as:

$$H_d = C \, (t_1 - t_2) A_r \qquad (19.8)$$

Where, H_d = Heat dissipated(J/s)

C = Coefficient of heat transfer (W/m²/°C)

For $(t_1 - t_2) = 40$ °C $C = 30 \text{W/m}^2/\text{°C}$

For $(t_1 - t_2) = 180$ °C $C = 40 \text{W/m}^2/\text{°C}$

t_1 = Temperature of rotating surface (°C)

t_2 = Temperature of surroundings (°C)

A_r = Surface area of rotating drum surface(m²)

Only 15 per cent to 20 per cent of this heat is dissipated to the surrounding while the rest increases the temperature Δt of drum and can be calculated from the following relation:

$$\Delta t = \frac{H_g}{m\,C_p} \tag{19.9}$$

Where, Δt = Temperature rise of brake drum (°C)

H_g = Heat generated during braking time (J)

m = Mass of rotating brake drum (kg)

C_p = Specific heat of the drum material (J/kg °C)

For aluminium	$C_p = 910$ J/kg °C
For cast steel	$C_p = 490$ J/kg °C
For cast iron	$C_p = 460$ J/kg °C

It is not possible to calculate the temperature rise exactly; hence to limit the temperature, a maximum pv value is fixed as described in the next section.

19.6 Lining Wear (pv Value)

If the temperature rise is more, lining starts wearing at higher speed, and hence life of lining decreases. The wear rate depends on normal pressure and rubbing velocity. Hence to limit the wear rate / temperature, a maximum pv value is recommended to design for a reasonable life. Recommended values are as under:

Table 19.2   Maximum pv Value

Type of Service	Heat Dissipation	Pv (N/mm s)*
Continuous	Poor	1.0×10^3
Intermittent	Poor	1.9×10^3
Continuous	Good	2.9×10^3

*p in N/mm² (MPa) and v in m/s.

Example 19.1

Heat generated in braking a hoisting drum with given load

A hoist carrying a load of 500 kg is lowered for a height of 10 m (Figure 19.S1). Outside diameter of hollow brake drum is 0.6 m and inside diameter 0.9 times outside diameter and weighs 150 kg. The hoist at its lowermost position is to be stopped in 10 s. Assuming the following values:

Coefficient of friction $\mu = 0.2$

Normal pressure = 0.4 N/mm²

Brake shoe angle substandard at the center = 60°

Specific heat of drum material C_p = 490 J/kg °C

Calculate:

a. Heat generated, while stopping.

b. Width of block.

c. Temperature rise of the drum, assuming that only 10 per cent of heat generated is dissipated to the surrounding.

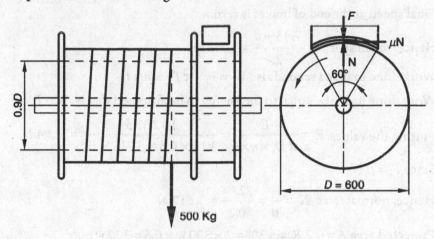

Figure 19.S1 A hoisting drum

Solution

Given $m = 500$ kg $h = 10$ m $2r = D = 0.6$ m $d = 0.9\,D$

$m_d = 150$ kg $\mu = 0.2$ $p = 0.4$ N/mm² $t = 10$ s

$\theta = 60°$ $C_p = 490$ J/kg °C

a. Calculate velocity attained after falling through a height of 10 m using equation:

$$v^2 - u^2 = 2\,g\,h \quad \text{or} \quad v^2 - 0 = 2 \times 9.81 \times 10 \quad \text{or} \quad v = 14\,\text{m/s}$$

Accelerating force = Mass × Acceleration = 500 × 9.81 = 4,905 N

Total load = Dead load + Accelerating force = (500 × 9.81) + 4,905 = 9,810 N

Rotational speed of the brake drum $\omega_1 = \dfrac{v}{R} = \dfrac{14}{0.3} = 46.67\,\text{rad/s (Because } v = w\,R]$

Radius of gyration for a hollow drum is $K = \dfrac{\sqrt{D^2 + d^2}}{8}$

Moment of inertia $I = m_d K^2 = m_d \times \dfrac{D^2 + d^2}{64} = 150 \times \dfrac{0.6^2 + 0.54^2}{64} = 1.5\,\text{kg m}^2$

Energy due to rotation of the drum: Moving body is completely stopped, that is, $\omega_2 = 0$

From Equation (19.1b): $E_R = \dfrac{1}{2}I\omega_1^2 = 0.5 \times 1.5 \times 46.67^2 = 1{,}633.5$ J

From Equation (19.3): Potential energy $E_p = m\,g\,h = 500 \times 9.81 \times 10 = 49{,}000$ J

Total energy $E = E_R + E_p = 1633.5 + 49{,}000 = 50{,}633.5$ J

Heat generated $H_g = E = 50{,}633.5$ J

b. Rotational speed of the drum, when brakes are applied $n = \dfrac{\omega}{2\pi} = \dfrac{46.67}{6.28} = 7.43$ rps

Final speed at the end of brakes is zero.

Hence, mean speed $= \dfrac{7.43 + 0}{2} = 3.715\,\text{rps}$

Work done in time t seconds is: $W = F_t \times \pi D_b \times n \times t$

Work done has to be equal to the energy to be absorbed, that is, $W = E$

Putting the values: $F_t = \dfrac{E}{\pi D_b \times n \times t} = \dfrac{50{,}633.5}{3.14 \times 0.6 \times 3.715 \times 10} = 723.4$ N

Since $F_t = \mu N$

Hence, normal force $N = \dfrac{F_t}{\mu} = \dfrac{723.4}{0.2} = 3{,}617$ N

Projected area $A = w\,2\,R_b \sin 30° = 2 \times 300\,w \times 0.5 = 300\,w$ mm^2

Normal force $N = pA = 0.4 \times 300\,w$

Substituting the value of N: $3{,}617 = 120\,w$ Hence $w = 30.2$ mm

c. Only 10 per cent heat is dissipated to surroundings, and hence rest 90 per cent is responsible for increase in temperature of the drum.

Heat generated per second $H_g' = \dfrac{H_g}{t} = \dfrac{50{,}633.5}{10} = 5{,}063.4$ J/s

From Equation (19.9), $\Delta t = \dfrac{H_g'}{mC_p} = \dfrac{0.9 \times 50{,}633.5}{150 \times 490} = 0.62\,°\text{C}$

19.7 Block Shoe Brakes

19.7.1 Fixed block shoe brakes

A fixed shoe brake is in which the brake block is fixed to the lever firmly. Another arrangement is pivoted block, in which the block can tilt and adjust according to the circumference of the brake drum. It is described in Section 19.7.5.

Fixed shoe brake is the simplest form of brake. It consists of a block curved with same radius as the brake drum. A friction lining is glued or riveted to the block Figure (19.2). The block is fixed at a distance b with a lever of length a, to increase the force at the drum.

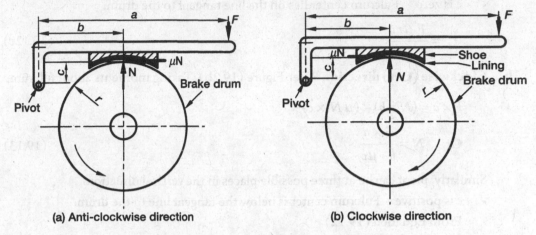

(a) Anti-clockwise direction   (b) Clockwise direction

Figure 19.2 Fixed pad shoe brake with bell crank lever

When a force F is applied at the end of the lever, it exerts an increased force $(F \times a / b)$, in the same direction due to leverage on the shoe, whose normal reaction N is equal and opposite to the direction of the force, that is,

$$N = F\frac{a}{b} \tag{19.10}$$

The frictional force F_f is μ times this reaction, that is, $F_f = \mu N$ and acts in the direction opposite to the direction of rotation, but its reaction on the shoe acts in the direction of rotation.

For drum of radius r, the frictional braking torque $T_f = F_f \times r = \mu N r$ (19.11)

a. **Anticlockwise direction**   Refer Figure (19.2a). Taking moments about fulcrum:

$$F \times a = (N \times b) + (\mu N \times c)$$

Or, $N = \dfrac{F \times a}{b + \mu c}$ (19.12)

The fulcrum or pivot can be at three possible places in the vertical direction.

- c is positive   Fulcrum center is below the tangent line to the drum.
  From Equation (19.12)

$$T_f = \mu N r = \frac{F \mu r a}{b + \mu c} \tag{19.12a}$$

- c is negative   Fulcrum center is above the tangent line to the drum.

$$T_f = \frac{F\,\mu r a}{b - \mu c} \tag{19.12b}$$

- c is zero   Fulcrum center lies on the line tangent to the drum.

$$T_f = \frac{F\,\mu r a}{b} \tag{19.12c}$$

b. Clockwise (CW) direction Refer Figure (19.2b). Taking moments about fulcrum:

$$F \times a = (N \times b) - (\mu N \times c)$$

Or, $$N = \frac{F \times a}{b - \mu c} \tag{19.13}$$

Similarly, pivot can be at three possible places in the vertical direction.

- c is positive   Fulcrum center is below the tangent line to the drum.
  From Equation (19.12)

$$T_f = \mu N r = \frac{F\,\mu r a}{b - \mu c} \tag{19.13a}$$

- c is negative   Fulcrum center is above the tangent line to the drum.

$$T_f = \frac{F\,\mu r a}{b + \mu c} \tag{19.13b}$$

- c is zero   Fulcrum center is on the tangent line to the drum.

$$T_f = \frac{F\,\mu r a}{b} \tag{19.13c}$$

From the above analysis, following can be noted:

- If $c = 0$, direction of rotation has no effect in braking torque.
- Braking torque for CW rotation is same as for anticlockwise rotation, if value of c is changed from positive to negative or vice versa.

Reaction at the pin   Reaction at the pivot has two components:

Horizontal force $F_H = \mu N$

Vertical force $F_V = F - N$

Resultant reaction at the pivot is: $R = \sqrt{F_H^2 + F_V^2}$

Width of block is generally taken 25 per cent to 50 per cent of the drum diameter. Size of block depends on allowable pressure p on lining and can be calculated from the relation:

$$N = p\,l\,w \tag{19.14}$$

Where, N = Normal force (N)

  p = Allowable pressure (MPa)

  l = Projected length of block (mm)

  w = Width of the block (mm)

19.7.2 Self-energizing brakes

Self-energizing means, that the frictional force is used to help brake application, in addition to applied force by the operator. It can be seen from Equation (19.13) that for CW direction:

$$N \times b = (F \times a) + (\mu N \times c)$$

Moment of frictional force $(\mu N \times c)$ helps in moment of actuation force $(F \times a)$. This phenomenon is called self-energizing.

19.7.3 Self-locking

In Equation (19.13) if $b = \mu c$, then the force required F for braking is zero, that is, brakes are applied even without force. Such a situation is called self-locking. It is not desirable as the operator should have a brake feel as to how much brake is being applied.

If $b < \mu c$, the force F is negative, which is also not desirable condition and results in grabbing of the drum causing uncontrolled braking. Hence b should be more than μc.

Example 19.2

Force on a fixed block brake with different directions of rotation

A brake drum of diameter 400 mm is braked with the help of a cast iron shoe with coefficient of friction 0.3. The force is applied at 800 mm away from center line of the drum as shown in Figure 19.S2. The pivot is 300 mm away from central axis on the other side and 10 mm below the surface of drum.

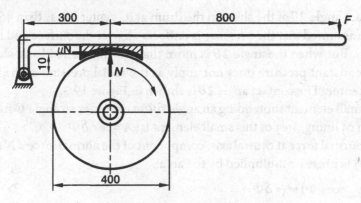

Figure 19.S2 Fixed block brake

For applying a braking torque of 500 Nm, calculate:

a. Force F for clockwise rotation.

b. Force F for anti-clockwise rotation.

c. For anti-clockwise rotation, distance y for brake to be self-locking.

Solution

Given $2r = 0.4$ m $\mu = 0.3$ $a = 1.1$ m $b = 0.3$ m

 $c = 0.01$ m $T_f = 500$ Nm

Frictional torque $T_f = \mu N r$ or $500 = 0.3 \times N \times 0.2$ or $N = 8{,}333$ N

a. Clockwise rotation   Taking moments about the pivot:

$$F \times 1.1 = -\mu N \times 0.01 + N \times 0.3 \quad \text{Substituting the values:}$$

$$1.1\,F = -(0.3 \times 8{,}333 \times 0.01) + (0.3 \times 8{,}333) \quad \text{or} \quad F = 2{,}250 \text{ N}$$

b. Anti-clockwise rotation   Taking moments about the pivot:

$$F \times 1.1 = \mu N \times 0.01 + N \times 0.3 = 0 \quad \text{Substituting the values:}$$

$$1.1\,F = (0.3 \times 8{,}333) + (0.3 \times 8{,}333 \times 0.01) \quad \text{or} \quad F = 2{,}295 \text{ N}$$

c. Self-locking

For brake to be self-locking, $F = 0$. Let y be the vertical distance between the pivots and drum surface. Taking moments about the pivot O:

$$0 - \mu N \times y + N \times 0.3 = 0 \text{ Substituting the values:}$$

$$0.3 \times 8{,}333 \times y = 0.3 \times 8{,}333 \quad \text{or} \quad y = 1 \text{ m}$$

19.7.4 Small / long shoe brake

When the contact angle 2θ of the shoe on the drum at its center is less than $45°$ (called **short shoe**), it can be assumed that the pressure is uniform all over the surface and hence wear will also be uniform. But when the angle 2θ is more than $45°$ it is called **long shoe**. The above assumption of constant pressure does not apply as the radial pressure at the ends is not the same as in the center. The contact angle 2θ is shown in Figure 19.3.

Consider a small element subtending an angle $\delta\Phi$ at center, at an angle Φ from central axis. If w is the width of lining, area of this small element is: $A = w\,r\,\delta\Phi$

If p_{max} is the normal force at central axis, component of the normal force dN on this element of length $(r\,\delta\Phi)$ is pressure multiplied by the area:

$$dN = (p_{max} \cos \Phi)\,w\,(r\,\delta\Phi)$$

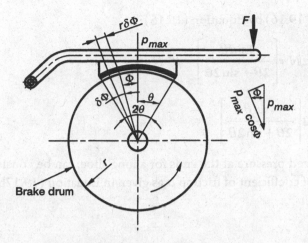

Figure 19.3 Effect of shoe angle

Total vertical force N can be got, by integrating above equation between limits $-\theta$ to θ.

$$N = \int dN \cos \Phi = w\, r \int_{-\theta}^{\theta} p_{max} \cos^2 \Phi \, d\Phi$$

Using trigonometric relation $\cos^2 \Phi = \dfrac{1}{2}\left(1 + \cos 2\,\Phi\right)$:

$$N = \frac{1}{2}\, w\, r\, p_{max} \int_{-\theta}^{\theta} \left(1 + \cos 2\,\Phi\right) d\Phi$$

$$\text{Or, } N = \frac{1}{2} w\, r\, p_{max}\left[\Phi + \frac{\sin 2\Phi}{2}\right]_{-\theta}^{\theta} = \frac{1}{4}\, w\, r\, p_{max}\left[2\Phi + \sin 2\Phi\right]_{-\theta}^{\theta}$$

$$\text{Or, } N = \frac{1}{4}\, w\, r\, p_{max}\left[\left(2\theta + 2\theta\right) + \left\{\sin 2\theta - \left(\sin -2\theta\right)\right\}\right]$$

$$\text{Or, } N = w\, r\, p_{max}\left[\frac{2\theta + \sin 2\theta}{2}\right] \tag{19.15}$$

Frictional torque $T_f = \mu\, r \times r \displaystyle\int_{-\theta}^{\theta} p \cos \Phi \, d\Phi \qquad$ Because $p = p_{max} \cos \Phi$

$$T_f = \mu\, r^2\, w\, p_{max} \int_{-\theta}^{\theta} \cos \Phi \, d\Phi$$

$$T_f = \mu\, r^2\, w\, p_{max}\left[\sin \Phi\right]_{-\theta}^{\theta} = 2\,\mu\, r^2\, w\, p_{max} \sin \theta \tag{19.16}$$

Dividing Equation (19.16) by Equation (19.15):

$$T_f = \mu N r \left[\frac{4\sin\theta}{2\theta + \sin 2\theta} \right] = \mu' N r \qquad (19.17a)$$

Where, $\mu' = \mu \left[\dfrac{4\sin\theta}{2\theta + \sin 2\theta} \right]$ $\qquad (19.17b)$

Thus effect of reduced pressure at the ends for a long shoe, can be considered by modifying the effective value of coefficient of friction μ as given in Equation (19.17b).

Example 19.3

Force required on long shoe brake for specified torque
A brake drum of 500 mm diameter is braked with the help of a cast iron shoe subtending an angle of 120° at the center of the drum with coefficient of friction 0.3. The force F is applied at 700 mm away from the center line of the drum as shown in Figure 19.S3. The pivot is 300 mm away from the central axis on the other side and 15 mm below the surface of the drum. Calculate the force F required for a brake torque of 450 Nm with clockwise and anti-clockwise rotation.

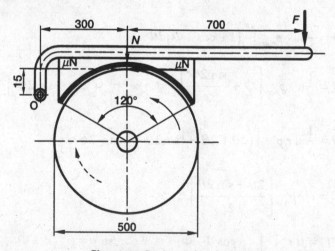

Figure 19.S3 Long shoe brakes

Solution

Given $2r = 0.5$ m $\mu = 0.3$ $a = 1.0$ m $b = 0.3$ m
$c = 0.015$ m $2\theta = 120°$ $T_f = 450$ Nm

For a long shoe the effective coefficient of friction $\mu' = \mu \left[\dfrac{4\sin\theta}{2\theta + \sin 2\theta} \right]$

$$\text{Contact arc } 120° = \frac{3.14 \times 120}{180} = 2.093 \text{ rad}$$

From Equation (19.17b):

$$\mu' = 0.3 \left[\frac{4 \sin 60°}{120° + \sin 120°} \right] = 0.3 \left[\frac{4 \times 0.866}{2.093 + 0.866} \right] = 0.351$$

Frictional torque $T_f = \mu' N r$ or $450 = 0.351 \times N \times 0.25$ or $N = 5{,}128$ N

Anti-clockwise rotation Taking moments about the pivot O:

$$F \times (0.7 + 0.3) - (\mu' N \times 0.015) - (N \times 0.3) = 0$$

Substituting the values:

$$F = (0.351 \times 5{,}128 \times 0.015) + (0.3 \times 5{,}128) = 1{,}565 \text{ N}$$

Clockwise rotation Taking moments about the pivot O:

$$F \times (0.7 + 0.3) + (\mu' N \times 0.015) - (N \times 0.3) = 0$$

Substituting the values:

$$F = (0.3 \times 5{,}128) - (0.351 \times 5{,}128 \times 0.015) = 1{,}538 - 27 = 1{,}511 \text{ N}$$

19.7.5 Pivoted block brakes

Frictional force between the lining and drum tries to move the shoe with respect to the lever. Solution to this problem is achieved by using pivoted block brakes. Such shoes are used in railways and hoists. The pivot is so located that the moment of frictional force about the pivot is zero. A pivoted block brake is shown in Figure 19.4.

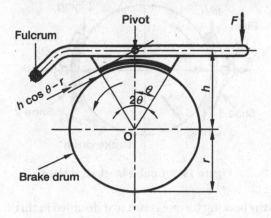

Figure 19.4 Pivoted block brake

Ideally normal reaction passes through this pivot point and hence no moment occurs on a shoe. For this condition, the pivot has to be very close to the drum surface. To obtain such a condition, vertical height h of pivot from the center of the brake drum should be:

$$h = \left[\frac{4r \sin \theta}{2\theta + \sin 2\theta} \right]$$

(19.18)

Where, r = Radius of brake drum

θ = Semiangle of the brake shoe over the drum

Frictional force μN causes a turning moment as $\mu N (h \cos \theta - r)$.

Advantages of pivoted block brakes are:

- Provides greater braking torque than fixed block brakes.
- Offers uniform wear of the lining.
- Pressure is more or less uniform.

Disadvantages are:

- Locating the pivot very near to the drum is difficult.
- Mounting is not at correct position of the pivot, as the lining wears out.

19.7.6 Double block shoe brakes

In a single shoe brake, normal force at the shoe causes bending of the shaft at drum location. To balance this force, double block shoe brakes are used. The force applied on the both shoes gets cancelled automatically, as these are in opposite directions. Figure 19.5 shows such a brake.

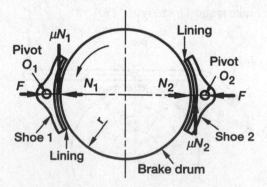

Figure 19.5 Double block shoe brake

It may be noted that the braking torque is not just doubled in this case. The braking torque by each shoe is different as demonstrated in Example 19.4.

Example 19.4

Force required on a double shoe brake with two levers

A double shoe brake with each having contact angle at the center of the brake drum 90°. The drum has diameter 300 mm and transmits a torque of 300 Nm rotating anti-clockwise at 500 rpm. Coefficient of friction can be taken as 0.3 and maximum pv value 1.7×10^6. A force F force is applied at the end of lever, 200 mm away from the central axis of the drum. Pivots are 200 mm apart at the other side of the drum as shown in Figure 19.S4. Calculate:

a. Force F b. Width of shoes

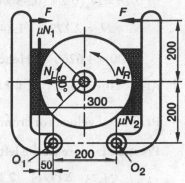

Figure 19.S4 Double shoe brake

Solution

Given $2r = 0.3$ m $\theta = 0.3$ $a = 0.4$ m $b = 0.2$ m

$c = 0.15 - 0.1 = 0.05$ m $2\theta = 90°$

$T_f = 300$ Nm $N = 500$ rpm $pv = 1.7 \times 10^6$

Since the contact angle 2θ is more than 45°, effective coefficient of friction μ' is calculated.

$$\text{Contact arc } 90° = \frac{3.14 \times 90}{180} = 1.57 \text{ rad}$$

$$\mu' = \mu \left[\frac{4\sin\theta}{2\theta + \sin 2\theta} \right] = 0.3 \left[\frac{4\sin 45°}{90° + \sin 90°} \right] = 0.3 \left[\frac{4 \times 0.707}{1.57 + 1} \right] = 0.333$$

Frictional torque T_f = Frictional force $F_f \times$ drum radius r

$$300 = F_f \times 0.15 \quad \text{or} \quad F_f = 2,000 \text{ N}$$

This total frictional force is sum of frictional force of both the shoes, that is, $F_f = F_{fL} + F_{fR}$

$$F_{fL} = \mu' N_L = 0.333 N_L \text{ and } F_{fR} = \mu' N_R = 0.333 N_R$$

Hence $F_f = 0.333(N_L + N_R)$ or $2,000 = 0.333 (N_L + N_R)$ or $N_L + N_R = 6,000$ N

Taking moments about pivot O_1 for left hand yoke:

$$F \times 0.4 + (\mu N_L \times 0.05) - (N_L \times 0.2) = 0 \quad \text{Substituting the values:}$$

$$0.4F = N_L[0.2 - (0.33 \times 0.05)] \quad \text{or} \quad N_L = 2.179 F$$

Taking moments about pivot O_2 for right hand yoke:

$$F \times 0.4 - \left(\mu N_R \times 0.05\right) - \left(N_R \times 0.2\right) = 0 \quad \text{Substituting the values:}$$

$$0.4F = N_R\left[0.2 + (0.33 \times 0.05)\right] \quad \text{or} \quad N_R = 1.847\,F$$

$$N_L + N_R = 2.179\,F + 1.847\,F = 4.026\,F$$

$$6{,}000 = 4.026\,F \quad \text{Hence } F = 1{,}490\ \text{N}$$

b. Area of the shoes $A = 2\,w\,r\sin\theta = 2\,w \times 0.15 \times \sin 45° = 0.212\,w\ \text{m}^2$

Peripheral velocity of drum $v = \dfrac{\pi D N}{60} = \dfrac{3.14 \times 0.3 \times 500}{60} = 7.85\ \text{m/s}$

Pressure is more for the left shoe hence:

$$p = \frac{N_L}{A} = \frac{2.179\,F}{A} = \frac{2.179 \times 1{,}490}{0.212\,w} = \frac{15{,}315}{w}\ \text{N/m}^2$$

$$p\,v = 1.7 \times 10^6 \quad \text{Substituting the values:}$$

$$\frac{15{,}315}{w} \times 7.85 = 1.7 \times 10^6 \quad \text{or} \quad w = 0.071\ \text{m} = 71\ \text{mm}$$

Example 19.5

Braking torque with screw and lever on a double shoe brake

A drum of diameter 200 mm is braked with two short shoes as shown in Figure 19.S5. Shoes are moved using a threaded shaft 130 mm above the center line of drum of diameter 25 mm having left and right hand threads on each side. Lead of the screw is 5 mm and coefficient of friction is 0.3 for shoes and 0.15 for the screw. Pivots are 200 mm apart and 130 mm below the center line. Find braking torque, if force F is 10 N on a lever of 400 mm length.

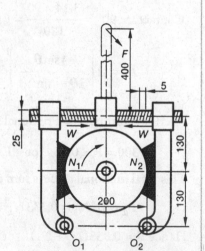

Figure 19.S5 Double shoe brakes

Solution

Given
$2r = 200\ \text{mm}$ $\quad \mu = 0.3$ $\quad \tan\Phi = 0.15$

$L = 400\ \text{mm}$ $\quad F = 10\ \text{N}$ $\quad l = 5\ \text{mm}$

$d = 25\ \text{mm}$ $\quad b = 130\ \text{mm}$ $\quad a = 260\ \text{mm}$

$c = 0$

For screw

Helix angle of the screw $\tan\alpha = \dfrac{\text{Lead}}{\pi d} = \dfrac{5}{\pi \times 25} = 0.0637$

Using Equation (18.1) of Chapter 18 of Volume 1, frictional torque on screw

$$T = F \times L = W \times 0.5d \frac{\tan\alpha + \tan\Phi}{1 - \tan\alpha \tan\Phi}$$

Substituting the values, $10 \times 400 = W \times 12.5 \dfrac{0.0637 + 0.15}{1 - (0.0637 \times 0.15)}$ or $W = 1,482$ N

Force on each side $= \dfrac{1,482}{2} = 7,41$ N

Taking moments about O_1: $W \times 260 = 130 \times N_1$ or $N_1 = 741 \times 2 = 1,482$ N

Taking moments about O_2: $W \times 260 = 130 \times N_2$ or $N_2 = 741 \times 2 = 1,482$ N

Total frictional torque $T_f = \mu(N_1 + N_2)r = 0.3 \times (1,482 + 1,482) \times 0.1 = 88.9$ Nm

19.8 Design Procedure for Block Shoe Brakes

1. Select brake drum diameter D, that is, $2r$ from the given data of load, speed, power, etc. Following Table 19.3 can help in deciding brake drum diameter.

    Table 19.3   Brake Drum Diameters and Power

Drum Diameter	Power (kW)		
D(mm)	600 rpm	760 rpm	960 rpm
160	7	7.5	8
200	10	11.5	12
250	15	16	18
320	24	27	30
400	39	45	50
500	65	75	84

2. For frictional torque T_p calculate frictional force $F_f = \dfrac{T_f}{r}$

3. Select material from Table 19.1 and find coefficient of friction for the combination of materials.

4. Fix lever dimensions based on the space available. By taking moments about pivot, find normal reaction N.

5. Get allowable pressure p for the selected material from Table 19.2, and calculate shoe area $A = N / p$.

6. Area of shoe = Width of shoe × projected length $= w \times 2r\sin\dfrac{\theta}{2}$, where, θ is contact arc angle. Width w can be assumed $0.25\,D$ to $0.33\,D$ (D is brake drum diameter)

7. Calculate heat generated H_g and heat dissipated H_d. Check that $H_d > H_g$.

19.9 Band Brakes

A band brake consists of a steel band with lining of leather, asbestos or wooden blocks fixed on a lever with pin joints as shown in picture on right side and also in Figure 19.6. Arrangement of fixing the band can be simple [Figure 19.6(a)], in which one of the bands is fixed at the fulcrum of the lever or differential in which both the ends are fixed away from the fulcrum at unequal distances [Figure 19.6(b)].

Band thickness is taken as 0.005 times the brake drum diameter D.

Empirically, thickness of band is taken as: $t_b = 0.005\,D$

Brake drum diameter	Band width w
Less than 1 m	100 mm
More than 1 m	100 mm to 150 mm

Width w and thickness t_b of band are given in Table 19.4.

Table 19.4 Width and Thickness of Lining (mm)

Width of Lining w	25–30	40–60	70–90	100–130	140–200
Thickness	3	3–4	4–6	5–6	7–10

19.9.1 Simple band brakes

In simple band brake, shown in Figure (19.6a), pin O_1 acts as fulcrum. When force F is applied on the end of lever, pin O_2 moves and the band is pulled applying brake over the drum. This causes tensions in the band in a similar way as in flat belts making one side tight side with tension T_1 and other slack side with tension T_2. Ratio of these tensions depends on the contact angle θ and is given as:

$$\frac{T_1}{T_2} = e^{\mu\theta} \tag{19.19}$$

Frictional torque $T_f = (T_1 - T_2)r$ (19.20)

Taking moments about the pivot O_1:

$$F \times a = T_2 \times b \quad \text{or} \quad F = T_2 \frac{b}{a} \tag{19.21}$$

Because the tensions are not constant in the band, the pressure is also not uniform. It is maximum at the tight side and minimum at slack side. Average pressure:

$$p = \frac{T_1 + T_2}{Dw} \tag{19.22}$$

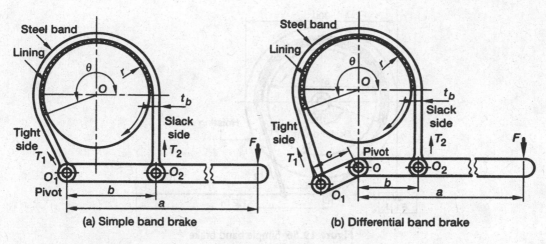

Figure 19.6 Band brakes

Where, D = Brake drum diameter

w = Width of band, hence $D\,w$ = Projected area

Maximum pressure $p_{max} = \dfrac{T_1}{D\,w}$ (19.23a)

Minimum pressure $p_{min} = \dfrac{T_2}{D\,w}$ (19.23b)

Tensile stress in the band can be taken between 50 MPa and 80 MPa. For a permissible tensile stress σ_t in the band, maximum tension, which it can take is:

$T_1 = \sigma_t\,w\,t_b$ (19.24)

Where, t_b = Thickness of band

Example 19.6

Force required for a simple band brake of a hoisting drum

A brake drum of diameter 0.6 m has a steel band with 5 mm thick lining for contact arc of 200° rotates clockwise. Coefficient of friction is 0.3. The band is pressed over the drum surface using a lever 1.2 m long, whose one end is at the pivot and the other end is fixed at a distance of 150 mm from the pivot as shown in Figure 19.S6. The brake drum is on the shaft of a hosting drum of 0.8 m diameter, with a dead load of 6 kN. Calculate:

a. Force required to support this vertical load of 6 kN.

b. Width of steel band, if tensile stress is not to exceed 50 MPa and thickness is 5 mm.

c. Force required, if direction is reversed.

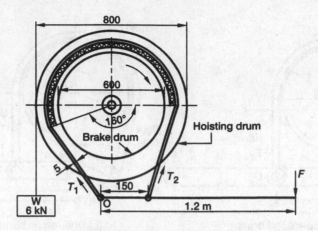

Figure 19.S6 Simple band brake

Solution

Given

$2r = 0.6$ m	$\mu = 0.3$	$a = 1.2$ m	$b = 0.15$ m
$t = 5$ mm	$\theta = 200°$	$W = 6$ kN	$D = 0.8$ m
$\sigma_t = 50$ MPa			

a. Contact angle $\theta = 360° - 160° = 200° = \dfrac{200 \times 3.14}{180} = 3.49$ rad

$$\frac{T_1}{T_2} = e^{\mu\theta} = e^{0.3 \times 3.49} = e^{1.046} = 2.85$$

Torque required for the brake drum: $T_f = W \times 0.5\,D = 6{,}000 \times 0.5 \times 0.8 = 2{,}400$ Nm

Net tension $\left(T_1 - T_2\right) = \dfrac{T_f}{r} = \dfrac{2{,}400}{0.3} = 8{,}000$

Substituting the value of T_1: $2.85\,T_2 - T_2 = 8{,}000$ or $T_2 = 4{,}324$ N

Hence Tension on tight side, $T_1 = 4{,}324 \times 2.85 = 12{,}324$ N

Taking moments about the pivot:

$\quad 1.2\,F = 4{,}324 \times 0.15$ or $F = 540.5$ N

b. Width of band $w = \dfrac{T_1}{t \times \sigma_t} = \dfrac{12{,}324}{5 \times 50} = 49.3$ mm say 50 mm

c. When direction of rotation is reversed, T_1 and T_2 get interchanged, hence:

$\quad 1.2\,F = 12{,}324 \times 0.15$ or $F = 1{,}540.5$ N

19.9.2 Differential band brakes

Refer Figure 19.6(b) for differential brake. When force F is applied at the end of lever, pin O_2 moves more than pin O_1 and hence the band is pulled applying brake over the drum. This causes tensions in the band in a similar way as in flat belts making one side tight side with tension T_1 and other slack side with tension T_2. Equations (19.19) and (19.20) can be used to calculate friction torque.

Taking moments about the pivot O:

$$F \times a = T_2 \times b - (T_1 \times c) \quad \text{Or,} \quad F = \frac{(T_2 \times b) - (T_1 \times c)}{a} \tag{19.25}$$

Rest of the procedure is same as for simple band brake.

Example 19.7

Force and size of belt for a differential band brake for the given torque

A differential band brake has contact angle of 225° to sustain a torque of 350 Nm over a cast iron drum of diameter 0.4 m as shown in Figure 19.S7. The band has lining having coefficient of friction 0.3. Calculate:

a. Force required at the end of lever 0.6 m from the pivot, if the band is fixed at a distance of 30 mm and 150 mm.

b. Width and thickness of brake band, if tensile stress is not to exceed 60 MPa.

c. Depth and thickness of brake lever with safe tensile stress 60 MPa.

d. Diameter of pin, if shear stress is limited to 40 MPa.

e. Size of rivets to fix the band with four rivets as shown in figure on right side.

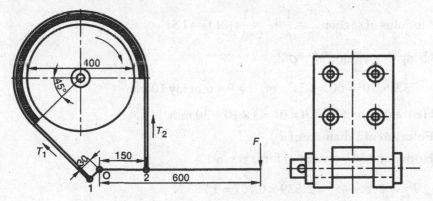

Figure 19.S7 Differential band brake

Solution

Given $D = 2r = 0.4 \text{ m}$ $T = 350 \text{ Nm}$ $\theta = 225°$ $\mu = 0.3$ $a = 0.6 \text{ m}$
$b = 0.15 \text{ m}$ $c = 0.03 \text{ m}$ $\sigma_t = 60 \text{ MPa}$ $\tau = 40 \text{ MPa}$ $p_b = 20 \text{ MPa}$

a. Braking force

Contact angle $\theta = 225° = \dfrac{225 \times 3.14}{180} = 3.925$ rad

$\dfrac{T_1}{T_2} = e^{\mu\theta} = e^{0.3 \times 3.925} = e^{1.1775} = 3.246$

Net tension $\left(T_1 - T_2\right) = \dfrac{T}{r} = \dfrac{350}{0.2} = 1{,}750$ N

Substituting the value of T_1: $3.246\,T_2 - T_2 = 1{,}750$ or $T_2 = 779$ N

Hence $T_1 = 779 \times 3.246 = 2{,}529$ N

Taking moments about the pivot: $0.6\,F + (2{,}529 \times 0.03) - (779 \times 0.15) = 0$

Or, $0.6\,F = 116.85 - 75.9$ or $F = 68.3$ N

b. Size of steel band

Thickness of band $t_b = 0.005\,D = 0.005 \times 0.4 = 0.002$ m $= 2$ mm

Width of band $w = \dfrac{T_1}{t_b\,\sigma_t} = \dfrac{2{,}529}{2 \times 60} = 21$ mm say 25 mm

c. Size of lever

Maximum bending moment M will be due to tight side pull, and hence lever is designed for that force.

$M = T_1 \times c = 2{,}529 \times 0.03 = 75.9$ Nm

Assume depth of lever h three times the thickness t, that is, $h = 3t$

Modulus of section $Z = \dfrac{1}{6}th^2 = \dfrac{1}{6}t(3t)^2 = 1.5t^3$

Using the relation $M = \sigma_t Z$

$75.9 \times 10^3 = 60 \times 1.5t^3$ or $t = 9.4$ mm say 10 mm

Hence depth of lever $h = 3t = 3 \times 10 = 30$ mm

d. Fulcrum pin diameter (d_p)

Horizontal component of force at pin 1:

$F_{1h} = T_1 \cos 45° = 2{,}529 \times 0.707 = 1{,}788$ N

Vertical component of force at pin 1:

$F_{1v} = T_1 \sin 45° = 2{,}529 \times 0.707 = 1{,}788$ N

Horizontal component of force at pin 2:

$$F_{2h} = T_2 \cos 90° = 779 \times 0 = 0 \text{ N}$$

Vertical component of force at pin 2:

$$F_{2v} = T_2 \sin 90° = 779 \times 1.0 = 779 \text{ N}$$

Resultant force at the pivot

$$R = \sqrt{(1,788 + 0)^2 + (1,788 + 779)^2} = 10^3 \sqrt{3.197 + 6.589} = 3,128 \text{ N}$$

The pin will be in double shear, hence $2\dfrac{\pi}{4}d_p^2 \times \tau = R$

Or, $1.57 \times 40 \times d_p^2 = 3,128$ or $d_p = 7 \text{ mm}$

Total width w as calculated above is 25 mm, hence length of pin taking the load is taken as half of 25, that is 12.5 say 13 mm as shown in figure below.

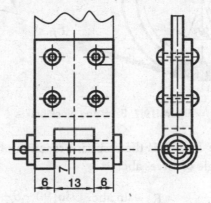

e. **Left and right rivet diameters**

Since the pull is more on the tight side, the rivet diameter to fix the band will be calculated on this basis. See right side of Figure 19.7 or figure above.

Assuming number of rivets $n = 4$:

Load on each rivet is $\dfrac{2,529}{4} = 632 \text{ N}$

Rivets are in double shear, hence $632 = 2 \times \dfrac{\pi}{4}d^2 \times \tau = 1.57d^2 \times 40$

$d = 3.2 \text{ mm}$ say 4 mm

19.9.3 Band and block brakes

In this type of brakes, wooden blocks are fixed to a steel band as friction material. Each block subtends and angle of 2θ at the center of the drum brake. Figure 19.7 shows such a brake.

Depending on the direction of rotation, one side of the band acts as a tight side and the other as slack side.

Let there be n blocks. Following nomenclature is being used for the tensions.

F_0 = Tension on slack side of the first block.

F_1 = Tension on tight side of the first block.

F_n = Tension on tight side of the nth block.

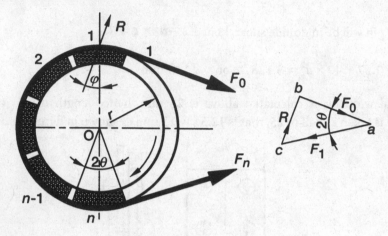

Figure 19.7 Band and block brakes

Refer right side of figure 19.7. R is the resultant reaction of F_0 and F_1 at angle Φ with vertical line. From the triangle of forces abc:

$$\frac{F_1}{\sin abc} = \frac{F_0}{\sin acb} \quad \text{or} \quad \frac{F_1}{F_0} = \frac{\sin abc}{\sin acb} = \frac{\sin\left(90 - \theta + \Phi\right)}{\sin\left(90 - \theta - \Phi\right)}$$

Or,
$$\frac{F_1}{F_0} = \frac{\cos\left(\theta - \Phi\right)}{\cos\left(\theta + \Phi\right)} = \frac{\cos\theta\cos\Phi + \sin\theta\sin\Phi}{\cos\theta\cos\Phi - \sin\theta\sin\Phi}$$

Dividing by $\cos\theta\,\cos\Phi$ in numerator and denominator:

$$\frac{F_1}{F_0} = \frac{1 + \tan\theta\tan\Phi}{1 - \tan\theta\tan\Phi} = \frac{1 + \mu\tan\theta}{1 - \mu\tan\theta} \tag{a}$$

Where, $\mu = \tan\Phi$

Tight side tension of the first block becomes the slack side tension of the second block.

Hence for the second block: $\dfrac{F_2}{F_1} = \dfrac{1 + \mu\tan\theta}{1 - \mu\tan\theta}$

Thus for the nth block: $\dfrac{F_n}{F_{n-1}} = \dfrac{1 + \mu \tan \theta}{1 - \mu \tan \theta}$

For all the blocks: $\dfrac{F_n}{F_0} = \dfrac{F_n}{F_{n-1}} \times \dfrac{F_{n-1}}{F_{n-2}} \times \ldots \times \dfrac{F_2}{F_1} \times \dfrac{F_1}{F_0}$

Or,

$$\dfrac{F_n}{F_0} = \left[\dfrac{1 + \mu \tan \theta}{1 - \mu \tan \theta} \right]^n \tag{19.26}$$

Frictional torque, $T_f = \left(F_n - F_o \right) \times r$ (19.27)

Example 19.8

Force required for a band and block brake for given power

A band and block brake is applied on a drum of diameter 800 mm transmitting 150 kW at 300 rpm. A steel band with 10 blocks of thickness 60 mm each subtends an angle of 20° at the center of the drum. A differential band brake as shown in Figure 19.S8 is used with force F acting at distance 600 mm from the pivot. Band is pinned at distance of 40 mm and 200 mm for tight and slack side, respectively. If coefficient of friction is 0.35, calculate the force F required to apply brake, so that drum does not rotate.

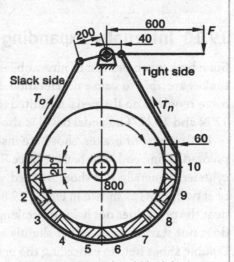

Figure 19.S8 Band and block brake

Solution

Given		
$D = 2r = 0.8$ m	$P = 150$ kW	
$N = 300$ rpm	$n = 10$	
$t = 60$ mm	$\theta = 20°$	
$\mu = 0.35$	$a = 0.6$ m	
$b = 0.04$ m	$c = 0.2$ m	

$$P = \dfrac{2 \pi N T}{60} \quad \text{or} \quad T = \dfrac{60 P}{2 \pi N} = \dfrac{60 \times 150 \times 1,000}{2 \times 3.14 \times 300} = 4{,}777 \text{ Nm}$$

$$T = \text{Net force} \times \text{radius} \quad \text{or} \quad 4{,}777 = \left(F_n - F_o \right) \times 0.4$$

Or, $\left(F_n - F_o \right) = \dfrac{4{,}777}{0.4} = 11{,}942$ N (a)

From Equation (19.26) ratio of forces for last and first block:

$$\frac{F_n}{F_o} = \left[\frac{1 + \mu\tan\theta}{1 - \mu\tan\theta}\right]^n = \left[\frac{1 + (0.35\times\tan 20°)}{1 - (0.35\times\tan 20°)}\right]^{10} = 3.46$$

Substituting the values in equation (a):

$$3.46\,F_o - F_o = 11{,}942\,\text{N} \quad\text{or}\quad F_o = 4{,}855\,\text{N}$$

Hence $F_n = 3.46\,F_o = 3.46 \times 4{,}855 = 16{,}798\,\text{N}$

Taking moments about the pivot:

$$0.6\,F + (16{,}798 \times 0.04) - (4{,}855 \times 0.2) = 0$$

Or, $\quad 0.6\,F = 971 - 671.9 \quad\text{or}\quad F = 498\,\text{N}$

19.10 Internally Expanding Shoe

Such brakes can be seen in automobiles. The brakes are said to be 100 per cent efficient, if the brakes are able to cause deceleration of the vehicle equal to g (i.e. 9.81 m/s²). The braking force required on the pedal is limited to maximum between 20 N and 30 N and with booster 12 N and 20 N. The pedal travel is also limited to 150 mm.

In this type of brakes, shoes are inside a hollow drum. It is of the shape of semi-circular, pivoted at one end. The braking force F is applied at the other free end using a cam or a hydraulic cylinder. Generally, two shoes are used, which can be mounted on a single pivot [Figure 19.8(a)] or at two pivots as shown in Figure 19.8(b) and picture on right side. The portion of the lining near the pivot does not help any braking as shown by dashed line, and hence the friction lining does not start from pivot but slightly away from it or pivot is shifted away from central axis. Double shoes help in cancelling the braking force and hence no force on the drum shaft.

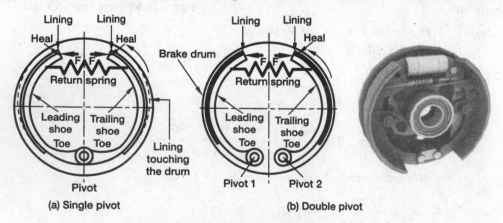

Figure 19.8 Internally expanding shoe brakes

It can be seen from Figure 19.8(a) that when the shoes expand (shown by dashed line) the lining near the pivot does not touch the drum as pivot is fixed and shoes move about the pivot. The shoes are returned, by a return tension spring between the shoes.

Analysis of these brakes is given in the next section and concluded that both the shoes do not give same braking torque. The leading shoe (in which the frictional force helps in applying brakes, i.e., self-energizing) offers greater force than the other shoe called trailing shoe.

Advantages of internal expanding shoe brakes are:

- Since the shoes are inside a drum, which provides protection from foreign particles.
- Simple construction.
- Less number of parts.
- Requires very less maintenance.
- Leading shoe offers self-energization, and hence less braking force.

Disadvantages of these brakes are:

- Heat dissipation is less as the shoes are inside the drum.

When the angle subtended by the lining at the center of the brake drum (2θ) is less than 45°, it is considered as a small shoe brake. For such a shoe, uniform pressure can be assumed over the entire surface of the shoe.

When the angle subtended is more than 45°, condition of uniform pressure no more exists. The force at the ends is same in the direction of force, but the force normal to the surface is reduced.

19.10.1 Analysis of internal shoe brakes

An internal double shoe brake is shown in Figure 19.9 with O as center and r as the radius of the brake drum. Centers of the pivot of the shoes are O_1 and O_2. Following nomenclature is being used:

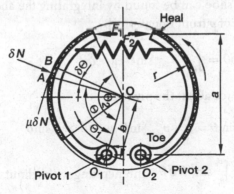

Figure 19.9 Forces on the shoe

Where, p_{max} = Maximum pressure
p = Normal pressure at any angle

θ = A general angle made by a small element AB from the pivot

$\delta\theta$ = Angle subtended by the small element of lining AB

θ_1 = Angle between the start of lining and pivot (See Figure 19.9).

θ_2 = Angle between the end of lining and pivot

b = Distance between drum center O and pivot center O_1.

a = Distance between the pivot center O_1 and the point of application of force

w = Width of the shoe

δN = Normal force on the element AB

$\mu\,\delta N$ = Tangential force on the element AB

It can be seen that the pressure p is proportional to the distance from the pivot center for any angle θ, that is, $p \propto b \sin\theta$. It will be maximum when the angle θ is 90° (As sin 90° = 1). When the sign of proportionality is converted to equal sign to a constant K appears.

Hence $p = K b \sin\theta$ $\hspace{3cm}$ (a)

Where, K = Brake constant (N/m³)

Moment due to normal force

Normal pressure acting on the element of area $A = w\,r\,\delta\theta$ is:

$$p = p_{max}\left(\frac{\sin\theta}{\sin 90}\right) = p_{max}\sin\theta \hspace{2cm} \text{(b)}$$

Normal force δN acting on the element of area $\delta A = w\,r\,\delta\theta$ is:

$$\delta N = p\,(w\,r\,\delta\theta)$$

Total force on the whole shoe can be found by integrating the above equation between the limits θ_1 and θ_2. Use value of p from Equation (b).

$$N = \int_{\theta_1}^{\theta_2} p w r \delta\theta = \int_{\theta_1}^{\theta_2}\left(p_{max}\sin\theta\right) w\,r\,\delta\theta$$

Moment due to normal force about the pivot O_1

$$M_N = N \times b \sin\theta = \int_{\theta_1}^{\theta_2}\left(p_{max}\sin\theta\right)w r \delta\theta \times b \sin\theta$$

Or, $\hspace{1cm} M_N = p_{max} w r b \int_{\theta_1}^{\theta_2}\sin^2\theta\,\delta\theta$ $\hspace{0.5cm}$ Intenerating within limits:

$$M_N = \frac{p_{max}\,wrb}{4}\left[2\left(\theta_2 - \theta_1\right) - \left(\sin 2\,\theta_2 - \sin 2\,\theta_1\right)\right] \hspace{1.5cm} (19.28)$$

Moment and torque due to tangential force

Tangential frictional force on this element $\delta F_f = \mu \, \delta N = \mu \, p_{max} \, w \, r \, \delta\theta \sin\theta$

Braking frictional torque due to this element force $\delta T_f = \delta F_f \times r = \mu \, p_{max} \, w \, r^2 \sin\theta \, \delta\theta$

Braking frictional torque on whole shoe about center O:

$$T_f = \mu \, p_{max} \, w r^2 \int_{\theta_1}^{\theta_2} \sin\theta \, \delta\theta \qquad \text{On integrating within the limits:}$$

$$T_f = \mu \, p_{max} = wr^2 \left(\cos\theta_1 - \cos\theta_2\right) \tag{19.29}$$

Taking moments of the frictional force on the element about the pivot O_1:

$$\delta M_f = \delta F_f \times (r - b\cos\theta) = (\mu \, p_{max} \, w \, r \, \delta\theta \sin\theta) \times (r - b\,\cos\theta)$$

On simplifying $\delta M_f = \mu \, p_{max} \, w r \, \delta\theta \, (r\sin\theta - 0.5b \sin 2\theta)$

Moment for the whole shoe is calculated by integrating the above equation:

$$M_f = \mu \, p_{max} \, wr \int_{\theta_1}^{\theta_2} \left(r\sin\theta - 0.5\,b\sin 2\,\theta\right) \delta\theta$$

On simplifying, $M_f = \mu \, p_{max} \, wr\left[r\left(\cos\theta_1 - \cos\theta_2\right) + 0.25\,b\left(\cos 2\theta_2 - \cos 2\theta_1\right)\right]$ (19.30)

For the anti-clockwise direction of rotation of the drum, the left shoe is leading shoe.

Taking moments about the pivot O_1:

$$F_1 \times a = M_N - M_f \quad \text{or} \quad F_1 = \frac{M_N - M_f}{a} \tag{19.31a}$$

Similarly for the trailing shoe, take moments about pivot O_2:

$$F_2 = \frac{M_N + M_f}{a} \tag{19.31b}$$

19.10.2 Shoe actuation

Shoes of the internally expanding brakes are expanded using either a cam or a hydraulic cylinder. When a cam is rotated, the shoes move outwards to apply brake. **Displacement** of both the shoes is equal [Figure 19.10(a)]. In hydraulic type as shown in Figure 19.10(b), oil pressure from the master cylinder supplies high pressure oil to all wheel cylinders. Piston moves outwards and **force** on each shoe is equal in this case.

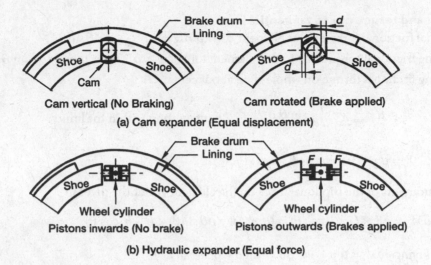

Figure 19.10 Shoe expanders

19.10.3 Shoe and brake factor

In Section 19.10.1, it is described that in internally expanding shoe drum brakes, a shoe may be leading or trailing. It is possible to take the advantage of self-energization for the leading shoe to increase tangential force. This reduces the force required at the shoe tip. Multiplication of force due to self-energization is called shoe factor. It is defined as:

$$\text{Shoe factor } S = \frac{\text{Tangential force on drum}}{\text{Force at shoe tip}}$$

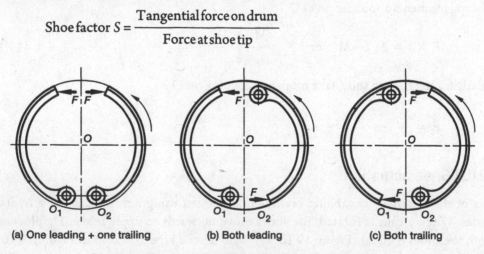

Figure 19.11 Leading and trailing shoe arrangements

In drum brakes, pivot position in the drum can make it leading or trailing. See Figure 19.11(a) in which the arrangement gives one leading and one trailing shoe. If pivot position is changed as shown in Figure 19.11(b), both the shoes become leading. Figure 19.11(c) shows with both shoes as trailing.

Brake factor depends on shoe arrangement: If S_L and S_T are the shoe factors for leading and trailing shoes, respectively, the brake factor B is given as under:

$B = S_L + S_T$ (F for one leading and one trailing shoe)

$B = 2S_L$ (F for both leading shoes)

$B = 2S_T$ (for both trailing shoes)

19.10.4 Maximum normal force for retarding wheel

When brakes are applied to the wheel of an automobile, the braking torque should be such that the wheels do not skid. The maximum torque T_w, which a wheel can take without slip, depends on load on wheel, road condition, and tyre surface giving adhesion.

$$T_w = \mu_a W r_w \qquad (19.32)$$

Where, μ_a = Adhesion friction coefficient between road and tyre

W = Vertical load on the wheel

r_w = Radius of wheel

The torque by the braking system: $T_B = \mu N r_d$ $\qquad (19.33)$

Where, μ = Friction coefficient between lining and drum

N = Normal radial force on lining

r_d = Radius of drum

For a condition of no slip, $T_w = T_B$ or $\mu_a W r_w = \mu N r_d$

Hence maximum normal force is: $N = \dfrac{\mu_a W r_w}{\mu \, r_d}$ $\qquad (19.34)$

Example 19.9

Braking torque by internally expanding leading and trailing shoes

In an internally expanding symmetrical shoe brakes, the pivots are arranged such that one shoe is leading and the other is trailing. Actuating force is acting at 140 mm from the center of the pivot of the shoes. Pivot is 75 mm away from drum center on the other side. Drum radius is 90 mm and coefficient of friction 0.36. Lining width is 40 mm and makes an angle of 5° and 130° from line of centers joining pivot and drum as shown in Figure 19.S9. Maximum allowable pressure on lining is 1.3 MPa. Calculate total braking torque with:

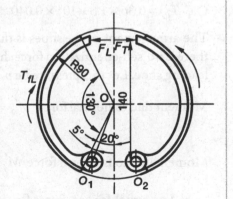

Figure 19.S9 Internally expanding shoe brakes

a. Equal force on both the shoes.

b. Equal displacement of the tip of shoes for the same total force.

Solution

Given $a = 140$ mm $b = 75$ mm $r = 90$ mm $\mu = 0.36$
           $w = 40$ mm $\theta_1 = 5°$ $\theta_2 = 130°$ $p_{max} = 1.3$ MPa

From Equation (19.28):

$$M_N = \frac{p_{max}\,w\,r\,b}{4}\left[2\left(\theta_2 - \theta_1\right) - \left(\sin 2\theta_2 - \sin 2\theta_1\right)\right]$$

Or,

$$M_N = \frac{1.3 \times 10^6 \times 0.040 \times 0.09 \times 0.075}{4}\left[2\left(130° - 5°\right)\frac{\pi}{180} - \left(\sin 260° - \sin 10°\right)\right] = 484\,\text{Nm}$$

From Equation (19.30):

$$M_f = \mu\, p_{max}\, w r\left[r\left(\cos\theta_1 - \cos\theta_2\right) + 0.25b\left(\cos 2\,\theta_2 - \cos 2\,\theta_1\right)\right]$$

$$M_f = \left(0.36 \times 1.3 \times 10^6 \times 0.040 \times 0.09\right) \times \left[\left\{0.09 \times \left(\cos 130° - \cos 5°\right)\frac{\pi}{180}\right\}\right.$$

$$\left. + \left\{0.25 \times 0.075\left(\cos 260° - \cos 10°\right)\right\}\right] = 212\,\text{Nm}$$

From Equation (19.31a): Force for leading shoe $F_L = \dfrac{M_N - M_f}{a} = \dfrac{484 - 212}{0.14} = 1{,}943$ N

From Equation (19.29): Braking torque by leading shoe $T_{fL} = \mu\, p_{max}\, w r^2\left(\cos\theta_1 - \cos\theta_2\right)$

Or, $T_{fL} = 0.36 \times 1.3 \times 10^6 \times 0.040 \times (0.09)^2 \times \left(\cos 10° - \cos 130°\right) = 248.5$ Nm

The arrangement of the shoes is that one is leading and the other is trailing on which there is no self-energization force; hence pressure on trailing shoe will not be same as on leading shoe. Let this pressure be p'. Prime is being used for trailing shoe.

Moment due to normal force $M_N' = \dfrac{M_N \times p'}{p_{max}} = \dfrac{484\,p'}{1.3 \times 10^6} = \dfrac{372.3\,p'}{10^6}$

Moment due to frictional force $M_f' = \dfrac{M_f \times p'}{p_{max}} = \dfrac{212\,p'}{1.3 \times 10^6} = \dfrac{163.1\,p'}{10^6}$

a. **For equal forces,** Force for trailing shoe = Force for leading shoe = 1,943 N

From Equation (19.31b): Force for trailing shoe $F_T = \dfrac{M_N + M_f}{a} = 1{,}943$ N

$$\frac{(372.3 + 163.1)p'}{0.14 \times 10^6} = 1943 \quad \text{or} \quad p' = 0.95 \text{ MPa}$$

Hence braking torque by trailing shoe: $T_{fT} = \dfrac{0.95 \times 248.5}{1.3} = 181.6$ Nm

Total braking torque $T_B = T_{fL} + T_{fT} = 248.5 + 181.6 = 430.1$ Nm

b. For equal displacement, forces F_L and F_T have to be in an inverse ratio of the torques obtained with equal forces. Total force = 1,943 + 1,943 = 3,886 N

$$\frac{F_T}{F_L} = \frac{248.5}{181.6} \quad \text{or} \quad F_T = 1.368 F_L$$

Total force $= F_T + F_L = 1.368 F_L + F_L = 3,886$

Or, $2.386 F_L = 3,886$ or $F_L = 1,641$ N

Hence $F_T = 1.368 F_L = 1.386 \times 1,641 = 2,245$

$$T_{fL} = 248.5 \times \frac{1,641}{1,943} = 210 \text{ Nm}$$

$$T_{fT} = 181.6 \times \frac{2,245}{1,943} = 210 \text{ Nm} \quad \text{Both shoes give same braking torque}$$

Example 19.10

Heat energy and shoe angle for a given stopping distance of a car

A car fitted with drum brakes has internally expanding shoes on all the four wheels. Total weight of the car with four passengers is 15 kN. 60 per cent of the weight is on the front wheels and rest on rear wheels. The car moving with a speed of 60 km/h on a downward gradient of 5° is to be stopped completely in a distance of 20 m. The data for the car are:

Wheel diameter = 0.7 m

Brake drum diameter = 0.3 m

Brake lining width = 50 mm

Coefficient of friction between drum and lining = 0.35

Average allowable pressure = 0.75 MPa

Calculate:

a. Heat generated

b. Shoe angle required for the front wheels. Assume equal force on each shoe.

Solution

Given

$$W_T = 15\,\text{kN} \qquad W_F = 0.6\,W_T \qquad U = 60\,\text{km/h} \qquad V = 0\,\text{km/h}$$
$$S = 20\,\text{m} \qquad \alpha = 5° \qquad 2r_w = 0.7\,\text{m} \qquad 2r_d = 0.3\,\text{m}$$
$$w = 50\,\text{mm} \qquad \mu = 0.35 \qquad p_{max} = 0.75\,\text{MPa}$$

a. Car speed $U = 60\,\text{km/h} = \dfrac{60 \times 1,000}{3,600} = 16.667\,\text{m/s}$

Retardation $f = \dfrac{V^2 - U^2}{2S} = \dfrac{0^2 - 16.667^2}{2 \times 20} = -6.95\,\text{m/s}^2$ (Minus means retardation)

Force due to retardation $F = \dfrac{W_T}{g}f + W_T\sin\alpha = W_T\left(\dfrac{f}{g} + \sin\alpha\right)$

Or, $F = 15,000\left(\dfrac{6.95}{9.81} + \sin 5°\right) = 11,943\,\text{N}$

Work done $W = F \times S = 11,943 \times 20 = 238,860\,\text{Nm}$

Heat generated = Work done = 238.860 kNm = 238.860 kJ

b. Braking force on front wheels $F_f = 0.6\,F = 0.6 \times 11,943 = 7,166\,\text{N}$

Braking force on each front wheels $F_{f1} = \dfrac{7,166}{2} = 3,583\,\text{N}$

Braking torque on each wheel $T_{f1} = F_{f1} \times r_w = 3,583 \times 0.35 = 1,254\,\text{Nm}$

Braking torque on each drum = Braking torque on each wheel = 1,254 Nm (a)

Also braking torque on drum = $2\,\mu\,N\,r_d$ (b)

Equating Equations (a) and (b):

Normal force $N = \dfrac{1,254}{2 \times 0.35 \times 0.15} = 11,943\,\text{N}$

Area of one side of lining A = Width × Radius × Arc in radians = $w \times r_d \times \theta$

Or, $A = 50 \times 150 \times (\theta \times \pi/180) = 131\,\theta\ \text{mm}^2$

Average pressure on lining $P_{av} = \dfrac{N}{A}$ or $0.75 = \dfrac{11,943}{131\theta}$ or $\theta = 121°$

19.11 Externally Contracting Brakes

These brakes are so called because the shoes move inwards towards center to touch the brake drum from outside (Figure 19.12). Cast iron shoes apply force on wheel, which acts as a drum in railways as shown in picture below. Their analysis is also similar, as given above in previous sections.

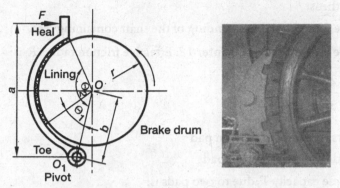

Figure 19.12 Externally contracting brakes

19.12 Disc Brakes

19.12.1 Arctual pads

In a disc brake, the drum is replaced by a circular disc. Two shoes of the shape of an arctual segment as shown in Figure 19.13 are used. One shoe is on one side of the disc and the other at the other side of the disc. When the shoes are pressed against the disc, brake is applied. These brakes can be seen in motor cycles and modern cars also as shown in picture on right side of Figure 19.13.

Drum brakes are being replaced by disc brakes in automobiles due to the following advantages:

- Pressure is uniform, and hence wear is also same for the whole surface.
- Since the rotating disc is open to atmosphere, cooling is better.
- Direction of rotation has no effect on braking torque.

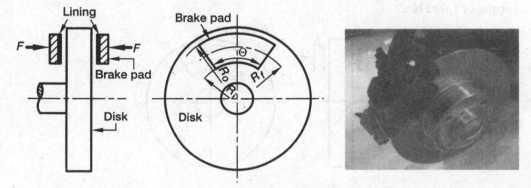

Figure 19.13 Disc brakes

- It will never be self-locking.
- Force of one shoe is cancelled by other shoe as they are in opposite directions. hence, no axial thrust
- Since the force is axial, no bending of the shaft containing disc.

Uniform pressure is discussed in Chapter 18. Effective friction radius R_f is:

$$R_f = \frac{2\left(R_o^3 - R_i^3\right)}{3\left(R_o^2 - R_i^2\right)}$$ (19.35)

Where, R_o = Outside radius of friction pad
R_i = Inside radius of friction pad

The braking torque capacity T_f due to two pads is:

$$T_f = 2\mu F R_f$$ (19.36)

Where, F = Brake actuation force at each pad
μ = Coefficient of friction

Area of an annular disc depends on outer and inner radii and is $\pi\left(R_o^2 - R_i^2\right)$.

For an arctual angle θ, area of pad A can be calculated using the relation:

$$A = \frac{\theta\pi\left(R_o^2 - R_i^2\right)}{2\pi} = \frac{\theta\left(R_o^2 - R_i^2\right)}{2}$$ (19.37)

Example 19.11

Size of arctual pad for given braking torque
A disc brake with two pads; one on each side has outer radius of pad 200 mm subtending contact angle of 60° is shown in Figure 19.S10. Assuming maximum average pressure 1.2 MPa and coefficient of friction 0.32, find the inner radius of the pad for a braking torque of 1,000 Nm.

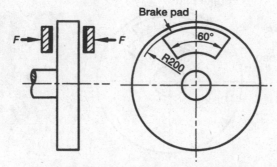

Figure 19.S10 Disc brakes with arctual pads

Solution

Given $R_o = 200$ mm $\theta = 60°$ $\mu = 0.32$ $p = 1.2$ MPa
                $T_f = 1,000$ Nm

Let the inner radius be x times the outer radius, that is, $R_i = x\,R_o$

Effective friction radius from Equation (19.35):

$$R_f = \frac{2\left(R_o^3 - R_i^3\right)}{3\left(R_o^2 - R_i^2\right)} = \frac{2R_o^3\left(1-x^3\right)}{3R_o^2\left(1-x^2\right)} = \frac{2 \times 200^3\left(1-x^3\right)}{3 \times 200^2\left(1-x^2\right)} = 133.3\left(1-x\right)$$

Angle of pad $\theta = 60° = \dfrac{60 \times 3.14}{180} = 1.047$ rad

From Equation (19.37), area of a pad:

$$A = \frac{\theta\left(R_o^2 - R_i^2\right)}{2}$$

$$A = \frac{1.047\left[200^2 - \left(x \times 200\right)^2\right]}{2} = \frac{1.047 \times 200^2\left[1-x^2\right]}{2} = 20,940\left(1-x^2\right)\,\text{mm}^2$$

Force on a pad $F = p \times A = 1.2 \times 20,940 \times (1 - x^2) = 25,128 \times (1 - x^2)$ N

Frictional torque for each pad $T_f = \dfrac{1,000}{2} = 500$ Nm

Also $T_f = \mu\,F\,R_f$ Substituting the values:

$$500 \times 10^3 = 0.32 \times 25,128 \times (1 - x^2) \times 133.3\,(1 - x)$$

Or, $(1 - x^2)(1 - x) = 0.466$

Solving by trial and error:

$x = 0.5$	LHS $= 0.375$
$x = 0.4$	LHS $= 0.505$
$x = 0.45$	LHS $= 0.447$
$x = 0.43$	LHS $= 0.464$; It is close to 0.466.

Hence $R_i = x\,R_o = 0.43 \times 200 = 86$ mm

19.12.2 Disc brakes with circular pads

Sometimes, an arctual pad is replaced by a simple circular pad of radius R. The effective radius R_f does not lie at the center of the circular pad placed at radius R_c as shown in Figure 19.14. Value of R_f is given in Table 19.5.

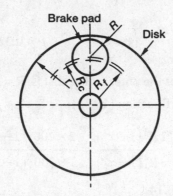

Figure 19.14 Disc brake with circular pad

Table 19.5 Effective Radius R_f for a Circular Pad of radius R

R / R_c	0.1	0.2	0.3	0.4	0.5	0.6
R_f	$0.983\ R_c$	$0.969\ R_c$	$0.957\ R_c$	$0.947\ R_c$	$0.937\ R_c$	$0.928\ R_c$

Normal force $F = p \times A$

The braking torque capacity due to two pads is given as:

$$T_f = 2 \mu F R_f \qquad (19.38)$$

Example 19.12

Braking torque with given size of circular pad
A disc brake with two circular pads of diameter 60 mm
at radius 100 mm as shown in Figure 19.S11. Assuming
maximum average pressure 1.5 MPa and coefficient of
friction 0.28, calculate braking torque.

Solution

Given $R_c = 100$ mm $2R = 60$ mm
               $\mu = 0.28$ $p = 1.5$ MPa

Area of pad $A = \pi R^2 = 3.14 \times 30^2 = 2{,}826$ mm^2

Normal force $F = p \times A = 1.5 \times 2{,}826 = 4{,}239$ N

$$\frac{R}{R_c} = \frac{60}{100} = 0.6$$

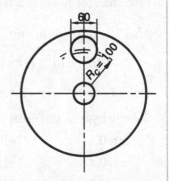

Figure 19.S11 Disc brakes
with circular pads

From Table 19.5, $R_f = 0.928\ R_c$

$R_f = 0.928\ R_c = 0.928 \times 100 = 92.8$ mm $= 0.928$ m

$R_f = 0.928\ R_c = 0.928 \times 100 = 92.8$ mm $= 0.928$ m

Frictional torque $T_f = 2 \mu F R_f = 2 \times 0.28 \times 4{,}239 \times 0.928 = 2{,}203$ Nm

Summary

Definition   A brake is a mechanical device, used to stop or slow down a moving part by absorbing its kinetic or potential energy by friction.

Functions   To stop or slow down a moving part or system, for example automobiles, trains, etc.

- To hold a part in position, for example, hand brake of a vehicle.
- To absorb potential energy released by moving parts like hoists.

Types of brakes   They can be classified as under:

a.   On the basis of its operation:

- *Mechanical brakes*   Operated by lever or pedal.
- *Hydraulic brakes*   Operated by oil pressure in a cylinder.
- *Pneumatic*   Operated by air pressure in a cylinder.
- *Magnetic*   Operated by magnetic forces, eddy currents, and magnetic particles.

b.   On the basis of type of shoes:

- *Fixed block* brakes or pivoted shoe as in railways.
- *Internally expanding* shoe brakes.
- *Externally contracting* shoe brakes and band brakes as in hoists.
- *Disc brakes:* Shoes are on the sides of a rotating disc. The force is applied axially.

Materials for brake lining: They should possess the following qualities;

- High coefficient of friction.
- Value of coefficient of friction should not change with temperature.
- Low wear rate.
- Withstand heat without losing properties.
- High mechanical strength.
- Not affected much by oil or moisture.

Materials with their properties for brake lining are tabulated in Table 19.1.

Energy absorbed by brakes I = Mass moment of inertia (kgm^2)

- *Pure rotation:* $E_R = \dfrac{1}{2}I\left(\omega_1^2 - \omega_2^2\right)$, ω_1 and ω_2 are angular velocities before and after braking.

- *Pure translation:* $E_T = \dfrac{1}{2}I\left(v_1^2 - v_2^2\right)$, v_1 and v_2 are linear velocities before and after braking.

- *Potential energy:* $E_P = mg(h_1 - h_2)$, m = Mass, h_1 and h_2 are initial height before and after braking, respectively.

- *Combined rotation and translation:* Energy absorbed due to combined effect is: $E = E_R + E_T + E_P$

Work done $W = F_t \pi D N t = E$, where F_t is force applied for brakes.

Energy absorbed in t seconds $E = F_f \pi D N t$ or   tangential force on the brake drum, $F_f = \dfrac{E}{\pi D N t}$

The frictional torque T_f is given as: $T_f = 0.5D\,F_f = r\,F_f$

Heat dissipated   Heat dissipated H_d depends on coefficient of friction, braking force, surface area of the drum, braking time, and heat dissipating capacity of the drum.

$H_d = C\,(t_1 - t_2)\,A_r$, where, t_1 and t_2 are temperatures of the rotating surface and of surroundings respectively; A_r = Area of rotating drum surface; and C = coefficient of heat transfer 30 W/m²/°C to 40 W/m²/°C. Only 15 per cent to 20 per cent of this heat is dissipated to the surrounding, while the rest increases the temperature of drum and lining.

Allowable temperatures   For the different lining materials, temperatures are: Wood = 60°C; fibre and leather = 70°C; asbestos without oil = 90°C; asbestos with oil = 105°C; and forodo = 180°C.

Heat generated per unit time is given as: $H_g = \mu\,Nv = \mu\,pAv$

Where, μ = Coefficient of friction N = Force on the brake shoe

           A = Projected area of contact surface p = Normal pressure and

           v = Peripheral velocity of drum

Temperature rise can be calculated from $\Delta t = \dfrac{H_g}{mC_p}$

   For Aluminium, C_p = 910 J/kg °C, for cast steel C_p = 490, and for cast iron C_p = 460.

Lining wear   The wear rate depends on normal pressure and rubbing velocity; hence to limit the wear rate/temperature, a maximum pv value given in Table 19.2. is recommended.

Block brakes   These are of following types: (a) Fixed block (b) Pivoted block

   a.   *Fixed block*   Brake block is fixed to the lever firmly. It consists of a block curved with same radius as the brake drum. When a force F is applied at the end of the lever, its normal reaction N is equal and opposite to the direction of the force, that is, $N = F\dfrac{a}{b}$

        Frictional force $F_f = \mu N$ and acts in the direction of rotation.

        Frictional braking torque $T_f = F_f \times r = \mu\,N\,r$

   b.   *Pivoted block*   If c is the distance of pivot from the drum surface. Taking moments about fulcrum:

$$F \times a = (N \times b) + (\mu N \times c) \quad \text{or} \quad N = \dfrac{F \times a}{b + \mu c}$$

   -    If fulcrum center is below the tangent line to the drum $= T_f = \mu N r = \dfrac{F\mu r a}{(b + \mu c)}$

   -    If fulcrum center is above the tangent line to the drum $T_f = \dfrac{F\mu r a}{b - \mu c}$

   -    If fulcrum center is on the tangent line to the drum $T_f = \dfrac{F\mu r a}{b}$

If c = 0, direction of rotation has no effect in braking torque.

Reaction at the pin   Horizontal force $F_H = \mu\,N$; Vertical force $F_V = F - N$

Resultant reaction at the pivot is: $R = \sqrt{F_H^2 + F_V^2}$

Width of block is generally taken 25 per cent to 50 per cent of the drum diameter.

Size of block depends on allowable pressure p on lining. $N = p\,l\,w$

N = Normal force, p = Allowable pressure, l = Projected length, and w = Width of block.

Self-energizing brakes   Frictional force helps in braking. $N \times b = (F \times a) + (\mu\,N \times c)$

Moment of frictional force $(\mu\,N \times c)$ helps in moment of actuation force.

Self-locking brakes   If $b = \mu\,c$, then the force required for braking is zero. Brake is applied even without force. If $b < \mu\,c$, F is negative; hence b should be more than $\mu\,c$.

Small / long shoe brake: When contact angle 2θ of the shoe on the drum at its center is less than 45° it is called short shoe. When angle 2θ is more than 45° it is called long shoe.

$$T_f = \mu' N r$$

The effect of reduced pressure at the ends for a long shoe can be considered by modifying the effective value of coefficient of friction as $\mu' = \mu\left[\dfrac{4\sin\theta}{2\theta + \sin 2\theta}\right]$

b. Pivoted block brakes   Block can tilt and adjust according to circumference of the drum.

The pivot is so located that the moment of frictional force about the pivot is zero. For this condition, the pivot has to be very close to the drum surface. Distance of pivot from the center of the brake drum:

$$h = \left[\dfrac{4r}{2\theta + 2\sin 2\theta}\right]$$

r = Radius of drum and θ = Semiangle of brake shoe.

Frictional force $\mu\,N$ causes a turning moment as $\mu\,N\,(h\cos\theta - r)$.

Advantages   Provides greater braking torque than fixed block brakes, offers uniform wear of the lining, and pressure is more or less uniform.

Disadvantages   Mounting is not in correct position of the pivot as the lining wears out.

Double block shoe brakes   The normal force at shoe causes bending of the shaft at drum cancels with the normal force of the other shoe. Normal reactions on both shoes are different as:

$$N_1 = \dfrac{aF}{b + c\mu} \quad\text{and}\quad N_2 = \dfrac{aF}{b - c\mu}$$

Design procedure for block shoe brakes

1. Select brake drum diameter from given load, speed, etc. (Table 19.3).
2. Calculate frictional force $F_f = T_f / r$ where, r is radius of brake drum
3. Select material from Table 19.1 and find coefficient of friction
4. Fix lever dimensions based on the space available. Find normal reaction N.
5. Assuming allowable pressure from Table 19.2. Calculate shoe area $A = N / p$.
6. Area of shoe = Width of shoe × Projected length = $w \times 2\,r\sin(0.5\theta)$, where, θ is contact arc angle.

Assume width $w = 0.25\ D$ to $0.33\ D$.

7. Calculate heat generated H_g and heat dissipated H_d. Check that H_d is more than H_g.

Band brakes These brakes have a steel band with lining of leather, asbestos, or wooden blocks fixed on a lever with pin joints. Arrangement of fixing the band can be simple or differential in which both the ends are fixed away from the fulcrum at unequal distances.

Band thickness t_b is taken as $0.005\ D$, width w, and thickness t_b of band is given in Table 19.4.

Width of lining $w = 100$ mm for $D < 1$ m and $w = 100$ mm to 150 mm for $D > 1$ m.

Simple band brake Ratio of these tensions is: $\dfrac{T_1}{T_2} = e^{\mu\theta}$; Frictional force $F = T_2\,\dfrac{b}{a}$; Frictional torque $T_f = (T_1 - T_2)\,r$

Because tensions are not constant in band, pressure is also not uniform. It is maximum at the tight side

$$p_{max} = \frac{T_1}{Dw}$$ and minimum at slack side $$p_{min} = \frac{T_2}{Dw}.$$ Average pressure: $$p = \frac{T_1 + T_2}{Dw}$$

Tensile stress in the band σ_t can be taken between 50 MPa and 80 MPa. For thickness of band t_b, maximum tension, which it can take for width w is: $T_1 = \sigma_t w\, t_b$

Differential band brake: If a, b and c are the distances of the force, slack side and tight side respectively from the fulcrum then: Frictional force $F = \dfrac{(T_2 \times b) - (T_1 \times c)}{a}$

Rest of the procedure is same as for simple band brake.

Band and block brakes Wooden blocks are fixed to a steel band as friction material. Each block subtends and angle of 2θ at the center of the drum brake.

Let there be n blocks. $F_0 =$ Tension on slack side of the first block, $F_1 =$ Tension on tight side of the first block, and $F_n =$ Tension on tight side of the nth block.

$$\frac{F_1}{F_0} = \frac{1 + \tan\theta \tan\varPhi}{1 - \tan\theta \tan\varPhi} = \frac{1 + \mu\tan\theta}{1 - \mu\tan\theta} \quad \text{where, } \mu = \tan\varPhi$$

Tight side tension of the first block becomes the slack side tension of the second block.

For all the blocks $\dfrac{F_n}{F_0} = \dfrac{F_n}{F_{n-1}} \times \dfrac{F_n - 1}{F_{n-2}} \times \ldots \times \dfrac{F_2}{F_1} \times \dfrac{F_1}{F_0} = \left[\dfrac{1 + \mu\tan\theta}{1 - \mu\tan\theta}\right]^n$

Frictional torque $T_f = (F_n - F_0) \times r$

Internally expanding shoe Automobiles use this type. Shoes of semi-circular shape are inside of a hollow drum. It is pivoted at one end. The braking force F is applied at the other free end using a cam or a hydraulic cylinder. The shoes are returned, by a return tension spring.

The leading shoe (in which the frictional force helps to apply brakes, i.e., self-energizing offers greater force than the other shoe called trailing shoe.

Advantages Shoes are inside a drum, which provides protection from foreign particles, simple construction, less number of parts, requires very less maintenance, leading shoe offers self-energization, and hence less braking force.

Disadvantages Heat dissipation is less as the shoes are inside the drum.

Analysis of internal shoe brakes

Pressure p on lining is proportional to the distance from the pivot center for any angle θ.

Hence $p = Kb \sin \theta$, Where, K = Brake constant (N/m³).

Let w be the width of lining, b = distance between drum and pivot center, θ_1 and θ_2 are the angles of lining from line of centers of pivot and drum, a = distance between pivot and point of application of force.

Moment due to normal force: $M_N = \dfrac{p_{max} \, w \, r \, b}{4} \left[2(\theta_2 - \theta_1) - (\sin 2\theta_2 - \sin 2\theta_1) \right]$

Braking torque due to friction: $T_f = \mu p_{max} \, w \, r^2 (\cos \theta_1 - \cos \theta_2)$

Moment due to friction: $M_f = \mu p_{max} \, w \, r \, [r (\cos \theta_1 - \cos \theta_2) + 0.25 \, b (\cos 2\theta_2 - \cos 2\theta_1)]$

Force on tip of leading shoe: $F_1 = \dfrac{M_N - M_f}{a}$ and force on trailing shoe: $F_2 = \dfrac{M_N + M_f}{a}$

Shoe actuation is done either by a cam or by a hydraulic cylinder. Cam gives equal displacement to both the shoes, while a hydraulic cylinder provides equal forces.

Shoe factor Multiplication of force due to self-energization is called shoe factor. Shoes can be both leading $(2L)$, both trailing $(2T)$. Generally, it is kept $(1L + 1T)$.

Brake factor depends on shoe arrangement: If S_L and S_T are the shoe factors for leading and trailing shoes, respectively, for one leading and one trailing shoes the brake factor $B = S_L + S_T$

$B = 2S_L$ for both leading shoes and $B = 2S_T$ for both trailing shoes.

Maximum torque T_w, which a retarding wheel of radius r_w can take without slip, depends on load W on wheel, and adhesions friction coefficient μ_a between road and tyre. $T_w = \mu_a W r_w$

The torque by the braking system $T_B = \mu N r_d$ for a condition of no slip

$T_w = T_B \quad$ or $\quad \mu_a W r_w = \mu N r_d$

Where, r_d = Radius of drum and N = Normal radial force on lining

Hence, $N = \dfrac{\mu_a W r_w}{\mu r_d}$

Externally contracting bakes These brakes are so called, because the shoes move inwards towards center to touch the brake drum. Cast iron shoes apply force on the drum.

Disc brakes In these brakes, drum is replaced by a circular disc. Shoes called pads can be of arctual shape or circular.

Arctual pads Two shoes of the shape of an arctual segment are used; one on one side of the disc and the other at other side of the disc. When the shoes are pressed against disc, brake is applied.

Advantages Pressure is uniform, cooling is better, direction of rotation has no effect on braking torque, never be self-lock, and no bending of the shaft. If R_o = Outside radius, R_i = Inside radius of pad, for uniform pressure effective friction radius $R_f = \dfrac{2\left(R_o^3 - R_i^3\right)}{3\left(R_o^2 - R_i^2\right)}$

For a force F, the braking torque capacity due to two pads is given as: $T_f = 2\,\mu\,F R_f$

Area of an annular disc depends on outer and inner radii and is $\pi\left(R_o^2 - R_i^2\right)$.

For an arctual angle θ, area of pad $A = \dfrac{\theta\pi\left(R_o^2 - R_i^2\right)}{2\pi} = \dfrac{\theta\left(R_o^2 - R_i^2\right)}{2}$

Disc brakes with circular pads Effective radius R_f does not lie at the center of circular pad, placed at radius R_c. Value of R_f is given in Table 19.5. Normal force $F = p \times A$

The braking torque capacity due to two pads is given as: $T_f = 2\,\mu\,F R_f$

Theory Questions

1. Write the functions of a braking system and types of brakes.
2. Compare a single shoe and double shoe brakes with the help of simple sketches.
3. What types of energies can be absorbed by brake? Write the governing equations for each.
4. Discuss the heat generated and heat dissipation in drum brakes.
5. Derive an expression for equivalent friction coefficient for a long shoe.
6. Describe a pivoted shoe brake with the help of a sketch. What are its advantages?
7. Draw a sketch of a band brake. What are the different types of band brake?
8. Differentiate between simple and differential band brakes with neat sketches.
9. Derive an expression for the ratio between tight and slack side tension of the band for a given number of blocks.
10. Describe the construction of internally expanding brakes with the help of a sketch.
11. Sketch the shoe arrangements for internally expanding shoes.
12. How are the shoes actuated in a double shoe drum brakes? Describe the various methods.
13. Discuss the braking torque given by leading and trailing shoes.
14. What is meant by shoe factor and brake factor?
15. What do you mean by self-energizing and self-locking brake?
16. Compare externally contracting and internally expanding drum brakes.
17. What is a disc brake? How does it differ from drum brake? Write its advantages and disadvantages over the drum brakes.
18. What is an effective radius of a brake pad for a disc brake? Discuss for arctual pad and circular pads.

Multiple Choice Questions

1. Brakes are used to:
   (a) Stop a rotating machine element
   (b) Stop a moving object
   (c) Slow down a moving or rotating object
   (d) All given in (a), (b), and (c)

2. Most of the cars use:
   (a) Hydraulic brakes
   (b) Pneumatic brakes
   (c) Magnetic brakes
   (d) None of the above

3. Type of brakes used in railways is:
   (a) Discbrakes
   (b) Fixed pad brakes
   (c) Externally contracting shoe brakes
   (d) Electric brakes

4. Energy absorbed by the brakes is proportional to the:
   (a) Linear velocity
   (b) Angular velocity
   (c) None of (a) and (b)
   (d) Square of linear or angular velocity

5. Condition assumed for the design of brakes is:
   (a) Uniform pressure
   (b) Uniform wear
   (c) Uniform heat generated
   (d) Either (a) or (b)

6. In internally expanding shoe brakes generally:
   (a) Both the shoes apply same braking effect
   (b) One shoe applies more braking then the other
   (c) Depends on speed and direction of rotation
   (d) None of the above

7. A self-locking brake is:
   (a) When the brake is applied without any effort
   (b) Drum gets locked in case of emergency
   (c) Both the shoes are self-energizing
   (d) None of the above

8. Disc brakes are becoming popular due to:
   (a) Their effectiveness
   (b) No shoe adjustment is needed
   (c) Better cooling then drum brakes
   (d) All given above

9. A caliper is used for:
   (a) Drum brakes
   (b) Disc brakes
   (c) Band brakes
   (d) None of the above

10. The force required on the brake pedal depends on:
    (a) Type of surface; dry or oily
    (b) Coefficient of friction
    (c) Material of lining
    (d) All given above

Answers to multiple choice questions

1. (d)   2. (a)   3. (c)   4. (d)   5. (d)   6. (b)   7. (a)   8. (d)   9. (b)   10. (d)

Unsolved Examples

1. A Shoe brake is used on a drum of 200 mm radius transmitting 200 Nm torque. Width is 25 per cent of the drum diameter. If coefficient of friction is 0.3, leverage is 2, and pressure should not exceed 1.2 MPa, calculate:

   (a) Actuation force.

   (b) Contact angle of the shoe. $[F = 3{,}333 \text{ N}, \ 2\theta = 64°]$

2. A single shoe block brake with 90° arc of contact on a drum of 300 mm diameter is braked with actuation force at a distance of 250 mm from the center line of the drum. The bell crank lever is pivoted 250 mm on the other side of the drum and the fulcrum is 100 mm below the surface of the drum as shown in Figure 19.P1. If coefficient of friction is 0.3, calculate the torque, which it can handle with a force $F = 800$ N. $[91.3 \text{ Nm}]$

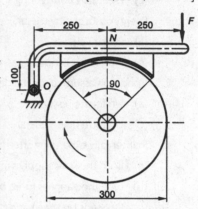

Figure 19.P1 Single shoe block brake

3. A double shoe brake with two blocks making an arc of 90° on a brake drum of 380 mm diameter. The two yokes are pivoted at 100 mm apart and 250 mm below the center line of the drum as shown in Figure 19.P2. Brakes are applied by a lever 600 mm long, rotating a sprocket of 60 mm diameter, which pull the yokes with two chains in opposite directions Coefficient of friction is 0.35, calculate force F for a torque of 200 Nm. $[62.3 \text{ N}]$

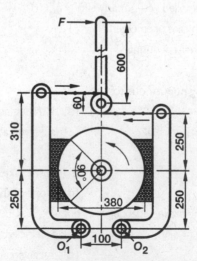

Figure 19.P2 Double shoe block brake with actuating chain and sprocket

4. A double shoe block brake is shown in Figure 19.P3. A force of 300 N is applied using a tension spring attached at the ends of yokes at a distance of 300 mm from the center line of the drum. The shoes make an arc of 110° at the center of drum. The drum of 400 mm rotates clockwise with coefficient of friction 0.34. Only one fulcrum for both the yokes is 250 mm below the center line of the drum. Find the torque, which it can absorb with the spring force without slipping.

[110.8 Nm]

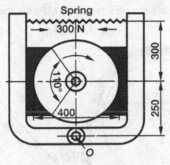

Figure 19.P3 Double shoe block brake with spring force

5. A single shoe brake acts on a drum of radius 200 mm to sustain a torque of 250 Nm. An actuation force F acts at the end of a lever at 300 mm from center line as shown in Figure 19.P4. The fulcrum is 100 mm away from center line of the drum and 40 mm above the line, tangent to the drum. Calculate:

(a) Actuation force for clockwise rotation of the drum.

(b) Actuation force for counter clockwise rotation of the drum.

(c) Distance x of the fulcrum instead of 40 mm, so that the brake is self-locking.

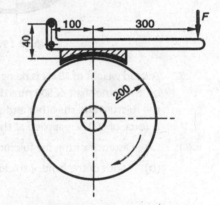

Figure 19.P4 A single shoe brake

[(a) F = 768 N; (b) F = 1,018 N; (c) x = 285 mm]

6. A band brake has to sustain a torque of 400 Nm with lining having coefficient of friction 0.3 covering the drum for 225° of diameter 400 mm rotating in counter clockwise rotation as shown in Figure 19.P5. One end of the band is at a distance of 40 mm from the fulcrum and other at 200 mm. Calculate:

(a) Tensions on tight and slack side.

(b) Actuation force F at a distance of 500 mm from the fulcrum.

(c) How the distance 200 be changed to make the brake self-locking.

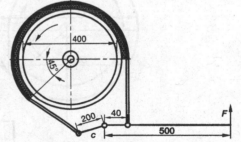

Figure 19.P5 A crane drum with band brake

[(a) T_1 = 2,890 N, T_2 = 890 N; (b) F = 125 N; (c) c = 130 mm]

7. A band brake has contact angle of 270° over a drum of 500 mm as shown in Figure 19.P6. Actuating force of 600 N acts at the end of a lever 600 mm long from the fulcrum. If coefficient of friction is 0.28, find:

   (a) Frictional torque, which it can sustain.

   (b) If thickness of the band is 2 mm, width of band, if safe tensile stress is 60 MPa.

   (c) Section of the lever, if safe tensile stress is 50 MPa for the lever material.

   [(a) T = 520 Nm; (b) w = 24 mm; (c) t = 17 mm, h = 50 mm]

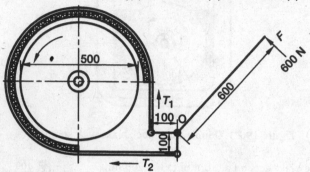

Figure 19.P6 A crane drum with band brake

8. A dead weight of 10 kN is being supported on a crane having drum of 600 mm with a brake drum on the same shaft of 500 mm diameter. The drum is braked with a band having one end fixed to the fulcrum and the other end at a distance of 60 mm from the fulcrum as shown in Figure 19.P7. A force of 300 N is applied at the end of a bell crank lever. If coefficient of friction is 0.3, calculate:

   (a) Distance x from the fulcrum, where this force is to be applied to support the load.

   (b) Width of the band, if thickness is 5 mm and safe tensile strength 50 MPa.

   [(a) x = 827 mm; (b) w = 65 mm]

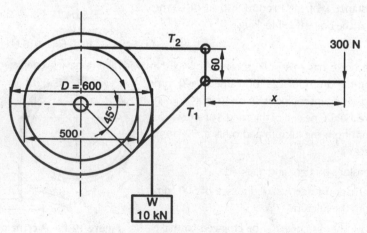

Figure 19.P7 A band brake on a crane drum

9. In a band and block brake, there are 10 blocks, each subtending an arc of 15° at the center of the drum. The blocks of thickness 20 mm are so spaced that they cover 215° over the drum of radius 0.4 m. It is required to absorb a power of 250 kW rotating at 250 rpm in clockwise direction. Assuming coefficient of friction 0.3, calculate the actuation force F at a distance of 800 mm from the fulcrum. One end of the band is connected at 30 mm as shown in Figure 19.P8 and other at 80 mm on the other side of the lever. [F = 500 N]

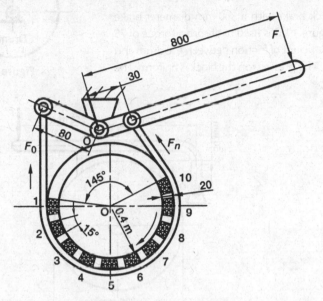

Figure 19.P8 A band and block brake

10. A passenger car of total weight 16 kN uses four wheel brakes. It is slowed down uniformly from 80 km/h to 40 km/h in a distance of 100 m. Gradient of road during braking is 1 in 20 downwards. Assuming equal weights on all wheels, calculate the mean lining pressure from the following data:

Brake lining width = 60 mm

Effective wheel diameter = 720 mm

Brake drum diameter = 350 mm

Lining angle = 130°

Coefficient of friction = 0.33   [p = 0.52 MPa]

Previous Competition Examination Questions

GATE

1. A four-wheel vehicle of mass 100 kg moves uniformly in a straight line with the wheels revolving at 10 rad/s. The wheels are identical, each with a radius of 0.2 m. Then a constant braking torque is applied to all the wheels and the vehicle experiences a uniform deceleration. For the vehicle to stop in 10 s, the breaking torque (Nm) on each wheel is_____

[Answer: 10 Nm] [GATE 2014]

2. A drum brake is shown in the Figure 19.E1 with dimensions in millimeters. The drum is rotating in anticlockwise direction. The coefficient of friction between drum and shoe is 0.2. The breaking torque (Nm) for the brake shoe is_____

    [Answer: 64 Nm] [GATE 2014]

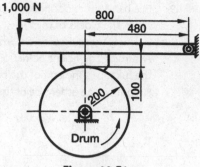

Figure 19.E1

3. A single block brake with a 300 mm diameter brake shown in Figure 19.E2 is used to absorb a torque of 75 Nm. The coefficient of friction between the drum and lining is 0.35. The pressure on the block is uniform. The force P will be:

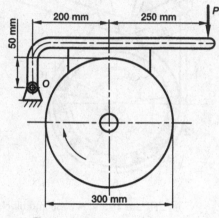

Figure 19.E2

(a) 480 N   (b) 540 N   (c) 580 N   (d) 640 N

[Answer: (c)] [GATE 2013]

4. A force of 400 N is applied to the brake drum of 0.5 m diameter in a band brake system shown in Figure 19.E3, where the wrapping angle is 180°. If the coefficient of friction between the drum and the band is 0.25, the braking torque applied in newton meters is:

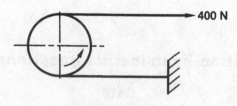

Figure 19.E3

(a) 100.6   (b) 54.4

(c) 22.1   (d) 15.7

[Answer: (b)] [GATE 2012]

5. A band brake having bandwidth of 80, drum diameter of 250 mm, coefficient of friction of 0.25, and angle of wrap 270° is required to exert a friction torque 1,000 Nm. The maximum tension (kN) developed in the band is:

    (a) 1.88   (b) 3.56

    (c) 6.12   (d) 11.56

    [Answer: (d)] [GATE 2010]

6. A block brake shown in Figure 19.E4 has a face width of 300 mm and a mean coefficient of friction 0.25. For an activating force of 400 N, the braking torque in newton meters is:

    (a) 30   (b) 40

    (c) 45   (d) 60

    [Answer: (c)] [GATE 2007]

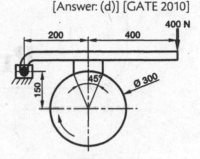

Figure 19.E4

Linked Answer Questions 7 and 8

A band brake shown in Figure 19.E5 consists of a lever attached to one end of the band. The other end of the band is fixed to the ground. The wheel has a radius of 200 mm and the wrap angle of the band is 270°. The braking force applied to the lever is limited to 100 N and the coefficient of friction between the band and the wheel is 0.5. No other information is given.

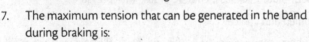

Figure 19.E5

7. The maximum tension that can be generated in the band during braking is:

    (a) 1,200 N   (b) 2,110 N

    (c) 3,224 N   (d) 4,420 N

    [Answer: (b)] [GATE 2005]

8. The maximum wheel torque that can be completely braked is:

    (a) 200 N-m   (b) 382 N-m

    (c) 604 N-m   (d) 844 N-m

    [Answer: (b)] [GATE 2005]

9. A simple band brake is to absorb 4.5 kW at 150 rpm. The coefficient of friction is 0.12 and maximum pressure between lining and the drum is to be 720 kPa. Determine the width of the band, if the drum is 300 mm and the angle of wrap is 215°.

    [Answer: w = 50.2 mm] [GATE 1993]

□□□

Pressure Vessels

20.1 Introduction and Applications

Pressure vessels are used as containers of fluid at high pressure as shown in picture on right side. These are used for thermal, chemical, nuclear plants, hydraulic cylinders, accumulators, etc. Vessels having internal pressure greater than 0.7 atmosphere fall under this category. The fluid inside may be in liquid or gaseous state or can change state

also from liquid to gas as in boiler. These have to be designed with great care and higher factor of safety as their failure can be injurious to life. These are generally made of ductile material like mild steel.

20.2 Classification

Pressure vessels can be classified in many ways:

a. **According to wall thickness**
   - Thin wall: If inner diameter to wall thickness ratio is more than 15.
   - Thick wall: If inner diameter to wall thickness ratio is less than 15.

b. **According to shape of vessel**
   - Cylindrical
   - Spherical

c. **According to end cover**
   - Flat
   - Ellipsoid
   - Hemispherical
   - Conical

20.3 Materials and Allowable Stresses

The materials used for pressure vessels are tabulated in Table 20.1 given below:

Table 20.1 Yield Strength S_{yt} of Material Used for Pressure Vessels

Material	C15	C15Mn75	22Si10S5P5	20Mo55	20Mn2	15Cr90Mo55
S_{yt} (MPa)	205	225	250	275	290	290

For operation under constant pressure and temperature up to 250 °C, S_{yt} varies from 200 MPa to 250 MPa. For high temperature applications, alloy steels are used. To calculate allowable stresses σ_t, a FOS is taken as under:

$$\sigma_t = \frac{S_{yt}}{1.5} \quad \text{or} \quad \sigma_t = \frac{S_{ut}}{3}$$

Three terms are used for the pressures in design calculations:

- **Working pressure**   Actual pressure inside the vessel (p).
- **Design pressure**   Generally taken same as working pressure but it can be taken as 5 per cent more than the working pressure, that is, $1.05\ p$.
- **Hydrostatic test pressure**   Pressure vessels are tested with an internal pressure, which is three times more than design pressure, that is, $3.15\ p$.

Stress σ_t is calculated on the basis of design pressure and should not exceed the safe stress.

20.4 Corrosion Allowance

Corrosion allowance is to be considered, if the pressure vessel is subjected to the following working conditions:

- Rusting, due to moisture.
- Chemical attacks, due to corrosive chemicals inside it.
- Scaling, due to high temperature application.
- Erosion, if the fluid in the pipe is flowing at high velocity.

This allowance varies from 1.5 mm to 3 mm, which is added to the thickness, calculated by the analysis given in the following sections.

In case of cylinders of engine a reboring allowance of 3 to 6 mm is added to the thickness of cylinder.

20.5 Class of Pressure Vessels

Class of a pressure vessel depends on the type of fluid it contains and the working conditions. There are three classes as given below:

Class 1 For any one or more than one working conditions:

- For poisonous gases and liquids like acids that are harmful to life.
- Applications, where temperature is more than 250 °C

Joints for this class are fully radiographed and have to be double butt joint or single butt with backing plate.

Class 2 Vessels, which do not come under class 1 or class 3 and the wall thickness is less than 38 mm, are categorized as class 2.

Type of joint is same as for class 1, but these are only spot radiographed and not fully radiographed.

Class 3 These vessels are where environmental conditions and fluids are not of class 1 and meant for light duties. Internal pressure is not more than 1.75 MPa and wall thickness is less than 16 mm. The joints are not radiographed.

20.6 Stresses due to Internal Pressure

When a pressure vessel is subjected to internal pressure, two types of stresses are developed; circumferential and longitudinal, which are described in the following sections. If the vessel is open at ends, longitudinal stresses do not appear.

20.6.1 Circumferential stresses

In a cylindrical vessel, pressure acts radially in all directions as shown in Figure 20.1(a). At right angle to a longitudinal section X–X, components of the pressure are shown in

Figure 20.1(b), which try to break the cylinder into two pieces. The force acting on this section is:

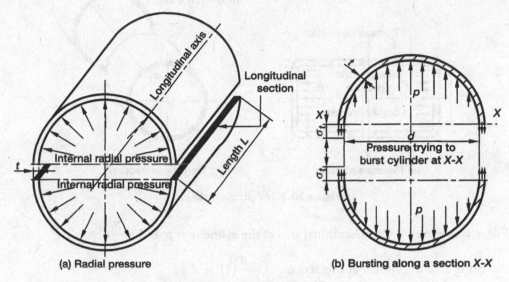

Figure 20.1 Circumferential stresses

Force F = Pressure × Projected area of the cylinder = $P_i \times L \times d$

This force is resisted by the cylinder walls at this section, whose area $A = 2t \times L$

For the cylinder to be safe, the stress should not exceed the allowable tensile stress, that is,

$$F = p \times L \times d = \sigma_{tc} \times 2\,t \times L$$

Or, Circumferential tangential stress (MPa): $\sigma_{tc} = \dfrac{pd}{2\,t}$ \hfill (20.1)

Or, Wall thickness $t = \dfrac{p \times d}{2\sigma_{tc}}$

Where, p = Internal pressure (MPa)

L = Length of cylinder (mm)

t = Wall thickness (mm)

20.6.2 Longitudinal stresses

In a cylindrical shell, the internal pressure at the ends shown in Figure 20.2(a) tries to break the cylinder into two pieces across the circular section as shown in Figure 20.2(b). This also causes tensile stress σ_{tl} across the circular section.

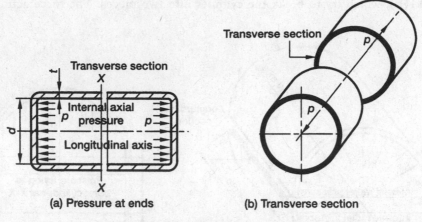

Figure 20.2 Longitudinal stresses

Force = Pressure × Cross-sectional area of the cylinder $= p \times A = \dfrac{\pi}{4}d^2 p$

Resisting area of cylinder at this section $A = \dfrac{\pi}{4}(D^2 - d^2)$

Where, d = Internal diameter (mm)

D = Outside diameter (mm)

Hence the longitudinal stress is: $\sigma_{tl} = \dfrac{\text{Force}}{\text{Resisting area}} = \dfrac{pd^2}{(D^2 - d^2)}$ \hfill (20.2a)

If wall thickness is less, resisting area can be approximately taken as: $A = \pi d t$

Hence $\sigma_{tl} = \dfrac{\text{Force}}{\text{Resisting area}} = \dfrac{p\,\pi d^2}{4(\pi dt)} = \dfrac{p\,d}{4t}$ \hfill (20.2b)

From Equations (20.1) and (20.2b), it can be seen that the longitudinal stress is half the circumferential stress.

Example 20.1

Longitudinal and circumferential stresses for a given pressure in a vessel
A closed-end cylindrical pressure vessel having wall thickness 15 mm and inside diameter 900 mm is filled with steam at pressure 1.5 MPa. Calculate longitudinal and circumferential stresses.

Solution

Given $\quad t = 15$ mm $\qquad d = 900$ mm $\qquad p = 1.5$ MPa

Ratio of $d/t = \dfrac{900}{15} = 60$

It is more than 15; hence it will be treated as a thin cylinder.

From Equation (20.1): Circumferential stress $\sigma_{tc} = \dfrac{pd}{2t} = \dfrac{1.5 \times 900}{2 \times 15} = 45$ MPa

From Equation (20.2): Longitudinal stress $\sigma_{tl} = \dfrac{pd}{4t} = \dfrac{1.5 \times 900}{4 \times 15} = 22.5$ MPa

20.6.3 Effect of pressure on size

When a closed cylindrical pressure vessel is subjected to internal pressure its diameter and length both increase.

Increase in diameter $\delta d = \dfrac{p \times d^2}{2t E}\left(1 - \dfrac{\mu}{2}\right)$ \hfill (20.3)

Increase in length $\delta L = \dfrac{p\, d\, L}{2t E}\left(1 - \dfrac{\mu}{2}\right)$ \hfill (20.4)

Due to this increase in diameter and length, its volume also increases.

Volume without pressure $V1 = \dfrac{\pi}{4} d^2 L$

Volume with pressure $V2 = \dfrac{\pi}{4}(d + \delta d)^2 (L + \delta L)$

Increase in volume $\delta V = (V2 - V1) = \left[\dfrac{\pi}{4}(d + \delta d)^2 (L + \delta L)\right] - \dfrac{\pi}{4} d^2 L$

Since δd is small its square will be still smaller and hence neglected.

Hence, $\delta V = \dfrac{\pi}{4}[d^2\, \delta L + (2dL\, \delta d)]$ \hfill (20.5)

Example 20.2

Wall thickness and increase in diameter and volume due to pressure
A cylindrical pressure vessel of inside diameter 200 mm and length 900 mm handles fluid at pressure 6 MPa. If Young's modulus of elasticity is 2×10^5 MPa and $\mu = 0.3$, calculate:

   a. Wall thickness, if tensile stress is not to exceed 50 MPa

   b. Increase in diameter

   c. Increase in volume

Solution

Given $d = 200$ mm $L = 900$ mm $p = 6$ MPa $\sigma_{tc} = 50$ MPa

            $E = 2 \times 10^5$ MPa

a. From Equation (20.1):

$$t = \frac{p \times d}{2\sigma_{tc}} = \frac{6 \times 200}{2 \times 50} = 12 \text{ mm}$$

d / t ratio $= \dfrac{200}{12} = 16.7$, is more than 15, and hence taken as thin wall vessel.

b. From Equation (20.3):

Increase in diameter $\delta d = \dfrac{p_i \times d^2}{2tE}\left(1 - \dfrac{\mu}{2}\right)$ Substituting the values:

$$\delta d = \frac{6 \times 200^2}{2 \times 12 \times 2 \times 10^5}\left(1 - \frac{\mu}{2}\right) = 0.05\left(1 - \frac{0.3}{2}\right) = 0.0425 \text{ mm}$$

c. From Equation (20.4): Increase in length $\delta L = \dfrac{p\,d\,L}{2tE}\left(1 - \dfrac{\mu}{2}\right)$

Or, $\delta L = \dfrac{6 \times 200 \times 900}{2 \times 12 \times 2 \times 10^5}\left(1 - \dfrac{0.3}{2}\right) = 0.19 \text{ mm}$

From Equation (20.5): Increase in volume $\delta V = \dfrac{\pi}{4}[d^2 \delta L + (2dL\,\delta d)]$

Or, $\delta V = 0.785[(200^2 \times 0.19) + (2 \times 200 \times 900 \times 0.0425)] = 18,016 \text{ mm}^3$

Alternately: Initial volume, $V1 = \dfrac{\pi}{4}d^2 L = 0.785 \times 200^2 \times 900 = 28,260,000 \text{ mm}^3$

Volume with pressure, $V2 = \dfrac{\pi}{4}(d + \delta d)^2 (L + \delta L)$

$$= 0.785 \times (200.0425)^2 \times (900.19) = 28,278,012 \text{ mm}^3$$

Increase in volume $\delta V = (V2 - V1) = 28,278,012 - 28,260,000 = 18,012 \text{ mm}^3$

Note   This value will be more exact, because higher order terms have not been neglected.

20.7 Thick Cylinders

The analysis given in Section 20.6 above is for thin cylinders. To withstand high pressure, wall thickness is to be increased and the cylinder becomes thick. In a thick cylinder, the tensile stress at inner surface is more than outside surface as shown in Figure 20.3 and picture on its right side. For thin cylinders, it was assumed uniform, that is, same for inside and outside.

In addition to radial stress [Figure 20.3(a)], there exists a tangential stress along the circumference also, which is maximum at inner radius and zero at outer radius as shown in Figure 20.3(b).

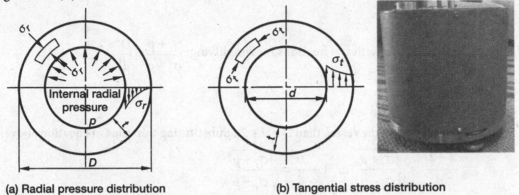

(a) Radial pressure distribution   (b) Tangential stress distribution

Figure 20.3 Stresses distribution in thick cylinder

Design of thick cylinder is done by any one of the following equations, depending on type of materials and ends, that is, whether open or closed:

For closed ends:
- **Lame's** equation: for brittle materials
- **Clavarino's** equation: for ductile materials

For open ends:
- **Birnie's** equation: for ductile materials-based on inside diameter
- **Barlow's** equation: for ductile materials-based on outside diameter

20.7.1 Lame's equation

This equation is used to find wall thickness of a pressure vessel made of brittle materials such as cast iron. According to this theory a vessel fails, if the principal stress exceeds the allowable stress of the material. Three principal stresses in a cylinder are:

$$\text{Radial stress } \sigma_r = -p \tag{20.6}$$

$$\text{Tangential stress } \sigma_t = \frac{p(D^2 + d^2)}{(D^2 - d^2)} \tag{20.7}$$

$$\text{Longitudinal stress } \sigma_l = \frac{pd^2}{(D^2 - d^2)} \tag{20.8}$$

Where, p = Internal pressure

D = Outside diameter

d = Inside diameter

Out of these three stresses, tangential stress is maximum, that is, $\sigma_t > \sigma_i > \sigma_r$, which is considered for design:

$$\frac{\sigma_t}{p} = \frac{D^2 + d^2}{D^2 - d^2}$$

Using mathematical properties of fractions for simplifying: $\dfrac{\sigma_t + p}{\sigma_t - p} = \dfrac{D^2}{d^2}$

Or, $\quad \dfrac{D}{d} = \sqrt{\dfrac{\sigma_t + p}{\sigma_t - p}}$

If t is the wall thickness of the vessel, then $D = d + 2t$. Substituting this value in equation above:

$$\frac{d + 2t}{d} = \sqrt{\frac{\sigma_t + p}{\sigma_t - p}} \quad \text{or} \quad 1 + \frac{2t}{d} = \sqrt{\frac{\sigma_t + p}{\sigma_t - p}}$$

Hence, wall thickness $t = \dfrac{d}{2}\left[\sqrt{\dfrac{\sigma_t + p}{\sigma_t - p}} - 1 \right]$ \hfill (20.9)

This equation to calculate wall thickness from inside diameter and internal pressure is called Lame's equation.

Example 20.3

Wall thickness of cast iron cylinder using Lame's equation

A cast iron cylinder of internal diameter 60 mm has ultimate strength of 220 MPa. It is supplied with internal pressure 8 MPa. Using an FOS = 4, calculate:
  a. Wall thickness
  b. What maximum axial force it can handle?

Solution

Given $\quad d = 60$ mm $\qquad\qquad S_{ut} = 220$ MPa $\qquad\qquad p = 8$ MPa $\qquad\qquad$ FOS = 4

  a. Safe tensile stress $\sigma_t = \dfrac{220}{4} = 55\,\text{MPa}$

  Since cast iron is a brittle material, Lame's equation (20.9) will be used

  Wall thickness $t = \dfrac{d}{2}\left[\sqrt{\dfrac{\sigma_t + p}{\sigma_t - p}} - 1 \right] \qquad$ Substituting the values:

  $$t = \frac{60}{2}\left[\sqrt{\frac{55 + 8}{55 - 8}} - 1 \right] = 30 \times 0.58 = 17.5 \text{ mm say 18 mm}$$

  b. Force = Pressure × Area $= 8 \times \dfrac{\pi}{4} 60^2 = 22{,}608$ N

20.7.2 Clavarino's equation

This equation is used when the cylinders are closed at both ends and are made of ductile materials. It assumes maximum strain theory of failure, that is, a material fails when the maximum strain is equal to the yield point of a simple tension test, that is:

$$\text{Strain } \varepsilon = \frac{\sigma}{E} = \frac{S_{yt}}{FOS \times E} \tag{a}$$

Where, S_{yt} = Tensile yield strength (MPa)

σ = Stress (MPa)

FOS = Factor of safety

E = Young's modulus of elasticity (MPa)

Also, strain is given as: $\varepsilon = \dfrac{1}{E}[\sigma_t - \mu(\sigma_l + \sigma_r)]$ \hfill (b)

Equating Equations (a) and (b):

$$\sigma = \sigma_t - \mu(\sigma_l + \sigma_r) \tag{20.10}$$

Substituting the values of the principal stresses from Equations (20.6, 20.7, and 20.8):

$$\sigma = \left[\frac{p(D^2 + d^2)}{D^2 - d^2} - \mu\left(\frac{pd^2}{(D^2 - d^2)} - p \right) \right]$$

Substituting value of $D = d + 2t$ and rearranging the terms:

$$t = \frac{d}{2}\left[\sqrt{\frac{\sigma + (1 - 2\mu)p}{\sigma - (1 + \mu)p}} - 1 \right] \tag{20.11}$$

Equation (20.11) is called Clavarino's equation.

Example 20.4

Wall thickness of pressure vessel using Clavarino's equation

A cylindrical pressure vessel of diameter 200 mm made of carbon steel having ultimate strength 360 MPa and Poisson's ratio 0.27 is filled with compressed air at a pressure of 14 MPa. Calculate its wall thickness assuming an $FOS = 4$.

Solution

Given $\quad d = 200 \text{ mm} \quad\quad S_{ut} = 360 \text{ MPa} \quad\quad \mu = 0.27 \quad\quad p = 14 \text{ MPa} \quad FOS = 4$

a. Safe tensile stress $\sigma_t = \dfrac{360}{4} = 90 \text{ MPa}$

Since carbon steel is a ductile material, Clavarino's equation (20.11) will be used. It is given as under:

$$t = \frac{d}{2}\left[\sqrt{\frac{\sigma + (1 - 2\mu)p}{\sigma - (1 + \mu)p}} - 1\right]$$

Substituting the values:

$$t = \frac{200}{2}\left[\sqrt{\frac{90 + (1 - 2 \times 0.27) \times 14}{90 - (1 + 0.27) \times 14}} - 1\right] = 100\left[\sqrt{\frac{96.44}{72.22}} - 1\right]$$

Or, $t = 100 \times 0.155 = 15.55$ mm say 16 mm

20.7.3 Birnie's equation

This equation is used when the cylinders are open ended and made of ductile material. When the cylinders are open ended, pressure in the axial direction gets released, and hence longitudinal stress becomes zero, that is, $\sigma_l = 0$. The stress σ in the shell is calculated only from radial and tangential stresses. Thus Equation (20.10) reduces to:

$$\sigma = \sigma_t - \mu\,\sigma_r$$

Substituting the values of principal stresses from Equations (20.6) and (20.7):

$$\sigma = \frac{p\,(D^2 + d^2)}{(D^2 - d^2)} - \mu p$$

Or, $\dfrac{\sigma}{p} = \left[\dfrac{(D^2 + d^2)}{(D^2 - d^2)} - \mu\right] = \dfrac{(1 + \mu)D^2 + (1 - \mu)d^2}{(D^2 - d^2)}$

Rearranging the terms: $\dfrac{D}{d} = \sqrt{\dfrac{\sigma + (1 - \mu)p}{\sigma - (1 + \mu)p}}$

Substituting $(D = d + 2t)$ and simplifying:

$$t = \frac{d}{2}\left[\sqrt{\frac{\sigma + (1 - \mu)p}{\sigma - (1 + \mu)p}} - 1\right] \tag{20.12}$$

This is called Birnie's equation.

Example 20.5

Wall thickness of a pipe using Birnie's equation

A steel pipe of diameter 75 mm is carrying a pressurized liquid at a pressure of 10 MPa. If its ultimate strength is 300 MPa and Poisson's ratio 0.27, assuming an FOS = 5, calculate its wall thickness using appropriate theory.

Solution

Given $d = 75$ mm $S_{ut} = 300$ MPa $\mu = 0.27$ $p = 10$ MPa $FOS = 5$

Safe tensile stress $\sigma_t = \dfrac{300}{5} = 60$ MPa

Since steel is a ductile material and pipe is open at both ends, Birnie's equation (20.12) will be used. It is given as under:

$$t = \frac{d}{2}\left[\sqrt{\frac{\sigma + (1-\mu)p}{\sigma - (1+\mu)p}} - 1\right] = \frac{75}{2}\left[\sqrt{\frac{60 + (1-0.27)\times 10}{60 - (1+0.27)\times 10}} - 1\right]$$

Or, $t = 37.5\left[\sqrt{\dfrac{67.3}{47.3}} - 1\right] = 7.23$ mm say 8 mm

20.7.4 Barlow's equation

Barlow's equation is used for high pressure oil and gas pipe lines. It is similar to thin cylinder, but uses outside diameter instead of inside diameter. It is given below:

$$t = \frac{p \times D}{2\sigma_t} \tag{20.13a}$$

Since $D - d + 2t$, substituting its value in above equation: $t = \dfrac{p \times (d + 2t)}{2\sigma_t}$

After simplification: $t = \dfrac{pd}{2(\sigma_t - p)}$ \hfill (20.13b)

Example 20.6

Pipe diameter using Barlow's equation

An iron pipe is to deliver water at rate of 2 m³/s at a velocity of 0.65 m/s and pressure 0.8 N/m². If tensile stress is 25 MPa, calculate:

a. Pipe diameter and

b. Wall thickness with corrosion allowance of 6 mm

Solution

Given $Q = 2$ m³/s $v = 0.65$ m/s $\sigma_t = 25$ MPa
            $p = $ N/m² $CA = 6$ mm

a. Discharge $Q = $ Area × velocity, hence $A = \dfrac{Q}{v} = \dfrac{2}{0.65} = 3.077$ m²

    Or, $\dfrac{\pi}{4}d^2 = 3.077$ or $d = 2$ m $= 2{,}000$ mm

b. Using Barlow's equation (20.13b):

$$t = \frac{p \times d}{2(\sigma_t - p)} = \frac{0.8 \times 2,000}{2(25 - 0.8)} = 33 \text{ mm}$$

Adding corrosion allowance $t = 33 + 6 = 39$ mm say 40 mm

20.8 Thin Spherical Vessels

20.8.1 Plate thickness

Sometimes, pressure vessels are made spherical also as shown in picture below on right side. The bursting force acts on an annular cross section, as shown in Figure 20.4.

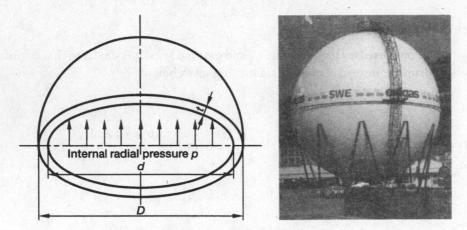

Figure 20.4 Spherical vessel

The bursting force can be calculated from the relation:

$$F = \frac{\pi}{4} d^2 \times p \qquad\qquad\qquad (a)$$

This force is resisted by

$$F = \sigma_t \times \text{Resisting area} = \sigma_t \times \pi d\, t \qquad\qquad\qquad (b)$$

Equating Equations (a) and (b): $\frac{\pi}{4} d^2 \times p = \sigma_t \times \pi d t$

Or, $$t = \frac{p \times d}{4\sigma_t} \qquad\qquad\qquad (20.14)$$

Example 20.7

Maximum pressure in a spherical vessels of given size and thickness

A spherical pressure of diameter 2 m is made of 15 mm thick plate. With $FOS = 4$, and ultimate tensile stress 320 MPa.

a. What maximum pressure it can handle?

b. What is mass per unit volume capacity of sphere, if specific gravity of the material is 7,650 kg/m³?

c. What thickness is required, if pressure is set at 1.5 MPa?

Solution

Given $d = 2$ m $S_{ut} = 320$ MPa $t = 15$ mm $FOS = 4$
            $\rho = 7,650$ kg/m³ $p = 1.5$ MPa

a. Safe tensile stress $\sigma_t = \dfrac{320}{4} = 80$ MPa

Using Equation (20.14) for spherical pressure vessels: $t = \dfrac{p \times d}{4\sigma_t}$

Or, $p = \dfrac{4t\sigma_t}{d} = \dfrac{4 \times 15 \times 80}{2,000} = 2.4$ MPa

b. Inner radius $r = \dfrac{2}{2} = 1$ m

Outside radius $R = r + t = 1.00 + 0.015 = 1.015$ m

Inside volume of sphere $V = \dfrac{4}{3}\pi r^3 = \dfrac{4}{3} \times 3.14 \times 1^3 = 4.187$ m³

Volume of material of hollow sphere

$$v = \dfrac{4}{3}\pi(R^3 - r^3) = \dfrac{4}{3} \times 3.14\,(1.015^3 - 1^3) = 0.191 \text{ m}^3$$

Mass of the spherical vessel $m = v \times \rho = 0.191 \times 7,650 = 1,461$ kg

Mass per unit volume $= \dfrac{m}{V} = \dfrac{1,461}{4.187} = 350$ kg/m³

c. Using Equation (20.14)

$$t = \dfrac{p \times d}{4\sigma_t} = \dfrac{1.5 \times 2,000}{4 \times 80} = 9.4 \text{ mm say } 10 \text{ mm}$$

20.8.2 Change in size of spherical pressure vessel with internal pressure

When a thin sphere is subjected to internal pressure, its diameter increases. The change in diameter is given by the relation:

Change in diameter, $\delta d = \dfrac{p \times d^2}{4 t E}(1 - \mu)$ (20.15)

Where, p = Internal pressure

d = Inside diameter of the shell

t = Thickness of the shell

E = Young's modulus of elasticity of the shell material and

μ = Poisson's ratio

Volume of a spherical vessel $V = \dfrac{4\pi}{3} r^3 = \dfrac{\pi}{6} d^3$

Increase in volume = Final volume after pressure – Initial volume without pressure

Increase in volume $\delta V = \dfrac{\pi}{6}(d + \delta d)^3 - \dfrac{\pi}{6} d^3$

If higher order terms are neglected, $\delta V = \dfrac{\pi}{6}(3d^2 \times \delta d)$

Substituting the value of δd from Equation (20.15):

$$\delta V = \frac{3 \pi d^2}{6}\left[\frac{p \times d^2}{4 t E}(1 - \mu)\right]$$

Or, $\delta V = \dfrac{\pi p \times d^4}{8 t E}(1 - \mu)$ (20.16)

Example 20.8

Change in capacity of spherical pressure vessel under pressure
A spherical pressure vessel having inside diameter 0.8 m and plate thickness 20 mm is subjected to an internal pressure of 2 MPa. E = 200 GPa and μ = 0.3. Calculate percentage increase in volume with pressure.

Solution

Given $d = 0.8$ m $t = 20$ mm $p = 2$ MPa $E = 200$ GPa

           $\mu = 0.3$

From Equation (20.16): $\delta V = \dfrac{\pi p d^4}{8t E}(1 - \mu)$

Or, $\delta V = \dfrac{3.14 \times 2 \times 800^4}{8 \times 20 \times 200 \times 10^3}(1 - 0.3) = 56.26 \times 10^3 \text{ mm}^3$

Initial volume without pressure $V = \dfrac{\pi}{6}d^3 = \dfrac{3.14}{6} \times 800^3 = 268 \times 10^6 \ \text{mm}^3$

Percentage increase in volume $= \dfrac{\delta V}{V} = \dfrac{56.26 \times 10^3 \times 100}{268 \times 10^6} = 0.021$ per cent

20.9 End Covers

End covers are used to cover the cylinder to form a closed vessel. Broadly, end covers are of three types:

- Flat [Figure 20.5(a)]
- Domed: These are further classified as:
  - Hemispherical [Figure 20.5(b)]
  - Semi-ellipsoid [Figure 20.5(c)]
  - Tori-spherical [Figure 20.5(d)]
- Conical [Figure 20.5(e)]

(a) Flat   (b) Hemispherical   (c) Semi ellipsoid   (d) Tori-spherical   (e) Conical

Figure 20.5 Types of end covers

End covers are neither supported freely nor rigidly fixed, and hence design is done mostly on the basis of empirical relations as given in subsequent sections for different types of end covers. It is assumed that the load acts uniformly over the surface.

In the design of shells and end covers containing corrosive fluids, a corrosion allowance of 2 mm–3 mm is added to the thickness calculated by the governing equations.

20.9.1 Flat circular

Flat covers are the easiest to make. Thickness of circular cover plate is calculated using the empirical relation given in Equation (20.17).

$$t_C = k_2 d \sqrt{\dfrac{p}{\sigma_t}} \qquad\qquad (20.17)$$

Where, t_C = Thickness of cover plate (mm)

       k_2 = A constant depending on material and end conditions for circular plate.
           See Table 20.2 for its value.

       σ_t = Safe tensile stress of cover material (N/mm² or MPa)

       d = Inside diameter of cylindrical pressure vessel (mm)

       p = Pressure inside the vessel

Table 20.2 Value of Constants

Shape of Flat End Cover	Constant	Material			
		Cast Iron		Mild Steel	
		Type of Support		Type of Support	
		Free	Fixed	Free	Fixed
Rectangular	k_1	0.75	0.62	0.6	0.49
Circular	k_2	0.54	0.44	0.42	0.35
Elliptical	K_3	1.5	1.2	1.2	0.9

Example 20.9

Flat circular cover plate

A cast iron cylindrical pressure vessel closed at both ends has flat circular end cover plate integral with the cylindrical shell of internal diameter 230 mm. It is filled with a fluid at a pressure of 10 N/mm². Safe tensile stress is 35 N/mm². Calculate:

a. Wall thickness of the shell

b. Thickness of end cover plate

Solution

Given $d = 230$ mm $p = 10$ MPa $\sigma_t = 35$ MPa

a. Using Lame's Equation (20.9):

$$t = \frac{d}{2}\left[\sqrt{\frac{\sigma_t + p}{\sigma_t - p}} - 1\right] = \frac{230}{2}\left[\sqrt{\frac{35 + 10}{35 - 10}} - 1\right] = 40 \text{ mm}$$

b. Using Equation (20.17): $t_C = k_2 d\sqrt{\dfrac{p}{\sigma_t}}$

End cover can be considered as fixed. Value of k_2 for cast iron from Table 20.2 is 0.44.

$$t_C = k_2 d\sqrt{\frac{p}{\sigma_t}} = 0.44 \times 230\sqrt{\frac{10}{35}} = 27 \text{ mm}$$

20.9.2 Flat rectangular

Sometimes, the pressure vessel may be of rectangular shape, for example, steam chest of a steam engine. Thickness of rectangular cover plate can be calculated using Equation (20.18):

$$t_c = a\,b\,k_1\sqrt{\frac{p}{\sigma_t(a^2 + b^2)}} \tag{20.18}$$

Where, t_c, p, σ_t are same as defined above

> a = Length of cover plate (mm)
>
> b = Width of cover plate (mm)
>
> k_1 = A constant depending on material and end conditions for rectangular cover plate. See Table 20.2 for its value.

Example 20.10

Wall thickness of a flat rectangular cover plate

A steam chest of internal dimensions 200 mm × 300 mm and pressure 1.4 MPa has a flat rectangular freely supported cast iron cover plate with allowable tensile stress 30 MPa. Calculate wall thickness of end cover plate

Solution

Given $a = 300$ mm $b = 200$ mm $p = 1.4$ MPa $\sigma_t = 30$ MPa

Using Equation (20.18): $t_c = a\, b\, k_1 \sqrt{\dfrac{p}{\sigma_t (a^2 + b^2)}}$

Value of k_1 from Table 20.2 is 0.75.

Substituting the values:

$$t_c = 300 \times 200 \times 0.75 \sqrt{\frac{1.4}{30 \times (300^2 + 200^2)}} = 27 \text{ mm}$$

20.9.3 Elliptical plate

Some pressure vessels are not cylindrical but of elliptical cross section. Such vessels will require elliptical cover plate. Ratio of major axis to minor axis is generally kept two. Thickness of the cover plate is given as:

$$t_c = a\, b\, k_3 \sqrt{\frac{p}{\sigma_t (a^2 + b^2)}} \tag{20.19}$$

Where, t_c, p, σ_t are same as defined above

> a = Major axis of cover plate (mm)
>
> b = Minor axis of cover plate (mm)
>
> k_3 = A constant depending on material and end conditions for ellipsoid cover plate. See Table 20.2 for its value.

Example 20.11

Wall thickness of an elliptical cover plate
An elliptical pressure vessel has major axis 900 mm and minor axis 400 mm. It contains fluid at a pressure of 2 MPa. It is covered rigidly with an elliptical flat plate. If safe tensile strength of the mild steel cover material is 60 MPa, calculate its thickness.

Solution

Given $\quad\quad a = 900$ mm $\quad\quad b = 400 \quad\quad p = 2$ MPa $\quad\quad \sigma_t = 60$ MPa

Using Equation (20.20): $t_c = a\, b\, k_3 \sqrt{\dfrac{p}{\sigma_t\,(a^2 + b^2)}}$

From Table 20.2, value of $k_3 = 0.9$. Substituting the values:

$$t_c = 900 \times 400 \times 0.9 \sqrt{\frac{2}{60\,(900^2 + 400^2)}} = 60 \text{ mm}$$

20.9.4 Hemispherical

It is just half of a sphere, having inside diameter same as the diameter of cylindrical pressure vessel. It offers minimum thickness of plate for the same strength, hence light and economical. Wall thickness of a hemispherical end cover, either integral or welded is given as:

$$t_c = \frac{p \times d}{4\sigma_t} \tag{20.20}$$

Example 20.12

Wall thickness of a hemispherical cover plate
A thin cylindrical pressure vessel is made of carbon steel with safe strength of 68 MPa has diameter 800 mm. It is closed at both the ends by hemispherical end covers. The internal pressure is 2.5 MPa. Calculate:
   a. Wall thickness of the shell
   b. Thickness of end cover plate

Solution

Given $\quad\quad d = 800$ mm $\quad\quad p = 2.5$ MPa $\quad\quad \sigma_t = 68$ MPa

Since it is given as thin, Equation (20.1) can be used to calculate wall thickness.

$$t = \frac{p \times d}{2\sigma_t} = \frac{2.5 \times 800}{2 \times 68} = 15 \text{ mm}$$

For hemispherical ends, using Equation (20.20)

$$t_c = \frac{p \times d}{4\sigma_t} = \frac{2.5 \times 800}{4 \times 68} = 7.5 \text{ mm}$$

Example 20.13

Vessel of required capacity with hemispherical cover plate

A pressure vessel having hemispherical ends is to contain corrosive fluid at 1.5 MPa with about 2.5 m³ capacity. Material has yield strength is 250 MPa and joint efficiency of the welded joint is 85 per cent, FOS is 1.5, and corrosion allowance 2.5 mm. Length should not exceed 3.5 m. Calculate:

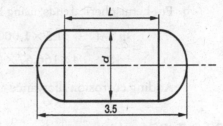

Figure 20.S1 Vessel with hemispherical cover plates

a. Shell thickness

b. Cover wall thickness

Solution

Given $p = 1.5$ MPa $V \cong 2.5$ m³ $S_{yt} = 250$ MPa $\eta = 85$ per cent

$FOS = 1.5$ $CA = 2.5$ mm $L < 3.5$ m

Length of hemispherical end is 0.5 d, and hence length of cylindrical shell is:

$$L = 3 - d$$

Volume of vessel V = Volume of cylindrical shell + Volume of spherical ends

Or, $V = \left(\dfrac{\pi}{4}d^2 \times L \right) + \dfrac{\pi}{6}d^3$

Substituting the values: $2.5 = \left[\dfrac{3.14}{4} \times d^2 \times (3-d) \right] + \dfrac{3.14}{6}d^3$

Simplifying: $d^3 - 10.5\,d^2 + 9.54 = 0$

Solving this cubical equation using Excel sheet or trial and error: $d = 1$ m

Hence length $L = 3.5 - 1 = 2.5$ m

Calculating exact volume:

$$V = \left(\dfrac{\pi}{4}d^2 \times L \right) + \dfrac{\pi}{6}d^3 = \left(\dfrac{3.14}{4}1^2 \times 2.5 \right) + \dfrac{3.14}{6}1^3 = 2.48 \text{ m}^3$$

It is almost same as 2.5 m³

a. Safe tensile stress $\sigma_t = \dfrac{S_{yt}}{1.5} = 166.67$ MPa

Pressure is not high so first considering it as thin cylinder and using Equation (20.1) to calculate wall thickness with joint efficiency as under.

$$t = \dfrac{p \times d}{2\sigma_t \eta} = \dfrac{1.5 \times 1,000}{2 \times 166.67 \times 0.85} = 5.3 \text{ mm}$$

Adding corrosion allowance $t = 5.3 + 2.5 = 7.8$ mm say 8 mm

b. For hemispherical ends, using Equation (20.20) with joint efficiency as 85 per cent.

$$t_c = \frac{p \times d}{4\sigma_t} = \frac{1.5 \times 1,000}{4 \times 166.67 \times 0.85} = 2.65 \text{ mm}$$

Adding corrosion allowance $t_c = 2.65 + 2.5 = 5.1$ mm say 5 mm

20.9.5 Dished

When the cover is not hemispherical but its frustum, it is called as dished. It could be either welded as shown in Figure 20.6(a) or riveted [Figure 20.6(b)] or bolted [picture 20.6(c)]. The camber c is less than the hemispherical covers.

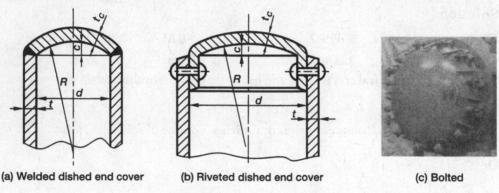

(a) Welded dished end cover   (b) Riveted dished end cover   (c) Bolted

Figure 20.6 Dished end covers

Formula to calculate wall thickness for such a riveted cover is:

$$t_c = \frac{k\,p\,R}{\sigma_u} \tag{20.21}$$

Where, t_c and p are same as defined above

σ_u = Ultimate stress of material

R = Inside radius of curvature of cover plate. (R should not be more than shell diameter).

k = A constant depending on material and end conditions for the cover

Without any opening or manhole $k = 4.16$

With an opening or manhole $k = 4.8$

For integral or welded dished heads:

$$t_c = \frac{p\,(d^2 + 4c^2)}{16c\sigma_t} \tag{20.22}$$

Where, t_c, p, and σ_t are same as defined above

c = Camber (mm)

Hemispherical is a dished head, where $c = 0.5\ d$. Substituting this value in Equation (20.22):

$$t_c = \frac{p \times d}{4\sigma_t} \text{ which is same as Equation (20.20) for hemispherical heads}$$

20.9.6 Semi-ellipsoid

For a semi-ellipsoid end cover, generally the major axis is double the length of the minor axis (Figure 20.7). The cover plate is of elliptical shape. It has a straight length $S = 3t$, which is about three times the wall thickness or 20 mm, whichever is more. Thickness for this cover plate required is maximum.

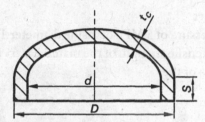

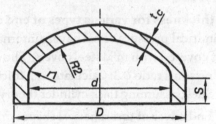

Figure 20.7 Semi-ellipsoid end cover   **Figure 20.8** Tori-spherical end cover

20.9.7 Tori-spherical end cover

This type of end cover has two radii as shown in Figure 20.8. It also has a straight length $S \simeq 0.3t$, which is about three times the wall thickness or 20 mm. Radius of spherical crown portion is $R2$. It should be less than outside shell diameter D, that is, $R2 < D$. Generally, it is kept 75 per cent to 80 per cent of D.

Crown radius $R2 = 0.8D$

Radius of knuckle $r1$ is about 6 per cent to 12 per cent of crown radius. It can be taken as 10 per cent of shell diameter, that is, $r1 = 0.1D$

It requires less forming than ellipsoid heads. Discontinuities between the shell and knuckle and knuckle and crown cause local stresses. Thickness of such an end cover is given by the relation:

$$t_c = \frac{0.885 p \times R_2}{\sigma_t} \tag{20.23}$$

Volume of such an end cover of one side can be approximated with the above-mentioned proportions as:

$$V_c = 0.123\ d^3 \tag{20.24}$$

20.9.8 Conical end covers

Conical end cover is shown in Figure 20.5(e) and picture on right side. Thickness of such a cover plate is given by equation:

$$t_c = \frac{p \times d}{2 \cos\theta\, \sigma_t} \qquad (20.25)$$

Where, θ = Semi-cone angle.

Example 20.14

Wall thickness for various types of end covers

A cylindrical pressure vessel having internal pressure of 5 MPa of inside diameter 1,200 mm is covered with mild steel covers. The safe tensile strength of the material is 65 MPa and Poisson's ratio 0.3. Calculate wall thickness of the:

  a. Shell, assuming thick cylinder theory

  b. End cover of tori-spherical shape

  c. End cover of conical shape of cone angle 60°.

  d. Dished end cover with camber as 25 per cent of diameter

  e. End cover of spherical shape

  f. Which type of cover has minimum and maximum thickness?

Solution

Given $p = 5$ MPa $d = 1,200$ mm $\sigma_t = 65$ MPa

$\mu = 0.3$ $c = 0.25\, d$ $2\theta = 60°$

  a. Since the material is ductile and closed at both the ends, Clavarino's equation (20.11) will be used:

$$t = \frac{d}{2}\left[\sqrt{\frac{\sigma + (1-2\mu)p}{\sigma - (1+\mu)p}} - 1\right]$$

$$t = \frac{1,200}{2}\left[\sqrt{\frac{65 + (1-2\times 0.3)\times 5}{65 - (1+0.3)\times 5}} - 1\right] = 600\left[\sqrt{\frac{67}{58.5}} - 1\right] = 42 \text{ mm}$$

  b. For tori-spherical end cover, use Equation (20.23):

$$t_c = \frac{0.885\, p \times R_2}{\sigma_t}$$

Taking crown radius R2 = 0.8d = 0.8 × 1,200 = 960 mm

Substituting the values: $t_c = \dfrac{0.885 \times 5 \times 960}{65} = 65$ mm

c. For conical end cover, Equation (20.25):

$$t_c = \frac{p \times d}{2\cos\theta\,\sigma_t} = \frac{5 \times 1{,}200}{2 \times \cos 30° \times 65} = 53.3 \text{ mm}$$

d. For dished end covers, use Equation (20.22):

$$t_c = \frac{p(d^2 + 4c^2)}{16c\,\sigma_t}$$

Camber $c = 0.25\,d = 0.25 \times 1{,}200 = 300$ mm

Substituting the values: $t_c = \dfrac{5 \times [1{,}200^2 + (4 \times 300^2)]}{16 \times 300 \times 65} = 29$ mm

e. For hemispherical vessels, use Equation (20.20):

$$t_c = \frac{p \times d}{4\sigma_t} = \frac{5 \times 1{,}200}{4 \times 65} = 23 \text{ mm}$$

f. For this example, minimum end cover thickness is with hemispherical cover and maximum is for tori-spherical. However, maximum thickness is for ellipsoid covers.

20.10 Fixing of End Covers

Ends of cylindrical pressure vessels are either integral or fixed with bolts and nuts or rivets or welded as shown in Figure 20.9.

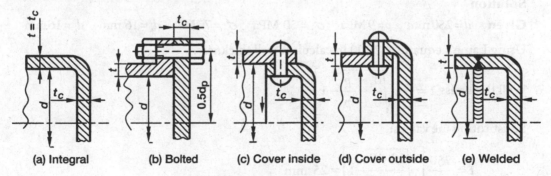

(a) Integral   (b) Bolted   (c) Cover inside   (d) Cover outside   (e) Welded

Figure 20.9 Joining of end covers

20.10.1 Integral

For integral covers, if c is the camber (Figure 20.6a) of dished end cover, its thickness is given same as Equation (20.22) as:

$$t_c = \frac{p\,(d^2 + 4\,c^2)}{16\,\sigma_t \times c}$$

20.10.2 Bolted

Thickness of cover plate, which is bolted to the pressure vessel, following empirical relation can be used:

$$\text{Thickness of cover plate } t_C = d_p\sqrt{\frac{k\,p}{\sigma_t}} \qquad (20.26)$$

Where, d_p = Pitch circle diameter of bolts [Figure 20.9(b)] = $d + 2\,t + 1.5 \times$ (Bolt diameter d_b)

k = 0.16 for bolted or riveted joint of cover plate.

k = 0.30 for flanged cover plate attached with a lap joint [Figure 20.9(c and d)]

k = 0.25 for welded cover plate [Figure 20.9(e)]

Example 20.15

Thickness of bolted cover plate
A cast iron cylinder of inner diameter 250 mm covered at both ends with flat cover plate has internal pressure 9 MPa. The hoop stress in cylinder is not to exceed 50 MPa. The cover plate made of steel having safe tensile stress 70 MPa is bolted to the cylinder with bolts of same strength having diameter 16 mm. Calculate:
   a. Wall thickness of the cylinder
   b. Thickness of cover plate

Solution
Given $d = 250$ mm $p = 9$ MPa $\sigma_{t1} = 50$ MPa $\sigma_{t2} = 70$ MPa $d_b = 16$ mm $d_b = 16$ mm
Using Lame's equation (20.1) to calculate wall thickness.

$$\text{Wall thickness } t = \frac{d}{2}\left[\sqrt{\frac{\sigma_t + p}{\sigma_t - p}} - 1\right]$$

Substituting the values:

$$t = \frac{250}{2}\left[\sqrt{\frac{50 + 9}{50 - 9}} - 1\right] = 25 \text{ mm}$$

Since the cover is bolted, hence using Equation (20.26) for thickness of cover plate, value of k for bolted joint is 0.16.

Pitch circle diameter $d_p = d + 2t + 1.5 \times$ (Bolt diameter d_b) $= 250 + 50 + 24 = 324$ mm

Thickness of cover plate $t_C = d_p \sqrt{\dfrac{k \times p}{\sigma_t}} = 324 \sqrt{\dfrac{0.16 \times 9}{70}} = 46.5$ say 47 mm

20.11 Welded Joints

The type of welding used for pressure vessels is fusion welding with a groove on one side or both sides. The types of joints are:

- Double welded [Figure 20.10(a)]
- Single welded without backing plate [Figure 20.10(b)]
- Single welded with backing plate [Figure 20.10(c)]

(a) Double welded **(b) Single welded withoutbacking plate** **(c) Single welded with backing plate**

Figure 20.10 Types of welded joints

Efficiency of the joint depends on the joint used. Table 20.3 gives the joint efficiency for different types of welds.

Table 20.3 Joint Efficiency of Welded Joints (Per cent)

Type of Joint	Double Welded	Single Welded	
		With Back Plate	Without Back Plate
Fully radio-graphed	100	90	70
Spot radio-graphed	85	80	65
Not radio-graphed	70	–	60

Welding is done in different locations to make a completely closed shell. Longitudinal welding along the axis of the shell is done to make a cylindrical or elliptical shell shown by (L) in Figure 20.11. Circumferential joints (C) are to increase length of the shell. Nozzles are welded at the opening shown by (N) and flanges are welded (F) to make the bolted connections.

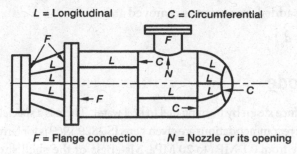

L = Longitudinal   C = Circumferential

F = Flange connection   N = Nozzle or its opening

Figure 20.11 Weld locations

20.12 Opening in Pressure Vessels

Openings are provided in a pressure vessel for the following reasons:

- Man holes of diameter about 400 mm, through which a man can enter inside the vessel to do repair or cleaning work.

- Small holes for mounting pressure gauge or a temperature gauge, if temperature of the fluid is high.

- For mounting safety valves so that the pressure does not exceed a preset value

The openings are generally circular, [Figure 20.12(a)] but elliptical openings are also seen. Owing to opening, some area of the shell is cut, which is called removed area (A_r). To compensate for the loss of this opening area, an additional reinforcement pad is provided at the opening as shown in Figure 20.12 (b). Figure 20.12(c) shows an opening fitted with a nozzle and reinforcement pad. The added pad area is called added area (A_a) and has to be equal to the removed area, that is, $A_a = A_r$. It may be noted that the compensation is done only for the area and not for the volume.

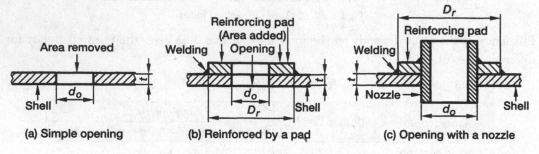

(a) Simple opening (b) Reinforced by a pad (c) Opening with a nozzle

Figure 20.12 Opening in a shell

Removed area for simple opening is:

$$A_r = t\, d_o \qquad\qquad (a)$$

Where, d_o = Diameter of opening

t = Thickness of the shell wall

Diameter of the reinforcing pad D_r is limited to twice the opening diameter $D_r = 2\, d_o$

Hence to have the added area same as removed area:

$$A_a = (D_r - d_o)\, t \qquad\qquad (b)$$

20.13 Boiler Code

A boiler is used to produce steam by burning fuel to boil water. There is a special code for the design of boiler. Some important recommendations are given here. IS 2829 1969 is for ferrous pressurized vessels having internal pressure from 0.1 MPa to 20 MPa. Materials of the shell should have minimum strength as under:

Tensile strength $\sigma_t = 385$ MPa
Compressive strength $\sigma_c = 665$ MPa
Shear strength $\tau = 308$ MPa

Thickness of the shell plate depends on its diameter and should not be less than the values given in Table 20.4:

Table 20.4 Diameter Versus Plate Thickness

Diameter of Shell (m)	Minimum Thickness of Shell Plate (mm)
0.9	6.35
0.94–1.37	8.00
1.4–1.8	9.5
More than 1.8	12.7

Containers having capacity less than 500 liters or diameter less than 150 mm are not in the scope of this code.

Example 20.16

Thickness of spherical tank in a pressurized system

A pump supplies oil at a pressure of 3 MPa to a spherical tank of diameter 1 m as shown in Figure 20.S2. The tank is welded with joint efficiency of 95 per cent. Safe tensile stress for the tank is 50 MPa.

a. Find thickness of the spherical tank.

b. Pressurized oil from the tank is supplied to a hydraulic cylinder with a pressure drop of 0.2 MPa to get a force of 15 kN. Assuming 10 per cent loss for force due to friction, find diameter of the cylinder and its wall thickness, if safe hoop tensile stress for the tank is 25 MPa.

c. If the piston moves 0.4 m in 10 s, find the power output in working cycle.

d. The work cycle repeats every minute. If hydraulic efficiency is 80 per cent and pump efficiency 65 per cent, calculate the power required for the motor to drive the pump.

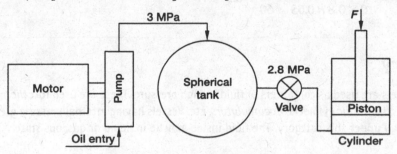

Figure 20.S2 Pressurized system

Solution

Given

a. $p_1 = 3$ MPa   b. $\Delta p = 0.2$ MPa   c. $L = 0.4$ m   d. $\eta_h = 80$ per cent

$d = 1$ m $F = 15$ kN $t_1 = 10$ s $\eta_p = 65$ per cent

$\eta_J = 95$ per cent $\eta_f = 90$ per cent $t_2 = 60$ s

$\sigma_{t1} = 50$ MPa $\sigma_{t2} = 25$ MPa

a. Using Equation (20.14) for spherical pressure vessels considering joint efficiency is:

$$t = \frac{pd}{4\sigma_t \eta_J} = \frac{3 \times 1000}{4 \times 50 \times 0.95} = 15.8 \text{ say } 16 \text{ mm}$$

b. Design force for the cylinder considering 10 per cent loss of friction:

$$F = 1.1F = 1.1 \times 15 = 16.5 \text{ kN}$$

Pressure after pressure drop $p_2 = p_1 - \Delta p = 3 - 0.2 = 2.8$ MPa

$$F' = p_2 \times \frac{\pi}{4} d_p^2 \quad \text{or} \quad d_p^2 = \frac{4F'}{\pi p_2} = \frac{4 \times 16,500}{3.14 \times 2.8}$$

Or, $(d_p)^2 = 7,506$ Hence, piston diameter $d_p = 86.7$ say 87 mm

Using Equation (20.1) for circumferential stress:

$$t = \frac{p \times d}{2\sigma_{t2}} = \frac{2.8 \times 87}{4 \times 25} = 4.8 \text{mm say } 5 \text{ mm}$$

c. Work done = Force × Distance travelled = $16.5 \times 0.4 = 6.6$ kNm

$$\text{Power delivered} = \frac{\text{Work done}}{\text{Time}} = \frac{6.6}{(60-10)} = 0.132 \text{ kW}$$

d. Power required from motor to lift 0.4 m in 10 s considering different efficiencies is:

$$P = \frac{16.5 \times 0.4}{10 \times 0.8 \times 0.65} \times \frac{10}{60} = 0.21 \text{ kW}$$

Summary

Pressure vessels are used as containers of fluid at high pressure. These are used for: *thermal, chemical, nuclear plants, hydraulic cylinders, accumulators*, etc. Vessels having internal pressure greater than 0.7 atmosphere fall under this category. The fluid inside may be in liquid or gaseous state.

Classification

a. *According to wall thickness*   Thin wall / Thick wall

b. *According to shape of vessel*   Cylindrical / Spherical

c. *According to end cover*   Flat / Ellipsoid / Hemispherical / Conical

Materials and allowable stresses   C15 - 205 MPa; C15 Mn 75 - 225 MPa; 20 Mn 2 - 290 MPa; 15 Cr 90 Mo 55 - 290 MPa.

Terms used   Working pressure: Actual pressure inside the vessel (p).

Design pressure   Generally taken as 5 per cent more than the working pressure, that is, 1.05 p.

Hydrostatic test pressure   Tested with internal pressure three times design pressure 3.15 p.

Corrosion allowance   Due to rusting, due to moisture, due to corrosive chemicals inside it, scaling, due to high temperature application, erosion, if the fluid in the pipe is flowing at high velocity. This allowance varies from 1.5 mm to 3 mm

Reboring allowance of 6 mm is added to thickness in case of engine *cylinders*.

Class of pressure vessels   It depends on type of fluid and working conditions.

Class 1   For poisonous gases or liquids like acids or temperature is more than 250 °C.

Class 2   Vessels that do not come under class 1 or class 3. Wall thickness < 38 mm.

Class 3   Meant for light duties. Internal pressure is < 1.75 MPa, wall thickness < 16 mm.

Stresses due to internal pressure   Two types of stresses are developed; *circumferential* and *longitudinal*. If the vessel is open at ends, longitudinal stress does not appear.

Circumferential stresses   In a cylindrical vessel, pressure acts radially in all directions. The force acting on a section is: F = Pressure × Projected area of cylinder = $P_i \times L \times d$

This force is resisted by the cylinder walls at this section, whose area $A = 2t \times L$

$$F = p \times L \times d = \sigma_{tc} \times 2t \times L \quad \text{or} \quad \sigma_{tc} = \frac{pd}{2t} \quad \text{or wall thickness } t = \frac{p \times d}{2\sigma_{tc}}$$

Longitudinal stresses   In a cylindrical shell, the internal pressure at the ends tries to break the cylinder in two pieces across an annular section of area (π d t) causing tensile stress σ_{tl} across this section. Hence

$$\sigma_{tl} = \frac{\text{Force}}{\text{Area}} = \frac{p\pi d^2}{4(\pi dt)} = \frac{pd}{4t}$$

Note   From above equations, note that longitudinal stress is half of the circumferential stress.

Effect of pressure on size   Increase in diameter $\delta d = \dfrac{p\,d^2}{2tE}\left(1 - \dfrac{\mu}{2}\right)$

Increase in length $\delta L = \dfrac{pdL}{2tE}\left(1 - \dfrac{\mu}{2}\right)$; Owing to this increase volume also increases.

Increase in volume $\delta V = V_2 - V_1 = \left[\dfrac{\pi}{4}(d + \delta d)^2\,(L + \delta L)\right] - \dfrac{\pi}{4}d^2$

$$= \text{Approximately } 0.785\,(d^2\,\delta L + 2\,dL\,\delta d)$$

Thick cylinders If inner diameter to wall thickness ratio is less than 15, it is considered thick cylinder. Design of thick cylinder is done by any one of the following equations depending on type of materials and ends whether closed or open:

Closed ends Lame's equation—for brittle materials

Clavarino's equation—for ductile materials

Open ends Birnie's equation—for ductile materials (based on inside diameter)

Barlow's equation—for ductile materials (based on outside diameter)

Lame's equation According to this theory, a vessel fails, if the *principal stress exceeds* the allowable stress of that material. Three principal stresses in a cylinder are:

Radial stress $\sigma_r = -p$

Tangential stress $\sigma_t = \dfrac{p(D^2 + d^2)}{(D^2 - d^2)}$

Longitudinal stress $\sigma_l = \dfrac{pd^2}{(D^2 - d^2)}$

Out of these three stresses, tangential stress is maximum, which is considered for design:

Wall thickness $t = \dfrac{d}{2}\left[\sqrt{\dfrac{\sigma_t + p}{\sigma_t - p}} - 1\right]$

Clavarino's equation It assumes maximum strain theory of failure. Strain $\varepsilon = \dfrac{S_{yt}}{FOS \times E}$

Wall thickness $t = \dfrac{d}{2}\left[\sqrt{\dfrac{\sigma + (1 - 2\mu)p}{\sigma - (1 + \mu)p}} - 1\right]$

Birnie's equation For open cylinders, no longitudinal stress, that is, $\sigma_l = 0$. Stress is: $\sigma = \sigma_t - \mu\sigma_r$

Wall thickness $t = \dfrac{d}{2}\left[\sqrt{\dfrac{\sigma + (1 - \mu)p}{\sigma - (1 + \mu)p}} - 1\right]$

Barlow's equation This equation is used for high-pressure oil and gas pipe lines.

Wall thickness $t = \dfrac{p \times D}{2\sigma_t}$

Since $D = d + 2t,\ t = \dfrac{p \times (d + 2t)}{2\sigma_t}$ or $t = \dfrac{pd}{2(\sigma_t - p)}$

Thin spherical vessels Wall thickness $t = \dfrac{pd}{4\sigma_t}$

Change in size of spherical vessel with internal pressure

Increase in diameter due to internal pressure: $\delta d = \dfrac{pd^2}{4tE}(1 - \mu)$

Volume of a spherical vessel $V = \dfrac{4\pi}{3}r^3 = \dfrac{\pi}{6}d^3$

Increase in volume $\delta V = \dfrac{\pi}{6}(d + \delta d)^3 - \dfrac{\pi}{6}d^3$ or $\delta V = \dfrac{\pi p d^4}{8tE}(1 - \mu)$

End Covers   These are of three types:

(a) *Flat* (b) *Domed*:   Semi-ellipsoid and tori-spherical (c) *Conical*

End covers are neither supported freely nor fixed rigidly and hence design is done empirically. Vessels containing corrosive fluids, a corrosion allowance of 2 mm–3 mm is added to the thickness.

Flat rectangular   Thickness of rectangular cover plate $t_c = a b k_1 \sqrt{\dfrac{p}{\sigma_t (a^2 + b^2)}}$

a = Length of cover plate, b = Width of cover plate, k_1 = A constant (Table 20.2)

Flat circular   Easiest to make. Thickness $t_c = k_2 d \sqrt{\dfrac{p}{\sigma_t}}$ Take value of k_2 from Table 20.2.

Elliptical plate   Ratio of major axis (a) to minor axis (b) is generally kept two.

Thickness of cover plate: $t_c = a b k_3 \sqrt{\dfrac{p}{\sigma_t (a^2 + b^2)}}$ Take k_3 from Table (20.2)

Hemispherical   It offers *minimum thickness* of plate for the same strength, hence light and economical.

Thickness is: $t_c = \dfrac{pd}{4\sigma_t}$

When the cover is not hemispherical but its frustum, it is called as dished. The camber c is less than the hemispherical covers. Thickness for a riveted cover is: $t_c = \dfrac{k p R}{\sigma_u}$

For *integral* or *welded* dished heads: $t_c = \dfrac{p(d^2 + 4c^2)}{16 c \sigma_t}$

c = Camber

Hemispherical is a dished head where c = 0.5 d. Hence $t_c = \dfrac{pd}{4\sigma_t}$

For a semi-ellipsoid end cover, generally the major axis is double the length of the minor axis. The cover plate is of elliptical shape. It has a straight length $S = 3 t_c$ or 20 mm, whichever is more. Thickness is maximum for this type.

Tori-spherical end cover   This type of end cover has two radii. It also has a straight length $S \cong 0.3$ t, or 20 mm. Radius of spherical crown portion is $R_2 = 0.8D$. Radius of knuckle $r1 = 0.1D$. Thickness of such a head is: $t_c = \dfrac{0.885 p R_2}{\sigma_t}$

Volume of such an end cover of one side can be approximated as: $V_c = 0.123\ d^3$

Conical end covers   Thickness of cover plate $t_c = \dfrac{pd}{2 \cos \theta\ \sigma_t}$; θ = Semicone angle

Fixing of end covers   Ends are either integral or fixed with bolts and nuts or rivets or welded. For integral covers, thickness of cover plate is $t_c = \dfrac{p(d^2 + 4c^2)}{16\ \sigma_t \times c}$

For bolted cover plate thickness of cover plate $t_c = d_p \sqrt{\dfrac{kp}{\sigma_t}}$ Where, d_p = Pitch circle diameter of bolts,

k = 0.16 for bolted / riveted, 0.30 for flanged, and 0.25 for welded.

Welded joints   Joints used are: Double welded or single welded with backing plate.

Efficiency of the joint depends on the joint used. See Table 20.3.

Welding is done in different locations to complete a closed shell. Longitudinal welding along axis of the shell is done to make a cylindrical or elliptical shell. Spherical ends are made by joining developed surfaces longitudinally of a hemisphere. Length of shell is increased by doing circumferential welding. Nozzles are welded at the opening and flanges.

Theory Questions

1.   Differentiate between a thin and thick cylinder.
2.   Discuss the various types of stresses coming in a cylindrical pressure vessel.
3.   List the theories used for the design of thick cylinders and their usage.
4.   Give equations to calculate principal stresses for cylindrical pressure vessel.
5.   Derive Lame's equation to calculate thickness of a thick cylinder.
6.   When do you use Clavarino's equation? Derive this equation.
7.   What are the theories for pressure vessels with open end? Differentiate between their uses. Derive any one to calculate thickness of cylinder wall.
8.   What are the different types of end covers? Give a neat sketch of each.
9.   Write the advantages and disadvantages of a hemispherical and semiellipsoid end covers.
10. Sketch a torispherical shape of an end cover. Describe its shape.
11. Why openings are provided in pressure vessels? How does the design changes with an opening?
12. What are the various classes of the pressure vessels?
13. What is a boiler code and what are its recommendations?

Multiple Choice Questions

1.   A pressure vessel is considered thick if:
     (a) It is very heavy
     (b) Its diameter is more than 1 m
     (c) Its diameter is less than 15 times wall thickness
     (d) Its outside diameter is 1.2 times the internal diameter
2.   A closed-end pressure vessels has:
     (a) Longitudinal stresses   (b) Circumferential stresses
     (c) Both given in (a) and (b)   (d) Hoop stresses

3. In an open-end pressure vessels the stresses are:

    (a) Hoop stresses

    (b) Circumferential stresses

    (c) Longitudinal stresses

    (d) Both given in (c) and (d)

4. In a thin cylinder of diameter d, if p is internal pressure and σ_t is tensile strength, thickness is calculated using the relation:

    (a) $\dfrac{p \times d}{2\sigma_t}$

    (b) $\dfrac{p \times d}{4\sigma_t}$

    (c) $\dfrac{p \times d^2}{4\sigma_t}$

    (d) None of given in (a), (b), or (c)

5. In a thick cylinder, radial stresses are:

    (a) Maximum inside and minimum outside

    (b) Maximum inside and zero outside

    (c) Maximum inside and half of maximum at outside

    (d) Depends on thickness

6. Lame's equation is for:

    (a) Thin cylinders

    (b) Thick cylinders with open ends and brittle materials

    (c) Thick cylinders with closed ends and brittle materials

    (d) Thick cylinders with closed ends and ductile materials

7. Thick cylinders with closed ends and ductile materials are designed using:

    (a) Clavarino's equation

    (b) Lame's equation

    (c) Birnie's equation

    (d) Barlow's equation

8. Theories that can be used for open-end pressure vessels are:

    (a) Lame's equation or Clavarino's equation

    (b) Clavarino's equation and Birnie's equation

    (c) Birnie's equation and Barlow's equation

    (d) None of the above

9. Thickness of a spherical shell is calculated using equation:

    (a) $t_c = \dfrac{p\,d}{2\sigma_t}$

    (b) $\dfrac{p \times d}{\sigma_t}$

    (c) $\dfrac{p \times d^2}{2.5\,\sigma_t}$

    (d) $\dfrac{p \times d^2}{1.25\,\sigma_t}$

10. Thickness of circular cover plate is calculated using equation:

    (a) $t_c = \dfrac{p\,d}{4\sigma_t}$

    (b) $t_c = \dfrac{k\,p\,R}{\sigma_u}$

    (c) $t_c = \dfrac{0.885\,p R_2}{\sigma_t}$

    (d) $t_c = k_1 d \sqrt{\dfrac{p}{\sigma_t}}$

11. Thickness of flat elliptical cover plate, with major and minor axes a and b, respectively, and k a constant is calculated using equation:

(a) $t_c = \dfrac{p(a^2 + b^2)}{4\sigma_t}$

(b) $t_c = \dfrac{p(a^2 + 4b^2)}{16\sigma_t}$

(c) $t_c = abk_1 \sqrt{\dfrac{p}{\sigma_t(a^2 + b^2)}}$

(d) $t_c = abk \sqrt{\dfrac{p}{\sigma_t(a + b)}}$

12. A welded long cylindrical pressure vessel has:

(a) Longitudinal welds

(b) Circumferential welds

(c) Both longitudinal welds and circumferential welds

(d) None of the above

13. Opening in a pressure vessel is made for:

(a) Man holes

(b) Mounting gauge

(c) Mounting a safety valve

(d) Any one or all given above

Answers to multiple choice questions

1. (c)   2. (c)   3. (b)   4. (a)   5. (b)   6. (c)   7. (a)   8. (c)   9. (a)   10. (d)
11. (c)   12. (c)   13. (d)

Design Problems

1. A boiler shell has diameter 2 m and pressure 0.9 MPa. Material used is mild steel having tensile strength of 480 MPa with *FOS* 5 and joint efficiency 75 per cent. Find:

   (a) Thickness of the shell.

   (b) Stress in shell plate.

   [(a) 12.7 mm; (b) 71 MPa;]

2. A hydraulic cylinder made of cast ion has a bore of 200 mm and subjected to internal pressure of 9 MPa. If tensile stress is not to exceed 40 MPa, calculate wall thickness.   [25.7 mm]

3. A cast iron pipe has to deliver water at the rate of 2.4 m³/s at a velocity of 0.6 m/s. Pressure in the pipe is 0.72.4 N/mm². Taking allowable strength of cast iron 25 N/mm², calculate diameter of the pipe and wall thickness.   [d = 2.25 m, t = 34 mm]

4. A steam engine has cast iron cylinder of 200 mm diameter. Maximum pressure is 1.2 MPa. Find the thickness of the cylinder if safe tensile strength is 15 MPa. Assume reboring allowance 6 mm.

   [t = 14 mm]

5. A cylindrical steel tank having diameter 150 mm and length 200 mm contains gas at a pressure of 12 N/mm². Safe tensile strength is 60 N/mm². Calculate:

(a) Wall thickness assuming it as a thin cylinder.

(b) Wall thickness according to Clavarino's equation, if poison's ratio is 0.3.

[(a) t = 15 mm; (b) t = 16 mm]

6. A hydraulic cylinder made of cast iron has a bore of 200 mm and is subjected to a pressure of 10 N/mm². Find its wall thickness, if tensile stress is not to exceed 35 N/mm². Take corrosion allowance as 6 mm. [t = 35 mm]

7. A diesel engine having diameter 240 mm has maximum pressure in the power stroke 2.8 MPa. Ultimate strength of the material is 240 MPa. Taking FOS 8 and reboring allowance 6 mm, calculate ts wall thickness. [t = 18 mm]

8. A cast iron cylindrical pressure vessel closed at both ends has circular-end cover plate integral with the cylindrical shell of internal diameter 180 mm. It is filled with a fluid at a pressure of 8 N/mm². Safe tensile stress is 30 N/mm². Calculate:

(a) Wall thickness of the shell

(b) Thickness of dished end cover plate with camber 10 % of diameter

[(a) t = 24 mm; (b) t_c = 31 mm]

9. A spherical pressure of diameter 1,500 mm is made of 12 mm thick plate with ultimate tensile stress 60 MPa.

(a) What maximum pressure it can handle?

(b) What thickness is required, if pressure is set at 1.5 MPa? [(a) p = 1.92 MPa; (b) t = 9.4 mm]

10. A spherical pressure vessel having inside diameter 0.6 m and plate thickness 15 mm is subjected to an internal pressure of 2 MPa. E = 200 GPa and μ = 0.3. Calculate increase in volume with pressure.

[23,738 mm³]

11. A steam chest of internal dimensions 400 mm × 250 mm and steam pressure 1.3 MPa has a flat rectangular freely supported cover plate made of carbon steel with allowable tensile stress 56 MPa. Calculate wall thickness of end cover plate. [t = 20 mm]

12. A thin cylindrical pressure vessel of diameter 600 mm made of carbon steel with yield strength 240 MPa. It is closed at both the ends by hemispherical end covers. The internal pressure is 3 MPa. Calculate:

(a) Wall thickness of the shell assuming FOS = 4

(b) Thickness of hemispherical end cover plate [(a) t = 15 mm; (b) t_c = 7.5 mm]

Competition Examination Questions

IES

1. The internal diameter of a hydraulic ram is 10 cm. Find the thickness required to withstand an internal pressure of 500 atm (1 atm = 98.07 kPa), if the yield point for the material (in tension as well as compression) is σ_y = 500 MPa. Use an FOS of 2. [Answer: t = 19.5 mm] [IES 2014]

2. A thin cylindrical shell with hemispherical ends is subjected to internal fluid pressure. For equal maximum stress to occur in both the cylindrical and the spherical portions, what would be the ratio of thicknesses of the spherical portion to that of the cylindrical portion?

[Answer: t_c / t = 0.5] [IES 2013]

3. What is the safe working pressure for a spherical pressure vessel 1.5 m internal diameter and 1.5 cm wall thickness, if the maximum allowable stress is 45 MPa?

    (a) 0.9 MPa   (b) 3.6 MPa   (c) 2.7 MPa   (d) 1.8 MPa

    [Answer: (d)] [IES 2013]

GATE

4. A cylindrical vessel with closed ends is filled with compressed air at a pressure of 500 kPa. The inner radius of the tank is 2 m and has its wall thickness of 10 mm. The magnitude of maximum in-plane shear stress (MPa) is _____

    [Answer: τ = 50 MPa] [GATE 2015]

5. A gas is stored in a cylindrical tank of inner radius 7 m and wall thickness 50 mm. The gauge pressure of the gas is 2 MPa. The maximum shear stress (MPa) in the wall is:

    (a) 35   (b) 70   (c) 140   (d) 280

    [Answer: (c)] [GATE 2015]

6. A cylindrical pressure vessel 200 cm in diameter and 350 cm in length is made of 1.3 cm thick plates. It is subjected to an internal pressure of 10 kg/cm². Calculate the longitudinal and circumferential stress developed in the pressure vessel.

    [Answer: σ_{fl} = 37.7 MPa, σ_{tc} = 75.4 MPa] [GATE 2015]

7. A thin gas cylinder with an internal radius of 100 mm is subjected to internal pressure of 10 MPa. The maximum permissible working stress is restricted to 100 MPa. The minimum wall thickness (mm) for safe design must be_____

    [Answer: t = 10 mm] [GATE 2014]

8. A long thin walled cylindrical shell closed at both ends is subjected to an internal pressure. The ratio of hoop stress (circumferential stress) to longitudinal stress developed in the shell is:

    (a) 0.5   (b) 1   (c) 2   (d) 4

    [Answer: (c)] [GATE 2014]

9. A thin gas cylinder with an internal radius of 100 mm is subjected to an internal pressure of 10 MPa. The maximum permissible working stress is restricted to 100 MPa. The minimum cylinder wall thickness (mm) for safe design must be _____

    [Answer: t = 10] [GATE 2014]

Common data questions 10 and 11

A spherical pressure vessel of 600 mm internal diameter is made of 3 mm thick cold drawn sheet steel with static strength properties of S_{ut} = 440 MPa and S_{yp} = 370 MPa. The pressure fluctuates between 0 and p_{max}, when yielding is not permitted (do not apply any FOS)

10. Maximum pressure for static yielding will be:

    (a) 5.4 MPa   (b) 6.4 MPa

    (c) 7.34 MPa   (d) 8.4 MPa

    [Answer: (c)] [GATE 2013]

11. Maximum pressure for eventual fatigue failure will be:

    (a) 5.5 MPa   (b) 6.5 MPa

    (c) 7.5 MPa   (d) 8.5 MPa

[Answer: (a)] [GATE 2013]

12. Bolts in the flanged end of pressure vessel are usually pretensioned. Indicate which of the following statement is not true?

    (a) Pretensioning helps seal the pressure vessel.

    (b) Pretensioning increases the fatigue life of the bolts.

    (c) Pretensioning reduces the maximum tensile stress in the bolts.

[Answer: (c)] [GATE 2013]

13. A long thin walled cylindrical shell, closed at both ends, is subjected to an internal pressure. The ratio of the hoop stress (circumferential stress) to longitudinal stress developed in the shell is

    (a) 0.5   (b) 1.0   (c) 2.0   (d) 4.0

[Answer: (c)] [GATE 2013]

Common data questions 14 and 15

A boiler shell of diameter 1.65 m is having steam at pressure 0.8 MPa with longitudinal riveted joint with double cover plate having joint efficiency 80 per cent. Assuming corrosion allowance 1 mm. Tensile strength of the plate is 92 MPa.

14. Thickness of the boiler shell should be:

    (a) 10 mm   (b) 12 mm

    (c) 15 mm   (d) 18 mm

[Answer: (a)] [GATE 2013]

15. Rivet diameter to be used should be:

    (a) 18 mm   (b) 19 mm

    (c) 20 mm   (d) 22 mm

[Answer: (b)] [GATE 2013]

16. A thin walled spherical shell is subjected to an internal pressure. If the radius of the shell is increased by 1 per cent and the thickness is reduced by 1 per cent, with the internal pressure remaining the same, the percentage change in the circumferential (hoop) stress is:

    (a) 0   (b) 1   (c) 1.08   (d) 2.02

[Answer: (d)] [GATE 2012]

17. If a thin cylinder of inner radius 500 mm and thickness 10 mm subjected to an internal pressure of 5 MPa, the average circumferential (hoop) stress in MPa is:

    (a) 100   (b) 250

    (c) 500   (d) 1,000

[Answer: (b)] [GATE 2012]

□□□

References

1. Shigley, J. E. 2004. *Mechanical Engineering Design*. McGraw Hill Book Company.

2. Khurmi, R. S., and J. K. Gupta. 2005. *A Text Book of Machine Design*. New Delhi: Eurasia Publishing House Pvt. Ltd. and S. Chand and Company Ltd. (Distributor).

3. Bhandari, V. B. 2010. *Design of Machine Elements*. New Delhi: Tata McGraw Hill Education Private Limited.

4. Sharma, P. C., and D. K. Aggarwal. 2006. *Machine Design*. New Delhi: S. K. Kataria and Sons.

5. Kulkarni, S. G. *Machine Design*. 2008. New Delhi: Tata McGraw Hill Education Private Limited.

6. Timothy, H., and P. E. Wentzell. 2004. *Machine Design*. New Delhi: Cengage Learning India Private Limited.

7. Jadon, V. K., and S. Verrma. 2010. *Analysis and Design of Machine Elements*. New Delhi: I. K. International Publishing House Pvt. Ltd.

8. Shariff, Abdullah. 2009. *Design of Machine Elements*. New Delhi: Dhanpat Rai and Sons.

9. Gupta, R. B. 1982. *Machine Design: Mechanical Engineering Design*. New Delhi: Satya Prakashan.

10. Madhavan, K, and Balaveera Reddy. 1987. *Design Data Hand Book*. New Delhi: CSB Publishers.

11. Faculty of PSG College of Technology. 1971. *PSG Design Data Book*. Coimbatore: PSG College of Technology.

12. Indian Standards from Bureau of Indian Standards.

Index

Printed in the United States
By Bookmasters